AF865767

Endothelial Cell Dysfunctions

Endothelial Cell Dysfunctions

Edited by
Nicolae Simionescu
and
Maya Simionescu
Institute of Cellular Biology and Pathology
Bucharest, Romania

Springer Science+Business Media, LLC

Library of Congress Cataloging-in-Publication Data

Endothelial cell dysfunctions / edited by Nicolae Simionescu and Maya Simionescu.
p. cm.
Includes bibliographical references and index.

1. Vascular endothelium--Pathophysiology. 2. Endothelium--Pathophysiology. I. Simionescu, N. (Nicolae) II. Simionescu, Maya.
[DNLM: 1. Endothelium--physiopathology. QS 532.5.E7 E5634]
RC691.4.E53 1991
616.1--dc20
DNLM/DLC
for Library of Congress 91-37285
CIP

ISBN 978-1-4899-0723-3 ISBN 978-1-4899-0721-9 (eBook)
DOI 10.1007/978-1-4899-0721-9

Originally published by Plenum Press, New York in 1992.
Softcover reprint of the hardcover 1st edition 1992

Contributors

J. S. Alexander • Department of Biomedical Engineering, Vanderbilt University, Nashville, Tennessee 37235

Ali Andalibi • Department of Microbiology, UCLA School of Medicine, Los Angeles, California 90024-1679

Robert Auerbach • Center for Developmental Biology, University of Wisconsin, Madison, Wisconsin 53706

Robert S. Bar • Diabetes and Endocrinology Research Center, University of Iowa, Iowa City, Iowa 52246

Craig T. Basson • Department of Pathology, Yale University School of Medicine, New Haven, Connecticut 06510

Leonard Bell • Department of Internal Medicine–Cardiology, Yale University School of Medicine, New Haven, Connecticut 06510

Paula N. Belloni • Department of Tumor Biology, The University of Texas M. D. Anderson Cancer Center, Houston, Texas 77030

Judith A. Berliner • Department of Pathology, UCLA School of Medicine, Los Angeles, California 90024-1732

Maria Luisa Brandi • Department of Physiopathology, University of Florence Medical School, Florence 50139, Italy

Jane M. Brinker • Connective Tissue Research Institute and Department of Medicine, University of Pennsylvania, and University City Science Center, Philadelphia, Pennsylvania 19104

L. Maximilian Buja • Department of Pathology, University of Texas Medical School, and Cardiology Research, Texas Heart Institute, Houston, Texas 77225–0345

J. P. Caen • Institut des Vaisseaux et du Sang, Hôpital Lariboisière, 75010 Paris, France

M. Clauss • Rover Research Laboratory, Department of Physiology and Cellular Biophysics, College of Physicians and Surgeons, Columbia University, New York, New York 10032

Ramzi S. Cotran • Department of Pathology, Brigham and Women's Hospital, and Harvard Medical School, Boston, Massachusetts 02115

Gregory Richard Criscuolo • Division of Neurosurgery, Yale University School of Medicine, New Haven, Connecticut 06510

H. F. Cuénoud • Department of Pathology, University of Massachusetts Medical School, Worcester, Massachusetts 01655

Myron I. Cybulsky • Vascular Research Division, Department of Pathology, Brigham and Women's Hospital, and Harvard Medical School, Boston, Massachusetts 02115

Elisabetta Dejana • Mario Negri Institute, 20157 Milan, Italy

G. V. Doern • Department of Medicine, University of Massachusetts Medical School, Worcester, Massachusetts 01655

Marwan E. El-Sabban • Cancer Biology Laboratories, Department of Pathology, College of Veterinary Medicine, Cornell University, Ithaca, New York 14853

Andrea Fattorossi • Immunology Unit, Medical Department, DASRS, Rome, Italy

A. P. Fishman • Cardiovascular–Pulmonary Division, Department of Medicine, University of Pennsylvania, Philadelphia, Pennsylvania 19104

Alan M. Fogelman • Department of Medicine, UCLA School of Medicine, Los Angeles, California 90024-1679

Joy S. Frank • Departments of Medicine and Physiology, UCLA School of Medicine, Los Angeles, California 90024-1679

H. Gerlach • Rover Research Laboratory, Department of Physiology and Cellular Biophysics, College of Physicians and Surgeons, Columbia University, New York, New York 10032

Michael A. Gimbrone, Jr • Vascular Research Division, Department of Pathology, Brigham and Women's Hospital, and Harvard Medical School, Boston, Massachusetts 02115

Paolo Golino • Division of Cardiology, University of Naples II School of Medicine, Naples 80131, Italy

John Gordon • British Bio-technology Limited, Cowley, Oxford OX4 5LY, United Kingdom

John M. Harlan • Department of Medicine, University of Washington, Seattle, Washington 98195

F. R. Haselton • Department of Biomedical Engineering, Vanderbilt University, Nashville, Tennessee 37235

Dorian O. Haskard • Department of Medicine (Rheumatology), RPMS, Hammersmith Hospital, London W12 0NN, United Kingdom

Fumitada Hazama • Department of Pathology, Shiga University of Medical Science, Seta, Otsu 520-21, Japan

Ada J. Huang • Department of Medicine, College of Physicians and Surgeons, Columbia University, New York, New York 10032

Susan Imes • Department of Medicine, UCLA School of Medicine, Los Angeles, California 90024-1679

Robert C. Johnson • Cancer Biology Laboratories, Department of Pathology, College of Veterinary Medicine, Cornell University, Ithaca, New York 14853

I. Joris • Department of Pathology, University of Massachusetts Medical School, Worcester, Massachusetts 01655

Nicholas A. Kefalides • Connective Tissue Research Institute and Department of Medicine, University of Pennsylvania, and University City Science Center, Philadelphia, Pennsylvania 19104

Olivier Kocher • Department of Pathology, Yale University School of Medicine, New Haven, Connecticut 06510

Volkhard Lindner • Department of Pathology, University of Washington, Seattle, Washington 98195

Thomas F. Lüscher • Department of Research, Laboratory of Vascular Research, and Department of Medicine, Divisions of Clinical Pharmacology and Cardiology, University Hospital, Basel, Switzerland

Aldons J. Lusis • Departments of Medicine and Microbiology, UCLA School of Medicine, Los Angeles, California 90024-1679

Bernadette Lyons-Giordano • Departments of Biochemistry and Biophysics, University of Pennsylvania, and University City Science Center, Philadelphia, Pennsylvania 19104

Joseph A. Madri • Department of Pathology, Yale University School of Medicine, New Haven, Connecticut 06510

G. Majno • Department of Pathology, University of Massachusetts Medical School, Worcester, Massachusetts 01655

June R. Merwin • Department of Pathology, Yale University School of Medicine, New Haven, Connecticut 06510

William J. Mileski • Department of Surgery, University of Washington, Seattle, Washington 98195

Steven A. Moore • Department of Pathology, University of Iowa College of Medicine, Iowa City, Iowa 52242

S. N. Mueller • Coriell Institute for Medical Research, Camden, New Jersey 08103

Mahamad Navab • Department of Medicine, UCLA School of Medicine, Los Angeles, California 90024-1679

Lindsey Needham • British Bio-technology Limited, Cowley, Oxford OX4 5LY, United Kingdom

Garth L. Nicolson • Department of Tumor Biology, The University of Texas M. D. Anderson Cancer Center, Houston, Texas 77030

S. Ogawa • Rover Research Laboratory, Department of Physiology and Cellular Biophysics, College of Physicians and Surgeons, Columbia University, New York, New York 10032

F. William Orr • Department of Pathology, McMaster University, Hamilton, Ontario, L8N 3Z5, Canada

William M. Pardridge • Department of Medicine, Division of Endocrinology, and Brain Research Institute, UCLA School of Medicine, Los Angeles, California 90024-1682

Bendicht U. Pauli • Cancer Biology Laboratories, Department of Pathology, College of Veterinary Medicine, Cornell University, Ithaca, New York 14853

Robin Perlmutter • Department of Pathology, Yale University School of Medicine, New Haven, Connecticut 06510

Jordan S. Pober • Department of Pathology, Brigham and Women's Hospital, and Harvard Medical School, Boston, Massachusetts 02115

Christian Prinz • Department of Pathology, Yale University School of Medicine, New Haven, Connecticut 06510

Tripathi B. Rajavashisth • Division of Medical Genetics, Harbor UCLA Medical Center, Torrance, California 90502-2064

Michael A. Reidy • Department of Pathology, University of Washington, Seattle, Washington 98195

Charles L. Rice • Department of Surgery, University of Washington, Seattle, Washington 98195

Russell Ross • Department of Pathology, University of Washington, Seattle, Washington 98195

Masakiyo Sasahara • Department of Pathology, Shiga University of Medical Science, Seta, Otsu 520-21, Japan

Samuel C. Silverstein • Department of Physiology and Cellular Biophysics, College of Physicians and Surgeons, Columbia University, New York, New York 10032

Maya Simionescu • Institute of Cellular Biology and Pathology, Bucharest 79691, Romania

Nicolae Simionescu • Institute of Cellular Biology and Pathology, Bucharest 79691, Romania

Arthur A. Spector • Department of Biochemistry, University of Iowa College of Medicine, Iowa City, Iowa 52242

D. M. Stern • Rover Research Laboratory, Department of Physiology and Cellular Biophysics, College of Physicians and Surgeons, Columbia University, New York, New York 10032

Mary C. Territo • Department of Medicine, UCLA School of Medicine, Los Angeles, California 90024-1679

J. M. Underwood • Department of Pathology, University of Massachusetts Medical School, Worcester, Massachusetts 01655

Paul M. Vanhoutte • Center for Experimental Therapeutics, Baylor College of Medicine, Houston, Texas 77030

Nicholas B. Vedder • Department of Surgery, University of Washington, Seattle, Washington 98195

D. Vilette • Institut des Vaisseaux et du Sang, Hôpital Lariboisière, 75010 Paris, France

J. L. Wautier • Institut des Vaisseaux et du Sang, Hôpital Lariboisière, 75010 Paris, France

Leonard Weiss • Department of Experimental Pathology, Roswell Park Memorial Institute, Buffalo, New York 14263

James T. Willerson • Department of Internal Medicine, University of Texas Medical School, and Cardiology Research, Texas Heart Institute, Houston, Texas 77225-0345

Stuart K. Williams • Department of Surgery, University of Arizona Health Sciences Center, Tucson, Arizona 85724

Robert K. Winn • Departments of Surgery and Physiology–Biophysics, University of Washington, Seattle, Washington 98195

Preface

It has taken biomedical research some four decades to reveal that vascular endothelium is one of the master systems of the body. We have witnessed the gradual discovery of its key roles in blood–tissue exchanges, in providing a compatible container for the blood cells, and in the active monitoring of homeostasis. That was a cumulative result of elegant observations by a number of investigators embarked on a search for understanding of a cellular system that is not only enormously complex but also central to life.

Technological advances in cellular and molecular biology have made it possible to explore in more depth the intimate mechanisms of endothelial functions not only in normal but, more recently, also in abnormal conditions. Thus, it has been firmly established that in pathological circumstances, endothelial cells are not only a target for various injuries, but also a source for response modifiers that enable the cells to adapt to and compensate for the effects of abnormal stimuli. Upon its interactions with injurious agents, endothelium can undergo functional and structural modulations which may either secure its recovery or cause it to evolve into a dysfunctional state. Along this line of responses, endothelial cells may suffer significant changes in their constitutive properties or may acquire new induced functions.

Conclusive evidence supports an active role for endothelium in the initiation, evolution, and outcome of several notorious diseases and conditions including inflammation, immune reactions, atherosclerosis, thrombosis, diabetes, viral infection, bleeding disorders, neoplasia and metastasis, and others. One should, however, admit that in several instances at present we are just handling an impressive mass of mostly descriptive data with rather little understanding, as yet, of the basic mechanism of endothelial participation in the pathogenesis of these disturbances. Although perhaps suggestive, a temporal association of some events does not prove a causal relationship. This leaves open a very exciting field of research ahead of us.

This book is a collection of selected authoritative accounts of endothelial cell dysfunction and injury detected in a number of major abnormalities. The research-oriented chapters are authored by some of those who have contributed more measurably to our present knowledge of endothelial cell biopathology.

The inventory of agents promoting endothelial dysfunction is incomplete, and the mechanisms by which they act are often still obscure. Thus, the elucidation of endothelial biochemical and reactive heterogeneity is largely needed, especially the identification of organ-specific cell surface antigens. This can be crucial not only for understanding the vascular involvement and manifestation in various diseases, but also for designing drugs that can specifically interact with the endothelial organotypic determinants. This may open new avenues for controlled drug delivery bearing on the endothelial cell molecular makeup expressed in a given disease. Thus, a new installment may be endothelium-mediated drug targeting.

Endothelial cell dysfunction, whether pathogenic or reactive, renders endotheliology a whole new ball game in which, at least for a while, we should stoically live with the uncertainty, as well as with the need to balance between exhilaration and frustration—these two ubiquitous and inevitable ingredients of research. As such, this book is above all a stimulus for other investigators to join the venture, because these days, witnessing or participating in discoveries in endothelial cell biopathology may be just a matter of being there.

Nicolae Simionescu
Maya Simionescu

Bucharest, Romania

Contents

I

General

1

Endothelial Cell Response to Normal and Abnormal Stimuli

Modulation, Dysfunction, Injury; Adaptation, Repair, Death

Maya Simionescu

What we do not understand,
we do not possess.
Goethe

I. CELL INTERACTIONS AND ADAPTATION TO THEIR ENVIRONMENT

Normal eukaryotic cells interact continuously and subtly with their surrounding milieu; variations in the nature, duration, and intensity of the extracellular stimuli generate a bidirectional interplay between the two. By virtue of their programmed genotype and the expressed metabolic, functional, and structural phenotype, cells are endowed with a finite capacity to adapt. In a homeostatic steady state, cells are able to handle normal physiologic demands. If the limits of their adaptive capacity are exceeded, as the result of excessive physiologic stimuli or pathologic stresses, the steady state can be altered. By their struggle to adapt, cells try to escape damage and premature death.

The stimuli acting on a cell can be either normal or abnormal. According to the nature (injurious or noninjurious) of the stimuli, the cell response may take a relatively wide spectrum of forms, in most cases states along a continuum which may span from minute metabolic modifications to irreversible cell damage leading to death.*

*(The terms used to describe the effects of an aggression on the cell structure and functioning are largely arbitrary. There is no unanimity as to how these words should be defined, and often they mean different things to different authors. Although in this case there is nothing to be gained by an exercise in semantics, hereafter the terms we

Maya Simionescu • Institute of Cellular Biology and Pathology, Bucharest 79691, Romania.

Endothelial Cell Dysfunctions, edited by Nicolae Simionescu and Maya Simionescu. Plenum Press, New York, 1992.

To a normal stimulus, a cell may respond by adjusting within a physiological range its constitutive functions with no significant perturbations of cell structure or interactions; this is a process of *modulation*, e.g., cell hypertrophy induced by augmented secretion of growth factors (Fig. 1).

An abnormal stimulus may impair the expression of constitutive functions or may induce new properties or functions reflected in perturbation of cell interactions and activities. This *dysfunction* may be reversible or may lead to more advanced injury, e.g., inflammation triggering the induction of new adhesive molecules on macrophages and leukocytes. Because the term "injury" may ambiguously describe alternatively either an adverse cause or a disturbing effect, we prefer to restrict it to the physical damage of a cell abolishing partially or fully its capacity to adapt. As such an injury can be *reversible* followed by cell *repair* or *irreversible* leading to *cell death*. Any change through which cells escape irreversible damage defines the *adaptation*. The efficiency of cell adaptation depends on the balance between the nature, duration, and severity of the aggression, and the cell intrinsic state and adaptive resources. Pathology may be thought of as a continuum with normal biology, on which the cell's struggle for adaptation is a dynamic, progressive, interrelated process leading to reversible or irreversible cellular alterations. In response to one or several adverse influences (e.g., genetic, nutritional, physical, chemical, infectious, immunologic), at the inception of cell reaction is a metabolic modification continued by more pronounced functional and structural derangements. Four intracellular systems appear to be particularly vulnerable: (1) maintenance of the integrity of cell membranes, (2) aerobic respiration, (3) synthesis of structural, export, and enzymatic proteins, (4) preservation of the integrity of genetic apparatus.[8] The complex mechanisms of a cell's successful or unsuccessful adaptation to various injurious stimuli are only partially understood. Some valuable descriptions are given in Refs. 3, 6–8, 10, 13, 15, 17, 23, 24, 35.

II. ENDOTHELIAL CELL RESPONSE AND ADAPTATION TO STIMULI

Vascular endothelium is critically interposed between a circulating tissue—the blood (of continuous turnover and changing composition)—and all other tissues of the body. By virtue of this strategic location, the metabolically active endothelial cells constitute not only a highly sophisticated monitor of plasma homeostasis, but also a source and a relay station for messages between the intra- and the extravascular compartments. There is substantial evidence that under a common theme, endothelium is a polarized tissue, highly heterogeneous, being morphologically and functionally differentiated as a defining component of the host organ, to the extent that one can consider one tissue = one endothelium. With an aggregate mass that equals that of the liver (~ 1000 g), vascular endothelium constitutes a

---

prefer to use are still imperfect and represent our own choice. They may be at variance with the connotation given to these words by other contributors to this book.)

In most dictionaries, the following general definitions are given for the terms used in this chapter. *Modulation*: the normal capacity of cell adaptability to its environment. *Dysfunction*: abnormal, impaired, or incomplete functioning of an organ or part of it. *Injury*: physical harm or damage to a thing (or person); or, impairment of body function or structure due to adverse influence. *Adaptation*: a change in structure, function, or form that produces adjustment, of a body to its environment.

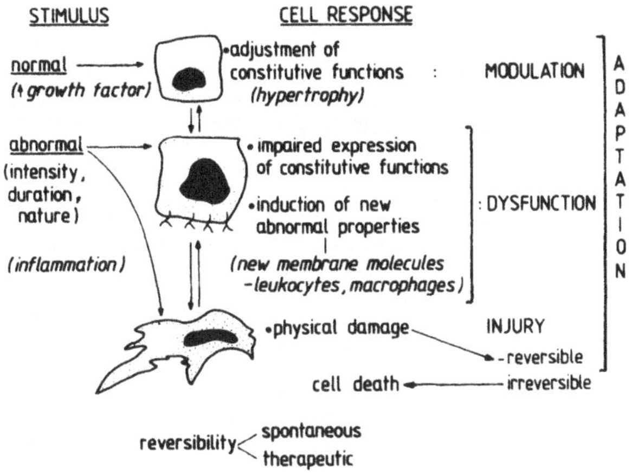

Figure 1. Schematic illustration of cell response to normal or abnormal stimuli from its environment. Depending on the nature, intensity, and duration of the injurious stimuli and on its functional state, the cell may react by modulation, dysfunction, or injury. Through these adaptive reactions, the cell tries to escape irreversible injury and death. A few examples are given in parentheses.

multifunctional organ with a broad spectrum of properties and activities. Among these constitutive functions of special significance are: permselective barrier, blood-compatible container, synthetic organ, integrator and transducer of blood-borne signal, regulator of the vascular growth, vascular tone, metabolism of hormones and vasoactive mediators.[12]

Bearing on its unique situation and its multiple properties, endothelium is crucially involved both in homeostasis and in a large variety of diseases, being able to undergo drastic functional and structural modifications in response to various physiopathologic stimuli. An impaired functioning of any of these interrelated compartments (i.e., blood, endothelium, tissues) may affect significantly the other compartments. There are a plethora of examples from the pathology and pathogenesis of diabetes, bleeding disorders, atherosclerosis, and so forth.

In addition to the broad scope of constitutive functions mentioned above, endothelium is also capable of responding to abnormal stimuli by expressing newly acquired, *induced properties and functions*. These are diverse (expression of adhesive molecules for leukocytes, such as ELAM-1,[6,7,12] release of mitogens and growth inhibitors, expression of cytokines IL-1, alpha and beta interferons, colony-stimulating factors, and so forth[5,6,9,12,22,38] and can markedly alter the properties of endothelium in ways that can be relevant for a certain disease.

As any other cell, the endothelial cell can respond to normal stimuli by adjusting some of its *constitutive* functions: examples of such *modulation* are represented by the increased secretion of endothelial-derived relaxing factor (EDRF) upon stimulation by acetylcholine, or endothelial adaptation according to the blood flow parameters[21,41] (Fig. 2).

Abnormal stimuli, such as those appearing in inflammation, immunity, and coagulation by augmenting the production of cytokines IL-1 and TNF can generate the *induction* of new properties of endothelium which is significantly altered. An example of this *dysfunction* is the "endothelial activation," conveniently but more comprehensively defined in relation to macrophage and lymphocyte activation. This phenomenon, initially described in *in vitro* systems, also occurs *in vivo*.[6] An endothelial dysfunction appears as a localized alteration of endothelium brought up temporarily to a nonadaptive functional state bearing on a reversible phenotypic change.[11,12] An "activated" endothelium exhibits an increased

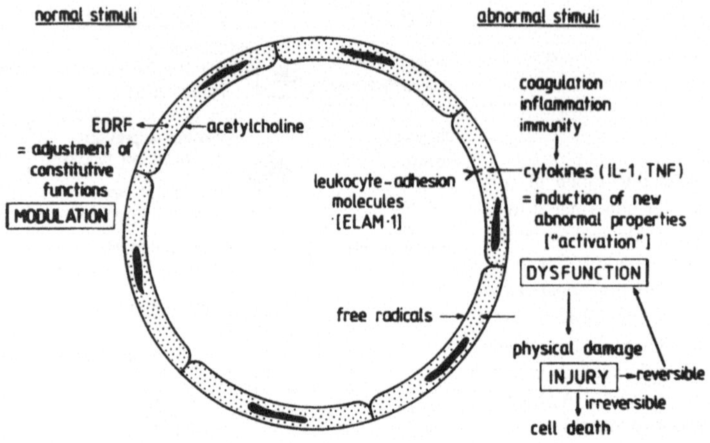

Figure 2. Diagrammatic illustration of the endothelial cell reaction to normal stimuli by modulation and to abnormal stimuli by dysfunction and/or injury. The latter can be either reversible or irreversible leading to cell death. The three modes of cell response are exemplified.

permeability to macromolecules, hypertrophic cuboidal cells with large ovoid nuclei and augmented number of biosynthetic organelles.[8]

Although the term and the state named "activation" needs further characterization regarding the stimulus, the signal transduction mechanisms, and the quantitation of the resulting phenotype,[31,38] the concept has been seminal for the cardiovascular research.[6,7]

Excessive adverse influences resulting in physical damage of endothelial cells characterize the altered state of *injury*, as largely documented in the case of alterations produced by exposure to free radicals[1,2,18,32,36,39,43] or in hypercholesterolemia.[33,34] Endothelial injury may be *reversible* either through a repair process taking place within the framework of the damaged cell, or by the replacement of a denuded endothelial area as a result of local tissue regeneration.[27,29] An *irreversible* injury may lead to *cell death* as in other tissue systems (Fig. 2).

By their nature and final physiologic efficiency, through modulation, dysfunction, and reversible injury the cell secures its *adaptation* to the new conditions imposed by the adverse influences. Obviously this adaptive capability is exceeded during an irreversible injury.

III. ENDOTHELIAL CELL DYSFUNCTION AND INJURY IN VARIOUS DISEASES

So far, no primary disease of the endothelium has been identified. Vascular endothelium rather participates in the pathogenesis of certain disorders or suffers secondarily from a variety of perturbations associated with some diseases.

Actually this is the topic of this book. An introductory installment is shown in Fig. 3, and more comprehensive accounts on this subject are given in the coming chapters.

Inflammation (6, 7, 22, 38)	Diabetes
Immune and autoimmune diseases (6, 7, 38)	Pulmonary hypertension/edema
	Respiratory distress syndrome
Neutrophil-mediated injury	Rheumatoid arthritis
Oxidant-mediated injury (1, 2, 10, 15, 16, 18, 28, 32, 36, 39, 43)	Lupus erythematosus
Bleeding disorders	Multiple sclerosis
Thrombosis (20)	Preeclampsia (30)
Hypercholesterolemia	Neoplasia-metastasis
Atherosclerosis (11, 14, 25, 33, 34, 37)	Viral infections (19)
Hypertension (25, 26)	AIDS/Kaposi's sarcoma
Coronary Disease (40)	Aging
Cyclosporin treatment (42)	
EDRF dysfunction	

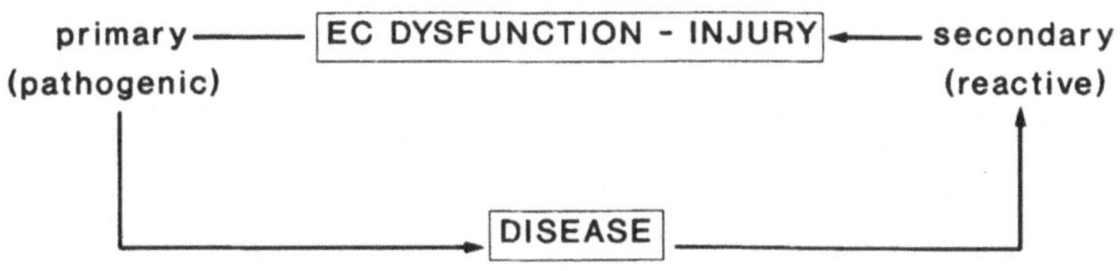

Figure 3. Tabulation of endothelial cell dysfunction and injury identified in some major pathologic conditions in humans. In addition to the reports given in other chapters of this book, for some disorders a few references are indicated by numbers.

The title of this book reflects the prevalence of the endothelial cell dysfunctions, a term that can also be taken under a broader connotation as expressing endothelial pathogenic involvement or secondary implication in several diseases.

Three decades of sustained endeavor and progressively elaborate investigation of endothelial cell biology and its participation in various diseases brought up the excitement of few though incomplete answers to some of the initial questions. Concomitantly, however, as always in science, a score of additional more difficult questions arose. That evokes the feeling so beautifully expressed in this old Zen saying:

> When one is very young and knows a little,
> mountains are mountains, water is water,
> and trees are trees.
>
> When he has studied and has become sophisticated,
> mountains are no longer mountains,
> water is no longer water, and trees
> are no longer trees.
>
> When he thoroughly understands,
> mountains are again mountains,
> water is water, and trees are trees.

REFERENCES

1. Andreoli, S. P., Mallett, C. P., and Bergstein, J. M., 1986, Role of glutathione in protecting endothelial cells against hydrogen peroxide oxidant injury, *J. Lab. Clin. Invest.* **108:**190–198.
2. Block, E. R., Patel, J. M., Angelides, K. J., Sheridan, N. P., and Garg, L. C., 1986, Hyperoxia reduces plasma membrane fluidity: A mechanism for endothelial cell dysfunction, *J. Appl. Physiol.* **60:**826–835.
3. Brandle, E. R., and Gabbiani, G., 1983, The role of cytoskeletal and cytocontractile elements in pathologic processes, *Am. J. Pathol.* **110:**361–370.
4. Byron, M. A., Allington, M. J., Chapel, H. M., Mowat, A. G., and Cedarholm-Williams, S. A., 1987, Indications of vascular endothelial cell dysfunction in systemic lupus erythematosus, *Ann. Rheum. Dis.* **46:**741–745.
5. Cavender, D. E., Edelbaum, D., and Ziff, M., 1989, Endothelial cell activation induced by tumor necrosis factor and lymphotoxin, *Am. J. Pathol.* **134:**551–560.
6. Cotran, R. S., 1987. New roles for the endothelium in inflammation and immunity, *Am. J. Pathol.* **129:** 407–413.
7. Cotran, R. S., and Pober, J. S., 1988, Endothelial activation: Its role in inflammatory and immune reactions, in: *Endothelial Cell Biology in Health and Disease* (N. Simionescu and M. Simionescu, eds.), Plenum Press, New York, pp. 335–347.
8. Cotran, R. S., Kumar, V., and Robbins, S. L. (eds.), 1989, *Pathologic Basis of Disease*, Saunders, Philadelphia, pp. 1–38.
9. Elices, M. J., Osborn, L., Takada, Y., Crouse, C., Luhovsky, S., Hemler, M. E., and Lobb, R. R., 1990, VCAM-1 on activated endothelium interacts with the leukocyte integrin VLA-4 at a site distinct from the VLA-4/fibronectin binding site, *Cell* **60:**577–584.
10. Freeman, B. A., and Crape, J. D., 1982, Biology of disease: Free radicals and tissue injury, *Lab. Invest.* **47:**412–420.
11. Gimbrone, M. A., Jr., 1989, Endothelial dysfunction and atherosclerosis, *J. Cardiac Surg.* **4:**180–183.
12. Gimbrone, M. A., Jr., and Bevilacqua, M. P., 1988, Vascular endothelium: Functional modulation at the blood interface, in: *Endothelial Cell Biology in Health and Disease* (N. Simionescu and M. Simionescu, eds.), Plenum Press, New York, pp. 255–273.
13. Gordon, J. L., and Pearson, J. D., 1982, Responses of endothelial cells to injury, in: *Pathobiology of the Endothelial Cell* (H. L. Nossel and H. J. Vogel, eds.), Academic Press, New York, pp. 433–454.
14. Hajjar, K. A., Gavish, D., Breslow, J. L., and Nachman, R. L., 1989, Lipoprotein (a) modulation of endothelial cell surface fibrinolysis and its potential role in atherosclerosis, *Nature* **339:**303–305.
15. Halliwell, B., 1987, Oxidants and human disease: Some new concepts, *FASEB J.* **1:**358–369.
16. Hennig, B., and Chow, C. K., 1988, Lipid peroxidation and endothelial cell injury: Implications in atherosclerosis, *Free Radical Biol. Med.* **4:**99–106.
17. Henson, P. M., and Johnson, R. B., Jr., 1987, Tissue injury in inflammation: Oxidants, proteases and cationic proteins, *J. Clin. Invest.* **79:**669–674.
18. Hirosumi, J., Ouchi, Y., Watanabe, M., Kusunoki, J., Nakamura, t., and Orimo, H., 1988, Effect of superoxide and lipid peroxide on cytosolic free calcium concentration in cultured pig aortic endothelial cells, *Biochem. Biophys. Res. Commun.* **152:**301–307.
19. Hirschberg, H., 1981, Presentation of viral antigens by human vascular endothelial cells in vitro, *Hum. Immunol.* **2:**235–246.
20. Jorgensen, L., Grothe, A. G., Larsen, T., Kinlough-Rathbone, R. L., and Mustard, J. F., 1986, Injury to cultured endothelial cells by thrombin-stimulated platelets, *Lab. Invest.* **54:**408–415.
21. Joris, I., Zand, T., and Majno, G., 1982, Hydrodynamic injury of the endothelium in acute aortic stenosis, *Am. J. Pathol.* **106:**394–408.
22. Jutila, M. A., Berg, E. L., Kishimoto, T. K., Picker, L. J., Bargatze, R. F., Bisho, D. K., Orosz, C. G., Wu, N. W., and Butcher, E. C., 1989, Inflammation-induced endothelial cell adhesion to lymphocytes, neutrophils and monocytes, *Transplantation* **48:**727–731.
23. Lindquist, S., 1986, The heat shock response, *Annu. Rev. Biochem.* **55:**1151–1191.
24. Majno, G., 1975, *The Healing Hand: Man and Wound in the Ancient World*, Harvard University Press, Cambridge, Mass., p. 43.
25. Majno, G., Zand, T., Nunnari, J. J., Kowala, M. C., and Joris, J., 1988, Intimal response to shear stress,

hypercholesterolemia and hypertension: Studies in the rat aorta, in: *Endothelial Cell Biology in Health and Disease* (N. Simionescu and M. Simionescu, eds.), Plenum Press, New York, pp. 349–367.

26. Nag, S., 1986, Cerebral endothelial plasma membrane alterations in acute hypertension, *Acta Neuropathol.* **70:**38–43.
27. Pederson, D. C., and Bowyer, D. E., 1985, Endothelial injury and healing in vitro. Studies using an organ culture system, *Am. J. Pathol.* **119:**264–272.
28. Poumay, Y., and Ronveau-Dupal, M. F., 1988, Incubation of endothelial cells in a superoxide-generating system: Impaired low density lipoprotein receptor-mediated endocytosis, *J. Cell. Physiol.* **136:**289–296.
29. Reidy, M. A., 1985, A reassessment of endothelial injury and arterial lesion formation, *Lab. Invest.* **53:** 513–526.
30. Roberts, J. M., Taylor, R. N., Musci, T. J., Rodgers, J. M., Hubel, C. A., and McLaughlin, M. K., 1989, Preeclampsia: An endothelial cell disorder, *Am. J. Obstet. Gynecol.* **161:**1200–1204.
31. Ryan, U. S., 1987, Endothelial cell activation responses, in: *Pulmonary Endothelium in Health and Disease* (U. S. Ryan, ed.), Dekker, New York, pp. 3–33.
32. Sasaguri, Y., Morimatsu, M., Kinoshita, T., Nakashima, T., Inagaki, T., and Yagi T., 1985, Difference in susceptibility to injury by linoleic acid hydroperoxide between endothelial and smooth muscle cells of arteries, *J. Appl. Biochem.* **7:**70–78.
33. Simionescu, N., 1988, Prelesional changes of arterial endothelium in hyperlipoproteinemic atherogenesis, in: *Endothelial Cell Biology in Health and Disease* (N. Simionescu and M. Simionescu, eds.), Plenum Press, New York, pp. 385–429.
34. Simionescu, N., Vasile, E., Lupu, F., Popescu, G., and Simionescu, M., 1986, Prelesional events in atherogenesis: Accumulation of extracellular cholesterol-rich liposomes in the arterial intima and cardiac valves of the hyperlipidemic rabbit, *Am. J. Pathol.* **123:**109–125.
35. Slater, T. F., 1984, Free radical mechanisms in tissue injury, *Biochem. J.* **222:**1–4.
36. Starkebaum, G., and Harlan, J. M., 1986, Endothelial cell injury due to copper-catalyzed hydrogen peroxide generation from homocysteine, *J. Clin. Invest.* **77:**1370–1376.
37. Walker, L. N., and Bowyer, D. E., 1984, Endothelial healing in rabbit aorta and the effect of risk factors for atherosclerosis: Hypercholesterolemia, *Arteriosclerosis* **4:**479–488.
38. Wallis, W. J., and Harlan, J. M., 1986, Effector functions of endothelium in inflammatory and immunologic reactions, *Pathol. Immunopathol. Res.* **5:**73–103.
39. Warren, J. S., and Ward, P. A., 1986, Review: Oxidative injury to the vascular endothelium, *Am. J. Med. Sci.* **292:**97–103.
40. Werns, S. W., Walton, J. A., Hsia, H. H., Nabel, E. G., Sanz, M. L., and Pitt, B., 1989, Evidence of endothelial dysfunction in angiographically normal coronary arteries of patients with coronary disease, *Circulation* **79:**287–291.
41. Zand, T., Nunnari, J. J., Hofman, A. H., Savilonis, B. J., MacWilliams, B., Majno, G., and Joris, I., 1988, Endothelial adaptations in aortic stenosis. Correlation with flow parameters, *Am J. Pathol.* **133:**407–418.
42. Zoja, C., Furci, L., Ghilardi, F., Zilio, P., Benigni, A., and Remuzzi, G., 1986, Cyclosporin-induced endothelial cell injury, *Lab. Invest.* **55:**455–462.
43. Zweier, J. L., Kuppusami, P., and Lutty, G. A., 1988, Measurement of endothelial cell free radical generation: Evidence for a central mechanism of free radical injury in postischemic tissues, *Proc. Natl. Acad. Sci. USA* **85:**4046–4050.

2

Interactions of Matrix Components and Soluble Factors in Vascular Responses to Injury

Modulation of Cell Phenotype

Joseph A. Madri, June R. Merwin, Leonard Bell, Craig T. Basson, Olivier Kocher, Robin Perlmutter, and Christian Prinz

I. INTRODUCTION

The vascular system is lined by mitotically quiescent but metabolically active endothelial cells, which in addition to having a broad range of metabolic activities, provide a nonthrombogenic surface for blood flow. Beneath the endothelium, smooth muscle cells are found in the media of large vessels, and pericytes are found in close association with the endothelial cells of microvascular beds. The smooth muscle cells (pericytes) are thought to play major roles in maintaining vessel wall integrity, being responsible for the maintenance fo the connective tissues of the vessel wall and in the control of vascular tone.[8] Vascular cells (large and small vessel derived endothelial, pericyte, and smooth muscle cells) have been found to respond to injury in specific ways, depending upon the vascular bed and the cell type(s) injured. For example, following denudation injury evoked by angioplasty, endarterectomy or autologous or synthetic grafting, large vessel endothelial cells bordering the affected area will exhibit rapid sheet migration over the exposed extracellular matrix and proliferate in an attempt to reconstitute the normal continuous endothelial cell lining.[15,20] The medial smooth muscle cells of large and medium-sized vessels respond to vessel injury by migrating into the intima, where they proliferate and synthesize matrix components,

---

Joseph A. Madri, June R. Merwin, Craig T. Basson, Olivier Kocher, Robin Perlmutter, and Christian Prinz • Department of Pathology, Yale University School of Medicine, New Haven, Connecticut 06510. *Leonard Bell* • Department of Internal Medicine–Cardiology, Yale University School of Medicine, New Haven, Connecticut 06510.

Endothelial Cell Dysfunctions, edited by Nicolae Simionescu and Maya Simionescu. Plenum Press, New York, 1992.

which results in the formation of a thickened intima which narrows the vessel lumen.[34] In contrast, following soft tissue injury or in response to a variety of angiogenic factors, microvascular endothelial cells respond by freeing themselves from the constraints of their investing basement membranes. Following this, they migrate and proliferate in the surrounding three-dimensional interstitial stroma and ultimately form new microvessels.[17] The role(s) of pericytes (smooth muscle cell analogues which are found surrounding microvessels following new vessel formation) in the response to injury has been less well studied and their origin(s) (endothelial, undifferentiated mesenchymal–fibroblastic, or vascular smooth muscle cell) is still a matter of controversy.[12,17]

In this chapter we will discuss the modulation of large vessel endothelial and smooth muscle cell responses to *in vivo* denudation and *in vitro* migration injury by selected soluble factors including platelet-derived growth factor (PDGF), transforming growth factors β1 and β2 (TGF-β1, TGF-β2). In addition, we will discuss the roles of these factors in modulating the behavior and differentiation of microvascular endothelial cells during the angiogenic response.

II. METHODS AND MATERIALS

Cells: Bovine aortic endothelial (BAEC), bovine aortic smooth muscle (BASMC), bovine adrenal cortical microvascular (BCE), and rat epididymal fat pad microvascular (RFC) cells were isolated, cultured, and characterized as described.[5,18,19]

Matrices: Collagen types I, III, IV, and V, laminin, and fibronectin were isolated, purified, and characterized as described.[14,20,21] Bacteriologic plastic culture dishes were coated with matrix components as described.[20]

Cell Proliferation: Proliferative rates were determined by cell counting using a Coulter counter and by quantitating cellular DNA as described.[22,26]

Migration Assay: BAEC and BASMC migration rates were quantitated using a previously described assay utilizing a Teflon fence to initiate migration.[30] Migration areas and migration rates were determined using a Macintosh Plus computer.

Angiogenesis Assays: *In vitro* angiogenesis (tube formation) was assayed in three-dimensional cultures morphologically, using either light microscopic (Hoffman Interference Optics) analyses of 8-μm tissue sections or standard electron microscopic techniques[22,26] and by standard electron microscopic methods.[26]

Immunofluorescence: Immunofluorescence of BAEC, BASMC, RFC, and BCE monolayer cultures and three-dimensional cultures were performed as described.[2,37]

Electron Microscopy: Electron microscopy of cells in two- and three-dimensional culture was performed using standard procedures.[26]

ELISA: Quantitative assay of matrix components was performed using an ELISA developed in the laboratory.[7,18,24]

Northern Blot Analyses: Quantitation of mRNA levels of selected matrix components and cell surface matrix binding proteins was performed by Northern blotting techniques as described.[12,24]

Growth Factors: TGF-β1 and TGF-β2 were generous gifts of Drs. Anita Roberts and Michael Sporn, Laboratory of Chemoprevention, NCI, NIH, Bethesda, Md. PDGF-AA, AB, and BB were purchased from Collaborative Research, Inc., Boston, Mass.

III. RESULTS AND DISCUSSION

A. The Effects of Extracellular Matrix Components and Soluble Factors on Bovine Aortic Endothelial Cell Migration in Vitro and in Vivo: Models of Deendothelialization and Repair

Mechanical, viral, or chemical injury to the endothelium, followed by platelet aggregation and release of growth factors and vasoactive agents, such as PDGF, TGF-β1, serotonin, norepinephrine, and histamine, which promote vascular smooth muscle cell chemotaxis, migration, and proliferation, has been proposed as an important event in the development of arteriosclerosis.[25] In addition, several of the currently available invasive treatments of large and medium vessel occlusive disease (angioplasty, autologous and synthetic grafting, endarterectomy, atherectomy, and laser ablation) result in significant deendothelialization and medial injury of vascular segments, and the resultant platelet adhesion, aggregation, and release of soluble factors and "activation" of local medial smooth muscle cells are thought to be associated with the development of posttherapy stenosis that is noted in up to 50% of the patient populations studied.[28] For several years we have investigated the role of matrix components and selected soluble factors in the modulation of endothelial and vascular smooth muscle cell attachment, proliferation, and migration.[4,21,23,25]

It is widely accepted that matrix components have a significant effect on BAEC migration *in vitro*. Using an *in vitro* migration assay that was developed in our laboratory, we have demonstrated that interstitial collagens elicit the most rapid migration rate, basement membrane components evoke intermediate migratory rates, and fibronectin elicits the lowest migration rate.[13,20,21,30,31] Additionally, we found that migratory behavior on these substrates could be modulated by altering the amounts of matrix coated on the culture dishes and migratory behavior could be correlated with the organizational patterns and dynamics of various cytoskeletal components including fodrin,[30] protein 4.1,[13] vinculin, and α-actinin.[23] BAEC cultured on substrates that elicit high to intermediate migratory rates, including collagen types I, III, IV and laminin, were also noted to reorganize the cortical cytoskeletal components fodrin and protein 4.1 following stimuli to migrate. In contrast, substrates which elicit low migratory rates, such as fibronectin, do not drive reorganization of these cytoskeletal components following stimuli to migrate.[13,30] These data are consistent with the concept that the existing and newly synthesized extracellular matrix modulates cellular behavior by affecting the organization and dynamics of the cytoskeleton presumably via cell surface matrix binding proteins.[20,21,23,25,30]

We have demonstrated that both integrin and nonintegrin matrix binding proteins participate in *in vitro* BAEC adhesion and migration events.[2,37] Immunoprecipitation and immunoblot analyses revealed that BAEC express at least two different integrin heterodimers, β1 and β3 class integrins. Attachment assays using RGD-containing peptides illustrated dose-dependent RGD sensitivity on fibrinogen, laminin, and fibronectin substrates. Immunofluorescence studies of BAEC integrins revealed rapid matrix protein-specific organization within one hour of plating. β_1 integrin molecules were observed to become arrayed in linear stress fiber-type patterns on fibronectin and laminin substrates. In contrast, β3 integrins were noted to organize into punctate patterns on a fibrinogen substratum at this early time point. The nonintegrin laminin binding protein LB69 was not observed to organize on any of these matrix components at this early time point (Fig. 1).[2] However, at later time points LB69 was noted to organize following plating (Fig. 2), suggesting a spatiotemporal segregation of integrin and nonintegrin binding proteins during adhesive events. Supporting the concept of a spatiotemporal segregation of integrin and nonintegrin binding proteins during adhesive events in BAEC are the findings that both antibodies directed against integrin and RGD-containing peptides inhibit BAEC attachment *and* spreading while antibodies directed against LB69 as well as YIGSR-containing peptides *only* affect BAEC spreading. In addition, selective disorganization of β1 integrins (but not LB69) is noted following RGD-containing peptide inhibition of BAEC sheet migration on a laminin substrate (Figs. 3 and 4). Recent studies have demonstrated that while the underlying matrix elicits specific patterns of integrin organization of BAEC, there are no changes in the sizes of cell surface pools of β1 or β3 integrins. However, in contrast, selected soluble factors (TGF-β1 and PDGF) were observed to modulate (increase) sizes of cell surface pools of BAEC β1 or β3 integrins and BASMC β3 integrins but did not alter integrin organization in these two cell types *in vitro*.[3] These data suggest a complex modulation of vascular cell behavior occurring, in part, through a coordination of soluble factor and extracellular matrix protein regulation of integrin expression and organization.[2,3,24]

In addition, we have found that in a rat carotid balloon deendothelialization model, in which there is incomplete reendothelialization of the deendothelialized area, there is increased fibronectin and TGF-β staining throughout the media and luminal surface of the chronically deendothelialized region of the vessel (Fig. 5).[24] These *in vivo* findings correlate well with *in vitro* studies in which BAEC migration on a type I collagen coating was *inhibited* and BASMC migration was *enhanced* by: (1) addition of soluble fibronectin to the culture media; (2) coatings of increasing fibronectin concentration; and (3) TGF-β1 treatment of migrating cells in which there is a significant increase in fibronectin mRNA and protein levels (Fig. 6).[25] Thus, both *in vivo* and *in vitro*, the composition of the existing and newly synthesized underlying extracellular matrix as well as the presence of soluble factors appear to have profound effects upon the migratory abilities of large vessel endothelial and medial smooth muscle cells. Further studies using this *in vivo* model have revealed distinct differences in the localization patterns of β1 or β3 integrins in the neointimal and luminal regions of both the chronically deendothelialized and reendothelialized areas.[3] In the chronically deendothelialized area, neointimal smooth muscle cells exhibited increased staining for β3 integrins but no changes in β1 integrins compared with normal and postinjury medial smooth muscle cells. This observation is consistent with the effects of TGF-β1 and PDGF on BASMC noted *in vitro*. In the reendothelialized area, the neointimal smooth muscle cells nearer the lumen display less intense staining for β3 integrins (as well as for fibrinogen) than those deeper in the neointima. These observations support the well-accepted concept of endothelial cell modulation of smooth muscle phenotype.

In addition to these matrix- and soluble factor-driven effects, BAEC have been found to be sensitive to a variety of soluble factors including those obtained from platelet releasates. In the past, several investigators have demonstrated effects on motility, proliferation, morphology, and surface protein expression when large vessel endothelial cells have been incubated with a variety of soluble factors including PDGF, TGF-β1, serotonin, histamine, and a variety of cytokines.[1,4,6,9,10,33,36] In our studies, the platelet factors histamine,

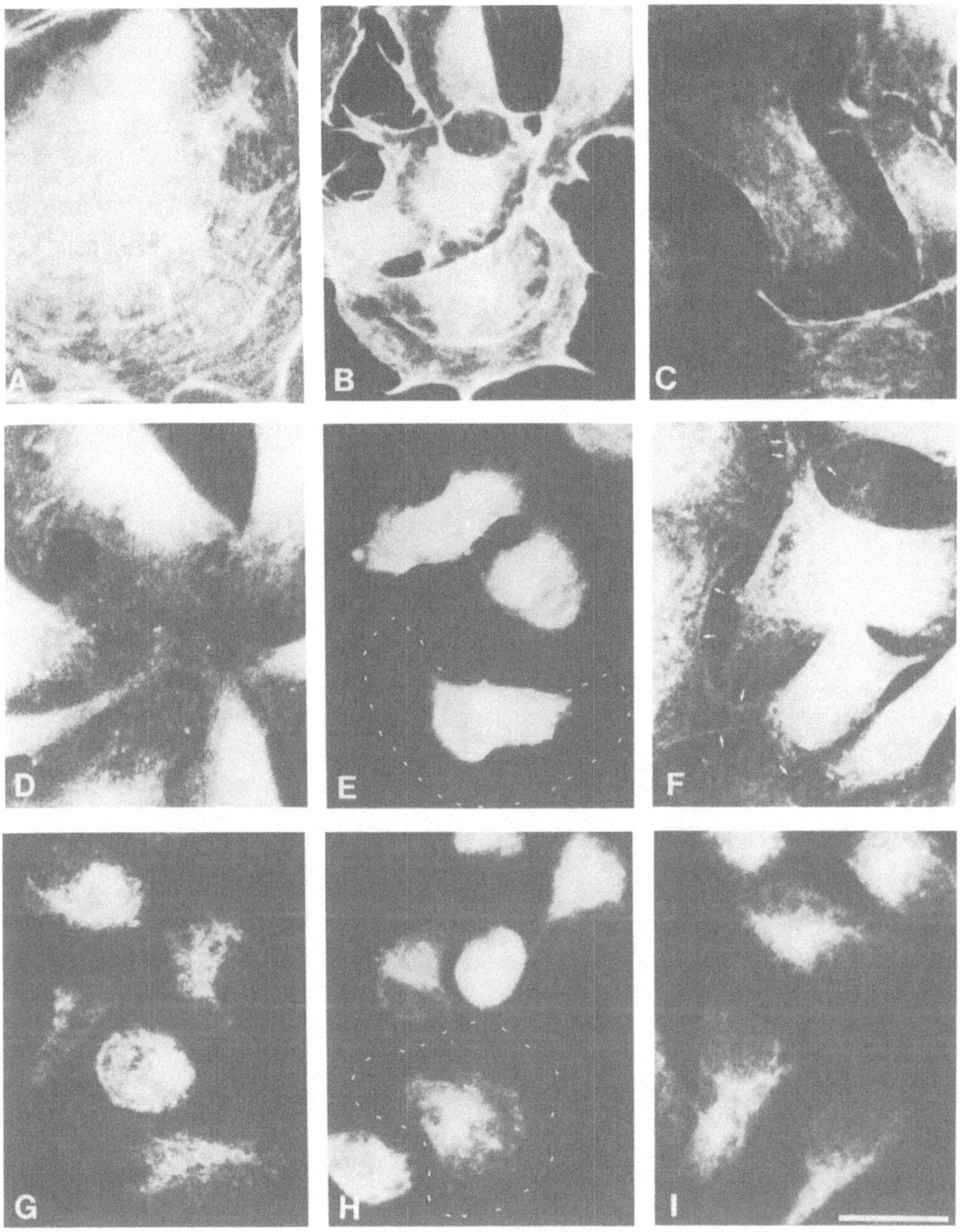

Figure 1. Immunofluorescence localization of matrix binding proteins in BAECs 1 hr after plating. BAECs were plated on fibronectin (A, D, G), laminin (B, E, H), or fibrinogen (C, F, I), fixed 1 hr later and stained with anti-β1/10P (A–C), anti-GPIIIa (D–F), or anti-LnR (G–I). Note the β1 integrin organization into concentric arcs when the cells are plated onto Fn (A) or Ln (B) and the β3 integrin localization to cell periphery (arrows) when BAECs are plated onto Fb (F). In all cases at this time point, LB69 (G–I) remained in a perinuclear distribution and did not exhibit cytoplasmic process localization even in BAECs plated on Ln (I). Dashed lines in E and H denote cell borders of these representative cells. Bar = 5 μm. Reproduced from Ref. 2 with permission.

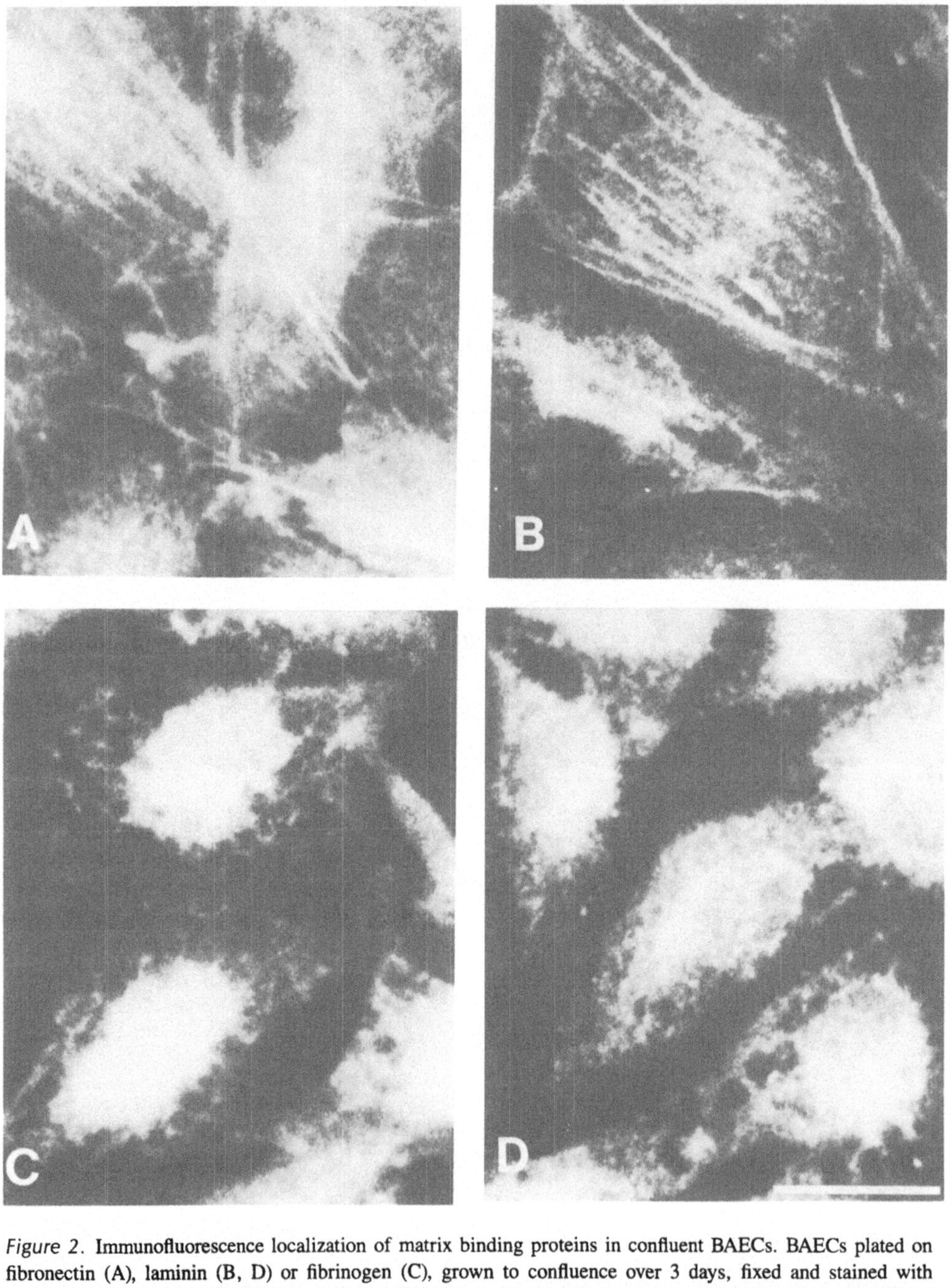

Figure 2. Immunofluorescence localization of matrix binding proteins in confluent BAECs. BAECs plated on fibronectin (A), laminin (B, D) or fibrinogen (C), grown to confluence over 3 days, fixed and stained with anti-β1/10P (A, B), anti-GPIIIa (C), or anti-LnR (D). β1 integrin organization in response to Fn (A) and Ln (B) and β3 integrin organization in response to Fb (C) are maintained. Note, however, the reorientation of β1 integrin fibrillar streaks (A, B). At this time point, LB69 now exhibits a diffuse punctate cytoplasmic pattern in BAECs plated on Ln (D). Bar = 5 μm. Reproduced from Ref. 2 with permission.

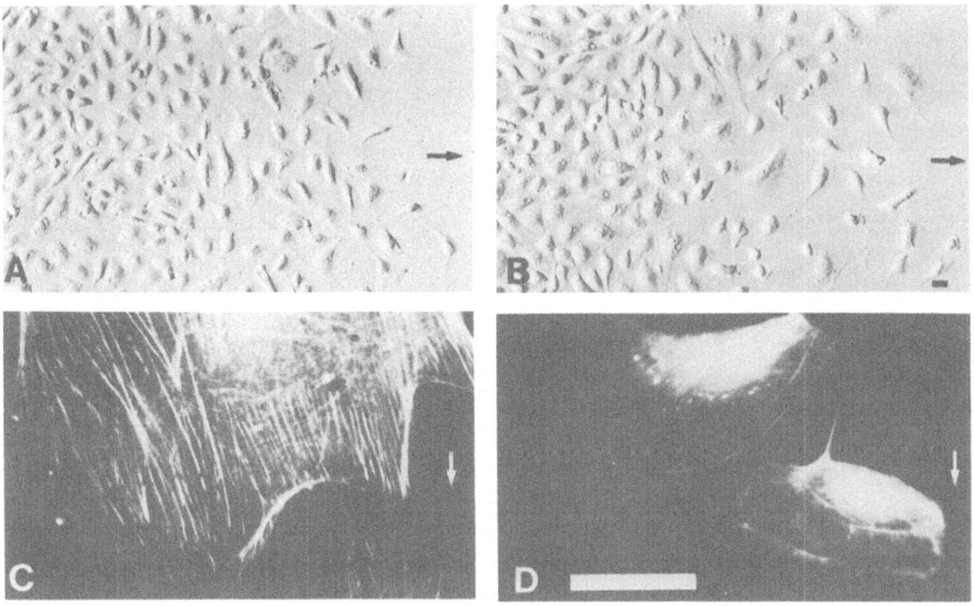

Figure 3. Hoffman Interference Microscopy and immunofluorescence localization of β1 integrin in BAECs migrating on laminin. BAECs were allowed to migrate over a laminin substrate in the absence (A, C) or presence (B, D) of YGRGDSP-BSA at a final peptide concentration of 5 μm. After 6 days, cultures were studied by Hoffman Interference Microscopy (A, B) and then fixed and stained with anti-β1/10P (C, D). Panels C and D show cells at the leading edge of the migrating front. Arrows indicate direction of BAEC migration. Note that the cells at the migrating front of RGD-treated cultures (B, D) are smaller and more densely packed than untreated cells (A, C). Furthermore, treatment with RGD causes a dramatic disorganization (D) of β1 integrins in these cells when compared with the prominent β1 integrin fibrillar streaks of untreated migrating cells (C). Bar = 25 μm. Reproduced from Ref. 2 with permission.

serotonin, norepinephrine, PDGF, and TGF-β1 were all found to decrease BAEC sheet migration while having markedly different effects on BAEC proliferation. In addition, inhibition of proliferation with mitomycin C treatment did not alter the effects of these added factors, supporting the concept that the processes of migration and proliferation are not linked.[4] It is important to recognize that platelets are not the sole cellular source for any of these factors and it is conceivable that *in vivo* the physiologic sources for these factors may be nonplatelet mesenchymal or circulating cells.

Ongoing studies investigating possible mechanisms by which extracellular matrix components and TGF-β1 mediate changes in BAEC migratory behavior have revealed that both matrix composition *and* soluble factors modulate levels of BAEC plasminogen activator and plasminogen activator inhibitor-1, and inhibitors of serine proteases decrease migration rates. These data suggest that changes in this protease–protease inhibitor system may also play a role in modulating migratory behavior. Specifically, plasminogen activator activity levels are *decreased* by TGF-β1 treatment. In contrast, plasminogen activator inhibitor-1 levels are *increased* by treatment with TGF-β1 (L. Bell and J. A. Madri, in preparation).

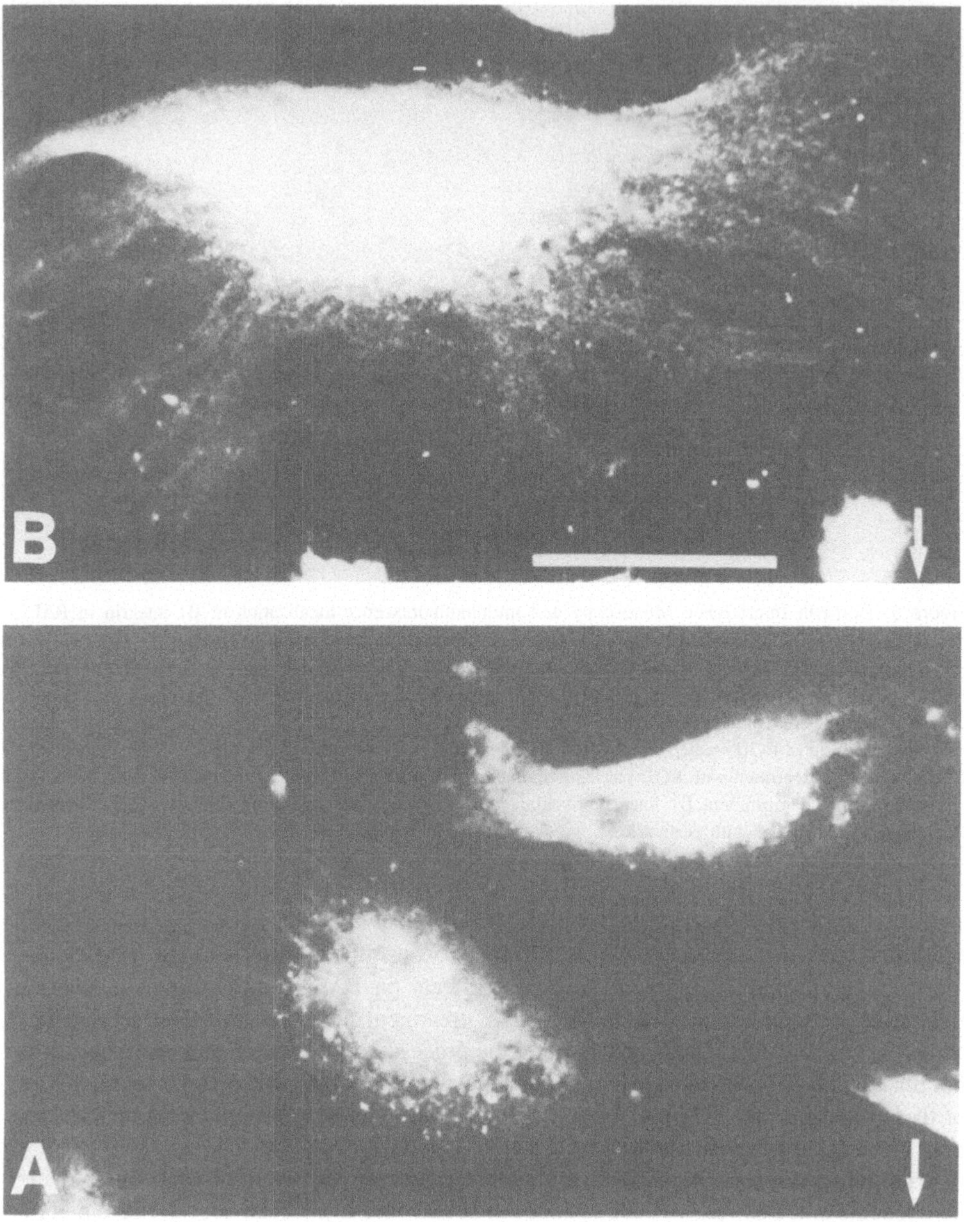

Figure 4. Immunofluorescence localization of β3 integrins and LB69 in BAECs migrating on laminin. Migrating BAEC cultures on laminin were established as in Fig. 3 in the presence or absence of YGRGDSP-BSA and processed for immunofluorescence staining with anti-GPIIIa (A) or anti-LnR (B). No change in staining pattern was seen upon addition of RGD peptide, and cells from the leading edge of untreated migrations are shown here. Arrows indicate direction of migration. The perinuclear localization of β3 integrins (A) is consistent with inability to bind laminin. Most notable is the lacy linear cytoplasmic distribution of nonintegrin LB69 in migrating BAECs (B). Bar = μm. Reproduced from Ref. 2 with permission.

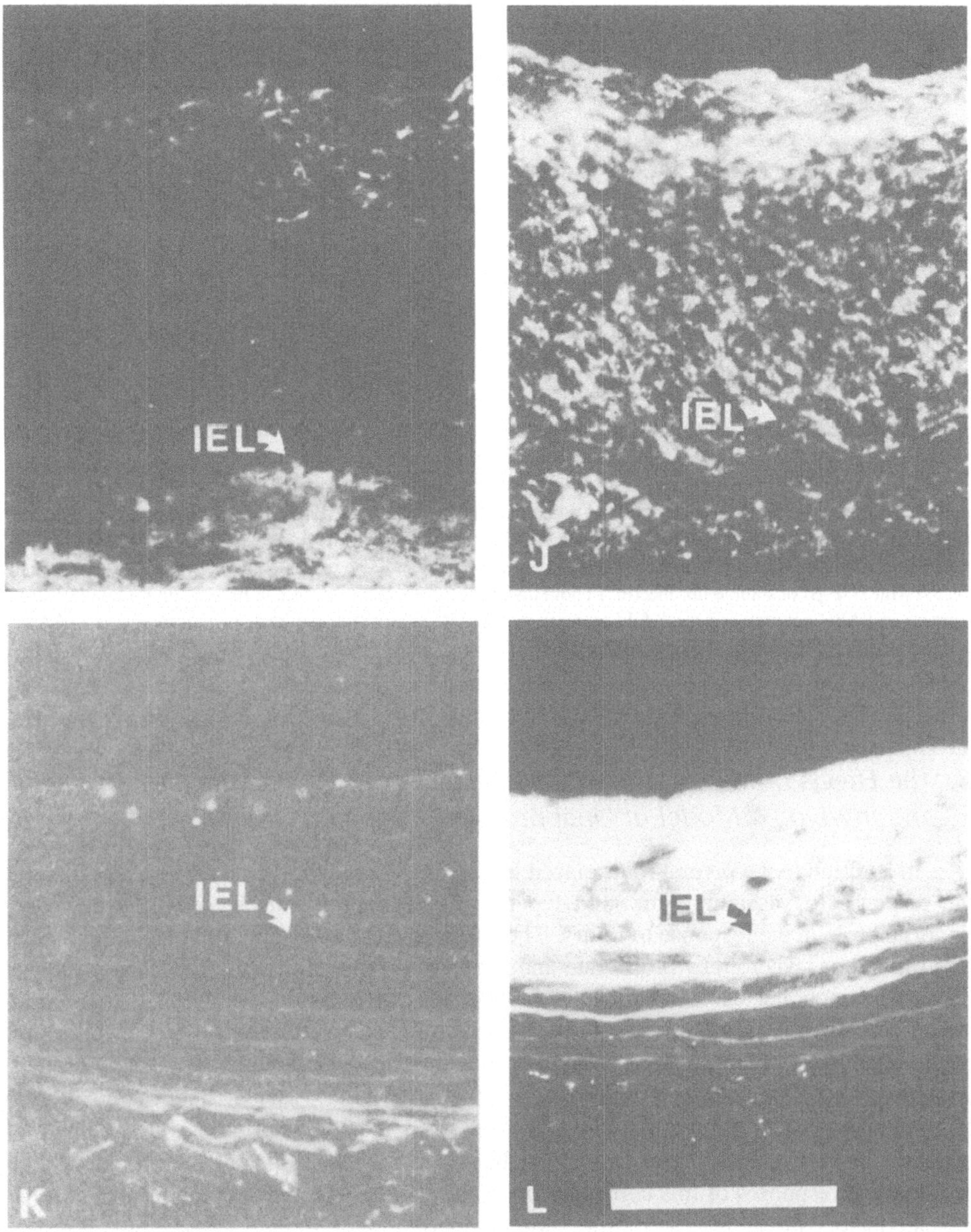

Figure 5. Immunofluorescence localization of fibronectin and TGF-β1 in rat balloon deendothelialized carotid artery. Immunofluorescence labeling of cross sections of a balloon deendothelialized carotid artery 10 weeks after injury at the reendothelialized zone (I, K) and the deendothelialized (luminal smooth muscle cell) zone (J, L) using antibodies against fibronectin (I, J) and TGF-β1 (C, D) at 10 weeks after balloon deendothelialization. IEL = internal elastic lamina. Bar = 100 μm. Reproduced from Ref. 26 with permission.

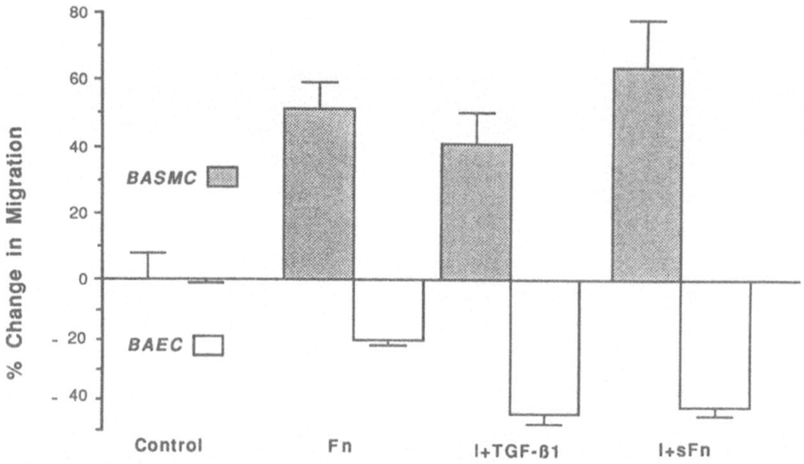

Figure 6. Relative changes in migration rates of bovine aortic endothelial cells (BAEC) and bovine aortic smooth muscle cells (BASMC) cultured on type I collagen (I) and treated with media alone (control), TGF-β1 (0.5 ng/ml/48 hr), or soluble fibronectin (sFn) (10 μg/48 hr) or cultured on fibronectin (Fn) over a 6-day period. BAEC exhibited a lower migration rate when cultured on fibronectin as compared to type I collagen and also exhibited low migration rates when treated with TGF-β1 or soluble fibronectin. In contrast, BASMC exhibited higher migration rates when grown on fibronectin as compared to type I collagen and also exhibited higher migration rates when treated with TGF-β1 or soluble fibronectin. In both instances there was an increase in fibronectin protein (2.2-fold increase in BAEC cultures and a 1.5-fold increase in BASMC cultures) and mRNA (4-fold increases in BAEC cultures and BASMC cultures) in response to TGF-β1.

B. The Effects of Matrix and Soluble Factors on BASMC Migration in Vitro and in Vivo: A Model of "Inappropriate" Repair

In addition to the pivotal roles played by the overlying endothelial cells in modulating the response to vascular injury and repair, medial smooth muscle cells also have an important role in the response to injury. Their response is one of migration from the media into the intima, where they proliferate and synthesize extracellular matrix. This response (driven by mononuclear cells, platelet factors, loss of overlying endothelial cells, and autocrine/paracrine factors produced by the smooth muscle cells themselves as well as by dysfunctional endothelium at or near the site of injury) is thought to lead to intimal thickening and ultimately atherosclerosis.[28,33] It is known that vascular smooth muscle cell migration and proliferation are modulated, in part, by the composition of the surrounding extracellular matrix and platelet factors (such as PDGF and TGF-β) and the migratory and proliferative responses of this cell type contribute to atherogenesis.[4,34] In our studies, the platelet factors histamine, serotonin, norepinephrine, PDGF, and TGF-β1 all increase bovine and rat aortic medial smooth muscle cell migration, while having markedly different effects on BASMC proliferation. In addition, as noted in experiments using BAEC, inhibition of proliferation with mitomycin C treatment did not alter the effects of these added factors, again supporting the concept that the processes of migration and proliferation are not linked.[4] Additional studies have revealed that in addition to increasing BASMC migratory rate, TGF-β1 treatment increases fibronectin mRNA and protein levels in confluent and migrating BASMC cultures and that TGF-β1 and PDGF treatment of cultured BASMC elicits an increase in the pool size of β3 integrins, supporting the concept that TGF-β1 may be eliciting its effects on migration, in part, via the selective modulation of matrix component(s) and cell surface matrix receptors including the integrins (Fig. 6).[3,25]

Thus, it appears that selected matrix components (fibronectin) and soluble factors (TGF-β1) elicit different effects in aortic endothelial and smooth muscle cells, enhancing migration in one case (BASMC) while inhibiting migration in the other (BAEC), which may, in part, lead to the development of atherosclerotic lesions. In addition, TGF-β1 appears to be having its effects on these vascular cells, in part, by modulating the matrix synthetic profile of the cells, eliciting increases in fibronectin mRNA and protein levels in both aortic endothelial and smooth muscle cells as well as by selectively modulating the sizes of surface pools of β1 and β3 integrins in these cell types. Furthermore, as noted for BAEC, both matrix composition *and* soluble factors (TGF-β1) modulate plasminogen activator and plasminogen activator inhibitor-1 levels in BASMC. In contrast to BAEC, BASMC plasminogen activator levels are *decreased* in cells plated and migrating on a fibronectin substrate compared to type I collagen and *increased* in response to TGF-β1. Conversely, BASMC levels of plasminogen activator inhibitor-1 are *increased* following TGF-β1 treatment (L. Bell and J. A. Madri, in preparation). These data, taken together with our observations in BAEC cultures, are consistent with the concept that modulation of vascular cell migratory behavior is complex and involves several mechanisms including protease–protease inhibitor systems, changes in selected matrix component synthesis and organization, changes in expression and organization of cell surface matrix binding proteins (integrins and nonintegrins), and cytoskeletal reorganization.[2,3,18,22–25]

C. Matrix Composition, Organization, and Transforming Growth Factors Beta: Modulators of Microvascular Endothelial Cell Differentiation in Vitro

In contrast to large vessel endothelial cells which undergo sheet migration in response to denudation injury, microvascular endothelial cells respond to injury by freeing themselves from the constraints of their investing basement membranes, migrating into and proliferating in a three-dimensional interstitial stroma, forming new microvessels, stabilizing them, and ultimately dismantling the newly formed microvascular bed following the conclusion of the repair response.[17]

In studying the complex process of angiogenesis and microvascular endothelial cell differentiation we have employed several *in vitro* culture systems,[16,18,22,29] including the culture of microvascular endothelial cells from the rat epididymal fat pad and bovine calf adrenal cortex *on* selected extracellular matrix components in two-dimensional culture, *on* the interstitial and basement membrane aspects of the amnion, and *in* three-dimensional type I collagen gels. Under all conditions studied these microvascular endothelial cells retain von Willebrand factor positivity. When microvascular endothelial cells are placed in culture, they are removed from a three-dimensional environment where they have a significant arc of curvature and intimate associations with specific matrix basement membrane components and pericytes and placed in a two-dimensional environment in which they have no arc of curvature, dramatically different associations with the substratum and no contacts or associations with pericytes. This change in environment may contribute, in part, to loss of differentiated phenotype in culture. Although still in a two-dimensional culture environment, placing these cells on basement membrane components or preexisting basement membranes (amnion) elicits a return of a more differentiated phenotype, namely, tube formation with tight junctions and luminal and abluminal plasma membrane specializations.[12,16–18,21–23,26,29,31] When placed on the interstitial aspect of the amnion, these cells exhibit a high proliferative rate, migrate *into* the three-dimensional stroma, and ultimately form tubelike structures having junctional complexes and abluminal basal lamina formation.[18,31] In contrast, when placed on interstitial collagen types I or III, these cells, while still retaining von Willebrand factor staining, lose their tight junctions and ZO1 staining and begin to express α-smooth muscle actin mRNA and protein, consistent with a modulation of phenotype driven, in part, by the composition and organization of the extracellular matrix.[12,18,23,25,26,31] Furthermore, when microvascular endothelial cells cultured on coatings of interstitial collagens are treated with TGF-β1 there is an eightfold induction of α-smooth muscle actin mRNA, further supporting the concept that under certain conditions microvascular endothelial cells can be induced to exhibit smooth muscle/pericyte phenotypic features.[12,23,25]

When grown on selected matrix components (IV, Ln, and Fn) in two-dimensional culture, microvascular endothelial cells (RFCs) exhibited dramatic changes in patterns of protein 4.1, vinculin, and actin staining in the absence and presence of TGF-β1. Cells cultured on type IV collagen and Ln substrates assumed spindle morphologies and the cytoskeletal components protein 4.1 and vinculin exhibited diffuse, punctate staining patterns, while cells cultured on a Fn substratum assumed a large, flattened polygonal morphology with linear, lacy staining of protein 4.1 and intense, short, interrupted linear staining patterns of vinculin.[25] In the presence of picogram amounts of TGF-β1, the microvascular endothelial cells assumed a large, spread, polygonal shape and exhibited a linear protein 4.1 staining pattern and an intense interrupted linear vinculin staining pattern on *all three* matrix components. In addition, Western blots of lysates from control and TGF-β1-treated cultures revealed changes in the intensity of several protein 4.1 isoforms. Namely, with the addition of TGF-β1, there was an increase in the staining intensity of the 80-kDa band, no change in the 120-kDa band, and the appearance of a 160-kDa band.[25] In double-labeling experiments, which examined the colocalization of vinculin and actin filaments in microvascular endothelial cells grown on Ln in the absence and presence of TGF-β1, vinculin organization was noted to change from a diffuse granular pattern containing occasional intense, short, interrupted punctate localizations to a pattern of predominantly intense, short, interrupted punctate staining at the periphery of the cells in the presence of TGF-β1. Actin filament staining changed from thin longitudinal filamentous staining along the long axis of the cells with occasional colocalization with vinculin at apparent terminations of filaments to thick, intense filamentous staining with extensive colocalization with vinculin and terminations of the thick filaments in the presence of TGF-β1. These studies suggest that TGF-β1 may be affecting changes in cell behavior (as demonstrated by degree of spreading and dramatic inhibition of proliferative rates) by directly modulating cortical and filamentous cytoskeletal elements or by indirectly modulating them by altering matrix and matrix receptor composition and organization, which then, in turn, would elicit changes in cytoskeletal organization.

When microvascular endothelial cells are dispersed and cultured *in* a three-dimensional type I collagen gel, their behavior, compared to that observed in two-dimensional culture systems, is quite different. In the presence of TGF-β1, TGF-β2, or bFGF, the cells organize into multicellular tubelike aggregates with lumina in a calcium-dependent process (Fig. 7), form tight junctions between cell processes, and deposit an organized basal lamina abluminally (Fig. 8).[27] Additionally, the cells no longer express α-smooth muscle actin mRNA or detectable protein and do not express β chain of the PDGF receptor or exhibit any proliferative response to any of the PDGF isoforms (see below), suggesting that the organization (architecture) of the surrounding matrix drives the cells to a differentiated endothelial phenotype having markedly different responses (and cell surface receptor repertoire) to cytokines produced during inflammatory, repair, and angiogenic processes (Fig. 8).

Recently, when we reexamined microvascular endothelial cells grown *on* collagen types I and IV in two-dimensional culture and *in* a three-dimensional type I collagen gel using an antibody probe directed against the β chain of the PDGF receptor antibody in Western blot analyses, we found that microvascular endothelial cells cultured in the two-dimensional environment expressed the β chain of the PDGF receptor while cells cultured in the three-dimensional environment did not. Large vessel smooth muscle cells expressed this PDGF receptor chain under the culture conditions used. In addition to finding this inducible receptor on the cell surface, we demonstrated that microvascular endothelial cells were responsive to selected isoforms of PDGF when cultured in two-dimensional but not in three-dimensional culture. Namely, PDGF-BB elicited a dramatic increase in proliferation while PDGF-AB caused a modest decrease in proliferative rate and PDGF-AA had no effect on proliferation. No PDGF isoform had any effect on proliferation when tested on three-dimensional cultures (R. Perlmutter and J. A. Madri, in preparation). These findings, together with our previous observations documenting the inducible presence of α-smooth muscle actin mRNA and protein in microvascular endothelial cells in two-but not three-dimensional cultures, lend further support to the concept that microvascular endothelial cells may have the capacity to give rise to pericytes during the angiogenic process in response to particular stimuli (Fig. 9).[12,23,25]

D. TGF-β Isoforms as Modulators of Vascular Cell Behavior

Transforming growth factors beta are thought to play important regulatory roles in many biological processes including organ and tissue development, wound healing, and arteriosclerosis. TGF-β1 is one member of a family of polypeptides including the heterodimer TGF-β1.2, the homodimer TGF-β2, and the more recently discovered isoforms TGF-β3, TGF-β4, and TGF-β5.[25] TGF-β1 and TGF-β2 are the most intensely studied members of this family and have been found to have similar, if not identical, activities in many mesenchymal cell types.[35] However, recent studies suggest that they elicit different responses in selected cell types.[32] Previous studies on large vessel endothelial cells revealed that TGF-β2 had no effect on proliferation.[11] In light of this finding and the observation that TGF-β1 and TGF-β2 are present in differing amounts in a variety of cell types,[35] it is likely that these TGF-β isoforms may prove to be important in modulating the behavior of vascular cells in the processes of wound healing, angiogenesis, and atherosclerosis.[27] In recent studies we have confirmed the observation that TGF-β2 has no effect on BAEC proliferation and demonstrated that TGF-β2 does not compete at all with TGF-β1 in functional assays (proliferation) and only modestly in migration assays or for receptor binding using

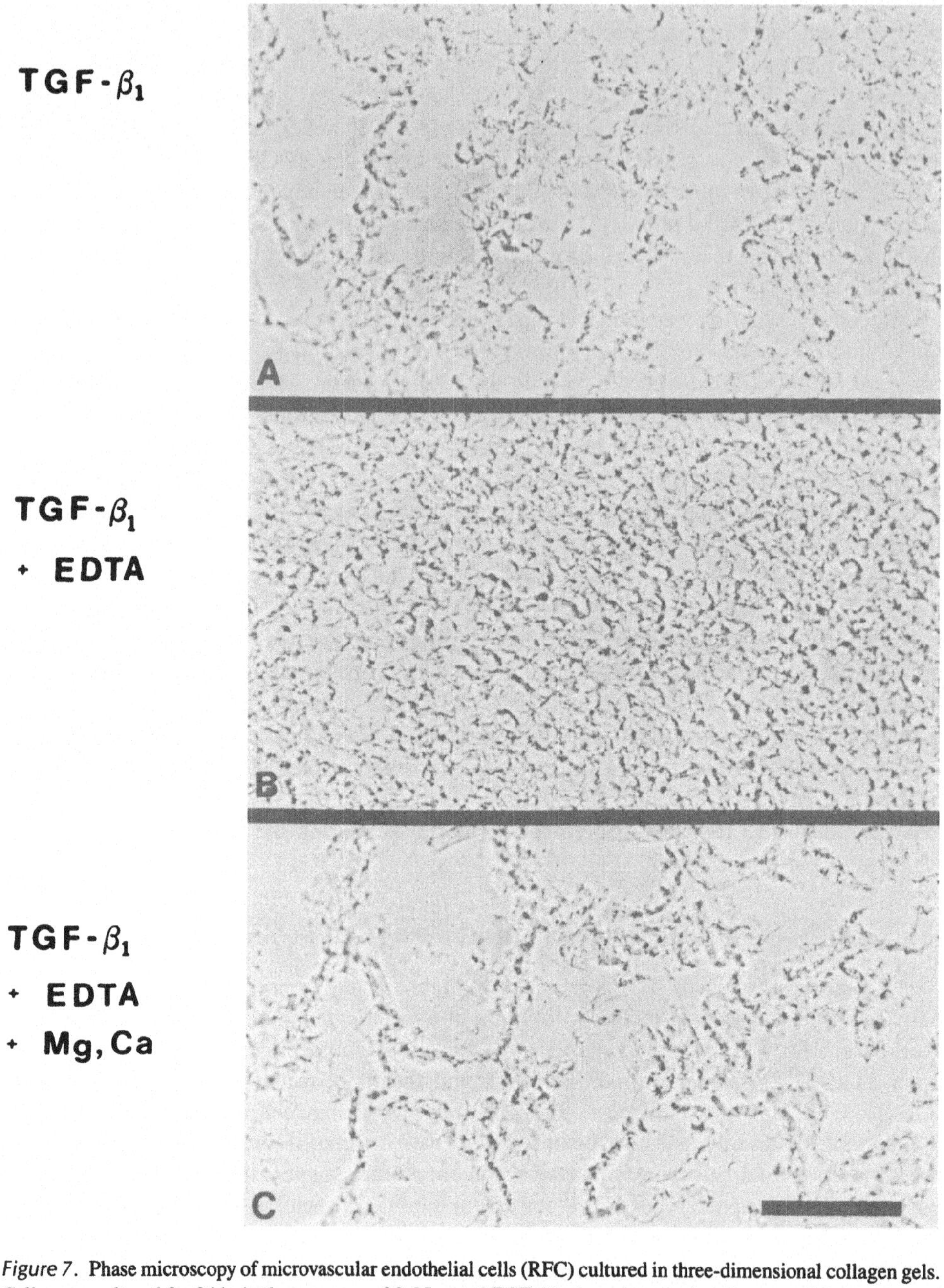

Figure 7. Phase microscopy of microvascular endothelial cells (RFC) cultured in three-dimensional collagen gels. Cells were cultured for 24 hr in the presence of 0.05 ng/ml TGF-β1, the gels were then frozen and 8-μm sections cut and analyzed using phase microscopy. In the presence of TGF-β1, multicellular aggregates forming tubelike structures were noted throughout the gel (A). In the presence of TGF-β1 and 5 mM EDTA, no tube formation was noted and the cells remained as isolated entities (B). When cultures were incubated with 0.05 ng/ml TGF-β1 and 5 mM EDTA for 24 hr and then 6 mM $MgCl_2$ and 6 mM $CaCl_2$ added and the cultures incubated for an additional 24 hr, tubelike structures were observed throughout the gels (C). Bar = 15 μm.

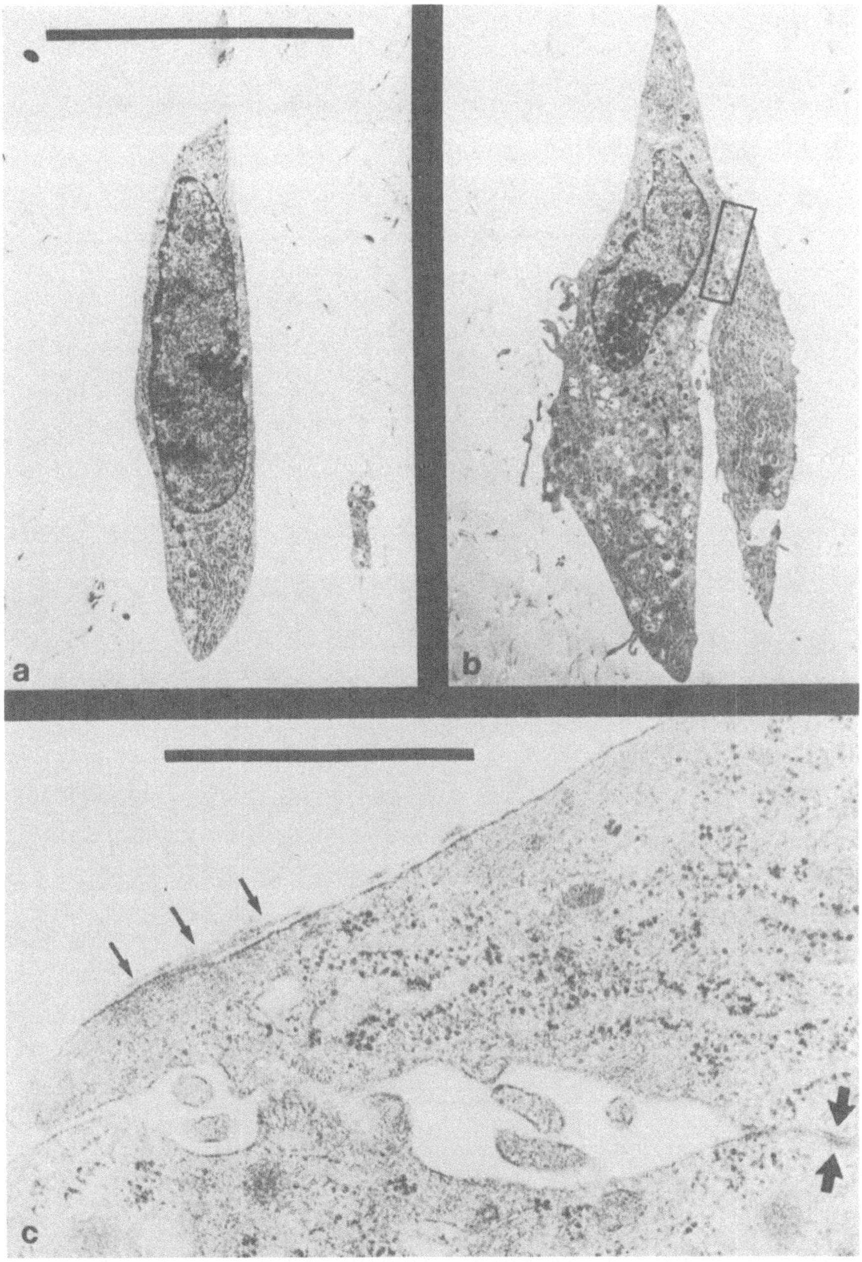

Figure 8. Electron microscopic analyses of microvascular endothelial cells cultured for 4 days in three-dimensional collagen gels. Isolated and quiescent control cell (a); TGF-β1-treated three-cell aggregate with extensive pseudopod formation, increased vacuoles and endoplasmic reticulum (b); the boxed area in (b) is enlarged, illustrating lumenlike formation closed off by junctional complex (large arrows) and deposition of basal lamina-like material abluminally (small arrows) (c); lumen formations sealed off by junctional complexes. Small arrows represent luminal plasmalemma membrane specializations (d, e). Bars = 10 μm in a and b; 1 μm in c and e; 4 μm in d. Reproduced from Ref. 28 with permission.

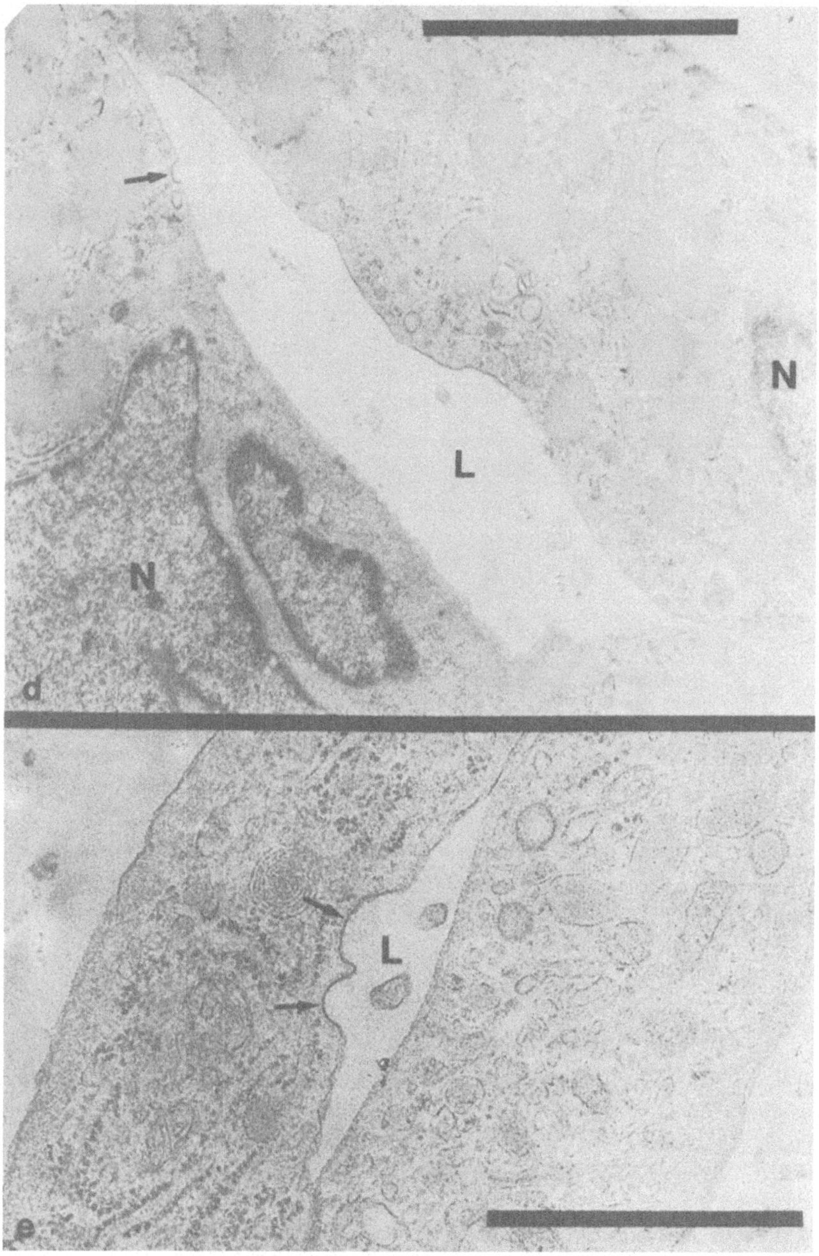

Figure 8. (*continued*)

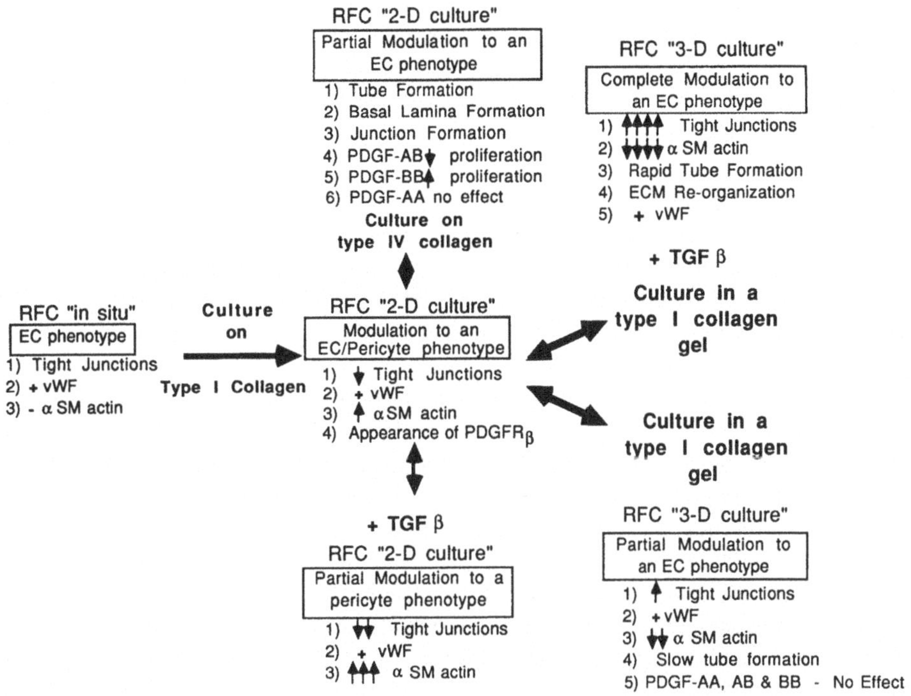

Figure 9. Schematic representation of the modulation of microvascular endothelial cell phenotype mediated by changes in extracellular matrix (ECM) composition, organization, and soluble factors (TGF-β1).

morphological assays (competitive binding with biotinylated TGF-β1 followed by avidin–gold localization), FACS analyses (competitive binding with biotinylated TGF-β1 followed by avidin-FITC fluorescence analyses), and radioligand binding assays (competitive binding with ^{125}I TGF-β1 followed by cross-linking. SDS-PAGE, and autoradiography).[27] In contrast, we have noted that microvascular endothelial cell proliferation in two-dimensional cultures is inhibited by TGF-β2 at all concentrations tested (0.05 to 5.0 ng/ml), but at levels lower than those observed with the same concentrations of TGF-β1. We have also observed that TGF-β2 enhances *in vitro* angiogenesis (tube formation) in three-dimensional cultures of microvascular endothelial cells, again only at concentrations tenfold higher than those observed for TGF-β1. Lastly, we have observed that the proliferation of cultured bovine aortic medial smooth muscle cells is inhibited by TGF-β2 to the same extent as TGF-β1, while smooth muscle cell migration was unaffected by TGF-β2 at all concentrations tested (0.05 to 5.0 ng/ml).[27] In addition, the ^{125}I TGF-β1 cross-linking analyses revealed differences in the type I to type II TGF-β receptor ratios in these three vascular cell types, namely BAEC 1:1, RFC 1.5:1, and BASMC 3:1. Whether these differences in receptor type ratio mediate the differences noted in the TGF-β1/TGF-β2 proliferative and migratory responses noted in these cell types or just correlate with these differences in cellular responses has yet to be determined. Nonetheless, these data suggest that vascular cells derived from diverse vascular beds exhibit individual differential sensitivities to various isoforms of TGF-β and this differential sensitivity and the broad range of effects elicited by them may have a significant role in modulating the regulation of inflammatory and healing responses observed throughout the vascular system.

IV. CONCLUSIONS

In this chapter we have presented evidence supporting the concept that *in vitro* and *in vivo* specific extracellular matrix proteins elicit *organization* of integrin and nonintegrin vascular cell surface matrix binding proteins and selected cytoskeletal elements while cytokines, such as PDGF and TGF-β1, modulate the size of surface membrane-associated vascular cell integrin *pools*, in addition to their effects on cytoskeletal organization and matrix synthesis and protease–protease inhibitor systems.

In an *in vivo* model of vascular endothelial and smooth muscle cell response to injury which results in incomplete reendothelialization and intimal thickening following balloon catheter denudation of rat carotid artery, the endothelial cells respond to the denudation injury initially by migration and proliferation. However, following an acute platelet response, plasma fibronectin deposition and synthesis by local vascular cells, secretion and activation of TGF-β1 by both endothelial and smooth muscle cells, endothelial cell migration/proliferation is inhibited, creating a chronically deendothelialized region of vessel wall. In addition, the deposition of plasma fibronectin (as well as other plasma components such as fibrinogen and fibrin split products), the development of a platelet releasate, and the synthesis, secretion, and activation of TGF-β1 by both endothelial and smooth muscle cells stimulate medial smooth muscle cells to migrate to the area of injury, proliferate, and synthesize, secrete, and deposit matrix components (such as fibronectin). These events further enhance smooth muscle cell migration into the injured area. Thus, the complex interactions of particular vascular cell populations with surrounding and newly synthesized matrix components and a variety of cytokines elicit expression of "dysfunctional" phenotypes in local endothelial and smooth muscle cell populations which favor the development of arteriosclerosis.

In contrast, the responses of microvascular endothelial cells to TGF-β1, TGF-β2, and the isoforms of PDGF depend, in part, upon the *organization* of the matrix. In three-dimensional culture these factors elicit an angiogenic response complete with tube formation. However, in two-dimensional culture these same cells express cell surface receptors and cytoskeletal components and have responses to these cytokines similar to those observed in vascular smooth muscle cells and pericytes.[23] Our *in vitro* data are consistent with the notion that the microvascular endothelial cell is capable of giving rise to pericytes and although our data suggest that the extracellular matrix may play an important role in this process *in vivo*, it is as yet undefined. Thus, like the large vessel endothelial and smooth muscle cells, the microvascular endothelial cell displays a "plastic" phenotype in response to injury, but rather than the "dysfunctional" phenotypes displayed by the large vessel cells, the microvascular endothelial cell displays a variety of phenotypes normally observed in the microvasculature.

Although still incomplete, our knowledge of how the extracellular matrix and soluble factors affect vascular cells is constantly growing. The mechanisms by which information is transduced across the plasma membrane are complex and most likely will involve a variety of matrix and soluble factor receptors, second messenger systems, and dynamic cytoskeletal organization.[23,25]

ACKNOWLEDGMENTS. Supported, in part, by USPHS Grants RO1-HL28373 and PO1-DK38979 (J.A.M.), a Swiss National Science Foundation Fellowship (O.K.), a Joshua Macy Pre-doctoral Fellowship (J.R.M.), MSTP award GM-07205 (C.T.B.), and NRSA Fellowship F32-HL-07442 (L.B.).

REFERENCES

1. Assoian, R. K., 1988, The role of growth factors in tissue repair IV: Type β transforming growth factor and stimulation of fibrosis, in: *The Molecular and Cellular Biology of Wound Repair* (R. F. Clark and P. Henson, eds.), Plenum Press, New York, pp. 273–280.
2. Basson, C. T., Knowles, W. J., Bell, L., Albelda, S. M., Castronovo, V., Liotta, L., and Madri, J. A., 1990, Spatiotemporal segregation of endothelial cell integrin and non-integrin extracellular matrix binding proteins during adhesion events, *J. Cell Biol.* **110:**789–802.
3. Basson, C. T., Asis, A., Reidy, M. A., and Madri, J. A., 1991, Extracellular matrix and soluble factors differentially modulate vascular cell integrins during bovine aortic cell adhesion in vitro and rat carotid atherogenesis in vivo, *Am. J. Pathol.* (under revision).
4. Bell, L., and Madri, J. A., 1989, Effect of platelet factors on migration of cultured bovine aortic endothelial and smooth muscle cells, *Circ. Res.* **65:**1057–1065.
5. Carley, W., Milici, A. J., and Madri, J. A., 1988, Extracellular matrix specificity for the differentiation of capillary endothelial cells, *Exp. Cell Res.* **178:**426–434.
6. Cotran, R. S., and Pober, J. S., 1988, Endothelial activation: Its role in inflammatory and immune reactions, in: *Endothelial Cell Biology in Health and Disease* (N. Simionescu and M. Simionescu, eds.), Plenum Press, New York, pp. 335–348.
7. Davis, B. H., Pratt, B. M., and Madri, J. A., 1987, Hepatic ito cell culture: Modulation of collagen phenotype and cellular retinol binding protein by retinol and extracellular collagen matrix, *J. Biol. Chem.* **262:**10280–10286.
8. Fishman, A. P. (ed.), 1982, *Endothelium*, The New York Academy of Sciences, New York.
9. Gimbrone, M. A., and Bevilacqua, M. P., 1988, Vascular endothelium: Functional modulation at the blood interface, in: *Endothelial Cell Biology in Health and Disease* (N. Simionescu and M. Simionescu, eds.), Plenum Press, New York, pp. 255–274.
10. Huang, J. S., Olsen, T. J., and Huang, S. H., 1988, The role of growth factors in tissue repair I: Platelet-derived growth factor, in: *The Molecular and Cellular Biology of Wound Repair* (R. F. Clark and P. Henson, eds.), Plenum Press, New York, pp. 243–252.
11. Jennings, J. C., Mohan, S., Linkhart, T. A., Widstrom, R., and Baylink, D. J., 1988, Comparison of the biological actions of TGF beta-1 and TGF beta-2: Differential activity in endothelial cells, *J. Cell. Physiol.* **137:**167–172.
12. Kocher, O., and Madri, J. A., 1989, Modulation of actin mRNAs in cultured vascular cells by matrix components and TGF-β1, *In Vitro* **25:**424–434.
13. Leto, T. L., Pratt, B. M., and Madri, J. A., 1986, Mechanisms of cytoskeletal regulation: Modulation of aortic endothelial cell protein band 4.1 by the extracellular matrix, *J. Cell. Physiol.* **127:**423–431.
14. Madri, J. A., 1982, The preparation of type V collagen, in: *The Immunochemistry of the Extracellular Matrix*, Volume I (H. Furthmayr, ed.), CRC Press, Boca Raton, Fl., pp. 75–90.
15. Madri, J. A., 1987, The extracellular matrix as a modulator of neovascularization, in: *Cardiovascular Disease: Molecular and Cellular Mechanisms, Prevention, Treatment* (L. Gallo, ed.), Plenum Press, New York, pp. 177–184.
16. Madri, J. A., and Pratt, B. M., 1986, Endothelial cell–matrix interactions: In vitro models of angiogenesis, *J. Histochem. Cytochem.* **34:**85–91.
17. Madri, J. A., and Pratt, B. M., 1988, Angiogenesis, in: *The Molecular and Cellular Biology of Wound Repair* (R. F. Clark and P. Henson, eds.), Plenum Press, New York, pp. 337–358.
18. Madri, J. A., and Williams, S. K., 1983, Capillary endothelial cell cultures: Phenotypic modulation by matrix components, *J. Cell Biol.* **97:**152–165.
19. Madri, J. A., Dreyer, B., Pitlick, F., and Furthmayr, H., 1980, The collagenous components of subendothelium: Correlation of structure and function, *Lab. Invest.* **43:**303–315.

20. Madri, J. A., Pratt, B. M., and Yannariello-Brown, J., 1988, Matrix-driven cell size changes modulate aortic endothelial cell proliferation and sheet migration, *Am. J. Pathol.* **132:**18–27.
21. Madri, J. A., Pratt, B. M., and Yannariello-Brown, J., 1988, Endothelial cell–extracellular matrix interactions: Matrix as a modulator of cell function, in: *Endothelial Cell Biology in Health and Disease* (N. Simionescu and M. Simionescu, eds.), Plenum Press, New York, pp. 167–190.
22. Madri, J. A., Pratt, B. M., and Tucker, A. M., 1988, Phenotypic modulation of endothelial cells by transforming growth factor-β depends upon the composition and organization of the extracellular matrix, *J. Cell Biol.* **106:**1375–1384.
23. Madri, J. A., Kocher, O., Merwin, J. R., Bell, L., and Yannariello-Brown, J., 1989, The interactions of vascular cells with solid phase (matrix) and soluble factors, *J. Cardiovasc. Pharmacol.* **14:**S70–S75.
24. Madri, J. A., Reidy, M. A., Kocher, O., and Bell, L., 1989, Endothelial cell behavior following denudation injury is modulated by TGF-β1 and fibronectin, *Lab. Invest.* **60:**755–765.
25. Madri, J. A., Kocher, O., Merwin, J. R., Bell, L., Tucker, A., and Basson, C. T., 1989, Interactions of vascular cells with transforming growth factors beta, *Ann. N.Y. Acad. Sci.* **593:**243–258.
26. Merwin, J. R., Anderson, J., Kocher, O., van Itallie, C., and Madri, J. A., 1990, Transforming growth factor β1 modulates extracellular matrix organization and cell–cell junctional complex formation during in vitro angiogenesis, *J. Cell. Physiol.* **142:**117–128.
27. Merwin, J. R., Newman, W., Beall, L. D., Tucker, A., and Madri, J. A., 1990, Vascular cells respond differentially to transforming growth factors-beta$_1$ and beta$_2$, *Am. J. Pathol.* **138:**37–51.
28. Munro, J. M., and Cotran, R. S., 1988, The pathogenesis of atherosclerosis: Atherogenesis and inflammation, *Lab. Invest.* **58:**249–253.
29. Nicosia, R. F., and Madri, J. A., 1987, The microvascular extracellular matrix: Developmental changes during angiogenesis in the aortic ring–plasma clot model, *Am. J. Pathol.* **128:**78–90.
30. Pratt, B. M., Harris, A. S., Morrow, J. S., and Madri, J. A., 1984, Mechanisms of cytoskeletal regulation: Modulation of aortic endothelial cell spectrin by the extracellular matrix, *Am. J. Pathol.* **117:**337–342.
31. Pratt, B. M., Form, D., and Madri, J. A., 1985, Endothelial cell–extracellular matrix interactions, *Ann. N.Y. Acad. Sci.* **460:**274–288.
32. Rizzino, A., Kazakoff, P., Ruff, E., Kuszynski, C., and Nebelsick, J., 1988, Regulatory effects of cell density on the binding of transforming growth factor beta, epidermal growth factor, platelet derived growth factor, and fibroblast growth factor, *Cancer Res.* **48:**4266–4271.
33. Ross, R., 1986, Medical progress: The pathogenesis of atherosclerosis—An update, *N. Engl. J. Med.* **314:**488–500.
34. Ross, R., 1988, Endothelial injury and atherosclerosis, in: *Endothelial Cell Biology in Health and Disease* (N. Simionescu and M. Simionescu, eds.), Plenum Press, New York, pp. 371–384.
35. Sporn, M. B., and Roberts, A. B., 1990, The transforming growth factors-beta: Past, present and future, *Ann. N.Y. Acad. Sci.* **593:**1–6.
36. Terkeltaub, R. A., and Ginsberg, M. H., 1988, Platelets and response to injury, in: *The Molecular and Cellular Biology of Wound Repair* (R. F. Clark and P. Henson, eds.), Plenum Press, New York, pp. 35–56.
37. Yannariello-Brown, J., Wewer, U., Liotta, L., and Madri, J. A., 1988, Distribution of a 69kD laminin binding protein in aortic and microvascular endothelial cells: Modulation during cell attachment, spreading and migration, *J. Cell Biol.* **106:**1773–1786.

3

Endothelial Cell Regrowth

Michael A. Reidy and Volkhard Lindner

I. INTRODUCTION

Endothelial cells, either *in vivo* or *in vitro*, grow as a monolayer and despite their key location in blood vessels and the important functions which they carry out, we know relatively little about the mechanisms which control their growth *in vivo*.[1–3] This is perhaps because most endothelial cells, at least *in vitro*, were found to grow readily and little attention was given to understanding what factors might influence this process. The widespread use of vascular grafting and the discovery that reendothelialization is often severely limited,[4–8] however, clearly demonstrate that endothelial growth and maintenance of an intact endothelial monolayer are not simple automatic processes. In this chapter we will discuss those experiments which have focused on understanding what factors are important in controlling endothelial cell replication and whether these factors can influence endothelial cell growth in the denuded blood vessel.

II. FACTORS THOUGHT TO INFLUENCE ENDOTHELIAL CELL GROWTH

Over the past several years there have been reports of both inhibitors and promoters of endothelial cell growth and it is not the intention of this brief chapter to review all these studies. Instead, we will concentrate on a few factors which by their nature are thought to be present in an injured vessel wall and so could be responsible for inhibiting endothelial cell regrowth.

It is now well recognized that the surface on which a cell grows can strongly influence its behavior. For example, several years ago Moscona and Folkman[9,10] showed that *in vitro* endothelial cell shape was strongly influenced by the chemical composition of the culture dish. Several investigators have continued with this approach and have asked how the various components of the extracellular matrix known to be present in the arterial wall can

Michael A. Reidy and Volkhard Lindner • Department of Pathology, University of Washington, Seattle, Washington 98195.

Endothelial Cell Dysfunctions, edited by Nicolae Simionescu and Maya Simionescu. Plenum Press, New York, 1992.

influence endothelial cell migration and replication.[11,12] Studies by Madri and co-workers have shown that endothelial cells from large arteries grow rapidly on laminin, collagen type I, III, and IV coated surfaces but a significant reduction in their growth was observed when the dishes were coated with fibronectin.[13] The significance of this finding to the *in vivo* state was unclear since fibronectin is known to be a major component of the subendothelial matrix. A recent study, however, has shown that after deendothelialization with a balloon catheter, intimal cells synthesize TGF-β.[14,15] TGF-β can cause an approximate threefold increase in fibronectin synthesis by endothelium.[16,17] This is of particular interest since TGF-β would be released from platelets at sites of endothelial cell injury as well as synthesized by the smooth muscle cells of an injured artery[14] and furthermore, TGF-β itself is *in vitro* an inhibitor of endothelial cell proliferation.[18–20] Thus, TGF-β might not only act directly on the endothelial cells but might also induce synthesis of a fibronectin which inhibits endothelial cell growth.

TGF-β has been shown by several investigators to be an inhibitor of endothelial cell growth in culture.[18–20] Both migration and proliferation of endothelium are influenced by TGF-β although their effects are not necessarily linked.[21] TGF-β has also been shown to inhibit the stimulatory effect of FGF on endothelium. Interestingly, one report has shown that the effects of TGF-β on endothelial cells are only transient and that within 48 hr after addition, no inhibition of endothelial cell growth could be observed.[19] Recently, we have asked if administration of TGF-β *in vivo* would influence endothelial cell growth immediately after a denuding injury. Using continuous infusion of TGF-β at doses which have been shown to be effective on smooth muscle cells, we detected no change in the replication rate of endothelium. In part, this finding agrees with anecdotal data from *in vivo* studies where the endothelium exhibits high replication rates immediately after denudation yet platelet interaction with the exposed subendothelium and therefore presumably release of TGF-β is most pronounced. One possibility is that TGF-β released from platelets is inactive[22,23] and some form of cell–cell association is required to activate this molecule. In arteries of experimental animals subjected to denuding injuries, smooth muscle cells migrate into the intima after several days and so move into close apposition to the endothelium.[24,25] As suggested by Antonelli-Orlidge *et al.*[26] this cellular contact will activate TGF-β which then inhibits endothelial cell growth.[27] This time frame would match the *in vivo* data showing that endothelial cell replication indeed starts to slow within the first 2 weeks after injury when smooth muscle cells are well established in the intima and so are in potential contact with the endothelium. Much more information, however, is required as to when and how activated TGF-β is formed before any firm conclusion can be made as to the role of this ubiquitous peptide with respect to endothelial cell growth.

Platelets have also been thought to play a role in controlling endothelial cell growth independently of TGF-β. There have been several reports of low-molecular-weight inhibitors released by platelets[28,29] and platelets have also been shown to possess an endothelial growth factor.[30] Several molecules have been partially isolated and characterized by different laboratories[31–33] but only one has been positively identified.[34] Miyazono *et al.*[35] have cloned a platelet-derived endothelial cell growth factor (PD-ECGF) and an antibody raised against PD-ECGF completely neutralized the endothelial cell mitogenic activity of platelet lysate,[36] thus indicating that PD-ECGF is perhaps the only mitogen in platelets. It therefore appears that platelets contain both one well-recognized endothelial cell inhibitor, namely TGF-β, and also an endothelial cell mitogen, PD-ECGF. Recent data have shown that several cells synthesize PD-ECGF.[35] Usuki and co-workers[37] have shown that certain human carcinoma cell lines as well as human fibroblasts not only express the mRNA for this mitogen but also synthesize the active protein. These workers have also shown that PD-ECGF lacks a hydrophobic signal sequence as do other endothelial cell mitogens, namely acidic[38] and basic FGF,[39] and therefore the mechanisms of secretion of PD-ECGF are not understood. One obvious possibility is that it is released from dying cells.[40–42] Unlike FGF, PD-ECGF is not thought to be made by endothelial cells.

Yet another platelet product, namely platelet-derived growth factor (PDGF), has also been suggested to influence endothelial cells.[43] It has been known for several years that endothelial cells synthesize and secrete PDGF and yet these cells neither bind nor respond to this mitogen.[44] This finding strongly suggested that PDGF synthesis by endothelium might act in a paracrine manner to stimulate the neighboring smooth muscle cells. Some doubt was cast upon this concept when Hermansson *et al.*[45] found that the proliferating microvascular endothelial cells of a human glioblastoma expressed the mRNA for both PDGF B chain and the PDGF receptor. Yet another recent report has shown that cultured microvascular endothelium bound PDGF and subsequently was stimulated to replicate,[43] while endothelial cells derived from large arteries did not bind or respond to PDGF. Bell and Madri,[21] however, demonstrated that PDGF decreases the migration rate and proliferation rate of aortic endothelial cells at concentrations which were shown to stimulate smooth muscle cells. Thus, no clear conclusion can be reached as to the effect of PDGF on vascular endothelium and one approach to resolve this issue will be to understand the expression of the PDGF receptors by these various endothelial cells.

In late 1989, a series of papers provided evidence of yet another endothelial cell mitogen which was referred to both as vascular permeability factor (VPF) and vascular endothelial growth factor (VEGF).[46–52] VPF was first identified as a secreted product of several human and rodent tumor cell lines which promoted fluid and protein extravasation.[53] Recent purification and cloning of this protein[48] led to experiments which showed that VPF would also stimulate the growth of endothelial cells *in vitro* and that this mitogenic effect was specific for endothelium. Two other groups isolated and characterized a similar protein from pituitary follicular stellate cells[49,51] which bound to heparin–Sepharose and was secreted with a molecular mass of 45,000 and named it vascular endothelial growth factor (VEGF). This factor was soon shown to be identical to VPF since the cDNA for both VPF and VEGF was found to code for the same polypeptide which was similar in structure to the B chain of PDGF.[48] In fact, the overall homology with PDGF-B was 18% and all eight of the cysteines in PDGF-B were found to be conserved in VPF which is an indication that the folding of these proteins are similar. There is little known about the biology of VPF/VEGF except that it is mitogenic for large vessel endothelium and angiogenic in a variety of systems[52] and thus far has not been found to be active on other cells. One important point is that this protein can be synthesized and secreted by a variety of transformed cells including a monocyte-like cell[50] and has been found to be active only on endothelium.

Perhaps the best known endothelial cell mitogen is FGF.[54–56] There are now at least six[57–62] members of the FGF family but to date most studies have focused on the effects of acidic and basic FGF on endothelial cell biology. These two molecules have a 53% homology in their sequence and possess many similarities in their mode of action. Both bind to heparin and would appear to act on the same receptor of numerous cell types.[63,64] They are mitogenic and chemotactic to many cells including endothelium and have been shown to be potent angiogenic factors.[65–67] Until a few years ago, it was not clear if endothelium required FGF for its growth since the sequence for both acidic and basic FGF have no consensus signal peptide and therefore are not thought to be secreted proteins.[38,39] Furthermore, FGF was not found in plasma or serum.[68,69] It was subsequently shown that endothelial cells express the mRNA for at least basic or acidic FGF[70,71] as do both smooth muscle cells[72,73] and macrophages.[74] Therefore, a variety of cells which come into contact with endothelium could provide a source of FGF. Another potential source of FGF is the basement membrane which can be thought of as a storage pool of FGF,[75–78] and it has been suggested that mechanical injury or proteolytic degradation of the basement membrane might release FGF which would then act on adjacent endothelial cells. The question of how FGF gets to be localized in the basement membrane in the first place is not understood.

III. CAPACITY FOR ENDOTHELIAL REGROWTH IN VIVO

The data from the preceding section illustrate the fact that endothelial cell growth can indeed be initiated by known molecules and in this section we will examine under what *in vivo* conditions endothelial cells are not capable of growth. Early work on endothelial cell growth had suggested that *in vitro* endothelial cell replication might occur in the absence of any known mitogens[79] and this agreed with *in vivo* data where endothelial cells would always replicate in response to any defects in the endothelial cell monolayer.[80–82] Studies by this laboratory have shown that *in vivo* the response of the endothelium to a denuding injury is immediate and that small defects in the monolayer are repaired within a few hours by cell movement into the wound.[83,84] Larger injuries necessitated cell replication which can be detected within the first 24 hr and seemed to continue at a significantly high rate until repair was complete.[85] It was, therefore, somewhat surprising that several years ago we observed that after total denudation of a rabbit aorta, the early rapid replication rate of the endothelial cells slowed markedly and by 2–4 weeks post injury, their regrowth had almost stopped despite the fact that large areas of the artery were not repopulated with new endothelium (Fig. 1).[86] This finding suggested that the growth of endothelium in a denuded artery does not always continue until cell contacts are reestablished and that presumably other factors

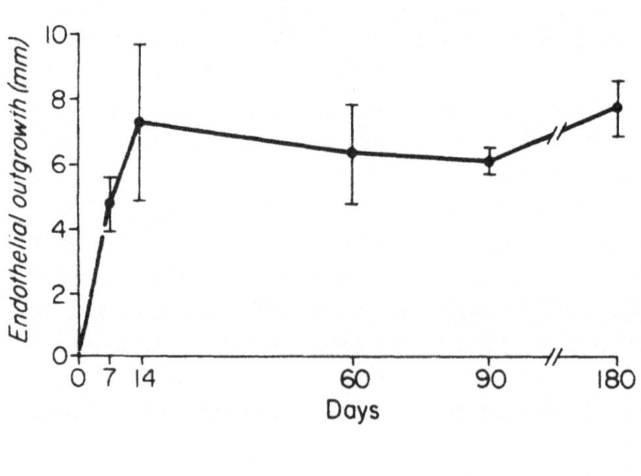

Figure 1. Outgrowth of endothelial cells in rabbit thoracic aortas after total denudation with a balloon catheter. Each point is the mean of total outgrowth from six individual intercostal arteries.

influence this process. The finding that endothelial cell growth *in vivo* is limited has been known by vascular surgeons for several years since synthetic vascular grafts in man rarely achieve a total endothelial cover[7] and more often the extent of reendothelialization is limited to those areas adjacent to anastomosis with the native artery.[6,8] Indeed, this single finding is thought by many to be responsible for the high frequency of failures with these prostheses.

Over the past 8 years we have begun to examine the process that controls endothelial cell growth *in vivo* and to understand why the ability of endothelial cells to repopulate a denuded surface is so limited. These studies were carried out *in vivo* because at the time little was known about endothelial cell mitogens and because this spontaneous cessation of endothelial cell growth was only found to occur in the intact artery. Even now it has not been possible to mimic this endothelial response *in vitro*. One early explanation for this lack of growth was that rapid and prolonged replication initiated after injury could ultimately lead to endothelial cell senescence. If this were true, then it might explain why endothelial cell growth slowed only after a prolonged period of high replication. To test this hypothesis, carotid arteries were totally denuded of endothelium and left for 6 weeks after which time endothelial regrowth had spontaneously stopped. These partially endothelialized arteries were then mechanically reinjured and endothelium replication assessed by [^{3}H]thymidine uptake and *en face* autoradiography (see Fig. 2). These injured endothelial cells showed a marked increase in their replication (23.5% versus 0.02%) and the speed of healing this wound was identical to that observed in previously unmanipulated arteries. Interestingly, despite this increase in cell growth, the endothelial monolayer as a whole was not restimulated to grow into the previously denuded zones of the artery. An important conclusion from this study was that "inhibited" endothelial cells are indeed capable of replicating but that injury including the fresh interaction of platelets with the denuded surfaces was not a sufficient stimulus to induce regrowth onto the previously denuded segments of the artery.

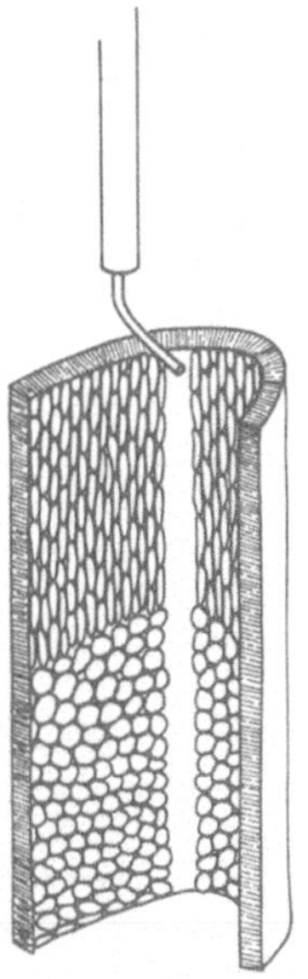

Figure 2. Illustration of experimental design to test whether endothelial cells senesced after period of rapid regrowth. Rat carotid arteries were totally denuded of endothelium and left for 6 weeks. During this time a limited regrowth occurred but had completely stopped after 6 weeks. Smooth muscle cells migrated to the luminal surface to form a pseudoendothelium in the still denuded zones (lower half). A small catheter was then inserted into the carotid artery to denude a small zone along the length of the vessel. [^{3}H]thymidine was given and cell replication assessed by autoradiography.

Since we and others had only observed inhibition of endothelial cell growth on an injured artery where smooth muscle formed the luminal surface (pseudoendothelium),[86,87] one possibility was that these cells somehow interfered with the regrowth of endothelium. In fact, one obvious possibility was that endothelial–smooth muscle cell contact was inhibiting endothelial growth in a similar manner as endothelial–endothelial cell contacts (Fig. 3). Therefore, an experiment was undertaken in which smooth muscle cell migration

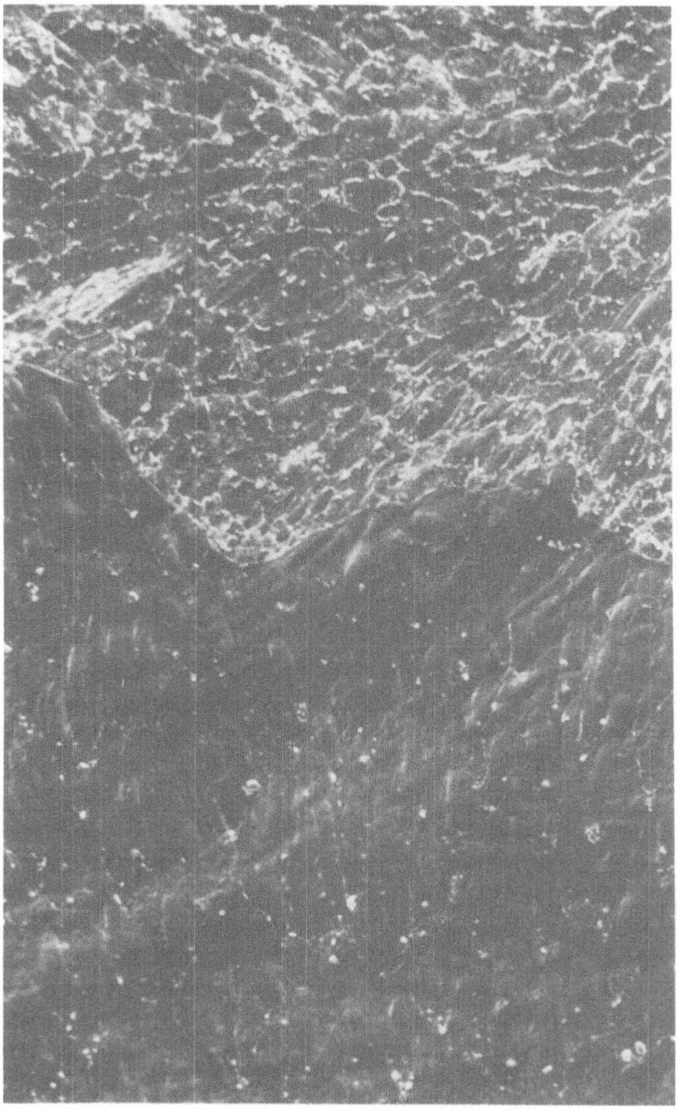

Figure 3. Interface between vascular endothelium and smooth muscle cells in a rabbit carotid artery 6 weeks after balloon catheter denudation. Note that the endothelial cells (lower half) directly abut the luminal smooth muscle cells. (×430.)

to the denuded surface was inhibited and endothelial cell regrowth was then examined under these conditions. In this experiment, totally inert vascular grafts were transplanted into the carotid arteries of rabbit. Thus, endothelial cells which grew from the cut end of the artery onto a graft surface would not encounter any smooth muscle cells on the luminal surface. The rate of endothelial cell replication and the total endothelial outgrowth onto these grafts, however, were found to be identical to those observed in an artery subjected to balloon injury with the exception that there was a slight increase in replication of the endothelial cells near the anastomosis (Table 1). Examination of these grafts by electron microscopy showed that no other cells except platelets were present on the luminal surface and that the ingrowing endothelium was not in contact with any cells on the luminal surface. This result led us to conclude that the progress of reendothelialization was not retarded by the presence of smooth muscle cells on the luminal surface. This finding appears to be at variance with the recent data from D'Amore and co-workers[88] who have shown that pericytes and smooth muscle cells when in contact with endothelial cells *in vitro* inhibit their growth and do so by activating TGF-β (see above). A key feature of their studies was that some form of contact had to be made between the endothelium and smooth muscle cells in order to activate TGF-β. At first sight, this study would seem to be similar in approach to that reported by us in that the focus was on interaction of smooth muscle cells with the endothelium. It should be remembered, however, that the aim of our studies *in vivo* was to determine whether the presence of any other cells on the luminal surface would interfere with endothelial regrowth. Endothelial cells, however, do form contacts with smooth muscle cells via their basal surface and we did observe that as the endothelial cells grew onto the graft, they were accompanied by underlying smooth muscle cells. Thus, even though there were no other cells present on the denuded surface, the regrowing endothelium was always in close approximation with underlying smooth muscle cells. The possibility therefore still exists that a physical connection between these two cell types can play a role in inhibiting endothelial cell growth.

Our data so far clearly illustrate that endothelial cells have a limited capacity to reheal denuded surfaces and yet surprisingly little has been published on the subject. One reason for this may be because the retardation of endothelial cell growth can only be observed in certain arteries and under certain conditions. This statement in no way implies that endothelial cells from different arteries have different growth potentials but simply that *in vivo* this retardation of endothelial cell growth can only be observed after prolonged reendothelialization. Therefore, if a denuded artery can be totally healed within a relatively short period of time (less than 2 weeks), then no retardation of endothelial cell growth can be detected. This point is well illustrated by comparing endothelial growth in the aorta and carotid artery of rats. If aortas are totally denuded of endothelium, total regrowth is achieved by approximately 3 weeks.[24] In this situation, rapid regrowth of endothelium was

Table 1. Endothelial Cell Outgrowth onto Rabbit Carotid Arteries

Time after surgery	Balloon catheter denudation	Aldehyde-fixed carotid graft	Autologous carotid graft
6 weeks	4.1 ± 0.6 mm	3.0 ± 0.6 mm	
12 weeks	5.3 ± 1.9 mm	2.9 ± 0.8 mm	2.9 ± 1.0 mm

achieved because the many branches of the aortas provide multiple sources of new endothelium. Thus, due to frequency of intercostal branches, which are only 3–4 mm apart, regrowth is rapid. In contrast, the rat carotid is a long vessel with no side branches. To achieve complete endothelial cell regrowth in these vessels, cells have to grow from the carotid bifurcation or from the aortic arch which is a distance of approximately 3.5 cm. In this situation, endothelial regrowth does stop when the artery is only partially healed. Therefore, depending upon the vessel studied and its anatomy with respect to its branching pattern, the fact that endothelial cell growth can be retarded might easily be overlooked. This observation has been more readily recognized in man because after implantation of vascular grafts there are only two sources of endothelium and the distance to be repopulated with new endothelium is often large.

IV. IN VIVO CONTROL OF ENDOTHELIAL CELL GROWTH

Until recently, we have had no success in stimulating complete endothelial regrowth over a large area, i.e., more than 2 cm in length. As discussed above, neither modifying the substrate nor stimulating the endothelial cells influenced the extent of endothelial cell outgrowth. One possible conclusion from this work is that endothelial cells have a finite ability to repopulate a denuded surface. We have now found this conclusion to be incorrect and that endothelial cells do have the ability to spontaneously repopulate large denuded areas. In a recent series of experiments, endothelial cells were removed from rat carotid arteries using a less traumatic device (gentle denudation) which does not disrupt the underlying smooth muscle cells.[89,14] As in balloon catheter-denuded carotids, endothelial regrowth occurred from either end of the vessel but these cells maintained a high proliferative rate (Fig. 4) until the vessel was totally repopulated with new endothelium (Fig. 5). At present, we do not fully understand why the growth rates of endothelium should

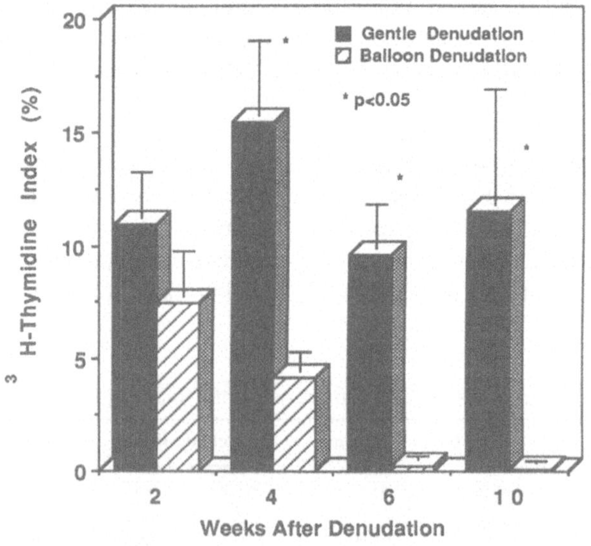

Figure 4. Endothelial cell replication after gentle and balloon denudation.

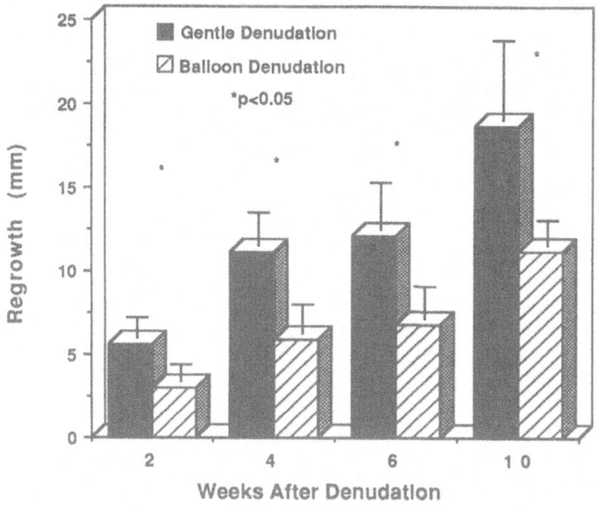

Figure 5. Total endothelial regrowth after gentle and balloon denudation.

be so different but this finding illustrates an important fact which is that endothelial cells do have the potential for continuous growth and that total regrowth of large denuded areas can be achieved under certain conditions.

It is still not clear what factors are involved in this process but one indication came from an experiment where we examined the endothelial cells for the presence of basic FGF. Our aim was to correlate the ability of those cells to replicate with the presence of this mitogen using a polyclonal antibody made against a known sequence of basic FGF. In rat carotid arteries, quiescent endothelium was found to stain only faintly with this antibody but regenerating endothelial cells stained strongly (Fig. 6). Of particular interest was the fact that 6 weeks after balloon catheter injury when endothelial cell growth had stopped, no cells stained with the basic FGF antibody, whereas the replicating endothelium in arteries denuded by the new "gentle technique" stained strongly for this mitogen. This led us to consider that basic FGF possibly synthesized by the endothelial cells themselves was important for their growth. To establish that basic FGF was effective on rat arterial endothelial cells, this mitogen was infused into animals whose carotid arteries had been previously denuded with a balloon catheter. We have shown that at this time the endothelial proliferation rate is close to background yet both continuous infusion and a single bolus injection of FGF caused a highly significant increase in endothelial cell replication (Fig. 7). To our knowledge, this is the first time that endothelial cell replication in large arteries has been restimulated after it has spontaneously stopped. One interesting fact was that the response of the endothelial cells in these animals was not uniform and only those endothelial cells in the still denuded artery showed this dramatic increase in their replication rates (see Fig. 7). The endothelium of adjacent untraumatized arteries, i.e., aorta or right carotid, showed an increased replication as compared to control animals but the increase was approximately 30-fold lower compared to the injured arteries. No data are currently available to explain why only certain cells respond so well to basic FGF but in order to answer this problem, future studies will need to address the availability of receptors on these

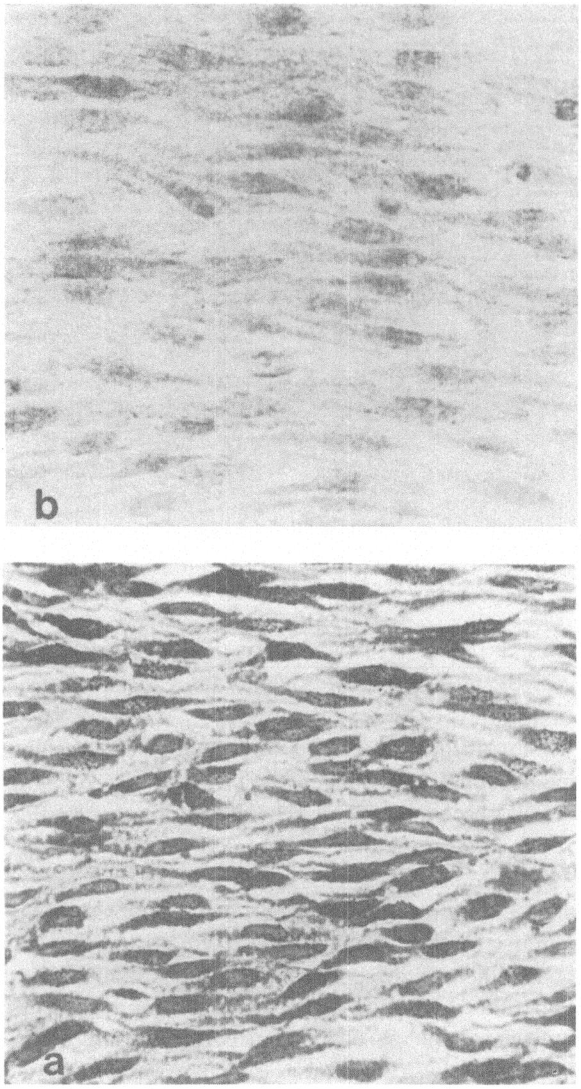

Figure 6. Luminal surface of rat aorta stained *en face* with an antibody to basic FGF. (a) Endothelial cells which are rapidly regenerating (adjacent to injured zone) stain strongly with the antibody. (b) Normal untraumatized endothelium shows little staining. (×400.)

cells and the presence of inhibitors thought to be present on confluent but not sparse endothelium.

An increase in cell replication might imply new endothelial growth onto a denuded surface but in previous work we have shown that replication can occur without any detectable outgrowth.[89,90] An experiment was therefore designed to test whether FGF would stimulate the growth of endothelial cells onto a denuded artery. Rat carotid arteries were denuded of endothelium with a balloon catheter and for a period of 8 weeks basic FGF (12 μg) was injected twice per week. The extent of endothelial growth onto this denuded surface was found to be significantly increased after FGF administration as compared to

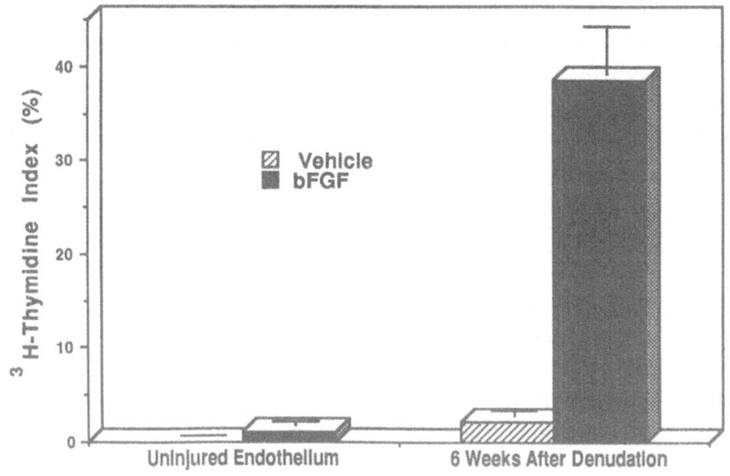

Figure 7. Endothelial replication after infusion of bFGF (120 μg over 8 hr).

animals receiving the vehicle alone (Fig. 8). It is apparent from this result that FGF is effective in stimulating endothelial cell growth in denuded arteries and perhaps this mitogen may have a role in promoting endothelial cell growth in vascular grafts.

V. CONSEQUENCE OF ENDOTHELIAL CELL LOSS

The importance of an intact endothelial cell monolayer might on first examination be obvious but over the past few years, a considerable body of data has accumulated which implies that loss of endothelium might have minimal or even no consequences to the arterial

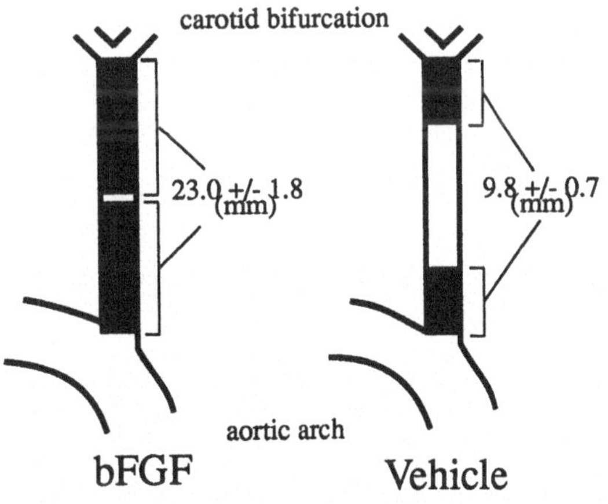

Figure 8. Endothelial outgrowth after injecting bFGF for 8 weeks.

wall. If this is true, then the importance of maintaining an intact endothelium might have minimal significance. The following section will briefly discuss the response of the arterial wall to loss of endothelium.

Loss of endothelium precipitates an immediate thrombotic response with platelets rapidly adhering to denuded areas.[91–93] While endothelial cells have unique antithrombotic and anticoagulation properties, arteries devoid of endothelium also rapidly become nonthrombogenic.[14] Furthermore, if smooth muscle cells migrate onto the luminal surface as commonly occurs, these cells adopt some of the properties of the endothelium in that their presence stimulates neither coagulation nor platelet adherence.[86,87] These cells have the ability to synthesize PGI_2 which in part explains this property.[94] Thus, the permanent loss of endothelium might only trigger the thrombotic pathways for only a short period of time and these denuded arterial surfaces are in essence nonthrombogenic.

The major thrust of work relating to loss of endothelium deals with the proliferative response of the underlying smooth muscle cells. One question which has received a great deal of attention over the past decade is whether endothelial loss is a key event in stimulating smooth muscle cell growth. There are two sets of data relating to this issue which are apparently contradictory. Denudation of arteries using devices such as a balloon catheter invariably stimulates widespread smooth muscle cell replication and the development of intimal lesions.[95–102] In contrast, removal of endothelium by nontraumatic approaches such as used in the laboratory does not stimulate smooth muscle cells or initiate lesion formation. With such opposing results, it has been difficult to come to any firm conclusion as to what importance the cell plays in intimal lesion development (Table 2). One possible reason for these marked differences is that the procedures used to remove endothelial cells have been shown to do far more than simply cause denudation. For example, a balloon catheter not only removes endothelial cells and initiates platelet adherence but also disrupts the medial cells and causes widespread cell death.[25] In the past few years, we have tried to focus on just one of the events and see what directly influences the growth of smooth muscle cells. In one experiment, we asked if platelets were a necessary component for the initiation of smooth muscle cell proliferation induced by injury. Rats were made thrombocytopenic and their carotid arteries subjected to conventional balloon catheter injury. Despite the almost total absence of platelets on these denuded surfaces, the initial replication rate of the smooth muscle cells was found to be identical to that observed in normal animals subjected to an identical denuding injury (Fig. 9).[103] The absence of platelets, therefore, appears to exert no effect on smooth muscle cell proliferation but interestingly, the presence of platelets did appear to strongly influence the ability of smooth muscle cells to migrate into the intima

Table 2. Response of Arterial Wall to Denudation

	Denudation with nylon filament	Denudation with balloon catheter	Denudation with nylon loop (gentle denudation)
Endothelial cell loss	3–15 cells wide	Total denudation	Total denudation
Impact to arterial media	Not detected	Acute loss of DNA	Not detected
Time for reendothelialization	< 96 hr	> 1 year	< 12 weeks
SMC proliferation (24–48 hr)	$< 1\%$	$> 13\%$	$< 2\%$
Intimal thickening	No	Yes	Yes (if still denuded)

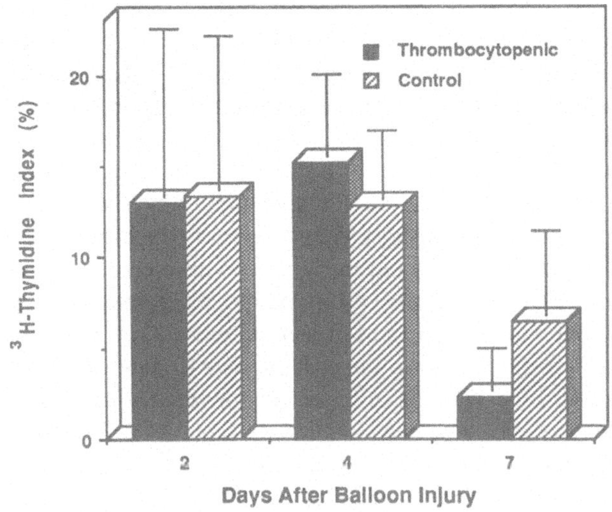

Figure 9. Replication of medial smooth muscle cells after platelet depletion.

(Fig. 10). Even the brief period of thrombocytopenia (approximately 40 hr) markedly reduced the ability of smooth muscle cells to migrate into the intima and form a lesion. The results from this study led us to conclude that smooth muscle proliferation can occur in the absence of platelets but still does not answer the question as to whether endothelial loss is a necessary event for smooth muscle cell proliferation.

A concern with all studies using mechanical techniques to remove endothelial cells is that they cause variable trauma to the underlying medial smooth muscle cells. For example, the balloon catheter causes an approximate 25% loss of DNA[25] and therefore the cellular

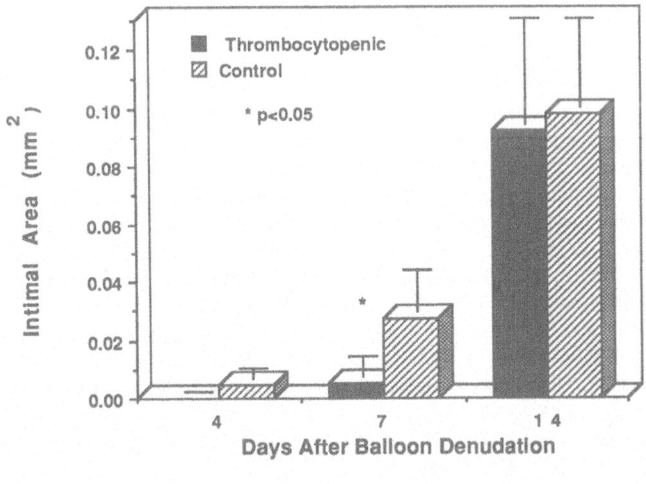

Figure 10. Intimal thickening after platelet depletion.

reaction could be in response to this trauma and not related to endothelial cell loss. To try to eliminate this problem, a new technique (gentle denudation) was used to remove the endothelium with no injury to the underlying medial cells.[14,89] Endothelial denudation by this procedure was found to cause a small increase in the replication rate of the smooth muscle cells but this value was significantly less than that induced by a balloon catheter injury (Table 2). Furthermore, the extent of intimal thickening in these animals was also markedly reduced. One interesting aspect of this study which sheds more light on the importance of endothelium was that those denuded areas which were rapidly (< 7 days) repopulated by new endothelial cells did not go on to develop intimal lesions despite a significant increase in smooth muscle cell proliferation. Yet those adjacent areas which had no endothelial cover but with similar rates of smooth muscle cell replication did develop intimal lesions. We feel that this finding makes it possible to suggest a unifying hypothesis which takes into account the very differing results previously obtained in a variety of experiments studying the effects of denudation (see Table 2). Loss of endothelium does cause an increase in the proliferation rate of vascular smooth muscle cells. However, the replication rate of these cells does not appear to be influenced by platelets. If the denuded artery is covered by new endothelium within a short time, approximately 7 days, then no intimal lesion forms. This would agree with previous data showing that when the area of denudation was limited and reendothelialization was rapid, no intimal lesion formed.[85] In contrast, denudation of a large area which is not repopulated with endothelium is associated with the migration of smooth muscle cells into the intima and development of intimal lesions. Therefore, endothelium would appear to be an important regulator of the migration of smooth muscle cells as well as influencing the proliferation of these cells. An area for further work is to understand the mechanism involved in smooth muscle cell movement and how the endothelium interferes with this process.

ACKNOWLEDGMENTS. This work was supported in part by grants HL42270 and HL41103 from NIH and 2,584 from the NIH Council for Tobacco Research.

REFERENCES

1. Abercrombie, M., 1970, Contact inhibition in tissue culture, *In Vitro* **6**:128–142.
2. Martz, E., and Steinberg, M. S., 1972, The role of cell–cell contact in "contact" inhibition of cell division: A review and new evidence, *J. Cell. Physiol.* **79**:189–210.
3. Heimark, R. L., and Schwartz, S. M., 1985, The role of membrane–membrane interactions in the regulation of endothelial cell growth, *J. Cell Biol.* **100**:1934–1940.
4. Goff, S. G., Wu, H. D., Sauvage, L. R., Usui, Y., Wechezak, A. R., Coan, D. E., Arnell, R. E., and Walker, M. W., 1988, Differences in reendothelialization after balloon catheter removal of endothelial cells, superficial endarterectomy, and deep endarterectomy, *J. Vasc. Surg.* **7**:119–129.
5. Reidy, M. A., Clowes, A. W., and Schwartz, S. M., 1983, Endothelial regeneration. V. Inhibition of endothelial regrowth in arteries of rat and rabbit, *Lab. Invest.* **49**:569–575.
6. Reidy, M. A., 1988, Endothelial regeneration. VIII. Interaction of smooth muscle cells with endothelial regrowth, *Lab. Invest.* **59**:63–43.
7. Berger, K., Sauvage, L. R., Rao, A. M., and Wood, S. J., 1972, Healing of arterial prostheses in man: Its incompleteness, *Ann. Surg.* **175**:118–127.
8. Sauvage, L. R., Berger, K. E., Wood, S. J., Yates, S. G., Smith, J. C., and Mansfield, P. B., 1974, Interspecies healing of porous arterial prostheses, *Arch. Surg.* **109**:698–705.

9. Folkman, J., and Moscona, A., 1978, Role of cell shape in growth control, *Nature* **273:**345–349.
10. Ingber, D. E., Madri, J. A., and Folkman, J., 1987, Endothelial growth factors and extracellular matrix regulate DNA synthesis through modulation of cell and nuclear expansion, *In Vitro Cell. Dev. Biol.* **23:** 387–394.
11. Herbst, T. J., McCarthy, J. B., Tsilibary, E. C., and Furcht, L. T., 1988, Differential effects of laminin intact type IV collagen, and specific domains of type IV collagen on endothelial cell adhesion and migration, *J. Cell Biol.* **106:**1365–1373.
12. Form, D. M., Pratt, D. M., and Madri, J. A., 1986, Endothelial cell proliferation during angiogenesis. In vitro modulation by basement membrane components, *Lab. Invest.* **55:**521–530.
13. Madri, J. A., Pratt, B. M., and Younariello-Brown, J., 1988, Matrix-driven cell size change modulates aortic endothelial cell proliferation and sheet migration, *Am. J. Pathol.* **132:**18–27.
14. Lindner, V., Reidy, M. A., and Fingerle, J., 1989, Regrowth of arterial endothelium. Denudation with minimal trauma leads to complete endothelial cell regrowth, *Lab. Invest.* **61:**556–563.
15. Majesky, M. W., Reidy, M. A., Twardzik, D. R., and Schwartz, S. M., 1989, Production of type-I transforming growth factor beta (TGF-beta) during repair of arterial injury, *FASEB J.* **3:**298.
16. Madri, J. A., Pratt, B. M., and Tucker, A. M., 1988, Phenotypic modulation of endothelial cells by transforming growth factor-beta depends upon the composition and organization of the extracellular matrix, *J. Cell Biol.* **106:**1375–1384.
17. Madri, J. A., Reidy, M. A., Kocher, O., and Bell, L., 1989, Endothelial cell behavior after denudation injury is modulated by transforming growth factor beta-1 and fibronectin, *Lab. Invest.* **60:**755–765.
18. Fràter-Schröder, M., Müller, G., Birchmeier, W., and Böhlen, P., 1986, Transforming growth factor beta inhibits endothelial cell proliferation, *Biochem. Biophys. Res. Commun.* **137:**295–302.
19. Heimark, R. L., Twardzik, D. R., and Schwartz, S. M., 1986, Inhibition of endothelial regeneration by type beta transforming growth factor from platelets, *Science* **233:**1078–1080.
20. Baird, D., and Durkin, T., 1986, Inhibition of endothelial cell proliferation by type beta-transforming growth factor: Interactions with acidic and basic fibroblast growth factors, *Biochem. Biophys. Res. Commun.* **138:**476–482.
21. Bell, L., and Madri, J. A., 1989, Effects of platelet factors on migration of cultured bovine aortic endothelial and smooth muscle cells, *Circ. Res.* **65:**1057–1065.
22. Miyazono, K., Hellmann, U., Wernstedt, C., and Heldin, C.-H., 1988, Latent high molecular weight complex of transforming growth factor beta-1. Purification from human platelets and structural characterization, *J. Biol. Chem.* **263:**6407–6415.
23. Wakefield, L. M., Smith, D. M., Flanders, K. C., and Sporn, M. B., 1988, Latent transforming growth factor-beta from human platelets. A high molecular weight complex containing precursor sequences, *J. Biol. Chem.* **263:**7646–7654.
24. Haudenschild, C. C., and Schwartz, S. M., 1979, Endothelial regeneration. II. Restitution of endothelial continuity, *Lab. Invest.* **41:**407–418.
25. Clowes, A. W., Reidy, M. A., and Clowes, M. M., 1983, Kinetics of cellular proliferation after arterial injury. I. Smooth muscle growth in the absence of endothelium, *Lab. Invest.* **49:**327–333.
26. Antonelli-Orlidge, A., Saunders, K. B., Smith, S. R., and D'Amore, P. A., 1989, An activated form of transforming growth factor beta is produced by co-culture of endothelial cells and pericytes, *Proc. Natl. Acad. Sci. USA* **986:**4544–4548.
27. Sato, Y., and Rifkin, D. B., 1989, Inhibition of endothelial cell movement by pericytes and smooth muscle cells: Activation of a latent transforming growth factor beta-1-like molecule by plasmin during co-culture, *J. Cell Biol.* **109:**309–315.
28. Fenselau A., Watt, S., and Mello, R. J., 1981, Tumor angiogenic factor, *J. Biol. Chem.* **256:**9605–9611.
29. Brown, M. T., and Clemmons, D. R., 1986, Platelets contain a peptide inhibitor of endothelial cell replication and growth, *Proc. Natl. Acad. Sci. USA* **83:**3321–3325.
30. Miyazono, K., Okabe, T., Urabe, A., Yamanaka, M., and Takafu, F., 1985, A platelet factor that stimulates the proliferation of vascular endothelial cells, *Biochem. Biophys. Res. Commun.* **126:**83–88.
31. Sipes, N. J., Meyskens, F. L., and Bregman, M. D., 1986, Biological characterization of an acid-sensitive growth factor from human platelets: Role in the proliferation of human melanoma and bovine endothelial cells, *Biochem. Biophys. Res. Commun.* **138:**795–802.
32. Clemmons, D. R., Isley, W. L., and Brown, M. T., 1983, Dialyzable factor in human serum of platelet origin stimulates endothelial cell replication and growth, *Proc. Natl. Acad. Sci. USA* **80:**1641–1645.

33. King, G.L., and Buchwald, S., 1984, Characterization and partial purification of an endothelial cell growth factor from human platelets, *J. Clin. Invest.* **73**:392.
34. Miyazono, K., Okabe, T., Urabe, A., Takaku, F., and Heldin, C.-H., 1987, Purification and properties of an endothelial cell growth factor from human platelets, *J. Biol. Chem.* **262**:4098–4103.
35. Ishikawa, F., Miyazono, K., Hellman, U., Drexler, H., Wernstedt, C., Hagiwara, K., Usuki, K., Takaku, F., Risau, W., and Heldin, C.-H., 1989, Identification of angiogenic activity and the cloning and expression of platelet-derived endothelial cell growth factor, *Nature* **338**:557–562.
36. Miyazono, K., and Heldin, C.-H., 1989, High-yield purification of platelet-derived endothelial cell growth factor: Structural characterization and establishment of a specific antiserum, *Biochemistry* **78**:1704.
37. Usuki, K., Heldin, N.-E., Miyazono, K., Ishikawa, F., Takaku, F., Westermark, B., and Heldin, C.-H., 1989, Production of platelet-derived endothelial cell growth factor by normal and transformed human cells in culture, *Proc. Natl. Acad. Sci. USA* **86**:7427–7431.
38. Jaye, M., Howk, R., Burgers, W., Ricca, G. A., Chiu, I. M., Ravera, M. W., O'Brien, S. J., Modi, W. S., Maciag, T., and Drohan, W. N., 1986, Human endothelial cell growth factor: Cloning, nucleotide sequence, and chromosome localization, *Science* **233**:541–545.
39. Abraham, J. A., Mergia, A., Whang, J. L., Tumolo, A., Friedman, J., Hjerrild, K. A., Gospodarowicz, D., and Fiddes, J. C., 1986, Nucleotide sequence of a bovine clone encoding the angiogenic protein, basic fibroblast growth factor, *Science* **233**:545–458.
40. Klagsbrun, M., and Vlodavsky, I., 1988, Biosynthesis and storage of basic fibroblast growth factor (bFGF) by endothelial cells: Implications for the mechanism of action of angiogenesis, in: *Growth Factors and Other Aspects of Wound Healing: Biological and Clinical Implications* (A. Barbul, E. Pines, M. Caldwell, and T. K. Hunt, eds.), Liss, New York, pp. 55–61.
41. McNeil, P. L., Muthukrishnan, L., Warder, E., and D'Amore, P. A., 1989, Growth factors are released by mechanically wounded endothelial cells, *J. Cell Biol.* **109**:811–822.
42. Gajdusek, C. M., and Carbon, S., 1989, Injury-induced release of basic fibroblast growth factor from bovine aortic endothelium, *J. Cell. Physiol.* **139**:570–579.
43. Bar, R. S., Boes, M., Booth, B. A., Dake, B. L., Henley, S., and Hart, M. N., 1989, The effects of platelet-derived growth factor in cultured microvessel endothelial cells, *Endocrinology* **124**:1841–1848.
44. Dicorleto, P. E., and Bowen-Pope, D. F., 1983, Cultured endothelial cells produce a platelet-derived growth factor-like protein, *Proc. Natl. Acad. Sci. USA* **80**:1919.
45. Hermansson, M., Nister, M., Betsholtz, C., Heldin, C.-H., Westermark, B., and Funa, K., 1988, Endothelial cell hyperplasia in human glioblastoma: Coexpression of mRNA for platelet-derived growth factor (PDGF) b chain and PDGF receptor suggests autocrine growth stimulation, *Proc. Natl., Acad. Sci. USA* **85**:7748–7752.
46. Connolly, D. T., Olander, J. V., Heuvelman, D., Moonsell, R., Siegel, N., Haymore, B. L., Leimgruber, R., and Feder, J., 1989, Human vascular permeability factor, *J. Biol. Chem.* **264**:20017–20024.
47. Connolly, D. T., Heuvelman, D., Nelson, R., Olander, J. V., Eppley, B. L., Delfino, J. J., Siegel, N. R., Leimgruber, R. M., and Feder, I., 1989, Tumor vascular permeability factor stimulates endothelial cell growth and angiogenesis, *J. Clin. Invest.* **84**:1470–1478.
48. Keck, P. J., Hauser, S. D., Krivi, G., Sanzo, K., Warren, T., Feder, J., and Connolly, D. T., 1989, Vascular permeability factor, an endothelial cell mitogen related to PDGF, *Science* **246**:1309–1312.
49. Henzel, W. J., and Ferrara, N., 1989, Pituitary follicular cells secrete a novel heparin-binding growth factor specific for vascular endothelial cells, *Biochem. Biophys. Res. Commun.* **161**:851–858.
50. Leung, D. W., Cachianes, G., Kuang, W.-J., Goeddel, D. V., and Ferrara, N., 1989, Vascular endothelial growth factor is a secreted angiogenic mitogen, *Science* **246**:1306–1309.
51. Gospodarowicz, D., Abraham, J. A., and Schilling, J., 1989, Isolation and characterization of a vascular endothelial cell mitogen produced by pituitary-derived folliculostellate cells, *Proc. Natl. Acad. Sci. USA* **86**:7311–7315.
52. Plouet, J., Schilling, J., and Gospodarowicz, D., 1989, Isolation and characterization of a newly identified endothelial cell mitogen produced by A+T−20 cells, *EMBO J.* **8**:3801–3806.
53. Senger, D. R., Peruzzi, C. A., Feder, J., and Dvorak, H. F., 1986, A highly conserved vascular permeability factor secreted by a variety of human and rodent tumor cell lines, *Cancer Res.* **46**:5629–5632.
54. Thomas, K. A., 1987, Fibroblast growth factors, *FASEB J.* **1**:434–440.
55. Baird, A., Esch, F., Mormede, P., Ueno, N., Ling, N., Böhlen, P., Ying, S.-Y., Wehrenberg, W. B., and Guillemin, R., 1986, Molecular characterization of fibroblast growth factor: Distribution and biological activities in various tissues, *Recent Prog. Horm. Res.* **42**:143–205.

56. Gospodarowicz, D., 1989, Fibroblast growth factor, *Crit. Rev. Oncogenesis* **1**:1–26.
57. Delli-Bovi, P., Curatola, A. M., Kern, F. G., Greco, A., Ittmann, M., and Basilico, C., 1987, An oncogene isolated by transfection of Kaposi sarcoma DNA encodes a growth factor that is a member of the FGF family, *Cell* **50**:729–737.
58. Taira, M., Yoshida, T., Miyagawa, K., Sakamoto, H., Terada, M., and Sugimura, T., 1987, cDNA sequence of human transforming gene hst and identification of the coding sequence required for transforming activities, *Proc. Natl. Acad. Sci. USA* **84**:2980–2984.
59. Dickson, C., Smith, R., Brookes, S., and Peters, G., 1984, Tumorogenesis by mouse mammary tumor virus: Proviral activation of a cellular gene in the common integration region, Int-2, *Cell* **37**:529–536.
60. Dickson, C., and Peters, G., 1987, Potential oncogene product related to growth factors, *Nature* **326**:833.
61. Zhan, X., Bates, B., Hu, X., and Goldfarb, M., 1988, The human FGF-5 oncogene encodes a novel protein related to fibroblast growth factors, *Mol. Cell. Biol.* **8**:3487–3495.
62. Finch, P. W., Rubin, J. S., Miki, T., Ron, D., and Aaronson, S. A., 1989, Human KGF is FGF-related with properties of a paracrine effector of epithelial cell growth, *Science* **245**:752–755.
63. Burrus, L. W., and Olwin, B. B., 1989, Isolation of a receptor for acidic and basic fibroblast growth factor from embryonic chick, *J. Biol. Chem.* **264**:18647–18653.
64. Lee, P., Johnson, D. E., Cousens, L. S., Fried, V. A., and Williams, L. T., 1989, Purification and complementary DNA cloning of a receptor for basic fibroblast growth factor, *Science* **245**:57–60.
65. Montesano, R., Vassali, J. D., Baird, A., Guillemin, R., and Orci, L., 1986, Basic fibroblast growth factor induces angiogenesis in vitro, *Proc. Natl. Acad. Sci. USA* **83**:7297–7301.
66. Hayek, A., Culler, F. L., Bearrie, G. M., Lopez, A. D., Cuevas, P., and Baird, A., 1987, In vivo model for study of the angiogenic effects of basic fibroblast growth factor, *Biochem. Biophys. Res. Commun.* **147**:876–880.
67. Folkman, J., and Klagsbrun, J., 1987, Angiogenic factors, *Science* **235**:442–447.
68. Mormède, P., Baird, A., and Pigeon, P., 1985, Immunoreactive fibroblast growth factor (FGF) in rat tissue: Molecular weight forms and effects of hypophysectomy, *Biochem. Biophys. Res. Commun.* **128**:1108–1113.
69. Gauthier, T., Maftouh, M., and Picard, C., 1987, Rapid enzymatic degradation of (^{125}I)[Tyr^{10}]FGF(1-10) by serum in vitro and involvement in the determination of circulating FGF by RIA, *Biochem. Biophys. Res. Commun.* **145**:775–781.
70. Schweigerer, L., Neufeld, G., Friedman, J., Abraham, J. A., Fiddes, J. C., and Gospodarowicz, D., 1987, Capillary endothelial cells express basic fibroblast growth factor, a mitogen that stimulates their own growth, *Nature* **325**:257–259.
71. Vlodavsky, I., Friedman, R., Sullivan, R., Sasse, J., and Klagsburn, M., 1987, Aortic endothelial cells synthesize basic fibroblast growth factor which remains cell-associated and platelet-derived growth factor-like protein which is secreted, *J. Cell. Physiol.* **131**:402–428.
72. Winkles, J. A., Friesel, R., Burgess, W. H., Mehlman, T., Weinstein, R., and Maciag, T., 1987, Human vascular smooth muscle cells both express and respond to heparin-binding growth factor I (endothelial cell growth factor), *Proc. Natl. Acad. Sci. USA* **84**:7124–7128.
73. Gospodarowicz, D., Ferrara, N., Haaparanta, T., and Neufeld, G., 1988, Basic fibroblast growth factor: Expression in cultured vascular smooth muscle cells, *Eur. J. Cell Biol.* **46**:144–151.
74. Baird, A., Mormède, P., and Böhlen, P., 1985, Immunoreactive fibroblast growth factor in cells of peritoneal exudate suggests its identity with macrophage-derived growth factor, *Biochem. Biophys. Res. Commun.* **126**:358–364.
75. Vlodavsky, I., Folkman, J., and Sullivan, R., 1987, Endothelial cell-derived basic fibroblast growth factor; synthesis and deposition into subendothelial extracellular matrix, *Proc. Natl. Acad. Sci. USA* **84**:2292–2296.
76. Baird, A., and Ling, N., 1987, Fibroblast growth factors are present in the extracellular matrix produced by endothelial cells in vitro: Implications for a role of heparinase-like enzymes in the neovascular response, *Biochem. Biophys. Res. Commun.* **142**:428–435.
77. Folkman, J., Klagsbrun, M., Sasse, J., Wadzinski, M., Ingber, D., and Vlodavsky, I., 1988, Heparin-binding angiogenic protein–basic fibroblast growth factor—is stored within basement membrane, *Am. J. Pathol.* **130**:393–400.
78. Bashkin, P., Doctrow, S., Klagsbrun, M., Svahn, C. M., Folkman, J., and Vlodavsky, I., 1989, Basic fibroblast growth factor binds to subendothelial extracellular matrix and is released by heparanase and heparin-like molecules, *Biochemistry* **28**:1737–1743.
79. Gajdusek, C. M., and Schwartz, S. M., 1982, Ability of endothelial cells to condition culture medium, *J. Cell. Physiol.* **110**:35–42.

80. Reidy, M. A., and Schwartz, S. M., 1983, Endothelial injury and regeneration. IV. Endotoxin: A nondenuding injury to aortic endothelium, *Lab. Invest.* **48:**25–34.
81. Reidy, M. A., 1985, Biology of disease. A reassessment of endothelial injury and arterial lesion formation, *Lab. Invest.* **53:**513–520.
82. Fishman, R. A., Ryan, G. B., and Karnovsky, M. J., 1975, Endothelial regeneration in the rat carotid artery and the significance of endothelial denudation in the pathogenesis of myointimal thickening, *Lab. Invest.* **32:**339–351.
83. Ramsay, M. M., Walker, L. N., and Bowyer, D. E., 1982, Narrow superficial injury to rabbit aortic endothelium: The healing process as observed by scanning electron microscopy, *Atherosclerosis* **43:**233.
84. Reidy, A. M., and Schwartz, S. M., 1981, Endothelial regeneration. III. Time course of intimal changes after small defined injury to rat aortic endothelium, *Lab. Invest.* **44:**301–308.
85. Reidy, M. A., and Silver, M., 1985, Endothelial regeneration. VII. Lack of intimal proliferation after defined injury to rat aorta, *Am. J. Pathol.* **118:**173–177.
86. Reidy, M. A., Standaert, D., and Schwartz, S. M., 1982, Inhibition of endothelial cell regrowth. Cessation of aortic endothelial cell replication after balloon catheter denudation, *Arteriosclerosis* **2:**216–220.
87. Schwartz, S. M., Stemerman, M. B., and Benditt, E. P., 1975, The aortic intima. II. Repair of the aortic lining after mechanical denudation, *Am. J. Pathol.* **81:**15–42.
88. Orlidge, A., and D'Amore, P. A., 1987, Inhibition of capillary endothelial cell growth by pericytes and smooth muscle cells, *J. Cell Biol.* **105:**1455–1462.
89. Reidy, M. A., Clowes, A. W., and Fingerle, J., 1989, Role of endothelium in the proliferation and migration of arterial smooth muscle cells, *Excerpta Med. Int. Congr. Ser.* **817:**559.
90. Reidy, M. A., Chopek, M., Chao, S., McDonald, T., and Schwartz, S. M., 1989, Injury induces increase in von Willebrand factor in rat endothelial cells, *Am. J. Pathol.* **134:**857–864.
91. Baumgartner, H. R., 1972, Platelet interaction with vascular structures, *Thromb. Diath. Haemorrh. Suppl.* **51:**161–176.
92. Groves, H. M., Kinlough-Rathbone, R. L., Richardson, M., Moore, S., and Mustard, J. F., 1979, Platelet interaction with damaged rabbit aorta, *Lab. Invest.* **40:**194–200.
93. Jørgensen, L., Grøthe, A. G., Groves, H. M., Kinlough-Rathbone, R. L., Richardson, M., and Mustard, J. F., 1988, Distribution of cellular responses in rabbit aorta following one and two injuries with a balloon catheter, *Br. J. Exp. Pathol.* **69:**351–364.
94. Eldor, A., Falcone, D. J., Hajjar, D. P., Minick, C. R., Weksler, B. B., 1981, Recovery of prostacyclin production by deendothelialized rabbit aorta. Critical role of neointimal smooth muscle cells, *J. Clin. Invest.* **67:**735–741.
95. Baumgartner, H. R., and Studer, A., 1966, Folgen des Fefässkatheterismus am normo-und hypercholesterinaemischen Kaninchen, *Pathol. Microbiol.* **29:**393.
96. Björkerud, S., 1969, Atherosclerosis initiated by mechanical trauma in normolipidemic rabbits, *J. Atheroscler. Res.* **9:**209.
97. Schwartz, S. M., Haudenschild, C. C., and Eddy, E. M., 1978, Endothelial regeneration. I. Quantitative analysis of intimal stages of intimal stages of endothelial regeneration in rat aortic intima, *Lab. Invest.* **38:**568–580.
98. Schwartz, S. M., Stemerman, M. B., and Benditt, E. P., 1975, The aortic intima. II. Repair of the aortic lining after mechanical denudation, *Am. J. Pathol.* **81;**15–42.
99. Spaet, T. H., Stemerman, M. B., Veith, F. J., and Lejnieks, I., 1975, Intimal injury and regrowth in the rabbit aorta: Medial smooth muscle cells as a source of neointima, *Circ. Res.* **36:**58–70.
100. Stemerman, M. B., Spaet, T. H., Pitlick, F., Cintron, J., Lejnieks, I., and Tiell, M. L., 1977, Intimal healing: The pattern of reendothelialization and intimal thickening, *Am. J. Pathol.* **87:**125–137.
101. Clowes, A. W., Collazzo, R. E., and Karnovsky, M. J., 1978, A morphologic and permeability study of luminal smooth muscle cells after arterial injury in the rat, *Lab. Invest.* **39:**141–150.
102. Hirsch, E. Z., and Robertson, A. L., 1977, Selective acute arterial injury and repair. I. Methodology and surface characteristics, *Atherosclerosis* **28:**271–287.
103. Fingerle, J., Johnson, R., Clowes, A. W., and Reidy, M. A., 1989, Role of platelets in smooth muscle cell proliferation and migration after vascular injury in rat carotid artery, *Proc. Natl. Acad. Sci. USA* **86:**8412–8416.

4

Modulation of Fibronectin and Thrombospondin Synthesis and mRNA Levels by Heparin in Human Endothelial and Smooth Muscle Cell Cultures

Bernadette Lyons-Giordano, Jane M. Brinker, and Nicholas A. Kefalides

I. INTRODUCTION

Since the initial observations that mast cells containing heparin accumulate at tumor sites prior to neovascularization,[1] numerous studies have underscored the potential importance of heparin in the regulation of new vessel formation. *In vitro* heparin potentiates human umbilical vein endothelial cell (EC)* chemotactic and growth responses to endothelial cell growth factor (ECGF), a member of the fibroblast growth factor (FGF) family of mitogens.[2,3] The growth and migration stimulatory activity of heparin is neutralized by protamine, an arginine-rich basic protein which is known to bind avidly to heparin.[2,4] *In vivo* local administration of protamine inhibits neovascularization induced by inflammatory agents or by immune reactions. Coadministration of heparin reverses the effect of protamine.[5]

The mechanism of action of heparin in the regulation of EC migration and growth has not been elucidated. Studies investigating the interaction of heparin with EC have shown

*Abbreviations used in this chapter: CCM, complete culture medium; EC, endothelial cells; ECGF, endothelial cell growth factor; FN, fibronectin; SMC, smooth muscle cells; SSC, standard saline citrate; TSP, thrombospondin.

Bernadette Lyons-Giordano • Departments of Biochemistry and Biophysics, University of Pennsylvania, and University City Science Center, Philadelphia, Pennsylvania 19104. *Jane M. Brinker* • Connective Tissue Research Institute and Department of Medicine, University of Pennsylvania, and University City Science Center, Philadelphia, Pennsylvania 19104. *Nicholas A. Kefalides* • Connective Tissue Research Institute and Department of Medicine, University of Pennsylvania, and University City Science Center, Philadelphia, Pennsylvania 19104.

Endothelial Cell Dysfunctions, edited by Nicolae Simionescu and Maya Simionescu. Plenum Press, New York, 1992.

that it is specifically bound to and internalized by these cells.[6,7] Heparin dissociation constants of 0.12 and 1.37 μM were observed for high- and low-affinity sites, respectively. It has been estimated that approximately 1.2×10^6 molecules of heparin bind per cell. Other glycosaminoglycans, such as chondroitin sulfate, dermatan sulfate, and intestinal mucosal heparan sulfate fraction I, are unable to compete for heparin binding. Heparan sulfate fraction IIa from intestinal mucosa competitively inhibits approximately 50% of the heparin binding. Heparan sulfate IIa has a higher degree of sulfation than fraction I which makes it more similar to heparin in charge density. Thus, the specificity of endothelial cell binding of heparin may reflect a requirement for highly anionic molecules.

While these data suggest that the mode of heparin regulation of EC behavior may be directly through plasma membrane receptors, there is evidence for an indirect action of heparin. In the absence of ECGF and at low serum concentrations, heparin inhibits the growth of these cells.[8] Likewise, it lacks chemotactic activity for endothelium in the absence of ECGF.[3] Moreover, heparin stabilizes ECGF and increases the apparent affinity of this mitogen for its receptor.[9,10] Since FGFs are known to be present in basement membrane,[11] the angiogenic activity of heparin *in vivo* may reflect its capacity to potentiate ECGF-induced EC migration and growth *in vitro*.

Peptides of the FGF family exhibit angiogenic activity *in vivo* and are potent mitogens for EC *in vitro*.[12] It has recently been reported that bovine aortic EC synthesize basic FGF.[13] While most of the endothelial-derived FGF remains cell-associated, up to 30% of the mitrogen is deposited in the subendothelial extracellular matrix. Studies of the binding of basic FGF to the extracellular matrix suggest that sequestration of basic FGF in the matrix occurs through interaction of the mitogen with heparinlike glycosaminoglycans in the extracellular matrix.[14] EC do synthesize heparan sulfates and these glycosaminoglycans are incorporated into their extracellular matrix.[15] Together these data favor the possibility that FGFs may be autocrine regulators of EC growth. Storage of FGFs in the extracellular matrix may regulate their availability to the cells. Release of these mitogens from the matrix through the action of heparan sulfate-degrading enzymes and proteases at sites of vessel injury may promote neovascularization.

In vitro studies of the effect of substrates prepared from purified extracellular matrix components on the proliferative response of bovine capillary endothelium to basic FGF provide additional insight into the possible significance of extracellular matrix in regulating the effects of this mitogen on the endothelium.[16] In the absence of serum or exogenous growth factors, bovine capillary EC attach to purified matrix components, but do not spread, and exhibit low levels of DNA synthesis. Addition of basic FGF dramatically increases DNA synthesis, independent of the substrate upon which the cells are plated. Based on the proliferative response of cells treated with basic FGF and attached to substrates of different composition, a hierarchy of matrix protein enhancement of basic FGF mitogenic effect was defined: basic FGF increases EC proliferation on type III collagen > gelatin > type IV collagen > fibronectin (FN) > type II collagen > laminin. In addition, the growth-modulating effects of the various extracellular matrix components relate directly to their capacity to modulate EC shape changes. In response to basic FGF, DNA synthesis increases exponentially in direct relation to increases in cell area. These findings not only extend previous studies which first correlated EC rate of proliferation to the shape-modulating effects of a synthetic substrate molecule,[17] but also suggest an alternate model for the interplay of extracellular matrix molecules and FGFs in regulating angiogenesis. In this paradigm, selective changes in extracellular matrix composition, through degradation and/or alterations in EC secretory phenotype, may enhance or restrict the proliferation of EC in the presence of mitogens important for angiogenesis. These growth regulatory effects of the extracellular matrix may be exerted through its modulation of cell shape.

In vivo studies have demonstrated qualitative changes in the subendothelial cell matrix during angiogenesis. These changes result from proteolytic degradation of the preexisting basal lamina prior to the migration stage of the repair process and subsequent alterations in the biosynthesis of matrix molecules by the EC.[18] *In vitro* studies have demonstrated the significance of the extracellular matrix composition and organization in the modulation of EC attachment,[19] migration,[20] proliferation,[21] and multicellular organization.[22] In culture, EC synthesize an extracellular matrix which contains proteoglycans,[15,23] collagens[24,25] and other glycoproteins including FN,[26–28] laminin,[29] and thrombospondin (TSP).[30] In addition, EC synthesize FGF which they sequester in their matrix.[13] The studies described herein were designed to elucidate the effects of heparin, in the presence of ECGF, on the growth of human umbilical vein EC and on their synthesis of FN and TSP. In addition, the results from preliminary studies of the effects of heparin on human vascular smooth muscle cell (SMC) synthesis of FN and TSP are included. Like EC, these cells are normally in a quiescent growth state *in vivo*, but will migrate and proliferate in response to vessel wall injury.[31] The growth of vascular SMC following EC injury may be important in the pathogenesis of atherosclerosis.[32] Heparin inhibits the growth of vascular SMC *in vivo* and *in vitro*.[33,34] The antiproliferative effect of heparin on SMC suggests a physiologic role for heparinlike molecules in the maintenance of vessel wall integrity. Given the proximity of vascular SMC and EC *in vivo*, the coordinate effects of heparin on the growth of secretory phenotype of each cell type may be important in the control of angiongenesis.

II. EXPERIMENTAL PROCEDURES

A. Cell Culture

Human umbilical vein EC were obtained as described previously.[35] EC cultures were grown in modified M199 growth medium containing 16.5% fetal calf serum, 2.0 mM glutamine, 15 mM Hepes, 1× antibiotics (10 μg/ml gentamycin, 0.4 μg/ml amphotericin B), and 30 μg/ml ECGF (Meloy Laboratories, Inc., Springfield, Va.) in the absence and presence of heparin from porcine intestinal mucosa (Sigma #H3125, St. Louis, Mo.). Preparation of SMC from human umbilical cord arteries was conducted essentially as described by Tumilowicz *et al.*[36] SMC were grown in Dulbecco's modified Eagle's medium (DMEM) with 10% fetal calf serum, 1 mM glutamine, 15 mM Hepes, and 1× antibiotics. For heparin-treated SMC, heparin (180 μg/ml) was added to the medium of subconfluent cultures. EC and SMC were maintained at 37°C in a humidified 5% CO_2 atmosphere.

B. Metabolic Labeling

For the 22- and 24-hr metabolic labeling studies, cultures were incubated in medium containing either radiolabeled proline or radiolabeled methionine. In studies using [^{14}C]proline labeling, subconfluent cultures were washed twice with phosphate-buffered saline and incubated at 37°C with 4 μCi/ml [^{14}C]proline (250 mCi/mmole; Amersham Corporation, Arlington Heights, Ill.) in MEM supplemented with 0.1% glucose, 1 mM glutamine, 50 μg/ml ascorbic acid, 50 μg/ml β-aminoproprionitrile fumerate, 15 mM Hepes, and in the absence and presence of various concentrations of heparin. For pulse-labeling studies of the effect of heparin on FN and TSP biosynthesis, cultures were incubated for 2 hr at 37°C with 40 μCi/ml [^{35}S]methionine (*trans* ^{35}S label; 1090 Ci/mmole; ICN Radiochemicals, Irvine, Calif.) in methionine-free DMEM supplemented as described for [^{14}C]proline labeling. For all analyses, medium and cell-matrix fraction were collected separately. Prior to analysis of radiolabeled proteins in the medium, protease inhibitors, *N*-ethylmaleimide (1 mM) and PMSF (1 mM), were added to the medium. After removal of the medium, the cell monolayers were washed three times with cold (4°C) PBS. The cell-matrix fraction was then extracted with 62.5 mM Tris-HCl (pH 6.8) containing 4 M urea and 2% SDS. Cell debris present in the medium was sedimented by centrifugation at 500*g* for 20 min at 4°C. The medium fractions were dialyzed exhaustively against a 0.5 M acetic acid at 4°C and then lyophilized prior to analysis by SDS-PAGE.

C. Protein Analyses

The radiolabeled proteins in the media, matrix and cell-matrix proteins were analyzed by SDS-PAGE according to Laemmli as described previously.[37] For quantitation of FN and TSP in the medium, cultures were incubated in MEM supplemented as described for metabolic labeling. After the medium was collected and centrifuged, it was dialyzed against PBS containing 0.05% Tween. FN and TSP levels in the medium were quantitated by indirect ELISA as described by Rennard *et al.*[38]

Immunoprecipitation of radiolabeled FN and TSP was conducted using protein A Sepharose beads as described previously by Lyons-Giordano *et al.*[39] The immunoprecipitates were then analyzed by SDS-PAGE. Human FN standard was purchased from Biomedical Technologies, Inc. (Cambridge, Mass.). Rabbit antibody against human FN was purchased from Calbiochem-Behring (San Diego, Calif.) TSP standard and rabbit anti-human TSP antibody were a gift from Dr. George Tuszynski, Lankenau Hospital, Philadelphia.

D. Nucleic Acid Analyses

For isolation of total RNA, the cells were homogenized with a glass/Teflon homogenizer and incubated in 10 mM Tris-HCl (pH 7.5), 5 mM EDTA, 0.5% SDS buffer containing 125 μg/ml proteinase K (Boehringer Mannheim Biochemicals, Indianapolis, Ind.) for 90 min at 42°C. Total cellular RNA was extracted with phenol/chloroform/isoamyl alcohol (25:24:1), selectively precipitated with 3 M sodium acetate, 5 mM EDTA, pH 7.0, and then washed with ethanol as adapted from the method described by Rowe *et al.*[40] RNA levels were quantitated by spectrophotometry at 260 nm.

For hybridization analyses, total cellular RNA (0.20–1.6 μg/slot) was applied to nitrocellulose using the Minifold II slot blot apparatus and procedure of Schleicher and Schuell, Inc., Keene, N.H.[41] The nitrocellulose filters were baked for 2 hr at 78°C under vacuum. The cDNA probes were nick-translated[42] with [^{32}P]dCTP and [^{32}P]dTTP (3000 Ci/mmole; Amersham) without excision from the plasmid to a specific activity of 5–8 × 10^8

cpm/μg. The filters were prehybridized overnight and then hybridized with the denatured labeled probes for 22 hr at 42°C in a solution containing 50% formamide, 5× standard saline citrate (SSC; 0.15 M sodium chloride, 0.015 M trisodium citrate), 1 mM EDTA, 2× Denhardt's solution, 100 μg/ml salmon sperm DNA, and 0.1% SDS. The filters were then washed with a final stringency of 0.5× SSC containing 0.1% SDS at 65°C. The filters were exposed to Kodak XAR-5 film with a DuPont Cronex Lightening Plus intensifying screen at −70°C for various lengths of time. Radioactivity hybridizing to RNA on the filters was quantitated by scanning densitometry of the autoradiograms.

The probes used include the cDNA clone pFH1 containing a 2.1-kb insert encoding about 1300 bp of the coding region plus the complete 3′ untranslated region of human FN mRNA, which was obtained from Dr. Francisco Baralle, University of Oxford.[43] The hybridization probe for TSP was a cDNA clone with a 1.9-kb insert which specifies the entire heparin binding domain and part of the type V collagen binding domain and was a gift from Dr. Vishva Dixit, University of Michigan.[44] Human fibroblast actin and tubulin cDNA clones were provided by Dr. Larry Kedes, Stanford University. The 1.8-kb actin insert includes sequences from the coding region up to amino acid 145 plus the complete 3′ untranslated region of the message. It hybridizes to both the β- and γ-actin mRNAs.[45] The β-tubulin cDNA clone pHF T-1, with its 2.1-kb insert, is virtually a full-length clone specific for β-tubulin mRNAs. This clone has the same sequences as the β-tubulin cDNA clone described by Hall *et al.*[46] The cDNA clones for human procollagen chains α2(I) and α1(III) were obtained from Dr. Jeanne Myers, University of Pennsylvania,[47] and Dr. Francesco Ramirez, State University of New York Health Science Center,[48] respectively.

III. RESULTS

A. Studies of the Effect of Heparin on EC Biosynthesis of FN

To assess the effects of heparin on EC synthesis of FN, cultures grown in the absence or presence of 90 μg/ml heparin were pulse-labeled for 2 hr with [^{35}S]methionine. Medium and cell-matrix fractions were combined for analysis by SDS-PAGE. As shown in Fig. 1, the level of radiolabeled FN was decreased in the medium–cell-matrix fraction from EC treated with heparin. By scanning densitometry, it was determined that heparin treatment resulted in a 50% reduction in radiolabeled FN. In contrast, the level of radiolabeled TSP was not affected by heparin treatment. Immunoprecipitation using antibodies specific for FN corroborated the apparent decrease in total radiolabeled FN in heparin-treated EC. Likewise, immunoprecipitation using antibodies specific for TSP showed no decrease in radiolabeled TSP in the combined medium–cell-matrix as a consequence of heparin treatment. Thus, using pulse-labeling conditions to minimize the possible effects of heparin on the rates of degradation of FN and TSP, the data demonstrate that heparin decreases EC synthesis of FN but not that of TSP.

The accumulation of FN and TSP in control cultures and EC grown in the presence of heparin (90 or 180 μg/ml) was determined utilizing ELISA techniques. As shown in Table 1, the effect of heparin on EC synthesis of FN was quantitatively reflected in the suppression of heparin-treated EC culture levels to 50% of that of control cultures. The levels of TSP in heparin-treated cultures did not vary significantly from that of control EC.

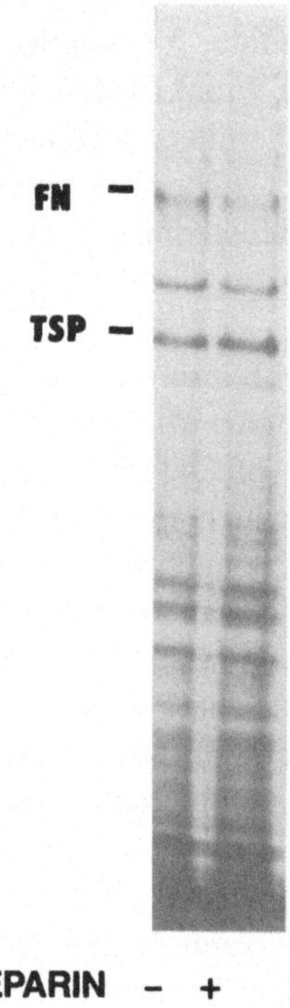

Figure 1. Heparin effects on EC biosynthesis of FN and TSP. The synthesis of FN and TSP by control and heparin-treated (90 μg/ml) EC was compared by pulse-labeling the cultures for 2 hr with 40 μCi/ml [^{35}S]methionine. For analysis, total radiolabeled proteins from control and heparin-treated cultures were resolved by SDS-PAGE. Each lane contains 10,000 cpm.

Table 1. Heparin Effects on EC Medium FN and TSP Levels[a]

Substrate	Heparin concentration (μg/ml)	FN (μg/10^6 cells)	TSP (μg/10^6 cells)
Gelatin	0	4.63 ± 0.40 (4)	39.3 ± 0.75 (2)
	90	1.93 ± 0.26 (4)	40.1 ± 2.70 (2)
	180	2.35 ± 0.11 (4)	37.7 ± 2.50 (3)
Plastic	0	9.32 ± 0.29 (2)	55.7 ± 5.70 (2)
	90	4.36 ± 0.07 (2)	59.2 ± 0.10 (2)
	180	4.51 ± 0.10 (2)	71.0 ± 3.60 (2)

[a]EC were grown on gelatin or plastic substrates in the absence or presence of heparin. For determination of FN and TSP levels in the medium, cells were incubated in serum-free medium for 22 hr. At the termination of incubation, the medium was removed and processed for competitive ELISA analysis as described in Materials and Methods. The cells were trypsinized and counted. The heparin concentrations shown were present both during the growth of the cells and in the serum-free incubation. The numbers in parentheses indicate the number of experiments. The data represent the mean ± standard deviation.

B. ECGF Modulation of the Heparin Effect

The preceding studies of the effect of heparin on FN synthesis were conducted using cells cultured in the presence of 30 μg/ml ECGF. Heparin is known to modulate the activity of ECGF. Specifically, heparin enhances the mitogenic efficacy of this peptide,[2] potentiates its chemotactic activity,[3] and increases the affinity of the peptide for its receptor.[9] To assess whether the effect of heparin on FN synthesis was modulated by ECGF, cells grown in medium lacking ECGF were treated with 90 μg/ml heparin in the absence and presence of 30 μg/ml ECGF for 24 hr. Control cells were grown in medium containing ECGF but lacking heparin. After 24 hr, the cells were labeled with [^{14}C]proline for 22 hr. Radiolabeled proteins recovered in the medium and cell-matrix fractions were resolved by SDS-PAGE and visualized by fluorography. Heparin and/or ECGF did not affect cell-matrix levels of FN (data not shown). The decrease in FN synthesis by cells cultured in the presence of heparin and ECGF was dependent on the presence of ECGF in the culture medium (Fig. 2). The difficulty in maintaining EC in culture medium without exogenous ECGF as well as the fact that EC synthesize ECGF[13] pose problems in assessing the effects of heparin, independent of ECGF, on EC biosynthetic phenotype.

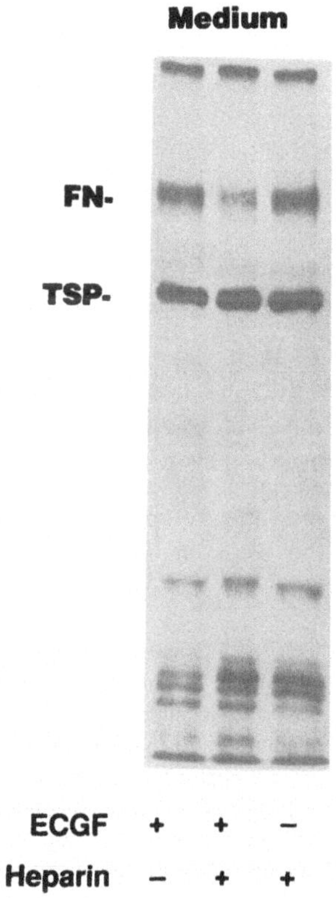

Figure 2. ECGF modulation of heparin effect. EC grown in CCM in the absence and presence of heparin (90 μg/ml) or grown in CCM lacking ECGF but supplemented with heparin (90 μg/ml) were labeled for 22 hr with [^{14}C]proline. Cells grown in the presence of heparin were labeled in the presence of heparin. ECGF was not present during the labeling of the cultures. Medium proteins visualized by SDS-PAGE and fluorography are shown. Each lane contains 10,000 cpm.

C. Dose Dependence of the Heparin Effect on EC FN Synthesis and Proliferation

EC synthesis of FN in response to different doses of heparin was quantitated using ELISA. EC were incubated for 12 hr in serum-free medium in the presence of heparin concentrations ranging from 0.09 to 90 μg/ml. FN levels per 10^6 cells in the heparin-treated cultures were normalized as percent inhibition relative to the levels in untreated cultures. The percent inhibition of FN synthesis observed with each heparin concentration was plotted as a function of heparin concentration as shown in Fig. 3. By interpolation, a half-maximal response would be expected with 2.5 μg/ml heparin. Maximal heparin-induced reduction of FN levels was observed with a heparin concentration of 90 μg/ml. No increase in suppression was evident in cultures treated with 180 μg/ml heparin as shown in Table 1. No effect on TSP levels was detected by ELISA over the same range of heparin concentrations.

The dose dependence of heparin-induced EC proliferation is also presented in Fig. 3. The concentration of heparin expected to yield half-maximal stimulation of proliferation was 2.5 μg/ml. This was the same concentration as required for half-maximal inhibition of FN synthesis. The correlation coefficient for heparin-induced stimulation of EC proliferation and inhibition of FN synthesis was determined to be −0.94 by correlation analysis.

D. Effect of Heparin on EC FN and TSP mRNA Levels

The rates of synthesis of extracellular matrix components may be regulated through changes in the steady-state level of their mRNAs, effects on the translatability of these mRNAs, or alterations in the rate of intracellular degradation of the newly synthesized

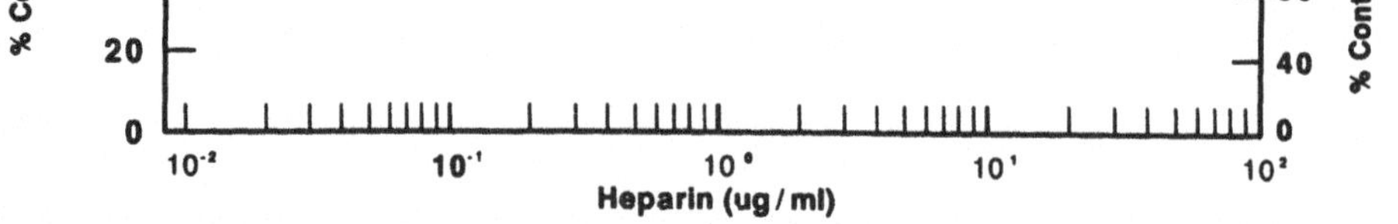

Figure 3. Concentration dependence of heparin effects on EC FN synthesis and proliferation. EC cultivated in the absence or presence of heparin concentrations ranging from 0.09 to 90 μg/ml were incubated for 12 hr in serum-free medium supplemented with the same concentration of heparin used to culture the cells. The FN level in the medium was then quantitated by ELISA and the cells counted. Values for FN levels in control and treated cultures were compared after normalization by the cell number in the respective cultures. To assay the concentration dependence of heparin-augmented proliferation, cells were cultured in the presence of heparin concentrations ranging from 0.09 to 90 μg/ml for 3 days and then trypsinized and counted. Data are presented as a percent of control cell proliferation and FN synthesis.

proteins. To elucidate the mechanism by which heparin reduced EC synthesis of FN, the effect of heparin on FN mRNA levels was assessed. Total RNA was prepared from subconfluent EC grown in the absence or presence of concentrations of heparin ranging from 0.09 to 90 μg/ml. For analysis, equivalent amounts of RNA for heparin-treated and control EC were applied to nitrocellulose using a slot blot apparatus and hybridized with a ^{32}P-labeled cDNA probe specific for FN. Concomitantly, hybridizations using radiolabeled cDNA probes for TSP, actin, and tubulin messages were performed to assess the specificity of the effects of heparin. Heparin caused a concentration-dependent decrease in FN and TSP message levels (Fig. 4). Maximal inhibition of both FN and TSP mRNA levels was observed with EC treated with 9.0 μg/ml heparin. Maximal heparin-induced suppression of FN mRNA was approximately 85% whereas that of TSP mRNA was 40%. By interpolation, the concentrations of heparin expected to yield half-maximal suppression of FN and TSP mRNA levels were the same (2.6 μg/ml). This concentration approximated that required for half-maximal inhibition of EC FN synthesis by heparin (Fig. 3). Heparin affected FN protein and mRNA levels in a coordinate manner. In contrast, heparin suppression of TSP mRNA levels was not accompanied by decreased TSP synthesis in EC cultures (Fig. 1). The effects of heparin on FN and TSP mRNA levels were not a consequence of a nonspecific suppression of RNA since approximately equal amounts of total RNA/cell were recovered from heparin-treated and control cultures. Also, actin and β-tubulin mRNA levels were not affected by treatment of EC with the same range of heparin concentrations used to assay the concentration dependence of heparin suppression of FN and TSP mRNA levels.

These data imply that heparin decreases EC expression of FN at least in part by decreasing the amount of FN mRNA available for translation. The failure of heparin to inhibit TSP expression, although it reduces TSP message levels, points to the possibility

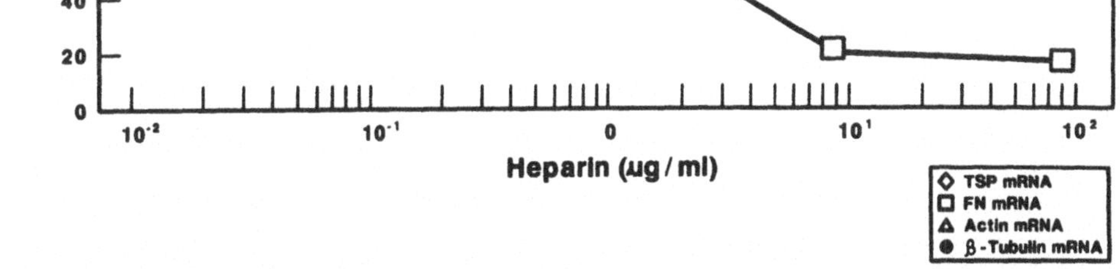

Figure 4. Quantitation of the concentration dependence of heparin suppression of FN and TSP mRNA levels in EC. The mRNA levels of FN, TSP, actin, and β-tubulin in EC grown in the presence of increasing concentrations of heparin were determined by scanning densitometry of autoradiograms following hybridization of total cellular RNA to the specific ^{32}P-labeled cDNA probes. The percent of each message relative to that in control cells is plotted as a function of heparin concentration present in the culture medium.

that the rate of EC TSP synthesis is regulated at the translational level. The coordinate effect of heparin on steady-state message levels for TSP and FN suggests that heparin may control the activity of cellular elements which regulate their mRNA levels in a concerted manner.

E. Effects of Heparin on SMC Synthesis of FN and TSP and Growth

The effects of heparin on human SMC biosynthesis of TSP and FN were studied using metabolic labeling and ELISA techniques. Cells, pretreated with heparin (180 μg/ml) for 24 hr, were labeled for 24 hr with medium containing [^{35}S] methionine and heparin. Control cells were incubated and labeled in the absence of heparin. As shown in Fig. 5, heparin caused an increase in the levels of both FN and TSP in the medium. No effects on cell-matrix-associated FN and TSP were observed, indicating no effect on compartmentalization of the proteins. These data were further substantiated by immunoprecipitation using rabbit antibodies specific for FN and TSP.

To quantify the effect of heparin on the synthesis of FN and TSP by human SMC, ELISA analyses were conducted. Heparin induced a 1.7-fold increase in FN levels/10^6 cells and a 10-fold increase in TSP levels/10^6 cells (Table 2). These data corroborated the increased FN and TSP apparent in heparin-treated SMC by metabolic labeling analysis.

To define the effects of heparin on human vascular SMC growth, cells seeded at low density were grown in the absence or presence of 180 μg/ml heparin. After 48 hr, the

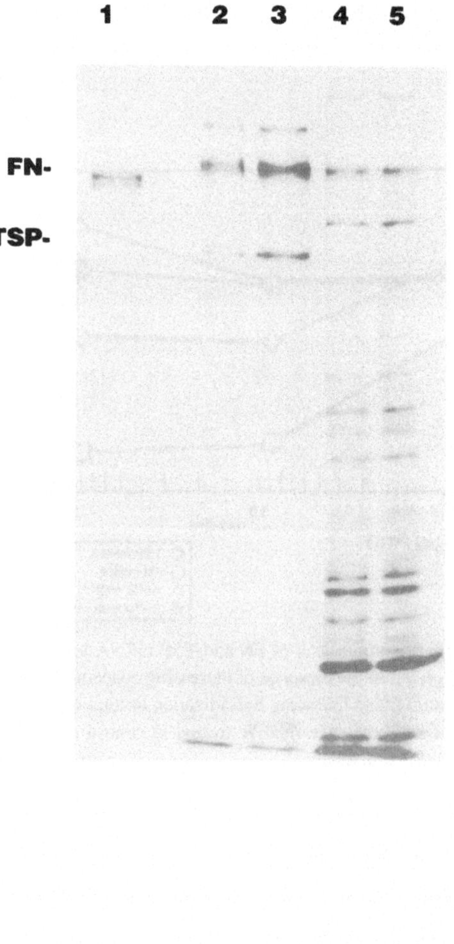

Figure 5. The effect of heparin on FN and TSP synthesis by human SMC. SMC were preincubated in growth medium containing 180 μg/ml heparin for 24 hr at 37°C with [^{35}S]methionine. Control SMC were preincubated and labeled in the absence of heparin. Radiolabeled proteins recovered in the medium and cell-matrix fractions of the heparin-treated and control cultures were fractionated by SDS-PAGE and visualized by autoradiography. Lane 1, FN standard; lane 2, medium from control SMC; lane 3, medium from heparin-treated SMC; lane 4, cell-matrix fraction from control SMC; lane 5, cell-matrix fraction from heparin-treated SMC.

Table 2. ELISA Analysis of the Effect of Heparin on Human Smooth Muscle Cell Synthesis of FN and TSP[a]

	μg FN/10[6] cells	μg TSP/10[6] cells
Control	2.22	1.21
Heparin-treated	3.81	12.4

[a]Heparin-treated human SMA were preincubated 24 hr at 37°C in growth medium containing 180 g/ml heparin. The cultures were then incubated 24 hr at 37°C in serum-free medium containing 180 μg/ml heparin. At the termination of the incubation, the medium was collected for ELISA and the cells were trypsinized and counted. Control cultures were preincubated in growth medium and incubated in the absence of heparin. Medium levels of FN and TSP quantitated by ELISA were normalized by cell number for comparison between control and heparin-treated cultures.

cultures were trypsinized and counted. The net growth of SMC treated with heparin was 55% less than that of control cells. Thus, similar to its effects on rat aortic SMC,[34] heparin is a potent inhibitor of human vascular SMC growth.

F. Effects of Heparin on SMC FN and TSP mRNA Levels

To elucidate the mechanism by which heparin increases human vascular SMC expression of FN and TSP, total RNA from control and heparin-treated cultures was isolated. The effects of heparin (180 μg/ml) on human vascular SMC steady-state levels of FN and TSP mRNA as well as those of procollagens α2(I) and α1(III) were quantitated by hybridizations to specific labeled cDNA probes to the RNA. Hybridizations utilizing a probe for actin verified the specificity of the effect of heparin on the message levels for the various extracellular matrix proteins. As summarized in Fig. 6, heparin treatment of SMC caused an increase in their steady-state mRNA levels for TSP, α2(I), and α1(III) procollagens. In contrast, no change in FN or actin mRNA levels was induced by heparin.

These data show that heparin coordinately increases SMC TSP mRNA and protein

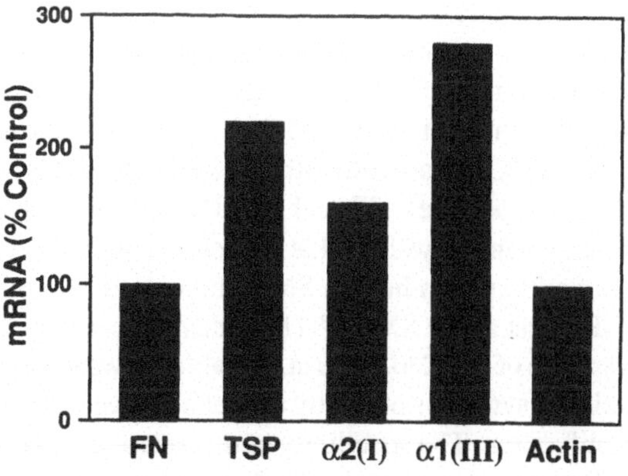

Figure 6. The steady-state levels of mRNAs from SMC grown in the absence or presence of heparin. The level of each mRNA in heparin-treated cultures relative to that in untreated cultures is plotted.

levels. The heparin-induced increase in SMC FN synthesis apparently reflects control at the translational or posttranslational level. The effects of heparin on SMC synthesis of type I and III procollagens warrant further investigation. The increased mRNA levels for procollagens α2(I) and α1(III) suggest that heparin may modulate SMC synthesis of collagens as well as FN and TSP.

IV. DISCUSSION

The kinetics of cellular proliferation following arterial injury have been extensively studied *in vivo*.[31] Upon denudation of a segment of the rat thoracic aorta, reendothelialization begins by the migration and proliferation of the adjacent EC. Intimal thickening, due to migration of SMC from the media to the intima where they proliferate, occurs in areas which are slowly and/or incompletely reendothelialized. Treatment with heparin diminishes the migration and proliferation of SMC without suppressing regeneration of the EC monolayer.[49]

Although it is widely accepted that FN and other extracellular matrix components play important roles in the regulation of cell migration and growth, limited data are available concerning the effects of angiogenic factors on EC synthesis of matrix components. The studies summarized herein demonstrate that heparin, in the presence of ECGF, suppresses EC synthesis of FN in a specific, dose-dependent manner. The suppression of EC synthesis of FN is reflected in a coordinate decrease in their steady-state levels of mRNA. EC synthesis of TSP, an extracellular matrix component believed to be important in the attachment of a variety of cells including those of mesenchymal origin,[50] is not affected by heparin treatment.

The inhibition of EC FN synthesis by heparin is modulated by ECGF. A requirement for ECGF has previously been reported for enhancement of EC migration and growth by heparin.[2,3] The possibility that the growth regulatory effects of heparin are related to its effects on EC synthesis of FN is further suggested by the high correlation between the dose dependence of the two phenomena. Additionally, it has been shown that transforming growth factor-β1, which reduces cultured EC sheet migration and growth, increases EC synthesis of FN.[50] These findings, together with our observations of the effects of heparin on EC, suggest that FN may function as an autocrine regulator of EC growth and migration.

The effects of heparin on SMC proliferation and synthesis of TSP and FN provide a striking contrast to its corresponding effects on EC. Heparin inhibits human vascular SMC growth and augments their expression of both FN and TSP. The antiproliferative effects of heparin on cultured SMC had been demonstrated using rat aortic cells.[34] Likewise, modulation of rat aortic SMC secretory phenotype by heparin has been reported.[51–53] In rat aortic SMC, treatment with heparin selectively induces the synthesis of two noncollagenous polypeptides (37 and 39 kDa) and modulates the expression of type III and type I procollagens as well as that of a short-chain (60 kDa) collagenous protein.[54] In contrast to our findings with human SMC, heparin does not change the net synthesis of either FN or TSP by rat aortic SMC.[53] These differences may reflect a species or tissue specificity in the response of SMC to heparin. Notable, however, is the observation that for both human SMC and EC, synthesis of FN by the cells is inversely related to heparin-induced changes in their proliferation. The specificity of this relationship is not the same for both cell types inasmuch as human vascular SMC TSP expression is modulated by heparin.

While the focus of our research has been on the modulation of vascular cell synthesis of matrix proteins by heparin, vascular cell migration and growth *in vivo* occur on a preexisting basal lamina. Endothelial-derived proteases, which break down components of the basal lamina, are important in facilitating EC migration and proliferation during neovascularization.[18] *In vitro* heparin, in the presence of ECGF, increases EC culture plasminogen activator activity by suppressing the synthesis of plasminogen activator inhibitor.[54] Thus, the facilitation of EC migration and growth as well as the concomitant remodeling of the basal lamina during vessel wound repair most likely reflect the capacity of angiogenic factors to change matrix protein catabolism as well as anabolism.

ACKNOWLEDGMENTS. This work was supported in part by NIH grants AR-20553, HL-29492, and HL-07502 and an award from the W. W. Smith Charitable Trust.

REFERENCES

1. Kessler, D. A., Langer, R. S., Pless, N. A., and Folkman, J., 1976, Mast cells and tumor angiogenesis, *Int. J. Cancer* **18**:703–709.
2. Thornton, S. C., Mueller, S. N., and Levine, E. M., 1983, Human endothelial cells: Use of heparin in cloning and long-term serial cultivation, *Science* **222**:623–625.
3. Terranova, V. P., DiFlorio, R. M., Hic, S., Friesel, R., Lyall, R. M., and Maciag, T., 1985, Human endothelial cells are chemotactic to endothelial cell growth factor and heparin, *J. Cell Biol.* **101**:2330–2334.
4. Azizkhan, R. G., Azizkhan, J. C., Zetter, B. R., and Folkman, J., 1980, Mast cell heparin stimulates migration of capillary endothelial cells *in vitro*, *J. Exp. Med.* **152**:931–944.
5. Taylor, S., and Folkman, J., 1982, Protamine is an inhibitor of angiogenesis, *Nature* **297**:307–311.
6. Glimelius, B., Busch, C., and Hook, M., 1978, Binding of heparin on the surface of cultured human endothelial cells, *Thromb. Res.* **12**:773–782.
7. Barzu, T., Molho, P., Tobelem, G., Petitou, M., and Caen, J., 1985, Binding and endocytosis of heparin by endothelial cells in culture, *Biochim. Biophys. Acta* **845**:196–203.
8. Rosenbaum, J., Tokelem, G., Molho, P., and Caen, J. P., 1986, Modulation of endothelial cell growth induced by heparin, *Cell Biol. Int. Rep.* **10**:437–446.
9. Schreiber, A. B., Kenney, J., Friesel, R., Mehlman, T., and Maciag, T., 1985, Interaction of endothelial cell growth factor with heparin. Characterization by receptor and antibody recognition, *Proc. Natl. Acad. Sci. USA* **82**:6138–6142.
10. Gospodarowicz, D., and Cheng, J., 1986, Heparin protects basic and acidic FGF from inactivation, *J. Cell. Physiol.* **128**:475–484.
11. Folkman, J., Klagsburn, M., Sasse, J., Wadzinski, M., Ingbee, D., and Vlodavsky, J., 1988, A heparin-binding angiogenic protein—basic fibroblast growth factor—is stored within basement membrane, *Am. J. Pathol.* **130**:393–400.
12. Gospodarowicz, D., Massoglia, S., Cheng, J., Lui, G.-L., and Böhlen, P., 1985, Isolation of bovine pituitary fibroblast growth factor purified by fast protein liquid chromatography (FPLC): Partial chemical and biological characterization, *J. Cell. Physiol.* **122**:323–333.
13. Vlodavsky, I., Folkman, J., Sullivan, R., Fridman, R., Ishai-Michaeli, R., Sasse, J., and Klagsburn, M., 1982, Endothelial cell derived basic fibroblast growth factor: Synthesis and deposition into subendothelial cell extracellular matrix, *Proc. Natl. Acad. Sci. USA* **84**:2292–2296.
14. Baird, A., and Ling, N., 1987, Fibroblast growth factors are present in the extracellular matrix produced by endothelial cells *in vitro*. Implications for a role of heparinase-like enzymes in the neovascular response, *Biochem. Biophys. Res. Commun.* **142**:428–435.
15. Kramer, R. H., Vogel, K. G., and Nicolson, G. L., 1982, Solubilization and degradation of subendothelial matrix glycoproteins and proteoglycans by metastatic tumor cells, *J. Biol. Chem.* **257**:2678–2686.
16. Ingber, D., Madri, J. A., and Folkman, J., 1987, Endothelial cell growth factors and extracellular matrix regulate DNA synthesis through modulation of cell and nuclear expansion, *In Vitro Cell. Dev. Biol.* **23**: 387–394.

17. Folkman, J., and Moscona, A., 1978, Role of cell shape in growth control, *Nature* **273**:345–349.
18. Glaser, B. M., Kalebic, T., Garbisa, S., Connor, T. B., and Liotta, L., 1983, Degradation of basement membrane components by vascular endothelial cells: Role in neovascularization, *Ciba Found. Symp.* **100**:150–162.
19. Macarak, E. J., and Howard, P. S., 1983, Adhesion of endothelial cells to extracellular matrix proteins, *J. Cell. Physiol.* **116**:76–86.
20. Young, W. C., and Herman, I. M., 1985, Extracellular matrix modulation of endothelial cell shape and motility following injury *in vitro*, *J. Cell Sci.* **73**:19–32.
21. Gospodarowicz, D., and Ill, C. R., 1981, Do plasma and serum have different abilities to promote cell growth? *Proc. Natl. Acad. Sci. USA* **77**:2726–2730.
22. Maciag, T., Kadish, L., Wilkins, M. B., Stemmerman, M. B., and Weinstein, R., 1982, Organizational behavior of human umbilical vein endothelial cells, *J. Cell Biol.* **94**:511–520.
23. Hassell, J. R., Gehron Robey, P., Barrach, H.-J., Wilzcek, J., Rennard, S. I., and Martin, G. R., 1980, Isolation of a heparan sulfate containing proteoglycan from basement membrane, *Proc. Natl. Acad. Sci. USA* **77**:4494–4498.
24. Howard, B. V., Macarak, E. J., Gunson, D., and Kefalides, N. A., 1976, Characterization of collagen synthesized by endothelial cells in culture, *Proc. Natl. Acad. Sci. USA* **73**:2361–2364.
25. Sage, H., Crouch, E., and Bornstein, P., 1979, Collagen synthesis by bovine aortic endothelial cells in culture, *Biochemistry* **18**:5433–5442.
26. Macarak, E. J., Kirby, E., Kirk, T., and Kefalides, N. A., 1978, Synthesis of cold-insoluble globulin by cultured calf endothelial cells, *Proc. Natl. Acad. Sci. USA* **75**:2621–2625.
27. Jaffe, E. A., and Mosher, D. F., 1978, Synthesis of fibronectin by cultured human endothelial cells, *J. Exp. Med.* **147**:1779–1791.
28. Birdwell, C. R., Gospodarowicz, D., and Nicolson, G. L., 1978, Identification, localization and role of fibronectin in cultured bovine endothelial cells, *Proc. Natl. Acad. Sci. USA* **75**:3273–3277.
29. Gospodarowicz, D., Greenburg, G., Foidart, J. M., and Savion, N., 1981, The production and localization of laminin in cultured vascular and corneal endothelial cells, *J. Cell. Physiol.* **107**:171–183.
30. Mosher, D. F., Doyle, M. J., and Jaffe, E. A., 1982, Synthesis and secretion of thrombospondin by cultured human endothelial cells, *J. Cell Biol.* **93**:343–348.
31. Clowes, A. W., Reidy, M. A., and Clowes, M. M., 1983, Kinetics of cellular proliferation after arterial injury. I. Smooth muscle growth in the absence of endothelium, *Lab. Invest.* **49**:322–333.
32. Ross, R., and Glomset, J. A., 1976, The pathogenesis of atherosclerosis, *N. Engl. J. Med.* **295**:369–377.
33. Clowes, A. W., and Karnovsky, M. J., 1977, Suppression by heparin of smooth muscle cell proliferation in injured arteries, *Nature* **265**:625–626.
34. Hoover, R. L., Rosenberg, R. D., Haering, W., and Karnovsky, M. J., 1980, Inhibition of rat arterial smooth muscle cell proliferation by heparin II: *In vitro* studies, *Circ. Res.* **47**:578–583.
35. Gimbrone, M. A., Cotran, R. S., and Folkman, J., 1974, Human vascular endothelial cells. Growth and DNA synthesis, *J. Cell Biol.* **60**:673–684.
36. Tumilowicz, J. J., Gawlik, M. E., Powell, B. B., and Trentin, J. J., 1985, Replication of cytomegalovirus in human arterial smooth muscle cells, *J. Virol.* **56**:839–845.
37. Laemmli, U. K., 1970, Cleavage of structural proteins during the assembly of the head of bacteriophage T4, *Nature* **227**:680–685.
38. Rennard, S. I., Berg, R., Martin, G. R., Foidart, J. M., and Gehron, P., 1980, Enzyme-linked immunoassay (ELISA) for connective tissue components, *Anal. Biochem.* **104**:205–214.
39. Lyons-Giordano, B., Brinker, J. M., and Kefalides, N. A., 1990, The effect of heparin on fibronectin and thrombospondin synthesis and mRNA levels in cultured human endothelial cells, *Exp. Cell. Res.* **186**:39–46.
40. Rowe, D. W., Moen, R. C., Davidson, J. M., Byers, P. H., Bornstein, P., and Palmiter, R. D., 1978, Correlation of procollagen mRNA levels in normal and transformed chick-embryo fibroblasts with different rates of procollagen synthesis, *Biochemistry* **17**:1581–1590.
41. Wahl, G., 1981, Schleicher and Schuell sequences: Application update No. 371, Schleicher and Schuell, Inc., Keene, N.H.
42. Maniatis, T., Fritsch, E. F., and Sambrook, J., 1982, *Molecular Cloning: A Laboratory Manual*, Cold Spring Harbor Laboratory, Cold Spring Harbor, N.Y.
43. Kornblihht, A. R., Vibe-Pederson, K., and Baralle, F. E., 1983, Isolation and characterization of cDNA clones for human and bovine fibronectins, *Proc. Natl. Acad. Sci. USA* **80**:3218–3222.

44. Dixit, V. M., Hennessey, S. W., Grant, G. A., Rotwein, P., and Frazier, W. A., 1986, Characterization of a cDNA encoding the heparin and collagen binding domains of thrombospondin, *Proc. Natl. Acad. Sci. USA* **83:**5449–5453.
45. Gunning, P., Ponte, P., Blau, H., Okayama, H., Engel, J., and Kedes, L., 1983, Human actin genes are single copy for a skeletal and a cardiac actin but multicopy for β and γ-cytoskeletal genes. 3′ untranslated regions are isotope specific but are conserved in evolution, *Mol. Cell. Biol.* **3:**854–862.
46. Hall, J. L., Dudley, L., Dobner, P. R., Lewis, S. A., and Gowan, N. J., 1983, Identification of two human β-tubulin isotypes, *Mol. Cell Biol.* **3:**787–795.
47. Myers, J. C., Chiu, M.-L., Fiaro, S. H., Clark, W. J., Prockop, D. J., and Ramirez, F., 1981, Cloning a cDNA for the pro α 2 chain of human type I collagen, *Proc. Natl. Acad. Sci. USA* **78:**3516–3520.
48. Chiu, M.-L., Weil, D., deWet, W., Bernard, M., Sippola, M., and Ramirez, F., 1985, Isolation of cDNA and genomic clones encoding human pro-α1(III) collagen. Partial characterization of the 3′ end region of the gene, *J. Biol. Chem.* **260:**4357–4363.
49. Clowes, A. W., and Clowes, M. M., 1985, Kinetics of cellular proliferation after arterial injury. II. Inhibition of smooth muscle growth by heparin, *Lab. Invest.* **52:**611–616.
50. Madri, J. A., Reidy, M. A., Kocher, O., and Bell, L., 1989, Endothelial cell behavior after denudation injury is modulated by transforming growth factor-β_1 and fibronectin, *Lab. Invest.* **60:**755–765.
51. Majack, R. A., and Bornstein, P., 1985, Heparin regulates the collagen phenotype of vascular smooth muscle cells: Induced synthesis of an M_r 60,000 collagen, *J. Cell Biol.* **100:**613–619.
52. Majack, R. A., and Bornstein, P., 1984, Heparin and related glycosaminoglycans modulate the secreting phenotype of vascular smooth muscle cells, *J. Cell Biol.* **99:**1688–1695.
53. Majack, R. A., Cook, S. C., and Bornstein, P., 1985, Platelet-derived growth factor and heparin like glycosaminoglycans regulate thrombospondin synthesis and deposition in the matrix by smooth muscle cells, *J. Cell Biol.* **101:**1059–1070.
54. Konkle, B. A., and Ginsburg, D., 1988, The addition of endothelial cell growth factor and heparin to human umbilical vein endothelial cell cultures decreases plasminogen activator inhibitor-1 expression, *J. Clin. Invest.* **82:**579–585.

5

Dysfunction of the Release of Endothelium-Derived Relaxing Factor

Thomas F. Lüscher and Paul M. Vanhoutte

I. INTRODUCTION

Due to its strategic anatomical position between the circulating blood and vascular smooth muscle cells, the endothelium is a primary target for injuries such as hypertension, hyperlipidemia, diabetes, and ischemia (Fig. 1).[1] Although it has been known for years that endothelial cells undergo morphological changes under these conditions, the functional alterations have only been characterized recently. As the endothelium plays a protective role in the circulation under physiological conditions by releasing endothelium-derived relaxing factor (which is a vasodilator and inhibitor of platelet function; for review see Refs. 1, 2), functional alterations of endothelial cells may be an important step in the pathogenesis of cardiovascular disease.

This review updates previous reports[1–3] and focuses on the dysfunction of the release of endothelium-derived relaxing factor occurring under several pathological conditions.

II. REGENERATED ENDOTHELIUM

In vivo denudation of the endothelium of porcine coronary arteries (by means of a balloon catheter) is followed by a rapid regeneration of the endothelial layer and recovery of endothelium-dependent relaxations to bradykinin and aggregating platelets within 1 week.[4] Four weeks after the procedure, however, the number of endothelial cells is increased (Fig. 2) and the endothelium-dependent relaxation to aggregating platelets is attenuated (Fig. 3). In contrast, the endothelium-dependent relaxations to ADP (Fig. 4), the Ca^{2+} ionophore A23187, bradykinin, thrombin, and platelet-activating factor are preserved. However, the

Thomas F. Lüscher • Department of Research, Laboratory of Vascular Research, and Department of Medicine, Divisions of Clinical Pharmacology and Cardiology, University Hospital, Basel, Switzerland. *Paul M. Vanhoutte* • Center for Experimental Therapeutics, Baylor College of Medicine, Houston, Texas 77030.

Endothelial Cell Dysfunctions, edited by Nicolae Simionescu and Maya Simionescu. Plenum Press, New York, 1992.

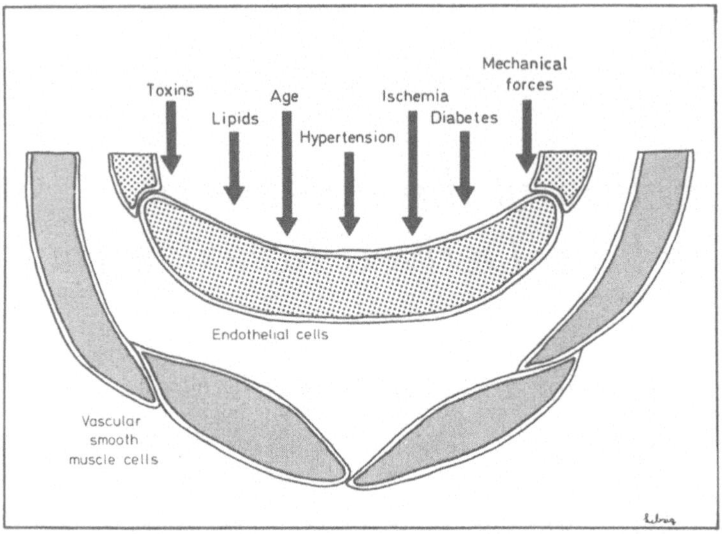

Figure 1. The endothelium as a target for vascular injuries. Various stimuli, such as toxins, lipids (in particular oxidized low-density lipoproteins), age, hypertension, ischemia, diabetes, and mechanical forces, do interfere with endothelium function. Thus, the cells may be important mediators of the development of functional alterations of the blood vessel wall.

endothelium-dependent relaxations to serotonin (Fig. 5) and the α_2-adrenergic agonist UK 14 304 are attenuated or lost, because of the loss of the pertussis toxin-sensitive component of the response (Fig. 6).[5] The lack of responsiveness to serotonin and α_2-adrenergic activation persists for 5 months at least, while in the same time period responses to ADP or bradykinin are only minimally affected. Thus, regenerated endothelial cells appear to lose the G-protein-mediated mechanism of release of endothelium-derived relaxing factor.[5]

In the regenerated state, the protective role of the coronary endothelium against vasoconstriction induced by aggregating platelets is impaired (Fig. 7). This explains why in the coronary circulation of the pig, platelet activation *in vivo* by intracoronary injection of serotonin or ADP can evoke coronary spasm at sites of previous endothelial injury.[6]

III. GRAFTED BLOOD VESSELS

Internal mammary arteries and saphenous veins are commonly used as bypass grafts to treat patients with coronary artery disease. Internal mammary artery grafts have a remarkably higher patency rate and patient outcome than saphenous vein grafts.[7] As this difference between arterial and venous grafts persists, if bypasses supplying the same vascular bed (for instance that supplied by the left anterior descending coronary artery) are compared, different biological properties of the internal mammary artery and saphenous vein have been implicated in that difference.[8]

Endothelium-dependent relaxations to acetylcholine are more pronounced in the internal mammary arteries than in the saphenous veins of patients undergoing coronary

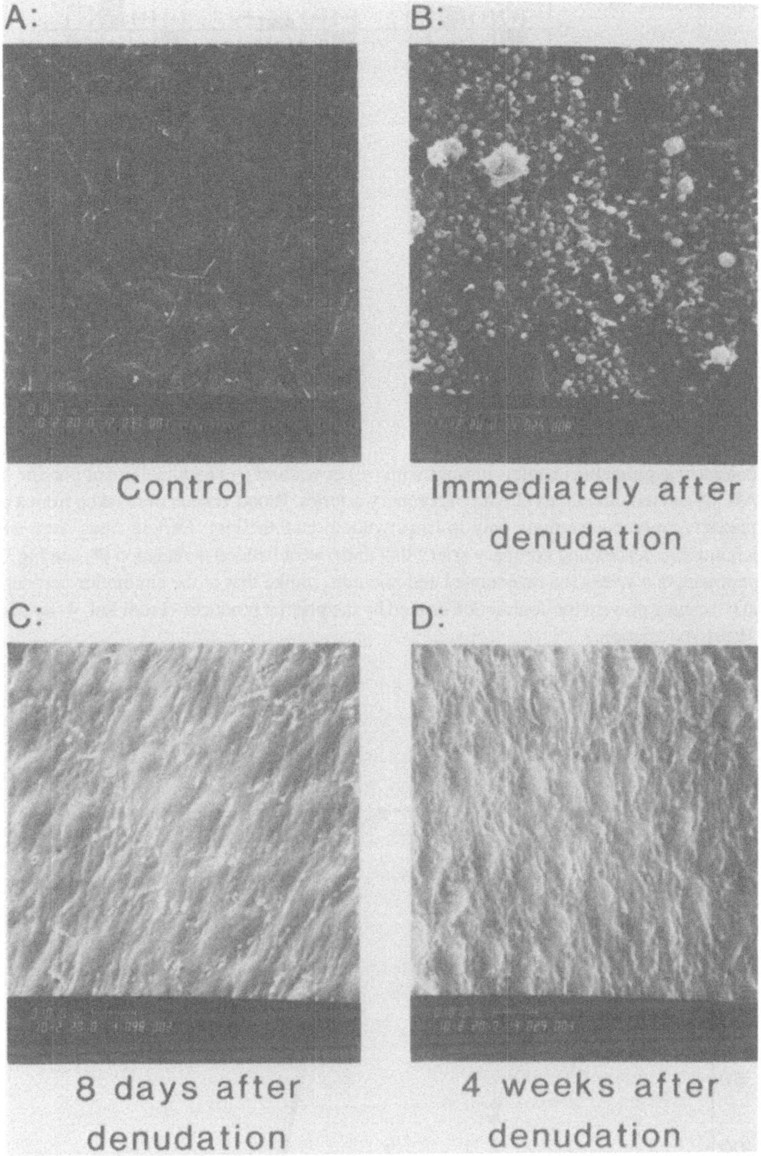

Figure 2. Scanning electron microscopy of porcine coronary arteries immediately after (B), 8 days (C) and 4 weeks (D) after the endothelium had been removed *in vivo* by means of a balloon catheter. Immediately after denudation, numerous platelets and a few leukocytes adhere to the lesion. Within 1 week a full endothelial lining has been regenerated. One month after denudation, the number of endothelial cells per area is larger (about twofold) than in control arteries. (From Ref. 4, by permission of the American Heart Association.)

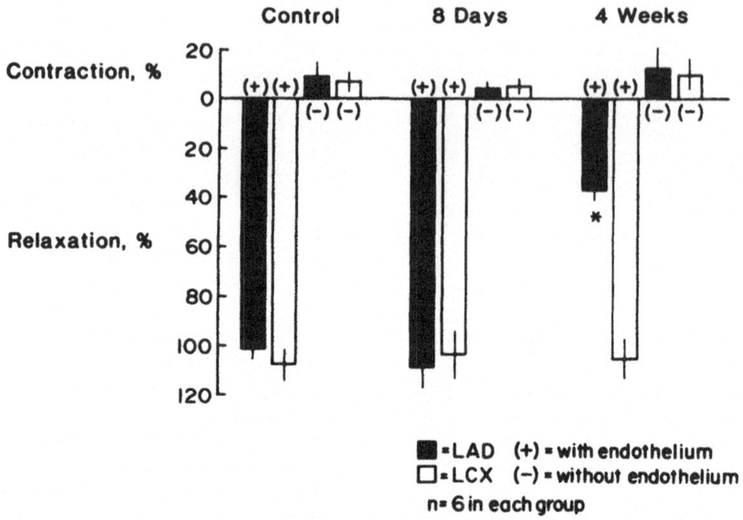

Figure 3. Responses to aggregating platelets in rings with (+) or without (−) endothelium of porcine left anterior descending (LAD; ■) and left circumflex (LCX; □) coronary arteries. Blood vessels were taken from a control pig; in both cases platelets evoke contractions only in rings without endothelium. Certain rings were taken from a portion of the left anterior descending coronary artery that underwent balloon denudation (■; see Fig. 2) 4 weeks prior to the experiment; at 4 weeks the regenerated endothelium, unlike that of the circumflex coronary artery of the same animal (□) cannot prevent the contraction evoked by the platelet products. (From Ref. 4, by permission of the American Heart Association.)

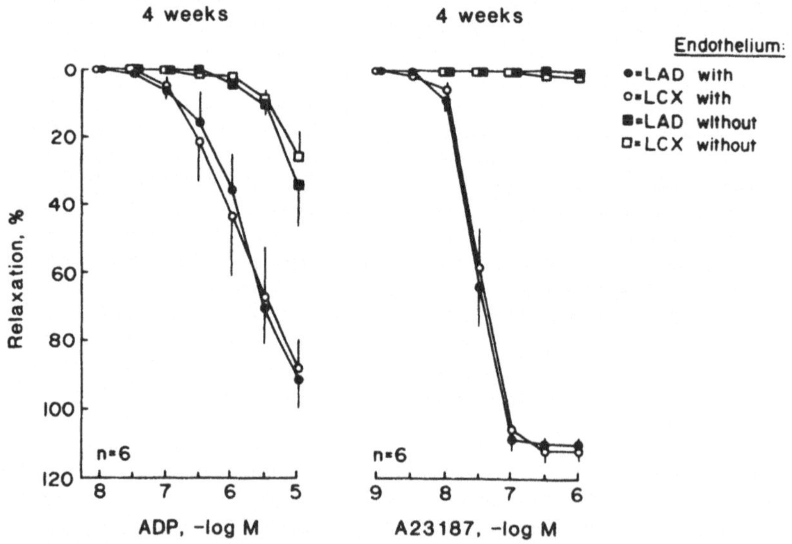

Figure 4. In rings of the left anterior descending coronary artery that underwent balloon denudation 4 weeks prior to the experiment, endothelium-dependent relaxations to ADP and the Ca^{2+} ionophore A23187 are comparable to control (left circumflex coronary artery), during contractions evoked by prostaglandin $F_{2\alpha}$. This demonstrates that the regenerated endothelium responds normally to these agents and is capable of releasing endothelium-dependent relaxing factor; this study also indirectly demonstrates that the sensitivity of the vascular smooth muscle for the factor has not changed. (From Ref. 4, by permission of the American Heart Association.)

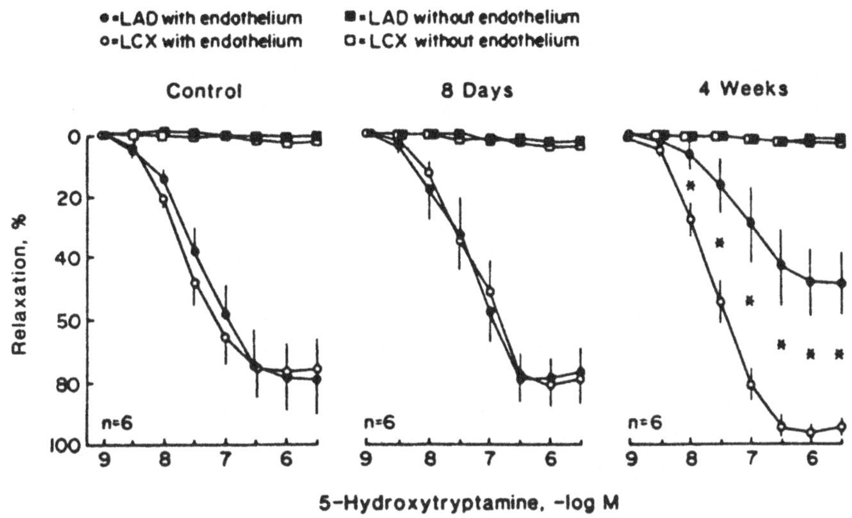

Figure 5. Endothelium-dependent relaxations to increasing concentrations of serotonin (5-hydroxytryptamine) in porcine left anterior descending (LAD) coronary arteries before (left) or after [8 days (center) or 4 weeks (right)] undergoing balloon denudation (see Fig. 2). As controls, the endothelium-dependent relaxations of the left circumflex (LCX) arteries of the same animals are shown. These experiments demonstrate that 4 weeks after denudation, the regenerated endothelium has a reduced responsiveness to serotonin. The experiments were performed in the presence of the $5HT_2$-serotonergic antagonist ketanserin to prevent activation of the vascular smooth muscle by the monoamine. (From Ref. 4, by permission of the American Heart Association.)

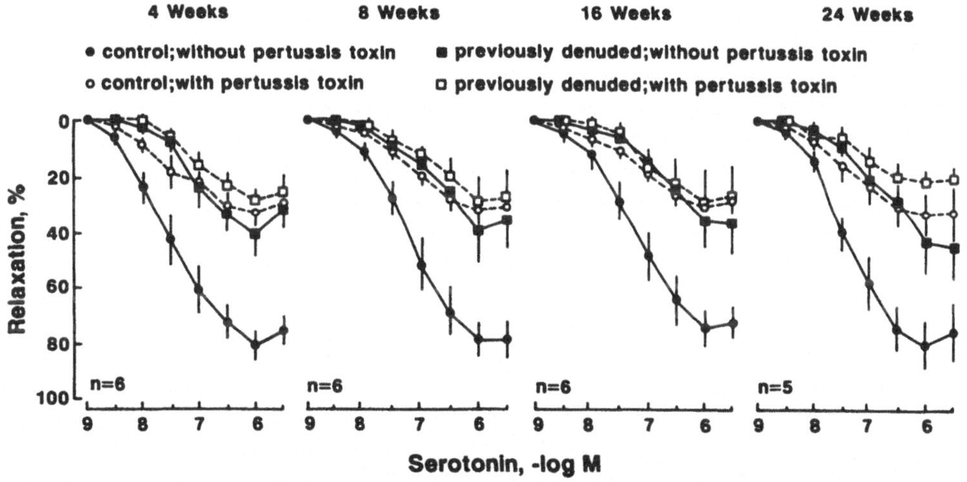

Figure 6. Endothelium-dependent relaxations to serotonin in control porcine coronary arteries and in arteries that underwent balloon denudation 4, 8, 16, or 24 weeks prior to the experiment. The effect of pertussis toxin (an inhibitor of G proteins) is shown also. Note that the degree of blunting of the response to serotonin observed in the previously denuded arteries is comparable to that obtained acutely in normal arteries with pertussis toxin, and that in the previously denuded arteries, the toxin causes no further inhibition of the relaxation to the monoamine. These observations strongly suggest that the regenerated endothelium has lost the pertussis toxin-sensitive mechanism of release of endothelium-dependent relaxing factor. In addition, this study illustrates that the abnormal response of the regenerated endothelium persists for at least 6 months. (From Ref. 5, by permission of the American Heart Association.)

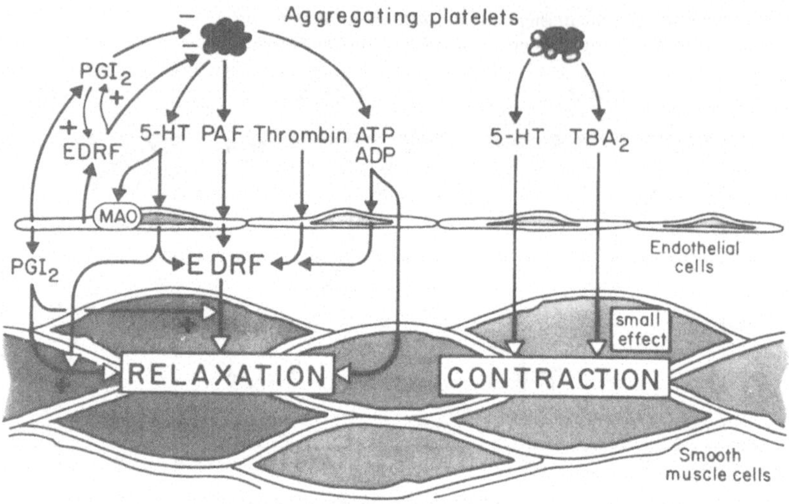

Figure 7. Endothelium-dependent responses under pathologic conditions. The endothelium is dysfunctional in the regenerated state, hypercholesterolemia, and atherosclerosis, releasing less endothelium-derived relaxing factor (EDRF), while the ability of the smooth muscle to contract is unaltered. As a result, the contractions predominate. In atherosclerosis, the production of both endothelium-dependent relaxing factor and prostacyclin (PGI_2) is reduced, and their synergistic actions against aggregating platelets may not occur. 5-HT = serotonin; MAO = monoamine oxidase; TBA_2 = thromboxane A_2; − = inhibition; + = synergism. (From Ref. 3, by permission of the American Heart Association.)

bypass surgery (Fig. 8).[9–13] Of clinical relevance is the fact that ADP, thrombin (Fig. 9),[9] and histamine (Fig. 10)[12] evoke endothelium-dependent relaxations of the internal mammary artery, while the responses are weak or absent in the saphenous vein.[9,12,13] Aggregating platelets release enough ATP and ADP to cause endothelium-dependent relaxations in the artery, while the veins contract when exposed to platelets.[14] As L-N^G-monomethyl arginine (L-NMMA; the inhibitor of nitric oxide formation from L-arginine) prevents endothelium-dependent relaxations to aggregating platelets in the internal mammary artery, nitric oxide must mediate the response.[14]

The weak endothelium-dependent relaxations to acetylcholine in saphenous veins must be related to a decreased production and/or to a less efficient release of endothelium-derived relaxing factor. Clearly, the phenomenon cannot be explained by an inability of venous vascular smooth muscle to respond to nitric oxide. Indeed, in the saphenous vein contracted with norepinephrine, exogenous nitric oxide and nitrovasodilators such as sodium nitroprusside and 3-morpholino-sydnonimine (SIN-1; the active metabolite of molsidomine) tend to be more effective than in the artery.[9,15] The difference also is unrelated to weak receptor-operated release mechanisms, as the Ca^{2+} ionophore A23187 also evokes only a half-maximal relaxation in the saphenous vein, but complete relaxation in the internal mammary artery.[9]

As in the internal mammary artery, the inhibitor of nitric oxide formation from L-arginine, L-NMMA, markedly reduces endothelium-dependent relaxations to acetylcholine and prevents those to histamine; nitric oxide must be the mediator of the response.[13] Similarly, the mediator of endothelium-dependent relaxations in the saphenous vein can be

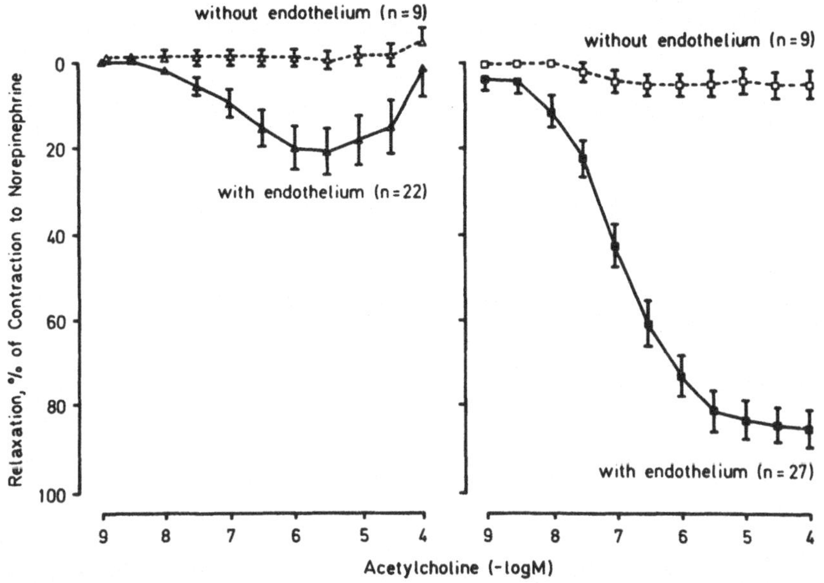

Figure 8. Effects of acetylcholine in the isolated human internal mammary artery (left) and saphenous vein (right). The muscarinic agonist evokes relaxations in preparations with (solid symbols), but not in those without endothelium (open symbols). Endothelium-dependent relaxations are more pronounced in the arteries than in the veins (From Ref. 9, by permission.)

identified as nitric oxide, since L-NMMA fully inhibits the relaxations to the muscarinic agonist.[13] In fact, under this conditions, endothelium-dependent contractions to acetylcholine are unmasked which can be prevented by inhibitors of cyclooxygenase or thromboxane synthetase. This suggests that the saphenous vein produces not only endothelium-derived nitric oxide, but also a cyclooxygenase-dependent endothelium-derived contracting factor ($EDCF_2$), most likely thromboxane A_2 (Fig. 11).[13] In contrast, although indomethacin unmasks endothelium-dependent relaxation to histamine in the saphenous vein, thromboxane receptor antagonists, but not inhibitors of thromboxane synthetase have a similar effect. Thus, histamine must elicit the release of an endothelium-derived cyclooxygenase product distinct from thromboxane A_2 which activates the thromboxane/endoperoxide receptor, most likely prostaglandin H_2 (Fig. 11).[13]

A reduced abluminal production of endothelium-derived relaxing factor in the saphenous vein as compared to the internal mammary artery would favor the occurrence of platelet-induced vasoconstriction in the former. Furthermore, if the luminal release of endothelium-derived relaxing factor were reduced also in the saphenous vein, platelet adhesion and aggregation would be inhibited to a lesser extent which would facilitate platelet disposition, plaque formation, and eventually occlusion of the graft.[2] The fact that the venous endothelium produces less prostacyclin than the internal mammary artery further would contribute to this phenomenon.[16] Thus, differences in endothelial function may importantly contribute to the function of coronary bypass grafts and may at least in part explain the more favorable outcome of patients receiving arterial as compared to venous grafts (Fig. 12).[8]

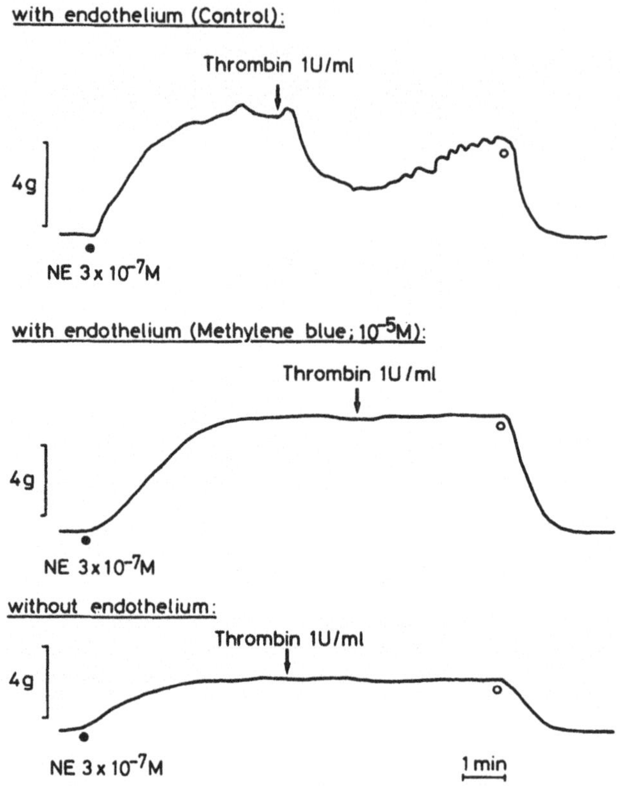

Figure 9. Effects of human thrombin (1 U/ml) in the isolated human internal mammary artery. In the ring with endothelium (control; top), but not in that without endothelium (bottom), the enzyme evokes a potent relaxation. Methylene blue prevents the response in a preparation with endothelium (middle), indicating that it is mediated by endothelium-derived relaxing factor. NE = norepinephrine. (From Ref. 9, by permission.)

The difference between arterial and venous endothelial cells probably persists after implantation of the grafts. Indeed, in the dog, grafting of the femoral vein into the arterial circulation does not enhance endothelium-dependent relaxations to acetylcholine, ADP (Fig. 13), and thrombin.[17] In contrast, at sites where myointimal thickening develops, mostly at the distal part of the venous grafts, endothelium-dependent relaxations tend to be reduced.[17] In chronically implanted venous grafts of patients with coronary artery disease, obtained during heart transplantation, the endothelium-dependent relaxation to acetylcholine is only slightly more pronounced than that reported in freshly harvested saphenous veins.[9,18] Arterialization of venous endothelium also does not enhance the luminal production of prostacyclin by the venous wall.[19] The reduced production of endothelium-derived relaxing factor and prostacyclin, particularly in areas of myointimal thickening, helps to explain why superimposed thrombi at these sites contribute to late closure of the grafts.[20]

IV. HYPERTENSION

A. Endothelium-Dependent Regulation of Blood Pressure

The systemic application of endothelium-dependent vasodilators such as acetylcholine, ADP, and substance P markedly decreases blood pressure in the rabbit.[21] Intra-

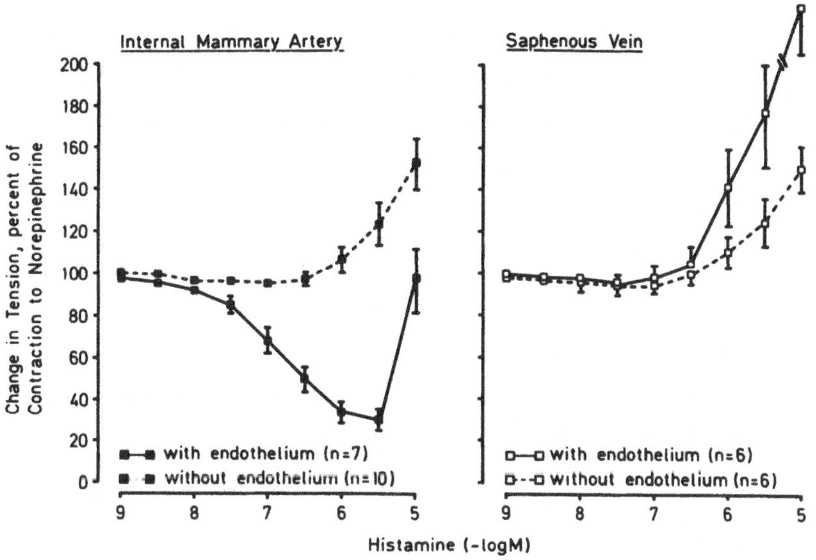

Figure 10. Effects of histamine in the isolated human internal mammary artery (left) and saphenous vein (right). Histamine evokes relaxations (via activation of an H_1-histaminergic receptor on the endothelium) in arterial preparations with (solid lines), but not in those without endothelium (dotted lines). In the absence of indomethacin, endothelium-dependent relaxations do not occur in the saphenous vein. Indeed, the presence of the endothelium facilitates the contractions to histamine. (From Ref. 12, by permission of the American Heart Association.)

venous infusion of L-NMMA, the inhibitor of the formation of nitric oxide from L-arginine, increases the arterial blood pressure in the rat and the rabbit (Fig. 14).[22] Similarly, infusion of hemoglobin leads to an increase in arterial blood pressure in healthy human subjects.[23] These observations imply that, in normotensive animals and subjects, the ongoing release of endothelium-derived relaxing factor participates in the overall regulation of peripheral vascular resistance. Thus, alterations in the release of endothelium-derived vasoactive factors could contribute to the increase in peripheral vascular resistance characteristic of hypertension.

B. Hypertension

Hypertension can interfere with both the basal and stimulated release of endothelium-derived relaxing factor. In perfused mesenteric resistance arteries, the presence of endothelial cells reduces the sensitivity and maximal response to norepinephrine (Fig. 15.[24] L-NMMA causes a comparable augmentation of the reactivity to the catecholamine as does endothelium removal, suggesting that nitric oxide in large part explains the inhibitory role of the endothelium against norepinephrine-induced contractions. As the responses to the α_1-adrenergic agonist phenylephrine are similarly inhibited as those to norepinephrine, the endothelium-dependent inhibition of the contractile response is most likely related to basal rather than stimulated release (through α_2-adrenergic receptors; see Ref. 2) of endothelium-derived nitric oxide.[24] In line with that interpretation, the basal release of endothelium-derived relaxing factor from the aorta of the spontaneously hypertensive rat is reduced

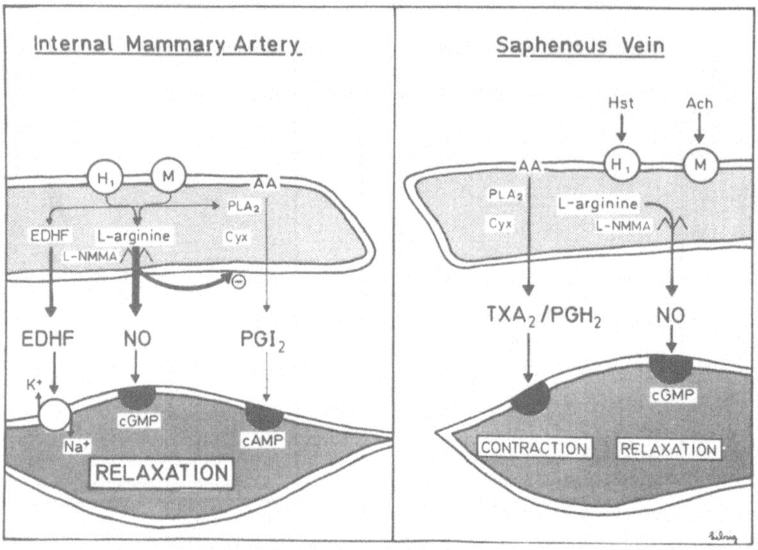

Figure 11. Endothelial mediators released from the human internal mammary artery and saphenous vein after stimulation with acetylcholine or histamine. In the artery, nitric oxide (NO) is the only or predominant mediator. It remains uncertain whether endothelium-derived hyperpolarizing factor (EDHF) contributes to the response. Prostacyclin (PGI_2) does not significantly contribute. In the saphenous vein, the relaxation is much weaker, but also caused by endothelium-derived NO. In addition, the venous endothelium produces endothelium-derived contracting factors originating from the cyclooxygenase pathway (i.e., thromboxane A_2, TXA_2; prostaglandin H_2, PGH_2). AA = arachidonic acid; cGMP = cyclic guanosine monophosphate; Cyx = cyclooxygenase; H_1 = histaminergic receptor; L-NMMA = L-N^G-monomethylarginine; M = muscarinic receptor.

under bioassay conditions.[25] Similarly, the basal levels of cyclic GMP are decreased in aortas with endothelium from rats with spontaneous, renal, DOCA salt hypertension and coarctation of the aorta.[26,27]

The endothelium-dependent relaxations to acetylcholine are attenuated in the thoracic aorta of rats with spontaneous hypertension (Fig. 16), renal hypertension, salt-induced hypertension, coarctation, and DOCA salt-induced hypertension.[28–37] Similar changes have been observed in the carotid artery of the spontaneously hypertensive rat.[38,39] In small mesenteric resistance arteries of the spontaneously hypertensive rat, the endothelium-dependent relaxations to acetylcholine are also impaired (Fig. 17).[40–44] In pressurized and perfused mesenteric resistance arteries, the intraluminal surface of the endothelial cells, which is most exposed to high blood pressure, exhibits the most pronounced alterations in function.[24] Thus, in 16- to 20-week-old rats, intraluminal, but not extraluminal application of the muscarinic agonist is associated with impaired endothelium-dependent relaxations.

Similar alterations in the response to endothelium-dependent agonists can be demonstrated in the intact organism.[45–47] In the cerebral microcirculation of the anesthetized, stroke-prone spontaneously hypertensive rat, the vasodilator responses to acetylcholine (but not those to nitroglycerin) are attenuated or converted into contraction.[46] Similarly, in the human forearm circulation, the vasodilatation evoked by acetylcholine is attenuated in hypertensive patients (Fig. 18).[48,49]

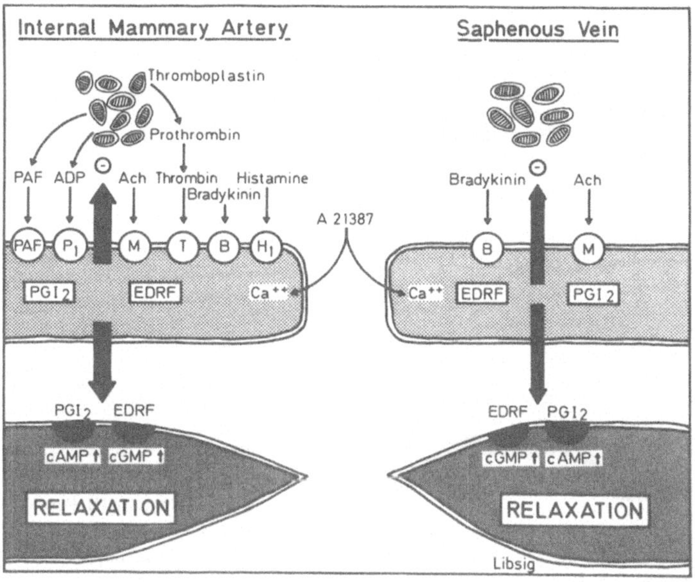

Figure 12. Platelet–vessel wall interaction in the human internal mammary artery and saphenous vein. In the artery platelet-derived mediators stimulate the release of endothelium-derived relaxing factor (EDRF), while in the vein this response is much less expressed. This difference may explain the more pronounced platelet adhesion and aggregation in venous as compared to arterial grafts. Ach = acetylcholine; B = bradykinin receptor; cAMP/cGMP = cyclic adenosine/guanosine monophosphate; H_1 = histaminergic receptor; P_1 = purinergic receptor; M = muscarinic receptor; PGI_2 = prostacyclin; T = thrombin receptor. (From Ref. 8, by permission.)

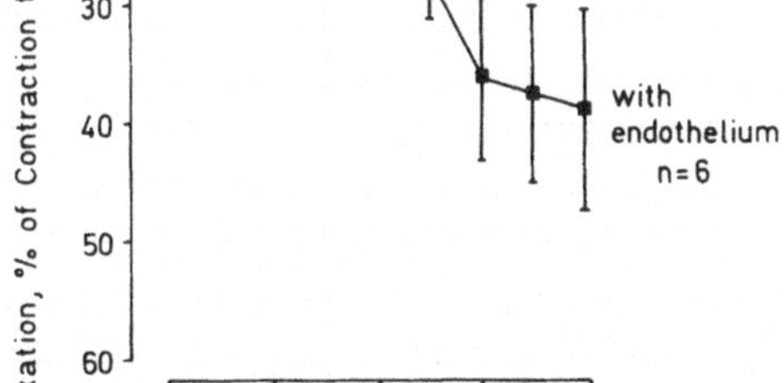

Figure 13. Endothelium-dependent relaxations to ADP in the human internal mammary artery. This response is weak or absent in the saphenous vein. (From Ref. 9, by permission.)

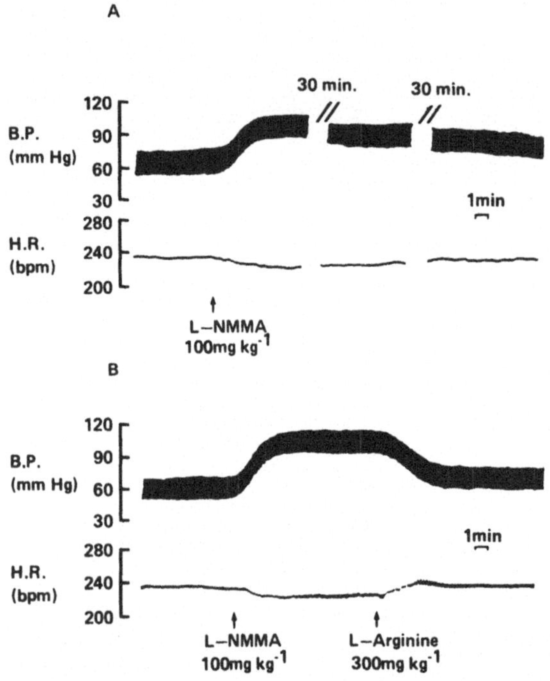

Figure 14. Effect of L-N^G-monomethyl arginine (L-NMMA; 100 mg/kg intravenously) on blood pressure (B.P.) and heart rate (H.R.) in the anesthetized rabbit. L-NMMA evokes a marked and sustained rise in blood pressure and a slight decease in heart rate (A). The infusion of L-arginine reverses the effects of L-NMMA (B). (From Ref. 22, by permission.)

In the aorta of the Dahl rat, the degree of impairment of endothelium-dependent relaxations to acetylcholine is directly related to the level of systolic blood pressure.[50] In hypertensive Dahl rats, successful therapy normalizes endothelium-dependent relaxations (Figs. 19 and 20).[51] In rats with coarctation of the aorta, endothelium-dependent relaxations are normal in the aortic segment distal to the stenosis.[29,32] These observations indicate that, at least in these secondary forms of hypertension, the reduced endothelium-dependent relaxations are a consequence rather than a cause of the high blood pressure. Likewise, in the spontaneously hypertensive rat, the endothelium-dependent relaxations to acetylcholine are unaltered in young animals, but they become progressively impaired as the blood pressure rises.[24,39,41,52] This suggests that the premature aging resulting from the high level of blood pressure plays a major role in the loss of endothelium-dependent responses. If this interpretation is correct, endothelium-dependent responses are not likely to play a primary role in the initiation of hypertension. However, at later stages of the disease, the endothelial dysfunction (particularly that occurring in small arteries[41–44,46–49]) would help to maintain or even augment further the increase in peripheral vascular resistance.

With regard to the complications of hypertension, such as myocardial infarction and stroke, interactions of platelets and coagulation products with the blood vessel wall, are of particular interest (Fig. 7; for review see Ref. 2). The endothelium-dependent relaxations to thrombin are reduced in Dahl hypertensive, but not in spontaneously hypertensive rats (Fig. 20).[30,53] By contrast, those to ADP are impaired in the aorta and in the carotid artery of both hypertensive strains.[30,38,53] Likewise, in the microcirculation of the stroke-prone spontaneously hypertensive rat, the dilatations induced by ADP and serotonin are re-

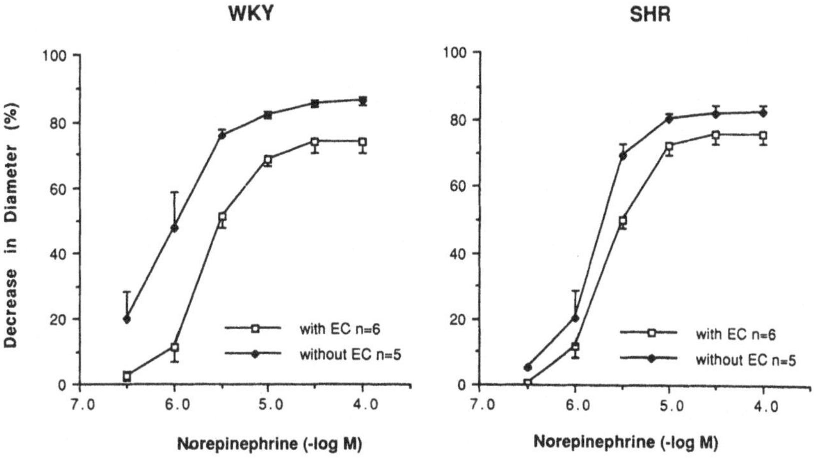

Figure 15. Effects of norepinephrine in perfused and pressurized mesenteric resistance arteries of Wistar–Kyoto (WKY; left) and spontaneously hypertensive rats (SHR; right). Note the augmented sensitivity and potency of the catecholamine after removal of the endothelium. (From Ref. 24, by permission of the American Heart Association.)

duced.[46,47] Serotonin and aggregating platelets do not evoke endothelium-dependent relaxations in the aorta or in the carotid artery of the rat, but in the aorta of normotensive rats, the contractions evoked by aggregating platelets are inhibited in the presence of the endothelium (Fig. 21).[38,53] By contrast, in spontaneously hypertensive rats the endothelium loses its inhibitory capacity.[53] The altered endothelial function, together with the greater aggregability of the platelets from hypertensives,[54,55] may favor the occurrence of vasoconstriction and thrombus formation in hypertensive blood vessels. Indeed, the uptake and content of serotonin of hypertensive platelets, particularly those of older hypertensive men, is reduced.[54,56,57] The thrombin-induced release of serotonin from platelets is increased in

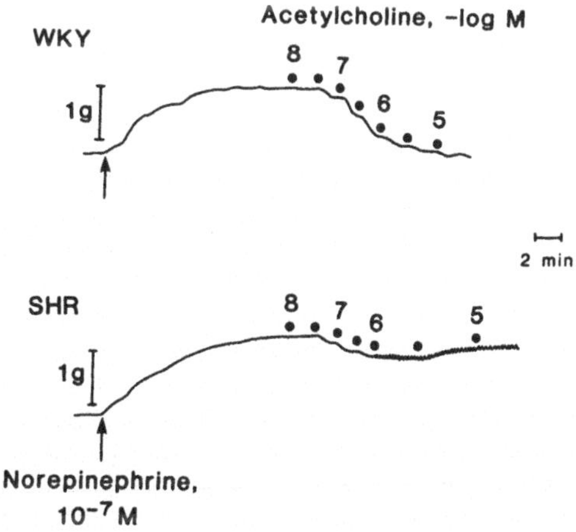

Figure 16. Relaxations to acetylcholine in aortic rings with endothelium obtained from a normotensive Wistar-Kyoto rat (WKY) and a spontaneously hypertensive rat (SHR). These experiments demonstrate that the relaxation to acetylcholine (concentration given in −log M) is reduced in the hypertensive rat. (From ref. 179, by permission.)

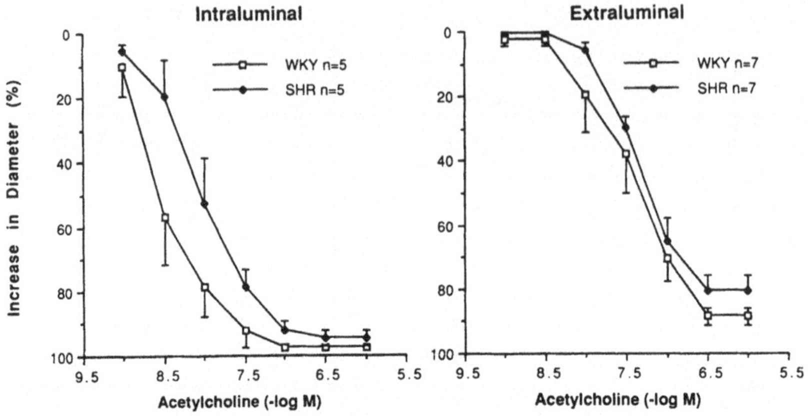

Figure 17. Effects of intraluminal (left) and extraluminal (right) acetylcholine in perfused and pressurized mesenteric resistance arteries of Wistar–Kyoto (WKY) and spontaneously hypertensive rats (SHR). Note the reduced endothelium-dependent relaxations with intraluminal, but not extraluminal infusion of the muscarinic agonist in the SHR. (From Ref. 24, by permission of the American Heart Association.)

spontaneously hypertensive rats and in patients with essential hypertension.[58,59] Platelet aggregation induced by ADP and serotonin is also enhanced in hypertensive patients.[60]

Although the endothelial function is altered in all types of hypertension, the reason underlying the dysfunction varies depending on the cause of the disease and the blood vessel studied. Indeed, in the aorta of rats with spontaneous, renal, and DOCA salt hypertension, the luminal release of endothelium-derived relaxing factor induced by acetylcholine and

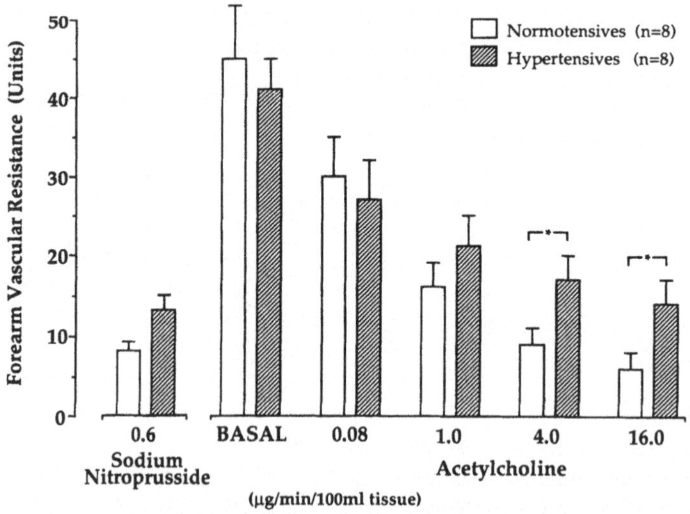

Figure 18. Effects of intraarterial infusion of acetylcholine in the forearm circulation of healthy volunteers (shaded bars) and patients with essential hypertension (open bars). In both groups, the muscarinic agonist markedly reduces forearm vascular resistance, but less so in hypertensive patients. In contrast, the response to sodium nitroprusside did not differ statistically. (From Ref. 48, by permission of the American Heart Association.)

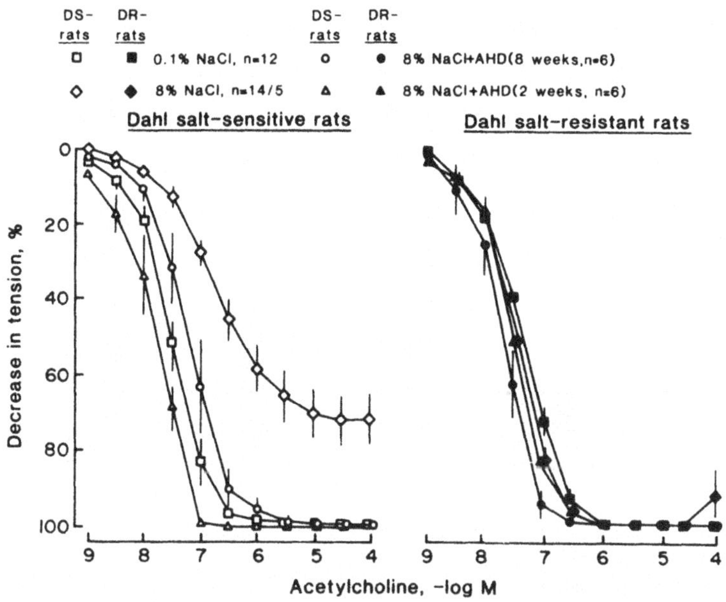

Figure 19. Demonstration that 2 and 8 weeks of successful antihypertensive therapy (AHD; a combination of reserpine plus hydrochlorothiazide plus hydralazine) restores or maintains a normal responsiveness to acetylcholine in Dahl-sensitive (DS) rats exposed to a high-salt diet, without affecting the response in the Dahl-resistant (DR) strain. (From Ref. 51, by permission of the American Heart Association.)

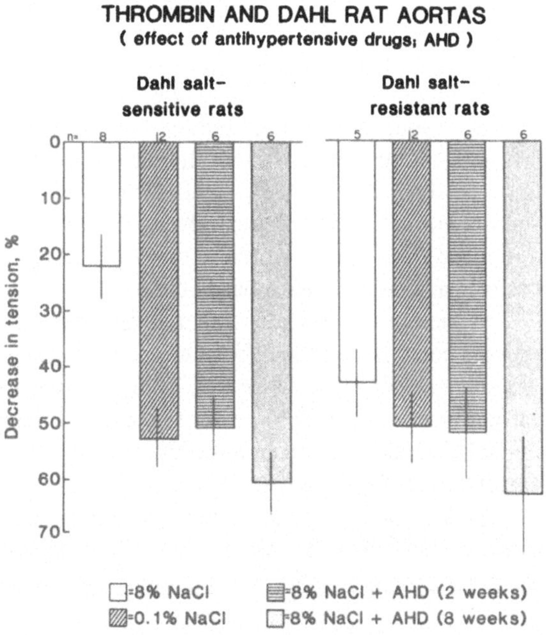

Figure 20. Demonstration that 2 and 8 weeks of successful antihypertensive therapy (AHD; a combination of reserpine plus hydrochlorothiazide plus hydralazine) restores or maintains the reduced endothelium-dependent relaxations to thrombin in Dahl-sensitive (DS) rats exposed to a high-salt diet, without affecting the response in the Dahl-resistant (DR) strain. (From Ref. 51, by permission of the American Heart Association.)

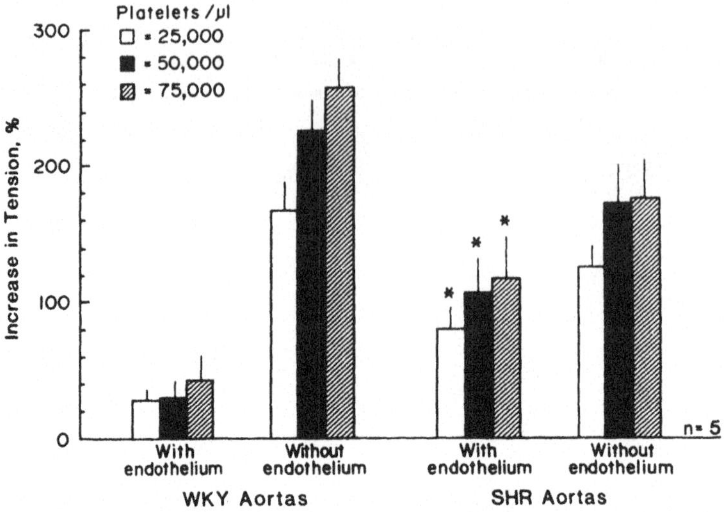

Figure 21. In the aorta of the normotensive Wistar–Kyoto rat (WKY), increasing numbers of aggregating platelets cause a large increase in tension in rings without, but a minimal reaction in rings with endothelium (left). By contrast, in the aorta of the spontaneously hypertensive rat (SHR), the contractions evoked by the aggregating platelets are not significantly different in rings with and without endothelium (right). These experiments demonstrate that the protective role of the endothelium in preventing platelet-induced vasoconstriction (see Fig. 7) is considerably curtailed in this model of hypertension. Asterisk indicates that the difference between aortas from hypertensive and normotensive rats is statistically significant. (From Ref. 53, by permission of the American Heart Association.)

histamine is normal.[25,36,61] In the aorta of these animal models of hypertension, an enhanced release of an endothelium-derived contracting factor ($EDCF_2$) derived from the cyclooxygenase pathway, most likely prostaglandin H_2, accounts in large part for the dysfunction of the endothelium.[62,63] In contrast, in the aorta of the Dahl salt-sensitive hypertensive rat, the reduced relaxation to acetylcholine is explained best by the combination of a reduced release of endothelium-derived relaxing factor and a decreased responsiveness of the vascular smooth muscle to the factor.[30] In the forearm circulation of hypertensive subjects, the vasodilator response to acetylcholine is reduced, while that to sodium nitroprusside is comparable to that obtained in normal subjects (Fig. 18).[48,49] Thus, a reduced release and/or production of endothelium-derived relaxing factor appears to be the primary defect under these conditions.

Although theoretically, subendothelial thickening occurring during the hypertensive process could impair the diffusion of substances released from the endothelium toward the vascular smooth muscle (see Ref. 2), it is unlikely to play an important role. Indeed, in the carotid artery of the rabbit, extensive intimal thickening occurs after mechanical denudation with a balloon catheter; after regeneration of the endothelium, however, the response to acetylcholine fully recovers.[64]

Functional alterations of the vascular smooth muscle (due to medial hypertrophy)[65] can contribute to the impaired endothelium-dependent responses in hypertension, but they are not a prerequisite for that phenomenon. Indeed, reduced, normal, or even enhanced relaxations to sodium nitroprusside have been reported.[29,30,41,42,44–46,66] Depressed endothelium-dependent relaxations, however, are observed in hypertensive arteries despite a normal or even an augmented response to the nitrovasodilators.[41,42,44]

V. DIABETES MELLITUS

The increased incidence of vascular disease in diabetes could at least in part be related to an early endothelial dysfunction.[67,68]

A. Basal Release of Relaxing Factor

It appears that in diabetes the spontaneous release of endothelium-derived relaxing factor can be normal, augmented, or reduced. Thus, in the aorta and in mesenteric arteries of alloxan- and streptotozin-diabetic rats, the sensitivity to norepinephrine and the selective α_1-adrenergic agonist methoxamine are increased.[69,70] The removal of the endothelium augments the response to both agonists, but to a lesser extent in arteries from diabetic than from control animals. By contrast, removal of the endothelium augments the maximal force generated by phenylephrine only in the femoral artery of diabetic, but not that of control dogs.[71,72] With α_2-adrenergic agonists, conflicting results have been reported also, ranging from a severely impaired to a facilitated inhibition by the diabetic endothelium.[69,70]

B. Stimulated Release of Relaxing Factor

In the aorta of spontaneously diabetic rats, the endothelium-dependent relaxations to acetylcholine and the Ca^{2+} ionophore A23187 are attenuated.[73] Since the relaxations to sodium nitroprusside and adenosine are unaltered, this suggests that either the release of endothelium-derived relaxing and contracting factors are reduced or augmented, respectively. The latter possibility is supported by experiments in rabbits with streptozotocin-induced diabetes where acetylcholine evokes endothelium-dependent contractions.[74] The contractions induced by the muscarinic agonist can be blocked by inhibitors of cyclooxygenase and antagonists of thromboxane receptors, suggesting that endothelium-derived thromboxane A_2 or endoperoxides mediate the response.

Contradictory results have been reported as regards endothelium-dependent relaxations to acetylcholine and to histamine in rats with alloxan- and streptotozin-induced diabetes.[69,70,75–78] In arteries of dogs and rabbits with diabetes, the endothelium-dependent relaxations to acetylcholine and A23187 are unaltered or slightly augmented.[71,72,79]

C. Microcirculation

Acetylcholine relaxes the microvessels of isolated penile corpora cavernosa. In tissue from diabetic men with impotence, the endothelium-dependent relaxations to acetylcholine, but not those to sodium nitroprusside and papaverine, are decreased, which suggests a reduced release of endothelium-derived relaxing factor (Fig. 22).[80] Likewise, in small mesenteric arteries of the diabetic rat studied *in situ*, the dilator responses to

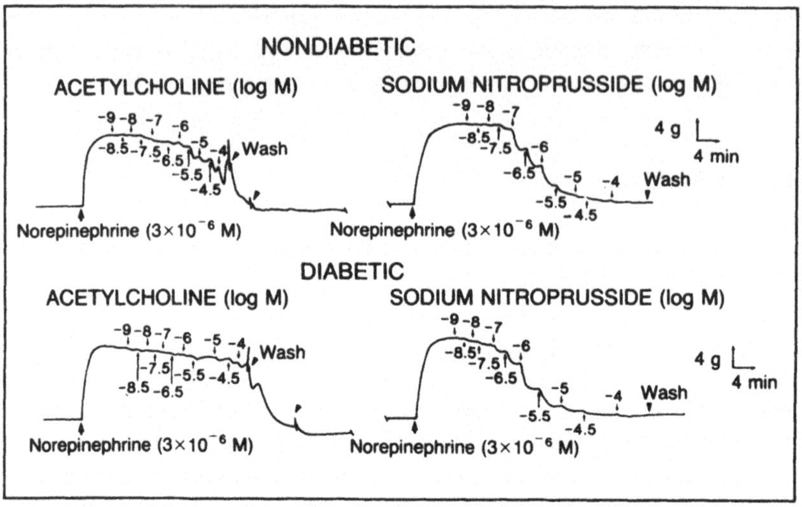

Figure 22. Effects of acetylcholine (left) and sodium nitroprusside (right) in isolated corpora cavernosa contracted with norepinephrine of nondiabetic patients (top) and patients with diabetes mellitus (bottom). These experiments demonstrate that sodium nitroprusside relaxes the preparations under both conditions, while the response to acetylcholine is reduced in the tissues from diabetic men. (From Ref. 80, by permission.)

histamine and bradykinin are blunted.[75,81] Since the response is unaffected by indomethacin[82] and is endothelium-dependent in isolated mesenteric resistance arteries,[41] this indicates a reduced capacity of histamine and bradykinin to release endothelium-derived relaxing factor. Similar conclusions have been reached in the cerebral microcirculation of anesthetized diabetic rats where the response to ADP (but not that to nitroglycerin) is severely impaired and the dilator response to serotonin converted into a contraction.[45] In the mesenteric microcirculation of the rat, the response to acetylcholine appears more resistant to the effects of diabetes as the response is only impaired 3 months after the onset of the disease.[81]

Previous administration of insulin normalizes the impaired responses to histamine and bradykinin.[82] The altered response to histamine and bradykinin is not linked to hyperglycemia, however, since it remains abnormal after restoring the plasma glucose levels by fasting. Similar functional defects as in diabetic animals can be induced in normal rats with 2-deoxyglucose, which causes intracellular glycopenia by inhibiting the utilization of the cellular sources of glucose.[82] This suggests that the endothelial effects of bradykinin and histamine (but not those of acetylcholine) require the intracellular utilization of glucose. In confirmation of this conclusion, the topical application of insulin normalizes the inhibitory effects of histamine and bradykinin in diabetic microvessels. By contrast, in normal animals given 2-deoxyglucose, insulin is ineffective in restoring the responses.[82]

In contrast to the mesenteric and cerebral microcirculation, the dilator response to acetylcholine (but not that to sodium nitroprusside) is augmented in perfused kidneys obtained from streptotozin-diabetic rats.[83] Since the response to the muscarinic agonist is blocked by hydroquinone, an augmented release of endothelium-derived relaxing factor may occur in the diabetic kidney.

VI. HYPERCHOLESTEROLEMIA

A. Lipoproteins

In the rabbit aorta contracted with either norepinephrine or serotonin, low-density lipoproteins (LDL), but not high-density lipoproteins (HDL), inhibit endothelium-dependent relaxations to acetylcholine.[84] The inhibitory concentration of the lipoprotein corresponds to that found in the plasma of patients with severe hyperlipidemia. In cultured porcine endothelial cells, native LDL inhibits the basal and receptor-operated release of endothelium-derived relaxing factor.[85] However, an effect is achieved only at very high concentrations (i.e., 2 mg/ml). In contrast, oxidized LDL markedly attenuates the release of the factor in bovine and porcine endothelial cells in culture at much lower concentrations.[85] In line with that observation, a predominant inhibitor effect of oxidized LDL on endothelium-dependent relaxations to acetylcholine also occurs in the rabbit aorta.[86,88]

B. Dietary Cholesterol

In the pig, a diet rich in cholesterol causes a moderate reduction of the endothelium-dependent relaxations of the coronary artery to serotonin (Fig. 23).[89,90] Hypercholesteremia *per se* also reduces endothelium-dependent relaxations to aggregating platelets, serotonin, and ADP in this and other arteries of the pig.[90–92] Similarly, in the porcine basilar artery, hypercholesteremia decreases the endothelium-dependent relaxations to aggregating platelets, but primarily because of a reduced response to platelet-derived ADP.[92] These results

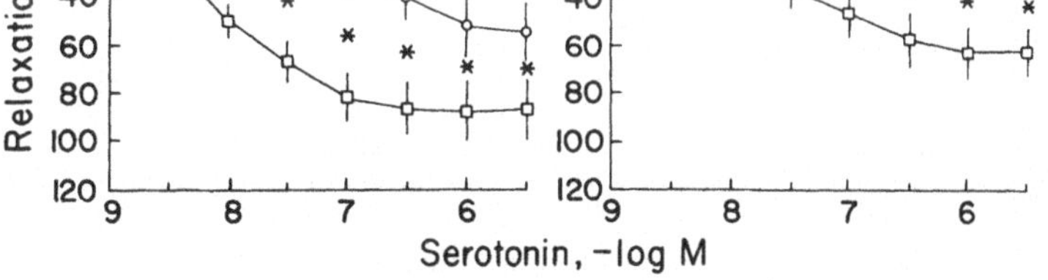

Figure 23. Endothelium-dependent relaxations to serotonin in porcine coronary arteries contracted with $F_{2\alpha}$ prostaglandin. In the right coronary artery (RCA; left panel) exposed to high cholesterol levels (hypercholesteremia), the response is moderately reduced as compared to control (see Fig. 4) and almost restored to normal by a diet enriched with fish oil. After previous endothelial denudation, hypercholesteremia leads to atherosclerotic changes in the left anterior descending coronary artery (LAD; right panel). The relaxations to serotonin are markedly reduced under these conditions, but much less so in arteries obtained from animals treated with a diet rich in fish oil. * $p < 0.05$. (From Ref. 90, by permission of the American Heart Association.)

contrast with the canine coronary artery, which appears relatively resistant to hypercholesteremia, as are the veins of the same species.[93]

C. Atherosclerosis

The endothelium-dependent relaxations to acetylcholine and ADP are reduced both *in vitro* and *in vivo* in the aorta of hypercholesteremic rabbits and in the common carotid and renal artery of Watanabe rabbits (Fig. 25).[94–100] The reduction of endothelium-dependent relaxations to these agonists is related to the amount of cholesterol in the diet, the duration of the diet, and the degree of fatty streak formation. In the aortic arch where atherosclerotic changes are the most severe, the endothelium-dependent relaxations to muscarinic and purinergic agonists are impaired more than in the abdominal aorta and the renal arteries.[100] In the hindlimb of atherosclerotic rabbits, the maximal relaxation, but not the sensitivity to acetylcholine, is reduced.[101] In diet-induced atherosclerosis of the rabbit, endothelium-dependent relaxations are impaired as early as 4 weeks and remain so for up to 16 weeks.[98,100]

In the coronary artery of the pig, the denudation of the endothelium followed by a hypercholesterolemic diet causes accelerated (within 4 weeks) atherosclerosis. In these blood vessels, the endothelium-dependent relaxations to aggregating platelets, serotonin (Fig. 24), ADP, thrombin, bradykinin, and substance P are markedly inhibited.[89–91] The release of endothelium-derived relaxing factor by serotonin is blunted in atherosclerotic porcine coronary arteries studied under bioassay conditions (Fig. 26).[90]

In the iliac artery of primates with diet-induced atherosclerosis, the endothelium-dependent relaxations to acetylcholine and thrombin are reduced.[93] In isolated atherosclerotic human coronary arteries, the endothelium-dependent relaxations to acetylcholine, bradykinin, substance P, ADP, and aggregating platelets are absent or smaller than in normal arteries (Fig. 27).[102–106] The dysfunction of the endothelium can also be demonstrated *in*

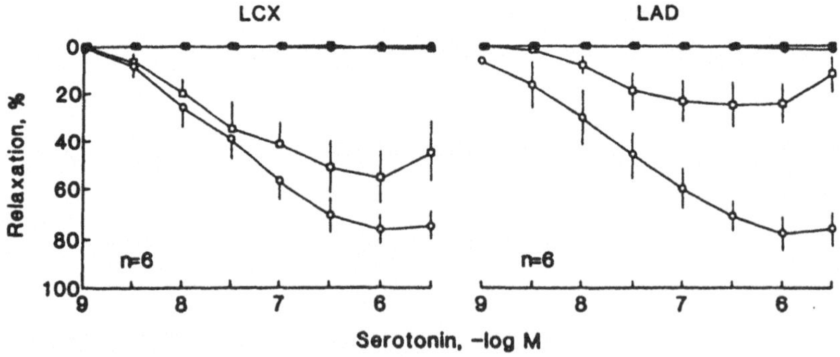

Figure 24. Endothelium-dependent relaxations to serotonin in the left circumflex (LCX; left) and the left anterior descending (LAD; right) coronary artery of pigs fed either with normal chow (○) or with a high-cholesterol (□) diet; the left anterior descending arteries had been denuded 4 weeks prior to the experiment, and became atherosclerotic. The cholesterol diet *per se* causes a modest blunting of the endothelium-dependent relaxation to serotonin; the inhibition was exacerbated in the atherosclerotic blood vessel, illustrating the marked reduction in endothelium-dependent relaxations caused by the atherosclerotic process. ●, ■ = rings without endothelium. (From Ref. 90, by permission of the American Heart Association.)

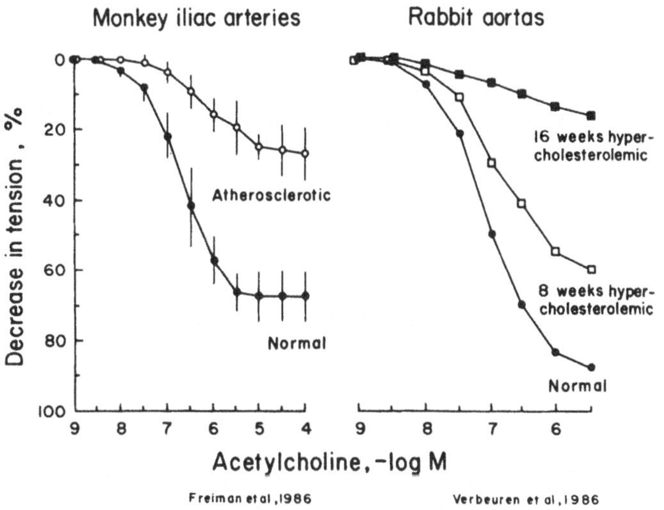

Figure 25. Endothelium-dependent relaxation in the atherosclerotic iliac arteries of primates (left) and aortas of the rabbit (right). Atherosclerosis was induced by a diet rich in cholesterol. These experiments demonstrate that the relaxation evoked by the muscarinic agonist is reduced in atherosclerotic arteries with endothelium. (Data from Refs. 93 and 100; figure from Ref. 180, by permission.)

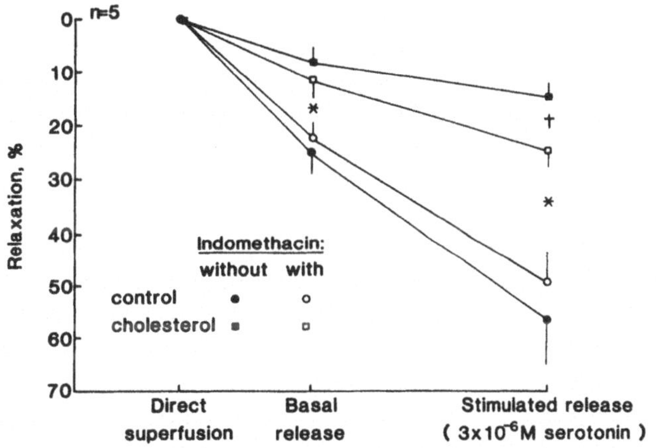

Figure 26. Bioassay of vasoactive factors (by means of a ring of left circumflex coronary artery without endothelium) released under basal conditions (basal release) and with exposure to serotonin (stimulated release) from control porcine left anterior descending coronary arteries (circles) and from such arteries which had been previously denuded and exposed to a high-cholesterol diet during the regeneration period (cholesterol group, squares). The experiments were performed both in the absence (solid symbols) and in the presence of the cyclooxygenase inhibitor, indomethacin (open symbols). The atherosclerotic arteries release significantly less (*) endothelium-dependent relaxing factor both under basal conditions and when stimulated with serotonin. Indomethacin does not affect the release of the factor in the control arteries but significantly (+) augments the relaxations of the bioassay tissue evoked by the perfusate from the atherosclerotic arteries. This suggests that the release of cyclooxygenase-dependent endothelium-derived contracting factor may have occurred. (From Ref. 90, by permission of the American Heart Association.)

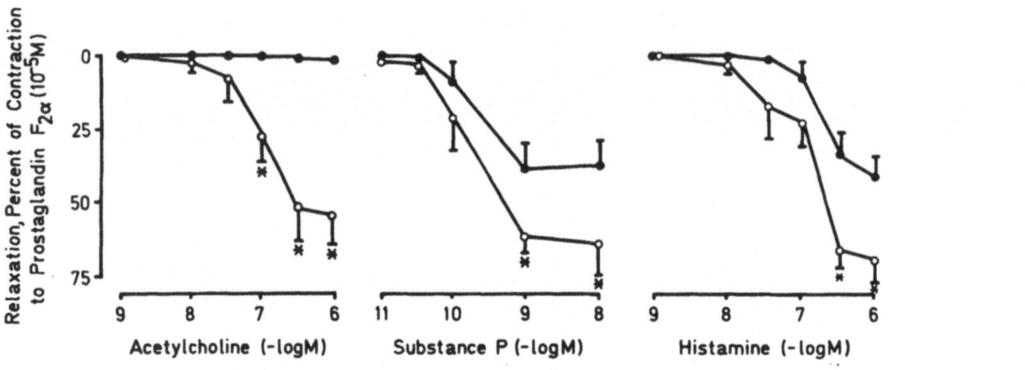

Figure 27. Endothelium-dependent relaxations to acetylcholine (left), substance P (middle), and histamine (right) of normal (open circles) and atherosclerotic (solid circles) human coronary arteries. The response is markedly reduced in patients with coronary artery disease. Asterisk indicates statistically significant difference between normal and atherosclerotic preparations. (Data from Ref. 105; Figure from Ref. 102, by permission.)

vivo, since in patients with coronary artery disease, the intracoronary infusion of acetylcholine into the left descending coronary artery evokes paradoxical contractions both in stenotic and in nonstenotic segments (Fig. 28).[107]

In atherosclerotic aortas of the rabbit and in human coronary arteries, endothelium-dependent relaxations to acetylcholine are impaired despite a maintained response to the Ca^{2+} ionophore A23187.[105] Other studies, however, reported impaired endothelium-dependent relaxations to all agonists tested, including A23187.[90,93,103–105] The latter studies suggest that the production of endothelium-derived relaxing factor is reduced in atherosclerotic arteries. The receptor-operated release of endothelium-derived relaxing factor may be defective at an earlier stage of the disease than the production of the factor. Indeed, although in established atherosclerosis the defect involves multiple receptors or a final common pathway for the release of endothelium-derived relaxing factor, early atherosclerosis and hypercholesteremia interfere primarily with the response to serotonin and ADP.[90,91]

As in hypertensive blood vessels, diffusion barriers (in particular, foam cells and fibrotic changes of the subendothelial layer) are unlikely to play an important role except at very late stages of the atherosclerotic process. Indeed, the fact that endothelium-dependent relaxations of atherosclerotic arteries are impaired most obviously at higher concentrations of the agonists argues against a major contribution of a diffusion barrier for the relaxing factor(s). In line with that interpretation, in the rabbit carotid artery, pronounced fibromuscular thickening does not reduce the endothelium-dependent relaxations.[64] The responsiveness of vascular smooth muscle cells to endothelium-derived relaxing factor appears to be maintained except in the very late stages of the atherosclerotic process.

Thus, in atherosclerotic iliac arteries of the monkey the response to nitroglycerine is maintained.[93] In the aorta of atherosclerotic rabbits, both the formation of cyclic GMP and the relaxation induced by nitroglycerine are also normal except in the most severely affected arterial segments.[100,102] In human coronary arteries, atherosclerosis slightly reduces the sensitivity to, but not the potency of nitroglycerine.[103]

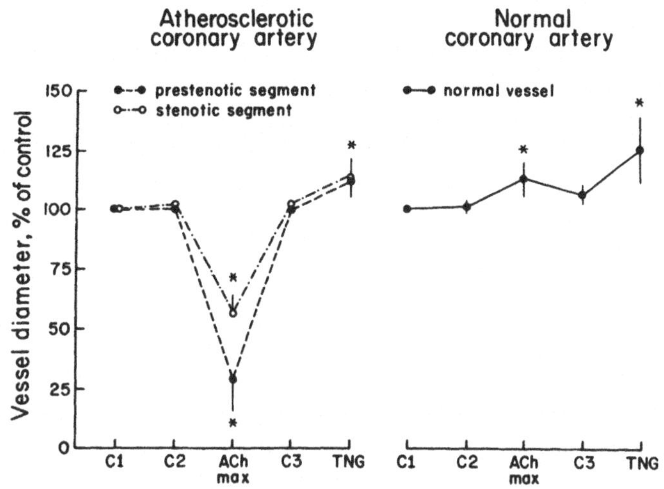

Figure 28. Effects of an intracoronary infusion of acetylcholine (ACh max) on the diameter of the left anterior descending human coronary artery in patients with (left) and without (right) evidence for coronary artery disease at angiography. These experiments demonstrate that acetylcholine dilates normal human coronary arteries *in vivo* (right), but contracts those of patients with coronary artery disease (left). C1, C2, C3 = control period, vehicle, and repeated control; TNG = nitroglycerine; * indicates a statistically significant difference from control. (Data from Ref. 107; Figure from Ref. 180, by permission.)

VII. CORONARY ARTERY DISEASE

A. Vasospasm

Increases in tone of epicardial coronary arteries contribute to myocardial ischemia in patients with coronary artery disease. However, vasospasm as seen in patients with variant angina rarely represents the primary cause of myocardial ischemia or infarction.[107–109] Vasospasm is a localized phenomenon that repeatedly affects the same coronary vascular segment. Thus, a local dysfunction of the blood vessel wall at the site of spasm is likely to be involved. It could be explained by a local hyperresponsiveness of the vascular smooth muscle. Adrenergic nerves are unlikely to play a role except if they were to accumulate serotonin at sites where platelets are activated and then would release the monoamine.[109,110]

In normal subjects, acetylcholine causes vasodilatation when infused into the left anterior descending coronary artery (Fig. 28).[107] In contrast, in patients with variant angina (and in those with coronary artery disease), coronary vasoconstriction can be provoked during angiography by methacholine and acetylcholine.[102,109] Thus, it is conceivable that local endothelial dysfunction may play a role in coronary vasospasm.

In patients with variant angina, ergonovine is the most reliable substance to provoke coronary spasm.[111] In the porcine coronary artery (contracted with prostaglandin $F_{2\alpha}$), ergonovine evokes endothelium-dependent relaxations.[112] The response is attenuated by the α_2-adrenergic antagonist rauwolscine and prevented by methiothepin (but not ketanserin), indicating that $5HT_1$-serotonergic receptors and in part α_2-adrenergic receptors are activated by the alkaloid. In the regenerated state, endothelium-dependent relaxations to ergonovine are reduced, most likely because of the loss of a pertussis toxin-sensitive G protein.[5,112] In addition, the endothelium-dependent inhibition of the contractions evoked by ergonovine are reduced in quiescent porcine coronary arteries with regenerated endothelium. Thus, at sites of previous endothelial injury, the blood vessel wall of epicardial coronary arteries may become hyperreactive to the vasoconstrictor effects of ergonovine and this may explain its capacity to elicit vasospasm in patients with variant angina. Atherosclerotic changes may further aggravate this phenomenon. Indeed, in the canine coronary artery, atherosclerosis is associated with a specific augmentation of the contractions induced by ergonovine.[113] In contrast to the pig, the $5HT_2$-serotonergic antagonist ketanserin does not prevent ergonovine-induced myocardial ischemia in patients with variant angina.[114] Depending on the species, however, different 5HT-serotonergic receptor subtypes may mediate the contractions of coronary smooth muscle to the monoamine. Thus, in the dog 5 HT_1-like while in the pig $5HT_2$-serotonergic receptors mediate serotonin-induced contraction of coronary arterial smooth muscle.[115]

In isolated human coronary arteries, histamine causes relaxations in rings with (Fig. 27), but contractions in rings without endothelium.[102,116] The intracoronary administration of histamine causes dilatation in humans, indicating that under physiological conditions its endothelium-dependent effect predominates.[117] Coronary arteries of patients with coronary artery disease contain an increased amount of the amine and they respond with an augmented vasoconstriction to exogenous histamine.[118] In the pig, vasospasm can be precipitated by histamine after denudation of the coronary artery followed by several weeks of an atherogenic diet.[119] An increased number of adventitial mast cells, which are a major source of histamine, has been reported in a patient with coronary spasm.[120] Thus, it is possible that in coronary vasospasm, histamine either fails to release endothelium-derived relaxing factor or that its direct vasoconstrictor effect predominates. The fact that histamine is effective in provoking coronary spasm only in about half of the patients with variant angina[111] demonstrates that it cannot be regarded as the only causal factor.

More recently, the potent vasoconstrictor peptide endothelin has been implicated in coronary vasospasm.[109] Indeed, the peptide provokes marked contractions of human coronary arteries in a much lower concentration range than that required with other known vasoconstrictor hormones.[121,122] In addition, threshold concentrations of the peptide potentiate contractions to norepinephrine and serotonin in human arteries (Fig. 29).[109,123] This is

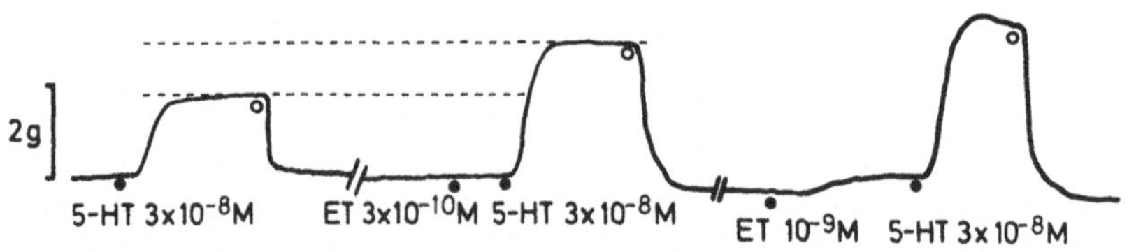

Figure 29. Potentiating effects of threshold and low concentrations of endothelin-1 in the human left anterior descending coronary artery. Serotonin causes significant contraction under control conditions (left), which is augmented after incubation of the blood vessel with a threshold concentration of endothelin-1 (middle). In the presence of a low concentration of endothelin, the contraction is further enhanced. (From Ref. 123, by permission of the American Heart Association.)

of particular interest in this context as the circulating levels of endothelin as well as those found in the coronary sinus are very low.[124] In patients with variant angina and provokable vasospasm during angiographic testing, the concentration of endothelin in the coronary sinus is elevated and decreases during the ischemic attack.[125] This may indicate that endothelin may act as a trigger mechanism for coronary vasospasm.

B. Hypoxia

As the production and release of endothelium-derived relaxing factor requires oxygen (for review see Ref. 2), the hypoxia and ischemia resulting from the vasospastic episodes may impair endothelial function further, particularly if the generation of superoxide anions is accelerated.[126] Hypoxia and ischemia increase coronary vasoconstrictor responses to serotonin and ergonovine, because it triggers the release of endothelium-derived contracting factor ($EDCF_1$).[127]

C. Myocardial Infarction

The major event in myocardial infarction is an acute occlusion of one or more epicardial coronary arteries. The event most commonly occurs in vascular segments with atherosclerotic changes. Fissuring of atherosclerotic plaques with intravascular hemorrhage and platelet adhesion, vasospasm, and thrombus formation are crucial factors in the development of the infarction (for review see Ref. 2). As a result of platelet activation, the urinary excretion of 2,3-dinor thromboxane B_2 (the metabolite of thromboxane A_2) is increased in unstable angina.[128,129] Fibrinopeptide A, a marker of thrombin generation, is also elevated under these conditions.[130] Thus, in evolving myocardial infarction, platelet-derived products such as ADP, serotonin, thrombin, platelet-derived growth factor, and thromboxane A_2 must be abundantly present. To judge from experiments in isolated human coronary arteries, these substances would promote coronary vasoconstriction and aggregation of platelets particularly at sites where the release of endothelium-derived relaxing factor and prostacyclin is reduced.[102–106] Intravascular bleeding during plaque rupture may liberate hemoglobin which would further inhibit endothelium-derived relaxing factor.[131]

D. Reperfusion

Ischemia can injury blood vessels and induce endothelial cell swelling[132,133] and obstruction of capillaries with neutrophils[134] which can lead to regional reperfusion defects ("no-reflow phenomenon").[132,135] Postischemic reperfusion is associated with an accelerated structural degeneration of endothelial cells and an increased adhesion and accumulation of neutrophils.[133,136,137]

Acute occlusion of a major coronary artery of the dog for 1 hr followed by reperfusion for 1 to 2 hr severely reduces endothelium-dependent relaxations to acetylcholine, thrombin, and aggregating platelets (Fig. 30), while the contractile response to potassium chloride, prostaglandin $F_{2\alpha}$ and the relaxations to sodium nitroprusside remain unaltered.[138–142] thus, a reduced release of endothelium derived relaxing factor(s) could contribute to the increased vascular tone and the tendency toward vasospasm occurring upon reperfusion following myocardial infarction.

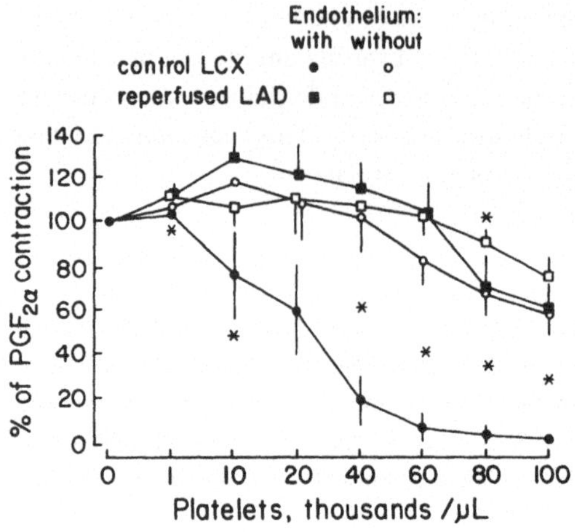

Figure 30. Isometric tension recording in rings with endothelium obtained from canine coronary arteries. Some arteries were occluded for 30 min *in vivo* and reperfused for 1 hr prior to the experiment (reperfused artery), while others were not (control arteries). This experiment demonstrates that the response to aggregating platelets is reduced up to 12 weeks in reperfused as compared to control arteries. (From Ref. 141, by permission of the American Heart Association.)

The injury to the endothelium caused by ischemia and/or reperfusion could be mediated by oxygen-derived free radicals produced in the reperfused myocardium or in the endothelial cells themselves.[143–146] Superoxide anions inactivate endothelium-derived relaxing factor and can activate vascular smooth muscle.[126,147,148] The effectiveness of superoxide dismutase in preventing the endothelial dysfunction after ischemia and reperfusion may be related to that effect.[149] In endothelial cells in culture, free radicals are formed during postanoxic reoxygenation in amounts sufficient to kill most of the cells.[150]

The impaired endothelium-dependent relaxations of occluded canine coronary arteries to aggregating platelets (Fig. 30), platelet-derived products (ADP and serotonin), and thrombin persist for up to 3 months, while those to acetylcholine recover.[141]

VIII. SUBARACHNOID HEMORRHAGE AND CEREBRAL VASOSPASM

A. Effects of Clot Products

The development and intensity of the vasospasm is related to the presence of a clot and the amount of blood in the subarachnoid space.[151] Vasoactive components present in the clot include platelet-derived substances, coagulation products, hemoglobin, and potassium.[151] The relatively high Ca^{2+} concentration in the subarachnoid space also could favor vascular contraction.[151]

Hemoglobin, fibrin degradation products, and plasmin—which are present in the cerebrospinal fluid after hemorrhage—can interfere with either the release and/or the action of endothelium-derived relaxing factor.[131,152,153] In the canine middle cerebral artery, hemolysate markedly attenuates endothelium-dependent relaxations to substance P and those to nitroglycerine.[154] In the canine basilar artery, hemoglobin and cerebrospinal fluid from patients with subarachnoid hemorrhage inhibit endothelium-dependent responses to the Ca^{2+} ionophore A23187.[155] In the perfused internal carotid artery of the dog, extraluminal hemoglobin potentiates the contractions induced by serotonin both in the presence and absence of the endothelium.[156] In contrast, even prolonged application (up to

3 hr) of extraluminal oxyhemoglobin does not prevent endothelium-dependent relaxations to intraluminally infused acetylcholine up to 3 hr after application.[156]

B. Experimental Hemorrhage

Two days after a single injection of blood into the cisterna magna, the hemoglobin-induced augmentation of the response to serotonin is lost, suggesting a reduced basal release of endothelium-derived relaxing factor.[157] The endothelium-dependent relaxations to ADP, but not those to acetylcholine, are decreased 2 days after the injection.[158] However, the impaired endothelium-dependent responses to ADP recover quickly. Two successive blood injections 2 days apart from each other cause more severe endothelial dysfunction than a single injection. Under these conditions, the relaxations to both acetylcholine and ADP are still impaired 4 days after the procedure.[159] While the relaxation to acetylcholine is normal 3 weeks after the injection, that to ADP remains attenuated.

In the canine basilar artery, endothelium-dependent relaxations to vasopressin are lost and those to thrombin markedly reduced 8 days after two consecutive intrathecal injections of blood (Fig. 31).[160] In contrast, endothelium-independent relaxations to papaverine are preserved and those to sodium nitroprusside are only minimally affected. The degree of vasospasm determined angiographically corresponds with the loss of endothelium-dependent relaxations in the isolated blood arteries. Bioassay experiments demonstrate that the luminal release of endothelium-derived relaxing factor by either vasopressin or bradykinin is normal 8 days after the subarachnoid hemorrhage (Fig. 32).[161] This indicates that a decreased diffusion and possibly in part also a responsiveness of endothelium-derived relaxing factor rather than a decreased liberation of the substance is responsible for the reduced relaxations occurring in subarachnoid hemorrhage.

Although endothelium-dependent relaxations are reduced, in experimental subarachnoid hemorrhage, endothelium-dependent contractions to arachidonic acid, acetylcholine, A23187, ADP, mechanical stretch, and hypoxia are preserved.[160,161] This imbalance of endothelium-derived relaxing and contracting factors may be important in the pathogenesis of cerebral vasospasm. Indeed, cerebrospinal fluid from patients with subarachnoid hemorrhage, but not that of normal subjects, inhibits endothelium-dependent relaxations to A23187 in isolated canine basilar arteries.[160,161]

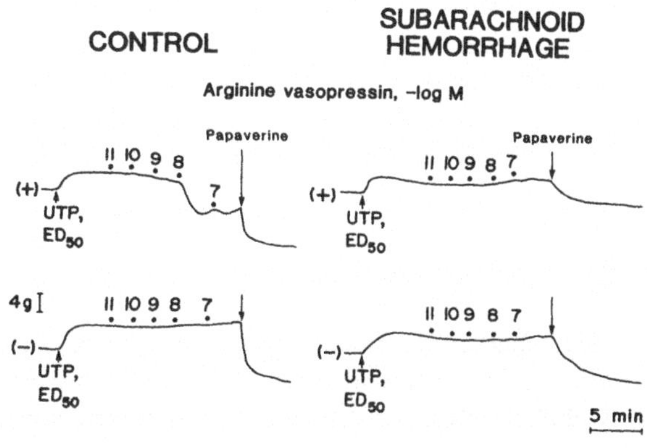

Figure 31. Effect of increasing concentrations of vasopressin on contractions (to uridine triphosphate, UTP) of canine basilar arteries with and without endothelium, taken from animals that underwent cerebral angiography only (control), or that were subjected to both angiography and two intracisteral injections of blood (SAH). Note that the subarachnoidal hemorrhages result in a loss of the endothelium-dependent relaxation to the hormone. (From Ref. 160, by permission.)

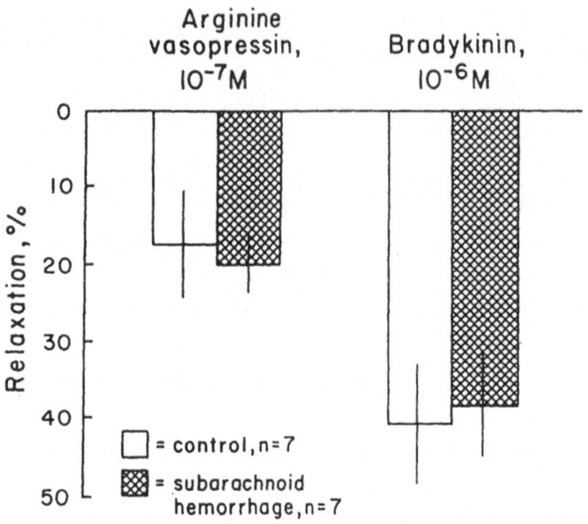

Figure 32. The relaxing activity of the perfusate flowing through basilar arteries with endothelium (taken from control dogs and from animals subjected to subarachnoid hemorrhage) was bioassayed with rings of coronary arteries without endothelium. The release of endothelium-derived relaxing factor evoked by vasopressin (left) and bradykinin (right) was comparable in the two groups. Thus, subarachnoidal hemorrhage does not reduce the ability of the endothelial cells to release relaxing factor, at least toward the lumen of the blood vessel. (From Ref. 161, by permission.)

IX. ACUTE RENAL FAILURE

A. Hemoglobin and Myoglobin

Of all causes of acute renal failure, rhabdomyolysis and hemorrhage with increased blood and urine levels of myoglobin and hemoglobin are most relevant to endothelium-dependent responses.[162,163] Indeed, myoglobin and hemoglobin are potent inhibitors of endothelium-dependent relaxations of isolated blood vessels.[131,152] Hemoglobin blocks the increase in cyclic GMP evoked by endothelium-derived relaxing factor in cultured mesangial cells.[164] Thus, high plasma levels of hemoglobin and/or myoglobin *in vivo* may, by inhibiting endothelium-derived relaxing factor, increase vascular tone. In the renal circulation, this may help to decrease renal blood flow and reduce glomerular filtration.[163]

B. Ischemic Acute Renal Failure

In ischemic acute renal failure of the rat, the renal vasodilatory response to a acetylcholine and bradykinin is lost and the decrease in blood flow in response to renal nerve stimulation is augmented.[165] This suggests that the production of endothelium-derived relaxing factor is decreased under these conditions; this could, at least in part, explain the increased sensitivity of vascular smooth muscle to constricting substances (released from adrenergic nerves or circulating in the blood) and the increase in the basal tone of the renal microvasculature observed under these conditions.

X. PULMONARY DISEASE

Endothelium-dependent relaxations have been demonstrated in large isolated pulmonary arteries from different species, including humans.[166,169] In the intact lung of the rabbit, acetylcholine reverses the increases in pulmonary artery pressure induced by a thromboxane analogue in the presence of indomethacin; since quinacrine and hemoglobin prevent that response, it is likely that endothelium-derived relaxing factor is released in the intact lung.[169]

A. Pulmonary Hypertension

In patients with various congenital cardiac shunts, pulmonary arterial pressure may rise to systemic levels and pulmonary vascular resistance may be markedly elevated, a situation described as Eisenmenger's syndrome. In these patients with secondary pulmonary hypertension, endothelium-dependent relaxations to acetylcholine are markedly attenuated as compared to pulmonary arteries obtained from patients with lung carcinoma and normal pulmonary pressures. In addition, at high concentrations of acetylcholine, paradoxical contractions occur in the blood vessels which were exposed to hypertension. Since the response to sodium nitroprusside is well-maintained in pulmonary arteries of patients with pulmonary hypertension, this disease must primarily interfere with endothelial function.[170]

B. Cystic Fibrosis

Cystic fibrosis is an autosomal recessive disease associated with progressive pulmonary dysfunction. In pulmonary arteries obtained from patients undergoing heart lung transplantation for cystic fibrosis, endothelium-dependent relaxations to acetylcholine and ADP are reduced. As all these patients do develop pulmonary hypertension, the altered endothelium-dependent relaxations occurring in cystic fibrosis may be related to elevated pressures in the pulmonary circulation and/or to respiratory failure.[171]

XI. HEART FAILURE

In the canine femoral artery studied *in vivo* in dogs with heart failure induced by prolonged rapid ventricular pacing, the vasodilator response to acetylcholine is depressed.[172] Blockade of cyclooxygenase with indomethacin does not depress the relaxation induced by acetylcholine in healthy dogs, but enhances the response to lower concentrations of the muscarinic agonist in dogs with heart failure, suggesting that a vasoconstrictor prostaglandin might be involved. Alterations in endothelium-dependent responses in heart failure may contribute to the increased peripheral vascular resistance observed in this disease.

XII. SHOCK

A. Tumor Necrosis Factor

Macrophages produce several mediators, including tumor necrosis factor (TNF), in response to endotoxin.[173,174] TNF reduces endothelium-dependent relaxations to acetylcholine in the cat carotid artery, without affecting that to acidified nitrate.[175] Since the inhibitory effect of the factor can be prevented by inhibitors of protein synthesis, it probably promotes the synthesis of proteins contributing to the damage of endothelial cells and reducing the release (or the transfer) of endothelium-derived relaxing factor.[175]

B. Endotoxin

Endotoxin reduces endothelium-dependent relaxations in rabbits previously exposed to the substance.[176] In addition, endotoxin activates the L-arginine pathway in the aorta of the rat.[177,178] Thus, contractions induced by norepinephrine are severely reduced in isolated aortas treated with the toxin. After incubation with the inhibitor of nitric oxide formation from L-arginine, L-NMMA, the contractions to norepinephrine can be normalized. As these effects of endotoxin are unrelated to the presence or absence of the endothelium, this suggests that the substance stimulates nitric oxide formation either in vascular smooth muscle cells or in other cells contained in the media, such as macrophages, leukocytes, or fibroblasts. This effect of endotoxin may importantly contribute to the known reduced effectiveness of catecholamines in patients with septic shock.

ACKNOWLEDGMENTS. The authors thank Robert R. Lorenz, Bernadette Libsig, and Sabine Bohnert for technical assistance. Original research reported in this chapter was made possible by grants from the Swiss National Research Foundation (Grant 32-25468.88), the National Institute of Health (NIH HL31183), the Swiss Cardiology Foundation, the Helmut Horten Foundation, and an educational grant from Hoechst Pharmaceutica, Paris, France. One of the authors (T.F.L.) is the recipient of a career development award from the Swiss National Research Foundation (SCORE-grant 3231-025150).

REFERENCES

1. Furchgott, R. F., and Vanhoutte, P. M., 1989, Endothelium-derived relaxing and contracting factors, *FASEB J.* **3**:2007–2018.
2. Lüscher, T. F., and Vanhoutte, P. M., 1990, *The Endothelium: Modulator of Cardiovascular Function*, CRC Press, Boca Raton, Fla.
3. Vanhoutte, P. M., and Shimokawa, H., 1989, Endothelium-derived relaxing factor and coronary vasospasm, *Circulation* **80**:1–9.
4. Shimokawa, H., Aarhus, L. L., and Vanhoutte, P. M., 1987, Porcine coronary arteries with regenerated endothelium have a reduced endothelium-dependent responsiveness to aggregating platelets and serotonin, *Circ. Res.* **61**:256–270.
5. Shimokawa, H., Flavahan, N. A., and Vanhoutte, P. M., 1989, Natural course of the impairment of endothelium-dependent relaxations after balloon endothelium-removal in porcine coronary arteries, *Circ. Res.* **65**:740–753.
6. Shimokawa, H., and Vanhoutte, P. M., unpublished observation.
7. Loop, F. D., Lytle, B. W., Cosgrove, D. M., Stewart, R. W., Goorastic, M., Williams, G. W., Golding, L. A. R., Gill, C. G., Taylor, P. C., Sheldon, W. C., and Proudfit, W. L., 1986, Influence of the internal mammary-artery graft on 10-year survival and other cardiac events, *N. Engl. J. Med.* **314**:1–6.
8. Yang, Z., and Lüscher, T. F., 1989, Endothelium-dependent regulatory mechanisms in human coronary bypass grafts: Possible clinical implications, *Z. Kardiol.* **78**(Suppl. 6):80–84.
9. Lüscher, T. F., Diederich, D., Siebenmann, R., Lehmann, K., Stulz, P., von Segesser, L., Yang, Z., Turina, M., Grädel, E., Weber, E., and Bühler, F. R., 1988, Difference between endothelium-dependent relaxation in arterial and in venous coronary bypass grafts, *N. Engl. J. Med.* **319**:462–467.
10. Schoeffter, P., Dion, R., and Godfraind, T., 1988, Modulatory role of the vascular endothelium in the contractility of human isolated internal mammary artery, *Br. J. Pharmacol.* **95**:531–543.

11. Thulesius, O., Ugaily-Thulesius, L., Neglen, P., and Shuhaiber, H., 1988, The role of the endothelium in the control of venous tone: Studies on isolated human veins, *Clin. Physiol.* **8:**359–366.
12. Yang, Z., Diederich, D., Schneider, K., Siebenmann, R., Stulz, P., von Segesser, L., Turina, M., Bühler, F. R., and Lüscher, T. F., 1989, Endothelium-derived relaxing factor and protection against contractions induced by histamine and serotonin in the internal mammary artery and saphenous vein, *Circulation* **80:**1041–1048.
13. Yang, Z., von Segesser, L., Bauer, E., Stulz, P., Turina, M., and Lüscher T. F., 1991, Different activation of the endothelial L-arginine and cyclooxygenase pathway in the human internal mammary artery and saphenous vein, *Circ. Res.* **68:**52–60.
14. Yang, Z., Stulz, P., von Segesser, L., Bauer, E., Turina, M., and Lüscher, T. F., 1991, Different platelet–vessel wall interaction in arterial and venous coronary bypass grafts, *Lancet* **337:**939–943.
15. Lüscher, T. F., Richard, V., and Yang, Z., 1990, Interaction between endothelium-derived nitric oxide and SIN-1 in human and porcine blood vessels, *J. Cardiovasc. Pharmacol.* **14**(Suppl. 11)**:**76–80.
16. Subramanian, V. A., Hermandez, Y., Tack-Goldman, K., Grabowski, E. F., and Weksler, B. B., 1986, Prostacyclin production by internal mammary artery as a factor in coronary artery bypass grafts, *Surgery* **100:**376–383.
17. Miller, V. M., Reigel, M. M., Hollier, L. H., and Vanhoutte, P. M., 1987, Endothelium-dependent responses in autogenous femoral veins grafted into the arterial circulation of the dog, *J. Clin. Invest.* **80:**1350–1357.
18. Ku, D. D., 1988, Personal communication, Copper Mountain FASEB Summer Conference on "Endothelium and Cardiovascular Function," Copper Mountain Co., USA.
19. Bush, H. L., Jr., Jakubowski, J. A., Curl, G. R., Deykin, D., and Nabseth, D. C., 1986, The natural history of endothelial structure and function in arterialized vein grafts, *J. Vasc. Surg.* **3:**204–215.
20. Fuster, V., and Chesebro, J. H., 1986, Role of platelets and platelet inhibitors in aortocoronary artery vein-graft disease, *Circulation* **73:**227–232.
21. Dudel, C., and Förstermann, U., 1988, Gossypol attenuates selectively the blood pressure lowering effect of endothelium-dependent vasodilators in the rabbit in vivo, *Eur. J. Pharmacol.* **145:**217–221.
22. Rees, D. D., Palmer, R. M. J., and Moncada, S., 1989, The role of endothelium-derived nitric oxide in the regulation of blood pressure, *Proc. Natl. Acad. Sci. USA* **86:**3375–3378.
23. Savitsky, J. P., Doczi, J., Black, J., and Arnold, J. D., 1978, A clinical safety trial of stroma-free hemoglobin, *Clin. Pharmacol. Ther.* **23:**73–80.
24. Dohi, Y., Thiel, M., Bühler, F. R., and Lüscher, T. F., 1990, Activation of endothelial L-arginine pathway in resistance arteries: Effect of age and hypertension, *Hypertension* **15:**170–179.
25. Lüscher, T. F., Romero, J. C., and Vanhoutte, P. M., 1986, Bioassay of endothelium-derived vasoactive substances in the aorta of normotensive and spontaneously hypertensive rats, *J. Hypertension* **4**(Suppl. 6)**:** 81–83.
26. Otsuko, Y., DiPiero, A., Hirt, E., Brennaman, B., and Lockette, W., 1988, Vascular relaxation and cGMP in hypertension, *Am. J. Physiol.* **254:**H163–H169.
27. Shirasaki, Y., Kolm, P., Nickols, G. A., and Lee, T. J.-F., 1988, Endothelial regulation of cyclic GMP and vascular responses in hypertension, *J. Pharmacol. Exp. Ther.* **245:**53–58.
28. Konishi, M., and Su, C., 1983, Role of endothelium in dilator responses of spontaneously hypertensive rat arteries, *Hypertension* **5:**881–886.
29. Lockette, W. E., Otsuha, Y., and Carretero, O. A., 1986, Endothelium-dependent relaxation in hypertension, *Hypertension* **8**(Suppl. II)**:**61–66.
30. Lüscher, T. F., Raij, L., and Vanhoutte, P. M., 1987, Endothelium-dependent responses in normotensive and hypertensive Dahl rats, *Hypertension* **9:**157–163.
31. Lüscher, T. F., Vanhoutte, P. M., 1986, Endothelium-dependent contractions to acetylcholine in the aorta of the spontaneously hypertensive rat, *Hypertension* **8:**344–348.
32. Miller, M. J. S., Pinto, A., and Mullane, K. M., 1987, Impaired endothelium-dependent relaxations in rabbits subjected to aortic coarctation hypertension, *Hypertension* **10:**164–170.
33. Sim, M. K., and Singh, M., 1987, Decreased responsiveness of the aortae of hypertensive rats to acetylcholine, histamine and noradrenaline, *Br. J. Pharmacol.* **90:**147–150.
34. Van de Voorde, J., Cuvelier, C., and Leusen, I., 1984, Endothelium-dependent relaxation effects in aorta from hypertensive rats, *Arch. Int. Physiol. Biochem.* **92:**P10–P11.
35. Van de Voorde, J., and Leusen, I., 1984, Endothelium-dependent and independent relaxation effects on aorta preparations of renal hypertensive rats, *Arch. Int. Physiol. Biochem.* **92:**P35–P36.

36. Van de Voorde, J., and Leusen, I., 1986, Endothelium-dependent and independent relaxation of aortic rings from hypertensive rats, *Am. J. Physiol.* **250**:H711–H717.
37. Winquist, R. J., Bunting, P. B., Baskin, E. P., and Wallace, A. A., 1984, Decreased endothelium-dependent relaxation in New Zealand genetic hypertensive rats, *J. Hypertension* **2**:536–541.
38. Lüscher, T. F., Diederich, D., Vanhoutte, P. M., Weber, E., and Bühler, F. R., 1988, Endothelium-dependent responses in the common carotid and renal artery of normotensive and spontaneously hypertensive rats, *Hypertension* **11**:573–578.
39. Hongo, K., Nakagomi, T., Kassell, N. F., Sasaki, T., Lehman, M., Vollmer, D. G., Tsukahara, T., Ogawa, H., and Torner, J., 1988, Effects of aging and hypertension on endothelium-dependent vascular relaxation in rat carotid artery, *Stroke* **19**:892–897.
40. Carvalho, M. H. C., Scivoletto, R., Fortes, Z. B., Nigro, D., and Cordellini, S., 1987, Reactivity of aorta and mesenteric microvessels to drugs in spontaneously hypertensive rats: Role of the endothelium, *J. Hypertension* **5**:377–382.
41. De Mey, J. G., and Gray, S. D., 1985, Endothelium-dependent reactivity in resistance vessels, *Prog. Appl. Microcirc.* **88**:181–187.
42. Diederich, D., Yang, Z., Bühler, F. R., and Lüscher, T. F., 1990, Impaired endothelium-dependent relaxations in hypertensive resistance arteries involve the cyclooxygenase pathway, *Am. J. Physiol.* **258**:H445–H451.
43. Lüscher, T. F., Aarhus, L. L., and Vanhoutte, P. M., 1990, Indomethacin enhances the impaired endothelium-dependent relaxations in small mesenteric arteries of the spontaneously hypertensive rat, *Am. J. Hypertension* **3**:55–58.
44. Tesfamariam, B., and Halpern, W., 1988, Endothelium-dependent and endothelium-independent vasodilation in resistance arteries from hypertensive rats, *Hypertension* **11**:440–444.
45. Mayhan, W. G., 1989, Impairment of endothelium-dependent dilatation of cerebral arterioles during diabetes mellitus, *Am. J. Physiol.* **256**:H621–H625.
46. Mayhan, W. G., Faraci, F. M., and Heistad, D. D., 1987, Impairment of endothelium-dependent responses of cerebral arterioles in chronic hypertension, *Am. J. Physiol.* **253**:H1435–H1440.
47. Mayhan, W. G., Faraci, F. M., and Heistad, D. D., 1989, Responses of cerebral arterioles to adenosine diphosphate, serotonin and the thromboxane analogue U-46619 during chronic hypertension, *Hypertension* **12**(Suppl. 6):556–561.
48. Linder, L., Kiowski, W., Bühler, F. R., and Lüscher, T. F., 1990, Indirect evidence for release of endothelium-derived relaxing factor in human forearm circulation in vivo, *Circulation* **81**:1762–1767.
49. Panza, J. A., Quyyumi, A. A., Brush, J. E., and Epstein, S. E., 1990, Abnormal endothelium-dependent vascular relaxation in patients with essential hypertension, *N. Engl. J. Med.* **323**:22.
50. Lüscher, T. F., Raij, L., and Vanhoutte, P. M., 1987, Effect of hypertension and its reversal on endothelium-dependent relaxations in the rat aorta, *J. Hypertension* **5**(Suppl. 5):153–155.
51. Lüscher, T. F., Vanhoutte, P. M., and Raij, L., 1987, Antihypertensive therapy normalizes endothelium-dependent relaxations in salt-induced hypertension of the rat, *Hypertension* **9**(Suppl. III):193–197.
52. Gray, S. D., and De Mey, J. G., 1985, Vascular reactivity in neonatal spontaneously hypertensive rats, *Prog. Appl. Microcirc.* **8**:173–180.
53. Lüscher, T. F., and Vanhoutte, P. M., 1986, Endothelium-dependent responses to aggregating platelets and serotonin in spontaneously hypertensive rats, *Hypertension* **8**(Suppl. II):55–60.
54. Amstein, R., Fetkovska, N., Lüscher, T. F., Kiowski, W., and Bühler, F. R., 1988, Age and the platelet serotonin vasoconstrictor axis in essential hypertension, *J. Cardiovasc. Pharmacol.* **11**(Suppl. 1):35–40.
55. De Clerck, F., 1986, Blood platelets in human essential hypertension, *Agents Actions* **18**:563–580.
56. Fetkovska, N., Amstein, R., Ferracin, F., Regenass, M., Pletscher, A., and Bühler, F. R., 1990, 5-Hydroxytryptamine kinetics and activation of blood platelets in patients with essential hypertension, *Hypertension* **15**:267–273.
57. Kamal, L. A., Quan-Bui, K. H. L., and Meyer, P., 1984, Decreased uptake of H^3-serotonin and endogenous content of serotonin in blood platelets in hypertensive patients, *Hypertension* **6**:568–573.
58. Baudouin-Legros, M., Dard, B., and Guicheney, P., 1986, Hyperreactivity of platelets from spontaneously hypertensive rats, *Hypertension* **8**:694–699.
59. Valtier, D., Guicheney, P., Baudoin-Legros, M., and Meyer, P., 1986, Platelets in human essential hypertension: In vitro hyperreactivity to thrombin, *J. Hypertension* **4**:551–555.

60. Nara, Y., Kihara, M., Mano, M., Horie, R., and Yamori, Y., 1984, Dietary effect on platelet aggregation in men with and without a family history of essential hypertension, *Hypertension* **6:**339–343.
61. Hoeffner, U., and Vanhoutte, P. M., 1990, Increases in flow reduce the release of endothelium-derived relaxing factor in the aorta of normotensive and spontaneously hypertensive rats, *Am. J. Physiol.* **256:**H828.
62. Kato, T., Iwama, Z., Okumura, K., Hashimoto, H., Ito, T., and Satake, T., 1990, Prostaglandin H^2 may be the endothelium-derived contracting factor released by acetylcholine in the aorta of the rat, *Hypertension* **15:**475–481.
63. Auch-Schwelk, W., Katusic, Z. S., and Vanhoutte, P. M., 1990, Thromboxane A_2-receptor antagonists inhibit endothelium-dependent contractions, *Hypertension* **15:**699–703.
64. Cocks, T. M., Manderson, J. A., Mosse, P. R. L., Campbell, G. R., and Angus, J. A., 1987, Development of a large fibromuscular intimal thickening does not impair endothelium-dependent relaxations in the rabbit carotid artery, *Blood Vessels* **24:**192–200.
65. Folkow, B., 1982, Physiological aspects of primary hypertension, *Physiol. Rev.* **62:**347–504.
66. Shirasaki, Y., Su, C., Lee, T. J.-F., Kolm, P., Cline, W. H., Jr., and Nickols, G. A., 1986, Endothelial modulation of vascular relaxation to nitrovasodilators in aging and hypertension, *J. Pharmacol. Exp. Ther.* **239:**861–866.
67. Arbogast, B. W., Berry, D. L., and Newell, C. L., 1984, Injury of arterial endothelial cells in diabetic, sucrose-fed and aged rats, *Atherosclerosis* **51:**31–45.
68. Steiner, G., 1981, Diabetes and atherosclerosis, *Diabetes* **30**(Suppl. 2)**:**1–7.
69. Harris, K. H., and MacLeod, K. M., 1988, Influence of the endothelium on contractile responses of arteries from diabetic rats, *Eur. J. Pharmacol.* **153:**55–64.
70. Wakabayashi, I., Hatake, K., Kimura, N., Kakishita, E., and Nagai, K., 1987, Modulation of vascular tonus by the endothelium in experimental diabetes, *Life Sci.* **40:**643–648.
71. Gebremedhin, D., Koltai, M. Z., Pogatsa, G., Magyar, K., and Hadhazy, P., 1987, Differential contractile responsiveness of femoral arteries from healthy and diabetic dogs: Role of the endothelium, *Arch. Int. Pharmacodyn.* **288:**100–108.
72. Gebremedhin, D., Koltai, M. Z., Pogatsa, G., Magyar, K., and Hadhazy, P., 1988, Influence of experimental diabetes on the mechanical responses of canine coronary arteries: Role of endothelium, *Cardiovasc. Res.* **22:**537–544.
73. Durante, W., Sen, A. K., and Sunahara, F. A., 1988, Impairment of endothelium-dependent relaxation in aortae from spontaneously diabetic rats, *Br. J. Pharmacol.* **94:**463–468.
74. Tesfamariam, B., Jakubowski, J. A., and Cohen, R. A., 1989, Contraction of diabetic rabbit aorta due to endothelium-derived PGH_2/TXA_2, *Am. J. Physiol.* **257:**H1327–H1333.
75. Fortes, Z. B., Leme, J. G., and Scivoletto, R., 1983, Influence of diabetes on the reactivity of mesenteric microvessels to histamine, bradykinin and acetylcholine, *Br. J. Pharmacol.* **78:**39–48.
76. Fortes, Z. B., Leme, J. G., and Scivoletto, R., 1983, Vascular reactivity in diabetes mellitus: Role of the endothelial cell, *Br. J. Pharmacol.* **79:**771–781.
77. Oyama, Y., Kawasaki, H., Hattori, Y., and Kanno, M., 1986, Attenuation of endothelium-dependent relaxation in aorta from diabetic rat, *Eur. J. Pharmacol.* **131:**75–78.
78. Pieper, G. M., and Gross, G. J., 1988, Oxygen free radicals abolish endothelium-dependent relaxation in diabetic rat aorta, *Am. J. Physiol.* **255:**H825–H833.
79. Cohen, R. A., and Zitnay, K. M., 1986, Augmented adrenergic responses of diabetic carotid arteries are dependent on the endothelium [Abstract], *Circulation* **74**(Suppl. II)**:**413.
80. De Tejada, I. S., Goldstein, I., Azadzoi, K., Krane, R. J., and Cohen, R. A., 1989, Impaired neurogenic and endothelium-mediated relaxation of penile smooth muscle from diabetic men with impotence, *N. Engl. J. Med.* **320:**1025–1030.
81. Takiguchi, Y., Satoh, N., Hashimoto, H., and Nakashima, M., 1988, Changes in vascular reactivity in experimental diabetic rats: Comparison with hypothyroid rats, *Blood Vessels* **25:**250–260.
82. Fortes, Z. B., Leme, J. G., and Scivoletto, R., 1984, Vascular reactivity in diabetes mellitus: Possible role of insulin on the endothelial cell, *Br. J. Pharmacol.* **83:**635–643.
83. Bhardwaj, R., and Moore, P. K., 1988, Increased vasodilator response to acetylcholine of renal blood vessels from diabetic rats, *J. Pharm. Pharmacol.* **40:**739–742.
84. Andrews, H. E., Bruckdorfer, K. R., Dunn, R. C., and Jacobs, M., 1987, Low-density lipoproteins inhibit endothelium-dependent relaxation in rabbit aorta, *Nature* **327:**237–239.

85. Boulanger, C., Bühler, F. R., and Lüscher, T. F., 1989, Low density lipoproteins impair the release of endothelium-derived relaxing factor from cultured porcine endothelial cells [Abstract] *Eur. Heart J.* **10:**331.
86. Kugiyama, K., Kerns, S. A., Morrisett, J. D., Roberts, R., and Henry, P. D., 1990, Impairment of endothelium-dependent arterial relaxation by lysolecithin in modified low-density lipoproteins, *Nature* **344:**160–162.
87. Simon, B. C., Cunningham, L. D., and Cohen, R. A., 1990, Oxidized low-density lipoproteins cause contraction and inhibit endothelium-dependent relaxation in the pig coronary artery, *J. Clin. Invest.* **86:** 75–79.
88. Tanner, F. C., Boulanger, C., and Lüscher, T. F., 1990, Oxidized low-density lipoproteins reduce relaxations to serotonin in porcine coronary arteries: Role of endothelium-derived nitric oxide, Proceedings of the Second Iuphar Satellite Meeting on Serotonin, Basel, p. 129.
89. Cohen, R. A., Zitnay, K. M., Haudenschild, C. C., and Cunningham, L. D., 1988, Loss of selective endothelial cell vasoactive functions caused by hypercholesterolemia in pig coronary arteries, *Circ. Res.* **63:**903–910.
90. Shimokawa, H., and Vanhoutte, P. M., 1989, Impaired endothelium-dependent relaxation to aggregating platelets and related vasoactive substances in porcine coronary arteries in hypercholesterolemia and atherosclerosis, *Circ. Res.* **64:**900–914.
91. Shimokawa, H., and Vanhoutte, P. M., 1989, Hypercholesterolemia causes generalized impairment of endothelium-dependent relaxation to aggregating platelets in porcine arteries, *J. Amer. Coll. Cardiol.* **13:**1402.
92. Shimokawa, H., Kim P., and Vanhoutte, P. M., 1988, Endothelium-dependent relaxation to aggregating platelets in isolated basilar arteries of control and hypercholesterolemic pigs, *Circ. Res.* **63:**604–612.
93. Freiman, P. C., Mitchell, G. G., Heistad, D. D., Armstrong, M. L., and Harrison, D. G., 1986, Atherosclerosis impairs endothelium-dependent vascular relaxation to acetylcholine and thrombin in primates, *Circ. Res.* **58:**783–789.
94. Bossaller, C., Yamamoto, H., Lichtlen, P. R., and Henry, P. D., 1987, Impaired cholinergic vasodilation in the cholesterol-fed rabbit in vivo, *Basic Res. Cardiol.* **82:**396–404.
95. Chappell, S. P., Griffith, T. M., Henderson, T. M., and Lewis, A. H., Influence of cholesterol feeding on endothelium-dependent vasomotor response in rabbit aortic strips [Abstract], *Br. J. Pharmacol.* **85:**266P.
96. Diederich, D., Bühler, F. R., Yang, Z., Weber, E., and Lüscher, T. F., 1988, Impact of atherosclerosis and hypertension on endothelium-dependent relaxations in the renal and carotid artery [Abstract], *Am. J. Hypertension* **1:**14A.
97. Habib, J. B., Bossaller, C., Wells, S., Williams, C., Morrisett, J. D., and Henry, P. D., 1986, Preservation of endothelium-dependent vascular relaxation in cholesterol-fed rabbit by treatment with the calcium blocker PN 200-110, *Circ. Res.* **58:**305–309.
98. Jayakody, L., Senaratne, M., Thomson, A., and Kappagoda, T., 1987, Endothelium-dependent relaxation in experimental atherosclerosis, *Circ. Res.* **60:**251–264.
99. Sreeharan, N., Jayakody, R. L., Senaratne, M. P. J., Thomson, A. B. R., and Kappagoda, C. T., 1986, Endothelium-dependent relaxation and experimental atherosclerosis in the rabbit aorta, *Can. J. Physiol. Pharmacol.* **64:**1451–1453.
100. Verbeuren, T. J., Jordaens, F. H., Zonnekeyn, L. L, VanHove, C. E., Coene, M.-C., and Herman, A. G., 1986, Effect of hypercholesteremia in vascular reactivity in the rabbit: I. Endothelium-dependent relaxations in isolated arteries of control and hypercholesteremic rabbits, *Circ. Res.* **58:**552–564.
101. Wright, C. E., and Angus, J. A., 1986, Effects of hypertension and hypercholesteremia on vasodilatation in the rabbit, *Hypertension* **8:**361–371.
102. Lüscher, T. F., Richard, V., Tschudi, M., Yang, Z., and Boulanger, C., 1989, Endothelial control of vascular tone in large and small coronary arteries, *J. Am. Coll. Cardiol.* **15:**519–527.
103. Förstermann, U., Mügge, A., Alheid, U., Haverich, A., and Frölich, J. C., 1988, Selective attenuation of endothelium-mediated vasodilation in atherosclerotic human coronary arteries, *Circ. Res.* **62:**185–190.
104. Förstermann, U., Mügge, A., Bode, S. M., and Frölich, J. C., 1988, Response of human coronary arteries to aggregating platelets: Importance of endothelium-derived relaxing factor and prostanoids, *Circ. Res.* **63:**306–312.
105. Förstermann, U., Mügge, A., and Frölich, J. C., 1986, Endothelium-dependent relaxation of human epicardial coronary arteries: Frequent lack of effect of acetylcholine, *Eur. J. Pharmacol.* **128:**277–281.
106. Thom, S., Hughes, A., and Sever, P. S., 1988, Endothelium dependent responses in human arteries, in:

Relaxing and Contracting Factors: Biological and Clinical Research (P. M. Vanhoutte, ed.), Humana Press, Clifton, N.J., pp. 511–528.
107. Ludmer, P. L., Selwyn, A. P., Shook, T. L., Wayne, R. R., Mudge, G. H., Alexander, R. W., and Ganz, P., 1986, Paradoxical vasoconstriction induced by acetylcholine in atherosclerotic coronary arteries, *N. Engl. J. Med.* **315**:1046–1051.
108. Maseri, A., L'Abbate, A., Baroldi, G., Chierchia, S., Marzilli, M., Ballestra, A. M., Severi, S., Parodi, O., Biagini, A., Distante, A., and Pesola, A., 1978, Coronary vasospasm as a possible cause of myocardial infarction: A conclusion derived from the study of "preinfarction" angina, *N. Engl. J. Med.* **299**:1271–1277.
109. Lüscher, T. F., 1990, Endothelin: Key to coronary spasm? *Circulation* **83**:701–706.
110. Cohen, R. A., Zitnay, K. M., and Weisbrod, R. M., 1987, Accumulation of 5-hydroxytryptamine leads to dysfunction of adrenergic nerves in canine coronary artery following intimal damage in vivo, *Circ. Res.* **61**:829–833.
111. Kaski, J. C., Crea, F., Meran, D., Rodriguez, L., Araulo, L., Chierchia, S., Davies, G., and Maseri, A., 1986, Local coronary supersensitivity to diverse vasoconstrictive stimuli in patients with variant angina, *Circulation* **74**:1255–1265.
112. Shimokawa, H., Flavahan, N. A., Shepherd, J. T., and Vanhoutte, P. M., 1989, Endothelium-dependent inhibition of ergonovine-induced contraction is impaired in porcine coronary arteries with regenerated endothelium, *Circulation* **80**:643–650.
113. Kawachi, Y., Tomoike, H., Maruoka, Y., Kikuchi, Y., Araki, H., Ishii, Y., Tanaka, K., and Nakamura, M., 1984, Selective hypercontraction caused by ergonovine in the canine coronary artery under conditions of induced atherosclerosis, *Circulation* **69**:441–450.
114. Freedman, S. B., Chierchia, S., Rodriquez-Plaza, L., Bugiardini, R., Smith, G., and Maseri, A., 1984, Ergonovine-induced myocardial ischemia: No role for serotonergic receptors? *Circulation* **70**:178–183.
115. Cohen, R. A., 1986, Contractions of isolated canine coronary arteries resistant to S_2-serotonergic blockade, *J. Pharmacol. Exp. Ther.* **237**:548–552.
116. Toda, N., 1987, Mechanism of histamine actions in human coronary arteries, *Circ. Res.* **61**:280–286.
117. Vigorito, C., Giordano, A., De Caprio, L., Vitale, D., Ferrara, N., Tuccillo, B., Maurea, N., Rispoli, M., and Rengo, F., 1986, Direct coronary vasodilator effects of intracoronary histamine administration in humans, *J. Cardiovasc. Pharmacol.* **6**:933–939.
118. Kalsner, S., and Richards, R., 1984, Coronary arteries of cardiac patients are hyperreactive and contain stores of amines: A mechanism for coronary spasm, *Science* **256**:H720–H725.
119. Shimokawa, H., Tomoike, H., Nabeyama, S., Yamamoto, H., Araki, H., Nakamura, M., Ishii, Y., and Tanaka, K., 1983, Coronary artery spasm induced in atherosclerotic miniature swine, *Science* **221**:560–562.
120. Forman, M. B., Oates, J. A., Robertson, D., Robertson, R. M., Roberts, L. J., and Virmani, R., 1985, Increased adventitial mast cells in a patient with coronary spasm, *N. Engl. J. Med.* **313**:1138–1141.
121. Yanagisawa, M., Kurihara, H., Kimura, S., Mitsui, Y., Kobayashi, M., Watanabe, T. X., and Masaki, T., 1988, A novel potent vasoconstrictor peptide produced by vascular endothelial cells, *Nature* **332**:411–415.
122. Lüscher, T. F., Yang, Z., Tschudi, M., von Segesser, L., Stulz, P., Boulanger, C., Siebenmann, R., Turina, M., and Bühler, F. R., 1990, Interaction between endothelin-1 and endothelium-derived relaxing factor in human arteries and veins, *Circ. Res.* **66**:1088–1094.
123. Yang, Z., Richard, V., von Segesser, L., Bauer, E., Stulz, P., Turina, M., and Lüscher, T. F., 1990, Threshold concentrations of endothelin-1 potentiate contractions to norepinephrine and serotonin in human arteries: A new mechanism of vasospasm? *Circulation* **82**:188–195.
124. Suzuki, H., Matsumoto, H., Kitada, C., Yanagisawa, M., Miyauchi, T., Masaki, T., and Fujino, M., 1989, Immunoreactive endothelin-1 in plasma detected by a sandwich-type enzyme immunoassay, *J. Cardiovasc. Pharmacol.* **13**(Suppl. 5):151–152.
125. Toyo-Oka, T., Aizawa, T., Suzuki, N., Hirata, Y., Miyauchi, T., Yanagisawa, M., Masaki, T., and Sugimoto, T., 1991, The increased plasma level of endothelin-1 and coronary spasm induction in patients with vasospastic angina pectoris, *Circulation* **83**:476–483.
126. Rubanyi, G. M., and Vanhoutte, P. M., 1986, Superoxide anions and hyperoxia inactivate endothelium-derived relaxing factor, *Am. J. Physiol.* **250**:H822–H827.
127. Rubanyi, G. M., and Vanhoutte, P. M., 1985, Hypoxia releases a vasoconstrictor substance from the canine vascular endothelium, *J. Physiol (London)* **364**:45–56.
128. Fitzgerald, D. J., Roy, L., Catella, F., and FitzGerald, G. A., 1986, Platelet activation in unstable coronary artery disease, *N. Engl. J. Med.* **315**:983–989.

129. Hirsh, P. D., Hillis, L. D., Campell, W. B., Firth, B. G., and Willerson, J. T., 1981, Release of prostaglandins and thromboxane into the coronary circulation in patients with ischemic heart disease, *N. Engl. J. Med.* **304:**687–691.
130. Theroux, P., Latour, J.-G., Leger-Gauthier, C., and De Lara, J., 1987, Fibrinopeptide A and platelet factor levels in unstable angina, *Circulation* **75:**156–162.
131. Martin, W., Villani, G. M., Jothianandan, D., and Furchgott, R. F., 1985, Blockade of endothelium-dependent and glyceryl trinitrate-induced relaxation of rabbit aorta by certain ferrous hemoproteins, *J. Pharmacol. Exp. Ther.* **233:**679–685.
132. Fishbein, M. C., Y-Rit, J., and Lando, U., 1980, The relationship of vascular injury to myocardial necrosis after reperfusion, *Circulation* **62:**1274–1280.
133. Kloner, R. A., Ganote, C. E., and Jennings, R. B., 1974, The "no-reflow" phenomenon after temporary coronary occlusion in the dog, *J. Clin. Invest.* **54:**1496–1502.
134. Engler, R. L., Schmid-Schonbein, G. W., and Pavelec, R. S., 1983, Leukocyte capillary plugging in myocardial ischemia and reperfusion in dogs, *Am. J. Pathol.* **111:**98–111.
135. Mullane, K. M., Read, N., Salmon, J. A., and Moncada, S., 1984, Role of leukocytes in acute myocardial infarction in anesthetized dogs: Relationship to myocardial salvage by anti-inflammatory drugs, *J. Pharmacol. Exp. Ther.* **228:**510–522.
136. Forman, M. B., Puett, D. W., Scott, B. A., Bingham, S. E., Virmani, R., Tantengo, M. V., Light, R. T., Bajaj, A., Price, R., and Friesinger, G., 1987, Preservation of endothelial cell structure and function by intracoronary perfluorochemical in a canine preparation of reperfusion, *Circulation* **76:**469–479.
137. Go, L. O., Murry, C. E., Richard, V. J., Weischedel, G. R., Jennings, R. B., and Reimer, K. A., 1988, Myocardial neutrophil accumulation during reperfusion after reversible and irreversible ischemic injury, *Am. J. Physiol.* **255:**H1188–H1198.
138. Ku, D. D., 1982, Coronary vascular reactivity after acute myocardial infarction, *Science* **218:**576–578.
139. Mehta, J. L., Nichols, W. W., Donnelly, W. H., Lawson, D. L., and Saldeen, T. G. P., 1989, Impaired canine coronary vasodilator response to acetylcholine and bradykinin after occlusion-reperfusion, *Circ. Res.* **64:** 43–54.
140. Nichols, W. W., Mehta, J. L., Donnelly, W. H., Lawson, D., Thompson, L., and ter Riet, M., 1988, Reduction in coronary vasodilator reserve following coronary occlusion and reperfusion in anesthetized dog: Role of endothelium-derived relaxing factor, myocardial neutrophil infiltration and prostaglandins, *J. Mol. Cell. Cardiol.* **20:**943–954.
141. Pearson, P. J., Schaff, H. V., and Vanhoutte, P. M., 1990, Long-term impairment of endothelium-dependent relaxations to aggregating platelets after reperfusion injury in canine coronary arteries, *Circulation* **81:**1921–1927.
142. VanBenthuysen, K. M., McMurtry, I. F., and Horwitz, L. D., 1987, Reperfusion after coronary occlusion in dogs impairs endothelium-dependent relaxation to acetylcholine and augments contractile reactivity in vitro, *J. Clin. Invest.* **79:**265–274.
143. Baker, J. E., Felix, C. C., Olinger, G. N., and Kalyanaraman, B., 1988, Myocardial ischemia and reperfusion: Direct evidence for free radical generation by electron spin resonance spectroscopy, *Proc. Natl. Acad. Sci. USA* **85:**2786–2789.
144. Simpson, P. J., and Lucchesi, B. R., 1987, Free radicals and myocardial ischemia and reperfusion injury, *J. Lab. Clin. Med.* **110:**13–30.
145. Zweier, J. L., Flaherty, J. T., and Weisfeldt, M. L., 1987, Direct measurement of free radical generation following reperfusion of ischemic myocardium, *Proc. Natl. Acad. Sci. USA* **84:**1404–1408.
146. Zweier, J. L., Kuppusamy, P., and Lutty, G. A., 1988, Measurement of endothelial cell free radical generation: Evidence for a central mechanism of free radical injury in postischemic tissues, *Proc. Natl. Acad. Sci. USA* **85:**4046–4050.
147. Gryglewski, R. J., Palmer, R. M. J., and Moncada, S., 1986, Superoxide anion is involved in the breakdown of endothelium-derived vascular relaxing factor, *Nature* **320:**454–456.
148. Vanhoutte, P. M., and Katusic, Z. S., 1988, Endothelium-derived contracting factor: Endothelin and/or superoxide anion? *Trends Pharm. Sci.* **9:**229–230.
149. Mehta, J. L., Nichols, W. W., Saldeen, T. P. G., and Franzini, D., 1988, Protection by superoxide dismutase from reperfusion-induced attenuation of coronary flow reserve [Abstract], *Clin. Res.* **36:**543A.
150. Schinetti, M. L., Sbarbati, R., and Scarlattini, M., 1989, Superoxide production by human umbilical vein endothelial cells in an anoxia–reoxygenation model, *Cardiovasc. Res.* **23:**76–80.

151. Kassell, N. F., Sasaki, T., Colohan, A. R. T., and Nazar, G. B., 1985, Cerebral vasospasm following aneurysmal subarachnoid hemorrhage, *Stroke* **16**:562–572.

152. Martin, W., Villani, G. M., Jothianandan, D., and Furchgott, R. F., 1985, Blockade of endothelium-dependent and glyceryl trinitrate-induced relaxation of rabbit aorta by certain ferrous hemoproteins, *J. Pharmacol. Exp. Ther.* **233**:679–685.

153. Watanabe, K., Ishida, T., Yoshitomi, F., and Tanaka, K., 1984, Fibrinogen degradation products influence PGI_2 synthesis by cultured porcine aortic endothelial and smooth muscle cells, *Atherosclerosis* **51**:151–161.

154. Toda, N., 1988, Hemolysate inhibits cerebral artery relaxation, *J. Cerebral Blood Flow Metab.* **8**:46–53.

155. Kanamaru, K., Waga, S., Kojima, T., Fujimoto, K., and Niwa, S., 1987, Endothelium-dependent relaxation of canine basilar arteries. Part 2: Inhibition by hemoglobin and cerebrospinal fluid from patients with aneurysmal subarachnoid hemorrhage, *Stroke* **18**:938–943.

156. Tanaka, Y., and Chiba, S., 1988, Relationship between extraluminal oxyhemoglobin to intraluminal 5-hydroxytryptamine in isolated canine internal carotid arteries, *Neurosurgery* **69**:263–268.

157. Hongo, K., Kassell, N. F., Nakagomi, T., Sasaki, T., Tsukahara, T., Ogawa, H., Vollmer, D. G., and Lehman, R. M., 1988, Subarachnoid hemorrhage inhibition of endothelium-derived relaxing factor in rabbit basilar artery, *J. Neurosurg.* **69**:247–253.

158. Nakagomi, T., Kassell, N. F., Sasaki, T., Fujiwara, S., Lehman, R. M., Johshita, H., Nazar, G. B., and Torner, J. C., 1987, Effect of subarachidonal hemorrhage on endothelium-dependent vasodilation, *J. Neurosurg.* **66**:915–923.

159. Nakagomi, T., Kassell, N. F., Sasaki, T., Fujiwara, S., Lehman, R. M., and Torner, J. C., 1987, Impairment of endothelium-dependent vasodilation induced by acetylcholine and adenosine triphosphate following experimental subarachnoid hemorrhage, *Stroke* **18**:482–489.

160. Kim, P., Sundt, T. M., and Vanhoutte, P. M., 1989, Alterations in endothelium-dependent responsiveness of the canine basilar artery, *J. Neurosurg.* **69**:239–246.

161. Kim, P., Lorenz, R. R., Sundt, T. M., and Vanhoutte, P. M., 1989, Release of endothelium-derived relaxing factor after subarachnoid hemorrhage, *J. Neurosurg.* **70**:108–114.

162. Levinsky, N. G., 1977, Pathophysiology of acute renal failure, *N. Engl. J. Med.* **296**:1453–1458.

163. Lüscher, T. F., Bock, A. H., Yang, Z., and Diederich, D., 1991, Endothelium-derived relaxing and contracting factors: Perspectives in nephrology, *Kidney Int.* **39**:575–590.

164. Schultz, P., Hartich, L., and Raij, L., 1988, The endothelium derived relaxing factor (EDRF) nitric oxide (NO) increases cGMP levels in rat mesangial cells [Abstract], *Am. J. Hypertension* **1**:75A.

165. Conger, J. D., Robinette, J. B., and Schrier, R. W., 1988, Smooth muscle calcium and endothelium-derived relaxing factor in the abnormal vascular responses of acute renal failure, *J. Clin. Invest.* **82**:532–537.

166. De Mey, J. G., and Vanhoutte, P. M., 1982, Heterogenous behavior of the canine arterial and venous wall: Importance of the endothelium, *Circ. Res.* **51**:439–447.

167. Greenberg, B., Rhoden, K., and Barnes, P., 1987, Endothelium-dependent relaxation of human pulmonary arteries, *Am. J. Physiol.* **252**:H434–H438.

168. Ignarro, L. J., Byrns, R. E., Buga, G. M., and Woods, K. S., 1987, Endothelium-derived relaxing factor from pulmonary artery and vein possesses pharmacological and chemical properties identical to those of nitric oxide radical, *Circ. Res.* **61**:866–879.

169. Cherry, P. D., and Gillis, C. N., 1987, Evidence for the role of endothelium-derived relaxing factor in acetylcholine-induced vasodilatation in the intact lung, *J. Pharmacol. Exp. Ther.* **241**:516–520.

170. Dinh, X. A. T., Higenbottam, T. W., Clelland, C. A., Pepke-Zaba, J., Cremona, G., Yazdani Butt, A., Large, S. R., Wells, F. C., and Wallwork, J., 1991, Impairment of endothelium-dependent pulmonary-artery relaxation in chronic obstructive lung disease, *N. Engl. J. Med.* **324**:1539–1547.

171. Dinh, X. A. T., Higenbottam, T. W., Pepke-Zaba, J., Clelland, C., and Wallwork, J., 1989, Reduced endothelium-dependent relaxation of cystic fibrosis pulmonary arteries, *Eur. J. Pharmacol.* **163**:401–403.

172. Kaiser, L., Spickard, R. C., and Olivier, N. B., 1989, Heart failure depresses endothelial cell dependent relaxation to acetylcholine in the canine femoral artery, *Am. J. Physiol.* **256**:H962–H967.

173. Tracey, K. J., Lowery, S. F., and Fahey, T. J., 1986, Cachectin/tumor necrosis factor induces lethal shock and stress hormone responses in the dog, *Gynecol. Obstet.* **164**:415–422.

174. Tracey, K. J., Beutler, B., Lowry, S. F., Merryweather, J., Wolpe, S., Milsark, I. W., Hariri, R. J., Fahey, T. J., III, Zentella, A., Albert, J. D., Shires, G. T., and Cerami, A., 1986, Shock and tissue injury induced by recombinant human cachectin, *Science* **234**:470–474.

175. Aoki, N., Siegfried, M., and Lefer, A. M., 1989, Anti-EDRF effect of tumor necrosis factor in isolated, perfused cat carotid arteries, *Am. J. Physiol.* **256**:H1509–H1512.
176. Vane, J. R., 1971, Inhibition of prostaglandin synthesis as a mechanism of action for aspirin-like drugs, *Nature* **231**:232–235.
177. Fleming, I., Gray, G. A., Julou-Schaeffer, G., Barratt, J. R., and Stoclet, J. C., 1990, Incubation with endotoxin activates the L-arginine pathway in vascular tissue, *Biochem. Biophys. Res. Commun.* **171**: 562–568.
178. Busse, R., Personal communication.
179. Lüscher, T. F., and Vanhoutte, P. M., 1988, Hypertension and endothelium-dependent responses, in: *Vasodilatation: Vascular Smooth Muscle, Peptides, Autonomic Nerves and Endothelium* (P. M. Vanhoutte, ed.), Raven Press, New York, pp. 523–529.
180. Lüscher, T. F., 1988, *Endothelial Vasoactive Substances and Cardiovascular Disease*, Karger, Basel, pp. 1–215.

6

Modulation of Endothelial Paracellular Permeability

A Mechanistic Approach

F. R. Haselton, J. S. Alexander, S. N. Mueller, and A. P. Fishman

I. INTRODUCTION AND SCOPE

Inflammatory edema is a significant element of many disease, e.g., adult respiratory distress syndrome, septicemia, and hypoxia/reperfusion syndrome. *In vivo* studies have sought to identify the mechanism(s) by which inflammatory mediators produce edema. However, this approach has met with limited success since edema often results from a combination of both pressure-induced filtration of plasma proteins, and changes in the permeability properties of the endothelial lining of the vascular wall.[1-3]

Experimental data suggest that the endothelial monolayer forms the main barrier to macromolecular passage from blood to tissues.[4-6] For example, luminally applied high-molecular-weight tracers are almost completely restricted by the endothelial cell layer and do not reach the abluminal compartment.[7] The accumulation of these tracers within interendothelial junctions also suggests that their transport is essentially junctional,[4,8] and that this is the region where most mass transfer occurs. Several studies support this view,[5,6] estimating that over 90% of transendothelial solute exchange occurs at the endothelial junction. Junction structure is apparently a major determinant of endothelial monolayer permeability.

Interendothelial junctions are complex integral membrane structures which associate with the network of cytoplasmic filaments,[14,15] i.e., the cytoskeleton. Cytoskeletal filaments are structurally regulated through signals transmitted via receptors and second messen-

F. R. Haselton and J. S. Alexander • Department of Biomedical Engineering, Vanderbilt University, Nashville, Tennessee 37235. *S. N. Mueller* • Coriell Institute for Medical Research, Camden, New Jersey 08103. *A. P. Fishman* • Cardiovascular–Pulmonary Division, Department of Medicine, University of Pennsylvania, Philadelphia, Pennsylvania 19104.

Endothelial Cell Dysfunctions, edited by Nicolae Simionescu and Maya Simionescu. Plenum Press, New York, 1992.

gers.[10,12,16,17] In addition, physiological agents which alter endothelial monolayer permeability *in vitro*[8–13] also modify the endothelial cytoskeleton.[10,14,15] Thus, structural regulation of the endothelial junction, and monolayer permeability, may be regulated by the cytoskeleton.

One mechanism by which autacoids could regulate endothelial monolayer permeability is illustrated in Fig. 1. There, binding of ligand (e.g., autacoid or hormone) to a cell surface receptor modifies levels of second messengers (e.g., diacylglycerol, calcium, or cAMP). These second messengers alter cell regulatory systems, e.g., kinases producing changes in cytoskeleton. Such resulting cytoskeletal changes affect transmembrane junctional components and regulate interendothelial junctional permeability.

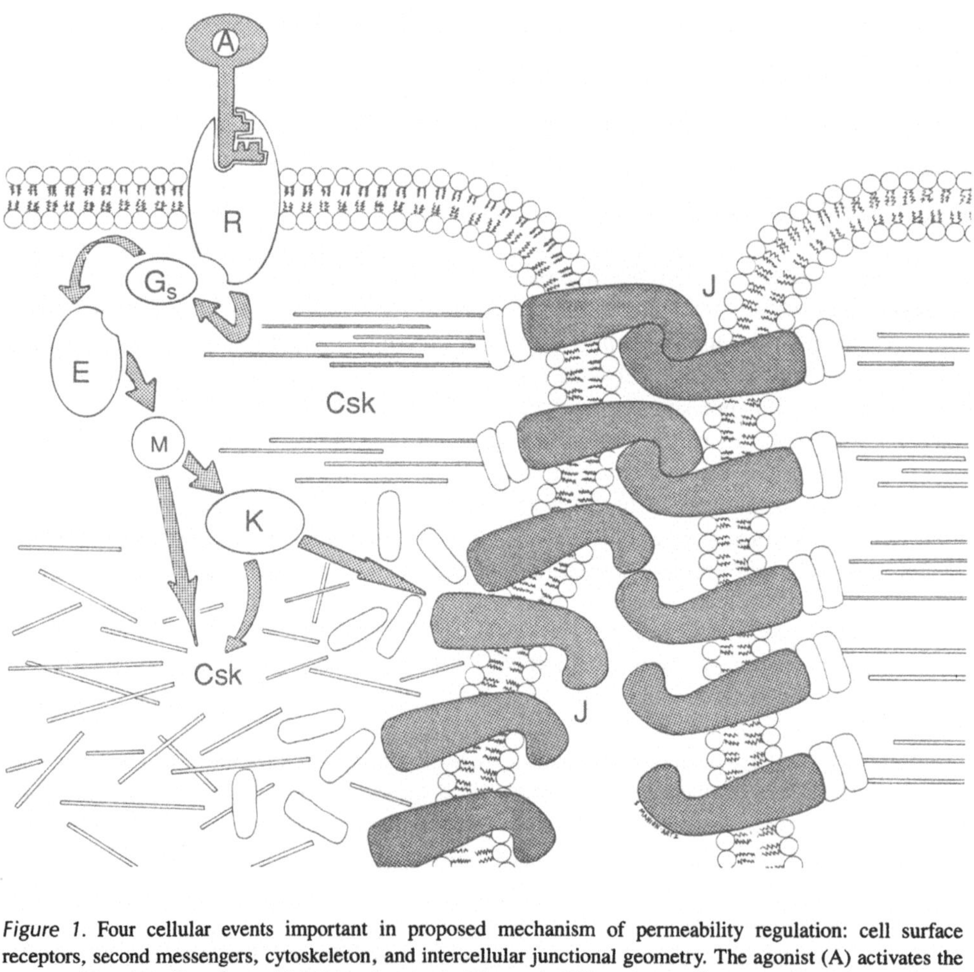

Figure 1. Four cellular events important in proposed mechanism of permeability regulation: cell surface receptors, second messengers, cytoskeleton, and intercellular junctional geometry. The agonist (A) activates the receptor (R) at the cell surface, which through a coupled G-protein (G_s), activates effector proteins (E) and changes levels of second messengers (M), which affect protein kinases (K) or cytoskeleton (C) directly, and leads to a change in interendothelial geometry (J).

In this chapter, we present two types of experimental evidence in support of the following components of this hypothesis: (1) permeability/cytoskeletal responses are receptor mediated, (2) these responses depend on second messengers, (3) specific cytoskeletal changes are associated with an increased/decreased permeability response, and (4) these responses are rapid and reversible and, therefore, may reflect normal physiological responses.

II. MATERIALS AND METHODS

A. Endothelial Monolayer Modulators

Agents were selected which activate different levels of the hypothetical pathway of barrier regulation shown in Fig. 1. These agents include cell surface agonists, second messenger mimetics, and agents targeting the cytoskeleton.

1. Receptor Agonists

Bradykinin, thrombin, ATP, adenosine, propranolol, and isoproterenol were purchased from Sigma (St. Louis, Mo.).

2. Second Messengers

Second messenger analogues and functional mimetics were used to directly alter intracellular levels of specific second messengers. Forskolin and dibutyryl cAMP were used to mimic elevated cell cAMP; A23187 and phorbol myristate acetate were used to mimic effects of phopholipase C path activation. These were purchased from Sigma.

3. Cytoskeletal Modulators

The microfilament structural modulators cytochalasin D and phalloidin were used. These were purchased from Sigma.

B. Cell Culture Methods

Well-established techniques were used to isolate and culture bovine endothelial cells; detailed procedures used for primary cell isolation, cell identification, monolayer culture subcultivation, cloning, and the determination of *in vitro* life span have been described previously.[18] Briefly, cultures of bovine adult and fetal aortic endothelial cells were obtained from thoracic aortas by means of collagenase treatment. Monolayer cultures were established in culture medium containing serum and were subcultured using trypsin–EDTA. Endothelial cell identity was verified by indirect immunofluorescent assay for factor VIII-related antigen. Both an uncloned cell line and a clonal strain derived from this line were used during the vigorous proliferative phase of their *in vitro* life span. The fetal bovine cell lines (AG-7680 and AG-7681) are available from the Cell Repository at the Coriell Institute for Medical Research (Copewood Street, Camden, N.J.).

C. Cell-Column Permeability Assay

This permeability assay provides a reliable and sensitive index of changes in endothelial intercellular permeability produced by stimulation of an endothelial monolayer. The method has three components: a chromatographic cell column made from cell-covered microcarrier beads, a multiple indicator column injection strategy, and a mathematical model which estimates monolayer permeability from the shape of the column elution profiles. These details of this method have been described previously,[9] and are briefly described in sections 1–3 below.

1. Microcarrier Culture Methods

Endothelial cells are cultured on microcarrier beads (Fig. 2) as previously described.[9,19,20] Endothelial cells are released from monolayer cultures using trypsin–EDTA and mixed with Cytodex-3 beads (Pharmacia, Sweden) to achieve a density of 2×10^4 cells per cm^2 of bead surface. The cell plus bead suspension is mixed intermittently overnight (2 min on–30 min off) at 60 rpm in a microcarrier stirrer apparatus (Techne, Cambridge,

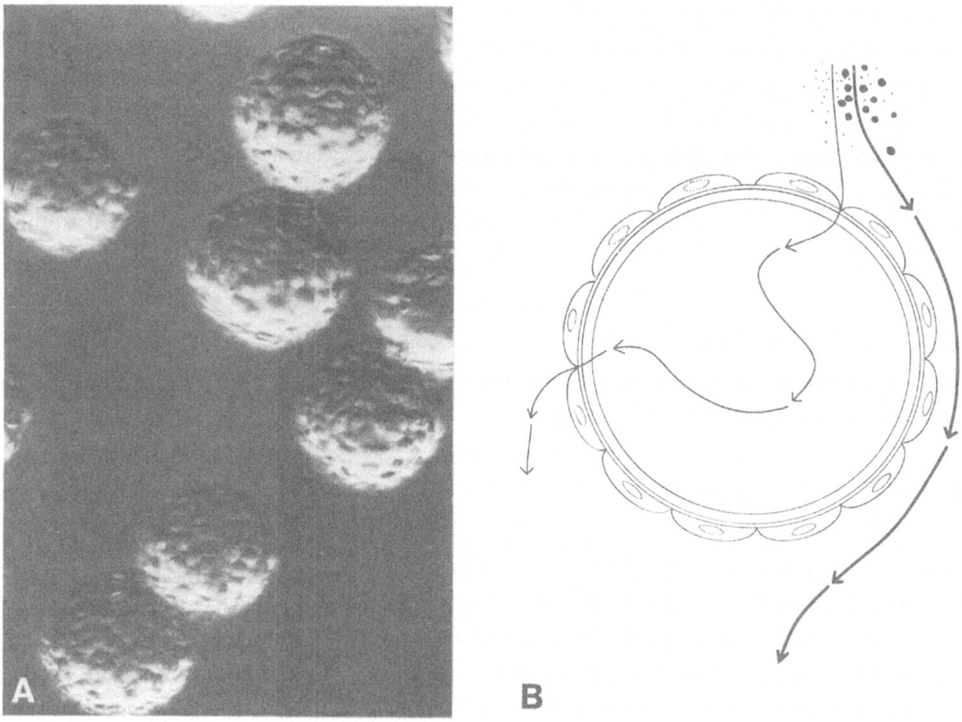

Figure 2. (A) Cytodex-3 microcarrier beads coated with a confluent layer of bovine aortic endothelial cells. Each bead is 175 μm in diameter and is covered with approximately 180 cells. (B) Schematic representation of the paths of two different molecular weight tracers through a column of cell-covered beads. The large tracer blue dextran (2×10^6 Da; large dots) does not cross the cell monolayer and rapidly passes through the column. The low-molecular weight tracer polyethylene glycol (900 Da; small dots) may cross one or more monolayers in its transit of the column and it therefore has a different elution pattern than that of blue dextran.

Mass.). Attachment of endothelial cells to the beads is monitored microscopically; the viability of the attached cells is nearly 100%, as judged by erythrosine B dye exclusion. Once the cells are attached to the beads, the cultures are continuously stirred at 60 rpm. The culture medium is replaced with fresh medium twice weekly and the cells grow from 2×10^4 cells/cm^2 to 1.5×10^5 cells/cm^2 (approximately 180 cells/bead) within 3 days. Endothelial cell microcarrier cultures are used for permeability experiments between 7 and 14 days following inoculation.

2. Cell-Column Chromatographic Methods

A chromatographic cell column consists of a 0.6-cm-diameter glass column (Rainin, Natick, Mass.) within a constant-temperature water jacket. Cell-covered beads are poured to a column height of 3 cm, representing 195 cm^2 of cell culture surface, or approximately 2.5×10^7 cells. The cell column is equilibrated in Hanks' balanced salt solution (HBSS) with 10% fetal bovine serum and gentamicin (50 μg/ml). The column is perfused at approximately 1 ml/min and 50 μl of tracer mixture is applied using a rotary injection valve (Rainin).

3. Multiple Indicator Strategy

Multiple tracer analysis was used to determine cell layer permeability from elution profiles of applied tracers. Using this approach, blue dextran (2000 kDa) was used as the flow tracer. The monolayer permeant tracer [^{3}H]polyethylene glycol (900 Da) (New England Nuclear) was applied simultaneously with the flow tracer. The paracellular penetration of polyethylene glycol through the cell monolayer and into the bead matrix causes the permeant tracer to be retained on the column for a longer time than the blue dextran which does not cross the cell layer (Fig. 2B). The effect of a cell monolayer on the chromatographic behavior of these tracers[9] is illustrated in Fig. 3. In addition to being small enough to pass between the cells, this permeant tracer is not metabolized or actively taken up by the cells.[4,7,21]

4. Mathematical Modeling

Estimates of permeability based on the column elution patterns of multiple tracers are an adaptation and extension of *in vivo* techniques used to assess capillary permeability.[9,22–27] To apply these techniques to these experiments, a mathematical model of tracer motion is constructed by assuming the physical picture of tracer behavior as shown in Fig. 2B. A modified Marquardt iteration scheme was used to estimate parameter values for permeability which best approximated the experimental data.[25] Best fit was determined by the minimization of the coefficient of variation between a computer generated prediction and the experimental data.

D. Rhodamine–Phalloidin Staining of Cultured Endothelial Cell Monolayers

Early passage (P3 to P6) endothelial cells were trypsinized in 0.25% trypsin/1 mM EDTA and seeded onto 1.2-cm-diameter glass coverslips at 2×10^4 cells/cm^2. Medium

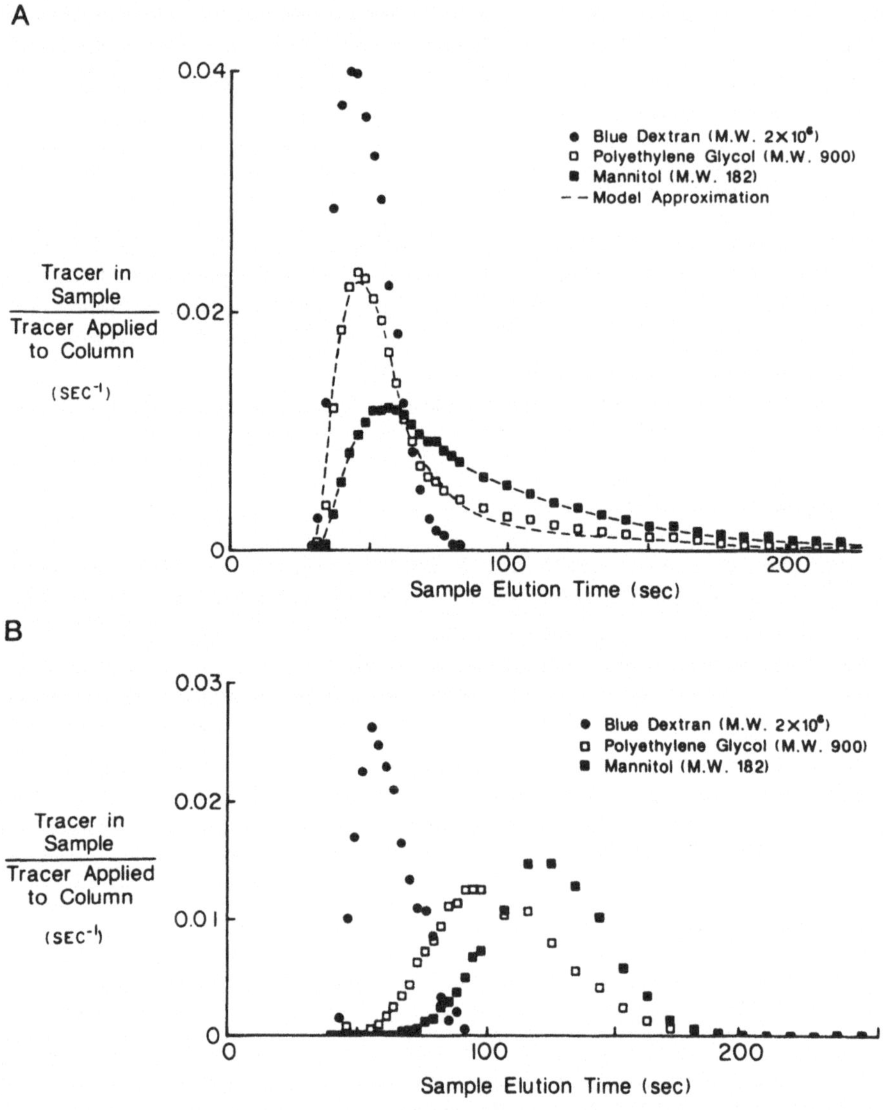

Figure 3. Comparison of elution patterns from a column of Cytodex-3 microcarrier beads covered with a confluent layer of endothelial cells (A) with the elution patterns from a column with the same number of uncovered beads (B). Both panels show the early elution of the high-molecular-mass tracer (blue dextran, 2×10^6 Da) followed by the elution of two lower-molecular-mass tracers ([^{3}H]polyethylene glycol, 900 Da; [^{14}C]mannitol, 182 Da) which have been retained on the column for a longer time because they are able to cross the cell monolayer and gain access to the interior of the microcarrier beads. The best fit of the Sangren–Sheppard model for the mannitol and polyethylene glycol data is shown as dashed lines in A. (From Ref. 9.)

consisted of DMEM plus 10% fetal calf serum with 100 U/ml penicillin, streptomycin, and amphotericin. Monolayers were used 5 days after establishing a confluent monolayer. Before the experimental manipulations, cells were incubated for 15 min in HBSS plus 0.5% bovine serum albumin (pH 7.35). Monolayers were exposed to test substances diluted in HBSS/0.5% BSA for 15 to 30 min. After treatment, monolayers were fixed in 3.7% phosphate-buffered formaldehyde (pH 7.35) for 10 min at 37°C. Monolayers were rinsed in phosphate-buffered saline (pH 7.35) and extracted in buffer (0.5% Triton X-100 in 1 mM PMSF, 0.6 M KCl, 1 mM $CaCl_2$, 1 mM $MgCl_2$) for 3 min. Cell monolayers were then washed 3 × with PBS and incubated with 1 U of rhodamine–phalloidin (Sigma Corp, St. Louis, Mo.) for 30 min. Cell monolayers were washed 3 × in PBS and mounted face down in 5 μl of glycerol/PBS + 0.1% *p*-phenylenediamine HCl (antiquenching agent); coverslips were sealed with nail polish and illuminated for fluorescence using an Olympus IMT-2 microscope. Fluorescent images were recorded for 20 seconds on T-MAX professional film, developed, and exposed identically.

Photographs of cell monolayers were qualitatively evaluated for five key features: density and regularity of the peripheral band structure (the junctionally associated microfilament complex), intercellular gaps (regions of cell retraction), central microfilaments ("aster"-like structures or central spanning microfilaments), and actin "paracrystals" or other nonfilamentous punctate F-actin staining.[28]

III. RESULTS

A. Physiological Agents that Increase Monolayer Permeability and Cause Retraction of the Actin Dense Peripheral Band

1. Bradykinin

Bradykinin, an inflammatory autacoid,[31] shifts the baseline elution profile of polyethylene glycol, lowering the peak of the polyethylene glycol profile and increasing the level of tracer in the tail of the curve (Fig. 4). This shape change begins to reverse within 30 min after the removal of bradykinin from the perfusate. The cell monolayer permeability during each of the three treatments given in Table 1 is based on analysis of the curves in Fig. 4 and those from similar experiments.

The effect of bradykinin begins after 5–6 min, peaks at 12–15 min, and returns to baseline values within 50 min (Fig. 5).

With respect to structural changes, bradykinin-treated monolayers (30 min at 10^{-6} M) show diminished dense peripheral band structure; many display retracted cell borders producing gaps (see Fig. 6B). Central microfilament "asters" also appear more often than in controls; actin "paracrystals"[29] appear less often; and intercellular junctions are less well-defined.

2. Thrombin

Thrombin, an inflammatory autacoid which figures prominently in blood coagulation, also increases cell-column permeability. Table 1 shows that, unlike bradykinin, the effect of thrombin does not reverse 60 min after removal of thrombin from the perfusate.

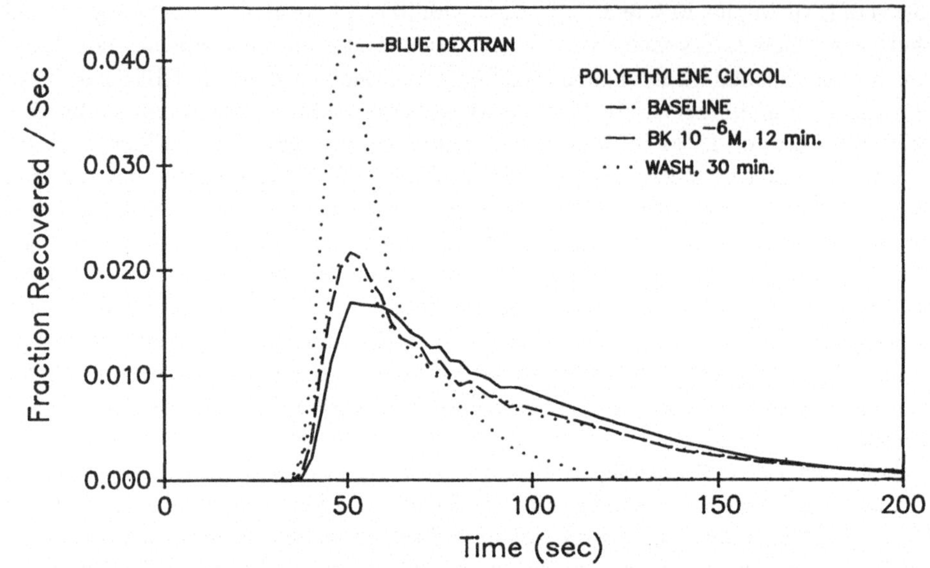

Figure 4. Effect of bradykinin on endothelial cell-column elution patterns. Composite from three measurements made during a single experiment comparing the elution pattern of blue dextran (labeled dotted line) with polyethylene glycol under three conditions: baseline (dashed line), baseline perfusate plus 10^{-6} M bradykinin after 12 min (solid line), and baseline perfusate 30 min after removal of bradykinin from the column perfusate (dotted line). Lines connect data points which have been omitted to simplify the figure. The elution of blue dextran remains constant; therefore, a single elution pattern is shown. Compared to baseline the addition of bradykinin lowers the peak of the elution pattern of polyethylene glycol and raises the concentration at later time points. Removing bradykinin results in an elution profile similar to baseline.

Table 1. Effect of Treatments on Endothelial Monolayer Permeability

Agent	Baseline	Treatment	Recovery
Bradykinin (10^{-6} M; 30′; n = 2)	5.78	7.66	5.84
Thrombin (2.5 U/ml; 30′; n = 3)	2.46	4.36*	4.49*
ATP (10^{-5} M; 15′; n = 2)	5.47	2.82	6.56
Adenosine (10^{-4} M; 15′; n = 5)	7.78	5.13*	14.8*
Isoproterenol (10^{-6} M; 15′; n = 4)	3.10	1.59*	4.25
Dibutyryl cAMP (10^{-3} M; 15′; n = 3)	3.83	2.03*	4.60
Forskolin (10^{-6} M; 15′; n = 2)	5.28	4.06	6.11
A23187 (10^{-6} M; 30′; n = 2)	8.09	14.9	18.2
Phorbol myristate acetate (10^{-6} M; 15′; n = 1)	5.01	18.7	18.9
Cytochalasin D (1 μg/ml; 15′; n = 5)	4.32	16.4*	6.27
Low calcium (0.14 mM; 10′; n = 5)	4.95	11.9*	6.88

*Significantly different from baseline, repeated measures ANOVA and Tukey's t-test.

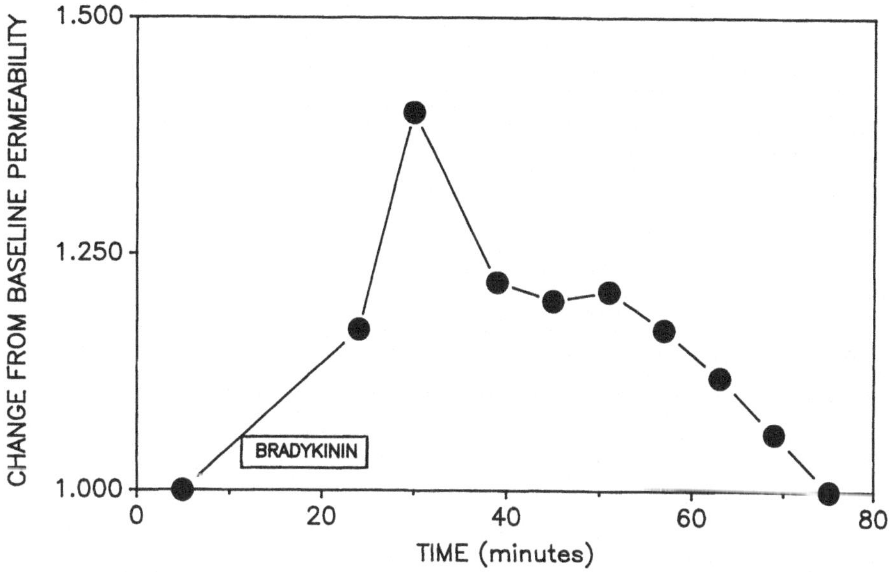

Figure 5. Time course of bradykinin response to treatment with 10^{-6} M bradykinin. Increase in permeability peaks after a 12-min exposure to bradykinin (box) and then declines to baseline following removal of bradykinin from the column perfusate.

Thrombin (30-min treatment at 2.5 U/ml) produces structural changes similar to those observed after bradykinin treatment (Fig. 6B,C). The dense peripheral microfilament structure is diminished compared to untreated controls, and central stress fibers are more abundant.

B. Physiological Agents that Decrease Permeability and Cause Protrusion of the Dense Peripheral Band

1. Isoproterenol and Other β-Adrenergic Agonists

Monolayers exposed to β-isoproterenol (10^{-6} M for 15 min), an anti-inflammatory β-adrenergic agonist, have a permeability to polyethylene glycol that is approximately 40% less than baseline. The addition of the β-adrenergic antagonist propranolol for 15 min reverses the permeability changes[9] (Table 1).

Endothelial microfilament structure is modified by exposure to β-isoproterenol (Fig. 7). Actin staining is greatly reduced and is often more distinct and tortuous than in controls. This suggests increased junctional bridging or overlap. Intercellular gaps are seen less often. In addition, central actin microfilaments are less pronounced.

2. Purinergic Agonists

Adenosine and ATP, which activate platelets and are associated with other cell functions, rapidly and reversibly decreased endothelial monolayer permeability.[30] ATP (10^{-6} M for 15 min) decreased polyethylene glycol permeability by 40%; permeability

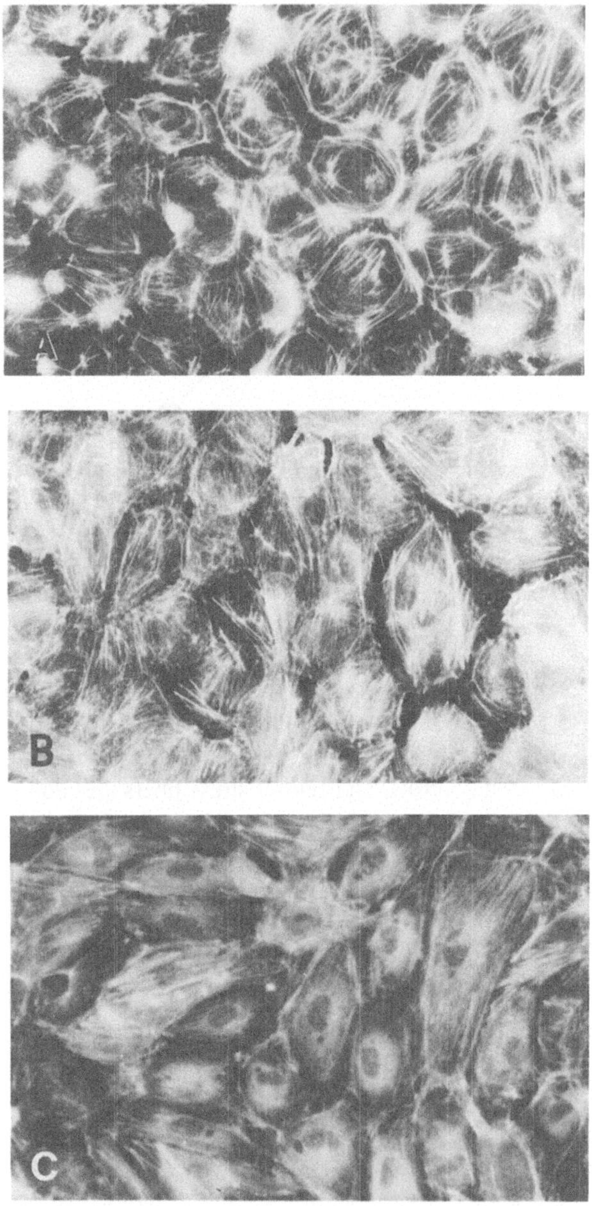

Figure 6. Patterns of actin staining resulting from physiological agents which increase permeability in cell columns: (A) no treatment, (B) bradykinin (10^{-6} M; 15 min), (C) thrombin (2.5 U/ml; 15 min). Note that in B and C the dense peripheral band outlining the cells in A is less pronounced. The diameter of each cell is approximately 10 μm.

returned to baseline after 15 min of ATP-free perfusate. Figure 8 illustrates the changes in polyethylene glycol elution produced by treatment such as adenosine which decreases permeability. Compared to the baseline elution profile, the curve is much more peaked and the concentration in the tail of the curve is lower. Analysis of these curves showed that adenosine (ADO) at 10^{-4} M decreases polyethylene glycol permeability by about 40%; permeability returns to baseline following 15 min of ADO-free perfusion (see Table 1).

Treatment of cultured monolayers with purinergic agonists (ATP 10^{-6} M and ADO

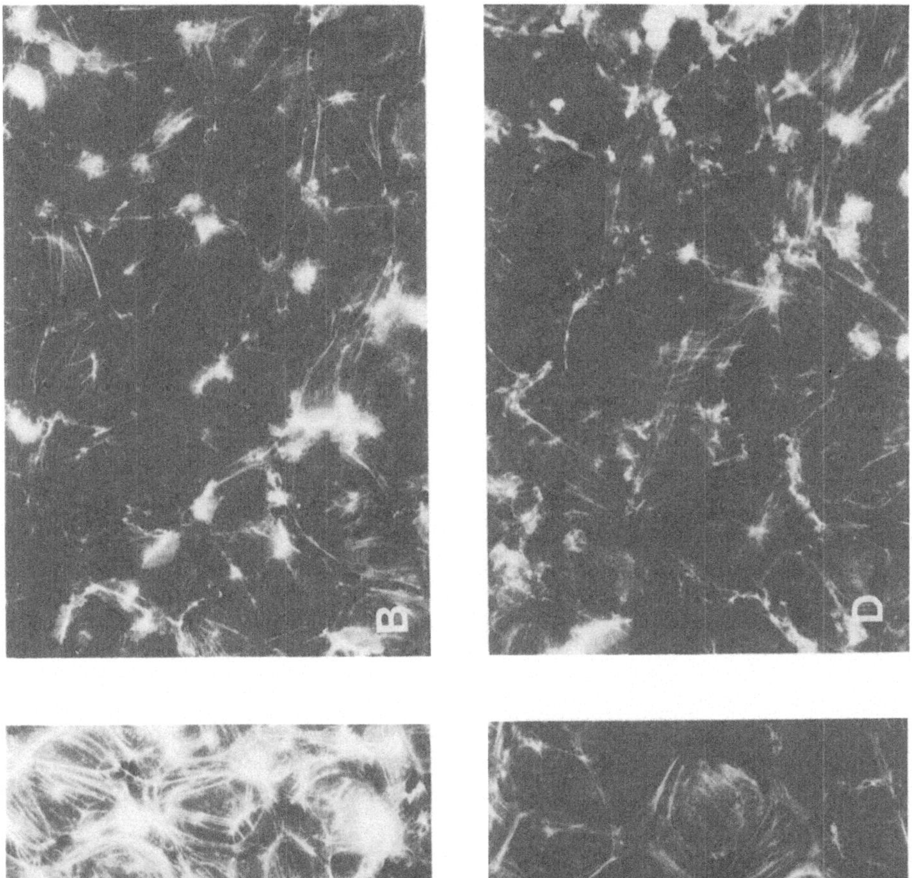

Figure 7. Patterns of actin staining resulting from agents which decrease permeability in cell columns: (A) no treatment, (B) isoproterenol (10^{-6} M; 15 min), (C) ATP (10^{-6} M; 15 min), and (D) adenosine (10^{-5} M; 15 min). Note the diminished central microfilaments following each of these treatments.

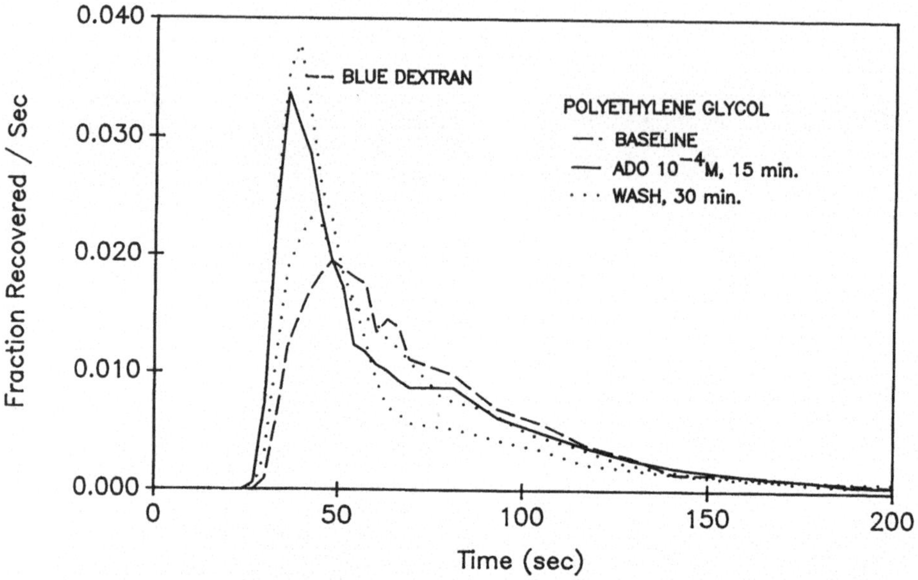

Figure 8. Effect of adenosine on endothelial cell-column elution patterns. Composite from three measurements made during a single experiment comparing the elution pattern of blue dextran (labeled dotted line) with polyethylene glycol under three conditions: baseline (dashed line), baseline perfusate plus 10^{-4} M adenosine after 15 min (solid line), and baseline perfusate 30 min after removal of adenosine from the column perfusate (dotted line). Compared to baseline the addition of adenosine raises the peak of the elution pattern of polyethylene glycol and lowers the concentration of later time points. Removing adenosine results in an elution profile similar to baseline.

10^{-4} M) produces alterations in microfilament appearance (Fig. 7C,D) similar to those produced by β-adrenergic agonism: there is a characteristic decrease in overall F-actin staining and much greater decrease in microfilament central staining.

C. Probes of the Second Messenger System

1. A23187

Calcium is an important intracellular second messenger which is increased in endothelium following exposure to inflammatory agents.[31] Similar changes in intracellular calcium are induced by the calcium ionophore A23187. Exposure of cell columns to A23187 at 10^{-6} M for 15 min produced an increase in monolayer permeability of almost 50%. These changes did not reverse 60 min after removal of the ionophore (see Table 1).

Actin cytoskeletal staining resulting from this treatment is shown in Fig. 9C. The changes produced by this agent are characterized by loss of regularity of the dense peripheral band, occasional gaps, central spanning microfilaments, microfilament "asters," and few if any "paracrystals."

2. Phorbol Myristate Acetate (PMA)

Another important second messenger is diacylglycerol, a potent activator of protein kinase C (PKC). Activation is associated with exposure to inflammatory mediators.

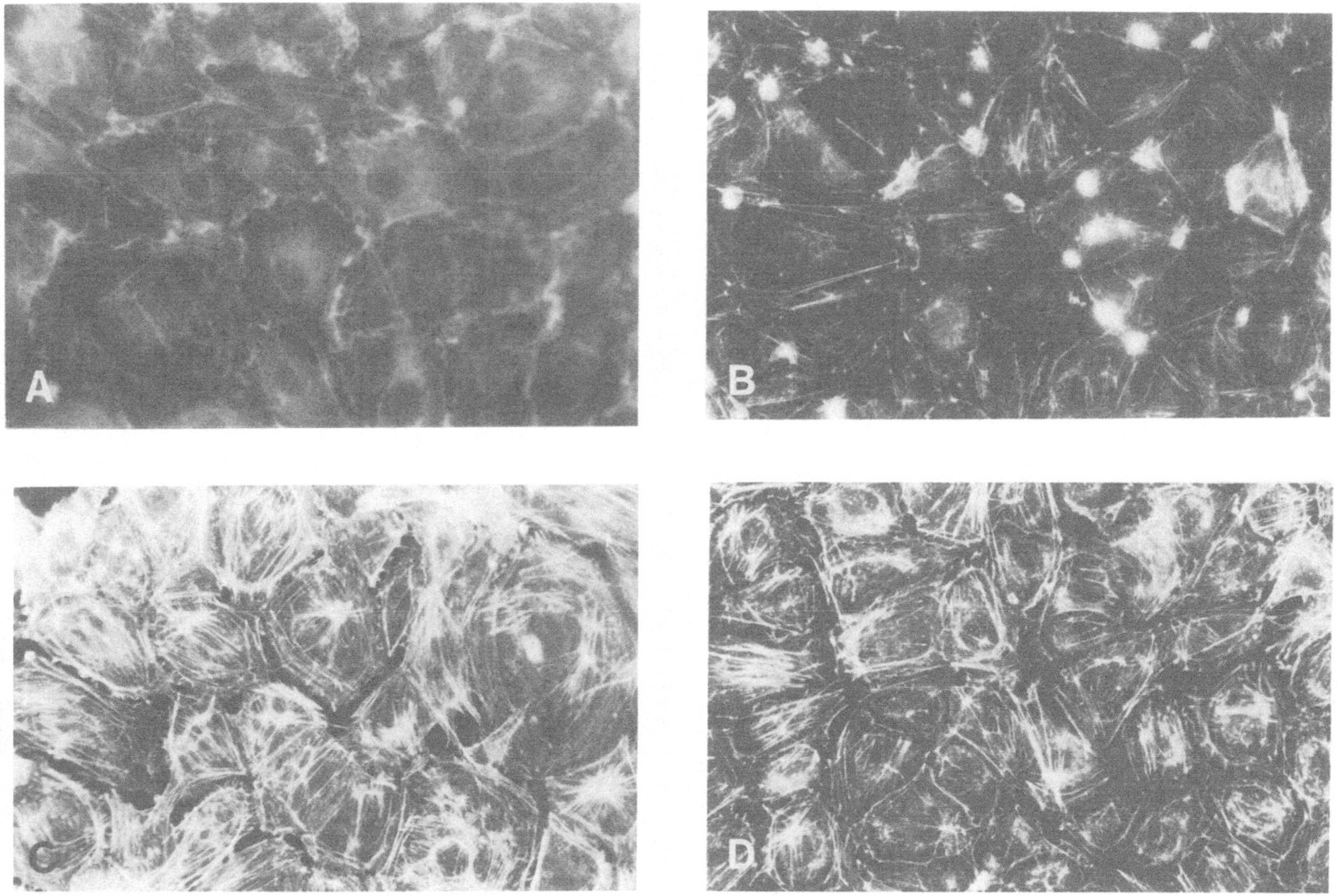

Figure 9. Patterns of actin staining resulting from experimental manipulations of second messenger systems which alter permeability in cell columns: (A) dibutyryl cAMP (10^{-3} M; 15 min), (B) forskolin (10^{-6} M; 2 min), (C) A23187 (10^{-6} M; 15 min), and (D) PMA (10^{-6} M; 10 min).

Activation of PKC by PMA (10^{-6} M for 15 min), a functional analogue for diacylglycerol, mimics the effect of endogenously generated diacylglycerol. In our study, the potent activation of PKC produced dramatic increases in endothelial monolayer permeability (Table 1).

PKC activation by PMA radically altered microfilament structure, producing extensive irregularities in the cell membrane, with intercellular gap formation; normal microfilament structure is replaced by small actin "asters" appeared frequently, and accounted for much of the microfilament F-actin staining (Fig. 9D).

3. *Dibutyryl cAMP and Forskolin*

Some cell surface receptors mobilize cAMP. The membrane-permeant dibutyryl analogue of cAMP effectively mimics the effects of receptor-mobilized cAMP. Endothelial permeability decreased within 15 min after exposure to 1 mM dibutyryl cAMP and reversed within 30 min of dibutyryl cAMP removal. Similarly, monolayer treatment with the specific adenylate cyclase activator, forskolin, decreased permeability (10^{-7} M for 15 min); permeability reverted to baseline levels within 30 min after removal of forskolin from the perfusate (see Table 1).

Both dibutyryl cAMP and forskolin enhanced dense peripheral band structure and regularity (Fig. 9A,B). Peripheral microfilament bands associated with the border and spanning microfilaments and "asters" were uncommon.

D. *Probes of the Cytoskeletal System and Intercellular Geometry*

1. *Cytochalasin D and Phalloidin*

Treatment with cytochalasin D for 15 min increases monolayer permeability by 200%; permeability reverses after removal of cytochalasin D (Table 1). Phalloidin was not tested for permeability effects in cell columns, but has been shown to decrease permeability.[10]

Cytochalasin D produced complete loss of the regular dense peripheral band (Fig. 11B), and extensive retraction of the intercellular border. Diffuse central F-actin and actin paracrystal staining replace most normal microfilaments. Phalloidin produced few changes in dense peripheral band staining, did not produce gaps, and did not affect "paracrystal" or "aster" formation (Fig. 11D).

2. *Low Extracellular Calcium*

Removal of 90% of the calcium from the cell-column perfusate for 10 min produced a 200% increase in permeability which returned to near baseline levels within 15 min after restoring normal calcium concentrations (see Fig. 10 and Table 1).

A decrease in the concentration of extracellular calcium caused the dense peripheral band (Fig. 11C) to disappear and, in general, the microfilament structure became more central and gaps occurred.

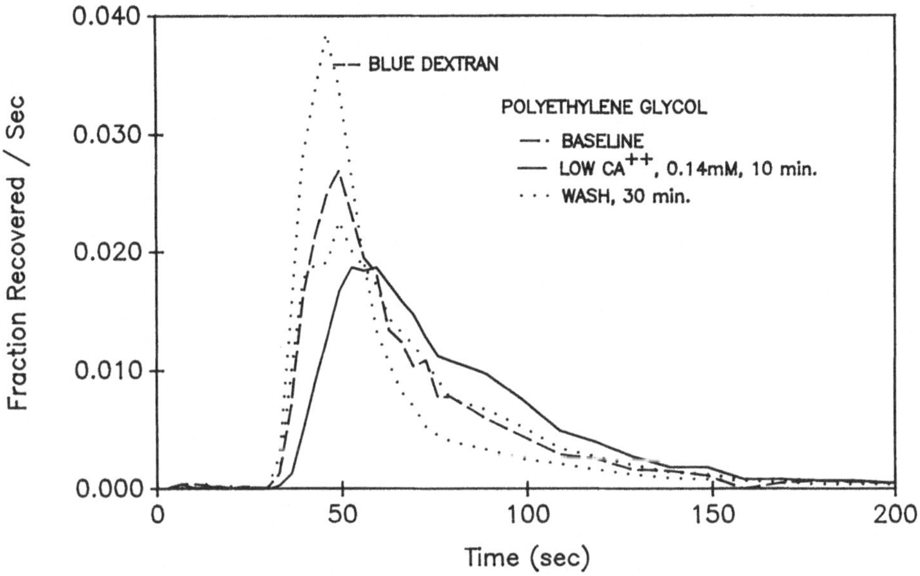

Figure 10. Effect of lowering extracellular calcium on endothelial cell-column elution patterns. Composite from three measurements made during a single experiment comparing the elution pattern of blue dextran (labeled dotted line) with polyethylene glycol under three conditions: baseline (1.4 mM Ca^{2+}, dashed line), low calcium (0.14 mM Ca^{2+} after 10 min, solid line), and baseline perfusate 30 min after a return to normal Ca^{2+} concentration (dotted line). Compared to baseline, 0.14 mM Ca^{2+} lowers the peak of the elution pattern of polyethylene glycol and raises the concentration at later time points. A return to normal calcium results in an elution profile similar to baseline.

IV. DISCUSSION

A. Permeability and Cytoarchitecture

Several studies suggest that there is a relationship between cytoskeletal disposition and numerous structure-related functions of the cell, such as shape change, locomotion, and monolayer permeability. Therefore, we have examined the effects of physiological and pharmacological agents on permeability and cytoskeleton using cell chromatography and rhodamine phalloidin staining for F-actin. Our results indicate that these agents affect both the endothelial permeability function and cytoskeletal structure.

We have demonstrated that the permeability and cytoskeleton of the endothelial monolayer rapidly and reversibly respond to stimulation. This behavior can only be attributed to agonist-induced changes in the endothelial monolayer behavior itself and to no other cells. These changes evidently are independent of changes in hydrostatic pressure and blood vessel recruitment.

B. Proposed Mechanism

In addition, our experiments have shown that diverse stimuli alter endothelial monolayer permeability and produce rearrangements of F-actin filaments. Based on these functional–structural sets of data, we propose a mechanistic model for the regulation of

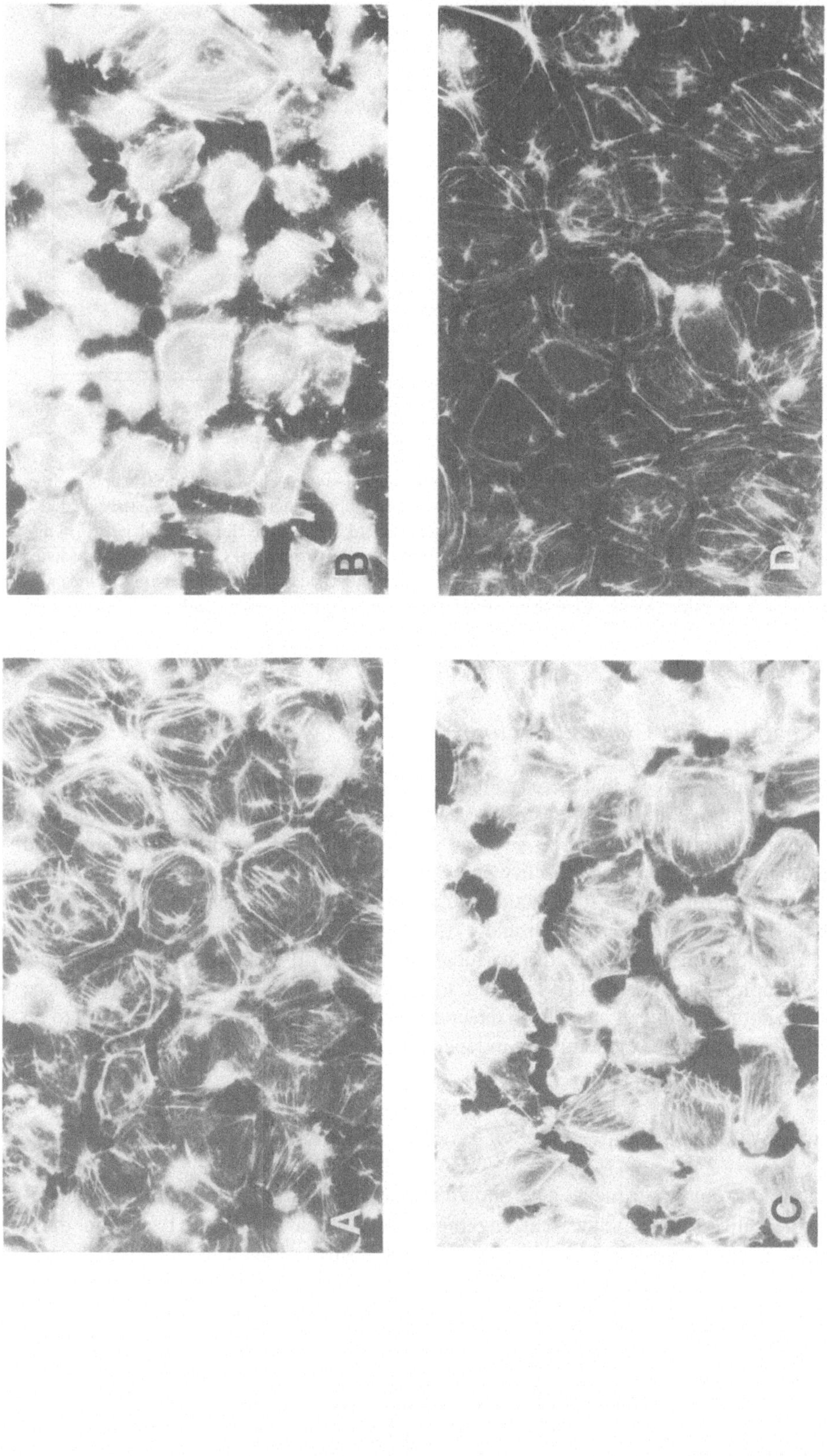

Figure 11. Patterns of actin staining resulting from experimental manipulations of cytoskeleton and extracellular Ca^{2+} which alter permeability in cell columns: (A) no treatment, (B) cytochalasin D (1 μg/ml; 15 min), (C) 0.14 mM extracellular calcium (10 min), and (D) phalloidin (10^{-5} M; 15 min).

intercellular junctional geometry, i.e., that change in permeability occurs through the specific chain of events described in Fig. 1. These events are: (1) cell surface receptor agonism, (2) signal amplification by second messengers, (3) cytoskeleton modification, and (4) subsequent alteration in the geometry of the interendothelial junction.

1. Receptors

It is well documented that the permeability/cytoskeletal response is linked to cell surface receptors. Endothelial cells have been found to express surface receptors which respond to extracellular signals: bradykinin,[31,32] thrombin,[13,33] purines,[34] and β-adrenergic agonists.[35–37] The permeability/cytoskeletal changes observed in the present study were produced by pharmacological concentrations of bradykinin, thrombin, adenosine, ATP, and β-isoproterenol.

The competitive antagonist propranolol blocked the decreased permeability response during treatment with β-isoproterenol. Such receptor-mediated responses occurred within 15 to 30 min, and with the exception of thrombin were reversed by the removal of the agent. Furthermore, we found that confluent monolayers did not respond to these agonists until 4 to 6 days postconfluence (data not shown). This time dependency is consistent with the notion that subcultivation using trypsin removes cell surface receptors,[33] which must be reexpressed to mediate the permeability/cytoskeletal responses of these agents.

2. Second Messengers

The second event in the proposed sequence leading to the permeability/cytoskeletal response is a change in the cytoplasmic concentrations of second messengers. The complex interactions of the various messengers are summarized schematically in Fig. 12. Our permeability/cytoskeletal responses are consistent with the action of second messengers observed in other cellular responses (secretion, shape change, motility, division). Figure 12 diagrams the effects of the inflammatory mediators thrombin and bradykinin through activation of the receptor labeled R_2. These autacoids increased permeability through an increase in intracellular calcium[38,39] by the pathway illustrated in the lower half of Fig. 12. As proposed by Berridge and Levine,[39] receptor-mobilized intracellular calcium requires phospholipase C (PLC) activation. Active PLC mediates production of diacylglycerol and IP3. These second messages release cell calcium stores, and lead to activation of cytoplasmic kinases like calmodulin-dependent protein kinases and protein kinase C (PKC).[40]

Direct evidence for the importance of Ca^{2+} in endothelial permeability/cytoskeletal regulation was found using the calcium ionophore A23187. An increase in cytoplasmic Ca^{2+} generated by A23187 was accompanied by an increase in endothelial monolayer permeability. The effect was rapid, and not reversible within 60 min. However, visual inspection of the cell-coated beads at the conclusion of ionophore exposure revealed that monolayers were intact.

Diacylglycerol (DG), another second messenger, alters cell shape and behavior via potent stimulation of PKC (Fig. 12). PMA mimics the action of DG. PMA, like DG, alters cell shape and behavior via potent stimulation of PKC.[40] PMA greatly increased cell-column endothelial permeability *in vitro* within 15 min and did not reverse within 60 min. Following PMA exposure, monolayers were intact.

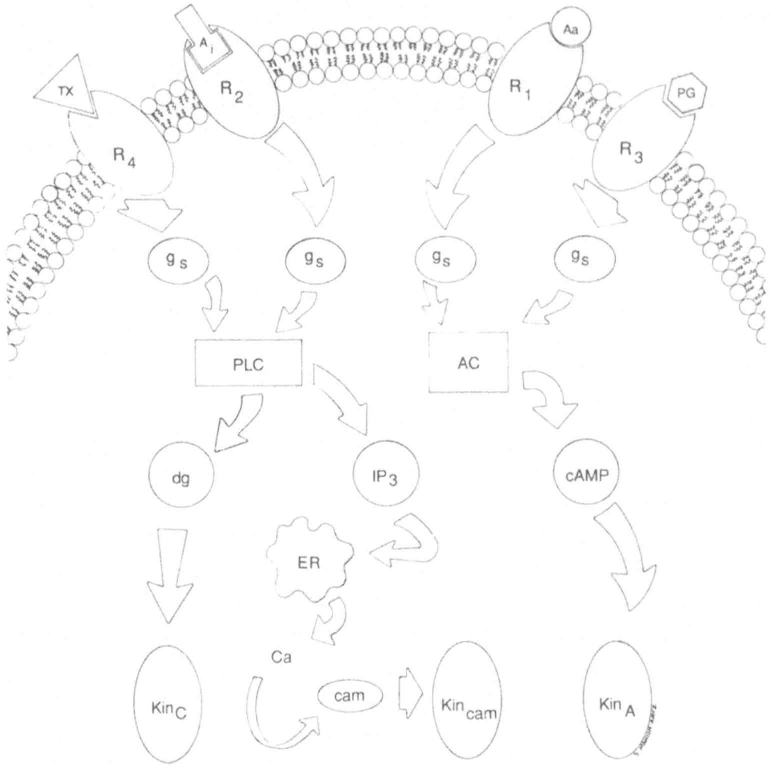

Figure 12. Inflammatory and anti-inflammatory second messenger pathways. The anti-inflammatory agonist (A_a) activates receptor R_1, stimulating adenylate cyclase (AC) through the action of a coupled stimulatory g-protein (g_s). This results in increased levels of cAMP which activate protein kinases (Kin_A). The inflammatory agonist (A_i) activates receptor R_2, stimulating phospholipase C (PLC) through the action of a coupled stimulatory g-protein (g_s). This results in increased levels of IP_3 or diacylglycerol which activate protein kinase C (Kin_C) or the release of calcium from the endoplasmic reticulum. Increased calcium is coupled through calmodulin to calmodulin-dependent kinases. Prostaglandins (PG) and thromboxanes (TX) may participate in feedback mechanisms which also mobilize these messenger pathways through surface receptors (R_3 and R_4).

The physiological mediators adenosine, ATP, and β-isoproterenol elevate cAMP levels by receptor-mediated stimulation of adenylate cyclase.[41,42] The pathway in Fig. 12 corresponding to these agonists begins with agonist stimulation of the receptor type labeled R_1. In the column assay all of these mediators decreased permeability. A similar effect was found when increased concentrations of the second messenger cAMP were mimicked through exposure of the monolayers to the membrane permeant analogue, dibutyryl cAMP. In addition, forskolin, a direct activator of adenylate cyclase[43] (Fig. 12), also decreased permeability.

3. *Regulation of Cytoarchitecture*

The third event in the proposed chain of events depicted in Fig. 1 is an alteration of the cytoarchitecture. Several investigations document receptor and second message modulation of cytoskeleton.[12,16,17] In confluent endothelial monolayers, microfilaments are organized as a "dense peripheral band" (DPB), the most conspicuous constituent of endothelial cytoarchitecture.[44] Microfilament assembly and cell shape are known to be sensitively regulated by cell calcium,[12,45] PKC,[46] and cAMP.[29] The edge-associated DPB has been shown to interact with junctional components[14,44] and may contribute to regulation of monolayer permeability.[10,15,47]

In general, we observed that agents which alter permeability modify F-actin staining patterns[16,17] (see Figs. 6, 7, 9, 11). Agents which increased permeability (bradykinin, thrombin, PMA, A23187, and cytochalasin D) produced characteristic changes in cytoarchitecture. The most dramatic, though infrequent, change in cytoarchitecture is retraction and opening of intercellular spaces as seen in the cytochalasin D and low Ca^{++} treated monolayers (Fig. 11B,C). Spaces were present in controls and were occasionally enhanced with other permeability-increasing agents. Visible gap formation may not be necessary for an increase in permeability to occur.

The structural change most often associated with agents that increase permeability is retraction and reorganization of the DPB (Figs. 6, 9C,D, 11B,C). A second common change was the formation of "aster" structures in the cytoplasm. These structures probably represent remodeled peripherally associated F-actin which retracts the cell border.

Agents which decreased permeability (ATP, adenosine, β-isoproterenol, dibutyryl cAMP, and forskolin) produced different characteristic changes in F-actin staining, often a distinct highlighting of the peripheral band after treatment indicating that the junctional region is preserved and possibly enhanced.[28]

Another common feature indicative of junction enhancement is decreased central staining and diminished microfilament spanning of the cell center. This was most evident for dibutyryl cAMP and forskolin treatment (Fig. 9A,B). One interpretation is that central parallel, or transcellular, actin filaments are involved in tension development; dissolution of these structures and junction enhancement may indicate redistribution of F-actin from central structures to junctional structures such as the DPB.

4. Direct Cytoskeletal Manipulation

Results obtained using physiological mediators and agents that mimic second messengers suggest that normal physiological modulation of the permeability of the endothelial barrier entails changes in the microfilament skeleton. The functional association between permeability and cytoskeleton is consistent with permeability effects of agents that directly alter the cytoarchitecture. For example, cytochalasins B and D are toxins which disturb actin microfilament structure; they have also been demonstrated to increase endothelial permeability *in vivo* and *in vitro*.[15,49] We found that permeability in the cell column was reversibly increased by treatment with cytochalasin D and that the DPB of the cytoskeleton was disintegrated. Perforations in the monolayer were common and the DPB of actin microfilaments was replaced by punctate aggregates of F-actin. Another cytoskeletal agent, phalloidin, stabilizes actin microfilaments against depolymerization.[48] Alexander *et al.*[10] demonstrated that phalloidin decreases permeability in cultured endothelial monolayers. We did not investigate phalloidin in the cell-column assay, but this agent does seem to enhance peripheral band structure (Fig. 11D). Direct modifications in the cytoarchitecture by cytochalasins and phalloidin suggest that an intact microfilament system is a requirement of the normal endothelial barrier.

5. Extracellular Calcium

Junctional stability requires adequate extracellular calcium. Many reports show that intercellular junctions are Ca^{2+} dependent.[30,50–52] Endothelial monolayers exposed to low concentrations (0.14 mM) of calcium exhibit increased permeability which reversed within 15 min after restoration of normal (1.4 mM) calcium concentrations. If extracellular Ca^{2+} levels were reduced below this critical level (0.14 mM), or if the monolayers were exposed to low Ca^{2+} for 30 min or more, normal barrier did not recover within the course of the experiment, suggesting that extracellular Ca^{2+} may be essential for the maintenance of the normal barrier function. Cytoskeletal staining shows that dissolution of the DPB results from this treatment (Fig. 11C). Thus, endothelial barrier maintenance and cytoskeleton are Ca^{2+} dependent.

6. Changes in Intercellular Geometry

The fourth event in the regulatory scheme is an alteration in the intercellular geometry as depicted in Fig. 13, which shows intercellular geometry changes as regulated by cytoplasmic features. Figure 13A shows the configuration of the key elements in regulating the structure of the barrier: actin microfilaments associate with, and organize, the endothe-

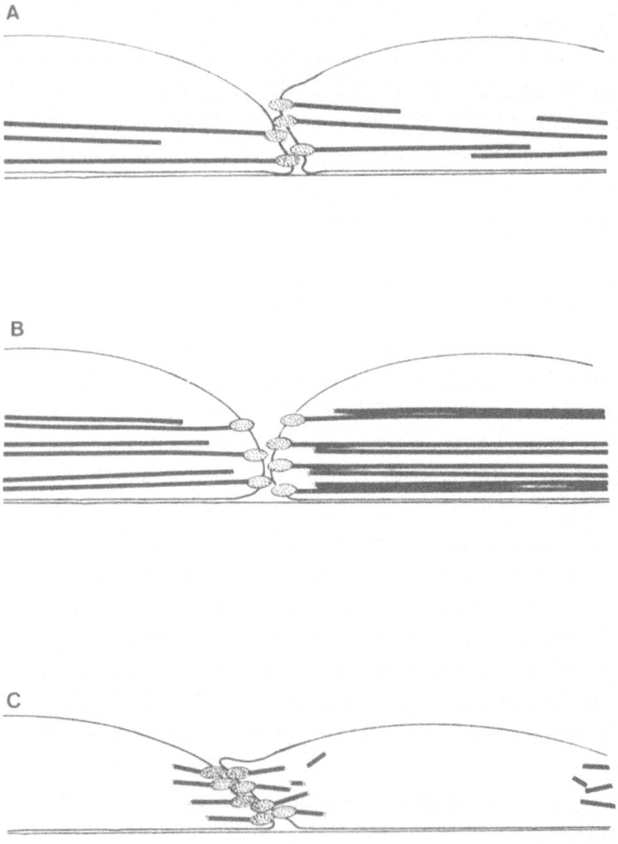

Figure 13. Schematic diagram of three states of the interendothelial junctions. A shows the configuration of the key elements important in regulating barrier. Actin microfilaments (heavy black lines) associate with and organize the endothelial junctional proteins (e.g., plakoglobins, uvomorulins, integrins, vinculin, talin; shown as speckled ovals). B shows the distributions of these elements during an inflammatory episode. Actin microfilaments retract to the center of the cell; junctional proteins also retract from each other, resulting in widening of the intercellular cleft. C shows the cytoplasmic regulatory elements during barrier promotion. Actin microfilaments associate with the junction, and enhance junctional protein organization.

lial junctional proteins (plakoglobins, integrins). Figure 13B shows the proposed distributions of these elements during an inflammatory episode: actin microfilaments retract to the center of the cell; junctional proteins also retract, resulting in widening of the intercellular cleft. Figure 13C shows the disposition of cytoplasmic regulatory elements during decreased permeability associated with the promotion of barrier. F-actin associates with the junction, and may enhance the organization of the junction. Transcellular tension is reduced, both by filament dissolution and by inactivation of the contractile apparatus. The combined effect of these two events is to reduce the intercellular space, thereby decreasing permeability.

The best evidence for the scheme proposed in Fig. 13 can be inferred from the changes in permeability produced by the agents discussed above. The most consistent physical interpretation of these permeability changes is an increase or decrease in the area available for the paracellular transport of tracers (e.g., polyethylene glycol) between the mobile phase of the cell column and the interior of the bead. Alternative explanations such as altered cellular uptake of polyethylene glycol are not consistent with the observation that within the time course of these measurements (3 min) no significant cellular uptake of polyethylene glycol occurs.[9] The most plausible explanation that combines the observations from both sets of data is the proposed change in interendothelial geometry depicted in Fig. 13.

The structural changes in DPB that occurred during challenge with mediators are consistent with a model in which DPB activation retracts cell borders, diminishes junctional integrity, and increases permeability. The rapid remodeling of the DPB observed during the change in permeability suggests that the DPB provides the necessary motile force for the structural rearrangement during the modulation of permeability postulated in Fig. 13. The close resemblance between actomyosin and the structure of sarcomeres strongly suggests functional homology with muscle. The development of microfilament-dependent tension in cultured endothelial cells has been demonstrated and has been shown to be receptor-regulated and second message-dependent.[53,54]

cAMP decreases monolayer permeability by reducing transcellular tension which may occur by both cAMP-mediated actomyosin inactivation and loss of DPB "sarcomere" organization. cAMP also produces centripetal DPB arrangement, paracrystal formation, and increased junctional apposition.

C. Another Cell Motile Phenomenon?

In many ways, the findings of the present study suggest that intercellular permeability regulation is another example of microfilament-mediated cell motility. Microfilaments participate in many other cell motile processes, including locomotion, shape change, cell division, and cytoplasmic circulation. Our study supports the observations of others showing that normal physiological modulation of endothelial monolayer permeability may be regulated by structural alteration or activation of the endothelial microfilament system.[10,16,17]

V. PERSPECTIVES

Our results show that endothelial cells regulate monolayer permeability independent of interstitial, adventitial, or luminal cells, and suggest that the cytoskeleton contributes to the regulation of normal monolayer permeability.

Many important details of the pathway proposed here have not been identified. For example, what is the link between the actin rearrangement and the physical change in intercellular geometry? Visualization studies directed against other intercellular and surface components may yield additional details of the molecular mechanism of this response.

In this report, we have only examined the microfilament-mediated contribution to monolayer barrier regulation in large vessel endothelium. Endothelial cells of varying histological and ontogenetic origins express other classes of junctional/cytoskeletal systems (zonula occludens and zonula adherens) in addition to endothelial adherens junctions.[14] Anticipated studies of microvascular permeability regulation could benefit from an examination of barrier regulation in these systems.

Since endothelial cell culture is a relatively recent innovation, not all culture conditions necessary to closely approximate *in vivo* conditions have been identified. The closer the *in vivo* and *in vitro* conditions approximate, the more confident we can be that the results of studies like these will apply *in vivo*.

Finally, the ability of the endothelial cell to respond to agents presented to its surface indicates continuing interaction with other cell types found in its local environment. The intercellular communication pathways need to be further identified. Now that the actions of many agents affecting microvascular permeability are accepted as endothelial cell specific, clinical and research approaches toward inflammation may focus increasingly on the endothelial component in inflammatory disease.

REFERENCES

1. Majno, G., Palade, G. E., and Schoefl, G. I., 1961, Studies on inflammation. II. The site of action of histamine and serotonin along the vascular tree: A topographic study, *J. Biophys. Biochem. Cytol.* **11:** 607–626.
2. Drenckhahn, D., and Franz, H., 1986, Identification of actin-, α-actin, and vinculin-containing plaques at the lateral membrane of epithelial cells, *J. Cell Biol.* **102:**1843–1852.
3. Meyrick, B., Perkett, E. A., Harris, T. R., and Brigham, K. L., 1987, Correlation of permeability with the structure of the endothelial layer of pulmonary artery intimal explants, *Fed. Proc.* **46:**2516–2520.
4. Renkin, E. M., 1952, Capillary permeability to lipid-soluble molecules, *Am. J. Physiol.* **168:**538–545.
5. Bottaro, D., Shepro, D., and Hechtman, H. B., 1986, Heterogeneity of intimal and microvessel endothelial cell barriers in vitro, *Microvasc. Res.* **32:**389–398.
6. Bundgaard, M., 1983, Capillary transport in capillary endothelium: Does it occur? *Fed. Proc.* **42:**2425–2430.
7. Simionescu, N., 1983, Cellular aspects of transcapillary exchange, *Physiol. Rev.* **63**(4):1536–1578.
8. Bottaro, D., Shepro, D., Peterson, S., and Hechtman, H. B., 1986, Serotonin, norepinephrine, and histamine mediation of endothelial cell barrier function in vitro, *J. Cell. Physiol.* **128:**189–194.
9. Haselton, F. R., Mueller, S. N., Howell, R. E., Levine, E. M., and Fishman, A. P., 1989, Chromatographic demonstration of reversible changes in endothelial permeability, *J. Appl. Physiol.* **67:**2032–2048.
10. Alexander, J. S., Hechtman, H. B., and Shepro, D., 1988, Phalloidin improves endothelial barrier function and protects against inflammatory agent mediated permeability in vitro, *Microvasc. Res.* **35:**308–315.
11. Casnocha, S. A., Eskin, S. G., Hall, E. R., and McIntire, L. V., 1989, Permeability of human endothelial monolayers: Effect of vasoactive agonists and cAMP, *J. Appl. Physiol.* **67**(5):1997–2005.
12. Rotrosen, D., and Gallin, J. I., 1986, Histamine type 1 receptor occupancy increases endothelial cytosolic calcium, reduces F-actin, and promotes albumin diffusion across cultured endothelial monolayers, *J. Cell Biol.* **103**(6, Pt. 1):2379–2387.
13. Minnear, F. L., Moon, D. G., and Hill, L. A., 1987, Thrombin-induced increase in pulmonary endothelial permeability in vitro: Protective effect of isoproterenol, *Ann. N.Y. Acad. Sci.* **485:**432–435.

14. Franke, W. W., Cowin, P., Grund, C., Kuhn, C., and Kapprell, H.-P., 1988, The endothelial junction: The plaque and its components, in *Endothelial Cell Biology in Health and Disease* (N. Simionescu and M. Simionescu, eds.), Plenum Press, New York, pp. 147–166.
15. Shasby, D. M., Shasby, S. S., Sullivan, M. J., and Petch, J. M., 1982, The role of the endothelial cell cytoskeleton in the modulation of endothelial barrier function, *Circ. Res.* **51**:657–661.
16. Welles, S. L., Shepro, D., and Hechtman, H. B., 1985, Vasoactive amines modulate actin cables (stress fibers) and surface area in cultured bovine endothelium, *J. Cell. Physiol.* **123**:337–342.
17. Welles, S. L., Shepro, D., and Hechtman, H. B., 1985, Eicosanoid modulation of stress fibers in cultured bovine aortic endothelium, *Inflammation* **9**(4):439–450.
18. Mueller, S. N., Rosen, E. M., and Levine, E. M., 1980, Cellular senescence in a cloned strain of bovine fetal aortic endothelial cells, *Science* **207**:889–891.
19. Ryan, U. S., Mortara, M., and Whitaker, C., 1980, Methods for microcarrier culture of bovine pulmonary artery endothelial cells avoiding the use of enzymes, *Tissue Cell* **12**:619–635.
20. Killackey, J. J., Johnston, M. G., and Movat, H. Z., 1986, Increased permeability of microcarrier-cultured endothelial monolayers in response to histamine and thrombin, *J. Pathol.* **122**:50–61.
21. Williams, S. K., Matthews, M. A., and Wagner, R. C., 1979, Metabolic studies on the micropinocytic process in endothelial cells, *Microvasc. Res.* **18**:175–184.
22. Crone, C., 1963, Permeability of capillaries in various organs as determined by use of the indicator diffusion method, *Acta Physiol. Scand.* **48**:292–305.
23. Sangren, W. C., and Sheppard, C. W., 1953, Mathematical derivation of the exchange of a labelled substance between a liquid flowing in a vessel and an external compartment, *Bull. Math. Biophys.* **15**:387–394.
24. Chinard, F. P., and Enns, T., 1954, Transcapillary pulmonary exchange of water in the dog, *Am. J. Physiol.* **178**:197.
25. Rowlett, R. D., and Harris, T. R., 1976, A comparative study of organ models and numerical techniques for evaluation of capillary permeability from multiple-indicator data, *Math. Biosci.* **29**:273–298.
26. Haselton, F. R., Roselli, R. J., Parker, R. E., and Harris, T. R., 1984, An effective diffusivity model of pulmonary capillary exchange: General theory, limiting cases and sensitivity analysis, *Math. Biosci.* **70**:237–263.
27. Haselton, F. R., Mueller, S. N., and Howell, R. E., 1987, Analysis of chromatography data from an endothelial cell column, *Proc. 13th N.E. Bioeng. Conf.*, Philadelphia.
28. Verderame, M., Alcorta, D., Egnor, M., Smith, K., and Pollack, T., 1980, Cytoskeletal F-actin patterns quantified with fluorescein isothiocyanate phalloidin in normal and transformed cells, *Proc. Natl. Acad. Sci. USA* **77**:6624–6628.
29. Osborn, M., and Weber, K., 1984, Actin paracrystal induction by forskolin and by dibutyryl cAMP in CHO cells, *Exp. Cell Res.* **150**:408–418.
30. Haselton, F. R., Alexander, J. S., and Blankenship, C., 1990, Permeability Response of Endothelial Monolayers to Changes in Extracellular Calcium, Microcirc. Soc., Washington, D.C.
31. Colden-Stanfeld, M., Schilling, W. P., Ritchie, A. K., Eskin, S. G., Navarro, L. T., and Kunze, D. L., 1987, Bradykinin induced increases in cytosolic Ca^{++} and ionic currents in cultured bovine aortic endothelial cells, *Circ. Res.* **61**:632–640.
32. Schilling, W., Ritchie, A. K., Navarro, L. T., and Eskin, S. G., 1987, Bradykinin stimulated Ca^{++} influx in cultured bovine aortic endothelial cells, *Circulation* **76**:52.
33. Jaffe, E. A., Grulich, J., Weksler, B. B., Hampel, G., and Watanabe, K., 1987, Correlation between thrombin induced prostacyclin production and inositol triphosphate and cytosolic free calcium levels in cultured human endothelial cells, *J. Biol. Chem.* **262**:8557–8565.
34. Luckhoff, A., and Busse, R., 1986, Increased free calcium in endothelial cells under stimulation with adenine nucleotides, *J. Cell. Physiol.* **126**:414–420.
35. Steinberg, S. F., Jaffe, E. A., and Bilezikian, J. P., 1984, Endothelial cells contain beta receptors, *Naunyn-Schmiedebergs Arch. Pharmacol.* **325**:310–313.
36. Howell, R. E., Albelda, S. M., Daise, M. L., and Levine, E. M., 1988, Characterization of beta adrenergic receptors in cultured human and bovine endothelial cells, *J. Appl. Physiol.* **65**:1251–1257.
37. Buonassisi, V., and Ventner, J. C., 1976, Hormone and neurotransmitter receptors in an established vascular endothelial cell line, *Proc. Natl. Acad. Sci. USA* **73**:1612–1616.
38. Ryan, U. S., Avdonin, P. V., Ya, E., Popov, E. G., Danilov, S. M., and Tkachuk, V. A., 1988, Influence of vasoactive agents on cytoplasmic free calcium in vascular endothelial cells, *J. Appl. Physiol.* **65**(5):2221–2227.

39. Berridge, M. J., and Levine, R. F., 1984, Inositol triphosphate: A novel second messenger in cellular communications, *Nature* **312**:315–321.
40. Kraft, A. S., and Anderson, W. B., 1983, Phorbol esters and protein kinase C, *Nature* **301**:621.
41. Clark, R. B., and Senney, N., 1976, Regulation of adenylate cyclase from cultured human cell lines by adenosine, *J. Biol. Chem.* **251**:4329–4336.
42. Makarski, J., 1981, Stimulation of cyclic AMP production by vasoactive agents in cultured bovine aortic and pulmonary artery endothelial cells, *In Vitro* **17**(50):450–457.
43. Seamon, K. B., and Daly, J. W., 1986, Forskolin: Its biological and chemical properties, *Adv. Cyclic Nucleotide Res.* **20**:3–138.
44. Wong, M. K. K., and Gottlieb, A. I., 1986, Endothelial cell monolayer integrity. I. Characterization of dense peripheral band of microfilaments, *Arteriosclerosis* **6**:212–219.
45. Europe-Finner, A. N., and Newell, P.C ., 1986, Inositol 1,4,5 triphosphate and calcium stimulation of actin polymerization in Dictyostelium discoideum, *J. Cell Sci.* **82**:41–51.
46. Schliwa, M., Nakamura, T., Porter, K. R., and Eutenever, U., 1984, A tumor promoter induces rapid and coordinated reorganization of actin and vinculin in cultured cells, *J. Cell Biol.* **99**:1045–1059.
47. Bentzel, C. J., Hainar, B., Ho, S., Hui, S. W., Adelman, A., Agnastopoules, T., and Bennedetti, K. L., 1980, Cytoplasmic regulation of tight junction permeability: Effect of plant cytokines, *Am. J. Physiol.* **239**: C75–C89.
48. Coluccio, L. M., and Tilney, L. G., 1984, Phalloidin enhances actin assembly by preventing monomer dissociation, *J. Cell Biol.* **99**:529–535.
49. Shasby, D. M., Shasby, S. S., and Peach, M. J., 1983, Granulocytes and phorbol myristate acetate increase permeability to albumin of cultured endothelial monolayers and isolated perfused lungs. Role of oxygen radicals and granulocyte adherence, *Am. Rev. Respir. Dis.* **127**(1):72–76.
50. Green, K. J., Geiger, B., Jones, J. C. R., Talian, J. C., and Goldman, R. D., 1987, The relationship between intermediate filaments and microfilaments before and during the formation of desmosomes and adherens-type junctions in mouse epidermal keratinocytes, *J. Cell Biol.* **104**:1389–1402.
51. Bologna, M., Allen, R., and Dulbecco, R., 1986, Organization of cytokeratin bundles by desmosomes in rat mammary cells, *J. Cell Biol.* **102**:560–567.
52. Shasby, D. M., and Shasby, S. S., 1986, Effects of calcium on transendothelial albumin transfer and electrical resistance, *J. Appl. Physiol.* **60**(1):71–79.
53. Kelly, C., D'Amore, P., Hechtman, H. B., and Shepro, D., 1988, Vasoactive hormones and cAMP affect pericyte contraction and stress fibers in vitro, *J. Musc. Res. Cell Motil.* **9**:184–194.
54. Morel, N., Dodge, A., Patton, W. F., and Shepro, D., 1990, Pulmonary microvascular endothelial cell contractility on silicone rubber substrates, *J. Cell. Physiol.* in press.

II

Endothelial Cell in Inflammation and Immunity

7

Endothelial–Leukocyte Adhesion Molecules in Acute Inflammation and Atherogenesis

Myron I. Cybulsky and Michael A. Gimbrone, Jr.

I. INTRODUCTION

Inflammation and atherosclerosis are pathophysiologic processes which share a common feature in their pathogenesis, the emigration of leukocytes from blood; yet, these processes involve different leukocytes, and have distinct locations, chronologies, and consequences. Acute inflammation is a rapid response of the host to local perturbations, such as trauma or bacterial infection, and is typically followed by resolution.[1] A prominent feature of acute inflammation is the emigration of neutrophil leukocytes from the blood of postcapillary venules and small veins into the extravascular space. Atherosclerosis, on the other hand, is a chronic process occurring in arteries, and initially involves the adherence and transmigration of circulating monocytes into the arterial intima.[2,3] The result is formation of atherosclerotic lesions that ultimately impinge on the arterial lumen and predispose to complications, such as thrombosis, hemorrhage, embolism, or dissection, which can have serious consequences for tissue perfusion.

Despite the differences in outcome of acute inflammation and atherosclerosis, an early event in both processes, leukocyte emigration, may share common mechanisms. The initial step in the emigration of leukocytes is their adherence to vascular endothelium. There is increasing evidence that vascular endothelial cells play an active role in this process by expressing on their surfaces endothelial–leukocyte adhesion molecules (ELAMs).[4,5] The localization of ELAM expression may account for venous versus arterial leukocyte emigration, and the leukocyte selectivity of a particular ELAM for neutrophil versus monocyte adhesion. In addition to supporting an adhesive interaction, the binding to an ELAM may transmit an activation signal to the leukocyte which may strengthen the

Myron I. Cybulsky and Michael A. Gimbrone, Jr. • Vascular Research Division, Department of Pathology, Brigham and Women's Hospital, and Harvard Medical School, Boston, Massachusetts 02115.

Endothelial Cell Dysfunctions, edited by Nicolae Simionescu and Maya Simionescu. Plenum Press, New York, 1992.

adhesive interaction, promote diapedesis, or induce the local production of cytokines. In this chapter, we will examine the experimental evidence for ELAM expression in both acute inflammation and atherosclerosis, and discuss the implications of this pathophysiologic mechanism.

II. ELAMs IN ACUTE INFLAMMATION

The adherence of neutrophils to vascular endothelium and their subsequent emigration is a hallmark of acute inflammation. This process is rapid and transient, its onset occurring within minutes and termination within hours after the initial inflammatory provocation.[6] In most tissues, neutrophil emigration occurs exclusively through the walls of postcapillary venules and small veins.[1,7,8] Although the usual purpose of neutrophil emigration, such as killing of invading microorganisms or phagocytosis of tissue debris, is beneficial to the host, this process potentially can result in microvascular dysfunction. A transient form of microvascular dysfunction is increased vascular permeability of postcapillary venules, which is neutrophil-dependent, unlike permeability changes induced by histamine or bradykinin.[9] Intense inflammatory reactions produce more severe injury to venules resulting in extravasation of erythrocytes,[10] and extensive plasma exudation with vascular stasis.[1]

Mediators of neutrophil emigration are produced in response to infection, tissue damage, or are liberated by microorganisms. Host-derived mediators include neutrophil chemoattractants (complement components, arachidonate products) and cytokines. Bacteria can liberate numerous biologically active products. In particular, gram-negative bacteria liberate endotoxins [lipopolysaccharides (LPS)] and the neutrophil chemoattractant *n*-formyl methionyl leucyl phenylalanine. In experimental models of dermal inflammation in the rabbit, these inflammatory stimuli appear to mediate neutrophil emigration by at least two different mechanisms.[11,12] Neutrophil chemoattractants elicit emigration with a rapid onset which cannot be suppressed by inhibiting local protein synthesis. In contrast, LPS or inflammatory cytokines such as interleukin 1 (IL-1) and tumor necrosis factor alpha (TNF) induce a protein synthesis-dependent neutrophil emigration with a delayed onset (30–60 min after intradermal injection). The biosynthesis and expression of leukocyte adhesion molecules by endothelium may account for this protein synthesis-dependent emigration.

For several years our laboratory has investigated endothelium-dependent mechanisms for leukocyte recruitment. We proposed that inducible, transient expression of ELAMs by the endothelium may contribute to the spatial and temporal patterns of leukocyte recruitment, and that a recognition mechanism involving leukocyte-specific ELAMs may add selectivity to this process. By utilizing *in vitro* models of leukocyte adhesion to cultured human umbilical vein endothelial cells (HUVEC), we and others have demonstrated that selective pretreatment of endothelium with LPS or a cytokine, such as IL-1 and TNF, upregulates the adhesion of neutrophils, monocytes, lymphocytes, and related cell lines in a time- and protein synthesis-dependent fashion.[13–18] These studies suggested that "activation" of vascular endothelium by certain inflammatory stimuli results in increased expression of cell surface adhesive molecules that can bind blood leukocytes. To better define cytokine-induced alterations in endothelial cell surface properties, our research group developed monoclonal antibodies to cytokine-activated endothelium. This strategy has yielded useful reagents for the immunochemical and functional characterization of putative

ELAMs *in vitro*, their molecular cloning, and investigation of their expression in human and animal tissues in pathophysiologic settings.

Studies by Bevilacqua and co-workers[19] led to the characterization of monoclonal antibodies H4/18 and H18/7, which recognize an inducible endothelial cell surface protein designated as "endothelial–leukocyte adhesion molecule 1 (ELAM-1)."[19] Its expression was upregulated from undetectable levels on unstimulated HUVEC by endotoxin and the same cytokine mediators (IL-1, TNF) that act on HUVEC to increase leukocyte adhesion.[20,21] Peak HUVEC expression occurred after approximately 4 hr of stimulation and was contemporaneous with peak expression of HUVEC adhesiveness for leukocytes. Other cytokines, including interferons and IL-2, did not induce ELAM-1 expression or stimulate leukocyte adhesion.

Monoclonal antibodies H4/18 and H18/7 were used to define the function and structure of ELAM-1. In leukocyte adhesion blocking assays, ELAM-1 was found to support neutrophil and HL-60 cells adhesion to IL-1-activated HUVEC monolayers.[22] In parallel studies, neither monoclonal antibody blocked adhesion of peripheral blood lymphocytes to HUVEC, under basal or cytokine-activated conditions. Immunoprecipitation studies determined that ELAM-1 is a glycoprotein which is biosynthesized only by activated endothelium.[19] Its cDNA was cloned from IL-1-activated HUVEC using a monoclonal antibody directed expression cloning strategy in COS cells.[23] The primary sequence of ELAM-1, predicted from its cDNA, indicated a transmembrane protein whose extracellular portion consisted of a complex mosaic of three different domain types: an amino-terminal lectinlike domain, an EGF domain, and six tandem repetitive motifs (about 60 amino acids each) related to those found in complement regulatory proteins. These unique structural characteristics, interestingly, were found to be shared by the recently cloned mouse protein MEL-14,[24,25] a lymphocyte cell surface homing receptor, and a human platelet granule membrane protein GMP-140.[26] The latter protein also has been recently described in endothelial cells, and in both cell types, upon activation by thrombin and other stimuli, appears to be rapidly mobilized from storage granules to the cell surface (hence, its alternative designation "PADGEM, platelet activation-dependent granule-external membrane protein"). The identification of these three cell surface molecules of similar complex domain structure thus appears to describe a new gene family, for which the name "vascular selectins" has been proposed.[27]

All members of this family have a defined role in leukocyte–endothelial adhesion. ELAM-1, expressed on activated endothelium, supports the adhesion of blood neutrophils and possibly monocytes.[19] MEL-14, a molecule expressed by neutrophils, monocytes, and lymphocyte subsets, mediates lymphocyte adhesion to specialized lymph node endothelial cells and may also participate in the neutrophil–endothelial interactions.[28] GMP-140 expression on the surface of thrombin-activated human platelets has been shown to support the adhesion of neutrophils and monocytes,[29] and when upregulated onto the endothelial surface it supports the adhesion of neutrophils.[30] Although stimuli such as thrombin and histamine can induce the translocation of GMP-140 from the endothelial storage pool to the cell surface, *in vivo* they do not elicit neutrophil emigration. Nevertheless, it is tempting to speculate that GMP-140 or other putative adhesive molecules with similar characteristics may contribute to the protein synthesis-independent neutrophil emigration in response to leukocyte chemoattractants.

In addition to ELAM-1, two other leukocyte adhesion molecules whose expression on the endothelial surface can be upregulated have been identified and cloned. They have been designated ICAM-1 (intracellular adhesion molecule-1)[31,32] and VCAM-1 (vascular cell adhesion molecule-1).[33] The expression of both molecules can be increased by LPS, IL-1, and TNF (ICAM-1 is also upregulated by interferon gamma); however, unlike ELAM-1, their elevated expression is not transient but rather is sustained over several days. The primary leukocyte ligand for ICAM-1 is the CD11a/CD18 heterodimer LFA-1, which is present on the surface of neutrophils, monocytes, and lymphocytes.[34] VLA-4 is a ligand for VCAM-1 and is present only on monocytes and lymphocytes.[35] Thus, ICAM-1 can support the adhesion of neutrophils,[36] lymphocytes,[37] and possibly monocytes, whereas VCAM-1, also identified as INCAM-110,[38,39] is mononuclear leukocyte-selective, and supports the adhesion of monocytes and lymphocytes, but not neutrophils.[33,39]

The distribution of leukocyte adhesion molecules in normal and diseased human tissues has been examined immunohistochemically with monoclonal antibodies. Human and baboon microvascular endothelium have been shown to express ELAM-1 transiently in certain pathological settings, in particular acute and chronic inflammatory processes in which cytokine generation is thought to occur.[40–42] Interestingly, ELAM-1 is expressed exclusively in endothelium of venules and is not expressed in uninflammed endothelial cells in various vessel types nor in other normal tissue components. The selective expression of ELAM-1 in endothelium of venules has been confirmed in explant studies using human foreskin treated *ex vivo* with recombinant cytokines.[43] Unlike ELAM-1, ICAM-1 is expressed constitutively in endothelium and other cell types including fibroblasts, epithelial cells, and cells of leukocyte origin,[44] but can be upregulated in various pathophysiologic settings. In inflammatory reactions, enhanced endothelial expression of ICAM-1 is limited predominantly to postcapillary venules and small veins[41,42]; however, it also has been detected in endothelium of normal arterioles.[45] VCAM-1/INCAM-110 is constitutively expressed by cells with dendritic morphology in lymphoid follicles and possibly normal endothelium in some tissues.[39] It is not expressed by endothelium in blood vessels of normal skin, but expression is induced in various inflammatory conditions.[39] The elucidation of temporal and anatomic patterns of leukocyte adhesion molecule expression in various pathophysiologic settings, including acute inflammatory reactions elicited by cytokines or by neutrophil chemoattractants, should provide further clues for understanding their roles *in vivo*. Ultimate proof for these roles will require blocking leukocyte adherence and emigration in experimental *in vivo* models.

III. LOCALIZED EXPRESSION OF AN ELAM BY VASCULAR ENDOTHELIUM DURING ATHEROGENESIS

A prominent component of atherosclerosis is the interaction of blood leukocytes with various cellular components of the arterial wall. Unlike acute inflammation, neutrophil adherence and emigration are not a feature of early atherogenesis. In atherosclerosis, one of the earliest detectable events is the adherence of monocytes to endothelium lining large arteries.[2,3] Subsequently, these cells migrate into the intima where they accumulate and become transformed into lipid-engorged "foam cells." Careful immunohistochemical, ultrastructural, and autoradiographic studies in various experimental animal models and in human tissues have established the relationship between adhering/migrating monocytes and initial foam cells, thus implicating this process in the initiation of atherosclerotic lesions. In addition, cell and molecular biological approaches have suggested multiple potential roles for this macrophage population in lesion progression, including: the localized generation of cytokines (e.g., IL-1, TNF), growth factors (e.g., PDGF, FGF), coagulation and fibrinolytic components, arachidonate-derived mediators, and reactive oxygen products.[46] It has also been proposed that macrophage foam cells can reenter the bloodstream and thus export lipids from lesions. Recently, T lymphocytes have been detected in advanced atherosclerotic plaques, and although their function in lesion pathogenesis remains to be defined, their presence may indicate an immune component.[47] Cytokines such as gamma interferon, released by activated lymphocytes, can modulate the functions of various cell types, including vascular endothelium and smooth muscle.[46] In particular, the inducible expression of class II MHC antigens (HLA-DR) in atherosclerotic plaques (but not in normal arteries)[47] indicates that this humoral cell–cell interaction can occur and suggests that the immunological status of these cells may be modified. Lymphocyte products similarly can influence the metabolism of lipoproteins by macrophages and other cells. All of these interactions are potentially relevant to the progression/regression of more advanced lesions.

We undertook to examine the mechanisms of the selective mononuclear leukocyte-endothelial interactions during atherogenesis. The hypothesis that we chose to test was that the localized nature of these interactions may be influenced by changes in the adhesive properties of the endothelial surface, involving the expression of "ATHERO-ELAMs," i.e., ELAMs which are selectively expressed in atherosclerotic lesions. Our first step in supporting this hypothesis was to identify an "ATHERO-ELAM." We reasoned that a candidate ATHERO-ELAM should: (1) support mononuclear leukocyte adhesion, (2) be inducibly expressed on the endothelial cell surface, and (3) be detectable in early atherosclerotic lesions or in areas with a predilection for developing lesions. The experimental search for molecules satisfying these criteria was undertaken in the rabbit, a species in which dietary and genetic models of atherosclerosis are well described.[48–50] We studied leukocyte–endothelial interactions utilizing standardized monolayer adhesion assays.[22] Treatment of cultured rabbit endothelial monolayers with LPS induced a hyperadhesive surface change which was demonstrable with U937 cells,[51] a human myelomonocytic cell line, and was similar to that reported by others with human peripheral blood leukocytes.[52] We found that U937 adhesion was time- and protein synthesis-dependent (inhibitable with cycloheximide). Adhesion increased over basal levels after 1–2 hr of LPS treatment, reached a maximum at 6–8 hr, and remained sustained through 96 hr (Fig. 1).

Monoclonal antibodies were produced to LPS-treated rabbit endothelium to identify putative ELAMs mediating the increased U937 adhesion. Three antibodies, designated Rb1/9, Rb2/3, and Rb2/4 (all IgG1 immunoglobulins), recognized LPS-inducible endothelial surface antigens. The time course of binding of Rb1/9 and Rb2/4 were virtually identical and were similar to the endothelial adhesive change for U937 cells (low in untreated endothelium, reaching a maximum in 6–8 hr, and sustained through 96 hr). Rb2/3, on the other hand, recognized a surface epitope with higher basal expression and a sustained LPS-induced elevation. We assessed the role of the inducible endothelial cell surface antigens recognized by the antibodies in supporting U937 adhesion. In adhesion blocking assays, pretreatment of endothelial monolayers with saturating concentrations of Rb1/9 significantly inhibited U937 adhesion to basal and 5 hr- or 24 hr-activated endothelial

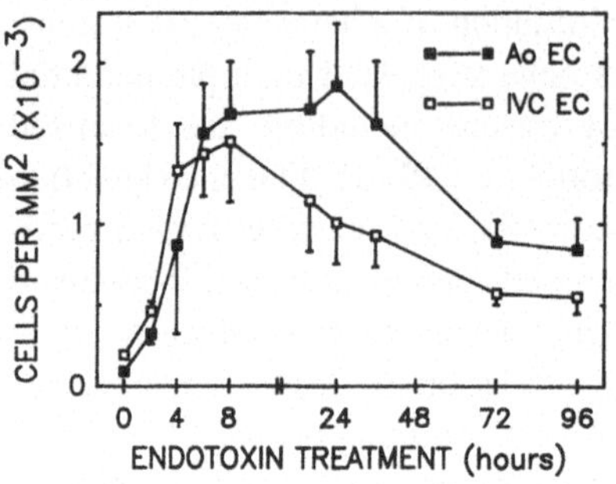

Figure 1. U937 adhesion to endothelial monolayers (passage 2), derived from the aorta (Ao) and inferior vena cava (IVC) of the same New Zealand White rabbit. Treatment of endothelium with LPS (*E. coli*, 1 μg/ml) increased U937 adhesion in a time-dependent fashion. The means and standard deviations of quadruplicate measurements are plotted.

monolayers (Fig. 2). In contrast, other monoclonal antibodies of the same immunoglobulin class (including Rb2/4, which recognizes the same molecule as Rb1/9,[51] and Rb2/3, which binds to basal and activated endothelial surfaces at higher densities than Rb1/9) did not inhibit U937 adhesion. These results established that the molecule recognized by Rb1/9 is an ELAM which appears to support mononuclear leukocyte adhesion. Immunochemical data indicated that this inducible ELAM is a newly synthesized protein, expressed on the endothelial surface in two forms (118 and 98 kDa, on reduced SDS-PAGE).[51]

Having identified an inducible rabbit ELAM which supported mononuclear leukocyte adhesion, we next examined its expression during atherogenesis by immunohistochemical staining. In rabbits fed a 1% cholesterol diet and in Watanabe heritable hyperlipidemic (WHHL) rabbits, specific staining with Rb1/9 was localized to aortic endothelium covering foam cell-rich intimal lesions.[51] Staining was evident at various stages of lesion development, ranging from focal regions with very small intimal accumulations of foam cells to near-circumferential lesions with abundant foam cells. Often, Rb1/9 staining of endothelium extended beyond the edges of intimal lesions (Fig. 3); however, in these same hypercholesterolemic animals, endothelium in adjacent uninvolved regions of the aorta did

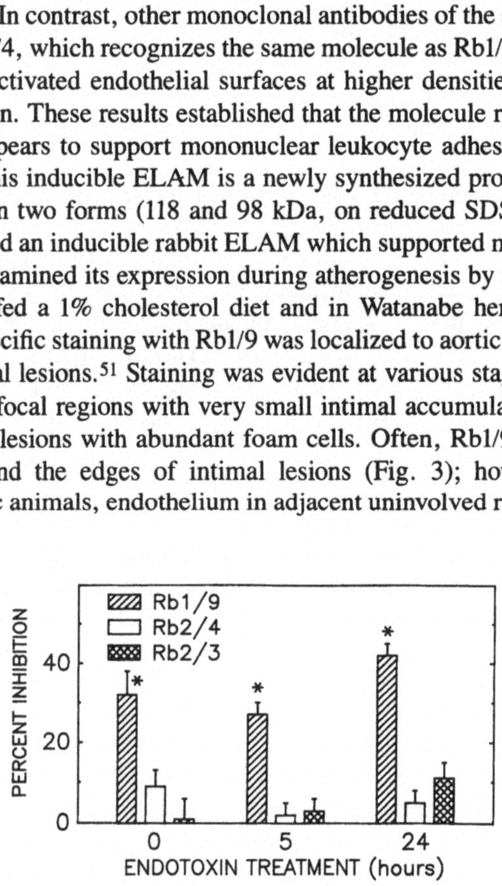

Figure 2. Monoclonal antibody inhibition of U937 adhesion to control and LPS-treated venous endothelial monolayers. In contrast to monoclonal antibodies Rb2/3 and Rb2/4, a 30-min preincubation (22°C) of monolayers with saturating concentrations of Rb1/9, inhibited significantly (*$p < 0.01$) U937 adhesion. Monoclonal antibodies were removed from wells prior to performing adhesion assays (10 min, 37°C), in which heat-aggregated rabbit IgG was included to block leukocyte Fc receptors. U937 adhesion to antibody-preincubated monolayers was compared to monolayers treated with control buffer (RPMI, 10% horse serum) and statistical significance was established using a paired *t* test. The means and standard errors of eight experiments with quadruplicate measurements are plotted.

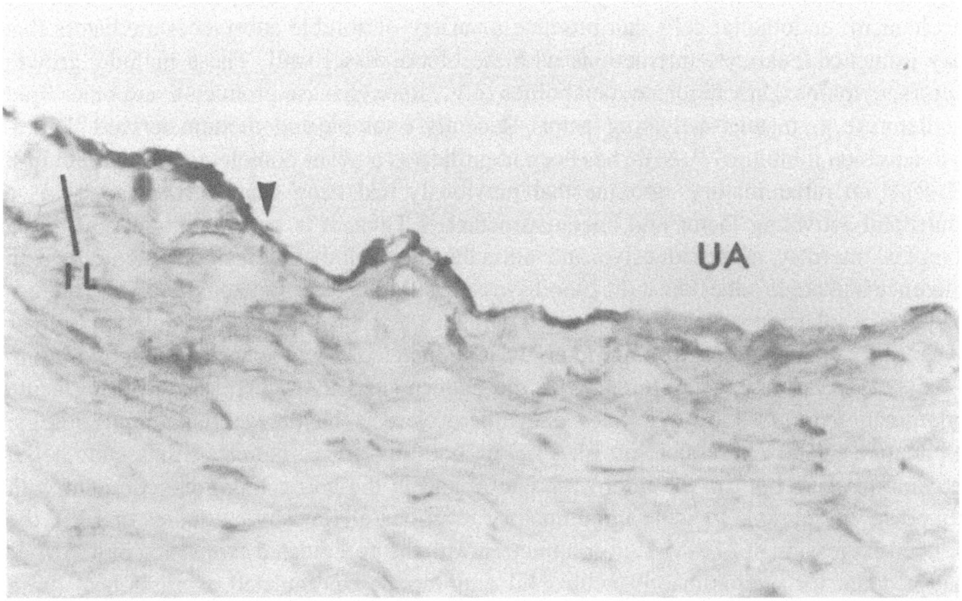

Figure 3. Immunohistochemical staining of an atherosclerotic lesion located in the descending thoracic aorta from a rabbit fed a 1% cholesterol diet for 9 weeks. Cryostat section stained for ATHERO-ELAM with Rbl/9 (undiluted culture supernatant, avidin–biotin peroxidase complex, 3-amino-9-ethylcarbazole). ATHERO-ELAM expression is detected in endothelium overlying a foam cell intimal lesion (IL) and extends for a short distance beyond the edge (arrowhead) of this lesion. Endothelium of the uninvolved aorta (UA) distant from the lesion failed to stain with Rbl/9. (Adapted from Ref. 51.)

not stain. Furthermore, endothelium in large arteries and veins of normal rabbits was uniformly unstained. Thus, the inducible ELAM recognized by Rbl/9 is an ATHERO-ELAM. These data are the first demonstration of lesion-selective ELAM expression in atherosclerosis, and this pattern of expression appears to differ from ELAM-1. To date, ELAM-1 has not been detected in the endothelium associated with human atherosclerotic plaques at various stages of progression (Munro, Pober, and Cotran, personal communication). The lack of ELAM-1 expression in human atherosclerosis (consistent with the absence of neutrophil emigration) suggests that the regulation of ELAM expression may be different in arteries than in microvessels.

IV. PATHOPHYSIOLOGIC IMPLICATIONS AND FUTURE DIRECTIONS

The adhesion and subsequent emigration of circulating leukocytes into extravascular tissues or into the intima of arteries plays a prominent role in the pathobiology of acute inflammation and atherosclerosis. This seemingly simple biologic phenomenon, in fact, may be regulated by several mechanisms. In this chapter, our focus has been on the upregulatable expression of ELAMs on the endothelial surface, a protein synthesis-dependent mechanism for leukocyte adhesion. In addition to this potential proadhesive mechanism, endothelial cells can produce a variety of soluble substances/mediators that may influence leukocyte interactions with the blood vessel wall. These include: growth factors, cytokines, arachidonate metabolites (e.g., lipoxygenase products), and other lipid mediators (e.g., platelet-activating factor). Recently, a soluble endothelium derived "leukocyte adhesion inhibitor" (LAI)[53] has been identified as a novel homologue of interleukin-8 (IL-8),[54] an inflammatory cytokine that previously had been characterized solely as a neutrophil-activating factor and chemoattractant.[55] Thus, it is becoming apparent that a complex interplay of proadhesive and antiadhesive endothelial factors potentially can influence leukocyte adhesion at the blood–vessel wall interface. Leukocyte factors, including surface adhesion molecules such as CD11/CD18, MEL-14, and the modulation of their levels or functional states, also are likely to influence leukocyte adhesion and emigration.

The similarities and differences in the patterns of ELAM expression during acute inflammation induced by cytokines and atherosclerosis have several pathophysiologic implications. The expression of ELAMs in both instances indicates that endothelial activation occurs both in the microvasculature and in the aortic lining concomitant with leukocyte emigration. In acute inflammatory reactions elicited by cytokines or LPS, the induced expression of ELAM-1, in conjunction with the upregulated expression of ICAM-1, may account for predominantly neutrophil emigration. Upregulated expression of these adhesion proteins on microvascular endothelium is localized to postcapillary venules, the site of neutrophil emigration. In rabbit models of atherosclerosis, we found arterial expression of an ATHERO-ELAM localized to endothelium overlying foam cell lesions, the site of ongoing mononuclear leukocyte adhesion and emigration. These observations are consistent with our hypothesis that ELAMs contribute to the anatomic location of leukocyte emigration.

The absence of ELAM-1 induction on arterial endothelium during inflammatory reactions and atherogenesis suggests that the regulation of ELAM expression may differ in arterial versus venous endothelium. In addition to this possibility, mediators produced in arteries, such as oxidized lipids[56,57] or hemodynamic factors, may selectively induce the expression of certain ELAMs. The absence of ELAM-1 induction in atherosclerotic lesions correlates with a lack of neutrophil recruitment at these sites. This observation is consistent with the notion that different patterns of ELAM expression may determine the specificity of leukocyte adhesion and emigration.

In normal arteries, an important role of the endothelium is to maintain homeostasis at the vessel–blood interface. Atherosclerosis or injury of the arterial wall can predispose the endothelium to dysfunction, i.e., modulation of the normally adaptive functional phenotype by pathophysiologic stimuli. Our data, demonstrating the expression of an ATHERO-ELAM, indicate that endothelial dysfunction may occur early in atherogenesis. This inducible molecular alteration of the endothelium may result in an abnormally hyperadhesive vascular luminal surface, and may promote mononuclear leukocyte recruitment independently, or in conjunction with other potential localizing mechanisms. These include lipid and peptide monocyte chemoattractants, which have been isolated from aortic tissue and from cultured endothelial or smooth muscle cells.[58–61]

The identification of an inducible ATHERO-ELAM suggests a number of experimental endeavors. Its role in leukocyte recruitment can be directly examined with blocking monoclonal antibodies in dietary and genetic models of atherosclerosis, thus helping to establish whether its expression is necessary for monocyte attachment. Mapping of its temporal and anatomic distribution will help to determine whether its expression precedes monocyte accumulation and, thus, whether it may have a role in the initiation of lesions. Further understanding of the selectivity for leukocyte recruitment in atherosclerosis (predominantly monocyte and lymphocyte versus neutrophil) will be enhanced by complete characterization of the leukocyte adhesive function of this ATHERO-ELAM and determining its primary structure. Molecular cloning will establish its relationship to other known vascular adhesion molecules and enable the identification of a human homologue. This and other potential ATHERO-ELAMs may find practical application as molecular markers in the diagnosis and treatment of early atherosclerotic disease. The definition of the molecular mechanisms regulating the expression of ELAMs in atherosclerosis, their relationship to known risk factors such as hypercholesterolemia,[61] and comparisons to the regulatory mechanisms operative in the microvasculature during inflammation should provide new insights into the pathogenesis of these complex disease processes.

ACKNOWLEDGMENTS. The authors acknowledge their colleagues and collaborators, especially Drs. M. P. Bevilacqua, R. S. Cotran, F. W. Luscinskas, J. M. Munro, and J. S. Pober, who actively contributed to various aspects of the studies of ELAM-1 summarized here. We also thank W. Atkinson, V. Davis, and K. Case for expert technical assistance. This research was supported primarily by grants from the National Institutes of Health. M.I.C. is a research fellow of the Medical Research Council of Canada.

REFERENCES

1. Movat, H. Z., 1985, *The Inflammatory Reaction*, Elsevier, Amsterdam.
2. Ross, R., 1986, The pathogenesis of atherosclerosis—An update, *N. Engl. J. Med.* **314**:488–500.
3. Munro, J. M., and Cotran, R. S., 1988, The pathogenesis of atherosclerosis: Atherogenesis and inflammation, *Lab. Invest.* **58**:249–261.
4. Gimbrone, M. A., Jr., and Bevilacqua, M. P., 1988, Vascular endothelium: Functional modulation at the blood interface, in: *Endothelial Cell Biology in Health and Disease* (N. Simionescu and M. Simionescu, eds.), Plenum Press, New York, pp. 255–273.
5. Stoolman, M. A., 1989, Adhesion molecules controlling lymphocyte migration, *Cell* **56**:907–910.
6. Colditz, I. G., and Movat, H. Z., 1984, Kinetics of neutrophil accumulation in acute inflammatory lesions induced by chemotaxins and chemotaxinigens, *J. Immunol.* **133**:2169–2173.
7. Clark, E. R., and Clark, E. L., 1935, Observations on changes in blood vascular endothelium in the living animal, *Am. J. Anat.* **57**:385–438.
8. Marchesi, V. T., 1961, The site of leukocyte emigration in inflammation, *Q. J. Exp. Physiol.* **46**:115–134.
9. Wedmore, C. V., and Williams, T. J., 1981, Control of vascular permeability by polymorphonuclear leukocytes in inflammation, *Nature* **289**:646–650.
10. Kopaniak, M. M., and Movat, H. Z., 1983, Kinetics of acute inflammation induced by *Escherichia coli* in rabbits: II. The effects of hyperimmunization, complement depletion and depletion of leukocytes, *Am. J. Pathol.* **110**:13–29.
11. Cybulsky, M. I., McComb, D. J., and Movat, H. Z., 1989, Protein synthesis dependent and independent mechanisms of neutrophil emigration. Different mechanisms of inflammation in rabbits induced by interleukin-1, tumor necrosis factor alpha or endotoxin versus leukocyte chemoattractants, *Am. J. Pathol.* **135**:227–237.
12. Rampart, M., and Williams, T. J., 1988, Evidence that neutrophil accumulation induced by interleukin-1 requires both protein biosynthesis and neutrophil CD18 antigen in vivo, *Br. J. Pharmacol.* **94**:1143–1148.

13. Bevilacqua, M. P., Pober, J. S., Wheeler, M. E., Cotran, R. S., and Gimbrone, M. A., Jr., 1985, Interleukin-1 acts on cultured human vascular endothelium to increase the adhesion of polymorphonuclear leukocytes, monocytes and related leukocyte cell lines, *J. Clin. Invest.* **76:**2003–2011.
14. Dunn, C. J., and Fleming, W. E., 1985, The role of interleukin-1 in the inflammatory response with particular reference to endothelial cell–leukocyte adhesion, in: *The Physiologic, Metabolic, and Immunologic Actions of Interleukin-1* (M. J. Kluger, J. J. Openheim, and M. C. Powanda, eds.), Liss, New York, pp. 45–54.
15. Schleimer, R. P., and Rutledge, B. K., 1986, Cultured human vascular endothelial cells acquire adhesiveness for neutrophils after stimulation with interleukin 1, endotoxin and tumor-promoting phorbol diesters, *J. Immunol.* **136:**649–654.
16. Gamble, R. J., Harlan, J. M., Klebanoff, S. J., Lopez, A. F., and Vadas, M. A., 1985, Stimulation of the adherence of neutrophils to umbilical vein endothelium by human recombinant tumor necrosis factor, *Proc. Natl. Acad. Sci. USA* **82:**8667–8671.
17. Pohlman, T. H., Stanness, K. A., Beatty, P. G., Ochs, H. D., and Harlan, J. M., 1986, An endothelial cell surface factor(s) induced *in vitro* by lipopolysaccharide, interleukin 1 and tumor necrosis factor-alpha increases neutrophil adherence by a CDw18-dependent mechanism, *J. Immunol.* **136:**4548–4553.
18. Cavender, D. E., Haskard, D. O., Joseph, B., and Ziff, M., 1986, Interleukin 1 increases the binding of human B and T lymphocytes to endothelial cell monolayers, *J. Immunol.* **136:**203–207.
19. Bevilacqua, M. P., Pober, J. S., Mendrick, D. L., Cotran, R. S., and Gimbrone, M. A., Jr., 1987, Identification of an inducible endothelial–leukocyte adhesion molecule, *Proc. Natl. Acad. Sci. USA* **84:**9238–9242.
20. Pober, J. S., Bevilacqua, M. P., Mendrick, D. L., Lapierre, L. A., Fiers, W., and Gimbrone, M. A., Jr., 1986, Two distinct monokines, interleukin 1 and tumor necrosis factor, each independently induce biosynthesis and transient expression of the same antigen on the surface of cultured human vascular endothelial cells, *J. Immunol.* **136:**1680–1687.
21. Pober, J. S., Lapierre, L. A., Stolpen, A. H., Brock, T. A., Springer, T. A., Fiers, W., Bevilacqua, M. P., Mendrick, D. L., and Gimbrone, M. A., Jr., 1987, Activation of cultured human endothelial cells by recombinant lymphotoxin: Comparison with tumor necrosis factor and interleukin 1 species, *J. Immunol.* **138:**3319–3324.
22. Luscinskas, F. W., Brock, A. F., Arnaout, M. A., and Gimbrone, M. A., Jr., 1989, Endothelial–leukocyte adhesion molecule-1-dependent and leukocyte (CD11/CD18)-dependent mechanisms contribute to polymorphonuclear leukocyte adhesion to cytokine-activated human vascular endothelium, *J. Immunol.* **142:**2257–2263.
23. Bevilacqua, M. P., Stengelin, S., Gimbrone, M. A., Jr., and Seed, B., 1989, Endothelial leukocyte adhesion molecule 1: An inducible receptor for neutrophils related to complement regulatory proteins and lectins, *Science* **243:**1160–1165.
24. Siegelman, M. H., van de Rijn, M., and Weissman, I. L., 1989, Mouse lymph node and homing receptor cDNA clone encodes a glycoprotein revealing tandem interaction domains, *Science* **243:**1165–1172.
25. Lasky, L. A., Singer, M. S., Yednock, T. A., Dowbenko, D., Fennie, C., Rodriguez, H., Nguyen, T., Stachel, S., and Rosen, S. D., 1989, Cloning of a lymphocyte homing receptor reveals a lectin domain, *Cell* **56:**1045–1055.
26. Johnston, G. I., Cook, R. G., and McEver, R. P., 1989, Cloning of GMP-140, a granule membrane protein of platelets and endothelium: Sequence similarity to proteins involved in cell adhesion and inflammation, *Cell* **56:**1033–1044.
27. Bevilacqua, M. P., 1989, Endothelial leukocyte adhesion molecule 1: An inducible receptor for neutrophils related to complement regulatory proteins and lectins, *Circulation* (Suppl.) **80**(4):II-I.
28. Butcher, E. C., 1990, Cellular and molecular mechanisms that direct leukocyte traffic, *Am. J. Pathol.* **136:** 3–11.
29. Larsen, E., Celi, A., Gilbert, G. E., Furie, B. C., Erban, J. K., Bonfanti, R., Wagner, D. D., and Furie, B., 1989, PADGEM protein: A receptor that mediates the interaction of activated platelets with neutrophils and monocytes, *Cell* **59:**305–312.
30. Geng, J.-G., Bevilacqua, M. P., Moore, K. L., McIntyre, T. M., Prescott, S. M., Kim, J. M., Bliss, G. A., Zimmerman, G. A., and McEver, R. P., 1990, Rapid neutrophil adhesion to activated endothelium mediated by GMP-140, *Nature* **343:**757–760.
31. Simmons, D., Makgoba, M. W., and Seed, B., 1988, ICAM, an adhesion ligand of LFA-1, is homologous to the neural cell adhesion molecule, *Nature* **331:**624–627.
32. Staunton, D. E., Marlin, S. D., Stratowa, C., Dustin, M. L., and Springer, T. A., 1988, Primary structure of

ICAM-1 demonstrates interaction between members of the immunoglobulin and integrin supergene families, *Cell* **52:**925–933.

33. Osborn, L., Hession, C., Tizard, R., Vassalio, C., Luhowskyj, S., Chi-Rosso, G., and Lobb, D., 1989, Direct expression cloning of vascular cell adhesion molecule 1, a cytokine-induced endothelial protein that binds to lymphocytes, *Cell* **59:**1203–1211.
34. Marlin, S. D., and Springer, T. A., 1987, Purified intracellular adhesion molecule-1 (ICAM-1) is a ligand for lymphocyte function-associated antigen 1 (LFA-1), *Cell* **51:**813–819.
35. Elices, M. J., Osborn, L., Takada, Y., Crouse, C., Luhowskyj, S., Hemler, M. E., and Lobb, R. R., 1990, VCAM-1 on activated endothelium interacts with the leukocyte integrin VLA-4/fibronectin binding site, *Cell* **60:**577–584.
36. Smith, C. W., Rothlein, R., Hughes, B., Mariscalco, M., Schmalsteig, F., and Anderson, D. C., 1988, Recognition of an endothelial determinant for CD 18-dependent human neutrophil adherence and transendothelial migration, *J. Clin. Invest.* **82:**1746–1756.
37. Haskard, D., Cavender, D. O., Beatty, P., Springer, T. A., and Ziff, M., 1986, T lymphocyte adhesion to endothelial cells: Mechanisms demonstrated by anti-LFA-1 monoclonal antibodies, *J. Immunol.* **137:**2901–2906.
38. Rice, G. E., and Bevilacqua, M. P., 1989, An inducible endothelial cell surface glycoprotein mediates melanoma adhesion, *Science* **246:**1303–1306.
39. Rice, G. E., Munro, J. M., and Bevilacqua, M. P., 1990, Inducible cell adhesion molecule 110 (INCAM-110) is an endothelial receptor for lymphocytes. A CD11/CD18-independent adhesion mechanism, *J. Exp. Med.* **171:**1369–1374.
40. Cotran, R. S., Gimbrone, M. A., Jr., Bevilacqua, M. P., Mendrick, D. L., and Pober, J. S., 1986, Induction and detection of human endothelial activation antigen in vivo, *J. Exp. Med.* **164:**661–666.
41. Cotran, R. S., Pober, J. S., Gimbrone, M. A., Jr., Springer, T. A., Wiebke, E. A., Gaspari, A. A., Rosenberg, S. A., and Lotze, M. T., 1988, Endothelial activation during interleukin 2 immunotherapy. A possible mechanism for the vascular leak syndrome, *J. Immunol.* **140:**1883–1886.
42. Munro, J. M., Pober, J. S., and Cotran, R. S., 1989, Tumor necrosis factor and interferon-gamma induce distinct patterns of endothelial activation and associated leukocyte accumulation in skin of *Papio Anubis*, *Am. J. Pathol.* **135:**121–133.
43. Messadi, D. V., Pober, J. S., Fiers, W., Gimbrone, M. A., Jr., and Murphy, G. F., 1987, Induction of an activation antigen on postcapillary venular endothelium in human skin organ culture, *J. Immunol.* **139:**1557–1562.
44. Dustin, M. L., Rothlein, R., Bhan, A. K., Dinarello, C. A., and Springer, T. A., 1986, Induction by IL 1 and interferon-gamma: Tissue distribution, biochemistry, and function of a natural adherence molecule (ICAM-1), *J. Immunol.* **137:**245–254.
45. Faull, R. J., and Russ, G. R., 1989, Tubular expression of intracellular adhesion molecule-1 during renal allograft rejection, *Transplantation* **48:**226–230.
46. Libby, P., 1987, The active roles of cells of the blood vessel wall in health and disease, in: *Molecular Aspects of Medicine*, Volume 9 (H. Baum, J. Gergely, and B. L. Fanburg, eds.), Pergamon Press, Elmsford, N.Y., pp. 499–567.
47. Hansson, G. K., Jonasson, L., Seifert, P. S., and Stemme, S., 1989, Immune mechanisms in atherosclerosis [Review], *Arteriosclerosis* **9:**567–578.
48. Buja, L. M., Kita, T., Goldstein, J. L., Watanabe, Y., and Brown, M. S., 1983, Cellular pathology of progressive atherosclerosis in the WHHL rabbit, *Arteriosclerosis* **3:**87–101.
49. Rosenfeld, M. E., Tsukada, T., Gown, A. M., and Ross, R., 1987, Fatty streak initiation in Watanabe heritable hyperlipemic and comparably hypercholesterolemic fat-fed rabbits, *Arteriosclerosis* **7:**9–23.
50. Havel, R. J., Yamada, N., and Shames, D. M., 1989, Watanabe heritable hyperlipidemic rabbit. Animal model for familial hypercholesterolemia, *Arteriosclerosis* Suppl. I **9:**I-33–I-38.
51. Cybulsky, M. I., and Gimbrone, M. A., Jr., 1991, Vascular endothelial cells express a monocyte adhesion molecule during atherogenesis, *Science* **251:**788–791.
52. Territo, M. C., Berliner, J. A., Almada, L., Ramirez, R., and Fogelman, A. M., 1989, Beta-very low density lipoprotein pretreatment of endothelial monolayers increases monocyte adhesion, *Arteriosclerosis* **9:**824–828.
53. Wheeler, M. E., Luscinskas, F. W., Bevilacqua, M. P., and Gimbrone, M. A., Jr., 1988, Cultured human endothelial cells stimulated with cytokines or endotoxin produce an inhibitor of leukocyte adhesion, *J. Clin. Invest.* **82:**1211–1218.

54. Gimbrone, M. A., Jr., Obin, M. S., Brock, A. F., Luis, E. A., Hass, P. E., Hebert, C. A., Yip, Y. K., Leung, D. W., Lowe, D. G., Kohr, W. J., Darbonne, W. C., Bechtol, K. B., and Baker, J. B., 1989, Endothelial-interleukin-8: A novel inhibitor of leukocyte–endothelial interactions, *Science* **246**:1601–1603.
55. Baggiolini, M., Walz, A., and Kunkel, S. L., 1989, Neutrophil-activating peptide-1/interleukin 8, a novel cytokine that activates neutrophils, *J. Clin. Invest.* **84**:1045–1049.
56. Steinberg, D., Parthasarathy, S., Carew, T. E., Khoo, J. C., and Witztum, J. L., 1989, Beyond cholesterol: Modifications of low-density lipoprotein that increase its atherogenicity, *N. Engl. J. Med.* **320**:915–924.
57. Berliner, J. A., Territo, M. C., Sevanian, A., Ramin, S., Kim, J. A., Bamshad, B., Esterson, M., and Fogelman, A. M., 1990, Minimally modified low density lipoprotein stimulates monocyte endothelial interactions, *J. Clin. Invest.* **85**:1260–1266.
58. Gerrity, R. G., Goss, J. A., and Soby, L., 1985, Control of monocyte recruitment by chemotactic factor(s) in lesion-prone areas of swine aorta, *Arteriosclerosis* **5**:55–66.
59. Berliner, J. A., Territo, M., Almada, L., Carter, A., Shafonsky, E., and Fogelman, A. M., 1986, Monocyte chemotactic factor produced by large vessel endothelial cells in vitro, *Arteriosclerosis* **6**:254–258.
60. Valente, A. J., Graves, D. T., Vialle-Valentin, C. E., Delgado, R., and Schwartz, C. J., 1988, Purification of a monocyte chemotactic factor secreted by non-human primate vascular cells in culture, *Biochemistry* **27**:4162–4168.
61. Quinn, M. T., Parthasarathy, S., and Steinberg, D., 1988, Lysophosphatidylcholine: A chemotactic factor for human monocytes and its potential role in atherogenesis, *Proc. Natl. Acad. Sci. USA* **85**:2805–2809.

8

Leukocyte-Mediated Endothelial Injury

Robert K. Winn, Charles L. Rice, Nicholas B. Vedder, William J. Mileski, and John M. Harlan

I. INTRODUCTION

The inflammatory response is characterized by leukocyte adherence to the vessel wall and subsequent infiltration of tissue by emigrated leukocytes.[1] Leukocytes contain or are capable of producing a number of products that can potentially cause cell injury or modify cell function (reviewed in Ref. 2). Consequently, many investigators have proposed a causal relationship between leukocyte accumulation at sites of inflammation or immune reaction and associated vascular and tissue injury. Leukocyte-mediated injury of vascular endothelium as an "innocent bystander" has been implicated in the pathogenesis of a wide variety of clinical disorders some of which are listed in Table 1.

Experimental evidence supporting a critical role of neutrophils in the pathogenesis of vascular injury comes from studies using animals depleted of circulating neutrophils by specific antibody or chemotherapeutic agents. Prior depletion of neutrophils attenuates vascular injury associated with the Arthus and Shwartzman reaction, serum sickness, nephrotoxic nephritis, acute arthritis, ischemia–reperfusion injury of the myocardium and intestine, and inflammatory skin lesions.[2] Several models of acute lung injury have also been shown to be attenuated by prior neutrophil depletion.[3]

Circulating leukocytes other than neutrophils may also contribute to vascular pathology. Mononuclear leukocytes rather than neutrophils are the predominant cell type at sites of chronic inflammation or immune reaction. Sites of allergic reaction exhibit a prominent infiltrate of eosinophils. Mediators released by these cell types during the process of adherence and emigration may also provoke vascular injury. However, because selective depletion or inhibition of these cell types *in vivo* is problematic, there is little direct experimental evidence for their role as mediators of vascular injury.

Robert K. Winn • Departments of Surgery and Physiology–Biophysics, University of Washington, Seattle, Washington 98195. *Charles L. Rice, Nicholas B. Vedder, and William J. Mileski* • Department of Surgery, University of Washington, Seattle, Washington 98195. *John M. Harlan* • Department of Medicine, University of Washington, Seattle, Washington 98195.

Endothelial Cell Dysfunctions, edited by Nicolae Simionescu and Maya Simionescu. Plenum Press, New York, 1992.

Table 1. Clinical Disorders Possibly Associated with Leukocyte-Mediated Vascular Injury

Adult respiratory distress syndrome
Multiple organ failure syndrome
Cerebral edema in bacterial meningitis
Inflammatory bowel disease
Inflammatory dermatoses
Vasculitis syndromes
Allograft rejection
Rheumatoid arthritis
Atherosclerosis
Vascular leak syndrome secondary to IL-2 therapy
Graft versus host disease
Asthma
Hypereosinophilic syndrome
Ischemia–reperfusion syndromes
Shock
Myocardial infarction
Stroke
Limb replantation
Organ transplantation

Obviously, interactions of leukocytes with endothelium during inflammatory or immune reactions must be carefully controlled; otherwise, leukocyte emigration in response to tissue damage, infection, or foreign antigen would result in vascular injury. Anti-inflammatory mechanisms serve to limit or prevent leukocyte-mediated damage to the vessel wall.[4] These include: chemotactic factor inhibitors; plasma protease inhibitors; antioxidants in plasma (e.g., methionine, ceruloplasmin, bilirubin); cellular antioxidants (e.g., catalase, superoxide dismutase, glutathione); cytokines (e.g., IL-8[5] and TGF-β[6]); and peptide inhibitors of IL-1 and TNF. Leukocyte-mediated injury likely occurs only when these endogenous regulatory mechanisms are deficient (congenitally or acquired) or they are overwhelmed by an exuberant inflammatory response.

Leukocyte adherence to endothelium appears to be necessary for leukocyte-mediated vascular injury to occur. Close approximation of the leukocyte to the endothelium allows short-lived mediators such as oxidants to act. Adherence also produces a protected microenvironment at the cell–cell interface that is inaccessible to plasma inhibitors. Inhibition of leukocyte adherence to endothelium by a mechanical barrier or a pharmacologic agent[7,8] is therefore effective in preventing or reducing endothelial cell injury by activated leukocytes *in vitro*.

Monoclonal antibodies to leukocyte or endothelial cell adhesion proteins that selectively inhibit leukocyte adherence to endothelium provide a powerful tool to assess the contribution of leukocytes to vascular injury in models of acute inflammation or immune reaction. Monoclonal antibodies directed to leukocyte membrane adhesion complex CD11/CD18 are particularly valuable in this regard. The CD11/CD18 complex consists of three heterodimeric subunits. Each subunit is composed of a distinct α chain (CD11a, CD11b, and CD11c) that is noncovalently associated with a common β chain (CD18). The CD11a/CD18

(LFA-1) subunit is expressed on all leukocytes, whereas the CD11b/CD18 (Mac-1, Mo-1) and CD11c/CD18 (p150, 95, Leu5) subunits are expressed on neutrophils, monocytes, and NK cells but not lymphocytes. The importance of the CD11/CD18 complex in neutrophil adherence and emigration in systemic vasculature is established by leukocyte adhesion deficiency (LAD), the congenital syndrome resulting from partial or complete deficiency of CD11/CD18. Patients with LAD suffer from recurrent bacterial infections due to a profound defect in neutrophil emigration as evidenced by absence of neutrophils at sites of infection. The defect in emigration appears to result from an inability of neutrophils to adhere to endothelium at sites of inflammation. Monoclonal antibodies directed to functional epitopes on CD11b/CD18 inhibit stimulated neutrophil adherence to endothelium *in vitro*[9] and block neutrophil emigration to sites of inflammation in the systemic circulation *in vivo*.[10–13]

As predicted from the *in vitro* studies utilizing a mechanical barrier or pharmacologic agents to prevent adherence, inhibition of neutrophil adherence to endothelium with MAbs to CD11b or CD18 also prevents endothelial injury by activated neutrophils *in vitro*.[14] Moreover, MAbs to CD11/CD18 prevent vascular injury in animal models previously shown to be neutrophil-dependent by depletion experiments.[10,15]

In this chapter we describe the use of the CD18 MAb 60.3[16] in several models of ischemia–reperfusion. We have examined: (1) the generalized ischemia–reperfusion injury following hemorrhagic shock and resuscitation in rabbits and primates, (2) ischemia–reperfusion injury to rabbit ears following vascular occlusion, and (3) ischemia–reperfusion injury following rewarming of previously frozen rabbit feet. The rationale for these studies is summarized in Fig. 1. We propose that an inflammatory stimulus is generated during ischemia or at the time of reperfusion that activates leukocytes and/or endothelial cells. The identity of the activating agent(s) is as yet unknown, but oxidants

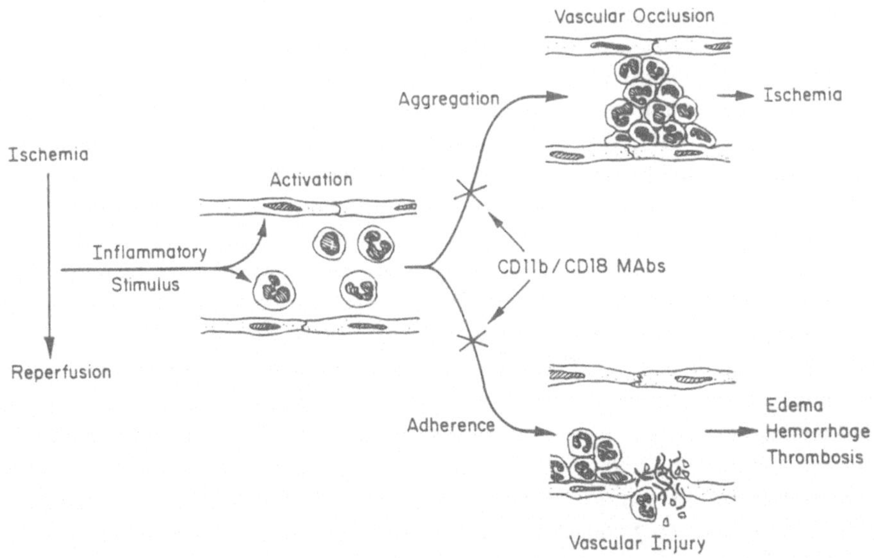

Figure 1. Proposed mechanisms of leukocyte-mediated vascular injury following ischemia–reperfusion.

have been implicated in some models.[17,18] As a consequence of activation, leukocytes adhere to endothelium and release mediators that provoke endothelial injury. Activation may also trigger intravascular leukocyte aggregation during reperfusion and result in microvascular occlusion. By inhibiting heterotypic or homotypic adhesion of leukocytes, CD11b or CD18 MAbs prevent vascular injury and reduce tissue damage.

II. EXPERIMENTAL PROCEDURES–METHODS

All experimental protocols were reviewed and approved by the Animal Care and Use Committee at the University of Washington. A protocol for the assessment of pain and discomfort was developed by the Department of Comparative Medicine, University of Washington. Parameters used to assess discomfort and pain included loss of weight (> 20% body weight), temperature, response to stimuli, and general appearance. In every case, evidence of mild to moderate pain or discomfort was treated with parenteral analgesics. If the animals showed evidence of moderate to severe pain or discomfort, the experiment was terminated and the animals were subjected to euthanasia with an overdose of pentobarbital.

A. Hemorrhagic Shock (Rabbits)

Rabbits were anesthetized with ketamine supplemented with local lidocaine and catheters were placed in the aorta and vena cava through the femoral artery and vein. The arterial catheter had one lumen for arterial pressure (P_a) monitoring and a thermistor at its tip for determination of cardiac output (Q_t) by thermal dilution. The venous catheter was used for injection of cold saline for measurement of Q_t. The arterial catheter was used to draw blood for measurement of arterial partial pressure of oxygen (PO_2), carbon dioxide (PCO_2), and pH. The venous catheter was used to withdraw blood in order to induce shock.

After surgical preparation, the rabbits were allowed to recover from anesthesia and were restrained in a standard rabbit restraint. Baseline measurements of Q_t, P_a, pH, PCO_2, and PO_2 were made for at least 1 hr; animals were then hemorrhaged by drawing blood into heparinized syringes until Q_t was approximately 30% of baseline. This corresponded to P_a of approximately 40 torr. These animals were kept at this Q_t by further withdrawal of blood as necessary for 2 hr. At the end of the period of hemorrhagic shock, rabbits were randomly assigned to receive a saline only (control) or to treatment with the CD18 MAb 60.3 (see below). The investigators were unaware of the treatment. Animals were then resuscitated with their shed blood and hemodynamic monitoring was continued for 3 hr. Lactated Ringers solution was given as needed during the 3-hr resuscitation phase to maintain cardiac output at baseline values. Following the resuscitation phase, catheters were removed and the animals were returned to their cages with free access to food and water. They were followed for 5 days with daily measurement of weight, PO_2, PCO_2, pH, and hematocrit. All animals were carefully monitored and euthanasia was performed if their clinical condition indicated significant discomfort or pain. Those animals killed were presumed to have irreversible injury and were considered deaths. Survivors were killed with an overdose of pentobarbital. Necropsy was performed on all animals and organs examined for signs of organ injury.

B. Hemorrhagic Shock (Primates)

Macaca mulatta monkeys were anesthetized with ketamine and catheters were placed in the aorta and vena cava via femoral artery and vein. Monitoring was carried out as described for rabbits. After establishing baseline values, hemorrhagic shock was induced by withdrawing blood to produce a Q_t of 30% of baseline which was maintained at that level for 1.5 hr. Animals were then randomly assigned to control (saline) and treatment (MAb 60.3) groups with the investigators unaware of the treatment. Resuscitation was begun by infusing the shed blood. A maintenance infusion of 4 ml/kg of lactated Ringers solution was started and increased as necessary to maintain Q_t at the preshock baseline value. The resuscitation phase continued for 24 hr. Catheters were then removed and the animals returned to their cage with free access to food and water. Measurement of weight and blood gas continued for 5 days. One pair of monkeys had electroencephalogram (EEG) leads placed on the scalp to measure global EEGs, and they were resuscitated with lactate Ringers solution alone (instead of shed blood). Initial fluid resuscitation was completed with three times the blood volume removed in these two monkeys. Endoesophagoscopy was performed on all animals prior to shock and at 24 hr after shock. The investigator performing the endoscopy had no knowledge of the treatment or fluid requirements. All animals were followed clinically by the staff of the Primate Center and euthanasia was performed when the clinical condition indicated that the injury was irreversible.

C. Ischemia–Reperfusion Rabbit Ear

Rabbits were anesthetized with ketamine and the ear transected with the exception of the central artery and vein. Transection of the nerves rendered the ear completely anesthetic. The ears were then reattached using sutures. Animals were divided into four groups. The sham group received no further treatment. The control group had artery and vein clamped for 10 hr and were treated with saline at the time of reperfusion. Two MAb 60.3-treated groups had the central artery and vein clamped for 10 hr. One group was treated at the time the artery and vein were clamped and the other was treated only at the time of reperfusion. The dose of MAb 60.3 was 2 mg/kg. Ear volume was measured daily on each animal for 6 days by displacement of water. All animals were killed with an overdose of pentobarbital.

D. Frostbite

Rabbits were anesthetized with ketamine. The left hind foot was shaved and frozen at −15°C for 30 min. Animals were randomly divided into two groups and their feet rewarmed in 39°C water for 30 min. Animals in the control group were treated with saline and those in the treatment group were treated with MAb 60.3 just prior to warming. Volume of the foot was monitored daily by displacement of water. All rabbits were treated with parenteral analgesics for the first 24 hr following frostbite. Freezing of the foot resulted in denervation rendering the foot anesthetic by 24 hr. All animals were killed with an overdose of pentobarbital at the end of the experiment.

E. Monoclonal Antibody

The MAb 60.3 used in these experiments is a murine IgG_{2a} antibody that recognizes a functional epitope on CD18. MAb 60.3 was prepared from mouse ascites fluid as previously described [16] and used at a concentration of 2 mg/ml in sterile saline. There was no detectable endotoxin in the MAb solution by limulus assay. This dose of MAb saturated neutrophil CD18 sites as determined by immunofluorescence flow cytometry. This dose also prevented neutrophil emigration toward *E. coli* endotoxin in skin lesions for up to 24 hr.[11]

III. RESULTS

A. Rabbit Hemorrhagic Shock[19]

There were no differences in the degree of shock between the two groups, measured by change in Q_t, calculated % blood removed, or P_a. During the resuscitation phase, there were no differences between the measured values for the two groups except P_a was reduced for control rabbits. This reduction occurred in spite of resuscitation resulting in a return to baseline value of Q_t.

All seven of the control rabbits died by the fifth day, whereas two of seven of the MAb 60.3-treated animals died. The control rabbits had a persistent acidosis that lasted more than 48 hr. The MAb 60.3-treated rabbits corrected their acidosis by 24 hr. Necropsy of these animals revealed gross injury to the stomach, liver, and lungs of control animals. The gastric mucosa of the stomach was particularly susceptible to injury with appearance of hemorrhagic gastritis in all of the control rabbits. Injury to the stomach and liver was markedly reduced in MAb 60.3-treated animals. Hematocrit of the control animals fell significantly (ΔHct ~ 10%) compared with treated rabbits (ΔHct ~ 3%). The lung injury in these two groups appeared to be unchanged by MAb 60.3 treatment.

B. Primate Hemorrhagic Shock[20]

Q_t, P_a, and blood removed per killigram body weight did not differ between control and MAb 60.3-treated animals. Two of five control animals died within the first 72 hr after hemorrhage; none of the MAb 60.3-treated animals died. There were no long-term deaths in either group at 2 months. Fluid requirement following resuscitation of the MAb 60.3-treated monkeys did not differ from maintenance requirements. In marked contrast, control animals required massive volumes of fluid to maintain cardiac output over the 24-hr period of resuscitation. Figure 2 shows the volume requirements for those animals resuscitated with their shed blood plus lactated Ringers solution at 6, 12, 18, and 24 hr. The control animals gained 0.7 kg (approximately 11% of body weight) and MAb 60.3-treated animals gained only 0.08 kg (approximately 1% of body weight) as a result of fluid resuscitation. There was no qualitative difference between the animals resuscitated with lactated Ringers solution only and those resuscitated with their shed blood except that the animals receiving fluid only required more volume as expected.

Esophagogastroscopy was normal in all animals prior to hemorrhage. None of the MAb 60.3-treated animals showed changes at 24 hr. The control animals all had signs of

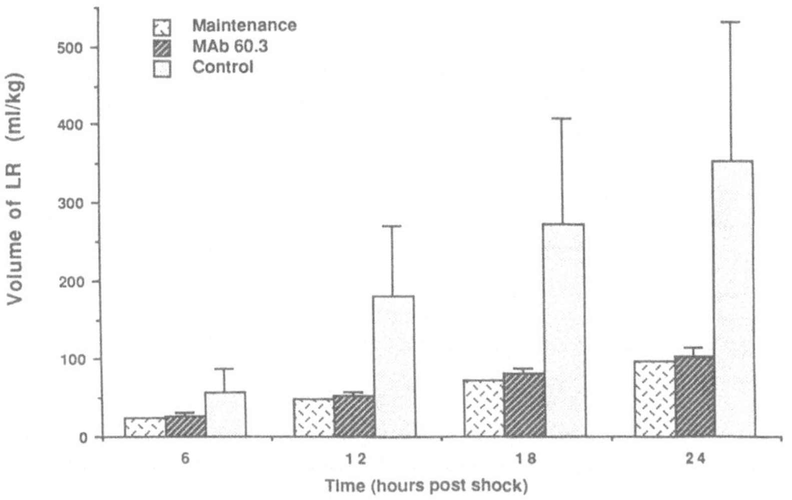

Figure 2. Fluid requirements following hemorrhagic shock and resuscitation in the primate. The cumulative volume (ml/kg) of lactated Ringers solution (LR) required over 24 hr to maintain cardiac output and mean arterial pressure is given for control and MAb 60.3-treated animals along with the calculated maintenance requirements (4 ml/kg/hr). Values represent the means (± S.D.) of five animals in each group. Differences between the control and MAb 60.3-treated animals were significant by analysis of variance and t test ($p < 0.05$).

gastritis either on endoscopy or at necropsy. Those animals that died also had evidence of bleeding lower in the intestinal tract.

There was clinical evidence of depressed brain function during and following hemorrhagic shock in both the rabbits and primates. CNS dysfunction was manifested by stupor that persisted throughout the resuscitation phase and by the reduced amount of ketamine anesthesia required at the time catheters were removed. The clinical symptoms of brain injury or depressed function were markedly reduced in MAb 60.3-treated animals. The EEG of a MAb 60.3-treated monkey showed a return to near baseline following resuscitation, whereas the control monkey remained obtunded with marked slowing of brain waves.

C. Rabbit Ear[21]

Figure 3 shows the ear volume (in ml of water displaced) for all four groups at baseline, 24 and 48 hr. The control rabbits (saline treatment) had grossly edematous ears with the swelling occurring in the first 24 hr and lasting for the entire 6 days of the experiments. Treatment with MAb 60.3 either prior to or following ischemia significantly reduced the degree of injury. In fact, the MAb 60.3-treated animals were not significantly different from the sham-operated group. Tissue necrosis as well as edema was markedly reduced by administration of MAb 60.3 either prior to or following ischemia.

D. Frostbite[22]

Figure 4 shows the volume of rabbit feet at baseline, 6 and 24 hr following freezing and rewarming for control animals and animals treated with MAb 60.3 at the time of rewarming.

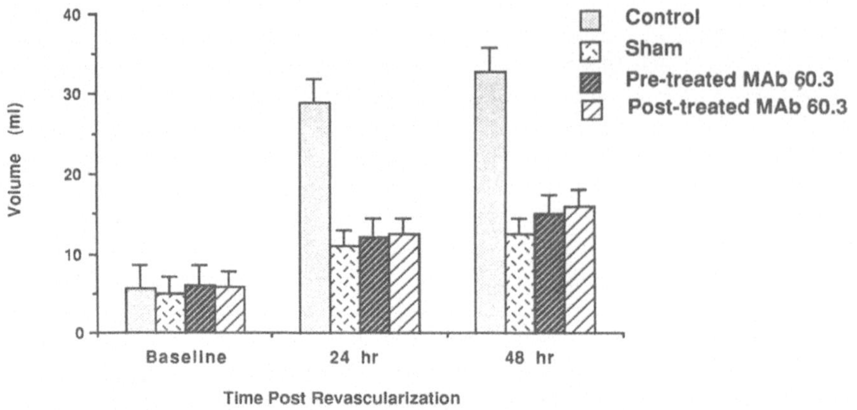

Figure 3. Rabbit ear edema following ischemia–reperfusion. Ear edema was determined by volumetric displacement following ischemia–reperfusion. Mean values (± S.D.) are given for four experimental groups: saline-treated (control, $n = 8$), sham-operated (no ischemia, $n = 3$), MAb 60.3-treated prior to ischemia–reperfusion (pretreated MAb 60.3, $n = 5$), and MAb 60.3 following ischemia and prior to reperfusion (posttreated MAb 60.3, $n = 5$).

The control group exhibited a marked increase in foot volume. This edema formation was essentially eliminated by treatment with MAb 60.3. Tissue necrosis was extensive in the control animals but was virtually eliminated by treatment with MAb 60.3.

IV. DISCUSSION

The dramatic efficacy of the CD18 MAb 60.3 in these three models of ischemia–reperfusion leads us to conclude that: (1) leukocytes mediate vascular injury following

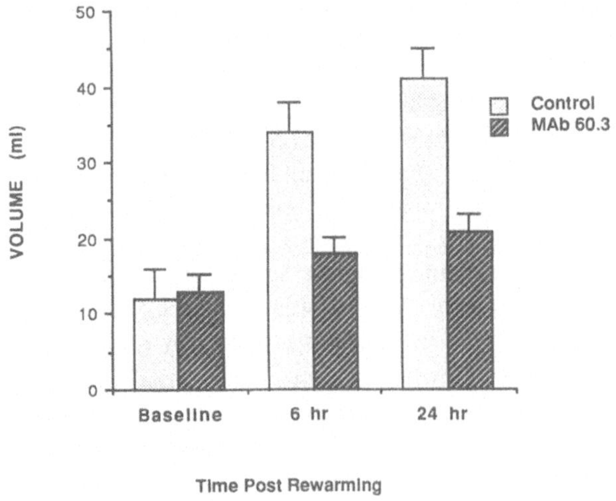

Figure 4. Limb edema following severe cold injury (frostbite). Limb volumes are presented for untreated (control) animals and animals treated with MAb 60.3 at the time of rewarming-reperfusion. Values represent the means (± S.D.) of six animals.

reperfusion of ischemic tissue and (2) in several conditions the major portion of tissue damage occurs during reperfusion rather than during ischemia. The first conclusion is based on the observation that administration of MAb 60.3 at the time of reperfusion significantly reduced tissue edema formation, a marker of endothelial dysfunction, in all three models, and prevented intestinal hemorrhage, an indication of vascular damage, in the shock model. The second conclusion is supported by the results in the rabbit ear and frostbite experiments. In both models, tissue damage was demonstrably reduced by MAb administration at the time of reperfusion. Measurement of edema formation in these two models provides a sensitive marker of a presumed increase in vascular permeability. Insignificant differences between the sham-operated group and the group receiving MAb 60.3 at the time of reperfusion of the ear suggest that injury was due to reperfusion and not ischemia. Also, the marked reduction in tissue edema in MAb 60.3-treated animals with frostbite suggests little or no ischemic injury in this model. Ischemic injury is more difficult to evaluate in hemorrhagic shock since measurement techniques are less sensitive. However, protection by treatment with MAb 60.3 at the time of reperfusion is consistent with a predominant reperfusion injury with little or no ischemic injury.

The observation that the level of consciousness in rabbits and monkeys was depressed to a greater degree in control animals compared with MAb 60.3-treated animals is of considerable interest. These results are consistent with either a reduction in brain blood flow due to microvessel occlusion by leukocyte aggregates or due to leukocyte-mediated cerebral vascular injury following resuscitation from hemorrhagic shock. A recent study in leukopenic dogs also supports a role of neutrophils in cerebral injury following ischemia–reperfusion.[23] Ischemia–reperfusion injury induced by injections of small amounts of air into one carotid artery was reduced by prior depletion of leukocytes with mechlorethamine.

Since the CD18 MAb 60.3 used in these studies recognizes the common β subunit present on all leukocytes (but not platelets or red cells), our results cannot determine which circulating leukocyte cell type mediates vascular injury following reperfusion. However, Simpson *et al.*[24] reported that the CD11b MAb designated 904 reduced infarct size in a canine model of myocardial ischemia–reperfusion indicating that neutrophils (and perhaps monocytes) rather than lymphocytes were involved. Interestingly, MAb 904 inhibits neutrophil aggregation but not adherence to the endothelium, suggesting that reperfusion injury in this model results from vascular occlusion by neutrophil aggregates rather than from endothelial injury by adherent neutrophils.

Not all tissue damage following ischemia–reperfusion is associated with significant leukocyte-mediated vascular injury. Depletion of neutrophils with antineutrophil serum failed to prevent an increased blood urea nitrogen and plasma creatine concentration following renal artery occlusion in rats. Likewise, administration of MAb 60.3 to rabbits provided no protection after ischemia–reperfusion of kidneys in rabbits.[25]

Nevertheless, our results and those of Simpson *et al.*[24] suggest that "antiadhesion" therapy using MAbs directed to leukocyte and endothelial adhesion proteins may prove useful in a number of clinical disorders involving ischemia and reperfusion. This novel approach may be applicable to a wide variety of disorders associated with leukocyte-mediated vascular injury (Table 1). For example, Tuomanen *et al.*[13] recently reported that a CD18 MAb prevented brain edema in a rabbit model of bacterial meningitis, presumably by preventing leukocyte-mediated damage to the blood–brain barrier. Also, Cosimi *et al.*[26] reported that inhibiting lymphocyte adherence to endothelium by a MAb to the endothelial adhesion protein ICAM-1 attenuated vascular damage and organ dysfunction in a primate model of renal allograft rejection.

V. PERSPECTIVES

Leukocyte–endothelial interactions are an important component of host defense and repair. Leukocytes must adhere to and then diapedese between endothelial cells in order to emigrate to extravascular sites of microbial invasion, foreign antigen, or tissue damage. Under some circumstances, however, the interactions of leukocytes with endothelium may provoke vascular injury. Mediators released by adherent leukocytes can induce endothelial cell dysfunction resulting in permeability edema, thrombosis, or hemorrhage. In addition, intravascular aggregation of leukocytes may lead to microvascular occlusion and tissue ischemia. Such leukocyte-mediated vascular injury may contribute to tissue damage in a variety of clinical disorders, ranging from the adult respiratory distress syndrome to allograft rejection. Surface proteins involved in the adhesive interaction of leukocytes and endothelium represent a target for therapy designed to prevent or limit leukocyte- mediated vascular injury. Monoclonal antibodies directed to functional epitopes on leukocyte or endothelial cell adhesion proteins have been demonstrated to reduce vascular and tissue injury in a number of animals models of inflammatory or immune disorders. These studies have not only established the importance of leukocytes in the pathogenesis of tissue damage in the experimental models, but have prompted interest in clinical trials of "antiadhesion therapy" with MAbs in myocardial infarction, hemorrhagic shock, and allograft rejection. Future trials of antiadhesion therapy in acute and subacute disorders will also likely involve peptide antagonists that inhibit the interaction of leukocyte receptors with endothelial ligands and cytokines that modify the expression or function of adhesion proteins. Careful studies in appropriate animal models are necessary to establish the safety as well as the efficacy of this approach, since even transient inhibition of leukocyte adherence to endothelium may impair host defense against extravascular infection. Some assurance in this regard comes from the observation that injection of two doses of CD18 MAb separated by 18 hr did not increase lethality in an animal model of sepsis.[27] Antiadhesion therapy with MAbs, peptides, or cytokines may not be practical for therapy of chronic inflammatory or immune disorders such as rheumatoid arthritis. It may be possible, however, to develop small molecules, i.e., drugs, that selectively impair the interaction of a class of leukocyte with endothelium by interfering with the signal transduction pathways involved in the expression or function of adhesion proteins or by preventing leukocyte receptors from binding to endothelial ligands.

In summary, antiadhesion therapy designed to prevent leukocyte-mediated vascular injury is presently a valuable experimental tool and may represent a novel approach to the therapy of inflammatory and immune clinical disorders.

ACKNOWLEDGMENTS. Research studies presented in this chapter were supported in part by USPHS grants GM07037, RR 05432, HL 30542, and a grant from the American Heart Association–Washington. J.M.H. is a recipient of an Established Investigatorship Award of the American Heart Association.

REFERENCES

1. Grant, L., 1973, The sticking and emigration of white blood cells in inflammation, in: *The Inflammatory Process*, Volume 2 (B. Zweifach, L. Grant, and L. McCluskey, eds.), Academic Press, New York, pp. 205–249.
2. Harlan, J. M., 1987, Consequences of leukocyte–vessel wall interactions in inflammatory and immune reactions, *Semin. Thromb. Hemostasis* **13:**434–444.
3. Tate, R. M., and Repine, J. E., 1983, Neutrophils and the adult respiratory distress syndrome, *Am. Rev. Respir. Dis.* **128:**552–559.
4. Harlan, J. M., 1987, Neutrophil-mediated vascular injury, *Acta Med. Scand. Suppl.* **715:**123–129.
5. Gimbrone, M. A., Jr., Obin, M. S., Brock, A. F., Luis, E. A., Hass, P. E., Hebert, C. A., Yip, Y. K., Leung, D. W., Lowe, D. G., Kohr, W. J., Darbonne, W. C., Bechtol, K. B., and Baker, J. B., 1989, Endothelial interleukin-8: A novel inhibitor of leukocyte–endothelial interactions, *Science* **246:**1601–1603.
6. Gamble, J. R., and Vadas, M. A., 1988, Endothelial adhesiveness for blood neutrophils is inhibited by transforming growth factor-β, *Science* **242:**97–99.
7. Sacks, T., Moldow, C. F., Craddock, P. R., Bowers, T. K., and Jacob, H. S., 1978, Oxygen radicals mediate endothelial cell damage by complement-stimulated granulocytes, *J. Clin. Invest.* **61:**1161–1167.
8. Shasby, D. M., Shasby, S. S., and Peach, M. H., 1983, Granulocytes and phorbol myristate acetate increase permeability to albumin of cultured endothelial monolayers and isolated perfused lungs. Role of oxygen radicals and granulocyte adherence, *Am. Rev. Respir. Dis.* **127:**72–76.
9. Wallis, W. J., Hickstein, D. D., Schwartz, B. R., June, C. H., Ochs, H. D., Beatty, P. G., Klebanoff, S. J., and Harlan, J. M., 1986, Monoclonal antibody-defined functional epitopes on the adhesion-promoting glycoprotein complex (CDw18) of human neutrophils, *Blood* **67:**1007–1013.
10. Arfors, K.-E., Lundberg, C., Lindbom, L., Lundberg, K., Beatty, P. G., and Harlan, J. M., 1987, A monoclonal antibody to the membrane glycoprotein complex CD18 inhibits polymorphonuclear leukocyte accumulation and plasma leakage *in vivo*, *Blood* **69:**338–340.
11. Price, T. H., Beatty, P. G., and Corpuz, S. R., 1987, In vivo inhibition of neutrophil function in the rabbit using monoclonal antibody to CD18, *J. Immunol.* **139:**4174–4177.
12. Rosen, H., and Gordon, S., 1987, Monoclonal antibody to the murine type 3 complement receptor inhibits adhesion of myelomonocytic cells *in vitro* and inflammatory cell recruitment *in vivo*, *J. Exp. Med.* **166:**1685–1701.
13. Tuomanen, E. I., Saukkonen, K., Sande, S., Cioffe, C., and Wright, S. D., 1989, Reduction of inflammation, tissue damage, and mortality in bacterial meningitis in rabbits treated with monoclonal antibodies against adhesion-promoting receptors of leukocytes, *J. Exp. Med.* **170:**959–968.
14. Diener, A. M., Beatty, P. G., Ochs, H. D., and Harlan, J. M., 1985, The role of neutrophil membrane glycoprotein 150 (GP-150) in neutrophil-mediated endothelial cell injury *in vitro*, *J. Immunol.* **135:**537–543.
15. Hernandez, L. A., Grisham, M. B., Twohig, B., Arfors, K.-E., Harlan, J. M., and Granger, D. N., 1987, Role of neutrophils in ischemia-reperfusion-induced microvascular injury, *Am. J. Physiol.* **253:**H699–H703.
16. Beatty, P. G., Ledbetter, J. A., Martin, P. J., Price, T. H., and Hansen, J. A., 1983, Definition of a common leukocyte cell-surface antigen (Lp 95-150) associated with diverse cell-mediated immune functions, *J. Immunol.* **131:**2913–2918.
17. Petrone, W. F., English, D. K., Wong, K., and McCord, J. M., 1980, Free radicals and inflammation: Superoxide-dependent activation of a neutrophil chemotactic factor in plasma, *Proc. Natl. Acad. Sci. USA* **77:**1159–1163.
18. Suzuki, M., Inauen, W., Kvietys, P. R., Grisham, M. B., Meininger, C., Schelling, C. M., Granger, H. J., and Granger, D. N., 1989, Superoxide mediates reperfusion-induced leukocyte– endothelial cell interactions, *Am. J. Physiol.* **257:**H1740–H1745.
19. Vedder, N. B., Fouty, B. W., Winn, R. K., Harlan, J. M., and Rice, C. L., 1989, Role of neutrophils in generalized reperfusion injury associated with resuscitation from shock, *Surgery* **106:**509–516.
20. Mileski, W. J., Winn, R. K., Pohlman, T. H., Vedder, N. B., Harlan, J. M., and Rice, C. L., 1990, Inhibition of CD18-dependent neutrophil adherence reduces organ injury after hemorrhagic shock in primates, *Surgery* **108:**206–212.
21. Vedder, N. B., Winn, R. K., Rice, C. L., Chi, E. Y., Arfors, K.-E., and Harlan, J. M., 1990, Inhibition of leukocyte adherence by anti-CD18 monoclonal antibody attenuates reperfusion injury in the rabbit ear, *Proc. Natl. Acad. Sci. USA* **87:**2643–2646.

22. Mileski, W., Winn, R., Harlan, J., Heimbach, D., and Rice, C., 1990, Inhibition of neutrophil (PMN) adherence with monoclonal antibody 60.3 reduces tissue loss following frostbite, *Proc. Amer. Burn Assoc.* **22**:164.
23. Dutka, A. J., Kochanek, P. M., and Hallenbeck, J. M., 1989, Influence of granulocytopenia on canine cerebral ischemia induced by air embolism, *Stroke* **20**:390–395.
24. Simpson, P. J., Todd, R. F., III, Fantone, J. C., Mickelson, J. K., Griffin, J. D., and Lucchesi, B. R., 1988, Reduction of experimental canine myocardial reperfusion injury by a monoclonal antibody (anti-Mo1, anti-CD11b) that inhibits leukocyte adhesion, *J. Clin. Invest.* **81**:624–629.
25. Thorton, M. A., Winn, R., Alpers, C. E., and Zager, R. A., 1989, An evaluation of the neutrophil as a mediator of *in vivo* renal ischemic-reperfusion injury, *Am. J. Pathol.* **135**:509–515.
26. Cosimi, A. B., Conti, D., Delmonico, F. L., Preffer, F. I., Wee, S.-L., Rothlein, R., Faanes, R., and Colvin, R. B., 1990, *In vivo* effects of monoclonal antibody to ICAM-1 (CD54) in nonhuman primates with renal allografts, *J. Immunol.* **144**:4604–4612.
27. Mileski, W. J., Winn, R. K., Harlan, J. M., and Rice, C. L., 1991, Transient inhibition of neutrophil adherence with the CD18 monoclonal antibody (MAb 60.3) does not increase mortality in abdominal sepsis, *Surgery* **109**:497–501.

9

Endothelial Cell Adhesive Interactions

Mechanisms and Consequences

Elisabetta Dejana, Lindsey Needham, and John Gordon

I. OVERVIEW

A. Introduction

Other chapters in this volume describe specific aspects of endothelial cell adhesive interactions in some detail—e.g., Chapters 7, 8, and 12. The aim of this chapter is to provide a broader perspective, and some additional information.

B. Classification

There are many potential ways in which endothelial cell adhesive interactions can be classified, but perhaps the simplest is to regard them as falling into three categories: interactions with the extracellular matrix; interactions with neighboring endothelial cells; and interactions with circulating cells in the blood.

C. Interactions with Matrix

The extracellular matrix of endothelial cells (EC), which constitutes the subendothelium, has prothrombotic properties—if exposed to blood, it can activate coagulation and induce deposition of platelets. Consequently, secure attachment of EC to the subendothelial matrix provides protection against thrombosis, by ensuring that the hemocompatible endothelial surface is a continuous lining throughout the vascular tree.

EC lay down their own extracellular matrix, and studies *in vitro* have shown that the composition and rates of formation can vary. An extreme example is provided by the EC from the cornea, which lay down a very thick matrix at a much faster rate than vascular EC.

Elisabetta Dejana • Mario Negri Institute, 20157 Milan, Italy. *Lindsey Needham and John Gordon* • British Bio-technology Limited, Cowley, Oxford OX4 5LY, United Kingdom.

Endothelial Cell Dysfunctions, edited by Nicolae Simionescu and Maya Simionescu. Plenum Press, New York, 1992.

Circulating cells from the bloodstream (such as leukocytes or tumor cells) can migrate between EC and temporarily reside in the subendothelium, before they emerge from the blood vessel to lodge in the extravascular tissue. Vascular smooth muscle cells, from the medial layers of the blood vessel wall, can migrate through the internal elastic lamina, to establish themselves in the subendothelium and proliferate there, thus contributing to the development of atherosclerotic plaques.

The formation of a matrix is important in neovascularization, because it contributes to vascular integrity and strength of new blood vessels, and because the movement of vascular cells along matrix components can provide an anatomical guide in the formation of new vessels. Interactions with the extracellular matrix also play a role in the differentiation of EC, during embryogenesis and in the adult, by helping to maintain the organ-specific characteristics of vascular endothelium.

D. Interactions with Circulating Cells

Initial adhesive interactions with circulating cells inevitably take place at the luminal surface of the endothelium. Interaction of circulating leukocytes with the endothelium is a normal, physiological event—a significant proportion (up to about 80%) of the intravascular neutrophil pool is marginated (i.e., adherent to endothelium) at any one time.[1] This marginated pool is in rapid equilibrium with the pool of freely circulating cells—marginated leukocytes can be released into the circulation by acute stimuli. These marginated leukocytes can emigrate through the blood vessel wall, but neutrophils, unlike lymphocytes, do not recirculate—most neutrophils are lost by traffic through to the intestinal and buccal mucosa. Lymphocytes emigrate from the vasculature mainly at specialized postcapillary venules in lymphoid tissues, where the endothelium usually has a characteristic plump morphology (see Chapter 12), returning to the blood via lymphatics after some hours.[2] There is also some emigration of lymphocytes (and other mononuclear leukocytes) in the systemic circulation, and the adhesion and emigration of all classes of leukocytes is greatly increased at sites of inflammation.[3,4] Most of this "inflammatory emigration" takes place in postcapillary venules.

Other classes of circulating cells can also interact with the vascular endothelium. For example, red blood cells infected with malaria parasites adhere to the microvascular endothelium, which leads to vascular occlusion and organ dysfunction, with particularly severe consequences when the cerebral circulation is affected.[5] Metastasis of tumors occurs when circulating tumor cells adhere to the vascular endothelium and emigrate to establish secondary growths in extravascular sites.[6]

E. Homotypic Endothelial Cell Interactions

Interactions between neighboring EC (which determine the nature of the intercellular junctions) are important in the maintenance of vascular integrity, in the regulation of vascular permeability (by limiting the egress of plasma proteins), and in the control of leukocyte traffic (see above).

The junctions between EC vary in different parts of the circulation: tight junctions predominate in specialized microcirculatory beds such as the brain and the retina, where edema is particularly dangerous, and, at the other extreme, there are regions of the circulation (e.g., in the bone marrow) where contacts between EC are intermittent, with large fenestrations. In most of the vasculature, the EC form a continuous monolayer lining the blood vessels, but with more gap junctions than tight junctions between the cells.

F. Adhesive Ligands

Several of the molecules on the endothelial surface which contribute to adhesive interactions have been identified.

Integrins play a major role in the interactions between endothelium and the extracellular matrix, and also in contacts between neighboring EC. The nature and function of integrins and their receptors are discussed later in this chapter. Molecules such as the ICAMs (intercellular adhesion molecules), which are members of the immunoglobulin superfamily, can contribute to homotypic and heterotypic intercellular interactions, as their name implies. Their structure and function are also discussed in this chapter, together with a brief account of other adhesive proteins. The molecule termed ELAM-1 (a member of the family called LEC-CAMS, or selectins), which plays an important role in leukocyte–endothelial interactions, is discussed at length in Chapter 7. ELAM-1 is an example of a protein which is not normally present on EC, but is synthesized and expressed on the endothelial surface when the cells are activated (e.g., by cytokines). Some proteins, such as ICAM-1, are present under normal conditions but can be up-regulated in response to activating stimuli. The expression of other surface proteins does not change during endothelial activation. Endothelial adhesive interactions are complex, and all of these categories of proteins can play important roles under different conditions.

II. INTEGRINS

A. General

Human EC adhere, spread, and organize their cytoskeleton on a variety of molecules of the extracellular matrix such as fibronectin, vitronectin, laminin, and collagen.[7,8] EC express receptor molecules which, on the outer side of the membrane, recognize and bind different components of the extracellular matrix and, on the cytoplasmic side, transmit intracellular signals and link a chain of proteins of the membrane–microfilament interaction complex involved in the mechanism of adhesion and cytoskeletal organization.[9]

Most of the EC receptors for extracellular matrix components belong to a recently discovered superfamily of adhesive membrane proteins termed "integrins."[10,11] These receptors have several structural and functional homologies so that it is believed that they differentiated from a common ancestral gene. They are all heterodimers of two noncovalently linked subunits. The larger subunit has been termed the "α chain" and the smaller subunit the "β chain." The name "integrins" arose because they are integral membrane proteins—i.e., each subunit has a transmembrane segment, a small C-terminal cytoplasmic domain, and a large N-terminal extracellular domain.[10,11] Some of these receptors specifically recognize in the ligand proteins a sequence of only three amino acids (arginine–glycine–aspartic acid, RGD). Many proteins contain this sequence but not all of them are recognized by an integrin receptor; however, this sequence is the cell recognition site of a large number of extracellular matrix and plasma proteins including fibrinogen, vitronectin, fibronectin, thrombospondin, and von Willebrand factor.[10] Despite the similarities in the cell binding sequences in ligand proteins, the cell can recognize them individually through specific and separate receptors. This suggests that other amino acids surrounding the RGD sequence, and the RGD steric conformation, confer specificity to the interactions with one integrin receptor or another.[10]

The number of the integrin family members is expanding. A subclassification of the family has been attempted based on the observation that some members of the integrin group have the same β chain but different α chains. This has resulted in the definition of three subfamilies: the β1 or VLA (very late antigens)[12]; the β2 or leu-CAM (leukocyte adhesion molecules)[13]; the β3 or cytoadhesins.[14] However, at least four additional and novel β chains have recently been described.[15–18] We will not go into the details of the structure and function of each integrin receptor but instead will focus on the receptors present on EC.

B. Structure and Function of Endothelial Integrins

Table 1 lists the members of the integrin family which to date have been identified in EC.[19,20] These cells possess at least four receptors belonging to the β1 subfamily. The VLA-2 or α2β1 integrin is identical to the platelet GpIa–IIa complex, which is the receptor for collagen in platelets and in other cells of hematopoietic and nonhematopoietic origin (for review see Ref. 21). In EC it behaves differently, acting as a major receptor for laminin and also binding (though rather less efficiently) to collagen and fibronectin.[22] VLA-3, or α3β1, which is expressed in lower amounts, appears to be a multifunctional receptor: it recognizes fibronectin, collagen, and laminin.[23] VLA-5, or α5β1, has been described as the fibronectin receptor in many types of cells including EC.[24] In platelets it was known as the GpIc–IIa complex (for review see Ref. 21). VLA-6, or α6β1, is present in very low amounts on EC. This molecule is the laminin receptor in platelets and possibly plays a similar role in EC (for review see Ref. 21).

EC express only one integrin belonging to the β3 or cytoadhesin subfamily, which was isolated and defined as the vitronectin receptor. This molecule has the same β chain as the platelet complex GpIIb–IIIa but a distinct α chain.[14] When it was isolated and reconstituted into artificial phospholipid membranes, it had a very low specificity and recognized (besides vitronectin) von Willebrand factor, thrombospondin, fibrinogen, and fibronectin.[25]

Table 1. Integrin Receptors in Endothelial Cells[a]

Receptor	Subunits	Ligand	RGD role
VLA-2	α2β1	lm, coll, fn	No
VLA-3	α3β1	lm, coll, fn	No
VLA-5	α5β1	fn	Yes
VLA-6	α6β1	lm	No
VNR	αvβ3	vn, fg, vW, fn, tsp	Yes
	αvβs	?	Yes

[a]lm, laminin; coll, collagen; fn, fibronectin; vn, vitronectin; vW, von Willebrand factor; tsp, thrombospondin; fg, fibrinogen; VNR, vitronectin receptor.

It behaves, however, in a different way in different cell types. In EC it maintains its multifunctional characteristics,[26] while in MG63 osteosarcoma cells or in vascular smooth muscle cells it recognizes only vitronectin.[27,28]

Finally, a novel integrin has been identified,[17] which has the same α chain as the vitronectin receptor but has a distinct β chain. The function of this molecule is still unknown but it has the property to become heavily phosphorylated when the cells are activated with phorbol esters.

Most of our data come from studies performed on cultured human umbilical vein or bovine thoracic aorta EC. However, when a comparison was attempted among EC from different parts of the vasculature, the integrin composition appeared to be essentially similar.[19] Also, EC integrins were not significantly modified by the time that the cells spent in culture, or by the passaging of the cells.[19]

C. Regulation of Integrin Expression and Activity

From what has been reported above, it appears that the same receptor (e.g., VLA-2 and the vitronectin receptor) might behave differently in EC, compared with other types of cells (e.g., platelets or tumor cells). This suggests that the diversity of the integrin system could be further augmented by a cell-specific type of regulation. Possible mechanisms of such regulation could include alternative mRNA splicing, posttranslational modification of the receptor, or association of the receptor with modifying components (e.g., gangliosides, glycosaminoglycans). However, no evidence is available documenting differences in the structure or in the synthetic pathways of endothelial integrins compared with other cells. Alternative splicings for integrin subunits have been described but these processes were present equally in all of the cell types studied, including EC.[29] A detailed study of synthesis of the vitronectin receptor shows that EC follow a pathway similar to that described for GpIIb–IIIa in megakaryocytes.[30]

Modulation of integrin synthesis and expression in EC is still a relatively unexplored area of research. Tarone *et al.*[20] reported that the combination of tumor necrosis factor and interferon induced a 50–70% decrease in vitronectin receptor number while no change was detected in the β1 subgroup of integrin molecules. This effect is particularly interesting considering that these cytokines are able to induce a dramatic modification of EC shape and matrix composition.[31]

Transforming growth factor β in different cell types elevates the expression of vitronectin receptor, LFA-1, and the β1 subgroup of integrins.[32] Whether this agent can also act on EC integrin expression is still unknown.

Cell adhesion to matrix or plasma proteins seems to be more efficiently regulated by modulation of the activity of integrin receptors than by changes in their number. This phenomenon has been widely studied for platelet GpIIb–IIIa[14] and leukocyte leu-CAM integrins.[33] Activation of these cells by aggregating agents or chemotactic stimuli can change integrin receptor conformation and in this way modify their binding capacity. Ionic concentration, and the phospholipid composition of the membrane, also modify integrin receptor affinity for ligand proteins in purified systems[25,34] and can change integrin receptor conformation. Finally, as with other types of receptors, phosphorylation could be another mechanism of regulation of integrin activity. Tyrosine phosphorylation of fibronectin receptors in oncogene-transformed cells has been described, and this was associated with an altered interaction with the cytoskeleton.[35] However, it is not known if this mechanism of regulation is present in normal cells, too.

D. Integrin-Mediated Cell Activation

Cell recognition of extracellular matrix proteins is a complex phenomenon which involves a variety of adhesive receptors and specific cell responses. It has been recognized for many years that cell interaction with the matrix is not simply a phenomenon of cell attachment but is followed by specific responses which lead to cell differentiation, migration, and growth. This implies the transmission of intracellular messages through specific receptor activation. Little is known about the biochemical signaling pathways that integrin receptors activate inside the cells. The platelet integrin GpIIb–IIIa regulates Ca^{2+} and Na^+/H^+ exchange[36,37] and tyrosine-specific protein phosphorylation.[38] However, we still do not have any evidence that these pathways are activated after integrin receptor occupancy in EC or other cell types.

EC express a variety of integrin molecules. An attractive possibility is that each receptor could induce specific cellular responses and/or that the engagement of more than one receptor is required for full cell activation. The same matrix protein can be bound by different integrins (see Table 1). For example, fibronectin is linked by VLA-2, VLA-3, VLA-5 and the vitronectin receptor, whereas laminin is recognized by VLA-2, VLA-3, and VLA-6. The receptors appear to bind to different domains of the molecules (i.e., RGD or other sequences), suggesting a multiple type of interaction between the ligand proteins and separate binding sites on the cell membrane.

It has been recently reported that laminin can induce EC differentiation into tubular structures and that this phenomenon requires cell attachment both to an RGD sequence via an integrin receptor and to a Tyr-Ile-Gly-Ser-Arg (YIGSR) sequence through a nonintegrin binding site.[39]

E. Interaction with the Cytoskeleton

In addition to chemical signaling, integrin molecules may convey regulatory information through interactions with cytoskeletal proteins. During attachment, the basal cell surface forms several types of contacts (known as focal contacts, or adhesion plaques) which represent the area of closest interaction between the substratum, the cell membrane, and the membrane insertion sites of actin microfilament bundles.[9] In EC[40] and other cells,[41] the fibronectin receptor (VLA-5, $\alpha5\beta1$) and the vitronectin receptor are clustered in focal contacts during cell adhesion. The organization of these receptors is strictly dependent on the specific molecules in the substratum. When EC are plated on vitronectin, only the vitronectin receptor is clustered, while when the cells are plated on fibronectin, only the fibronectin receptor is organized in focal contacts. When EC are seeded on fibrinogen and von Willebrand factor, both receptors are organized in adhesion plaques although only the vitronectin receptor is able to recognize these substrata. This paradox can be explained by the release of endogenous matrix proteins (mainly fibronectin) by EC, which causes binding and clustering of the fibronectin receptor.[19,28,42]

The mechanism of integrin receptor clustering and its consequences are still unknown. An attractive possibility is that integrin receptor clustering (as with other types of receptors) involves signal transmission across the membrane, which induces assembly of the cytoskeletal proteins vinculin and talin, which in turn mediate actin microfilament organization. It has been suggested that fibronectin receptors could be interconnected with actin fibers by directly binding to talin.[43]

The biological meaning of integrin clustering and cytoskeletal organization is still a matter of debate. It is probably essential for the maintenance of cell adhesion, cell shape, and motility but not for the first phases of cell attachment. When cytoskeletal protein assembly was inhibited by increasing cAMP in EC,[44] cell adhesion to different substrata was not modified. In addition, when recombinant fibronectin receptor lacking the cytoplasmic domain (and therefore the possibility to bind cytoskeletal proteins) was inserted into cells, there was no clustering or cytoskeletal organization, although the cells could still bind to fibronectin.[45]

III. INTERCELLULAR ADHESION MOLECULE-1

A. Expression and Structure

ICAM-1 is an integral membrane glycoprotein of heterogenous molecular mass (range 76–114 kDa) that is involved in both the homotypic and heterotypic cellular interactions that regulate the immune response.

ICAM-1 is constitutively expressed on a variety of hematopoietic and mesenchymal cells. In general, expression on nonhematopoietic cells is variable, ranging from low to moderate, but it can be up-regulated by cytokines and is often enhanced on tissues at sites of inflammation *in vivo*.[46–51] Peripheral blood leukocytes and some related promyelomonocytic cell lines have relatively low levels of ICAM-1 expression but this can be increased in response to activation or (in the case of promyelomonocytic cells) by inducing them to differentiate using agents such as phorbol ester.[46,52,53]

Sequence information obtained from ICAM-1 cDNA cloning has classified the molecule as belonging to the immunoglobulin superfamily.[54,55] ICAM-2 which possesses only two of the five homologous immunoglobulin domains of ICAM-1 has recently been defined by functional cDNA cloning. To date, this molecule shares greatest sequence homology with ICAM-1 and has also been shown to interact with LFA-1 but unlike ICAM-1 its expression cannot be regulated by cytokines.[56] Other members of the immunoglobulin superfamily such as neural cell adhesion molecule (NCAM) and Myelin-associated glycoprotein (MAG) share 20–25% homology with ICAM-1 whereas others, which perform more closely related functions such as CD2 and LFA-3, are unrelated.[54,57] A functional characteristic common to several members of the immunoglobulin superfamily is their ability to act as viral receptors[58,59]; this has been substantiated by the recent observation that ICAM-1 has been identified as the major rhinovirus receptor.[60,61]

B. Functions

A functional role for ICAM-1 was originally defined using monoclonal antibodies such as RR1 and LB2 which were shown to inhibit phorbol ester-induced homotypic aggregation of B lymphoblastoid cell lines.[62,63]

Previous experiments had shown this process to be dependent on the leukocyte cell surface glycoprotein LFA-1, a member of the CD18 leukocyte integrin family involved in many of the cell–cell adhesion reactions that regulate antigen-dependent and -independent leukocyte function.[64–66] On the basis of this it was suggested that intercellular adhesion may be mediated by the receptor–ligand pairing of LFA-1 and ICAM-1 and recently this has been confirmed in direct binding studies using the immunoaffinity-purified ligands reconstituted in planar lipid bilayers.[67,68]

The details of the molecular nature of ICAM-1–LFA-1 interactions have not been fully defined. Sequencing data have revealed that the conserved Arg-Gly-Asp (RGD) peptide sequence known to be present in the binding domains of other integrin ligands, is not apparent in either ICAM-1 or ICAM-2.[69,70] Biochemical characterization of ICAM-1–LFA-1 interactions has revealed that cell blinding, in common with other integrin-mediated processes, is magnesium dependent and requires metabolic energy, an intact cytoskeleton, and is temperature dependent.[67] Consistent with these observations is a recent finding that transient increases in LFA-1 avidity for ICAM-1 can occur in response to lymphocyte activation.[68] The molecular mechanisms responsible are initiated in response to signal transduction and are thought to involve conformational changes in the LFA-1 molecule. These changes are perhaps similar to those described for Macl (CD11b)—another member of the same integrin subclass, which in addition to its recognized ligand C3bi, can also interact with others such as ICAM-1 in response to certain chemoattractants.[68,70–72]

In addition to its immunoregulatory role in mediating lymphoid cell–cell interactions,[73–75] ICAM-1 has been implicated in regulation of leukocyte extravasation. Cytokine-inducible expression of ICAM-1 on endothelium and the fact that its ligand, LFA-1, is a member of the CD18 integrin family which are known to play an important role in the leukocyte response to inflammation[76,77] make it an attractive hypothesis. This has been substantiated by *in vitro* observations of a cytokine-inducible ICAM-1-dependent component to the adhesion of lymphoid[78] and myeloid cells to endothelium.[79] These findings, along with the demonstration of ICAM-1-dependent lymphocyte adhesion to other cell types such as dermal fibroblasts[46] and epidermal keratinocytes,[48] have also suggested a role for this molecule in the pathogenesis of several immunologically mediated disorders.[80,81] In the future, molecular studies involving mutation and epitope mapping of ICAM-1 will greatly facilitate the development of novel therapeutics for the treatment of a variety of immunological and inflammatory conditions in which this molecule could be involved.

IV. VASCULAR ADDRESSINS AND LYMPHOCYTE CIRCULATION

A. Tissue-Specific Lymphocyte Homing

For effective immune surveillance, an efficient and specific means of lymphocyte recirculation is required. Specificity of lymphocyte extravasation from the blood to the lymphoid organs is controlled by their selective migration through specialized postcapillary venules found in lymphoid tissue.[82] These "high endothelial" venules (HEV) are named after the phenotypically distinct form of endothelium that lines them which is often, but not exclusively, characterized by its plump cuboidal/columnar appearance.[83] Although this characteristic morphology is not directly responsible for lymphocyte adhesion, it has enabled the development of an *in vitro* assay based on the binding of lymphocytes to these easily identifiable cells in HEV of frozen lymph node sections.[84] This "Stamper–Woodruff" assay has been used extensively to define the molecular characteristics of lymphocyte–HEV interactions.

Murine and human lymphomas which interact in a lymphoid tissue-specific manner have established that lymphocyte–HEV interactions exhibit tissue specificity.[85] Control of this specificity appears to reside at the level of both the lymphoid-specific HEV and the lymphocyte. In the case of the latter, special receptors termed "homing receptors" have been functionally defined using monoclonal antibodies in humans and in several species of rodent model systems[86–88] and have established that at least three recognition systems involving specificity for HEV of peripheral lymph node, mucosal lymphoid tissue, and inflamed synovium exist. Some of these homing receptors, notably the murine (Mel 14) and human homologue (Leu 8) of the receptors involved in peripheral lymph node homing, have been cloned and shown to belong to the same family as the endothelial adhesion molecules ELAM-1 and GMP140 (see Refs. 89–92). Antibodies to these receptors selectively inhibit lymphocyte homing to peripheral lymph nodes *in vitro* and *in vivo* and have established that these receptors are confined to lymphoid and myeloid cells.[93,94] In the human, the 85- to 95- and 200-kDa Hermes/CD44 family of membrane antigens have been implicated in lymphocyte homing.[95–97] cDNA cloning has classified them as belonging to a new family that show substantial N-terminal sequence homology with the cartilage link proteins, indicating possible functional interactions of this class of molecules with the extracellular matrix and cytoskeleton.[98–100] The wider distribution of Hermes/CD44 variants on nonhematopoietic cells[101–103] coupled with the non-tissue-selective ability of certain polyclonal antibodies to inhibit lymphocyte homing[96] suggest a more generalized role for this family of antigens in regulating lymphocyte adhesion, perhaps similar to the accessory role described for the leukocyte integrin LFA-1.[104,105] Other homing receptors belonging to the VLA integrin superfamily have been functionally implicated in lymphocyte–mucosal HEV binding in the mouse and human.[106] There is still much to be learned about the molecules controlling lymphocyte homing receptor specificity, in particular what controls their developmental expression on lymphocyte subsets and what are the identities of their corresponding ligands on the HEV.

B. Vascular Addressins

Some progress in the mouse system has been made in defining two functionally and antigenically distinct HEV ligands that mediate tissue-specific lymphocyte homing to peripheral and mucosal lymphoid tissue. Monoclonal antibody MECA 367 specifically inhibits lymphocyte homing to mucosal lymphoid tissue both *in vitro* and *in vivo*.[107] Conclusive evidence that the 58- to 66-kDa polypeptide recognized by MECA 367 can function as an adhesion ligand for mucosal lymphoid tissue homing was obtained from binding studies on immunoaffinity-purified antigen reconstituted into lipid bilayers.[108] The HEV ligand involved in peripheral lymph node–lymphocyte interactions in the mouse is recognized by monoclonal antibody MECA 79.[109] The antibody recognizes a 92-kDa glycoprotein which possesses lectin binding activity and is sensitive to *N*-glycanase digestion; this tends to substantiate earlier reports on the importance of carbohydrate–lectin interactions in peripheral lymph node homing.[110,111]

Immunolocalization of MECA 367 and MECA 79 has revealed that in addition to the obvious respective tissue specificities for mucosal and peripheral lymph node HEV, both antigens are coexpressed in subsets of the HEV of mesenteric lymph nodes.[107] This implies that in addition to developmental tissue-specific signals, other humoral factors, perhaps present in afferent lymph, may be involved in regulating the expression of these so-called "vascular addressins." Cytokines generated *in situ*, at the site of an immune response, are possible candidates for humoral mediators of the HEV phenotype. *In vitro* experiments have already indicated that the cytokine γ-interferon can induce an antigen characteristic of the HEV phenotype on microvascular endothelium,[112] and vessels possessing HEV characteristics have been detected at sites of chronic inflammation.[113–115] Similar experiments have also identified several leukocyte adhesion molecules expressed on cytokine-treated endothelium[116–117] that are involved in the extravasation of leukocytes to nonlymphoid tissues at sites of inflammation.

Characterization of other differentiation-dependent antigens, and additional ligands involved in lymphocyte recirculation, remains a high priority for future research.

V. CONCLUSIONS

In this chapter, we have given a perspective on endothelial adhesive interactions, and outlined current knowledge of some of the ligands involved.

In some instances, individual molecules responsible for cell adhesion have been identified, and these are discussed in this chapter or in others in this volume. However, additional ligands still remain to be identified; in particular, there are relatively few cases where both partners in a pair of interacting molecules are known. For example, although it is known that the endothelial surface molecule ICAM-1 interacts with the leukocyte surface molecule known as LFA-1, the molecule on granulocytes which binds to the ELAM-1 molecule on endothelial cells has not been characterized.

Intercellular adhesive interactions, or cell–matrix interactions, are complex events, in which several different ligands are usually involved. The relative importance of these different ligands can vary in different circumstances. One of the reasons for this variable contribution of different ligands is that the expression of adhesive molecules on the cell surface can be up-regulated and down-regulated by different stimuli or environmental conditions. In addition, the same stimulus can induce the up-regulation of different cell surface molecules with different kinetics; for example, activation of endothelial cells *in vitro* by cytokines such as IL-1 or TNF results in the appearance of ELAM-1 on the surface, with expression peaking at 4–6 hr and declining over 12–24 hr, whereas the expression of ICAM-1 increases while ELAM-1 expression is declining.

This kinetic dimension is of great potential importance, and it is likely that there are more sophisticated variations than the simple appearance and disappearance of individual protein molecules; e.g., conformational changes in surface proteins, resulting in the exposure of new epitopes with different functions. This is especially likely in multidomain proteins such as ELAM-1, which has a lectinlike domain, an EGF-like domain, and six sequential, identical domains that resemble segments of complement proteins.

Future research on endothelial adhesive interactions will rely heavily on the techniques that have been developed in the past few years for the characterization of those surface adhesive proteins whose structure and function are now known. Monoclonal antibodies will confirm, by histopathological analysis, the distribution of such proteins in human tissues under different physiological and pathological conditions. Antibodies with cross-species specificity can be used for animal studies, to explore the regulation of such proteins under experimental conditions. Cloning, sequencing, and domain analysis of the genes for these proteins will eventually reveal not only which proteins are responsible for such adhesive interactions but also how the process of adhesion is mediated. Cell adhesion phenomena in general, and endothelial cell adhesion in particular, have been active fields of research in the recent past and there is certain to be considerable activity in these fields for the foreseeable future.

ACKNOWLEDGMENTS. Supported by CNR Special projects "Biotecnologie e Biostrumentazione" and "Tecnologie Biomediche e Sanitarie" and by the Associazione Italiana per la Ricerca sul Cancro.

REFERENCES

1. Cartwright, G. E., Athens, J. W., and Wintrobe, M. M., 1964, Kinetics of granulopoiesis in normal man, *Blood* **24:**780–803.
2. Ford, W. L., 1975, Lymphocyte migration and immune responses, *Prog. Allergy* **19:**1–59.
3. Gowans, J. L., 1970, Lymphocytes, *Harvey Lect.* **64:**87–119.
4. Graham, R. C., and Shannon, S. L., 1972, Peroxidase arthritis part 2—Lymphoid cell endothelial interactions during a developing immunologic inflammatory response, *Am. J. Pathol.* **69:**7–24.
5. Berendt, A. R., Simmons, D. L., Tansey, J., Newbold, C., and Marsh, K., 1989, Intercellular adhesion molecule-1 is an endothelial cell adhesion receptor for Plasmodium falciparum, *Nature* **341:**57–59.
6. Nicolson, G. L., and Winkelmake, J. L., 1975, Organ specificity of blood-borne tumour metastasis determined by cell adhesion, *Nature* **255:**230–231.
7. Form, D. M., Pratt, B. M., and Madri, J. A., 1983, Endothelial cell proliferation during angiogenesis. *In vitro* modulation by basement membrane components, *Lab. Invest.* **55:**521–525.
8. Ingber, D. E., and Folkman, J., 1989, How does extracellular matrix control capillary morphogenesis? *Cell* **58:**803–805.
9. Burridge, K., 1986, Substrate adhesion in normal and transformed fibroblasts: Organization and regulation of cytoskeletal, membrane and extracellular matrix components at focal contacts, *Cancer Rev.* **4:**18–78.
10. Ruoslahti, E., and Pierschbacher, M. D., 1987, New perspectives in cell adhesion: RGD and integrins, *Science* **238:**491–497.
11. Hynes, R. O., 1986, Integrins: A family of cell surface receptors, *Cell* **48:**549–554.
12. Hemler, M. E., Huang, C., and Schwarz, L., 1987, The VLA protein family: Characterization of five different surface heterodimers each with a common 130,000 Mr subunit, *J. Biol. Chem.* **262:**3300–3309.
13. Sanchez-Madrid, F., Nagy, J. A., Robbins, E., Simon, P., and Springer, T., 1983, A human leukocyte differentiation antigen family with distinct α-subunits and a common β-subunit. The lymphocyte function-associated antigen (LFA-1), the C3bi complement receptor (OKMl/Mac-1), and the Gp 150,95 molecule, *J. Exp. Med.* **158:**1785–1803.
14. Ginsberg, M. H., Loftus, J. C., and Plow, E. F., 1988, Cytoadhesins, integrins and platelets, *Thromb. Haemost.* **59:**1–20.
15. Kajiji, S., Tamura, R. N., and Quaranta, V., 1989, A novel integrin (αEβ4) from human epithelial cells suggests a fourth family of integrin adhesion receptors, *EMBO J.* **8:**673–680.
16. Cheresh, D. A., Smith, J. W., Cooper, H. M., and Quaranta, V., 1989, A novel vitronectin receptor integrin (αvβx) is responsible for distinct adhesive properties of carcinoma cells, *Cell* **57:**59–69.
17. Freed, E., Gailit, J., van der Geer, P., Ruoslahti, E., and Hunter, T., 1989, A novel integrin β subunit is associated with vitronectin receptor α subunit (αv) in a human osteosarcoma cell line and is a substrate for protein kinase C, *EMBO J.* **8:**2955–2965.

18. Kramer, R. H., McDonald, K. A., and Vu, M. P., 1989, Human melanoma cells express a novel integrin receptor for laminin, *J. Biol. Chem.* **264:**15642–15649.
19. Albelda, S. M., Daise, M., Levine, E. M., and Buck, C. A., 1989, Identification and characterisation of cell–substratum adhesion receptors on cultured human endothelial cells, *J. Clin. Invest.* **83:**1992–2002.
20. De Filippi, P., Truffa, G., Stefanuto, G., Altruda, F., Silengo, L., and Tarone, G., 1991, Tumor necrosis factor α and interferon γ modulate the expression of the vitronectin receptor (integrin β_3) in human endothelial cells, *J. Biol. Chem.* **266:**7638–7645.
21. Dejana, E., and Lauri, D., 1990, Biochemical and functional characteristics of integrins: A new family of adhesive receptors present in hemapoietic cells, *Haematologica* **75:**1–6.
22. Languino, L. R., Gehlsen, K. R., Wayner, E., Carter, W. G., Engvall, E., and Ruoslahti, E., 1989, Endothelial cells use $\alpha 2\beta 1$ integrin as a laminin receptor, *J. Cell Biol.* **109:**2455–2462.
23. Wayner, E. A., and Carter, W. G., 1987, Identification of multiple cell adhesion receptors for collagen and fibronectin in human fibrosarcoma cells possessing unique α and common β subunits, *J. Cell Biol.* **105:** 1873–1884.
24. Conforti, G., Zanetti, A., Colella, S., Abbadini, M., Marchisio, P. C., Pytela, R., Giancotti, F., Tarone, G., Languino, L. R., and Dejana, E., 1989, Interaction of fibronectin with cultured human endothelial cells. Characterization of the specific receptor, *Blood* **73:**1576–1585.
25. Conforti, G., Zanetti, A., Pasquali-Ronchetti, I., Quaglino Jr., D., Neyroz, P., and Dejana, E., 1990, Modulation of vitronectin receptor binding by membrane lipid composition, *J. Biol. Chem.* **265:**4011–4019.
26. Cheresh, D., 1987, Human endothelial cells synthesize and express an Arg-Gly-Asp-directed adhesion receptor involved in attachment to fibrinogen and von Willebrand factor, *Proc. Natl. Acad. Sci. USA* **84:**6471–6475.
27. Dejana, E., Colella, S., Languino, L. R., and Marchisio, P. C., 1986, Fibrinogen induces adhesion, spreading and microfilament organization of human endothelial cells *in vitro*, *J. Cell Biol.* **104:**1403–1411.
28. Dejana, E., Lampugnani, M. G., Giorgi, M., Gaboli, M., Federici, A. B., Ruggeri, Z. M., and Marchisio, P. C., 1989, Von Willebrand factor promotes endothelial cell adhesion via an Arg-Gly-Asp dependent mechanism, *J. Cell Biol.* **109:**367–375.
29. Van Kuppevelt, T., Languino, L. R., Gailit, J. O., and Ruoslahti, E., 1989, An alternative cytoplasmic domain of the integrin β_3 subunit, *Proc. Natl. Acad. Sci. USA* **86:**5415–5418.
30. Polack, B., Duperray, A., Troesch, A., Berthier, R., and Marguerie, G., 1989, Biogenesis of the vitronectin receptor in human endothelial cell: Evidence that the vitronectin receptor and GpIIb–IIIa are synthesized by a common mechanism, *Blood* **73:**1519–1524.
31. Stolpen, A. H., Eva, A. M., Guinan, C., Fiers, W., and Pober, J. S., 1986, Recombinant tumour necrosis factor and immune interferon act singly and in combination to reorganize human vascular endothelial cell monolayers, *Am. J. Pathol.* **123:**16–24.
32. Ignotz, R. A., Heino, J., and Messague, J., 1989, Regulation of cell adhesion receptors by transforming growth factor β. Regulation of vitronectin receptor and LFA-1, *J. Biol. Chem.* **264:**389–392.
33. Vedder, N. B., and Harlan, J. M., 1988, Increased surface expression of CD11b/CD18 (Mac-1) is not required for stimulated neutrophil adherence to cultured endothelium, *J. Clin. Invest.* **81:**676–682.
34. Gailit, J., and Ruoslahti, E., 1988, Regulation of the fibronectin receptor affinity by divalent cations, *J. Biol. Chem.* **263:**2065–2067.
35. Hirst, R., Horwitz, A., Buck, C., and Rohrschneider, L., 1986, Phosphorylation of the fibronectin receptor complex in cells transformed by oncogenes that encode tyrosine kinases, *Proc. Natl. Acad. Sci. USA* **83:**6470–6474.
36. Brass, L. F., 1985, Ca^{++} transport across the platelet plasma membrane. A role for membrane glycoproteins II–IIIa, *J. Biol. Chem.* **260:**2231–2238.
37. Banga, H. S., Simons, E., and Brass, L. F., 1986, Activation of phospholipases A and C in human platelets exposed to epinephrine: Role of glycoproteins IIb–IIIa and dual role of epinephrine, *Proc. Natl. Acad. Sci. USA* **83:**9197–9201.
38. Farrell, J. E., and Martin, G. S., 1989, Tyrosine specific protein phosphorylation is regulated by glycoprotein IIb–IIIa in platelets, *Proc. Natl. Acad. Sci. USA* **86:**2234–2238.
39. Grant, D. S., Tashiro, K. I., Segul-Real, B., Yamada, Y., Martin, G. R., and Kleinman, H. K., 1989, Two different laminin domains mediate the differentiation of human endothelial cells into capillary-like structures *in vitro*, *Cell* **58:**933–943.
40. Dejana, E., Colella, S., Conforti, G., Abbadini, M., Gaboli, M., and Marchisio, P. C., 1988, Fibronectin

and vitronectin regulate the organization of their respective Arg-Gly-Asp adhesion receptors in cultured human endothelial cells, *J. Cell Biol.* **107**:1215–1223.

41. Singer, I., Scott, S., Kawka, D. W., Kazazis, D. M., Gailit, J., and Ruoslahti, E., 1988, Cell surface distribution of fibronectin and vitronectin receptors depends on substrate composition and extracellular matrix accumulation, *J. Cell Biol.* **106**:2171–2182.
42. Dejana, E., Lampugnani, M. G., Giorgi, M., Gaboli, M., and Marchisio, P. C., 1990, Fibrinogen induces endothelial cell adhesion and spreading via the release of endogenous matrix proteins and the recruitment of more than one integrin receptor, *Blood* **75**:1509–1517.
43. Horwitz, A., Duggan, K., Buck, C., Beckerle, M. C., and Burridge, K., 1986, Interaction of plasma membrane fibronectin receptor with talin a transmembrane linkage, *Nature* **320**:531–533.
44. Lampugnani, M. G., Giorgi, M., Gaboli, M., Dejana, E., and Marchisio, P. C., 1990, Endothelial cell motility, integrin receptor clustering and microfilament organization are inhibited by agents that increase intracellular cAMP, *Lab. Invest.* **63**:521–531.
45. Solowska, J., Guan, J.-L., Marcantonio, E. E., Trevithick, J. E., Buck, C. A., and Hynes, R. O., 1989, Expression of normal and mutant avian integrin subunits in rodent cells, *J. Cell Biol.* **109**:853–861.
46. Dustin, M. L., Rothlein, R., and Bhan, A. K., 1986, Induction by IL1 and interferon γ: Tissue distribution, biochemistry and function of a natural adherence molecule (ICAM-1), *J. Immunol.* **137**:245–254.
47. Pober, J. S., Lapierre, L. A., Stolpen, A. H., Brock, T. A., Springer, T. A., Fiers, W., Bevilacqua, M. P., Mendrick, D., and Gimbrone, M. A. Jr., 1987, Activation of cultured human endothelial cells by recombinant lymphotoxin: Comparison with tumour necrosis factor and interleukin 1, *J. Immunol.* **138**:3319–3324.
48. Dustin, M. L., Singer, K. H., Tuck, D. T., and Springer, T., 1988, Adhesion of T lymphoblasts to epidermal keratinocytes is regulated by interferon γ and is mediated by intercellular adhesion molecule-1 (ICAM-1), *J. Exp. Med.* **167**:1323–1340.
49. Mentzer, S. J., Rothlein, R., Springer, T. A., and Faller, D., 1988, Intercellular adhesion molecule-1 (ICAM-1) is involved in the cytolytic T lymphocyte interaction with a human synovial cell line, *J. Cell. Physiol.* **137**:173–178.
50. Rothlein, R., Czajkowski, M., O'Neill, M. M., Marlin, S. D., Mainolfi, E., and Merluzzi, V. J., 1988, Induction of intercellular adhesion molecule-1 on primary and continuous cell lines by pro-inflammatory cytokines, *J. Immunol.* **141**:1665–1669.
51. Munro, J. M., Pober, J. S., and Cotran, R. S., 1989, Tumour necrosis factor and interferon γ induce distinct patterns of endothelial activation and associated leukocyte accumulation in skin of Papio anubis, *Am. J. Pathol.* **135**:121–133.
52. Clark, E. A., Ledbetter, J. A., Holly, R. C., Dinndorf, P. A., and Shu, G., 1986, Polypeptides on human β lymphocytes associated with cell activation, *Hum. Immunol.* **16**:100–113.
53. Dougherty, G. J., Murdoch, S., and Hogg, N., 1988, The function of human intercellular adhesion molecule-1 (ICAM-1) in the generation of an immune response, *Eur. J. Immunol.* **18**:35–39.
54. Simmons, D., Makgoba, M. W., and Seed, B., 1988, ICAM, an adhesion ligand of LFA-1 is homologous to the neural cell adhesion molecule NCAM, *Nature* **331**:624–627.
55. Staunton, D. E., Marlin, S. D., Stratowa, C., Dustin, M. L., and Springer, T. A., 1988, Primary structure of ICAM-1 demonstrates interaction between members of the immunoglobin and integrin supergene families, *Cell* **52**:925–933.
56. Staunton, D. E., Dustin, M. L., and Springer, T. A., 1989, Functional cloning of ICAM-2 a cell adhesion ligand for LFA-1 homologous to ICAM-1, *Nature* **339**:61–64.
57. Dustin, M. L., Sanders, M. E., Shaw, S., and Springer, T. A., 1987, Purified lymphocyte function associated antigen 3 binds to CD2 and mediates T lymphocyte adhesion, *J. Exp. Med.* **165**:677–692.
58. Maddon, P. J., Dalgleish, A. G., McDougal, J. S., Clapham, P. R., Weiss, R. A., and Axel, R., 1986, The T_4 gene encodes the AIDS virus receptor and is expressed in the immune system and the brain, *Cell* **47**:333–348.
59. Mendelsohn, C. L., Wimmer, E., and Racaniello, V. R., 1989, Cellular receptor for poliovirus: Molecular cloning, nucleotide sequence, and expression of a new member of the immunoglobulin superfamily, *Cell* **56**:855–865.
60. Greve, J. M., Davis, G., Meyer, A. M., Forte, C. P., Connolly Yost, S., Marlor, C. W., Kamarck, M. E., and McClelland, A., 1989, The major human rhinovirus receptor is ICAM-1, *Cell* **56**:839–847.
61. Staunton, D. E., Merluzzi, V. J., Rothlein, R., Barton, R., Marlin, S. D., and Springer, T. A., 1989, A cell adhesion molecule, ICAM-1, is the major surface receptor for rhinoviruses, *Cell* **56**:849–853.

62. Rothlein, R., Dustin, M. L., Marlin, S. D., and Springer, T. A., 1986, A human intercellular adhesion molecule (ICAM-1) distinct from LFA-1, *J. Immunol.* **137**:1270–1274.
63. Patarroyo, M., Clark, E. A., Prieto, J., Kantor, C., and Gahmberg, C. G., 1987, Identification of a novel adhesion molecule in human leukocytes by monoclonal antibody LB-2, *FEBS Lett.* **210**:127–131.
64. Springer, T. A., 1985, The LFA-1, Mac-1 glycoprotein family and its deficiency in an inherited disease, *Fed. Proc.* **44**:2660–2663.
65. Rothlein, R., and Springer, T. A., 1986, The requirement for lymphocyte function-associated antigen 1 in homotypic leukocyte adhesion stimulated by phorbol ester, *J. Exp. Med.* **162**:1132–1149.
66. Hynes, R. O., 1987, Integrins: A family of cell surface receptors, *Cell* **48**:549–554.
67. Marlin, S. D., and Springer, T. A., 1987, Purified intercellular adhesion molecule-1 (ICAM-1) is a ligand for lymphocyte function associated antigen 1 (LFA-1), *Cell* **51**:813–819.
68. Dustin, M. L., and Springer, T. A., 1989, T-cell receptor cross-linking transiently stimulates adhesiveness through LFA-1, *Nature* **341**:619–624.
69. Ruoslahti, E., and Pierschbacher, M. D., 1986, Arg-Gly-Asp, a versatile cell recognition signal, *Cell* **44**:517–518.
70. Detmer, P. A., Wright, S. D., Olsen, E., Kimball, B., and Cohn, Z. A., 1987, Aggregation of complement receptors on human neutrophils in the absence of ligand, *J. Cell Biol.* **105**:1137–1145.
71. Corbi, A. L., Kishimoto, T. K., Miller, L. J., and Springer, T. A., 1988, The human leukocyte adhesion glycoprotein Mac-1 (complement receptor type 3, CD11b) α subunit, *J. Biol. Chem.* **263**:12403–12411.
72. Smith, C. W., Marlin, S. D., Rothlein, R., Toman, C., and Anderson, D. C., 1989, Co-operative interactions of LFA-1 and Mac-1 with intercellular adhesion molecule-1 in facilitating adherence and transendothelial migration of human neutrophils *in vitro*, *J. Clin. Invest.* **83**:2008–2017.
73. Boyd, A. W., Wawryk, S. O., Burns, G. F., and Fecondo, J. V., 1988, Intercellular adhesion molecule 1 (ICAM-1) has a central role in cell–cell contact mediated immune mechanisms, *Proc. Natl. Acad. Sci. USA* **85**:3095–3099.
74. Makgoba, M. W., Sanders, M. E., Ginther Luce, G. E., Dustin, M. L., Springer, T. A., Clark, E. A., Mannoni, P., and Shaw, S., 1988, ICAM-1 a ligand for LFA-1 dependent adhesion of B, T and myeloid cells, *Nature* **331**:86–88.
75. Altmann, D. M., Hogg, N., Trowsdale, J., and Wilkinson, D., 1989, Co-transfection of ICAM-1 and HLA-DR reconstitutes human antigen presenting cell function in mouse L cells, *Nature* **338**:512–514.
76. Anderson, D. C., and Springer, T. A., 1987, Leukocyte adhesion deficiency: An inherited defect in the Mac1, LFA-1 and p150,95 glycoproteins, *Annu. Rev. Med.* **38**:175–194.
77. Pober, J. S., 1988, Cytokine mediated activation of vascular endothelium, *Am. J. Pathol.* **133**:426–433.
78. Dustin, M. L., and Springer, T. A., 1988, Lymphocyte function associated antigen-1 (LFA-1) interaction with intercellular adhesion molecule-1 (ICAM-1) is one of at least three mechanisms for lymphocyte adhesion to cultured endothelial cells, *J. Cell Biol.* **107**:321–331.
79. Smith, C. W., Rothlein, R., Hughes, B. J., Mariscalco, M. M., Rudloff, H. E., Schmalstieg, F. C., and Anderson, D. C., 1988, Recognition of an endothelial determinant for CD18-dependent human neutrophil adherence and transendothelial migration, *J. Clin. Invest.* **82**:1746–1756.
80. Griffiths, C. E. M., Voorhees, J. J., and Nickoloff, B. J., 1989, Characterisation of intercellular adhesion molecule 1 and HLA-DR expression in normal and inflamed skin. Modulation by recombinant gamma interferon and tumor necrosis factor, *J. Am. Acad. Dermatol.* **20**:617–629.
81. Adams, D. H., Shaw, J., Hubscher, S. G., and Rothlein, R., 1989, Intercellular adhesion molecule 1 on liver allografts during infection, *Lancet* **2**:1122–1124.
82. Gowans, J. L., and Knight, E. J., 1964, The route of recirculation of lymphocytes in rat, *Proc. R. Soc. London Ser. B* **159**:257–282.
83. Woodruff, J. J., Clarke, L. M., and Chin, Y. H., 1987, Specific cell-adhesion mechanisms determining migration pathways of recirculating lymphocytes, *Annu. Rev. Immunol.* **5**:201–222.
84. Stamper, H. B., and Woodruff, J. J., 1976, Lymphocyte homing into lymph nodes: *In vitro* demonstration of the selective affinity of recirculating lymphocytes for high endothelial venules, *J. Exp. Med.* **144**: 828–833.
85. Butcher, E. C., Scollay, R. G., and Weissman, I. L., 1980, Organ specificity of lymphocyte migration: Mediation by highly selective lymphocyte interactions with organ specific determinants, *Eur. J. Immunol.* **10**:556–561.

86. Gallatin, W. M., St. John, T. P., Siegelman, M., Reichert, R., Butcher, E. C., and Weissman, I. L., 1986, Lymphocyte homing receptors, *Cell* **44**:673–680.
87. Jalkanen, S., Reichert, R. A., Gallatin, W. M., Bargatze, R. F., Weissman, I. L., and Butcher, E. C., 1986a, Homing receptors and the control of lymphocyte migration, *Immunol. Rev.* **91**:39–60.
88. Jalkanen, S., Steere, A., Fox, R., and Butcher, E. C., 1986b, A distinct endothelial cell recognition system that controls lymphocyte traffic into inflamed synovium, *Science* **233**:556–558.
89. Camerini, D., James, S. P., Stamenkovic, I., and Seed, B., 1989, Leu8/TQ1 is the human equivalent of the Mel14 lymph node homing receptor, *Nature* **342**:78–82.
90. Lasky, L. A., Singer, M. S., Yednock, T. A., Dowbenko, D., Fennie, C., Rodriguez, H., Nguyen, T., Stachel, S., and Rosen, S. D., 1989, Cloning of a lymphocyte homing receptor reveals a lectin domain, *Cell* **56**:1045–1055.
91. Siegelman, M. H., Van de Rijn, M., and Weissman, I. L., 1989, Mouse lymph node homing receptor cDNA clone encodes a glycoprotein revealing tandem interaction domains, *Science* **243**:1165–1172.
92. Tedder, T. T., Isaacs, C. M., Ernst, T. J., Demetri, G. D., Adler, D. A., and Disteche, C. M., 1989, Isolation and chromosomal localization of cDNAs encoding a novel human lymphocyte cell surface molecule LAM-1, *J. Exp. Med.* **170**:123–133.
93. Gallatin, W. M., Weissman, I. L., and Butcher, E. C., 1983, A cell surface molecule involved in organ specific homing of lymphocytes, *Nature* **304**:30–34.
94. Lewinsohn, D. M., Bargatze, R. F., and Butcher, E. C., 1987, Leukocyte–endothelial cell recognition: Evidence of a common molecular mechanism shared by neutrophils, lymphocytes and other leukocytes, *J. Immunol.* **138**:4313–4321.
95. Jalkanen, S. T., Bargatze, R. F., Herron, L. R., and Butcher, E. C., 1986c, A lymphoid cell surface glycoprotein involved in endothelial cell recognition and lymphocyte homing in man, *Eur. J. Immunol.* **16**:1195–1202.
96. Jalkanen, S., Bargatze, R. F., De Los Toyos, J., and Butcher, E. C., 1987, Lymphocyte recognition of high endothelium: Antibodies to distinct epitopes of an 85–95 kDa glycoprotein antigen differentially inhibit lymphocyte binding to lymph node, mucosal or synovial endothelial cells, *J. Cell Biol.* **105**:983–990.
97. Jalkanen, S., Jalkanen, M., Bargatze, R., Tammi, M., and Butcher, E. C., 1988, Biochemical properties of glycoproteins involved in lymphocyte recognition of high endothelial venules in man, *J. Immunol.* **141**: 1615–1623.
98. Gallatin, W. M., Wayner, E. A., Hoffman, P. A., St. John, T., Butcher, E. C., and Carter, W. G., 1989, Structural homology between lymphocyte receptors for high endothelium and class III extracellular matrix receptor, *Proc. Natl. Acad. Sci. USA* **86**:4654–4658.
99. Goldstein, L. A., Zhou, D. F. H., Picker, L. J., Minty, C. N., Bargatze, R. F., Ding, J. F., and Butcher, E. C., 1989, A human lymphocyte homing receptor the Hermes antigen is related to cartilage proteoglycan core and link proteins, *Cell* **56**:1063–1072.
100. Stamenkovic, I., Amiot, M., Pesando, J. M., and Seed, B., 1989, A lymphocyte molecule implicated in lymph node homing is a member of the cartilage link protein family, *Cell* **56**:1057–1062.
101. Butcher, E. C., Lewinsohn, D., Duijvestijn, D., Bargatze, R. F., Wu, N., and Jalkanen, S., 1986, Interactions between endothelial cells and leukocytes, *J. Cell Biochem.* **30**:121–131.
102. Pals, S. T., Hogervorst, F., Keizer, G. D., Thepen, T., Horst, E., and Figdor, C. C., 1989, Identification of a widely distributed 90 kDa glycoprotein that is homologous to the Hermes-1 human lymphocyte homing receptor, *J. Immunol.* **143**:851–857.
103. Picker, L. J., De Los Toyos, J., Telen, M. J., Haynes, B. F., and Butcher, E. C., 1989, Monoclonal antibodies against the CD44 [In(Lu)-related p80], and Pgp-1 antigens in man recognise the Hermes class of lymphocyte homing receptors, *J. Immunol.* **142**:2046–2051.
104. Hamann, A., Jablonski-Westrich, D., Duijvestijn, A., Butcher, E. C., Baisch, H., Harder, R., and Thiele, H.-G., 1988, Evidence of an accessory role of LFA-1 in lymphocyte high endothelium interaction during homing, *J. Immunol.* **140**:693–699.
105. Pals, S. T., Denotter, A., and Miedema, F., 1988, Evidence that leukocyte function associated antigen-1 is involved in recirculation and homing of human lymphocytes via high endothelial venules, *J. Immunol.* **140**:1851–1853.
106. Holzmann, B., and Weissman, I. L., 1989, Integrin molecules involved in lymphocyte homing to Peyers patches, *Immunol. Rev.* **108**:45–61.

107. Streeter, P. R., Berg, E. L., Rouse, B. T. N., and Butcher, E. C., 1988, A tissue specific endothelial cell molecule involved in lymphocyte homing, *Nature* **331**:41–46.
108. Nakache, M., Berg, E. L., and Streeter, P. R., 1989, The mucosal vascular addressin is a tissue specific endothelial cell adhesion molecule for circulating lymphocytes, *Nature* **337**:179–181.
109. Streeter, P. R., Rouse, B. T. N., and Butcher, E. C., 1988, Immunohistologic and functional characterization of a vascular addressin involved in lymphocyte homing into peripheral lymph nodes, *J. Cell Biol.* **107**:1853–1862.
110. Stoolman, L. M., Tenforde, T. S., and Rosen, S. D., 1984, Phosphomannosyl receptors may participate in the adhesive interaction between lymphocytes and high endothelial venules, *J. Cell Biol.* **99**:1535–1540.
111. Rosen, S. D., Chi, S. I., True, D. D., Singer, M. S., and Yednock, T. A., 1989, Intravenously injected sialidase inactivates attachment sites for lymphocytes on high endothelial venules, *J. Immunol.* **142**:1895–1902.
112. Duijvestijn, A. M., Schreiber, A. S., and Butcher, E. C., 1986, Interferon gamma regulates an antigen specific for endothelial cells involved in lymphocyte traffic, *Proc. Natl. Acad. Sci. USA* **83**:9114–9118.
113. Freemont, A. J., and Ford, W. L., 1985, Functional and morphological changes in post capillary venules in relation to lymphocytic infiltration into BCG induced granulomata in rat skin, *J. Pathol.* **147**:1–12.
114. Duijvestijn, A. M., Horst, E., Pals, S. T., Rouse, B. T. N., Steere, A. C., Picker, L. J., Meijer, C. J. L. M., and Butcher, E. C., 1988, High endothelial differentiation in human lymphoid and inflammatory tissues defined by monoclonal antibody HECA-452, *Am. J. Pathol.* **130**:147–155.
115. Jutila, M. A., Berg, E. L., Kishimoto, T. K., Picker, L. J., Bargatze, R. F., Bishop, D. K., Orosz, C. G., Wu, N. W., and Butcher, E. C., 1989, Inflammation induced endothelial cell adhesion to lymphocytes, neutrophils and monocytes, *Transplantation* **48**:727–731.
116. Pober, J. S., 1988, Cytokine mediated activation of vascular endothelium, *Am. J. Pathol.* **133**:426–433.
117. Johnston, G. I., Cook, R. G., and McEver, R. P., 1989, Cloning of GMP-140 a granule membrane protein of platelets and endothelium: Sequence similarity to proteins involved in cell adhesion and inflammation, *Cell* **56**:1033–1044.

10

Modulation of Endothelial Cell Proliferation by Monocyte-Derived Cytokines

J. L. Wautier, D. Vilette, and J. P. Caen

I. INTRODUCTION

Vascular endothelium, the cellular interface between blood and tissue, is a quiescent population of cells *in vivo*. Indeed, angiogenesis, the process of new blood vessel formation by endothelial cells, rare under normal physiological conditions. Labeling studies have shown that vascular endothelial cell turnover is low, 0.01 to 0.1% of the cells being labeled after a 24-hr [^{3}H]thymidine pulse.[1–3] However, during a few physiological settings, including wound healing or menstruation, angiogenesis does occur,[4] but even during these processes, neovascularization is strongly regulated (i.e., is brief and strictly delimited). By contrast, uncontrolled angiogenesis is involved in serious diseases. For instance, vascularization is an absolute requirement for solid tumor growth,[4,5] and in diabetic retinopathy, vascularization of the retina often leads to blindness.[4]

The multistep neovascularization process involves controlled proliferation and migration of endothelial cells.[6–10] Thus, the characterization of modulators of endothelial proliferation and/or migration is critical to the understanding of the biochemical events that underlie angiogenesis. The availability of cultured endothelial cells[11,12] has led to the characterization of several regulators of endothelial cell growth. Heparin binding growth factors (acidic and basic FGF) were the first endothelial mitogens to be purified by taking advantage of their heparin affinity.[13–15] Epidermal growth factor (EGF),[16] transforming growth factor type α (TGF-α),[17] platelet-derived endothelial cell growth factor (PD-ECGF),[18] vasculotropin,[19] platelet-[20,21] and macrophage-derived[22,23] growth factors were shown to promote endothelial cell proliferation *in vitro*. On the other hand, activated monocytes/macrophages produce endothelial cell growth inhibitors. TGF-β inhibits *in vitro*

J. L. Wautier, D. Vilette, and J. P. Caen • Institut des Vaisseaux et du Sang, Hôpital Lariboisière, 75010 Paris, France.

Endothelial Cell Dysfunctions, edited by Nicolae Simionescu and Maya Simionescu. Plenum Press, New York, 1992.

endothelial cell proliferation[24–28] as does tumor necrosis factor type α (TNF-α).[29–31] Interferon γ (INF-γ) has also been shown to exert antiproliferative effects on endothelial growth.[32]

Beside these factors, we and others have suggested that normal unactivated monocytes produce an activity that inhibits the growth of human vascular endothelial cells.[33,34] Recently we have begun the characterization of a monocyte-derived endothelial cell inhibitory factor (MECIF).[35] In this study we have compared MECIF properties to the known cytokines which inhibit endothelial cell proliferation such as TNF-α, TGF-β, INF-γ.

II. MATERIALS AND METHODS

A. Reagents

Indomethacin, cycloheximide, formyl-methionyl leucine phenylalanine, endotoxin-LPS (*E. coli*), porcine heparin, type XIII TPCK-treated trypsin and soybean trypsin inhibitor were from Sigma (St. Louis, Mo.). ECGF was purchased from Meloy Laboratories (Springfield, Va.), TGF-β from Calbiochem (La Jolla, Calif.), INF-γ kindly provided by Unicet Schering (Levallois Perret, France), recombinant TNF-α and polyclonal antibodies specific for INF-γ from Boehringer Mannheim (Germany). Polyclonal antibodies specific for TGF-β were from R & D System (Minneapolis, Minn.).

B. Monocyte Isolation and Culture Supernatant Preparation

Citrated blood samples were collected from healthy normal donors. Mononuclear cells were separated by density gradient centrifugation and the monocytes were isolated either by adhesion to culture plastic or by flow elutriation using a Beckman J-6 MIE centrifuge (Beckman Instruments Inc., Palo Alto, Calif.).[35,36] Cell fractions were analyzed using a Coulter counter S+6 (Coulter Electronics Inc., Hialeah, Fla.).

Mononuclear cells (10^8) were incubated for 2 hr on culture dishes (50 mm, Nunc Rosklide, Denmark), nonadherent cells were removed, and adherent cells (95 ± 3% monocytes) were incubated in 2.5 ml of M199 culture medium supplemented with 2.5% FCS (Boehringer Mannheim, Germany) for up to 24 hr. The culture supernatants were collected and filter-sterilized (0.2 μm).

C. Cell Culture and Proliferation Assay

Human vascular endothelial cells (HEC) were collected from umbilical cords and cultured essentially as described[11] with modifications.[37] For most studies, primary cultures of human and porcine endothelial cells were used. Proliferation of human skin fibroblasts (up to 13th passage) and human smooth muscle cells (up to 8th passage) was also tested in the presence of MECIF.

MECIF activity was assayed in triplicate by incorporation of [*methyl*-^{3}H]thymidine (5 Ci/mmole, Commissariat à l'Energie Atomique, France) into acid-insoluble DNA of cultured vascular cells. Cells (15,000) in 200 μl of M199 culture medium supplemented with 20% FCS were seeded into 96-well microtiter plates (Nunc, Denmark) precoated with human purified fibronectin (10 μg/cm^2, Centre Régional de Transfusion Sanguine, Lille, France). When the endothelial cells were adherent (6 hr), the medium was removed and replaced by 200 μl of culture medium containing 10% FCS and the test fractions (5 to 40 μl). After 24 hr, the cells were pulsed for 16 hr with [*methyl*-^{3}H]thymidine (1 μCi/well). At the end of this period, the medium was removed, the cells were exposed to collagenase, and were harvested on glass fiber filters using a multiple cell harvester (Skatron, Lies, Norway).

D. Chromatographies of Crude Monocyte Supernatant

Monocyte supernatant (100 ml) was ten-fold concentrated using an ultrafiltration membrane (YM10 Amicon, Danvers, Mass.; molecular mass cutoff: 10,000 kDa) and fractionated by gel filtration chromatography on a Sepharose CL-6B column (Pharmacia, Uppsala, Sweden) equilibrated with 10 mM Tris-HCl pH 8.0 at a flow rate of 30 ml/hr. Typically, 100–110 fractions (2 ml per fraction) were collected and 10 μl of each was tested for inhibitory activity. Protein concentration was determined according to a dye fixation method[38] with a Bio-Rad Kit (Richmond, Calif.).

To estimate the molecular weight corresponding to MECIF activity, analytical quantities of crude monocyte supernatant were submitted to a gel-filtration HPLC. Samples were injected onto a TSK-250 column (600 × 7.5 mm, Bio-Sil) and the material was eluted with Hanks (without phenol red) at a flow rate of 1.0 ml/min.

E. Normal Rat Kidney (NRK) Fibroblast Proliferation Assay

Mitogen assay of NRK fibroblasts was performed essentially as described elsewhere.[39]

III. RESULTS

A. Production of MECIF Activity by Cultured Normal Human Monocytes

Isolated human blood monocytes were found to release an activity designated MECIF which can inhibit the proliferation of human umbilical vein endothelial cells as assessed by [*methyl*-^{3}H]thymidine incorporation. As shown in Fig. 1A, the antiproliferative effect of crude monocyte supernatant is dose dependent and this effect is half-maximal at 7 μg/ml corresponding to a 1:50 dilution. MECIF inhibitory effect was maximal after 24 hr incubation with endothelial cells. However, its action was already detectable after 6 hr incubation (64% of the value obtained after 24 hr incubation). Monocytes cultured in serum-free M199 medium produced a similar level of MECIF activity.

To exclude the possibility that contaminating cells could be the origin of MECIF activity, medium of cultured platelets or lymphocytes at concentrations normally found in the monocyte preparations, was tested on endothelial cell growth. These media induced a slight increased in [*methyl*-^{3}H]thymidine incorporation (122% for the platelets; 132–170% for the lymphocytes), thus suggesting that the generation of MECIF activity was directly attributable to the monocytes. To further confirm the monocyte origin of MECIF, mono-

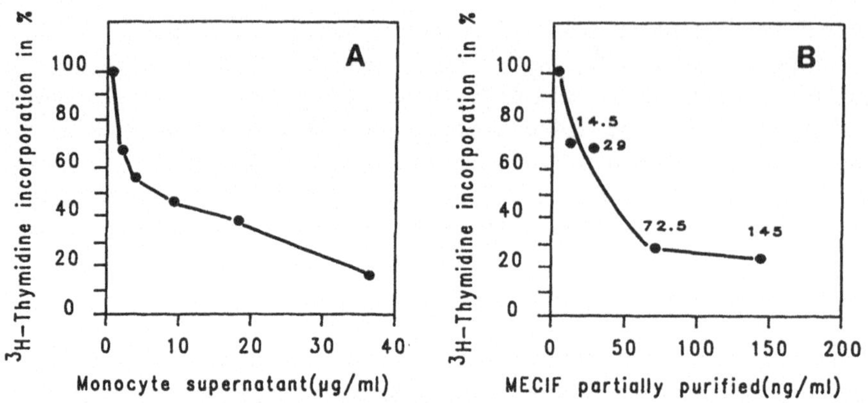

Figure 1. Inhibition of DNA synthesis in human endothelial cells by human normal monocyte conditioned-medium (A) and by fractionated MECIF (B). Monocyte supernatant (A) or fractionated MECIF (after Sepharose CL-6B) (B) was added at various concentrations to the endothelial cell culture medium and the results expressed in percentage of incorporation of [*methyl*-^{3}H]thymidine. The numbers by the experimental points are the protein concentrations in ng/ml. Protein concentration in crude monocyte supernatant was 1.8 mg/ml. The conditioned medium was diluted fivefold in M199 culture medium supplemented with 2.5% FCS and from 1 μl to 20 μl was added in a total volume of 200 μl per well. Thus, monocyte supernatant protein concentration was from 1.8 μg/ml to 36 μg/ml per well. Protein concentration of fractionated MECIF was 2.9 μg/ml. From 1 μl to 10 μl was added in a total volume of 200 μl per well. This resulted in fractionated MECIF concentration from 14.5 ng/ml to 145 ng/ml.

cytes from the same donor were purified either by counterflow elutriation or by selective adhesion. Both produced similar inhibitory activity (88 and 82% of inhibition, respectively).

To estimate the molecular weight of MECIF activity, analytical quantities of monocyte conditioned medium were subjected to high-performance gel filtration chromatography on a TSK-250 column. The inhibitory activity was eluted as a single peak corresponding to a molecular mass of 50–70 kDa (Fig. 2).

B. Production of MECIF Activity of Monocytes

The release of MECIF activity by monocytes was shown to be time dependent as demonstrated by the following experiment: monocytes were incubated for various periods of time (from 0.5 to 30 hr) and then the conditioned medium (0.2 μl) was tested for MECIF activity (Fig. 3). MECIF could already be detected after 3 hr of monocyte incubation and represented 40% of the maximal value observed after 24 hr incubation. The kinetics of MECIF activity production progressively increased after 6 hr (46%), and was higher after 18 hr (55%). The apparent augmentation of MECIF activity after 18 hr was observed consistently in four independent experiments. However, its biological significance is unclear.

Treatment of cultured monocytes by cycloheximide abrogated the release of MECIF activity to culture medium. On the other hand, treatment with indomethacin, a cyclooxygenase inhibitor, did not alter the expression of MECIF. Thus, it is suggested that protein synthesis, but not the prostaglandin endoperoxide production, is required for the secretion of MECIF. Activation of monocytes by endotoxin (LPS, *Escherichia coli*, 10 μg/ml) or FMLP (formyl-methionyl leucine phenylalanine, 0.1 μM) resulted in an increased expres-

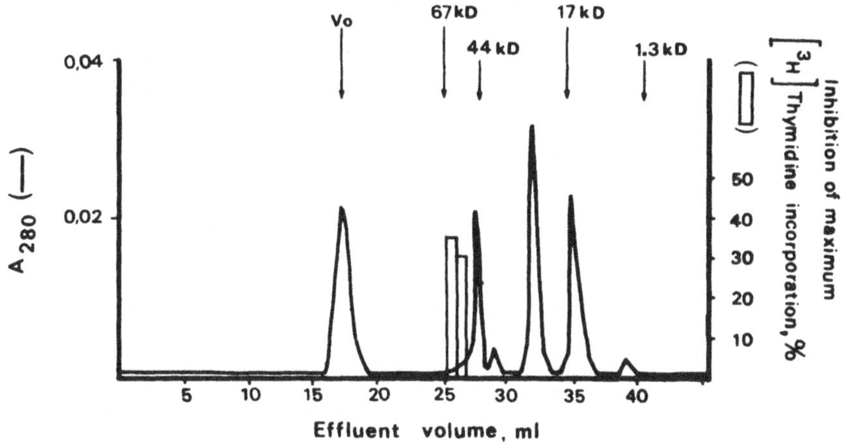

Figure 2. High-performance gel filtration chromatography of MECIF activity in crude monocyte supernatant TSK-250. Two hundred and fifty microliters of a tenfold diluted monocyte supernatant was injected and eluted with Hanks buffer at 1 ml/min. Forty microliters of each fraction was then tested for inhibitory activity. (Reprinted with permission of *Exp. Cell Res.*)

sion of procoagulant activity (tissue factor) as previously reported[40] but did not enhance the production of MECIF activity. Addition of Aprotinin in culture medium (10^{-8}–10^{-6} M) did not influence MECIF activity. These data are summarized in Table 1.

C. Chromatographically Fractionated Monocyte Conditioned Medium

One hundred milliliters of crude monocyte supernatant (180 mg of protein) was 10-fold concentrated with Amicon ultrafiltration and fractionated by chromatography on a Sepharose CL-6B column equilibrated with 10 mM Tris-HCl pH 8.0. From 5 to 10 μl of each

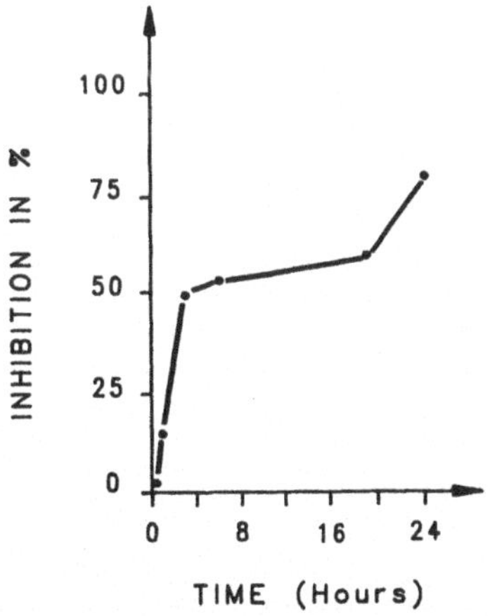

Figure 3. Production of MECIF activity by monocytes. Monocyte supernatant (corresponding to 0.2 μl of native supernatant per well) was tested for its inhibitory activity on [*methyl*-^{3}H]thymidine incorporation in endothelial cells.

Table 1. Modulation of MECIF Production by Monocytes[a]

Chemicals		% inhibitory activity[b]
Cycloheximide	10 μg/ml	0
Indomethacin	3×10^{-7} M	83–110
	3×10^{-8} M	80–110
f-MLP	0.2×10^{-7} M	83
	0.5×10^{-7} M	73
	1×10^{-7} M	66
LPS	2.5 μg/ml	91
	5 μg/ml	75
	10 μg/ml	82

[a]Adherent monocytes were treated with various chemicals for 4 hr and then incubated in fresh M199 + 2.5% FCS for 24 hr to allow the expression of MECIF activity.
[b]As compared to the nontreated monocytes.

fraction was tested for inhibitory activity on HEC growth and the profiles typically obtained are shown in Fig. 4. The fractions indicated with the bar were pooled and used for bioassays described below. As shown in Fig. 1, the half-maximal inhibition of [*methyl*-^{3}H]thymidine incorporation in endothelial cells was obtained at 7 μg/ml crude monocyte supernatant (Fig. 1A) and at 40 ng/ml after CL-6B chromatography (Fig. 1B). This suggests a 175-fold purification and the recovery was 12%.

We also verified that the inhibition of [*methyl*-^{3}H]thymidine incorporation was parallel to a reduction in the cell number as determined by cell counts. For this purpose, endothelial cells were incubated as described in Materials and Methods except that they were grown in

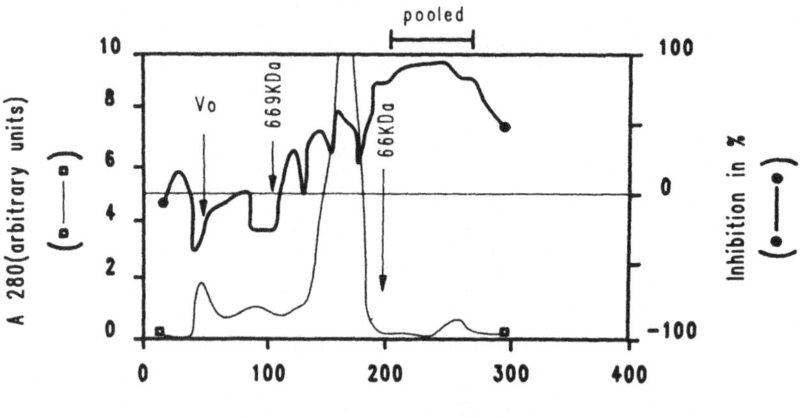

Figure 4. Chromatography on Sepharose CL-6B. Monocyte supernatant was concentrated tenfold by ultrafiltration and applied to a Sepharose CL-6B column equilibrated with 10 mM Tris-HCl pH 8.0. Aliquots (10 μl) of each fraction were then tested for their cell proliferation inhibitory activity (i.e., as determined by [*methyl*-^{3}H]-thymidine incorporation in human umbilical vein endothelial cells). (Reprinted with permission of *Exp. Cell Res.*)

Table 2. Inhibitory Effect of MECIF on Endothelial Cell Proliferation Estimated by Cell Count or [Methyl-³H]-Thymidine Incorporation

	Cell number	Radioactivity (cpm)
Control	28,866 ± 2100	31,178 ± 810
MECIF	10,800 ± 346	4,732 ± 298

the presence (or absence) of MECIF activity for 48 hr instead of 24 hr. At the end of this period, endothelial cells were either pulse-labeled (for the determination of the inhibition of [*methyl*-^{3}H]thymidine incorporation) or harvested (for determination of cell numbers). The results are shown in Table 2 and indicate that MECIF-induced inhibition of [*methyl*-^{3}H]-thymidine incorporation is associated with the inhibition of cellular proliferation. All of the biological properties described below have been established with fractionated MECIF obtained after Sepharose CL-6B chromatography.

D. Physical, Chemical, and Enzymatic Treatments of Fractionated MECIF

The biological activity of fractionated MECIF was tested under different conditions and data are summarized in Table 3. The MECIF activity was stable at 60°C for 10 min but heating at 80°C or incubation with trypsin resulted in a complete loss of MECIF activity. The inhibitory activity was stable in 1 M acetic acid for 30 hr. Freezing fractionated MECIF (but not crude monocyte supernatant) resulted in a dramatic loss in inhibitory activity. However, fractionated MECIF is stable for at least 2 months when stored at 4°C.

E. Effects of MECIF Exposure on Endothelial Cells

MECIF did not seem to exert its bioactivity on the endothelial cells through direct cytotoxicity as it did not provoke the release of ^{51}Cr from prelabeled endothelial cells. Furthermore, the antiproliferative effect of MECIF was reversible. Endothelial cells were first cultured for 24 hr in the presence or in the absence (control) of MECIF. This resulted in

Table 3. Sensitivity of MECIF Activity to Various Treatments

Treatment	% inhibitory activity[a]
Trypsin (2½ hr, 37°)[b]	0
60°, 10 min	82
80°, 10 min	2
Acetic acid 1M, 30 hr	100

[a]As compared to the nontreated activity.

[b]As control, trypsin was inactivated by incubating 50 μg of trypsin with 100 μg of soybean trypsin inhibitor for 2½ hr at room temperature prior to addition of fractionated MECIF. Inactivated trypsin did not affect MECIF activity.

a 40% inhibition of MECIF-treated cultures. Control and MECIF-treated cultures were then harvested, 15,000 cells were seeded in new wells in the presence of M199 culture medium supplemented with 20% FCS, 25 μg/ml ECGF, and 90 μg/ml heparin and allowed to grow for 5 days. Incorporations of [*methyl*-^{3}H]thymidine were determined at days 3, 4, and 6. As shown in Fig. 5, incorporation of [*methyl*-^{3}H]thymidine of MECIF-treated cultures was restored starting the fourth day.

The inhibition of proliferation was accompanied by a morphological alteration of endothelial cells. The presence of MECIF in the culture medium caused the cells to become elongated. However, as demonstrated by immunofluorescence studies and observations under transmission electron microscopy, the cells still retained their specific markers, i.e., the presence of von Willebrand factor and Weibel–Palade bodies.

We determined whether MECIF could inhibit ECGF-induced proliferation. Endothelial cells (after the second passage) were pulse-labeled in the presence of 10% FCS or in the presence of 10% FCS plus ECGF (25μg/ml) plus heparin (90 μg/ml). In these conditions, ECGF induced a 600% stimulation of [*methyl*-^{3}H]thymidine incorporation. Two concentrations of MECIF were tested and inhibitory effects were found to be similar in the presence (48 and 69% of inhibition) or in the absence of ECGF plus heparin (43 and 73% of inhibition). From these results, it appears that MECIF could inhibit ECGF-induced proliferation. We also found that subcultured endothelial cells did respond to MECIF activity at least up to six passages.

F. Specificity of MECIF Activity

MECIF activity displays cell type specificity which further supports the interpretation that the antiproliferative effect on endothelial cells is not due to a toxic contaminant in our preparations (Fig. 6).

Fibroblasts of normal human skin, human artery smooth muscle cells, and porcine

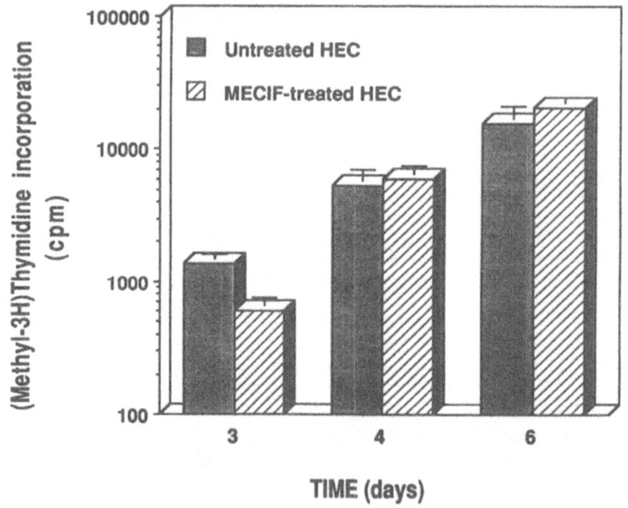

Figure 5. Reversibility of the inhibitory effect of MECIF activity. The endothelial cells were first cultured for 24 hr in the presence or in the absence (control) of MECIF. Cells were then harvested and replated in fresh medium containing M199 culture medium and 20% FCS plus ECGF (25 μg/ml) plus heparin (90 μg/ml) and allowed to grow for 5 days (from day 1 to day 6). Cells were pulse-labeled and [*methyl*-^{3}H]thymidine incorporation determined at days 3, 4, and 6.

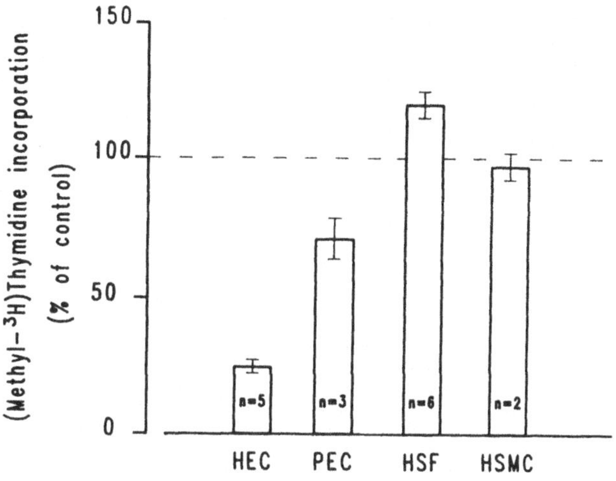

Figure 6. Cellular specificity of MECIF effect. Fractionated MECIF was added at the same concentration (100 ng/ml) to the cells of different species, namely human umbilical endothelial cells (HEC), porcine aortic endothelial cells (PEC), human skin fibroblasts (HSF), human smooth muscle cells (HSMC). The results are presented as the mean ± S.E.M. of the percentage of [*methyl*-^{3}H]thymidine incorporation that was obtained in the presence of the control culture medium. (Reprinted with permission of *Exp. Cell Res.*)

endothelial cells were tested in the absence or in the presence of 100 ng/ml MECIF and results expressed as a percentage of the control (i.e., untreated cells). Under these conditions, MECIF induced a strong growth inhibition of human umbilical vein endothelial cells (75 ± 2%, five experiments), while [*methyl*-^{3}H]thymidine incorporation was not inhibited in fibroblasts of normal skin (121 ± 4%, six experiments, nonsignificant stimulation) or in human artery smooth muscle cells (98 ± 3%, three experiments). Further evidence of the specificity of MECIF was shown as it inhibited the growth of human umbilical vein endothelial cells (HEC) more efficiently than arterial porcine endothelial cells (PEC) (27 ± 7% inhibition, three experiments, inhibition statistically significant: $p = 0.05$).

C. Inhibition of HEC Proliferation by MECIF, TGF-β, and INF-γ

TGF-β, TNF-α, and INF-γ are well-characterized growth factors that have been reported to inhibit endothelial cell growth *in vitro* and to be produced by activated monocytes.[24–32] TGF-β also elicits a strong mitogenic response on growth-arrested NRK monolayers.[39] To determine whether MECIF could be related to TGF-β, we thus examined the action of MECIF on NRK cells. For this purpose, NRK cells were grown, growth-arrested, incubated with either TGF-β or MECIF, and then pulse labeled overnight with [*methyl*-^{3}H]thymidine. As shown in Table 4, 2 ng/ml TGF-β and 40 ng/ml partially purified MECIF induced comparable biological effects on endothelial cell growth, i.e., 50% inhibition. However, TGF-β induced a strong mitogenic response on NRK cells (249 ± 24%) whereas MECIF did not (0 ± 3.5%). To further examine whether MECIF could be related to TGF-β, we used polyclonal antibodies specific for TGF-β. As shown in Table 4, preincubation of MECIF with polyclonal antibodies specific for TGF-β did not inhibit MECIF activity while in similar conditions TGF-β was neutralized. We performed a similar experiment with polyclonal antibodies specific for INF-γ which has been reported to inhibit endothelial cell growth.[32] Likewise, preincubation of MECIF with these antibodies did not reverse its inhibitory effect while INF-γ was neutralized (Table 4).

Table 4. Comparison of MECIF with Other "in Vitro" Endothelial Cell Growth Inhibitors[a]

	Incorporation of [*methyl*-^{3}H]thymidine in endothelial cells
TGF-β (2 ng/ml)	50 ± 8%
MECIF (40 ng/ml)	50 ± 3%
TGF-β (20 ng/ml) + anti-TGF-β (100 μg/ml)	121 ± 6%
MECIF (40 ng/ml) + anti-TGF-β (100 μg/ml)	33 ± 6%
INF-γ (30 U/ml)	44 ± 6.3%
MECIF (30 ng/ml)	40 ± 5.7%
INF-γ + anti-INF-γ (30 U/ml)	85 ± 2.5%
MECIF + anti-INF-γ (30 U/ml)	32 ± 3.2%

[a]Polyclonal antibodies were incubated for 1 hr with cytokines before being tested for their inhibitory activity on HEC growth. All values are the mean ± S.E.M. of triplicate determinations.

TNF-α has been reported to be a potent inhibitor of endothelial cell proliferation in culture.[29–31] Radioimmunometric analysis revealed low levels of TNF-α in our MECIF preparations, insufficient to promote the observed growth inhibitory effect. A MECIF sample inducing 35% inhibition was found to contain 0.05 ng/ml (3 u/ml) TNF-α. By comparison, the minimal concentration of TNF-α inducing 50% inhibition was reported to be 1 ng/ml.[31]

IV. DISCUSSION

Angiogenesis is a multistep process where endothelial cells have a fundamental role. They degrade the basement membrane,[41] are induced to proliferate, and then migrate into the tissue being vascularized.[6–10] Several growth factors have been shown to modulate endothelial cell growth *in vitro* and hence to be potential regulators of angiogenesis *in vivo*. Numerous factors (FGF, EGF, TGF-α, PD-ECGF, vasculotropin, macrophage- and platelet-derived growth factors) stimulate endothelial cell proliferation[13–23] while some (TGF-β, TNF-α, INF-γ) are inhibitors of endothelial cell growth.[24–32]

We have investigated the possible role of normal unstimulated human monocytes in regulating endothelial cell growth and have reported the initial characterization of a MECIF secreted by human monocytes and that inhibits human endothelial cell proliferation *in vitro*.[35] In our study we found that MECIF, TGF-β, and INF-γ consistently inhibited endothelial cell proliferation while the concentration of TNF-α required to inhibit endothelial cell proliferation seems more variable.

This protease-sensitive activity seems to be different from TGF-β, TNF-α, or INF-γ. These cytokines are preferentially released by activated monocytes while we found MECIF activity to be released under basal conditions by monocytes. MECIF activity is clearly distinguished from TGF-β by its failure to stimulate DNA synthesis in growth-arrested NRK cells. Furthermore, polyclonal antibodies specific for TGF-β did not inhibit MECIF activity. Likewise, preincubation of polyclonal antibodies specific for INF-γ did not reverse MECIF activity. Radioimmunometric analysis of our MECIF preparations revealed low levels of TNF-α insufficient to explain the observed inhibitory activity.

The monocyte origin of MECIF was established by the fact that the possible contaminant cells (platelets and lymphocytes) did not produce any inhibition of endothelial cell growth. Furthermore, monocytes obtained by flow-elutriation and hence free of significant platelet and lymphocyte levels produced similar level of MECIF activity. The production of MECIF activity in the monocyte culture involved protein synthesis and was time dependent, the maximum being reached after 24 hr incubation. The inhibitory effect on endothelial cell growth was not exerted by cytotoxic effects but rather by inhibiting DNA synthesis. MECIF activity was reversible and appeared to be cell specific since the growth of human smooth muscle cells, human skin fibroblasts, or NRK cells was not affected.

V. PERSPECTIVES

We do not know whether fractionated MECIF can modulate angiogenesis *in vivo*. Angiogenesis is a complex multistep process and factors can act via different mechanisms. For example, acidic and basic FGF promote endothelial cell proliferation and migration *in vitro*[13–15,42] and both are angiogenic *in vivo* in the developing chick chorioallantoic membranes and the corneal implants.[43,44] The same findings were described for PD-ECGF.[18,45] Their angiogenic properties probably work through the migration and proliferation stimulation of endothelial cells. By contrast, TGF-β actually inhibits endothelial cell growth *in vitro*[24–28] but is angiogenic in newborn mice.[46] Finally, other factors, including angiogenin,[47] or prostaglandins[48,49] are not endothelial mitogens while being angiogenic. These findings can be explained by the following mechanisms. First, some angiogenic factors might act primarily on endothelial cells by stimulating proliferation and mobility. Some of them can also affect endothelial mobility alone. Second, factors might be angiogenic indirectly, by mobilizing secondary cells to produce angiogenic factors or by releasing angiogenic factors stored in tissues or in basement membranes.

If, as expected, MECIF actually modulates neovascularization *in vivo*, it could play a role in the angiogenesis that accompanies tissue repair processes, diabetic retinopathy, and tumor-induced neovascularization.

REFERENCES

1. Schwartz, S. M., and Benditt, E., 1977, Aortic endothelial cell replication. I. Effects of age and hypertension in the rat, *Circ. Res.* **41:**248.
2. Hobson, B., and Deneckamp, J., 1984, Endothelial proliferation in tumours and normal tissues; continuous labelling studies, *Br. J. Cancer* **49:**405.
3. Engerman, R. L., Pfaffenbach, D., and Davis, M. D., 1967, Cell turnover of capillaries, *Lab. Invest.* **17:**738.
4. Folkman, J., and Klagsburn, M., 1987, Angiogenic factors, *Science* **235:**442–447.
5. Folkman, J., Merler, E., Abernaty, C., and Williams, G., 1971, Isolation of a tumor factor responsible for angiogenesis, *J. Exp. Med.* **133:**275–288.
6. Birdwell, C. R., Gospodarowicz, D., and Nicolson, G. L., 1977, Factors from 3T3 cells stimulate proliferation of cultured vascular endothelial cells, *Nature* **268:**528–531.
7. Gospodarowicz, D., Moran, J. S., and Braun, D. L., 1977, Control of proliferation of bovine vascular endothelial cells, *J. Cell. Physiol.* **91:**377–385.
8. Folkman, J., Haudenschild, C. C., and Zetter, E. R., 1979, Long-term culture of capillary endothelial cells, *Proc. Natl. Acad. Sci. USA* **76:**5217–5221.

9. Zetter, B. R., 1980, Migration of capillary endothelial cells is stimulated by tumor-derived factors, *Nature* **285:**41–43.
10. Azizkhan, R. G., Azizkhan, J. C., Zetter, B. R., and Folkman, J., 1980, Mast cell heparin stimulates migration of capillary endothelial cells in vitro, *J. Exp. Med.* **152:**931–944.
11. Jaffe, E. A., Nachman, R. L., Becker, C. G., and Minick, C. R., 1973, Culture of human endothelial cells derived from umbilical veins. Identification by morphologic and immunologic criteria, *J. Clin. Invest.* **52:**2745–2756.
12. Gimbrone, M. A., Cotran, R. S., and Folkman, J., 1974, Human vascular endothelial cells in culture. Growth and DNA synthesis, *J. Cell Biol.* **60:**673–684.
13. Gospodarowicz, D., Bialecki, H., and Greenburg, G., 1978, Purification of the fibroblast growth factor activity from bovine brain, *J. Biol. Chem.* **253:**3736–3743.
14. Maciag, T., Cerundolo, J., Ilsley, S., Kelly, P. R., and Forand, R., 1979, An endothelial cell growth factor from bovine hypothalamus: Identification and partial purification, *Proc. Natl. Acad. Sci. USA* **76:**5674–5678.
15. Shing, Y., Folkman, J., Sullivan, R., Butterfield, C., Murray, J., and Klagsbrun, M., 1984, Heparin affinity: Purification of a tumor-derived capillary endothelial cell growth factor, *Science* **223:**1296–1299.
16. Gospodarowicz, D., Brown, K. D., Birdwell, C. R., and Zetter, B. R., 1978, Control of proliferation of human vascular endothelial cells: Characterization of the response of human umbilical vein endothelial cells to fibroblast growth factor, epidermal growth factor, and thrombin, *J. Cell Biol.* **77:**774–788.
17. Schreiber, A. B., Winkler, M. E., and Derynck, R., 1986, Transforming growth factor-alpha: A more potent angiogenic mediator than epidermal growth factor, *Science* **232:**1250–1253.
18. Miyazono, K., Okabe, T., Urabe, A., Takaku, F., and Heldin, C. H., 1987, Purification and properties of an endothelial cell growth factor from human platelets, *J. Biol. Chem.* **262:**4098–4103.
19. Plouët, J., Schilling, J., and Gospodarowicz, D., 1989, Isolation and characterization of a newly identified endothelial cell mitogen produced by ALT-20 cells, *EMBO J.* **8:**3801–3806.
20. Clemmons, D. R., Isley, W. L., and Brown, M. T., 1983, Dialyzable factor in human serum of platelet origin stimulates endothelial cell replication and growth, *Proc. Natl. Acad. Sci. USA* **80:**1641–1645.
21. King, G. L., and Buchwald, S., 1984, Characterization and partial purification of an endothelial cell growth factor from human platelets, *J. Clin. Invest.* **73:**392–396.
22. Okabe, T., and Takaku, F., 1986, A macrophage factor that stimulates the proliferation of vascular endothelial cells, *Biochem. Biophys. Res. Commun.* **134:**344–350.
23. Polverini, P. J., and Leibovich, S. J., 1984, Induction of neovascularization in vivo and endothelial cell proliferation in vitro by tumor-associated macrophages, *Lab. Invest.* **51:**635–642.
24. Fräter-Schröder, M., Müller, G., Birchmeier, W., and Böhlem, P., 1986, Transforming growth factor-beta inhibits endothelial cell proliferation, *Biochem. Biophys. Res. Commun.* **137:**295–302.
25. Baird, A., and Durkin, T., 1986, Inhibition of endothelial cell proliferation by beta-transforming growth factor: Interactions with acidic and basic fibroblast growth factors, *Biochem. Biophys. Res. Commun.* **138:**476–482.
26. Heimark, R. L., Twardzik, D. R., and Schwartz, S. M., 1986, Inhibition of endothelial regeneration by type-beta transforming growth factor from platelets, *Science* **223:**1078–1080.
27. Müller, G., Behrens, J., Nussbaumer, U., Böhlem, P., and Birchmeier, W., 1987, Inhibitory action of transforming growth factor β on endothelial cells, *Proc. Natl. Acad. Sci. USA* **84:**5600–5604.
28. Takehara, K., Carwile LeRoy, E., and Grotendorst, G. R., 1987, TGF-β inhibition of endothelial cell proliferation: Alteration of EGF binding and EGF-induced growth-regulatory (competence) gene expression, *Cell* **49:**415–422.
29. Stolpen, A. H., Guinan, E. C., Fiers, W., and Pober, J. S., 1986, Recombinant tumor necrosis factor and immune interferon act singly and in combination to reorganize human vascular endothelial cell monolayers, *Am. J. Pathol.* **123:**16–24.
30. Sato, N., Goto, T., Haranaka, K., Satomi, N., Nariuchi, H., Mano-Hirano, Y., and Sawasaki, Y., 1986, Actions of tumor necrosis factor on cultured vascular endothelial cells: Morphologic modulation, growth inhibition and cytotoxicity, *J. Natl. Cancer Inst.* **76:**1113–1121.
31. Fräter-Schröder, M., Risau, W., Hallmann, R., Gautschi, P., and Böhlem, P., 1987, Tumor necrosis factor type α, a potent inhibitor of endothelial cell growth in vitro, is angiogenic in vivo, *Proc. Natl. Acad. Sci. USA* **84:**5277–5281.
32. Friesel, R., Komoriya, A., and Maciag, T., 1987, Inhibition of endothelial cell proliferation by gamma-interferon, *J. Cell Biol.* **104:**689–696.

33. Kahaleh, M. B., De Lustro, F., Bock, W., and LeRoy, E. C., 1986, Human monocyte modulation of endothelial cells and fibroblast growth: Possible mechanism for fibrosis, *Clin. Immunol. Immunopathol.* **39:**242–255.
34. Courillon-Mallet, A., Wautier, J. L., and Wautier, M. P., 1986, Monocyte-derived endothelial cell inhibitory factor (MECIF), Sixth Inter-Washington Spring Symposium (L. L. Gallo, ed.), p. 160a (Abstr).
35. Vilette, D., Setiadi, H., Wautier, M. P., Caen, J., and Wautier, J. L., 1990, Identification of an endothelial cell growth inhibitory activity produced by human monocytes, *Exp. Cell Res.* **188:**219–255.
36. Setiadi, H., Wautier, J. L., Courillon Mallet, A., Passa, P., and Caen, J., 1987, Increased adhesion to fibronectin and MO-1 expression by diabetic monocytes, *J. Immunol.* **138:**3230–3234.
37. Wautier, J. L., Pintigny, D., Maclouf, J., Wautier, M. P., Corvazier, E., and Caen, J., 1986, Release of prostacyclin after erythrocyte adhesion to cultured vascular endothelium, *J. Lab. Clin. Med.* **107:**210–215.
38. Bradford, M. M., 1976, A rapid and sensitive method for the quantitation of microgram quantities of protein utilizing the principle of protein-dye binding, *Anal. Biochem.* **72:**248–254.
39. Massague, J., 1984, Type β transforming growth factor from feline sarcoma virus-transformed rat cells, *J. Biol. Chem.* **259:**9756–9861.
40. Courillon Mallet, A., Bevilacqua, M. P., Wautier, J. L., Dervichian, M., Cattan, D., and Caen, J., 1986, Increased procoagulant response of monocytes from patients with familial Mediterranean fever, *Thromb. Haemostas.* **56:**211–213.
41. Gross, J. L., Moscatelli, D., and Rifkin, D. B., 1983, Increased capillary endothelial cell protease activity in response to angiogenic stimuli in vitro, *Proc. Natl. Acad. Sci. USA* **80:**2623–2627.
42. Montesano, R., Vassalli, J. D., Baird, A., Guillemin, R., and Orci, L., 1986, Basic fibroblast growth factor induces angiogenesis in vitro, *Proc. Natl. Acad. Sci. USA* **83:**7297–7301.
43. Shing, Y., Folkman, J., Haudenschild, C., Lund, D., Crum, R., and Klagsburn, M., 1985, Angiogenesis is stimulated by a tumor-derived endothelial cell growth factor, *J. Cell Biochem.* **29:**275–287.
44. Lobb, R. R., and Fett, J. W., 1984, Purification of two distinct growth factors from bovine neural tissue by heparin affinity chromatography, *Biochemistry* **23:**6295–6299.
45. Ishikawa, F., Miyazono, K., Hellman, U., Drexler, H., Takaku, F., Risau, W., Helden, C. H., Wernstedt, C., Hagiwenner, K., and Usuki, K., 1989, Identification of angiogenic activity and the cloning and expression of platelet-derived endothelial cell growth factor, *Nature* **338:**557–562.
46. Roberts, A. B., Sporn, M. B., Assoian, R. K., Smith, J. M., Roche, N. S., Wakefield, L. M., Heine, U. I., Liotta, L. A., Falanga, V., and Kehrl, J. H., 1986, Transforming growth factor type beta: Rapid induction of fibrosis and angiogenesis in vivo and stimulation of collagen formation in vitro, *Proc. Natl. Acad. Sci. USA* **83:**4167–4171.
47. Fett, J. W., Strydom, D. J., Lobb, R. R., Alderman, E. M., Bethune, J. L., Riordan, J. F., and Vallee, B. L., 1985, Isolation and characterization of angiogenin, an angiogenic protein from human carcinoma cells, *Biochemistry* **24:**5480–5486.
48. Form, D. M., and Auerbach, R., 1983, PGE2 and angiogenesis, *Proc. Soc. Exp. Biol. Med.* **172:**214–218.
49. Ziche, M., Jones, J., and Gullino, P., 1982, Role of prostaglandin E1 and copper in angiogenesis, *J. Natl. Cancer Inst.* **69:**475–482.

11

Recent Insights into the Mechanisms of Vascular Injury

Implications for the Pathogenesis of Vasculitis

Ramzi S. Cotran and Jordan S. Pober

Vascular injury and inflammation are the most important features of a heterogeneous group of disorders commonly referred to as the systemic necrotizing vasculitides.[1] Previous studies have established two pathogenetic mechanisms for vasculitis[2,3]: deposition of circulating immune complexes with subsequent complement activation and inflammation; and interactions between circulating antibodies with blood vessel wall antigens *in situ*, such as occurs in Goodpasture's syndrome, in which autoantibodies interact with glomerular and pulmonary capillary basement membranes. These mechanisms, however, cannot account for many forms of inflammatory vasculitis and several additional or alternative pathways have recently been explored. In this chapter we will review three settings of vascular injury which may shed light on the pathogenesis of vasculitis. First, we will examine the role of cytokine-induced activation[4] of endothelial cells in causing endothelial and vascular injury and the possible role of these effects in one form of vasculitis—Kawasaki's disease; second, we will describe the details of vascular injury occurring during transplantation reactions, and how these may be relevant to other forms of vasculitis; and third, we will review recent work associating the presence of antineutrophil cytoplasmic antibodies with certain forms of vasculitis, such as Wegener's granulomatosis.

I. CYTOKINE-INDUCED ENDOTHELIAL ACTIVATION AND VASCULITIS

As reviewed elsewhere in this book, and in a recent comprehensive survey, a large body of recent work has shown that cytokines induce a variety of structural and functional alterations in vascular endothelial cells, collectively referred to as "endothelial activation."

Ramzi S. Cotran and Jordan S. Pober • Department of Pathology, Brigham and Women's Hospital, and Harvard Medical School, Boston, Massachusetts, 02115.

Endothelial Cell Dysfunctions, edited by Nicolae Simionescu and Maya Simionescu. Plenum Press, New York, 1992.

Three of these alterations are particularly relevant to vasculitis. First, γ interferon regularly induces the synthesis and surface expression of class II histocompatibility antigens on endothelial cells[5]; second, interleukin 1 (IL-1) and tumor necrosis factor (TNF) induce endothelial cells to express increased surface procoagulant activity and decreased anti-coagulant properties, rendering the endothelial surface thrombogenic[5]; and third, these same cytokines cause increased adhesivity of the endothelial surface to neutrophils, lymphocytes, and monocytes by inducing the surface expression of endothelial adhesion molecules, such as ELAM-1 (endothelial leukocyte adhesion molecule 1), ICAM-1 (intercellular adhesion molecule 1), and INCAM-110/VCAM-1 (intercellular molecule 110 or vascular cell adhesion molecule 1) (reviewed in Refs. 6 and 7). Lymphocyte adhesion can also be induced by interleukin 4 (IL-4),[8] perhaps by increasing expression of INCAM-110/VCAM-1.

Although cytokine-exposed endothelial cells do not show signs of injury and cell death *in vitro*, there are several ways by which cytokines may ultimately result in loss of endothelial integrity. First, there is evidence that cytokines directly produce vascular leakiness. IL-1, TNF, and γ interferon cause rearrangements of endothelial cytoskeleton leading to long-lived increases in permeability.[9,10] The TNF effect, which begins over the course of 4 to 6 hr, is sensitive to pertussis toxin and thus may involve a G protein mediator.[10] Second, by initiating surface fibrin deposition, and favoring intravascular coagulation, endothelial injury may eventually result from the effect of fibrin-related proteins that are themselves toxic. Third, there is evidence that exposure of endothelial cells to cytokines, such as TNF and IL-1, renders these cells significantly more susceptible to both neutrophil-mediated and direct cellular injury.[11] And finally, immune injury to endothelium can be triggered by cytokine activation, as is suggested by the studies in Kawasaki's disease.

The possibility that immune injury to endothelium may play a role in vasculitis has been the subject of a great deal of speculation, and several groups have reported the presence of circulating antiendothelial antibodies in connective tissue diseases and in certain forms of systemic necrotizing vasculitis.[12] Cerilli and collaborators, for example, have identified circulating cytotoxic antibodies that reacted with monocyte and endothelial cells in a variety of vasculitides.[13] In 1986, Leung and his associates reported that children in the acute phase of Kawasaki's disease possessed in their sera at least two types of circulating antibodies which lyse cytokine-activated, but not normal endothelial cells.[14,15] The first set lyses IL-1- or TNF-treated endothelial cells and the second, γ interferon-treated endothelial cells, and the two sets recognize completely different antigens as shown by cross-adsorption studies. This work suggested a scenario by which these cytokines, which are known to be present in elevated concentrations in the sera of patients with this condition,[16–18] result in endothelial activation and induction of new endothelial surface antigens. These antigens elicit an antibody response and the antibodies in turn then cause cell lysis. We have recently tested parts of this hypothesis by examining for the presence of activation antigens in skin biopsies of patients with Kawasaki's disease, before and after treatment with intravenous gamma globulin, a therapy which is known to prevent the development of coronary artery aneurysms and attenuate the symptoms of the disease.[19] We found that ELAM-1 and ICAM-1 were present or accentuated in all biopsies prior to therapy, but disappeared or were markedly reduced in biopsies of patients who responded to therapy.[19] In addition, it was shown that peripheral blood mononuclear cells from patients with acute Kawasaki's disease spontaneously produced IL-1 but that cells isolated from gamma globulin-treated patients produced markedly lesser amounts of IL-1 than did such cells before therapy. These studies suggest that immune-mediated injury to cytokine-induced endothelial activation antigens exists in patients with Kawasaki's disease and that this may initiate the global vasculitis which ensues. The nature of the endothelial activation antigens in this setting, as well as the pathway leading from endothelial cell injury to necrotizing vasculitis, are unknown and deserve further study.

II. VASCULAR INJURY IN TRANSPLANTATION

Transplantation of vascularized organ grafts often results in immune-mediated injury of blood vessels because the donor vascular cells are foreign to the transplant recipient's immune system. Three distinct patterns of vascular injury have been described: hyperacute rejection; acute vascular rejection; and transplant-associated accelerated arteriosclerosis (TxAA).

Hyperacute rejection is caused by the presence of preexisting antibodies, usually of the IgM isotype, reactive with endothelial cells. Since ABO blood group incompatibilities are now avoided, hyperacute rejection has become distinctly uncommon in allotransplantation. However, the presence of natural antibodies reactive with xenogeneic cells constitutes the most formidable barrier to xenotransplantation, an otherwise attractive possibility for solving shortages of available organ grafts.

The histopathology of hyperacute rejection is characterized by extensive intravascular coagulation. As the name implies, this process can be initiated within minutes of vascular anastomosis and exposure of the graft to recipient blood. The endothelial lining normally functions to prevent intravascular coagulation by two active mechanisms: (1) through expression of cell surface heparan sulfate with anticoagulant properties, endothelial cells catalyze antithrombin III-mediated inactivation of thrombin; and (2) through expression of thrombomodulin, endothelial cells bind thrombin and alter its specificity so that thrombin catalyzes the generation of anticoagulant-activated protein C.[20] Inflammatory cytokines, such as IL-1 and TNF, inhibit thrombomodulin synthesis by endothelium and thereby block the generation of activated protein C, but this effect requires hours to occur. It is not known if heparan sulfate anticoagulant is affected by cytokines. Recently, it has been reported that exposure of porcine endothelial cells to human serum causes release of heparan sulfate, beginning in minutes.[21] This process depends upon activation of complement by antiendothelial IgM antibody, but does not necessarily cause endothelial cell death. Release of heparan proceeds via enzymatic cleavage, probably of the glycosaminoglycan core protein. It is not certain whether the relevant enzymes are provided by endothelial cells or by the serum. It is also not known whether loss of heparan depends only upon early complement components or requires assembly of membrane attack complex. Membrane attack complex may further contribute to a local hypercoagulable state by catalyzing formation of a prothrombinase[22] and by releasing unusually high-molecular-weight von Willebrand factor polymers stored in endothelial cell Weibel–Palade bodies.[23] These actions also occur in minutes. The heparan sulfate loss, prothrombinase assembly, and von Willebrand factor release may combine to produce local thrombosis in hyperacute rejection.

The histopathology of acute vascular rejection appears as vasculitis, often with fibrinoid necrosis. Several early studies emphasized the deposition of antibodies reactive with vascular cell alloantigens and referred to this lesion as acute humoral rejection. However, subsequent work has pointed out the presence of alloreactive lymphocytes in the injured vessels, leading to the more general term used here. Pathogenetically, acute vascular rejection is caused by an acquired immune reaction to graft alloantigens. MHC molecules are major targets of this response, but other studies have pointed to the prevalence of antibodies reactive with an alloantigen system shared by endothelial cells and monocytes (EM antigens) or with alloantigens unique to endothelium (VE antigens).[24] The biochemical nature of these antigens is not known. Acute vascular rejection is not widely appreciated as a major cause of rejection in organs other than kidney, but a recent study suggests that this entity may have been overlooked (and underdiagnosed) in cardiac transplantation.[25]

Perhaps the most clinically significant form of vascular injury in current transplantation is transplant-associated arteriosclerosis. This lesion, first noted in renal transplantation as a variant form of chronic rejection, is the major cause of late cardiac graft failure.[26] The histopathologic appearance of concentric bland intimal hyperplasia of both extramural and intramural coronary arteries without evidence of vascular cell necrosis or disruption of the internal elastic lamina, is a marker of classic vasculitis.

Immunohistochemistry has been used by Libby and co-workers to further analyze the lesions.[27] In the early phases, there is a subluminal accumulation of T lymphocytes, including approximately equal numbers of the $CD4^+$ and $CD8^+$ T cell subsets. Many of these lymphocytes appear in contact with the basal surface of the overlying vascular endothelium. The luminal endothelial cells of the lesional artery express both class I and class II MHC molecules, although class II MHC molecules are not normally found on arterial endothelium. Numerous macrophages appear adjacent to and beneath the subluminal band of T cells; scattered T cells are also present among these macrophages. The deepest layer of the neointima is comprised largely of smooth muscle cells and is poor in leukocytes. In contrast to the endothelium, these smooth muscle cells do not, in general, express class II MHC molecules. In some specimens, adventitial mononuclear cell infiltrates are also noted.

These immunohistochemical findings are most suggestive of a variant form of delayed hypersensitivity response.[28] In brief, it has been proposed that class I and class II molecules on endothelial cells are recognized by allospecific $CD8^+$ and $CD4^+$ T cells, respectively. The $CD8^+$ T cells may initially recognize class I molecules and produce the cytokines that induce endothelial cells to express class II molecules. Class II-expressing arterial endothelial cells can cause $CD4^+$ T cells to become activated and proliferate. Such T cells could also release cytokines which serve to recruit and activate macrophages. The macrophages, in turn, would produce other cytokines and growth factors which recruit and stimulate proliferation and matrix synthesis by smooth muscle cells. This proposed sequence of events differs from classic delayed hypersensitivity only in that medial smooth muscle cells, rather than interstitial fibroblasts, are induced by activated macrophages to proliferate and produce matrix. Further characterization of the resident leukocytes and vessel wall cells will be needed to test this interpretation.

Many transplant patients who develop transplant arteriosclerosis appear to be infected with cytomegalovirus.[29] Although there may be a positive correlation of transplant arteriosclerosis, with CMV infection in some studies, this may reflect coincidence of two common complications of cardiac transplantation. Alternatively, CMV infection of arterial wall cells could contribute to the delayed hypersensitivity response proposed above.

III. ANTINEUTROPHIL CYTOPLASMIC ANTIBODIES IN VASCULITIS

Another mechanism for the initiation of vasculitis is suggested by studies on the possible role of antineutrophil cytoplasmic autoantibodies (ANCA) in Wegener's granulomatosis.[12,30] This is a form of systemic vasculitis associated with chronic upper airway and sinus inflammation, necrotizing, often crescentic glomerulonephritis, and myelomonocyte proliferation.[31] The autoantibodies were originally defined by immunofluorescence microscopy, utilizing alcohol- or formaldehyde-fixed neutrophils as substrates, and for a time there was controversy and confusion regarding the specificity of these autoantibodies to Wegener's granulomatosis.[12] However, two reactive cytoplasmic antigens have recently been identified, and ELISA procedures are currently available to quantitate these autoantibodies; one of these antigens is myeloperoxidase, present in the primary granules of neutrophils.[32,33] Immunofluorescent staining for autoantibodies to this antigen is sometimes seen as nuclear or perinuclear staining (P-ANCA), because of artifactual redistribution of antigen during ethanol fixation.[32,33] The second antigen is a 29-kDa serine protease which has homology with, but is distinct from, leukocyte elastase and cathepsin.[34–36] This antigen appears to be identical to proteinase 3, a recently described proteinase of neutrophils which produces emphysema in experimental models.[37] Autoantibodies to this antigen always cause a cytoplasmic pattern of staining by immunofluorescence microscopy (C-ANCA) and are detected in the majority of patients with Wegener's granulomatosis and infrequently in other vasculitides and in patients with isolated glomerulonephritis.[31]

A number of studies have shown that disease activity and autoantibody titers are closely correlated and, thus, autoantibodies are thought to be an important factor in the pathogenesis of the disease. But how autoantibodies to these enzymes cause vascular injury is unknown. A recent study shows that these antibodies activate neutrophils *in vitro*, causing degranulation of neutrophils and a respiratory burst.[38] This activation is augmented by cytokines, such as TNF, resulting in the release of oxygen free radicals and lytic enzymes from neutrophils, neutrophil-mediated killing of endothelial cells (and possibly smooth muscle cells), and potentiating vasculitis. However, human proteinase 3 and other closely related neutral serine proteases may have other functions in the regulation of cell proliferation and terminal differentiation in hematopoietic cells, and it may thus be that the autoantibodies against proteinase 3 may have other effects in the immune evolution of these vasculitides.[39]

IV. SUMMARY AND CONCLUSIONS

Our increased understanding of the interactions between circulating leukocytes and vascular wall cells, and the influence of cytokines in these interactions has provided new insights on the possible mechanisms by which vascular injury can occur. Studies on cytokine-induced endothelial activation have shown that such activated cells are more susceptible to neutrophil-mediated as well as direct endothelial cell killing; further, in certain settings, such as in Kawasaki's disease, autoantibodies to activation antigens may cause endothelial lysis and perhaps initiate global vasculitis. Work on the mechanisms of vascular injury in transplantation has uncovered two relevant determinants of such injury: in acute vascular rejection, MHC molecules are major targets of the acquired immune response but antibodies to an alloantigen system on monocytes and/or endothelial cells also play an important role. In transplant arteriosclerosis, the evidence suggests that the mechanism of vascular injury is a delayed hypersensitivity reaction initiated by recognition of class I and class II MHC molecules on endothelial cells by allospecific T cells. This leads to a sequence of events involving T-cell proliferation, cytokine production, macrophage activation, and the production of growth factors that induce smooth muscle cell proliferation. Finally, evidence was reviewed which shows that autoantibodies to cytoplasmic enzymes of leukocytes are present in certain forms of vasculitis, and raises the possibility that interactions between cytokines, autoantibodies, and these leukocytes may trigger the release of these enzymes, causing vascular injury.

REFERENCES

1. Cupps, T. R., and Fauci, A. S., 1981, *The Vasculitides*, Saunders, Philadelphia.
2. Savage, C. O. S., and Ng, Y. C., 1986, The etiology and pathogenesis of major systemic vasculitis, *Postgrad. Med. J.* **62**:623.
3. Wilson, C. B., 1990, The renal response to immunological injury, in: *The Kidney*, 4th ed. (B. M. Brenner and R. C. Rector, eds.), Saunders, Philadelphia.
4. Pober, J. S., and Cotran, R. S., 1990, Cytokines and endothelial cell biology, *Physiol. Rev.* **70**:427.
5. Pober, J. S., Gimbrone, M. A., Jr., Cotran, R. S., Reiss, C. S., Burakoff, S. J., Fiers, W., and Ault, K. A., 1983, Ia expression by vascular endothelium is inducible by activated T-cells and by human γ interferon, *J. Exp. Med.* **157**:1339.
6. Cotran, R. S., and Pober, J. S., 1990, Cytokine–endothelial interactions in inflammation, immunity and vascular injury, *J. Am. Soc. Nephrol.* **1**:225–235.
7. Osborn, J, 1990, Leukocyte adhesion to endothelium in inflammation, *Cell* **62**:3.
8. Thornhill, M. H., Kyan-Aung, U., and Haskard, D. O., 1990, IL-4 increases human endothelial cell adhesiveness for T-cells but not for neutrophils, *J. Immunol.* **144**:3060.
9. Stolpen, A. H., Guinan, E. C., Fiers, W., and Pober, J. S., 1986, Recombinant tumor necrosis factor and immune interferon act singly and in combination to reorganize human vascular endothelial cell monolayers, *Am. J. Pathol.* **123**:16.
10. Brett, J., Gerlack, H., Nawroth, P., Steinberg, S., Godman, G., and Stern, D., 1989, Tumor necrosis factor/cachectin increases permeability of endothelial cell monolayers by a mechanism involving regulatory G-proteins, *J. Exp. Med.* **169**:1977.
11. Ward, P. A., and Varani, J., 1990, Mechanisms of neutrophil-mediated killing of endothelial cells, *J. Leukocyte Biol.* **48**:97.
12. Cotran, R. S., 1990, Pathogenesis of vasculitis: An update, in: *Advances in Pathology*, Vol. III, pp. 301–310.
13. Brasile, L., Kremer, J. M., Clarke, J. L., and Cerilli, J., 1989, Identification of an autoantibody to vascular endothelial cell-specific antigens in systemic vasculitis, *Am. J. Med.* **87**:74.
14. Leung, D. Y. M., Geha, R. S., Newburger, J. W., Burns, J. C., Fiers, W., Lapierre, L. A., and Pober, J. S., 1986, Two monokines, interleukin α and tumor necrosis factor, render cultured vascular endothelial cells susceptible to lysis by antibodies circulating during Kawasaki syndrome, *J. Exp. Med.* **164**:1958.
15. Leung, D. Y. M., Collins, T., Lapierre, L. A., Geha, R. S., and Pober, J. S., 1986, Immunoglobulin M antibodies present in the acute phase of Kawasaki syndrome lyse cultured vascular endothelial cells stimulated by γ-interferon, *J. Clin. Invest.* **77**:1428.
16. Maury, C. P. J., Salo, E., and Pelkonen, P., 1988, Circulating interleukin-1 in patients with Kawasaki disease, *N. Engl. J. Med.* **319**:1670.
17. Kurukawa, S., Matsubara, T., Jujoh, K., Yone, K., Sugawara, T., Sasai, K., Kato, H., and Yabuta, K., 1988, Peripheral blood monocyte/macrophage and serum tumor necrosis factor in Kawasaki disease, *Clin. Immunol. Immunopathol.* **48**:247.
18. Rowley, A. H., Shulman, S. T., Preble, O. T., Poiesz, B. J., Ehrlich, G. D., and Sullivan, J. R., 1988, Serum interferon concentrations and retroviral serology in Kawasaki syndrome, *Pediatr. Inf. Dis. J.* **7**:663.

19. Leung, D. Y. M., Cotran, R. S., Kurt-Jones, E. A., Burns, J. C., Newburger, J. W., and Pober, J. S., 1989, Endothelial activation and high interleukin-1 secretion in the pathogenesis of acute Kawasaki disease, *Lancet* **2:**1298.
20. Esmon, C., 1987, The regulation of natural anticoagulant pathways, *Science* **235:**1348.
21. Platt, J. L., Vercellotti, G. M., Lindman, B. J., Oegema, T. R. Jr., Bach, F. H., and Dalmasso, A. P., 1990, Release of heparan sulfate from endothelial cells: Implications for pathogenesis of hyperacute rejection, *J. Exp. Med.* **171;**1363.
22. Hamilton, K. K., Hattori, R., Esmon, C. T., and Sims, P. J., 1990, Complement proteins C5b-9 induce vesiculation of the endothelial plasma membrane and expose catalytic surface for assembly of the prothrombinase enzyme complex, *J. Biol. Chem.* **265:**3809.
23. Hattori, R., Hamilton, K. K., McEver, R. P., and Sims, P. J., 1989, Complement proteins C5b-9 induce secretion of high molecular weight multimers of endothelial von Willebrand factor and translocation of granule membrane protein GMP-140 to one cell surface, *J. Biol. Chem.* **264:**9053.
24. Cerilli, J., and Brasile, L., 1988, Endothelial cell alloantigens, *Transplant. Proc.* **12:**37.
25. Yowell, R. L., Hammond, E. H., Brislow, M., Watson, F. S., Renlund, D. G., and O'Connell, J. B., 1988, Acute vascular rejection involving the major coronary arteries of cardiac allograft, *Heart Transplant.* **7:**191.
26. Feogh, M. L., 1990, Chronic rejection-graft arteriosclerosis, *Transplant. Proc.* **22:**119.
27. Salomon, R. N., Hughes, C. C. W., Schoen, F. J., Payne, D. D., Pober, J. S., Libby, P., 1991, Human coronary transplantation-associated arteriosclerosis: Evidence for a chronic immune reaction to activated graft endothelial cells, *Am. J. Path.* **138:**791–798.
28. Libby, P., Salomon, R. N., Payne, D. D., Schoen, F. J., and Pober, J. S., 1989, Functions of vascular wall cells related to development of transplantation-associated coronary arteriosclerosis, *Transplant. Proc.* **21:**3677.
29. Grattan, M. T., Moreno-Cabral, C. E., Starnes, V. A., Oyer, P. E., Stinson, E. B., and Shumway, N. E., 1989, Cytomegalovirus infection is associated with cardiac allograft rejection and atherosclerosis, *J. Am. Med. Assoc.* **261:**3651.
30. Nölle, B., Specks, U., Ludemann, J., Rohrbach, M. S., DeRemee, R. A., and Gross, W. L., 1989, Anticytoplasmic autoantibodies: Their immunodiagnostic value in Wegener's granulomatosis, *Ann. Intern. Med.* **111:**28.
31. Feinberg, R., 1989, Morphologic and immunohistologic study of the evolution of necrotizing palisading granuloma of pathergic granulomatosis, *Semin. Respir. Med.* **10:**126.
32. Falk, R. J., and Jennette, J. C., 1988, Anti-neutrophil cytoplasmic autoantibodies with specificity for myeloperoxidase in patients with systemic vasculitis and idiopathic necrotizing and crescentic glomerulonephritis, *N. Engl. J. Med.* **318:**1651.
33. Jennette, J. C., Wilkman, A. S., and Falk, R. J., 1989, Anti-neutrophil cytoplasmic autoantibody-associated glomerulonephritis and vasculitis, *Am. J. Pathol.* **135:**921.
34. Goldschmeding, R., van der Schoot, C. E., Ten Bokkel Huinick, D., Hack, C. E., Van den Ende, M. E., Kallenberg, C. G., and Von dem Borne, A. E., 1989, Wegener's granulomatosis autoantibodies identify a novel DFP-binding protein in the lysosomes of normal human neutrophils, *J. Clin. Invest.* **84:**1577.
35. Niles, J. L., McCluskey, R. T., Ahmad, M. F., and Arnaout, M. A., 1989, Wegener's granulomatosis autoantigen is a novel neutrophil serine protease, *Blood* **74:**1988.
36. Ludemann, B. U., Utecht, B., and Gross, W. L., 1990, Antineutrophil cytoplasmic antibodies in Wegener's granulomatosis recognize an elastic enzyme, *J. Exp. Med.* **171:**357.
37. Kao, R. C., Wehner, N. G., Skubitz, K. M., Gray, B. H., and Hoidal, J. R., 1988, Proteinase 3: A distinct polymorphonuclear leukocyte proteinase that produces emphysema in hamsters, *J. Clin. Invest.* **82:**1963.
38. Falk, R. J., Terrell, R. S., Charles, L. A., and Jennette, J. C., 1990, Anti-neutrophil cytoplasmic autoantibodies induce neutrophils to degranulate and produce oxygen radicals in vitro, *Proc. Natl. Acad. Sci. USA* **87:**4115.
39. Jenne Tshopp, J., Ludemann, J., Uthatb, and Gross, W. L., 1990, Wegener's autoantigen decoded, *Nature* **346:**520.

12

The Role of Endothelium in the Pathophysiology of Rheumatoid Inflammation

Dorian O. Haskard

I. INTRODUCTION

Rheumatoid arthritis (RA) is a systemic inflammatory disease characterized by chronic destructive arthritis and various extraarticular manifestations such as rheumatoid nodules and vasculitis. This chapter is limited to a discussion of the role of endothelium in the pathogenesis of the articular features of the disease.

II. PATHOPHYSIOLOGY

Detailed reviews of the pathology and pathogenesis of rheumatoid synovitis can be found in standard textbooks.[1] Briefly, the pathology of RA consists of a thickening of the synovial lining of the joint cavity, an infiltration of the sublining layer of the synovial membrane with lymphocytes and monocytes, and the destruction of cartilage by an ingrowth of connective tissue "pannus" tissue.

The synovial membrane has many of the properties of an immunologically stimulated secondary lymphoid organ,[2] including (1) a characteristic cytoarchitecture with regions composed predominantly of T cells or plasma cells,[3] (2) the presence of HLA class II-positive macrophages and interdigitating reticulum cells in relation to T lymphocytes, suggesting the processing and presentation of antigen and subsequent T cell activation, (3) the presence of venules with cuboidal or columnar endothelium resembling the "high endothelial venules" (HEV) through which lymphocytes traffic into lymphoid parenchyma,[4,5] and (4) the local synthesis and secretion of immunoglobulins and cytokines.

Dorian O. Haskard • Department of Medicine (Rheumatology), RPMS, Hammersmith Hospital, London W12 0NN, United Kingdom.

Endothelial Cell Dysfunctions, edited by Nicolae Simionescu and Maya Simionescu. Plenum Press, New York, 1992.

The central importance of lymphocytes to the pathogenesis of RA is further supported by the therapeutic effects of measures that alter lymphocyte function, including thoracic duct drainage, leukocytopheresis, total lymphoid irradiation, and immunosuppression.

Immune interactions within synovial tissue have many pathophysiological consequences including the stimulation of B cells to secrete rheumatoid factors which self-associate into immune complexes, the development of locally invasive connective tissue "pannus" tissue, and the cytokine-mediated destruction of cartilage and bone. Superimposed upon this chronic process are episodic acute "flares" in disease activity that are largely attributable to the periodic entry of neutrophils into the joint. Neutrophils pass rapidly into the exudative synovial fluid within the joint cavity and are rarely seen in large numbers in the synovial tissue itself. Their activation by immune complexes and other factors, with the subsequent release of prostaglandins, leukotrienes, oxygen free radicals, and lysosomal enzymes, account for much of the pain and joint swelling that are witnessed clinically.

Synovial tissue is richly supplied with blood vessels which form the avenue through which lymphocytes, monocytes, and neutrophils reach the joint. This chapter will concentrate on the first step in the emigration of leukocytes from the blood, that of the adhesion of various types of leukocytes to the luminal surface of endothelium. If the number of leukocytes entering the tissues is rate-limiting for inflammatory responses, the control of leukocyte–endothelial cell adhesion could play an early and central role in both the acute and chronic phases of rheumatoid inflammation. However, it should be remembered that endothelial cells (EC) also contribute to the inflammatory response in other ways such as through the secretion of cytokines and growth factors,[6] through permeability changes leading to the exudation of protein from the blood into synovial tissue and fluid,[7] and through the vascularization of invasive "pannus" tissue.[8]

III. METHODS AND RESULTS

A. The Role of Cytokines in Lymphocyte Adhesion to Endothelium

It seemed possible that the recruitment of lymphocytes into inflammatory lesions might be attributable to an alteration in the adhesive capacity of EC in response to cytokines. In order to examine this possibility, experiments were performed in which human umbilical vein EC were preincubated with conditioned medium from mononuclear cell cultures that had been activated by allogeneic or mitogenic stimulation. These conditioned media were found to enhance the adhesion of ^{51}Cr-labeled T cells by an action on EC adhesiveness with no detectable effect on the adhesiveness of the lymphocytes.[9]

Further experiments have shown that purified or recombinant cytokines are indeed capable of enhancing EC adhesiveness for T cells. These fall into two pairs: interleukin 1 (IL-1)[10] and tumor necrosis factor (TNF)[11] which have a peak action on EC within 10 hr and the lymphokines interferon γ (IFNγ)[9] and interleukin 4 (IL-4)[12] which have peak actions later from 18 to 24 hr following stimulation of EC. The implication of these *in vitro* experiments is that cytokines might have an important feedback effect on endothelium in synovial tissue, resulting in enhanced lymphocyte recruitment and hence migration into the joint.

B. Selective Leukocyte Adhesion to EC

1. Lymphocytes and Neutrophils

As several laboratories documented enhanced adhesiveness of IL-1- and TNF-stimulated EC for different leukocyte types,[13–17] the question arose as to whether one mechanism could account for enhanced EC adhesiveness for leukocytes generally or whether different mechanisms needed to be considered for each leukocyte type. To address this question, experiments were conducted in which the adhesion of T cells and neutrophils were directly compared.[18]

Unstimulated adhesion of T cells to EC was found to be significantly greater than that of neutrophils, and this relatively greater adhesion of T cells was maintained when EC were stimulated with suboptimal concentrations of IL-1 or TNF. In contrast, neutrophils adhere significantly better than T cells to EC stimulated with higher concentrations of IL-1 or TNF. We also found that differences existed in the kinetics of IL-1- or TNF-stimulated enhanced EC adhesiveness for T cells and neutrophils. The enhanced adhesiveness for PMN was maximal 4–6 hr after stimulation and thereafter fell to baseline values by 72 hr. In contrast, T cell adhesion was maximal a little later at 8–10 hr and was more long-lasting, being still significantly elevated at 72 hr. These differences suggest that while the effect of IL-1 and TNF on vascular endothelium is to promote the adhesion and subsequent migration of a mixed population of leukocytes, neutrophil adhesion might be relatively favored at early time points following stimulation, and that as the stimulus persists there is a gradual selection for T cell adhesion. We have not yet compared monocyte with neutrophil or T cell adhesion in this way.

In contrast to the actions of IL-1 and TNF, IFNγ and IL-4 had no detectable effect on neutrophil–EC adhesion in the adhesion assay employed.[12,18] The release of IFNγ and IL-4 by activated lymphocytes may therefore be particularly important in the selective recruitment of T cells in an immune-mediated chronic inflammatory reaction.

2. Lymphocyte Subsets

A feature of the T–EC monolayer adhesion assay is that the proportion of T cells that adhere to EC under any one condition is always the same, regardless of the absolute number of T cells added to the EC monolayers, suggesting that there may be variation between different lymphocyte phenotypes in ability to adhere to EC.[19] We have used two methods to explore whether differences in capacity to adhere to EC might contribute to the profile of lymphocyte phenotypes found in synovial tissue and fluid in rheumatoid inflammation. The first has separated T cells of differing adhesiveness for EC by sequential "panning" of T cells on unstimulated and cytokine-stimulated EC monolayers, thereby allowing the separation of T cells into fractions whose phenotype can be analyzed using mAb staining and FACS analysis.[19] The second approach was to separate T cell subpopulations by flow cytometry and to assay adhesion to EC monolayers directly.

Initial experiments using lymphocytes separated from peripheral blood by sheep erythrocyte rosetting or by passage over nylon wool showed that there was a modest depletion of CD8 cells in the lymphocyte fraction that failed to adhere to EC, but indicated that differences in adhesive capacity between CD4 and CD8 cells were unlikely to fully explain why some lymphocytes bound well and some poorly to EC monolayers. Lymphocytes that adhered well to EC were, however, found to be enriched with cells bearing receptors for complement type 3 (CR3).[19] These cells, which copurify with true T cells, are heterogeneous and include large granular lymphocytes (LGL) and natural killer cells.[20] The finding that $CR3^+$ lymphocytes are highly adhesive to EC is consistent with the observation that synovial fluid lymphocytes, which presumably have emigrated from the blood in the course of a chronic inflammatory reaction, contain a significantly higher proportion of $CR3^+$ lymphocytes than peripheral blood lymphocytes.[21] In view of the highly adhesive properties of $CR3^+$ lymphocytes, it is tempting to speculate that these cells may be among the earliest lymphocytes to enter foci of inflammation.

More recently, the phenotype of rheumatoid synovial lymphocytes has been analyzed using monoclonal antibodies considered to identify naive and memory stages of differentiation. Whereas T cells within peripheral blood are approximately evenly divided between naive and memory cells, T cells within rheumatoid synovial fluid and tissue are almost entirely of memory cell phenotype.[22,23] This preferential accumulation of memory T cells is also found in other examples of chronic inflammation.[24]

Perhaps the most important observation, therefore, in these studies on lymphocyte subsets was the finding that memory T cells showed greater adhesion to both unstimulated and IL-1-stimulated EC than naive T cells.[24] Differences in adhesive capacity between these reciprocal T cell subsets may therefore be fundamental in determining the selective accumulation of memory T cells that is observed in the rheumatoid joint and other examples of chronic inflammation, and conversely in restricting the migration of unsensitized T cells to the pool of lymphocytes that recirculates through the fixed lymphoid system.

C. Molecular Basis of Leukocyte–EC Adhesion

The effects of cytokines such as IL-1 and TNF in enhancing EC adhesiveness for leukocytes require *de novo* protein synthesis[25] and are stable to fixation of EC with paraformaldehyde–lysine–periodate,[10] suggesting the induction or upregulation of adhesion molecules on the EC surface. Judging from differences in the degree and kinetics of increased IL-1- and TNF-stimulated EC adhesiveness for neutrophils and T cells,[18] and the selective actions of IFNγ and IL-4,[12,18] different molecules are likely to mediate the adhesion of neutrophils and T cells to cytokine-activated EC. Such an arrangement could allow for the separate control of lymphocyte and neutrophil migration into the inflamed synovium, as might be expected from the chronic and periodic "acute" phases of the disease. At this point, the molecules involved in the adhesion of T cells and neutrophils to cultured EC are incompletely defined. However, two cytokine-inducible EC molecules have been characterized and are present on EC in the rheumatoid synovial vasculature.

1. ELAM-1

ELAM-1 is an IL-1- or TNF-inducible adhesion molecule that selectively mediates the adhesion of neutrophils to EC *in vitro*.[26,27] We have generated a mAb, designated 1.2B6, that reacts with ELAM-1,[28] and have identified ELAM-1 on synovial EC *in situ*, using an immunoperoxidase staining technique on frozen sections of rheumatoid synovium.[29] The distribution of ELAM-1 on EC in blood vessels of rheumatoid subsynovial lining tissue was found to be widespread throughout areas with focal and diffuse collections of mononuclear cells (Fig. 1). Furthermore, while venules with flat endothelium did not invariably express ELAM-1, venules with "high endothelial" morphology were usually ELAM-1 positive.

Judging from the role of ELAM-1 in supporting IL-1- or TNF-stimulated adhesion to EC *in vitro*, it is reasonable to assume that its presence on blood vessels in rheumatoid synovium is important for the migration of neutrophils into the inflamed joint. As neutrophils tend to accumulate in synovial fluid and are seldom seen in synovial tissue, these observations on the location of ELAM-1-positive vessels are of interest as they provide insight as to the vessels through which neutrophil extravasation occurs. Furthermore, the positive staining of HEV-like vessels for ELAM-1 suggests that these vessels are involved in the traffic of neutrophils as well as lymphocytes.

We have recently found that the proportion of synovial blood vessels expressing ELAM-1 on EC drops during the first few weeks of gold therapy,[29] suggesting that gold may exert some of its rheumatoid disease-modifying actions by reducing the number of neutrophils entering inflamed joints, perhaps as a consequence of reduced local levels of cytokines.

2. ICAM-1

ICAM-1 (CD54) is a single-chain surface glycoprotein that is constitutively expressed on EC. The increased expression of ICAM-1 on IL-1- or TNF-stimulated EC is more long-lasting than that of ELAM-1.[30] Furthermore, ICAM-1 is also upregulated on EC by IFNγ.[30] ICAM-1 is known to act as a ligand for the leukocyte integrin LFA-1 (CD11a/CD18)[31,32] and

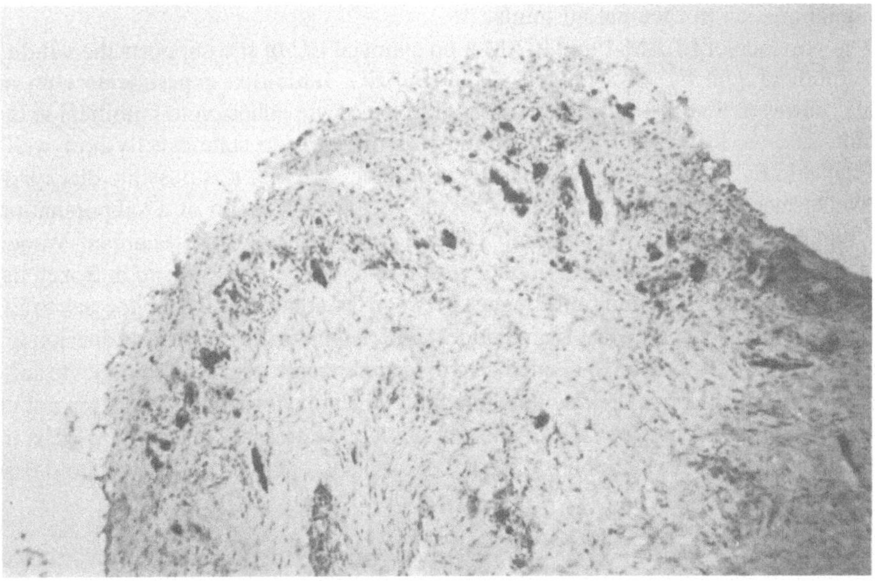

Figure 1. Immunoperoxidase staining of blood vessels in rheumatoid synovium. Vessels positive for ELAM-1 (mAb 1.2B6) are shown darkly stained throughout the subsynovium.

possibly also Mac-1 (CD11b/CD18).[33] Consistent with this, anti-ICAM-1 mAb partially inhibit the adhesion of neutrophils[33,34] and lymphocytes[35] to cytokine-stimulated umbilical vein EC monolayers. The presence of ICAM-1 on EC in blood vessels throughout the synovium suggests that it also has a role in leukocyte migration *in vivo*.[29,36]

IV. DISCUSSION

IL-1 and TNF are readily detected in rheumatoid synovial fluid and their mRNA can be found in cells derived from synovial tissue.[37–39] Support for a role for these cytokines in promoting leukocyte migration into inflamed joints comes from the demonstration of leukocyte infiltration following their injection into rabbit knees.[40,41] T cell clones derived from rheumatoid synovial tissue secrete IFNγ,[42] although only low levels of IFNγ have been found in rheumatoid synovial fluid or the supernatants of rheumatoid synovial explants.[43] While it is very likely that cytokines alter synovial EC adhesiveness for leukocytes in a similar way to that observed with cultured EC, the relative roles of these cytokines *in vivo* therefore require further study. It is of course also possible that important interactions might take place between the cytokines in their actions on EC, and that the list of cytokines which alter EC adhesiveness for leukocytes is not yet complete.

Although their number is seldom as great as in the acute synovial effusions due to sepsis or crystal arthritis, significant quantities of neutrophils do migrate into rheumatoid synovial fluid. This, together with the ready detection of ELAM-1 in rheumatoid synovium, suggests that as yet unidentified mechanisms exist that counteract ELAM-1 downregulation in chronic inflammatory tissues. It is possible, therefore, that the opposing tendencies of induction and downregulation of ELAM-1 are central to the relapsing-remitting course of acute inflammation in rheumatoid joints.

The presence of ELAM-1 and ICAM-1 on synovial EC *in situ* supports the validity of using umbilical vein EC as a model of synovial EC. Inhibition experiments with mAb indicate, however, that the molecules involved in leukocyte adhesion to umbilical vein EC have not yet been fully described[24,35,44] and it is too early to state exactly how well the umbilical vein EC model fits the *in vivo* situation. For example, it is possible that synovial EC have surface "addressins" that allow tissue-specific migration of a subpopulation of blood lymphocytes into the synovium. Using a modification of the Stamper–Woodruff assay for studying the adhesion of lymphocytes to microvascular EC *in situ* in frozen tissue sections, Jalkanen *et al.*[45] found differences between the adhesion of lymphocytes to EC in peripheral lymph nodes, mucosal lymph nodes, and rheumatoid synovium. Adhesion to EC in these three tissues appeared to be mediated by different epitopes on a lymphocyte surface molecule designated Hermes,[46] and now known to be equivalent to CD44.[47] It is not clear whether the interaction between CD44 and its putative "addressin" on synovial EC is truly specific for synovium or whether it is a feature of chronically inflamed nonlymphoid tissues more generally.

V. PERSPECTIVES

If separate mechanisms control the adhesion of neutrophils and lymphocytes to EC, and if the ability to adhere to EC determines the subsets of lymphocytes that migrate into synovial tissue, it may be possible pharmacologically to selectively manipulate the acute and chronic phases of rheumatoid inflammation.

REFERENCES

1. Harris, E. D., 1985, Pathogenesis of rheumatoid arthritis, in: *Textbook of Rheumatology*, 2nd ed. (W. N. Kelley, E. D. Harris, S. Ruddy, and C. B. Sledge, eds.), Saunders, Philadelphia, pp. 886–915.
2. Janossy, G., Panayi, G., Duke, O., Bofill, M., Poulter, L. W., and Goldstein, G., 1981, Rheumatoid arthritis: A disease of T-lymphocyte/macrophage immunoregulation, *Lancet* **2**:839–842.
3. Kurosaka, M., and Ziff, M., 1983, Immunoelectron microscopic study of the distribution of T cell subsets in rheumatoid arthritis, *J. Exp. Med.* **158**:1191–1210.
4. Freemont, A. J., Jones, C. J. P., Bromley, M., and Andrews, P., 1983, Changes in vascular endothelium related to lymphocyte collections in diseased synovia, *Arthritis Rheum.* **26**:1427–1433.
5. Iguchi, T., and Ziff, M., 1986, Electron microscopic study of rheumatoid synovial vasculature, *J. Clin. Invest.* **77**:355–361.
6. Haskard, D. O., 1989, Interleukin-1 and the vascular endothelial cell in inflammation, in: *Interleukin-1, Inflammation and Disease* (R. Bomford and B. Henderson, eds.), Elsevier, Amsterdam, pp. 123–142.
7. Simkin, P. A., and Wallis, W. J., 1985, Microvascular physiology of the rheumatoid synovium, in: *Rheumatoid Arthritis* (P. D. Utsinger, N. J., Zvaifler, and G. E. Ehrlich, eds.), Lippincott, Philadelphia, pp. 181–191.
8. Folkman, J., Ausprunk, D., and Langer, R., 1985, Connective tissue: Small blood vessels and capillaries, in *Textbook of Rheumatology*, 2nd ed. (W. N. Kelley, E. D. Harris, S. Ruddy, and C. B. Sledge, eds.), Saunders, Philadelphia, pp. 197–210.
9. Yu, C.-L., Haskard, D. O., Cavender, D. E., Johnson, A. R., and Ziff, M., 1985, Human gamma interferon increases the binding of T lymphocytes to endothelial cells, *Clin. Exp. Immunol.* **62**:554–561.
10. Cavender, D. E., Haskard, D. O., Joseph, B., and Ziff, M., 1986, Interleukin-1 increases the binding of human B and T lymphocytes to endothelial cell monolayers, *J. Immunol.* **136**:203–207.
11. Cavender, D. E., Saegusa, Y., and Ziff, M., 1987, Stimulation of endothelial cell binding of lymphocytes by tumour necrosis factor, *J. Immunol.* **139**:1855–1860.
12. Thornhill, M., Kyan-Aung, U., and Haskard, D. O., 1990, IL-4 increases human endothelial cell adhesiveness for T cells but not neutrophils, *J. Immunol.* **144**:3060–3065.
13. Dunn, C. J., and Fleming, W. E., 1984, Increased adhesion of polymorphonuclear leukocytes to vascular endothelium by specific interaction of endogenous (interleukin-1) and exogenous (lipopolysaccharide) substances with endothelial cells "in vitro," *Eur. J. Rheum. Inflam.* **7**:80–86.
14. Bevilacqua, M. P., Pober, J. S., Wheeler, M. E., Cotran, R. S., and Gimbrone, M. A. Jr., 1985, Interleukin 1 acts on cultured human vascular endothelium to increase the adhesion of polymorphonuclear leukocytes, monocytes, and related cell lines, *J. Clin. Invest.* **76**:2003–2011.
15. Gamble, J. R., Harlan, J. M., Klebanoff, S. J., and Vadas, M. A., 1985, Stimulation of the adherence of neutrophils to umbilical vein endothelium by human recombinant tumor necrosis factor, *Proc. Natl. Acad. Sci. USA* **82**:8667–8671.
16. Schleimer, R. P., and Rutledge, B. K., 1986, Cultured human endothelial cells acquire adhesiveness for neutrophils after stimulation with interleukin 1, endotoxin, and tumor-promoting phorbol esters, *J. Immunol.* **136**:649–654.
17. Goerdt, S., Zwadlo, G., Schlegel, R., Hagemeier, U.-H., and Sorg, C., 1987, Characterization and expression kinetics of an endothelial cell activation antigen in vivo present only in acute inflammatory diseases, *Exp. Cell Biol.* **55**:117–126.
18. Thornhill, M., Kyan-Aung, U., Lee, T. L., and Haskard, D. O., 1990, T cells and neutrophils exhibit differential adhesion to cytokine stimulated endothelial cells, *Immunology* **69**:287–292.
19. Cavender, D. E., Haskard, D. O., Maliakkal, D., and Ziff, M., 1988, Separation and characterization of human T lymphocytes with varying adhesiveness for endothelial cells, *Cell. Immunol.* **117**:111–126.
20. Bakke, A. C., Gray, J. D., Abo, W., Quismorio, F. P., Lash, A., Cooper, S. M., and Horwitz, D. A., 1986, Studies on human blood lymphocytes with iC3b (type 3) complement receptors, *J. Immunol.* **136**:1253–1259.
21. Fox, R. I., Fong, S., Tsoukas, C., and Vaughan, J. H., 1984, Characterization of recirculating lymphocytes in rheumatoid arthritis patients: Selective deficiency of natural killer cells in thoracic duct lymph, *J. Immunol.* **132**:2883–2887.

22. Duke, O., Bofill, M., Poulter, L. W., Panayi, G. S., and L. Janossy, G., 1987, Evidence of a deficiency of the suppressor-inducer and predominance of the helper-inducer T cell subsets in the rheumatoid synovial membrane, *Arthritis Rheum.* **30:**S39.
23. Pitzalis, C., Kingsley, G., Murphy, J., and Panayi, G. S., 1987, Abnormal distribution of the helper-inducer and suppressor-inducer T-lymphocyte subsets in the rheumatoid joint, *Clin. Immunol. Immunopathol.* **45:**252–258.
24. Pitzalis, C., Kingsley, G., Haskard, D. O., and Panayi, G. S., 1988, The preferential accumulation of helper-inducer T lymphocytes in inflammatory lesions: Evidence for regulation by selective endothelial and homotypic adhesion, *Eur. J. Immunol.* **18:**1397–1404.
25. Cavender, D. E., Haskard, D. O., Foster, N., and Ziff, M., 1987, Superinduction of T lymphocyte–endothelial cell binding by treatment of EC with interleukin-1 and protein synthesis inhibitors, *J. Immunol.* **138:**2149–2154.
26. Bevilacqua, M. P., Pober, J. S., Mendrick, D. L., Cotran, R. S., and Gimbrone, M. A., 1987, Identification of an inducible endothelial–leukocyte adhesion molecule, *Proc. Natl. Acad. Sci. USA* **84:**9238–9242.
27. Bevilacqua, M. P., and Gimbrone, M. A., 1988, Inducible endothelial functions in inflammation and coagulation, *Semin. Thromb. Hemost.***13:**425–433.
28. Wellicome, S., Thornhill, M., Pitzalis, C., Thomas, D. S., Lanchbury, J. S. S., Panayi, G. S., and Haskard, D. O., 1990, A monoclonal antibody that detects a novel antigen on endothelial cells that is induced by TNF, IL-1 or LPS, *J. Immunol.* **144:**2558.
29. Corkill, M. M., Kirkham, B. W., Barbatis, C., Haskard, D. O., Gibson, T. S., and Panayi, G. S., 1989, Changes in synovial membrane endothelial adhesion receptor expression following treatment of rheumatoid arthritis, *Br. J. Rheum.* **28:**66.
30. Pober, J. S., Gimbrone, M. A., Lapierre, L. A., Mendrick, D. L., Fiers, W., Rothlein, R., and Springer, T. A., 1986, Overlapping patterns of activation of human endothelial cells by interleukin 1, tumor necrosis factor, and immune interferon, *J. Immunol.* **137:**1893–1896.
31. Springer, T. A., Dustin, M. L., Kishimoto, T. K., and Marlin, S. D., 1987, The lymphocyte function-associated LFA-1, CD2, and LFA-3 molecules: Cell adhesion receptors of the immune system, *Annu. Rev. Immunol.* **5:**223–252.
32. Marlin, S. D., and Springer, T. A., 1987, Purified intercellular adhesion molecule 1 (ICAM-1) is a ligand for lymphocyte function associated antigen 1 (LFA-1), *Cell* **51:**813–819.
33. Smith, C. W., Marlin, S. D., Rothlein, R., Toman, C., and Anderson, D. C., 1989, Cooperative interactions of LFA-1 and Mac-1 with intercellular adhesion molecule-1 in facilitating adherence and transendothelial migration of human neutrophils in vitro, *J. Clin. Invest.* **83:**2008–2017.
34. Smith, C. W., Rothlein, R., Hughes, B., Mariscalco, M., Schmalstieg, F., and Anderson, D. C., 1988, Recognition of an endothelial determinant for CD18-dependent neutrophil adherence and transendothelial migration, *J. Clin. Invest.* **82:**1746–1756.
35. Dustin, M. L., and Springer, T. A., 1988, Lymphocyte function-associated antigen-1 (LFA-1) interaction with inter-cellular adhesion molecule-1 (ICAM-1) is one of at least three mechanisms for lymphocyte adhesion to cultured endothelial cells, *J. Cell Biol.* **107:**321–331.
36. Hale, L. P., Martin, M. E., and McCollum, D. E., 1989, Immunohistologic analysis of the distribution of cell adhesion molecules within the inflammatory synovial microenvironment, *Arthritis Rheum.* **32:**22–30.
37. Fontana, A., Hengartner, H., Weber, E., Fehr, K., Grob, P. J., and Cohen, G., 1982, Interleukin-1 activity in the synovial fluid of patients with rheumatoid arthritis, *Rheumatol. Int.* **2:**49–53.
38. Hopkins, S. J., and Meager, A., 1988, Cytokines in synovial fluid: II. The presence of tumor necrosis factor and interferon, *Clin. Exp. Immunol.* **73:**88–92.
39. Buchan, G., Barrett, K., Turner, M., Chantry, D., Maini, R. N., and Feldman, M., 1988, Interleukin-1 and tumour necrosis factor mRNA expression in rheumatoid arthritis: Prolonged production of IL1 alpha, *Clin. Exp. Immunol.* **73:**449–455.
40. Pettifer, E. R., Higgs, G. A., and Henderson, B., 1986, Interleukin 1 induces leukocytic infiltration and cartilage proteoglycan degradation in the synovial joint, *Proc. Natl. Acad. Sci. USA* **83:**8749–8753.
41. Henderson, B., and Pettifer, E. R., 1989, Arthritogenic actions of recombinant IL-1 and tumour necrosis factor alpha in the rabbit: Evidence for synergistic interactions between cytokines in vivo, *Clin. Exp. Immunol.* **75:**306–310.
42. Turner, M., Londei, M., and Feldman, M., 1987, Human T cells from autoimmune and normal individuals can produce tumour necrosis factor, *Eur. J. Immunol.* **17:**1807–1814.

43. Firestein, G. S., and Zvaifler, N. J., 1987, Peripheral blood and synovial fluid monocyte activation in inflammatory arthritis. II Low levels of synovial fluid and synovial tissue interferon suggest that interferon is not the primary macrophage activating centre, *Arthritis Rheum.* **30:**864–871.
44. Haskard, D. O., Cavender, D. E., Beatty, P., Springer, T. A., and Ziff, M., 1986, T cell adhesion to endothelial cells: Mechanisms demonstrated by anti-LFA-1 monoclonal antibodies, *J. Immunol.* **137:**2901–2906.
45. Jalkanen, S., Steere, A. C., Fox, R. I., and Butcher, E. C., 1986, A distinct endothelial recognition system that controls lymphocyte traffic into inflamed synovium, *Science* **233:**556–558.
46. Jalkanen, S., Bargatze, R. F., Toyos, J. de los, and Butcher, E. C., 1987, Lymphocyte recognition of high endothelium: Antibodies to distinct epitopes of an 85–95 kD glycoprotein antigen differentially inhibit lymphocyte binding to lymph node, mucosal, or synovial endothelial cells, *J. Cell Biol.* **105:**983–990.
47. Picker, L. J., Toyos, J. de los, Telen, M. J., Haynes, B. F., and Butcher, E. C., 1989, Monoclonal antibodies against the CD44 [In(Lu)-related p80], and Pgp-1 antigens in man recognize the Hermes class of lymphocyte homing receptors, *J. Immunol.* **142:**2046–2051.

13

Mechanisms of Neutrophil Migration across Endothelium

Ada J. Huang and Samuel C. Silverstein

I. INTRODUCTION

Polymorphonuclear leukocytes (PMN) arise in the bone marrow and travel within the vascular compartment. Their final destination is in the extravascular tissues where they perform most of their effector functions. A critical and precisely regulated step in an inflammatory response is the movement of PMN across the endothelium. It is now clear that there are at least two distinct mechanisms by which PMN cross the endothelium in response to inflammatory stimuli, a PMN-initiated mechanism and an endothelial cell (EC)-initiated mechanism.

In the PMN-initiated pathway, depicted in Fig. 1A, PMN emigrate from the blood into the tissues in response to soluble chemoattractants. This response occurs immediately upon contact of a PMN with a chemoattractant and is the major pathway for the entry of PMN into tissues within minutes of the introduction of an inflammatory stimulus.

In the EC-initiated pathway, depicted in Fig. 1B, EC play the dominant role in initiating PMN extravasation. They respond to an inflammatory stimulus by altering their surface properties, thereby promoting PMN–EC adhesion and subsequent PMN migration through the EC monolayer. Some of these changes in EC surface properties, such as the surface expression of platelet activating factor, can occur within minutes.[1] Other changes, such as the expression of new EC surface proteins, require at least 2 hr to reach detectable levels and do not promote maximal PMN–EC adhesion until 4 hr.[2] This slowly developing EC-initiated mechanism of PMN diapedesis may account for the continuous entry of leukocytes into an inflammatory site over a period of hours or days. Under physiologic circumstances, both of these pathways play a role in mediating the exodus of PMN from the

Ada J. Huang • Department of Medicine, College of Physicians and Surgeons, Columbia University, New York, New York 10032. *Samuel C. Silverstein* • Department of Physiology and Cellular Biophysics, College of Physicians and Surgeons, Columbia University, New York, New York 10032.

Endothelial Cell Dysfunctions, edited by Nicolae Simionescu and Maya Simionescu. Plenum Press, New York, 1992.

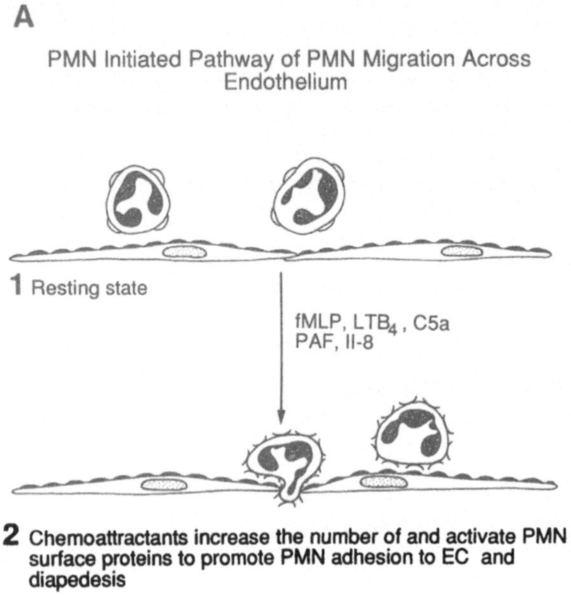

Figure 1A. PMN-initiated mechanism of PMN migration across endothelium. Solid semicircles on EC surface represent constitutively expressed EC surface proteins, such as ICAM-1, which mediate PMN binding to EC.

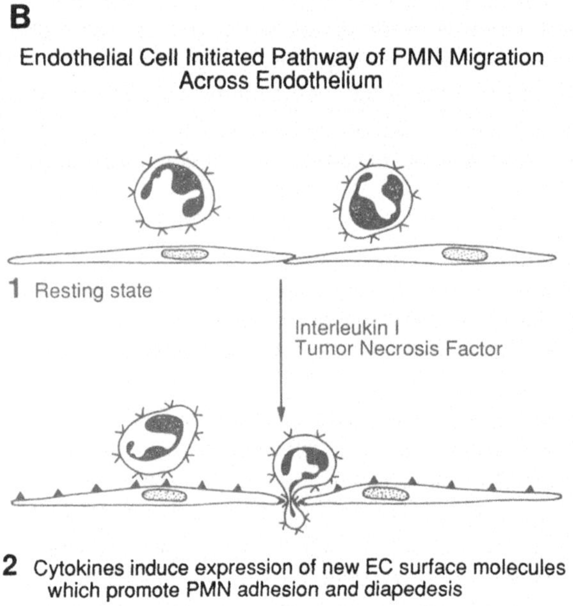

Figure 1B. EC-initiated mechanism of PMN migration across endothelium.

blood into the tissues. The two pathways complement one another and may overlap temporally. Here we review the cellular and molecular mechanisms by which leukocytes emigrate across an EC monolayer and the role of PMN in mediating the changes in vascular permeability to fluids and macromolecules which frequently accompany PMN extravasation.

II. DIFFICULTIES WITH IN VIVO STUDIES

Efforts to study the mechanism(s) by which PMN breach an EC barrier or by which PMN may regulate the permeability of an EC barrier during inflammation have been hindered by the multiplicity of factors involved in these processes *in vivo*. One complication is the heterogeneity of cells in the blood (PMN, monocytes, lymphocytes, and platelets) and connective tissue (mast cells, smooth muscle cells, and fibroblasts). Another consideration is the heterogeneity of EC. Even within a single vessel, there are at least three types of EC,

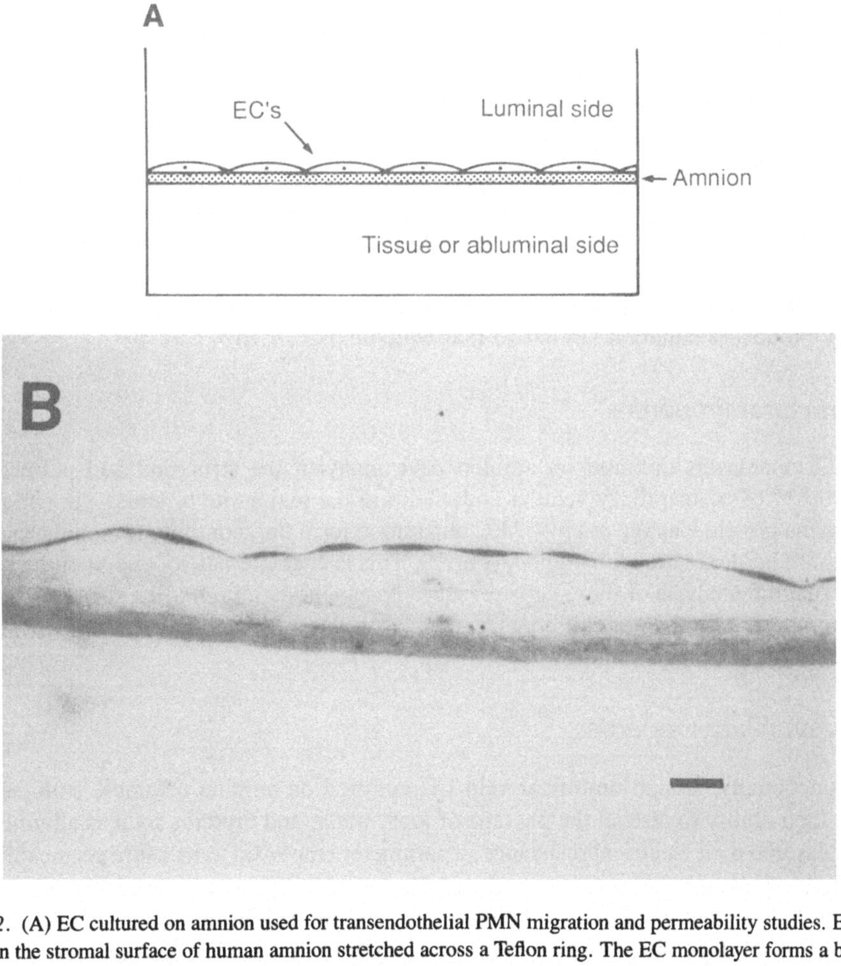

Figure 2. (A) EC cultured on amnion used for transendothelial PMN migration and permeability studies. EC are plated on the stromal surface of human amnion stretched across a Teflon ring. The EC monolayer forms a barrier between an upper (luminal) and lower (tissue or abluminal) compartment. For PMN migration studies, PMN are placed in the upper compartment and a chemoattractant (e.g., fMLP or leukotriene B_4) is placed in the lower compartment. For permeability studies, the flux of ions, water, or albumin across such EC monolayers is measured. (B) Light micrograph of a cross section of human umbilical vein EC cultured on amnion. Cells form a flat, continuous monolayer. Bar = 10 μm.

namely arteriolar EC, capillary EC, and venular EC, each of which may differ in their surface and permeability properties. For example, histamine receptors are localized to the surface of postcapillary venular EC[3] and venular endothelium is characteristically more permeable to ions and macromolecules than arteriolar endothelium of the same microvessel.[4,5] A third complication is the possible activation of one or more of the plasma effector systems such as the coagulation or complement cascades. Other factors are alterations in hydrostatic and/or oncotic pressure which may occur during an inflammatory response. As a result of these multiple variables, all of which are difficult to control in an *in vivo* setting, the mechanisms of PMN diapedesis have been difficult to analyze *in vivo*. For these reasons, many investigators have turned to *in vitro* models to examine PMN–EC interactions at a cellular and molecular level.

III. EC CULTURES ON AMNION—AN IN VITRO SYSTEM

Several laboratories have developed systems in which EC are cultured on synthetic porous substrates such as nitrocellulose or polycarbonate filters[6] or dialysis membranes.[7] We have developed a system in which bovine microvascular or human umbilical vein EC are cultured on a substrate derived from human amniotic membrane (Fig. 2A).[8,9] The advantages of using porous as opposed to solid substrates include: Porous substrates allow one to assess the movement of water, proteins, and cells across the EC barrier. They also allow one to selectively manipulate the environment on both sides of the EC monolayer and thereby produce a situation similar to that which exists *in vivo*.

A. Structural Properties

EC monolayers cultured on amnion have many of the structural and permeability properties[8,9] of postcapillary venular endothelium, the major site of leukocyte emigration and plasma protein leakage *in vivo*.[10] EC on amnion form flat, continuous monolayers (Fig. 2B), whose borders stain with $AgNO_3$ (Fig. 3). This is characteristic of endothelium *in situ*. Ultrastructural analysis of these cells reveals a basement membrane-like structure beneath the monolayer. At points of cell contact, they have plasma membrane densities which are characteristic of tight junctions.

B. Permeability Properties

Functionally, human umbilical vein EC cultured on amnion resemble endothelia *in vivo* in their ability to restrict the passage of ions, water, and proteins such as albumin. We have measured their electrical resistance, a parameter employed to measure permeability of epithelial monolayers to ions,[11] to evaluate the permeability of there monolayers. Monolayers of human umbilical vein EC cultured on amnion exhibit 6–12 $\Omega \cdot cm^2$ of electrical resistance 5–21 days after reaching confluence. Although these EC are confluent by day 2 after plating, they require 3–4 days before exhibiting any resistance (average of 1–4 $\Omega \cdot cm^2$) and 8–11 days to exhibit maximum resistance (average 8–9 $\Omega \cdot cm^2$). Once the EC have reached confluence there is no net change in their number but the percentage of EC in the

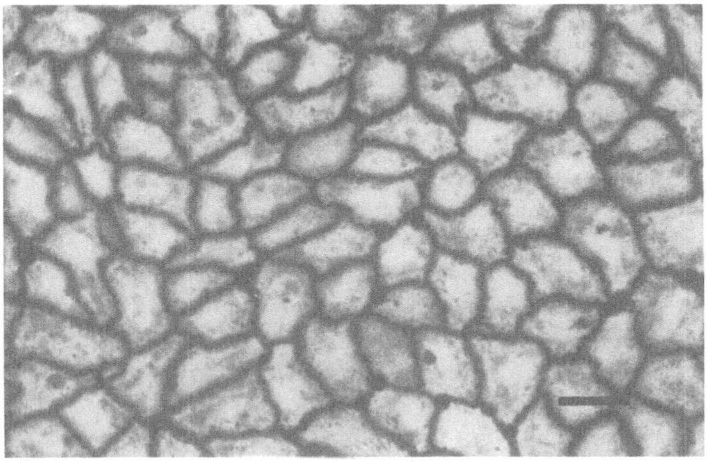

Figure 3. Light micrograph of human umbilical vein EC grown on amnion and stained with $AgNO_3$. EC are outlined by deposits of silver granules. Bar = 50 μm.

monolayer which incorporate [^{3}H]thymidine decreases from 5 ± 2% on day 3 after plating to 1 ± 0.5% on day 8 after plating.[9] Thus, it appears that the expression of electrical resistance by EC monolayers is related to a quantitative change in the number of cells in S phase of the cell cycle and a qualitative change in the contacts between individual EC. The magnitude of the transendothelial electrical resistance *in vitro* is in good agreement with the electrical resistance observed across continuous, nonfenestrated capillary endothelium *in vivo*. The latter values range from 1.8 $\Omega\cdot cm^2$ in mesenteric vessels[12] to 23–24 $\Omega\cdot cm^2$ in skeletal muscle and skin capillaries[4,13] (Table 1). Electrical resistance measurements across monolayers of rabbit aortic endothelium,[14] and porcine pulmonary endothelium[15] are in agreement with those noted for bovine microvessel and human umbilical vein EC (Table 1). It is of interest to note that brain endothelium can be distinguished from other types of

Table 1. Electrical Resistance across Vascular Endothelium

Type of endothelium	$\Omega\cdot cm^2$	Reference
In vitro measurements		
Human umbilical vein	6–12	(9)
Rabbit aortic	10–20	(14)
Porcine pulmonary aortic	6–7	(15)
Bovine microvascular	2–6	(18)
Bovine brain	157–783	(17)
In vivo measurements		
Mesentery	1.8	(12)
Skeletal muscle	23	(4)
Skin	24	(13)
Brain	1870	(16)

endothelium in this regard; frog brain endothelium has an average electrical resistance of 1870 $\Omega\cdot cm^2$ *in vivo*[16] (100–500-fold less permeable to ions than endothelium from other organ beds) and bovine brain endothelium has an electrical resistance of 157–783 $\Omega\cdot cm^2$ *in vitro*[17] (Table 1). Therefore, the relative permeability of EC monolayers to ions *in vitro* resembles that found *in vivo* and is a characteristic property of EC monolayers derived from most sources.

Human umbilical vein EC cultured on amnion restrict the transendothelial flux of albumin.[9] However, it is difficult to make quantitative comparisons between the albumin permeability of these cultured EC monolayers and the albumin permeability of capillaries *in vivo* since *in vivo* measurements are affected by flow and hydrostatic pressure, and by uncertainties with respect to the surface area available for albumin diffusion.[19]

Hydraulic conductivity (L_p), a measure of the relationship between water flow across a barrier and the driving pressure, is another property we have used a compare the permeability of endothelium *in vitro*[20] and *in vitro*. We have applied a method devised by Fischbarg[21] for measuring water movement across epithelia to measure fluid movement across EC monolayers on amnion.[22] Using this method, bovine aortic, bovine microvascular, and human umbilical vein EC monolayers cultured on amnion have hydraulic conductivities (L_p values) of $1.14 \pm 0.8 \times 10^{-6}$, $1.67 \pm 1.49 \times 10^{-6}$, and $2.9 \pm 0.85 \times 10^{-6}$ $cm\cdot sec^{-1}\cdot cm\ H_2O^{-1}$, respectively. These values are in rough agreement with hydraulic conductivities measured in single capillaries *in vivo* which range from 3×10^{-6} $cm\cdot sec^{-1}\cdot cm\ H_2O^{-1}$ in rabbit omentum[23] to 1×10^{-7} $cm\cdot sec^{-1}\cdot cm\ H_2O^{-1}$ in rat skeletal muscle.[24] Bovine aortic and bovine microvascular EC monolayers also sustain an osmotically driven transendothelial flow of water in response to an oncotic pressure difference across the monolayer.[22]

C. Response to Physiologic Agonists

Another parameter which has been used to evaluate the physiologic relevance of monolayers of cultured EC is the ability of the endothelium to respond to agonists (listed in Table 2) known to promote the leakage of proteins and fluids from postcapillary venules *in vivo*. Histamine and thrombin cause an increase in permeability of EC monolayers to albumin, low-density lipoprotein, and Evans blue (Table 8). Histamine, leukotriene C_4, and serotonin produce a 50% decrease in the electrical resistance of monolayers of human

Table 2. Substances which Increase Vascular Permeability in Vivo

	Reference
Histamine	(25)
Thrombin	(26)
Leukotriene C_4	(27)
Serotonin	(25)
Platelet activating factor	(28)
Tumor necrosis factor	(29)
Bradykinin	(30)

umbilical vein EC on amnion (A. J. Huang, unpublished observations). This is equivalent to a twofold increase in the permeability of these monolayers to ions. In agreement with this, histamine and serotonin also result in a two- to fourfold increase in the movement of water across bovine microvascular EC monolayers.[22] The role of most of the other substances listed in Table 2 in mediating increased fluid and macromolecule flux across EC monolayers continues to arouse controversy and is discussed below under vascular permeability.

The above studies demonstrate that monolayers of EC from different vascular sources cultured on amnion qualitatively exhibit the barrier functions of endothelium *in vivo*. These studies also suggest that bovine microvascular and human umbilical vein EC monolayers most closely resemble postcapillary venular endothelium in the extent to which they restrict the passage of ions and their responses to certain physiologic agonists.

IV. LEUKOCYTE MARGINATION AND ELECTRON MICROSCOPIC STUDIES

At any given moment, about half of the leukocytes in the blood are in a marginated pool, browsing along the surface of the endothelium.[31] This phenomenon occurs primarily in the microvasculature and is explained by the physical properties of blood cells and fluid flow through small-caliber vessels.[32] In tubes of small diameter, such as microvessels, flow rates and therefore wall shear rates are decreased. Under these conditions, erythrocytes tend to aggregate and form rouleaux. In contrast, leukocytes do not aggregate and are displaced by the rouleaux from the center of the vessel toward the periphery of its lumen. The lower flow rates in microvessels also permit the physical adhesive forces that exist between leukocytes and EC to prevail over shear forces that propel leukocytes along the length of the vessel.[33] Both of these factors contribute to the characteristic distribution of leukocytes at the periphery of the vessel lumen throughout the microvasculature.

Leukocytes are therefore in an excellent position to detect changes in EC surface properties or substances that diffuse into the blood vessels from the tissues. In response to such stimuli, PMN migrate across the endothelium. Electron microscopic studies of PMN diapedesis, both *in vivo*[34] and *in vitro*,[18] reveal the following sequence of events: (1) When appropriately stimulated, PMN adhere to the luminal or apical membrane of the EC (Fig. 4A). This occurs at random locations in monolayers of cultured EC *in vitro*, but primarily in postcapillary venules *in vivo*. PMN bind equally to the central and peripheral portions of the surfaces of EC in monolayers maintained *in vitro*. We are unaware of studies that have evaluated whether PMN bind randomly over the surfaces of EC *in vivo*. (2) PMN extend pseudopods and insert them into the intercellular cleft between two EC (Fig. 4B). Although no structural connections are seen between PMN and EC, their plasma membranes are very closely apposed ($\leq$150 Å) throughout their transit of an EC monolayer.[9] (3) PMN then move their cell bodies into and through the space between adjacent EC. (4) Once PMN have traversed the endothelium, the membranes of adjacent EC close over the PMN and become reapposed to one another. Continuity of the EC barrier is maintained throughout the period of PMN migration by close apposition between migrating PMN and EC (Fig. 4C). (5) Upon reaching the undersurface of the EC, the PMN dissect between the basal or abluminal surface of the EC and the EC basement membrane (Fig. 4C). (6) Finally, the PMN migrates out of the subendothelial cell space (Fig. 4D) and moves into the

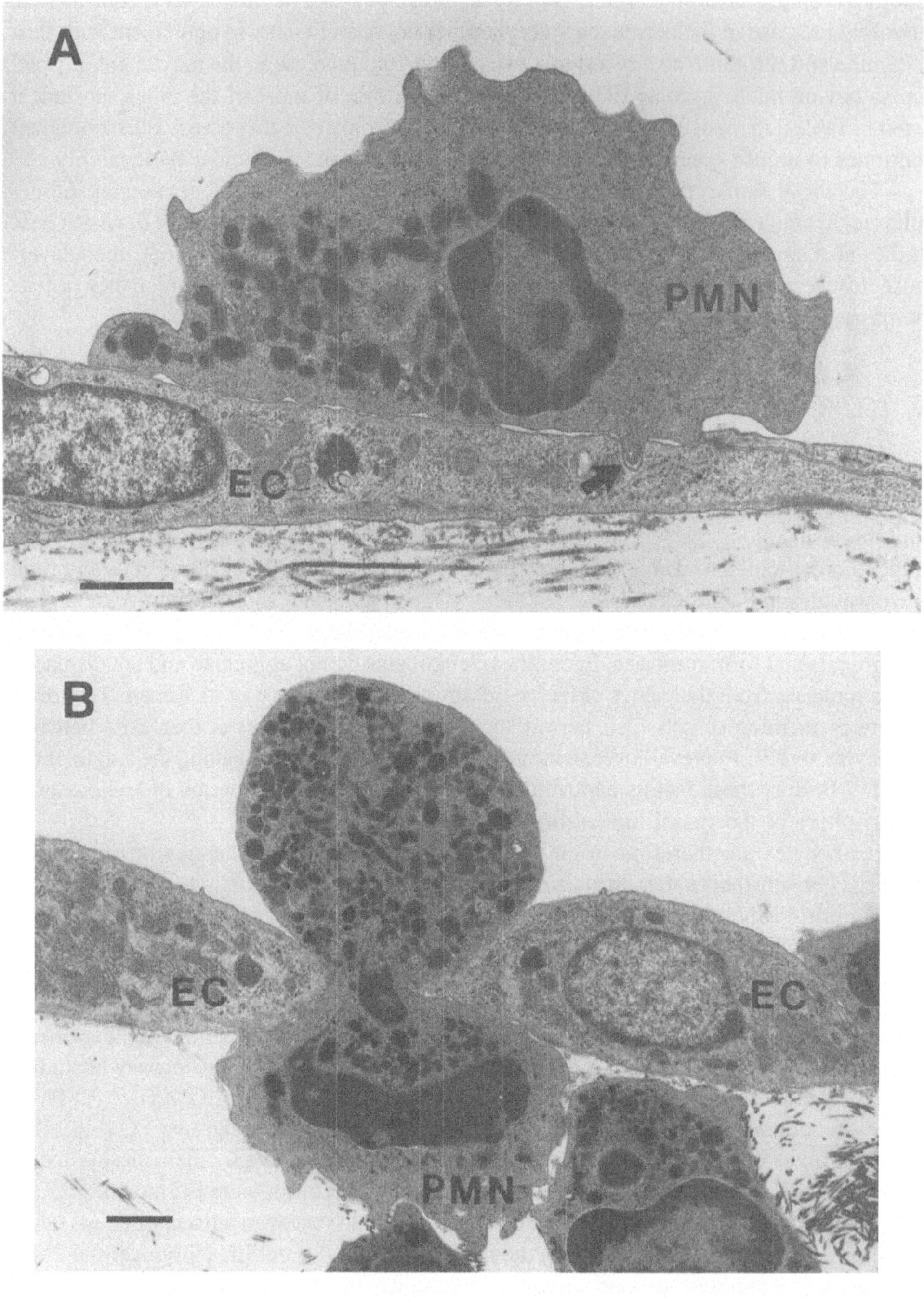

Figure 4. Transmission electron micrographs of PMN migrating across monolayers of EC grown on amnion. (A) A PMN adheres to the apical surface and indents the apical plasma membrane (arrow) of an EC following exposure to a chemoattractant placed below the EC monolayer. (B) A PMN migrating between two EC. The plasma membranes of PMN and EC are very closely apposed throughout PMN transit across an EC monolayer.

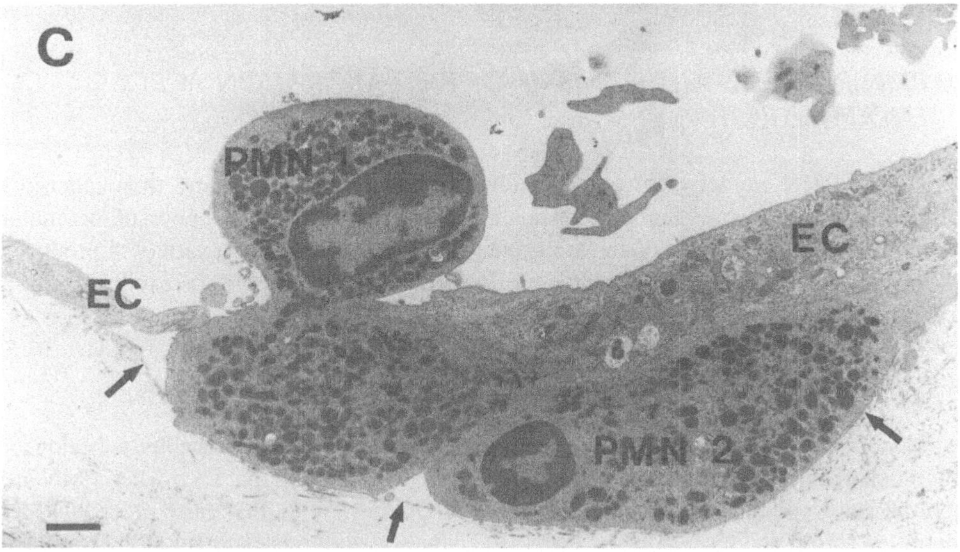

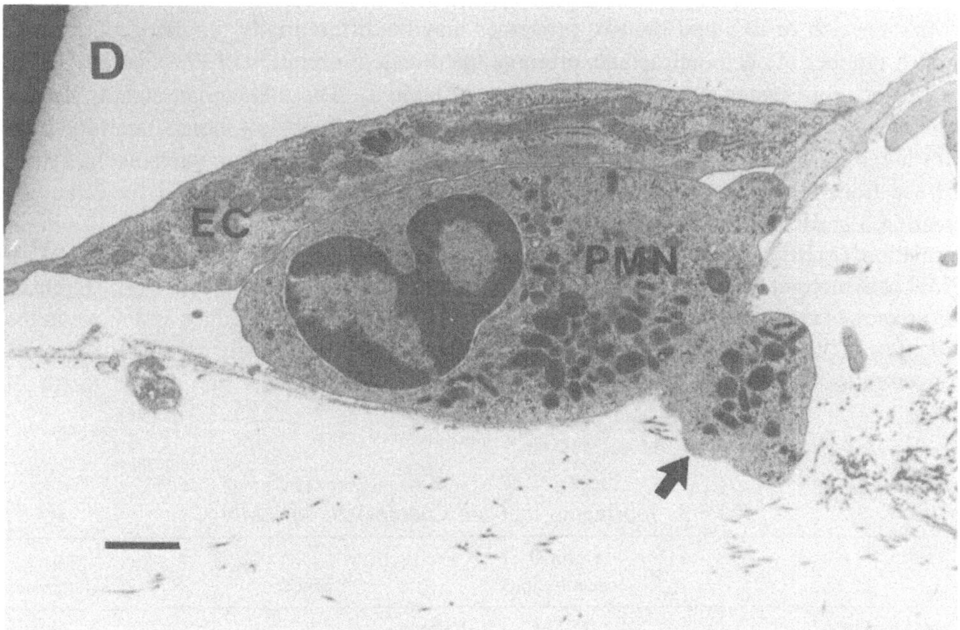

Figure 4. (*continued*) (C) One PMN migrating between two EC (PMN 1) and another (PMN 2) which has completed migration and dissected between the basal surface of the EC and its underlying basement membrane (arrows). (D) A PMN which has completed migration and is between the EC and its basement membrane. It is beginning to migrate through the basement membrane with a pseudopod (arrow). Bars = 1 μm.

connective tissue toward its final destination. By this time the endothelium is again structurally intact and appears as it did prior to PMN migration.

V. PMN-INITIATED MECHANISMS OF PMN MIGRATION ACROSS ENDOTHELIUM

When PMN are activated by a stimulus such as a chemoattractant, they undergo a number of structural and functional responses. These include the assumption of locomotor morphology, the modulation of surface proteins, and the secretion of a variety of products. All of these events can occur simultaneously and may play a role in the exodus of PMN from the vasculature.

A. Chemotactic Factors

Under resting or unstimulated conditions, there is little spontaneous adhesion or migration of PMN to or across an endothelium in comparison to that seen when PMN are appropriately stimulated.[9,18,35,36] Although PMN adhesion to EC must precede PMN migration across an EC monolayer, PMN–EC adhesion and transendothelial PMN migration are two distinct phenomena. Transendothelial migration does not necessarily follow PMN adhesion to EC, and the two processes may be differentially regulated.

A number of chemoattractants promote the directed migration of PMN across an EC barrier or in a Boyden chamber type of assay (Table 3). These chemoattractants share a number of features: (1) they are all soluble molecules of relatively low molecular mass (from 300 daltons for leukotriene B_4 to 30,000 daltons for platelet-derived growth factor); (2) they diffuse from the site at which they are generated into the surrounding tissues, thereby creating a gradient in which the concentration of chemoattractant is highest at its site of formation; (3) they each bind to a different but specific high-affinity receptor on PMN.[37–39] PMN are thought to sense the direction of a chemoattractant gradient by differential occupancy of chemoattractant receptors at their leading and trailing edges and move in the direction of highest receptor occupancy.[40] PMN respond maximally to chemoattractant concentrations of 10^{-9}–10^{-7} M (Table 3); these concentrations are in the range of

Table 3. Substances that Are Chemotactic for PMN

	Optimal concentration[a]	Source	Reference
f-Met-Leu-Phe	10^{-7} M	Bacteria	(43)
Leukotriene B_4	10^{-7}–10^{-8} M	PMN	(45)
C5a	10^{-9}–10^{-10} M	Complement proteins	(46)
Fibrinopeptide B	10^{-8} M	Fibrinogen	(47)
Neutrophil activating peptide-1/Il-8	10^{-8} M	Monocytes/EC/fibroblasts	(50)
Platelet activating factor	10^{-7} M	Granulocytes/EC/macrophages/platelets	(53)
Platelet derived growth factor	10^{-11} M	Platelets/EC/macrophages	(54)

[a]Concentration at which PMN exhibit the maximal chemotactic response in a Boyden chamber-type assay.

the dissociation constants reported for some of the chemoattractant receptors.[41] At these concentrations, PMN can sense a 2% difference in the concentration of chemoattractant across their 10-μm length. PMN exhibit locomotor morphology and random movement in the presence of a uniform concentration of chemoattractant; however, they do not move in a sustained fashion in any single direction.[42]

When a chemoattractant such as fMLP[43] is added to the basal or abluminal side of an EC monolayer, it diffuses across the monolayer and thus establishes a chemotactic gradient. In response to fMLP (10^{-7} M) placed on the abluminal side of an EC monolayer, 20% of PMN added to the apical or luminal side of such an EC monolayer adhere to and migrate across it. PMN migration across the EC monolayer occurs synchronously (within 10 min) and at random locations throughout the monolayer.[9,18] It is unclear why only 20% of PMN migrate in response to such a chemotactic gradient. This limitation in PMN response to chemoattractants has been observed in studies of PMN migration across filters as well.[44]

In contrast to fMLP, which is produced by bacteria, most other chemoattractants originate from endogenous sources (e.g., the host's own cells or circulating plasma proteins). Leukotriene B_4 is an example of an endogenously produced chemoattractant. It is generated by activated PMN[45] and may act to amplify recruitment of PMN to an inflammatory focus. The magnitude and time course of PMN migration across cultured EC monolayers in response to leukotriene B_4 (10^{-7} M) is similar to that described for fMLP.[9,18]

Other substances which are chemotactic for PMN (see Table 3) are C5a[46] and fibrinopeptide B.[47] C5a, a cleavage product of the fifth component of complement, is generated during the activation of the complement cascade and mediates PMN extravasation *in vivo*[48] and PMN adhesion to EC monolayers *in vitro*.[49] Fibrinopeptide B is a 14-amino-acid peptide produced by the action of thrombin on fibrinogen. In contrast to other chemoattractants, fibrinopeptide B induces PMN chemotaxis without stimulating release of lysosomal enzymes or superoxide anion.[47] The role of C5a and fibrinopeptide B in mediating directed PMN migration across cultured EC monolayers has not been reported.

Neutrophil activating peptide or interleukin 8 (Il-8), a 72-amino-acid peptide, is secreted by a variety of cells including monocytes,[50] EC,[51] and fibroblasts,[52] when these cells are stimulated by cytokines such as tumor necrosis factor (TNF) or interleukin 1 (Il-1). Il-8 is chemotactic for PMN.[50] Platelet activating factor (PAF)[53] and platelet-derived growth factor (PDGF)[54] are secretory products of activated platelets, EC, and macrophages, which also induce directed migration of PMN. Since these last three substances are secreted by EC, the generation of such products may represent a mechanism by which EC regulate PMN binding and migration either in a positive or a negative fashion. Recent reports indicate that Il-8 both stimulates PMN migration across EC monolayers (J. R., Gamble, personal communication) and inhibits PMN adhesion to EC.[55] There is at present no explanation for these two seemingly contradictory effects of Il-8. The effect of PDGF on PMN migration across EC monolayers has not been studied. Thus, there are a number of chemotactic substances which exert similar effects on PMN. In the present context the most important of these effects are PMN binding to and migration across vascular endothelium.

At present, we have no explanation for the diversity of receptors expressed by PMN for distinct chemoattractants. While it is possible that different chemoattractants will be shown to exert qualitatively distinct effects on PMN–EC interactions, it seems more likely that multiple chemoattractants evolved because PMN diapedesis and directed migration play so vital a role in host defense.

Not all mediators which promote PMN–EC interactions are chemoattractants for PMN. Il-1 and thrombin exert effects directly on EC as will be discussed below.[56,57] Il-1 and thrombin also may act indirectly to promote PMN–EC interactions; Il-1 enhances secretion of IL-8 from macrophages[58] and thrombin stimulates release of PDGF from platelets.[59] TNF is not itself chemotactic for PMN[60]; it promotes PMN–EC interactions by its direct and immediate effect on PMN adhesiveness and by its effect on the surface properties of EC.

B. PMN Surface Proteins

PMN have on their surfaces three glycoproteins (the CD11/CD18 family) which play an essential role in mediating white blood cell adhesion to EC (see Ref. 61 for review). Each of these proteins is a heterodimer consisting of a distinctive α subunit (CD11a, CD11b, or CD11c) and a common β subunit (CD18) (Table 4). This CD11/CD18 family of proteins belongs to a larger family of cell surface proteins, the integrins, all of which are involved in cell adhesion.[62] PMN and monocyte/macrophages express all three members of the CD11/CD18 family while lymphocytes primarily express CD11a/CD18.[61] In this review we focus on the role of the CD11/CD18 proteins in mediating PMN–EC interactions; however, each of these proteins has other functions as well (Table 4).

Infusion of antibodies to CD18 blocks PMN adhesion to vascular endothelium *in vivo*.[63] PMN deficient in CD11/CD18 proteins[64] and PMN which have been treated with antibodies to CD18[65] are markedly impaired in their ability to adhere to or migrate across EC monolayers *in vitro*. Together these findings provide convincing evidence that the CD11/CD18 proteins on PMN mediate the adhesion and migration of PMN to and across EC.

Studies of the effects of monoclonal antibodies against each of the distinctive α subunits of CD11/CD18 proteins demonstrate that both CD11a/CD18, and CD11b/CD18 are involved in adhesion of PMN to EC (Table 5). Antibodies to CD11a inhibit adhesion of chemoattractant-stimulated PMN to EC by 40–60%[66,67] whereas antibodies to CD11b inhibit such PMN–EC adhesion by 50–70%.[66–68] Antibodies against CD11a or CD11b are not as effective in inhibiting adhesion of PMN to EC as antibodies against CD18,

Table 4. CD11/CD18 Family of Leukocyte Surface Glycoproteins

Glycoprotein receptor α β	Cell distribution	Ligand	Function
CD11a/CD18	Granulocyte/monocyte/macrophage/ lymphocyte	ICAM-1	Adhesion between WBC
		ICAM-1	WBC–EC adhesion
		Endotoxin (LPS)	Bacterial cell adhesion
CD11b/CD18	Granulocyte/monocyte/macrophage	?	PMN–EC adhesion
		C3bi	Adhesion and phagocytosis
		Fibrinogen	Adhesion
		Endotoxin (LPS)	Bacterial cell adhesion
CD11c/CD18	Granulocyte/monocyte/macrophage	Fibrinogen	Adhesion
		Endotoxin (LPS)	Bacterial cell adhesion

Table 5. Inhibition of PMN-Initiated PMN–EC Adhesion by Monoclonal Antibodies against CD11/CD18 Glycoproteins

Antibody	% inhibition	References
Anti-CD18	80–90%	(65,69)
Anti-CD11a	50–70%	(66,67)
Anti-CD11b	40–60%	(66–68)
Anti-CD11c	No effect	(66)

presumably because PMN adhesion to EC is mediated by both CD11a/CD18 and CD11b/CD18. Antibodies against CD11c do not have any effect on PMN–EC adhesion.[66]

The relative importance of CD11a/CD18 and CD11b/CD18 in PMN–EC adhesion is dependent upon whether adhesion is initiated by PMN or EC (Table 6). CD11a/CD18 and CD11b/CD18 both participate when PMN–EC adhesion is initiated by PMN in response to a chemoattractant (fMLP,[67,68] C5a,[66] or TNF[66,68]) (Fig. 1A). In contrast, CD11a/CD18 but not CD11b/CD18 plays a role in PMN–EC adhesion initiated by Il-1-treated EC.[67]

Chemoattractants promote both quantitative (i.e., a 5- to 10-fold increase in their surface expression) and qualitative (i.e., phosphorylation of their cytoplasmic domains) changes in CD11/CD18 proteins on PMN. Several lines of evidence indicate that the qualitative changes in the CD11/CD18 proteins are principally responsible for the increased adhesion of chemoattractant-stimulated PMN to EC. First, agonists such as fMLP, C5a, and TNF increase the expression of CD11b/CD18 on PMN (Table 7) but there is no consistent relationship between the increase in the number of CD11b/CD18 on PMN surfaces and the number of PMN that bind to EC. Moreover, these agonists do not alter the quantity of CD11a/CD18 on the PMN surface. Second, PMN cytoplasts, which have been depleted of their intracellular storage pools of CD11/CD18 proteins,[66,78] and PMN treated with the anion channel blocking agent, DIDS,[78] which are unable to increase their surface expression of CD11b/CD18 and cD11c/CD18, adhere as well as untreated PMN to EC in response to agonists. Third, PMN adhesion to EC is a transient phenomenon even in the continuous

Table 6. Inhibition of PMN–EC Adhesion by Monoclonal Antibodies against CD11a and CD11b

Agonist	Anti-CD11a	Anti-CD11b	References
PMN-initiated PMN–EC adhesion			
Tumor necrosis factor	+	+	(66,68)
C5a	+	+	(66)
f-Met-Leu-Phe	+	+	(67,68)
Phorbol esters	+	+	(66,68)
Leukotriene B_4	?	+	(68)
Platelet activating factor	?	+	(68)
EC-initiated PMN–EC adhesion			
Interleukin 1	+	−	(67)

Table 7. Effects of Agonists on Expression of CD11/CD18 on PMN

Agonist	CD11a	CD11b	CD11c	CD18	References
f-Met-Leu-Phe	—[a]	↑ 5–10×	↑ 5–10×		(70,71)
C5a	—	↑ 2–10×	↑ 5–10×		(66,70)
Tumor necrosis factor	—	↑ 2–3×	?		(66,72,73)
Phorbol esters	—	↑ 2–3×	?		(66)
Interleukin 1	?	—	?	↑ 3–5×	(74,75)
Interleukin 8	?	↑ 2×	?		(76)
Granulocyte macrophage colony stimulating factor	?	↑ 2×	?		(77)

[a]No change in expression from that under resting conditions.

presence of an agonist.[66] PMN adhesion to EC declines to baseline levels after 30 min of agonist stimulation. However, increased PMN expression of CD11b/CD18 persists at this time. Fourth, T lymphocytes express CD11a/CD18, but not CD11b/CD18 or CD11c/CD18. Activation of these lymphocytes causes no increase in CD11a/CD18 expression but promotes their adhesion to the cell surface protein, intercellular adhesion molecule 1 (ICAM-1).[79] These findings are consistent with the hypothesis that qualitative changes in PMN surface CD11/CD18 proteins are principally responsible for the enhanced adhesion of agonist-stimulated PMN to EC. These qualitative changes could be alterations in the conformation of individual CD11/CD18 proteins in response to phosphorylation of their cytoplasmic domains,[80,81] and/or the redistribution (clustering) of these proteins in PMN surface membranes.[82]

VI. EC-INITIATED MECHANISMS OF PMN MIGRATION ACROSS ENDOTHELIUM

In the last 5 years a second mechanism by which PMN migration across an endothelium occurs has been identified. EC, rather than PMN, initiate this process.

A. EC Response to Cytokines

EC treated with either Il-1[56] or TNF[72] promote PMN–EC adhesion and transendothelial migration of PMN commensurate with that induced by chemotactic agents such as fMLP.[75,83] Despite the apparent lack of a gradient of soluble chemoattractant, PMN travel from the luminal or apical side of Il-1- or TNF-treated EC monolayers toward the abluminal or basal side, but not in the opposite direction.[75] Il-1 is secreted by a wide variety of cells which may interact with EC including macrophages and glial cells as well as by EC themselves.[84,85] In addition to promoting PMN–EC interactions, Il-1 induces a number of other changes in EC including the expression of procoagulant activity[85,86] and the secretion of a number of biologically active extracellular products.[87] TNF exerts many of the same effects as Il-1 on EC but is synthesized primarily by macrophages[88,89] and T lymphocytes.[90]

B. EC Surface Proteins

EC exhibit qualitative and quantitative changes in expression of surface proteins in response to stimulation by Il-1 and/or TNF. A qualitative change is the *de novo* synthesis and expression of endothelial leukocyte adhesion molecule 1 (ELAM-1).[2] A quantitative change is increased expression of ICAM-1.[91] Both molecules mediate PMN–EC adhesion.

ELAM-1 is a glycoprotein whose molecular mass varies from 78,000 to 115,000 daltons depending on its degree of glycosylation. It has a single membrane-spanning domain. Its extracellular segment contains three domains: an outermost lectinlike domain, a middle region homologous to epidermal growth factor, and an innermost region homologous to the repetitive sequences found in the external domain of complement receptors 1 and 2.[92] ELAM-1 is similar in structure to the lymph node homing receptor found on T and B lymphocytes that is recognized by the monoclonal antibody MEL-14,[93] and to granule membrane protein 140 found in secretory granules in platelets and in Weibel–Palade bodies in EC.[94] ELAM-1, the lymph node homing receptor, and granule membrane protein 140 are members of a new family of phylogenically related proteins. These three proteins bear no resemblance to the CD11/CD18 family of leukocyte surface proteins, although all of these proteins mediate adhesion between cells.

ELAM-1 is not found on unstimulated EC; its expression is induced by treatment of EC with Il-1, TNF, or endotoxin. ELAM-1 first becomes detectable on the surfaces of EC 30 min after addition of the inducing agent,[2] peaks at 4–6 hr, and returns to baseline after 24 hr even when the EC are maintained continuously in the presence of Il-1 or TNF.[95] PMN adhesion to EC correlates with ELAM-1 expression on the EC surface.[2] Surface expression of ELAM-1 by COS cells transfected with DNA encoding ELAM-1 also correlates with the ability of PMN to bind to the transfected cells.[92] The surface molecule on PMN which interacts with ELAM-1 on EC has not been identified. We refer to this molecule as "PMN_X" for the purposes of this discussion.

As noted above, ELAM-1 mediates PMN adhesion to EC. However, antibodies directed against ELAM-1 inhibit only 50% of the binding of PMN to Il-1- or TNF-treated EC.[2] These data suggest that EC surface protein(s) other than ELAM-1 also are involved in the attachment of PMN to Il-1- or TNF-treated EC.

Another EC surface protein which mediates PMN–EC adhesion is ICAM-1. ICAM-1 on the EC surface binds CD11a/CD18 on lymphocyte surfaces[96] and most likely to CD11a/CD18 on PMN surfaces as well. It also serves as a binding site for malaria (*P. falciparum*)-infected erythrocytes.[97] ICAM-1 is a member of the immunoglobulin family of proteins and therefore represents a third class of surface proteins which mediate cell adhesion between PMN and EC. Unlike the CD11/CD18 family of glycoproteins which is restricted to leukocytes, and ELAM-1 which has been identified only on EC, ICAM-1 has a wide tissue distribution. In addition to EC and leukocytes, it is expressed on fibroblasts and epithelial cells.

ICAM-1 is constitutively expressed on EC but its density on the cell surface varies according to the location and size of the vessel.[96,98] In response to Il-1, TNF, or interferon γ, ICAM-1 expression on EC increases within 1 hr, reaches a peak of 30- to 40-fold above baseline values at 8 hr, and remains elevated for at least 7 days.[91,96] Antibodies directed against ICAM-1, like antibodies directed against ELAM-1, inhibit adhesion of PMN to EC.

However, the extent of this inhibition is dependent on the mechanism used to initiate PMN–EC adhesion, and is discussed more fully below. These findings indicate that ICAM-1 plays a central role in the attachment of PMN to EC.

The observations that interferon γ increases EC ICAM-1 expression but not PMN adhesion to EC, and that PMN adhesion to EC declines to baseline at a time when ICAM-1 expression remains elevated, confirm that binding of PMN to EC is not a straightforward function of the number of ICAM-1 on the EC surface. Efforts to sort out the relationship between ICAM-1 expression and PMN adhesion are complicated by the fact that at least two different EC surface proteins, ELAM-1 and ICAM-1, mediate PMN adhesion to EC and the expression of both ELAM-1 and ICAM-1 on EC is regulated by common agonists, namely Il-1 and TNF.

PMN migration across an EC monolayer can be initiated by at least two different pathways: by PMN which have been stimulated with chemoattractants or by EC which have been stimulated with cytokines. The pathway by which PMN–EC interaction is initiated determines the EC surface proteins to which the PMN bind. When PMN–EC adhesion is induced by the PMN-initiated pathway, antibodies directed against ICAM-1 block PMN adhesion to EC almost completely ($\geq$99%), indicating that PMN receptors bind exclusively to ICAM-1 on the EC under these circumstances.[74] That ICAM-1 on the EC surface is the principal binding site for CD11/CD18 proteins on PMN is supported by two observations. First, antibodies directed against CD18 on PMN or antibodies directed against ICAM-1 on EC each result in virtually complete ($\geq$90%) inhibition of PMN–EC adhesion. Second, antibodies directed against ICAM-1 and CD18 together do not result in any greater inhibition of adhesion than either antibody alone.[74]

In contrast, when PMN attachment to EC is initiated by the EC pathway, antibodies directed against either ELAM-1[2] or ICAM-1[74] block PMN adhesion only partially ($\approx$50%) (The effect of treating EC with antibodies against ELAM-1 and ICAM-1 in combination on PMN–EC adhesion has not been reported.) Thus, PMN appear to interact with both ELAM-1 and ICAM-1 on cytokine-treated EC.

When PMN–EC adhesion is initiated by PMN and there is no direct stimulus to the EC, ELAM-1 is not expressed on EC. Under these conditions, CD11/CD18 proteins on the PMN surface are activated by chemoattractants and bind to constitutively expressed ICAM-1 on the EC surface. In situations where PMN–EC adhesion is initiated by EC, however, ELAM-1 expression on the EC surface is induced. Binding of ELAM-1 to "PMN_X" may in turn activate PMN CD11/CD18 proteins to bind to EC ICAM-1. Such a mechanism has been described in which ligation of the T cell receptor transiently increases the adhesiveness of CD11a/CD18 on these cells for ICAM-1.[79] Il-1 and TNF also may directly increase the adhesiveness of EC ICAM-1 for CD11a/CD18. Thus, both ELAM-1–"PMN_X" and ICAM-1–CD11a/CD18 systems participate in EC-initiated PMN–EC adhesion.

C. EC Surface-Associated PAF

In addition to the EC surface molecules ELAM-1 and ICAM-1, another EC surface-associated molecule which mediates PMN–EC adhesion is EC-derived PAF. Treatment of EC with thrombin,[57] leukotriene C_4, and leukotriene D_4[99] results in PMN–EC adhesion by a mechanism distinct from that mediated by the cytokines, Il-1 or TNF, described in the preceding section. Thrombin-, leukotriene C_4-, and leukotriene D_4-mediated PMN–EC adhesion occurs almost immediately and peaks within 5 min after exposure of EC to each of these agonists. This rapid-onset PMN–EC binding is mediated by PAF which is synthesized by and transiently expressed on the surfaces of EC treated with these agonists.[1] Whether EC surface-associated PAF promotes PMN transit across EC monolayers in addition to promoting PMN adhesion has not been reported.

Thus, there are at least two mechanisms by which EC initiate PMN–EC adhesion: (1) a cytokine-initiated mechanism which requires EC protein synthesis, involves the EC surface proteins, ELAM-1 and ICAM-1, and occurs 2–4 hr after EC exposure to cytokines; (2) a thrombin-, leukotriene C_4-, or leukotriene D-initiated mechanism which involves an EC surface-associated lipid, PAF, and occurs within minutes after EC stimulation.

D. EC Secretory Products

When appropriately stimulated (e.g., by endotoxin), EC are also capable of secreting a number of substances which act in an autocrine fashion to regulate PMN adhesion and migration. Among these is Il-1, which promotes PMN adhesion to and migration across endothelia as described above. Other EC secretory products which regulate PMN–EC interactions include Il-8,[51] PDGF,[100] and PAF,[101] all of which are chemotactic for PMN as discussed previously.

VII. PERMEABILITY STUDIES

Leukocyte extravasation during an inflammatory response is generally accompanied by an increase in vascular permeability to fluids and macromolecules.[10] The role of PMN in mediating these permeability changes is controversial. Some investigators report that edema formation does not occur in animals depleted of PMN[102,103] while others report contradictory findings.[29] Yet others report that PMN emigration from the vasculature occurs without evidence of tissue injury[104,105] or increased albumin leakage.[106,107]

A. Difficulties with Studies of EC Monolayer Permeability

Studies employing monolayers of EC have not entirely resolved the controversy regarding the role of PMN in mediating changes in permeability of endothelia for the following reasons (Table 8): (1) Different tracers have been used to assess EC monolayer permeability including water,[22] ions,[9] albumin,[108] Evans blue,[109] inulin, sorbitol,[110] and low-density lipoproteins.[111] The pathway(s) by which these molecules cross EC monolayers are not entirely clear. Some of these markers (e.g., ions) may passively diffuse through the inter-endothelial cell junctions[112]; others (e.g., low-density lipoproteins, albumin) may bind to EC surface receptors and be transcytosed across the endothelium.[113–115] Others (e.g., transferrin) may be actively transported by EC or utilize more than one of these pathways. Therefore, investigators utilizing different markers may in fact be measuring different phenomena. Agonists may have differential effects on each of these processes. In support of this concept, monocyte migration across EC monolayers is reported to increase low-density lipoprotein but not ion flux across these monolayers.[116] (2) Different types of

Table 8. Effect of Agonists on the Permeability of Endothelial Monolayers

Agonist	Type of EC/substrate	Effect on flux	Marker	References
Histamine	HUVEC[a]/filters	↑	Albumin	(118)
	HUVEC/amnion	↑	Ions	(9)
	HUVEC/beads	↑	Evans blue	(109)
	HUAEC[b]/filters	↑	LDL[c]	(111)
	BMEC[d]/amnion	↑	H_2O	(22)
	BAEC[e]/amnion	↑	Albumin	(119)
	BAEC/beads	↑	Albumin	(120)
Thrombin	HUVEC/beads	↑	Evans blue	(109)
	HUAEC/filters	↑	Ions, LDL	(111)
	BAEC/filters	↑	Albumin	(121)
	BPAEC[f]/filters	↑	Albumin	(26)
Leukotriene C_4	HUVEC/amnion	↑	Ions	(122)
	BAEC/filters	No effect	Albumin	(108)
	BPAEC/filters	No effect	Albumin	(123)
Serotonin	HUVEC/amnion	↑	Ions	(122)
	HUVEC/beads	No effect	Evans blue	(109)
	BMEC/amnion	↑	H_2O	(22)
	BAEC/amnion	↓	Albumin	(119)
	BAEC/beads	↓	Albumin	(120)
Platelet activating factor	HUVEC/filters	↑	Albumin	(124)
	BAEC/filters	No effect	Albumin	(108,125)
	BPAEC/filters	No effect	Albumin	(126,127)
Tumor necrosis factor	HUVEC/amnion	No effect	Ions	(83)
	HUVEC/filters	No effect	Albumin	(75)
	BAEC/filters	↑	Albumin, inulin, sorbitol	(110)
Bradykinin	HUVEC/beads	No effect	Evans blue	(109)
	HUAEC/filters	No effect	Ions, LDL	(111)

[a]Human umbilical vein endothelial cells.
[b]Human umbilical artery endothelial cells.
[c]Low-density lipoprotein.
[d]Bovine microvascular endothelial cells.
[e]Bovine aortic endothelial cells.
[f]Bovine pulmonary artery endothelial cells.

EC have been used in these permeability studies. Since permeability changes occur primarily in post capillary venules *in vivo*,[10] it may be that the capability to respond to vasoactive substances is restricted to EC originating from a particular vascular segment. This capability may be conferred by the presence on EC of receptors for specific agonists, the ability of EC from different vascular segments to secrete compounds which act in an autocrine manner to increase permeability of EC monolayers, or the expression of the cellular machinery to effect such permeability changes by certain types of EC. Although the vasoactive substances listed in Table 8 are found in most animal species, their ability to mediate changes in vascular permeability may vary in different species. For example, *in vitro* studies utilizing human EC monolayers fail to demonstrate a response to bradykinin. Perhaps this is not surprising since the role of this substance in mediating increases in vascular permeability in humans is not well documented. (3) A variety of substrates have been used to support EC monolayers used for permeability studies. Since an extracellular matrix can influence EC phenotype,[8,117] it may affect the ability of a particular type of EC to respond to agonists.

These difficulties with EC permeability studies are illustrated by the lack of consensus regarding the ability of soluble substances such as leukotriene C_4, serotonin, PAF, TNF, or bradykinin to increase the permeability of cultured EC monolayers (Table 8). All of these compounds are reported to increase vascular permeability *in vivo* (Table 2). Until more is known about the mechanisms or routes by which different markers of permeability traverse EC monolayers, the properties which distinguish EC originating from different vascular segments, and the effect of extracellular matrices on EC phenotype, it will be difficult to definitively establish the role of soluble substances or PMN in regulating EC monolayer permeability *in vivo*.

Increased vascular permeability also must be differentiated from vasodilatation. Although these two phenomena frequently accompany one another, they are differentially regulated. Substances such as nitric oxide, also known as endothelium-derived relaxing factor,[128] and prostaglandin E_2[129] mediate vasodilatation but there is no evidence to date that these latter substances directly mediate an increase in leakage of water and proteins from the vasculature.

B. Role of PMN in Mediating Increases in EC Monolayer Permeability

Despite the difficulties discussed above, *in vitro* studies demonstrate that PMN are not an absolute requirement for increases in permeability across an EC monolayer since substances such as histamine and thrombin (Table 8) increase permeability of EC monolayers in the absence of PMN. Although PMN may not be required to effect an increase in vascular permeability, under certain conditions they may mediate such permeability increases. Studies utilizing monolayers of cultured EC indicate that PMN can mediate an increase in albumin flux across an EC monolayer when they are stimulated by agonists placed on the same side of the EC monolayer.[108,130] Under such conditions, large numbers of PMN aggregate on the apical surface of the EC monolayer resulting in physical disruption of the monolayer. Ultrastructural studies of directed migration of PMN across endothelia *in vivo*[34] and *in vitro*[18] indicate that this does not necessarily occur. Studies of EC permeability *in vitro* confirm that increased ion and protein flux across an EC monolayer are not necessary consequences of directed PMN migration across such a monolayer but can result if PMN are present in sufficiently large numbers (PMN:EC ratio $\geq$ 25:1).[9] The small caliber of microvessels, where PMN extravasation occurs *in vivo*, makes it unlikely that such high PMN:EC ratio exist under physiologic conditions.

We measured the effect of transendothelial PMN migration in response to a chemoattractant gradient on the permeability of EC monolayers. PMN, at a PMN:EC ratio of 5:1, were added to the luminal side of EC monolayers and allowed to migrate in response to the chemoattractants fMLP or leukotriene B_4. We measured transendothelial electrical resistance (ionic permeability) or radiolabeled albumin flux during this process. Under these conditions, PMN migration was synchronous and rapid, resulted in one PMN per EC traversing the monolayer, but induced no change in the monolayer's permeability to ions or albumin. Thus, increased permeability of an EC monolayer is not a necessary consequence of PMN migration across such a monolayer. The increased permeability of endothelia which accompanies PMN–EC interactions *in vivo* may require the presence of additional factors such as changes in hydrostatic pressure, other cell types, or soluble mediators.

Since electrical resistance of a cellular monolayer is a measure of ions traversing the paracellular pathway and reflects the size of this pathway,[112] we studied the effect of PMN migration on the dimensions of the paracellular pathway. We measured the transit time and diameter of PMN as they traversed EC monolayers, as well as the closeness of apposition between PMN and the EC. Based on this information, we calculated the changes in size of the paracellular pathway expected to occur during transendothelial PMN migration. We used these data to generate a series of curves that predicted the change in electrical resistance that should occur as PMN traverse an EC monolayer (Fig. 5). Electron micrographs show very close apposition (≤150 Å) between PMN and EC during transendothelial PMN migration. Based on this measurement and the short time interval required for PMN to traverse EC monolayers (average of about 1 min), the size of the paracellular pathway should increase very little (≤0.22%). A change in resistance due to the presence of PMN in the EC intercellular spaces (ratio of 1 PMN per EC) would thus not be predicted. Only if ≥4 times the number of PMN than were actually observed are present in the intercellular spaces at a given time (i.e., 75% of the intercellular spaces occupied by PMN) or if the distance separating the PMN and EC is ≥300 Å (i.e., ≥2 times the distance observed), would a decrease in transendothelial cell resistance be predicted.

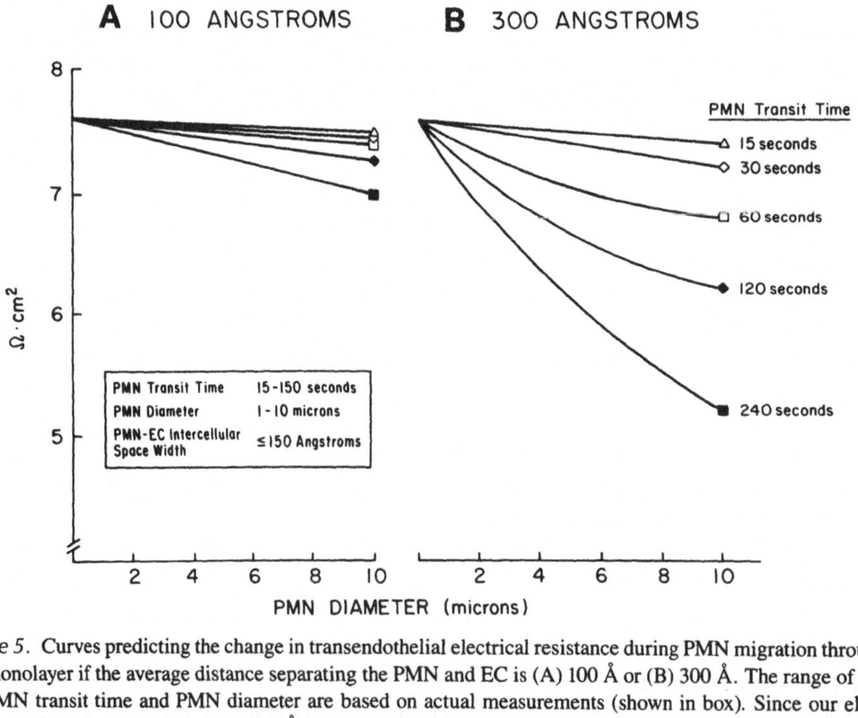

Figure 5. Curves predicting the change in transendothelial electrical resistance during PMN migration through an EC monolayer if the average distance separating the PMN and EC is (A) 100 Å or (B) 300 Å. The range of values for PMN transit time and PMN diameter are based on actual measurements (shown in box). Since our electron micrographs indicate a distance ≤150 Å between PMN and EC, no measurable change in resistance would be predicted (A). Only if this distance is ≥300 Å or twice the distance observed (B), or if ≥4 times the number of PMN than were actually observed are in the intercellular spaces at a given time (■), would a decrease in transendothelial electrical resistance be predicted.

A comparable number of PMN migrating across epithelial cell monolayers that exhibit high electrical resistance ($\geqslant$500 $\Omega \cdot cm^2$) has been reported to decrease, but not to abolish, the electrical resistance of these monolayers.[131,132] As observed in studies of EC monolayers, such permeability changes are not detected during PMN migration across epithelial monolayers which exhibit low electrical resistance ($\leqslant$150 $\Omega \cdot cm^2$).[131,133] These observations are consistent with our analysis of the dimensions of the paracellular pathway during transendothelial PMN migration. High-resistance epithelial monolayers are 100-fold less permeable to ions in their resting state than EC monolayers ($\geqslant$500 versus 6–12 $\Omega \cdot cm^2$ resistance, respectively). The high electrical resistance of these epithelial monolayers is due to the presence of complex occluding junctions. These "tight" junctions are disrupted transiently during PMN migration. The close apposition of PMN and epithelial cells under these circumstances cannot compensate for the disruption of occluding junctions between epithelial cells induced by the migrating PMN, so resistance falls. However, resistance does not fall to zero because the close apposition between migrating PMN and epithelial cells remains a significant barrier to the free flow of ions. In contrast, during PMN migration through monolayers of relatively low resistance, close apposition between migrating PMN and epithelial cells is sufficient to maintain the original permeability of the monolayer.

VIII. MECHANISM BY WHICH PMN OPEN INTER-EC JUNCTIONS

Although there is a great deal of information regarding the molecules that mediate PMN adherence to EC, the mechanism(s) by which PMN open the junctions between epithelial or EC cells remain unresolved.

A. PMN Secretory Products

Chemoattractant-stimulated PMN secrete oxygen radicals, lipid metabolites, and proteases.[134] It is evident that reactive metabolites of oxygen are not required to effect PMN extravasation. PMN from patients with chronic granulomatous disease are unable to make these metabolites but nonetheless form pus in response to infections. EC treated with hydrogen peroxide bind more PMN than control EC,[135] but the concentrations of hydrogen peroxide which result in enhanced PMN–EC adhesion cause lethal EC injury.[135,136] There is no evidence of such EC injury either ultrastructurally[18,34] or functionally (no change in permeability)[9] during PMN migration across an endothelium.

Lipid products of stimulated PMN, such as leukotriene B_4 and PAF, are themselves chemoattractants. Their release by PMN may be an amplification mechanism for attracting additional PMN to an inflammatory focus. Whether other PMN lipid metabolites (e.g., thromboxanes or leukotriene A_4) play any role in PMN extravasation is not clear.

PMN elastase activity is not required for PMN adhesion or migration. We have directly measured the elastase activity of PMN chemotaxing across filters. AAPVCK, a specific low-molecular-weight inhibitor of elastase, inhibits $\geqslant$99% of the elastase activity of these PMN but has no effect on PMN migration.[137] AAPVCK also has no effect on PMN migration across EC monolayers.[18]

The plasminogen activator–plasmin system also is not required for transendothelial PMN migration. PMN cross EC monolayers equally well in the presence of inhibitors of

PMN plasminogen activator (urokinase) and in serum depleted of plasminogen and plasmin. Plasmin has been shown to promote PMN–EC adhesion, but it does so by virtue of its lysine binding (negatively charged) sites and not by its proteolytic activity.[138]

While we cannot exclude the possibility that proteases released by PMN, other than elastase and plasminogen activator, degrade junctions between EC, thereby facilitating PMN diapedesis, their participation in this way seems unlikely for the following reasons. First, monolayers of epithelial cells whose intercellular junctions have been degraded by proteases do not reexpress their junctions or regain electrical resistance for 18–24 hr.[139] Second, cathepsin G, a cationic protease released from azurophilic granules of stimulated PMN, induces significant gap formation between EC comprising a monolayer and a 2- to 5-fold increase in albumin flux across such EC monolayers.[140] In contrast to these observations, the permeability of EC monolayers is unaffected by PMN migration across them.[9]

B. A Unifying Hypothesis

While it is not possible to rule out a role for PMN secretory products or plasma proteins in the transendothelial movement of PMN from the data presented above, it is clear that EC in monolayers can respond to soluble mediators by retracting from one another, thereby increasing the permeability of the monolayer to water and macromolecules (see Table 8). This response requires binding of the soluble mediator, whether it be histamine or thrombin, to a corresponding EC surface receptor. Ligation of histamine receptors on EC monolayers results in increases in cytosolic free calcium ($[Ca^{2+}]_i$) and F-actin on the EC, increased transendothelial albumin flux,[118] and decreased transendothelial electrical resistance.[9] Morphologic studies *in vivo* demonstrate that changes in EC shape and opening of inter-endothelial cell junctions accompany histamine-induced increases in vascular permeability to fluid and macromolecules.[141]

PMN binding to, and migration across, endothelia induces a sequence of events which resemble those that occur following histamine treatment of EC in many respects. PMN adhesion to EC involves the ligation of specific EC surface molecules (e.g., ICAM-1), by PMN surface proteins (e.g., CD11a/CD18). Morphologic studies demonstrate that changes in EC shape also accompany PMN adhesion to and migration across an endothelium.[18,34] Therefore, we hypothesized that PMN–EC interactions may initiate intracellular signals similar to those which are initiated when histamine binds to its receptors on EC, namely the generation of an EC $[Ca^{2+}]_i$ transient and the opening of junctions between EC. This would allow leukocytes to insert their pseudopods between EC, thereby initiating their passage across the endothelium.

To address these issues, we have measured $[Ca^{2+}]_i$ in single EC maintained in intact monolayers. PMN migration across these EC monolayers in response to a chemoattractant induces a 2- to 5-fold increase in EC $[Ca^{2+}]_i$.[142] This increase in EC $[Ca^{2+}]_i$ temporally corresponds with the period of PMN migration across the EC monolayer; it begins shortly after addition of the chemoattractant and returns to resting levels within 15 min, a time at which PMN transendothelial migration is complete. Chemoattractant-stimulated PMN are required to induce this increase in $[Ca^{2+}]_i$; neither unstimulated PMN, nor fMLP, in the absence of PMN induce a significant change in EC $[Ca^{2+}]_i$. These findings indicate that EC respond to contact with PMN surface proteins and/or PMN secretory products. They suggest that EC are active participants in the movement of PMN across endothelia.

IX. SUMMARY AND CONCLUSIONS

Leukocyte extravasation is of sufficient importance to the survival of mammals that multiple agonists (Table 9) and at least two temporally and mechanistically distinct pathways have evolved for effecting it. Leukocyte extravasation can be initiated by PMN or by EC, as illustrated in Fig. 6. Under physiologic circumstances, it is likely that regardless of the pathway by which leukocyte extravasation first occurs, both PMN- and EC-initiated pathways ultimately participate in the exodus of PMN from the blood into the extravascular tissues.

In the PMN-initiated pathway (Fig. 6A), chemotactic factors which are generated rapidly (seconds to minutes) by exogenous pathogens (e.g., fMLP) or endogenously by the animal's own cells (e.g, leukotriene B_4) or plasma (e.g., C5a) stimulate PMN to migrate across postcapillary venular endothelia. Binding of PMN to EC is mediated, at least in part, by a qualitative change in two PMN surface proteins, CD11a/CD18 and CD11b/CD18. At present, the role of PMN secretory products in mediating PMN diapedesis is not clear.

PMN traffic across an endothelium also can be regulated by an EC-initiated pathway (Fig. 6B). Cytokines, such as Il-1 and TNF, stimulate PMN binding to EC by inducing expression of an EC surface protein, ELAM-1, and by enhancing the adhesion between ICAM-1 on EC and CD11a/CD18 on PMN. EC may also regulate PMN traffic by secreting substances such as PAF which enhance PMN adhesion to EC and/or PMN migration across endothelia.

Increased permeability of EC monolayers can occur in the absence of PMN since substances such as histamine and thrombin increase the permeability of cultured EC monolayers. Increased permeability of EC monolayers to ions or albumin also is not a necessary consequence of PMN migration across such a monolayer. However, at very high ratios of chemoattractant-stimulated PMN:EC, increased permeability of EC monolayers does occur. Unresolved issues include whether at physiological PMN:EC ratios that occur *in vivo*, additional factors such as changes in hydrostatic pressure, the presence of other cell types, or soluble mediators must accompany PMN emigration to alter the permeability of endothelia.

Interaction of chemoattractant-stimulated PMN with EC induces a transient increase in EC $[Ca^{2+}]_i$. Since both PMN-initiated and EC-initiated mechanisms of PMN migration across an EC monolayer involve the ligation of specific EC surface receptors, we suggest

Table 9. Principal Cellular Target[a] of Biological Mediators of PMN–EC Interactions

	EC	PMN
Interleukin 1, thrombin	+	−
Tumor necrosis factor	+	+
f-Met-Leu-Phe, C5a, leukotriene B_4	−	+
Interleukin 8, platelet activating factor	?	+

[a]Cell having a receptor for the mediator.

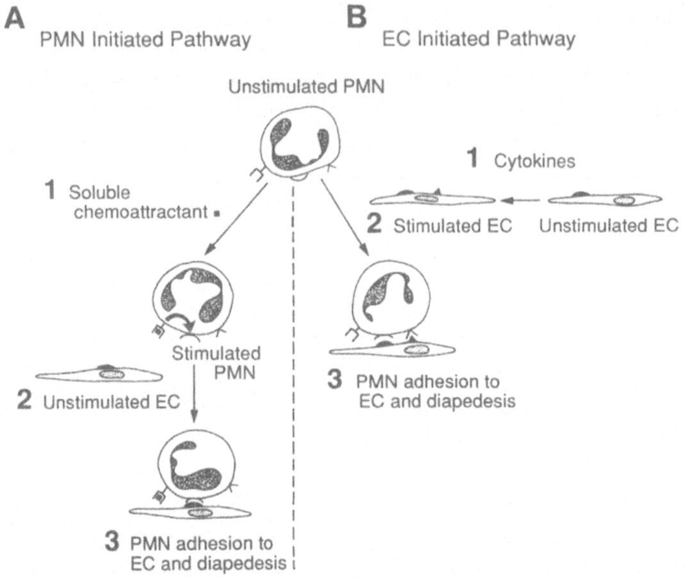

Figure 6. PMN- and EC-initiated pathways of PMN migration across EC. (A) In the PMN-initiated pathway, soluble chemoattractants bind to specific receptors on PMN, activate PMN surface proteins (e.g., CD11a/CD18) to bind to constitutively expressed proteins on the EC surface (e.g., ICAM-1) (solid semicircles), and promote PMN adhesion to EC and diapedesis. (B) In the EC-initiated pathway, cytokines such as Il-1 and TNF induce the expression of a new EC surface molecule (e.g., ELAM-1) (solid triangles). Binding of this induced EC surface molecule to its cognate PMN surface molecule (Λ) may activate a second PMN surface molecule (e.g., CD11a/CD18) to bind its corresponding EC surface protein (e.g., ICAM-1). Cytokines also may directly activate EC surface proteins (e.g., ICAM-1) to bind to proteins on the PMN surface (CD11a/CD18). Thus, when initiated by EC, at least two sets of PMN–EC surface proteins participate in PMN adhesion to EC and diapedesis.

that PMN binding to EC mediated by the EC-initiated pathway also induces an increase in EC $[Ca^{2+}]_i$. We hypothesize that this change in EC $[Ca^{2+}]_i$ reflects one of several signals that migrating PMN transmit to EC to induce the EC to change shape and to open their intercellular junctions, thereby facilitating PMN passage across endothelia. According to this view, EC play a major role in regulating not only the attachment of leukocytes to their surfaces, but also the movement of leukocytes across endothelia.

The hypothesis we have proposed for the mechanism by which migrating PMN induce the local and selective opening of junctions between EC is not restricted to PMN and EC. We think it may be applicable generally. The same mechanism may be used by PMN and other leukocytes as they traverse epithelia, by embryonic cells as they migrate through cellular layers to their ultimate anatomic destination, and by tumor cells as they spread throughout the body.

REFERENCES

1. Zimmerman, G. A., McIntyre, T. M., Mehra, M., and Prescott, S. M., 1990, Endothelial cell-associated platelet-activating factor: A novel mechanism for signaling intercellular adhesion, *J. Cell Biol.* **110:**529–540.

2. Bevilacqua, M. P., Pober, J. S., Mendrick, D. L., Cotran, R. S., and Gimbrone, M. A., 1987, Identification of an inducible endothelial–leukocyte adhesion molecule, *Proc. Natl. Acad. Sci. USA* **84:**9238–9242.
3. Heltianu, C., Simionescu, M., and Simionescu, N., 1982, Histamine receptors of the microvascular endothelium revealed in situ with a histamine–ferritin conjugate: Characteristic high affinity binding sites in venules, *J. Cell Biol.* **93:**357–364.
4. Olesen, S. P., and Crone, C., 1983, Electrical resistance of muscle capillary endothelium, *Biophys. J.* **42:** 31–41.
5. Palade, G. E., Simionescu, M., and Simionescu, N., 1979, Structural aspects of the permeability of the microvascular endothelium, *Acta Physiol. Scand. Suppl.* **463:**11–32.
6. Taylor, R. F., Price, T. H., Schwartz, S. M., and Dale, D. C., 1981, Neutrophil–endothelial cell interactions on endothelial monolayers grown on micropore filters, *J. Clin. Invest.* **67:**584–587.
7. King, G. L., and Johnson, S. M., 1985, Receptor-mediated transport of insulin across endothelial cells, *Science* **227:**1583–1586.
8. Furie, M. B., Cramer, E. B., Naprstek, B. L., and Silverstein, S. C., 1984, Cultured endothelial monolayers that restrict the transendothelial passage of macromolecules and electrical current, *J. Cell Biol.* **98:**1033–1041.
9. Huang, A. J., Furie, M. B., Nicholson, S. C., Fischbarg, J., Leibovitch, L. S., and Silverstein, S. C., 1988, Effects of human neutrophil chemotaxis across human endothelial cell monolayers on the permeability of these monolayers to ions and macromolecules, *J. Cell. Physiol.* **135:**355–366.
10. Zweifach, B. W., 1973, Microvascular aspects of tissue injury, in: *The Inflammatory Process*, Vol. 2 (B. W. Zweifach, L. Grant, and R. T. McCluskey, eds.), Academic Press, New York, pp. 3–46.
11. Cereijido, M., Robbins, E. S., Dolan, W. J., Rotunno, C. A., and Sabatini, D. D., 1978, Polarized monolayers formed by epithelial cells on a permeable and translucent support, *J. Cell Biol.* **77:**853–880.
12. Crone, C., and Christensen, O., 1981, Electrical resistance of a capillary endothelium, *J. Gen. Physiol.* **77:**349–371.
13. Olesen, S. P., Saint-Aubain, M. L. de, and Bundgaard, M., 1984, Permeabilities of single arterioles and venules in the frog skin: A morphologic study, *Microvasc. Res.* **28:**1–22.
14. Territo, M., Berliner, J. A., and Fogelman, A. M., 1984, Effect of monocyte migration on low density lipoprotein transport across aortic endothelial cell monolayers, *J. Clin. Invest.* **74:**2279–2284.
15. Shasby, D. M., and Shasby, S. S., 1986, Effects of calcium on transendothelial albumin transfer and electrical resistance, *J. Appl. Physiol.* **60:**71–79.
16. Crone, C., and Olesen, S. P., 1982, Electrical resistance of brain microvascular endothelium, *Brain Res.* **241:**49–55.
17. Rutten, M. J., Hoover, R. L., and Karnovsky, M. J., 1987, Electrical resistance and macromolecular permeability of brian endothelial monolayer cultures, *Brain Res.* **425:**301–310.
18. Furie, M. B., Naprstek, B. L., and Silverstein, S. C., 1987, Migration of neutrophils across monolayers of cultured microvascular endothelial cells: An in vitro model of leukocyte extravasation, *J. Cell Sci.* **88:** 161–175.
19. Taylor, A. E., and Granger, D. N., 1984, Exchange of macromolecules across the microcirculation, in: *Handbook of Physiology: The Cardiovascular System*, Vol. 4 (E. M. Renkin and C. C. Michel, eds.), Williams & Wilkins, Baltimore, pp. 467–520.
20. Michel, C. C., 1984, Fluid movements through capillary walls, in: *Handbook of Physiology: The Cardiovascular System*, Vol. 4 (E. M. Renkin and C. C. Michel, eds.), American Physiological Society, Bethesda, pp. 375–409.
21. Fischbarg, J., Warshavsky, C. R., and Lim, J. J., 1977, Pathways for hydraulically and osmotically induced water flows across epithelium, *Nature* **266:**71–74.
22. Luckett, P. M., Fischbarg, J., Bhattacharya, J., and Silverstein, S. C., 1989, Hydraulic conductivity of endothelial cell monolayers cultured on human amnion, *Am. J. Physiol.* **256:**H1675–H1683.
23. Lee, J. S., Smaje, L. H., and Zweifach, B. W., 1971, Fluid movement in occluded single capillaries of rabbit omentum, *Circ. Res.* **28:**358–370.
24. Shasby, D. M., and Peterson, W. P., 1987, Effects of albumin concentration on endothelial albumin transport in vitro, *Am. J. Physiol.* **253:**H654–H661.
25. Majno, G., and Palade, G. E., 1961, Studies on inflammation I—The effect of histamine and serotonin on vascular permeability: An electron microscopic study, *J. Biophys. Biochem. Cytol.* **11:**571–605.

26. Garcia, J. G. N., Siflinger-Birnboim, A., Bizios, R., DelVecchio, P. J., Fenton, J. W. II, and Malik, A. B., 1986, Thrombin-induced increase in albumin permeability across the endothelium, *J. Cell. Physiol.* **128:** 96–104.
27. Dahlen, S. E., Bjork, J., Hedqvist, P., Arfors, K. E., Hammarstrom, S., Lindgren, J. A., and Sammuelson, B., 1981, Leukotrienes promote plasma leakage and leukocyte adhesion in post-capillary venules: In vivo effects with relevance to the acute inflammatory response, *Proc. Natl. Acad. Sci. USA* **78:**3887–3891.
28. Humphrey, D. M., McManus, L. M., Satouchi, K., and Hanahan, D. J., 1982, Vasoactive properties of acetyl glycerol ether phosphorylcholine and analogues, *Lab. Invest.* **46:**422–427.
29. Horvath, C. J., Ferro, T. J., Jesmok, G., and Malik, A. B., 1988, Recombinant tumor necrosis factor increases pulmonary vascular permeability independent of neutrophils, *Proc. Natl. Acad. Sci. USA* **85:** 9219–9223.
30. Gallin, J. I., Goldstein, I. M., and Snyderman, R. (eds), 1988, *Inflammation: Basic Principles and Clinical Correlates*, Raven Press, New York, p. 111.
31. Wintrobe, M. M., 1962, *Clinical Hematology*, Lea & Febiger, Philadelphia, p. 242.
32. Chien, S., Usami, S., and Skalak, R., 1984, Blood flow in small tubes, in: *Handbook of Physiology: The Cardiovascular System*, Vol. 4 (E. M. Renkin, C. C. Michel, and S. R. Geiger, eds.), American Physiological Society, Bethesda, pp. 237–238.
33. Firrell, J. C., and Lipowsky, H. H., 1989, Leukocyte margination and deformation in mesenteric venules of rat, *Am. J. Physiol.* **256:**H1667–H1674.
34. Marchesi, V. T., and Florey, H. W ., 1960, Electron micrographic observations on the emigration of leukocytes, *Q. J. Exp. Physiol.* **45:**343–348.
35. Lackie, J. M., and DeBono, D., 1977, Interactions of neutrophil granulocytes (PMN's) and endothelium in vitro, *Microvasc. Res.* **13:**107–112.
36. Hoover, R. L., Briggs, R. T., and Karnovsky, M. J., 1978, The adhesive interaction between polymorphonuclear leukocytes and endothelial cells in vitro, *Cell* **14:**423–428.
37. Zigmond, S. H., 1974, Mechanisms of sensing chemical gradients by polymorphonuclear leukocytes, *Nature* **249:**450–452.
38. Wilkinson, P. C. (ed.), 1982, *Chemotaxis and Inflammation*, Churchill Livingstone, Edinburgh.
39. Zigmond, S. H., and Devreotes, P. N., 1988, Chemotaxis in eukaryotic cells: A focus on leukocytes and *Dictyostelium*, *Annu. Rev. Cell Biol.* **4:**649–686.
40. Zigmond, S. H., 1981, Consequences of chemotactic peptide receptor modulation for leukocyte orientation, *J. Cell Biol.* **88:**644–647.
41. Snyderman, R., and Pike, M. C., 1984, Chemoattractant receptors on phagocytic cells, *Annu. Rev. Immunol.* **2:**257–281.
42. Tranquillo, R. T., Lauffenburger, D. A., and Zigmond, S. H., 1988, A stochastic model for leukocyte random motility and chemotaxis based on receptor binding fluctuations, *J. Cell Biol.* **106:**303–309.
43. Robbins, S. L., Cotran, R. S., and Kumar, V. (eds.), 1984, *Pathologic Basis of Disease*, Saunders, Philadelphia, p. 48.
44. Harvath, L., and Leonard, E. J., 1982, Two neutrophil populations in human blood with different chemotactic activities: Separation and chemoattractant binding, *Infect. Immun.* **36:**443–449.
45. Ford-Hutchinson, A. W., Bray, M. A., Doig, M. V., Shipley, M. E., and Smith, M. J. H., 1980, Leukotriene B_4, potent chemokinetic and aggregating substance released from polymorphonuclear leukocytes, *Nature* **286:**264–265.
46. Fernandez, H. N., Henson, P. M., Otani, A., and Hugli, T. E., 1978, Chemotactic response to human C3a and C5a anaphylatoxins. I. Evaluation of C3a and C5a leukotaxis in vitro and under stimulated in vivo conditions, *J. Immunol.* **120:**109–115.
47. Senior, R. M., Skogen, W. F., Griffin, G. L., and Wilner, G. D., 1986, Effects of fibrinogen derivatives upon the inflammatory response, *J. Clin. Invest.* **77:**1014–1019.
48. Shaw, J. O., Henson, P. M., Henson, J., and Webster, R. O., 1980, Lung inflammation induced by complement-derived chemotactic fragments in the alveolus, *Lab. Invest.* **42:**547–558.
49. Tonnesen, M. G., Smedley, L. A., and Henson, P. M., 1984, Neutrophil–endothelial cell interactions—Modulation of neutrophil adhesiveness induced by complement fragments C5a, and C5a des arg and fMLP in vitro, *J. Clin. Invest.* **74:**1581–1592.
50. Yoshimura T., Matsuchima, K., Tanaka, S., Robinson, E. A., Appella, E., Oppenheim, J. J., and Leonard, E. J., 1987, Purification of a human monocyte-derived neutrophil chemotactic factor that has peptide

sequence similarity to other host defense cytokines, *Proc. Natl. Acad. Sci. USA* **84:**9233–9237.

51. Strieter, R. M., Kunkel, S. L., Showell, H. J., Remick, D. G., Phan, S. H., Ward, P. A., and Marks, R. M., 1989, Endothelial cell gene expression of a neutrophil chemotactic factor by TNF-alpha, LPS, and IL-1 beta, *Science* **243:**1467–1469.
52. Strieter, R. M., Phan, S. H., Showell, H. J., Remick, D. G., Lynch, J. P., Genard, M., Raiford, C., Eskandari, M., Marks, R. M., and Kunkel, S. L., 1989, Monokine-induced neutrophil chemotactic factor gene expression in human fibroblasts, *J. Biol. Chem.* **264:**10621–10626.
53. Hopkins, N. K., Schaub, R. G., and Gorman, R. R., 1984, Acetyl glycerol ether phosphorylcholine (pafacether) and leukotriene B_4-mediated neutrophil chemotaxis through an intact endothelial cell monolayer, *Biochim. Biophys. Acta* **805:**30–36.
54. Deuel, T. F., Senior, R. M., Huang, J. S., and Griffin, G. L., 1982, Chemotaxis of monocytes and neutrophils to platelet-derived growth factor, *J. Clin. Invest.* **69:**1046–1049.
55. Gimbrone, M. A., Obin, M. S., Brock, A. F., Luis, E. A., Hass, P. E., Hebert, C. A., Yip, Y. K., Leung, D. W., Lowe, D. G., Kohr, W. J., Darbonne, W. C., Bechtol, K. B., and Baker, J. B., 1989, Endothelial interleukin-8: A novel inhibitor of leukocyte–endothelial interactions, *Science* **246:**1601–1603.
56. Bevilacqua, M. P., Pober, J. S., Wheeler, M. E., Cotran, R. S., and Gimbrone, M. A., Jr., 1985, Interleukin 1 acts on cultured human vascular endothelium to increase the adhesion of polymorphonuclear leukocytes, monocytes, and related leukocyte cell lines, *J. Clin. Invest.* **76:**2003–2011.
57. Zimmerman, G. A., McIntyre, T. M., and Prescott, S. M., 1985, Thrombin stimulates the adherence of neutrophils to human endothelial cells in vitro, *J. Clin. Invest.* **76:**2235–2246.
58. Baggiolini, M., Walz, A., and Kunkel, S. L., 1989, Neutrophil-activating peptide-1/interleukin 8, a novel cytokine that activates neutrophils, *J. Clin. Invest.* **84:**1045–1049.
59. Ross, R., Raines, E. W., and Bowen-Pope, D. F., 1986, The biology of platelet-derived growth factor, *Cell* **46:**155–169.
60. Shalaby, M. R., Palladino, M. A., Jr., Hirabayashi, S. E., Eesalu, T. E., Lewis, G. D., Shepard, H. M., and Aggarwal, B. B., 1987, Receptor binding and activation of polymorphonuclear neutrophils by tumor necrosis factor-alpha, *J. Leukoc. Biol.* **41:**196–204.
61. Anderson, D. C., and Springer, T. A., 1987, Leukocyte adhesion deficiency: An inherited defect in the Mac-1, LFA-1, and p150,95 glycoproteins, *Annu. Rev. Med.* **38:**175–194.
62. Hynes, R. O., 1987, Integrins: A family of cell surface receptors, *Cell* **48:**549–554.
63. Arfors, K. E., Lundberg, C., Lindbom, L., Lundberg, K., Beatty, P. G., and Harlan, J. M., 1987, A monoclonal antibody to the membrane glycoprotein complex CD18 inhibits polymorphonuclear leukocyte accumulation and plasma leakage in vivo, *Blood* **69:**338–340.
64. Beatty, P. G., Harlan, J. M., Rosen, H., Hansen, J. A., Ochs, H. D., Price, T. H., Taylor, R. F., and Klebanoff, S. J., 1984, Absence of monoclonal-antibody-defined protein complex in boy with abnormal leukocyte function, *Lancet* **1:**535–537.
65. Harlan, J. M., Killen, P. D., Senecal, F. M., Schwartz, B. R., Yee, E. K., Taylor, R. F., and Klebanoff, S. J., 1985, The role of neutrophil membrane glycoprotein gp-150 in neutrophil adherence to endothelium in vitro, *Blood* **66:**167–178.
66. Lo, S. K., Detmers, P. A., Levin, S. M., and Wright, S. D., 1989, Transient adhesion of neutrophils to endothelium, *J. Exp. Med.* **169:**1779–1793.
67. Smith, C. W., Marlin, S. D., Rothlein, R., Toman, C., and Anderson, D. C., 1989, Cooperative interactions of LFA-1 and Mac-1 with intercellular adhesion molecule-1 in facilitating adherence and transendothelial migration of human neutrophils in vitro, *J. Clin. Invest.* **83:**2008–2017.
68. Zimmerman, G. A., and McIntyre, T. M., 1988, Neutrophil adherence to human endothelium in vitro occurs by CDw18 (Mol, MAC-1/LFA-1/GP 150,95) glycoprotein-dependent and independent mechanisms, *J. Clin. Invest.* **81:**531–537.
69. Tonnesen, M. G., Anderson, D. C., Springer, T. A., Knedler, A., Avdi, N., and Henson, P. M., 1989, Adherence of neutrophils to cultured human microvascular endothelial cells: Stimulation by chemotactic peptides and lipid mediators and dependence upon the Mac-1, FLA-1, p150,95 glycoprotein family, *J. Clin. Invest.* **83:**637–646.
70. Springer, T. A., Thompson, W. S., Miller, L. J., Schmalsteig, F. C., and Anderson, D. C., 1984, Inherited deficiency of the Mac-1, LFA-1, p150.95 glycoprotein family and its molecular basis, *J. Exp. Med.* **160:**1901–1918.
71. Berger, M., O'Shea, J., Cross, A. S., Folks, T. M., Chused, T. M., Brown, E. J., and Frank, M. M., 1984,

Human neutrophils increase expression of C3bi as well as C3b receptors upon activation, *J. Clin. Invest.* **74:**1566–1571.

72. Gamble, J. R., Harlan, J. M., Klebanoff, S. J., and Vadas, M. A., 1985, Stimulation of the adherence of neutrophils to umbilical vein endothelium by human recombinant tumor necrosis factor, *Proc. Natl. Acad. Sci. USA* **82:**8667–8671.
73. Berger, M., Wetzler, E. M., and Wallis, R. S., 1988, Tumor necrosis factor is the major monocyte product that increases complement receptor expression on human neutrophils, *Blood* **71:**151–158.
74. Smith, C. W., Rothlein, R., Hughes, B. J., Mariscalco, M. M., Rudloff, H. E., Schmalsteig, F. C., and Anderson, D. C., 1988, Recognition of an endothelial determinant for CD18-dependent human neutrophil adherence and transendothelial migration, *J. Clin. Invest.* **82:**1746–1756.
75. Moser, R., Schleiffenbaum, B., Groscurth, P., and Fehr, J., 1989, Interleukin 1 and tumor necrosis factor stimulate human vascular endothelial cells to promote transendothelial neutrophil passage, *J. Clin. Invest.* **83:**444–455.
76. Detmers, P. A., Olesen-Egbert, E., Lo, S. K., Olsen-Egbert, E., Walz, A., Baggiolini, M., and Cohn, Z. A., 1990, Neutrophil activating protein/interleukin 8 stimulates the binding activity of the leukocyte adhesion receptor CD11b/CD18 on human neutrophils, *J. Exp. Med.* **171:**1155–1162.
77. Lopez, A. F., Williamson, D. J., Gamble, J. R., Begley, C. G., Harlan, J. M., Klebanoff, S. J., Waltersdorph, A., Wong, G., Clark, S. C., and Vadas, M. A., 1986, Recombinant human granulocyte–macrophage colony-stimulating factor stimulates in vitro mature human neutrophil and eosinophil function, surface receptor expression, and survival, *J. Clin. Invest.* **78:**1220–1228.
78. Vedder, N. B., and Harlan, J. M., 1988, Increased surface expression of CD11b/CD18(Mac-1) is not required for stimulated neutrophil adherence to cultured endothelium, *J. Clin. Invest.* **81:**676–682.
79. Dustin, M. L., and Springer, T. A., 1989, T-cell receptor cross-linking transiently stimulates adhesiveness through LFA-1, *Nature* **341:**619–624.
80. Chatila, T. A., Geha, R. S., and Arnaout, A., 1989, Constitutive and stimulus-induced phosphorylation of CD11/CD18 leukocyte adhesion molecules, *J. Cell Biol.* **109:**3435–3444.
81. Buyon, J. P., Slade, S. G., Reibman, J., Abramson, S. B., Philips, M. R., Weissman, G., and Winchester, R., 1990, Constitutive and induced phosphorylation of the α- and β-chains of the CD11/CD18 leukocyte integrin family, *J. Immunol.* **144:**191–197.
82. Detmers, P. A., Wright, S. D., Olsen, E., Kimball, B., and Cohn, Z. A., 1987, Aggregation of complement receptors on human neutrophils in the absence of ligand, *J. Cell Biol.* **105:**1137–1145.
83. Furie, M. B., and McHugh, D. D., 1989, Migration of neutrophils across endothelial monolayers is stimulated by treatment of the monolayers with interleukin-1 or tumor necrosis factor-α, *J. Immunol.* **143:**3309–3317.
84. Gallin, J. I., Goldstein, I. M., and Snyderman, R. (eds.), 1988, *Inflammation: Basic Principles and Clinical Correlates*, Raven Press, New York, pp. 204–205.
85. Stern, D. M., Bank, I., Nawroth, P. P., Cassimeris, J., Kisiel, W., Fenton, J. W. II, Dinarello, C., Chess, L., and Jaffe, E. A., 1985, Self-regulation of procoagulant events on the endothelial cell surface, *J. Exp. Med.* **162:**1223–1235.
86. Bevilacqua, M. P., Pober, J. S., Majeau, G. R., Cotran, R. S., and Gimbrone, M. A., 1984, Interleukin 1 (Il-1) induces biosynthesis and cell surface expression of procoagulant activity in human vascular endothelial cells, *J. Exp. Med.* **160:**618–623.
87. Gallin, J. I., Goldstein, I. M., and Snyderman, R. (eds.), 1988, *Inflammation: Basic Principles and Clinical Correlates*, Raven Press, New York, pp. 198–199.
88. Pennica, D., Nedwin, G. E., Hayflick, J. S., Seeburg, P. H., Derynck, R., Palladino, M. A., Mohr, W. J., Aggarwal, B. B., and Goeddel, D. V., 1984, Human tumor necrosis factor: Precursor structure, expression and homology to lymphotoxin, *Nature* **312:**724–729.
89. Beutler, B., Mahoney, J., Le Trang, N., Pekala, P., and Cerami, A., 1985, Purification of cachectin, a lipoprotein lipase-suppressing hormone secreted by endotoxin-induced RAW 264.7 cells, *J. Exp. Med.* **161:**984–995.
90. Gray, P. W., Aggarwal, B. B., Benton, C. V., Bringman, T. S., Henzel, W. J., Jarrett, J. A., Leung, D. W., Maffatt, B., Ng, P., and Sverdersky, L. P., 1984, Cloning and expressing cDNA for human lymphotoxin, a lymphokine with tumor necrosis activity, *Nature* **312:**721–724.
91. Pober, J. S., Gimbrone, M. A., Lapierre, L. A., Mendrick, D. L., Fiers, W., Rothlein, R., and Springer,

T. A., 1986, Overlapping patterns of activation of human endothelial cells by interleukin 1, tumor necrosis factor, and immune interferon, *J. Immunol.* **137**:1893– 1896.
92. Bevilacqua, M. P., Stengelin, S., Gimbrone, M. A., and Seed, B., 1989, Endothelial leukocyte adhesion molecule 1: An inducible receptor for neutrophils related to complement regulatory proteins and lectins, *Science* **243**:1160–1165.
93. Siegelman, M. H., van de Rijn, M., and Weissman, I. L., 1989, Mouse lymph node homing receptor cDNA clone encodes a glycoprotein revealing tandem interaction domains, *Science* **243**:1165–1172.
94. Johnston, G. I., Cook, R. G., and McEver, R. P., 1989, Cloning of GMP-140, a granule membrane protein of platelets and endothelium: Sequence similarity to proteins involved in cell adhesion and inflammation, *Cell* **56**:1033–1044.
95. Pober, J. S., Bevilacqua, M. P., Mendrick, D. L., Lapierre, L. A., Fiers, W., and Gimbrone, M. A., 1986, Two distinct monokines, interleukin 1 and tumor necrosis factor, each independently induce biosynthesis and transient expression of the same antigen on the surface of cultured human vascular endothelial cells, *J. Immunol.* **136**:1680–1687.
96. Dustin, M. L., and Springer, T. A., 1988, Lymphocyte function-associated antigen-1 (LFA-1) interaction with intercellular adhesion molecule-1 (ICAM-1) is one of at least three mechanisms for lymphocyte adhesion to cultured endothelial cells, *J. Cell Biol.* **107**:321–331.
97. Berendt, A. R., Simmons, D. L., Tansey, J., Newbold, C. I., and Marsh, K., 1989, Intercellular adhesion molecule-1 is an endothelial cell adhesion receptor for Plasmodium falciparum, *Nature* **341**:57–59.
98. Dustin, M. L., Rothlein, R., Bhan, A. K., Dinarello, C. A., and Springer, T. A., 1986, Induction by IL 1 and interferon-gamma: Tissue distribution, biochemistry, and function of a natural adherence molecule (ICAM-1), *J. Immunol.* **137**:245–254.
99. McIntyre, T. M., Zimmerman, G. A., and Prescott, S. M., 1986, Leukotrienes C_4 and D_4 stimulate human endothelial cells to synthesize platelet-activating factor and bind neutrophils *Proc. Natl. Acad. Sci. USA* **83**:2204–2208.
100. DiCorleto, P. E., and Bowen-Pope, D. F., 1983, Cultured endothelial cells produce a platelet-derived growth factor-like protein, *Proc. Natl. Acad. Sci. USA* **80**:1919–1923.
101. Camussi, G., Aglietta, M., Malavasi, F., Tetta, C., Piacibello, W., Sanavio, F., and Bussolino, F., 1983, The release of platelet-activating factor from human endothelial cells in culture, *J. Immunol.* **131**:2397–2403.
102. Wedmore, C., and Williams, T. J., 1981, Control of vascular permeability by polymorphonuclear leukocytes in inflammation, *Nature* **289**:646–650.
103. Heflin, A. C., Jr., and Brigham, K. L., 1981, Prevention by granulocyte depletion of increased vascular permeability of sheep lung following endotoxemia, *J. Clin. Invest.* **68**:1253–1260.
104. Shaw, J. O., and Henson, P. M., 1982, Pulmonary intravascular sequestration of activated neutrophils—Failure to induce light microscopic evidence of lung injury in rabbits, *Am. J. Pathol.* **108**:17–23.
105. Meyrick, B., Hoffman, L. H., and Brigham, K. L., 1984, Chemotaxis of granulocytes across bovine pulmonary artery intimal explants without endothelial cell injury, *Tissue Cell* **16**:1–16.
106. Webster, R. O., Larsen, G. L., Mitchell, B. C., Goins, A. J., and Henson, P. M., 1982, Absence of inflammatory lung injury in rabbits challenged intravascularly with complement-derived chemotactic factors, *Am. Rev. Respir. Dis.* **125**:335–340.
107. Martin, T. R., Pistorese, B. P., and Chi, E. Y., Goodman, R. B., and Matthay, M. A., 1989, Effects of leukotriene B_4 in the human lung: Recruitment of neutrophils into the alveolar spaces without a change in protein permeability, *J. Clin. Invest.* **84**:1609–1619.
108. Harlan, J. M., Schwartz, B. R., Reidy, M. A., Schwartz, S. M., Ochs, H. D., and Harker, L. A., 1985, Activated neutrophils disrupt endothelial monolayer integrity by an oxygen radical-independent mechanism, *Lab. Invest.* **52**:141–150.
109. Killackey, J. J., Johnston, M. G., and Movat, H. Z., 1986, Increased permeability of microcarrier-cultured endothelial monolayers in response to histamine and thrombin: A model for the in vitro study of increased vasopermeability, *Am. J. Pathol.* **122**:50–61.
110. Brett, J., Gerlach, H., Nawroth, P., Steinberg, S., Godman, G., and Stern, D., 1989, Tumor necrosis factor/cachectin increases permeability of endothelial cell monolayers by a mechanism involving regulatory G proteins, *J. Exp. Med.* **169**:1977–1991.
111. Langeler, E. G., Snelting,-Havinga, I., and Van Hinsbergh, V. W. M., 1989, Passage of low density

lipoproteins through monolayers of human arterial endothelial cells: Effects of vasoactive substances in an in vitro model, *Arteriosclerosis* **9:**550–559.
112. Fromter, E., 1972, The rout of passive ion movement through the epithelium of Necturus gallbladder, *J. Membr. Biol.* **8:**259–301.
113. Vasile, E., Simionescu, M., and Simionescu, N., 1983, Visualization of the binding, endocytosis, and transcytosis of low-density lipoprotein in the arterial endothelium in situ, *J. Cell Biol.* **96:**1677–1689.
114. Ghitescu, L., Fixman, A., Simionescu, M., and Simionescu, N., 1986, Specific binding sites for albumin restricted to plasmalemmal vesicles of continuous capillary endothelium: Receptor-mediated transcytosis, *J. Cell Biol.* **102:**1304–1311.
115. Schnitzer, J. E., Carley, W. W., and Palade, G. E., 1988, Albumin interacts specifically with a 60-kDa microvascular endothelial glycoprotein, *Proc. Natl. Acad. Sci. USA* **85:**6773–6777.
116. Navab, M., Hough, G. P., Stevenson, L. W., Drinkwater, D. C., Laks, H., and Fogelman, A. M., 1988, Monocyte migration into the subendothelial space of a coculture of adult human aortic endothelial and smooth muscle cells, *J. Clin. Invest.* **82:**1853–1863.
117. Madri, J. A., and William, S. K., 1983, Capillary endothelial cell cultures: Phenotypic modulation by matrix components, *J. Cell Biol.* **97:**153–165.
118. Rotrosen, D., and Gallin, J. I., 1986, Histamine type I receptor occupancy increases endothelial cytosolic calcium, reduces F-actin, and promotes albumin diffusion across cultured endothelial monolayers, *J. Cell Biol.* **103:**2379–2387.
119. Doukas, J., Shepro, D., and Hechtman, H. B., 1987, Vasoactive amines directly modify endothelial cells to affect polymorphonuclear leukocyte diapedesis in vitro, *Blood* **69:**1563–1569.
120. Bottaro, D., Shepro, D., Peterson, S., and Hechtman, H. B., 1986, Serotonin, norepinephrine, and histamine mediation of endothelial cell barrier function in vitro, *J. Cell. Physiol.* **128:**189–194.
121. Baron, D. A., Lofton, C. E., Newman, W. H., et al., 1989, Atriopeptin inhibition of thrombin-mediated changes in the morphology and permeability of endothelial monolayers, *Proc. Natl. Acad. Sci. USA* **86:**3394–3398.
122. Huang, A. J., Unpublished observations.
123. Noonan, T. C., Selig, W. M., Kern, D. F., and Malik, A. B., 1986, Mechanism of peptidoleukotriene-induced increases in pulmonary transvascular fluid filtration, *J. Appl. Physiol.* **61:**1928–1934.
124. Bussolino, F., Camussi, G., Aglietta, M., Braquet, P., Bosia, A., Pescarmona, G., Sanavio, F., D'Urso, N., and Marchisia, P. C., 1987, Human endothelial cells are target for platelet-activating factor, *J. Immunol.* **139:**2439–2446.
125. Stoll, L., and Spector, A. A., 1989, Interaction of platelet-activating factor with endothelial and vascular smooth muscle cells in coculture, *J. Cell. Physiol.* **139:**253–261.
126. Garcia, J. G., Azghani, A., Callahan, K. S., and Johnson, A. R., 1988, Effect of platelet activating factor on leukocyte–endothelial cell interactions, *Thromb. Res.* **51:**83–96.
127. Burhop, K. E., Garcia, J. G., Selig, W. M., Lo, S. K., Van der zee, H., Kaplan, J. E., and Malik, A. B., 1986, Platelet-activating factor increases lung vascular permeability to protein, *J. Appl. Physiol.* **61:**2210–2217.
128. Palmer, R. M., Ferrige, A. G., and Moncada, S., 1987, Nitric oxide release accounts for the biological activity of endothelium-derived relaxing factor, *Nature* **327:**524–526.
129. Miller, V. M., and Vanhoutte, P. M., 1985, Endothelium-dependent contractions to arachidonic acid are mediated by products of cyclooxygenase, *Am. J. Physiol.* **248:**H432–H437.
130. Shasby, D. M., Shasby, S. S., and Peach, M. J., 1985, Polymorphonuclear leukocyte: Arachidonate edema, *J. Appl. Physiol.* **59:**47–55.
131. Evans, C. W., Taylor, J. E., Walker, J. D., and Simmons, N. L., 1983, Transepithelial chemotaxis of rat peritoneal exudate cells, *Br. J. Exp. Pathol.* **64:**644–654.
132. Nash, S., Stafford, J., and Madara, J. L., 1987, Effects of polymorphonuclear leukocyte transmigration on the barrier function of cultured intestinal epithelial monolayers, *J. Clin. Invest.* **80:**1104–1113.
133. Milks, L., and Cramer, E., 1984, Transepithelial electrical resistance studies during in vitro neutrophil (PMN) migration, *Fed. Proc. Abstr.* **43**(3)**:**777.
134. Robbins, S. L., Cotran, R. S., and Kumar, V. (eds.), *Pathologic Basis of Disease*, Saunders, Philadelphia, p. 51.
135. Lewis, M. S., Whatley, R. E., Cain, P., McIntyre, T. M., Prescott, S. M., and Zimmerman, G. A., 1988,

Hydrogen peroxide stimulates the synthesis of platelet-activating factor by endothelial cell-dependent neutrophil adhesion, *J. Clin. Invest.* **82**:2045–2055.

136. Ager, A., and Gordon, J. L., 1984, Differential effects of hydrogen peroxide on indices of endothelial cell function, *J. Exp. Med.* **159**:592–603.
137. Weitz, J. I., Huang, A. J., Landman, S. L., Nicholson, S. C., and Silverstein, S. C., 1987, Elastase-mediated fibrinogenolysis by chemoattractant-stimulated neutrophils occurs in the presence of physiologic concentrations of antiproteinases, *J. Exp. Med.* **166**:1836–1850.
138. Lo, S. K., Ryan, T. J., Gilboa, N., Lai, L., and Malik, A. B., 1989, Role of catalytic and lysine-binding sites in plasmin-induced neutrophil adherence to endothelium, *J. Clin. Invest.* **84**:793–801.
139. Griepp, E. B., Dolan, W. J., Robbins, E. S., and Sabatini, D. D., 1983, Participation of plasma membrane proteins in the formation of tight junctions by cultured epithelial cells, *J. Cell Biol.* **96**:693–702.
140. Peterson, M. W., Stone, P., and Shasby, D. M., 1987, Cationic neutrophil proteins increase transendothelial albumin movement, *J. Appl. Physiol.* **62**:1521–1530.
141. Majno, G., Shea, S. M., and Leventhal, M., 1969, Endothelial contraction induced by histamine type mediators, *J. Cell Biol.* **42**:617–672.
142. Huang, A. J., Nicholson, S. C., and Silverstein, S. C., 1989, Neutrophil migration across monolayers of human endothelial cells induces changes in endothelial cell cytosolic free calcium, *J. Cell Biol.* **109**:313a.

14

Endothelial Permeability in Inflammation
The Role of Capillaries versus Venules

I. Joris, H. F. Cuénoud, G. V. Doern, J. M. Underwood, and G. Majno

I. INTRODUCTION

The electron microscopic studies of N. Simionescu and collaborators[17–19] have clearly shown that the endothelium in the three basic segments of the microcirculation—arterioles, capillaries, and venules—differs in terms of ultrastructure and permeability. Segmental differences are obvious also in the inflammatory response. Indeed, those chemical mediators that increase microvascular permeability do so by affecting selectively the permeability of the venules; this was established by using a method that selectively marks leaky segments of the microcirculation, namely *vascular labeling*.[13] A clue to this selectivity for the venules was provided by Heltianu *et al.*[5] who showed that the endothelium of these vessels is especially rich in histamine receptors. Another difference between capillary and venular endothelium is that the latter is able to contract *in vivo* under the stimulus of inflammatory mediators[7]; rather than reducing the caliber of the vessel, this contraction produces gaps between the endothelial cells, allowing the escape of fluid.[7,14]

From this body of information it may be concluded that a large part of the inflammatory exudate (that part which is induced by permeability-increasing mediators) derives from the venules. Could the capillaries also contribute to the exudate? Three possibilities come to mind: (1) *Arteriolar vasodilatation increases the blood pressure* in the microcirculation, and this should lead to an increased level of transudation (i.e., ultrafiltration) from all vascular segments, including the capillaries, without an increase in permeability; this protein-poor transudate should then mix with the inflammatory exudate. (2) *Loss of whole plasma* from capillaries could occur as a result of direct injury.[2] (3) A third possibility—*capillary leakage* caused by inflammatory mediators—seems unlikely, because no inflammatory

I. Joris, H. F. Cuénoud, J. M. Underwood, and G. Majno • Department of Pathology, University of Massachusetts Medical School, Worcester, Massachusetts 01655. G. V. Doern • Department of Medicine, University of Massachusetts Medical School, Worcester, Massachusetts 01655.

Endothelial Cell Dysfunctions, edited by Nicolae Simionescu and Maya Simionescu. Plenum Press, New York, 1992.

mediator is known to increase capillary permeability. The purpose of this chapter is to describe a hitherto unsuspected form of capillary leakage, observed in a simple, "new" experimental model of inflammation.

Generally speaking, studies on vascular leakage in the microcirculation have been performed with single chemical mediators injected locally. In real life, however, vascular leakage occurs in situations that are much more complicated, such as infarcts or abscesses. To our knowledge, the pattern of vascular leakage under such conditions has not been examined. We therefore decided to test the pattern of vascular leakage that occurs during the evolution of an aseptic infarct. This raised a technical problem: if we produced a partial infarct of the kidney, by occluding a branch of the renal artery, vascular leakage would undoubtedly occur at the interface between live and dead tissue; however, the pathogenesis of this vascular malfunction would remain unclear. Three mechanisms would be possible, alone or in combination, namely: (1) direct vascular damage by ischemic injury, (2) secondary vascular damage induced by the inflammatory response, and (3) a functional response of intact vessels to mediators arising from the necrotic tissue. We therefore devised an experimental model that would simplify the interpretation. To this effect, the simplest procedure was to choose an appropriate target organ rich in blood vessels (striated muscle) and place against it a mass of aseptic, dying tissue such as liver or kidney. Basically, this model represents a *pseudoinfarct* (Fig. 1).

As a target organ we chose the cremaster muscle of the rat, which forms on each side an easily accessible bag, open to the abdominal cavity. The testis can be eased back into the peritoneal cavity, and a suitable fragment of tissue can be introduced into the cremasteric recess through a small laparotomy incision. At a selected time after implantation, the rat is given carbon black intravenously, so as to label those vessels that are leaking at that time; an hour later the animal is sacrificed and the cremaster is excised and examined by transillumination. This is an extremely convenient method for studying vascular labeling, because the excised cremaster is about 0.5 mm in thickness, and can be cleared and mounted between slide and coverslip for studying the intact vascular tree.

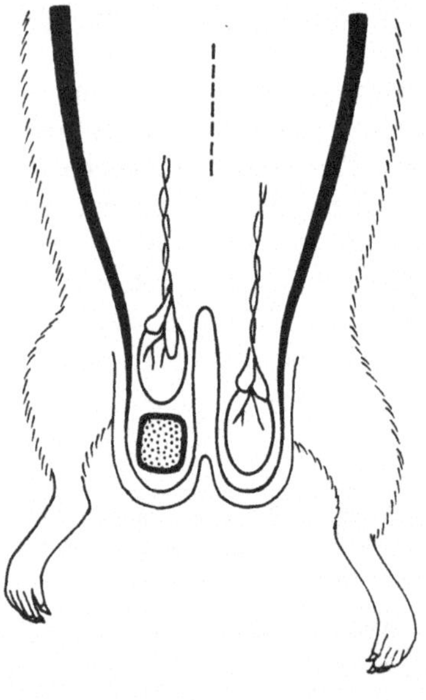

Figure 1. The experimental model of "pseudoinfarct": a fragment of tissue (represented as a square) is introduced into the scrotal cavity through a laparotomy incision (dashed line), after the testis has been displaced into the peritoneal cavity.

When we initiated this series of experiments, we anticipated that the cremaster exposed to dying tissue would soon develop intense venular labeling, and that this would fade away as the inflammatory response became chronic. The results were more complex.

II. MATERIALS AND METHODS

Full details are published elsewhere.[8] In essence: we used 139 Wistar rats, operated on both sides. Large rats were chosen as donors (380–650 g); *recipient rats* were in the 400–500 g range: with smaller rats under 300 g the implant was often lost into the peritoneal cavity, due to retraction of the cremaster. All animals were maintained and tested in accordance with recommendations in the *Guide for the Care and Use of Laboratory Animals* (DHHS, NIH publication 85-23, 1985) and the guidelines of the Animal Care Advisory Committee of the University of Massachusetts Medical School.

Prior to sacrifice under deep ether anesthesia, the donor rats were injected intravenously with 0.1 cm^3 of penicillin G solution (100,000 units/ml); this was followed by 0.1 cm^3 of a solution of streptomycin sulfate (200 mg/ml). Streptomycin was given in the second place, because it caused respiratory arrest within 5–15 sec. The purpose of these injections was to make sure that organs to be used as implants would be perfused with antibiotics. The organs were then harvested aseptically (liver, kidney, testis, heart, epididymal fat body, and abdominal muscle). The kidneys were decapsulated and used either whole or as two separate halves. From each liver four samples were cut as close as possible to the shape of a cube, and weighing 1.1–2 g.

All implants were introduced aseptically through a midline laparotomy incision (Fig. 1). Before an implant could be placed in the cremasteric sac, the testis was pushed into the peritoneal cavity by gentle external manipulation.

Control experiments. (a) *Boiled tissues*: Liver or kidney specimens were boiled in distilled water for about 10 min and implanted as usual. (b) *Teflon spheres*, ½ inch in diameter, were obtained from Small Parts, Inc., Miami, Fla. (#BT-8, weight 2.28 g). (c) *Hollow glass spheres* with a diameter of 12 mm were blown for us by Radnotti Glass Technology, Inc., Monrovia, Calif. These spheres weighed 1 g and had either a glossy or a frosted surface. (d) *Sham operation*: under ether anesthesia, a forceps bearing a tissue sample was introduced as usual into the cremasteric sac through a laparotomy incision, and retrieved immediately thereafter. (e) *Two normal controls* were also used.

Vascular labeling of the cremasters. The animals were sacrificed at intervals varying from 2 to 48 hr, and the cremasters were prepared as follows. Approximately 1 hr before sacrifice, the rat was anesthetized with ether, and given intravenously a 0.3-cm^3 injection of carbon black. The rats were sacrificed under heavy ether anesthesia by clamping the heart through a thoracotomy incision. The cremasters were then removed, stretched on dental wax, fixed in 10% neutral formalin, cleared in glycerin, and mounted in glycerin jelly as previously described.[7,13]

For electron microscopy all tissues were fixed in 3% glutaraldehyde, dripped on as well as injected into the cremasteric sac. Tissues were then postfixed in 1% OsO_4 in cacodylate buffer, and further processed for electron microscopy by routine methods.[7]

Bacterial cultures were performed on 14 implants according to the method of Krizek

2and Rabson.[9] When bacterial growth occurred, colony counts were made and the organisms were identified according to routine methods.[10]

Grading of the vascular labeling. Each cremaster muscle was transilluminated, examined blindly by two observers, and graded in a semiquantitative fashion. Grades were assigned to separate parameters: (1) *Overall intensity of the labeling.* This was estimated with the naked eye, according to an arbitrary scale of 0–5: 0 = no labeling, 1 = microscopic traces of labeling, 2 = a few labeled vessels barely visible by the naked eye, 3 = a distinct small patch of labeling, 4 = a large patch of labeling, 5 = extreme labeling throughout most of the cremaster. For grades 0 and 1, an examination under a 10× objective was sometimes performed. (2) *Percentage of venular versus capillary labeling.* To assign this grade, the entire labeled area was scanned with a 10× objective, and the result was expressed as V/C (venous versus capillary); for example, 8/2 would refer to 80% of the labeling being venular, 20% capillary.

III. RESULTS

Bacteriological tests showed that contamination had occurred in two implants only, one from an 18- and one from a 48-hr implant. The few colonies that developed were interpreted as due to skin contaminants (*Staphylococcus aureus* and *Proteus mirabilis*). Both the type and the amount of contamination suggested that a few microorganisms had been gathered during the final sampling process. These results ruled out the possibility that our findings could be attributed to bacterial infection.

Gross observations. About one of every five implants was lost to the study, because it had been expelled into the peritoneum. Up to a stage of 24 hr the implants were free in the cremasteric cavity, but at 48 hr they had become adherent to the inner surface of the cremasteric muscle by mild fibrinous adhesions. Adhesions between the implant and the testis never formed (although they did form between the implant and the epididymis or the mesotestis; this we cannot explain). *Edema* was apparent in the cremaster at 24 hr, and thereafter it also spread across the muscle to the subcutaneous tissue of the scrotum.

Vascular labeling. When vascular labeling occurred, it was maximal in the cremaster where it made contact with the implant; labeling occurred also in the epididymis and mesotestis, but never in the superficial vessels of the testis itself. *Labeling of venules only* was visible after 4 hr; the venules were easily recognized due to their branching pattern (Fig. 2). *Capillary labeling* began to appear at 8 hr, as venular labeling decreased; thereafter the proportion of capillary versus venular labeling continued to increase (Fig. 3), so that by 48 hr the labeling was almost entirely capillary (Figs. 4 and 5). Capillary labeling was recognized by the fact that the labeled blood vessels were all parallel, quite unlike the branching pattern of the venules; the parallel pattern is of course due to the fact that most capillaries run along the striated muscle fibers. In rats killed after 3 days, the implants were firmly attached to the muscle; both capillary and venular labeling had disappeared, and were replaced by a punctate pattern. This stage was not included in the present study.

The overall intensity of the vascular labeling was strongest in liver and kidney implants; there was no significant difference between the labeling produced by liver and kidney implants, except that maximal intensity of the labeling was reached at 8 hr with liver implants, somewhat later with kidney implants.

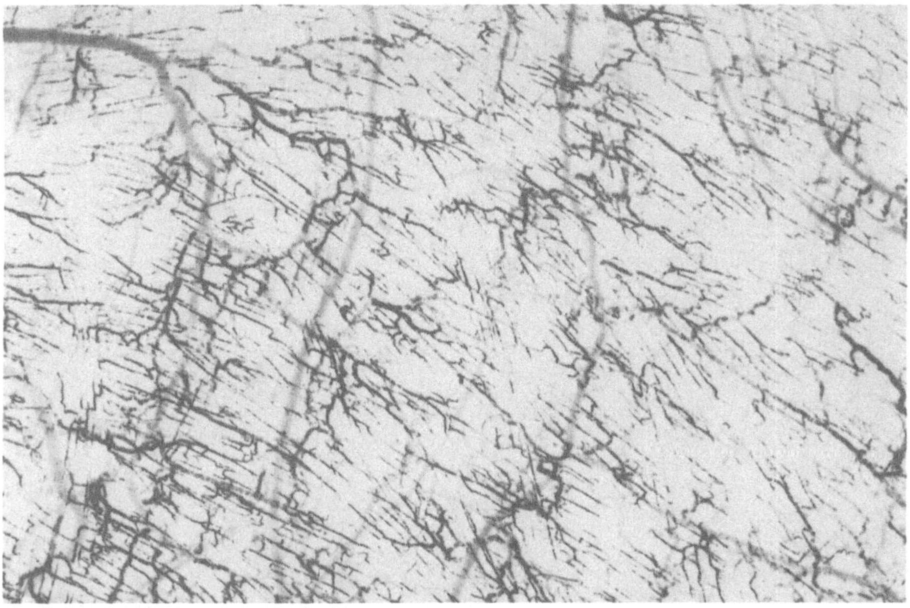

Figure 2. Rat cremaster cleared in glycerin and viewed by transillumination. The animal was sacrificed 1 hr after an intravenous injection of carbon black and a local injection of histamine. Note the characteristic labeling of the venules, represented by the black branching structures. (15×)

Control implants were all tested at 48 hr: *boiled tissues, Teflon, smooth and frosted glass beads produced the same effect: a relatively small area of capillary labeling, of mild intensity, and mixed (venular and capillary).*

Histology. Sections of the cremaster showed an acute inflammatory infiltrate, beginning at 2 hr and consisting mainly of neutrophils, a varying proportion of monocytes–macrophages, and a sprinkling of eosinophils. The cellular composition changed with time: neutrophils predominated up to 8 hr; between 12 and 18 hr the proportion of mononuclear cells increased, and by 24–48 hr they represented roughly 90% of the inflammatory cell population. The vascular labeling was, as usual, much easier to see in transilluminated whole mounts than in histologic cross sections.

Electron microscopy showed, as expected, that the accumulations of carbon black along the vessels indeed represented true labeling.[12,13] Interendothelial gaps were found in capillaries as well as in venules. Mitoses were found in the endothelium, pericytes, and fibroblasts. At 24 and 48 hr many endothelial cells were thickened and contained large numbers of free ribosomes; they could be best described as "activated."

IV. DISCUSSION

The essence of our findings in "pseudoinfarcts" can be summarized as follows: the inflammatory reaction induced by dying and dead tissue begins with a phase of venular leakage, which merges after 8–12 hr into a phase of capillary leakage lasting up to 48 hr.

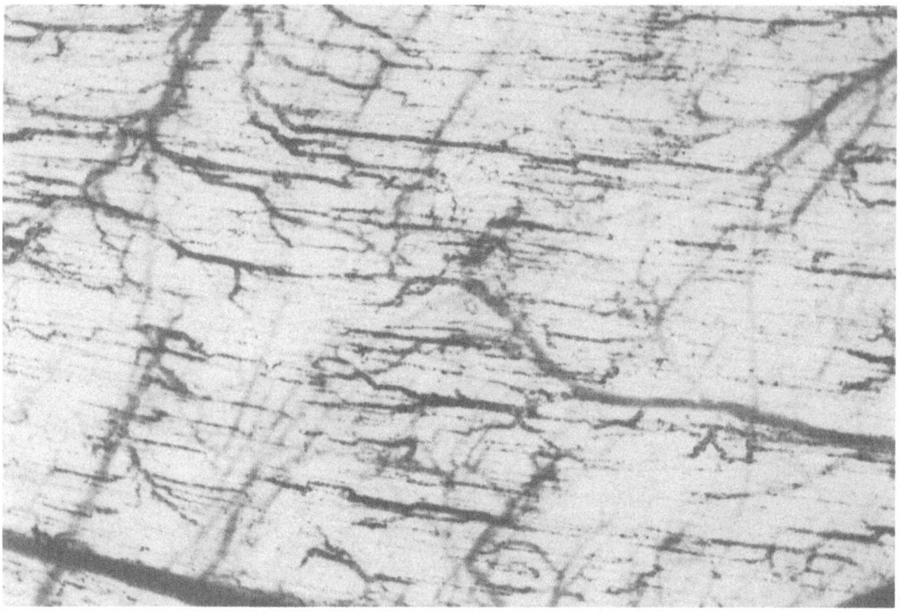

Figure 3. Rat cremaster, cleared and viewed by transillumination, *18 hr* after exposure to a "pseudoinfarct" (liver) and 1 hr after intravenous injection of carbon black. Note the venular labeling, admixed with some traces of capillary labeling. The latter is represented by grains of carbon in parallel lines because the capillaries run parallel to the muscle fibers. (15×)

We were not surprised by the early venular leakage: all the inflammatory mediators that have been tested have this effect, and a variety of such mediators should be produced within and around necrotic tissue. On the other hand, the later phase of capillary leakage was totally unexpected. To this day, not a single inflammatory mediator is known to have this effect.

This is not to say that capillary leakage has not been observed before. When tissues are physically injured, such as by mechanical trauma or a severe burn, all blood vessels will become leaky including the capillaries.[2] There are also several examples of a more specific, selective capillary leakage; this has been observed after a variety of different insults which have in common the production of a *delayed phase* of vascular leakage (reviewed in Ref. 20). Such injuries include mild heat,[3] toxins,[1,6] chronic sterile irritation,[4] and delayed hypersensitivity reactions.[21] The mechanism of the specific capillary response in these situations has not been explained; the prevailing opinion has been that the endothelium is injured, but if so, the nature of this damage and its specificity for the capillaries remain unknown.

It might be argued that the capillary leakage observed in our experiments could be produced by some toxic material arising from the dying or dead tissue, and that it could therefore be just a variant of the toxin-related "delayed capillary leakage" mentioned above. However, our experiments indicate that the presence of dying or dead tissue is not necessary: capillary leakage was observed also, at 48 hr, around the control implants that

Figure 4. Rat cremaster, cleared and transilluminated, *48 hr* after exposure to a "pseudoinfarct" (liver) and 1 hr after an intravenous injection of carbon black. The black parallel lines represent carbon labeling of capillaries; almost no venules are labeled. Note that the enlargement is identical to that of the previous figure. (15×)

consisted of boiled tissues, Teflon spheres, and smooth as well as frosted glass spheres. We used these implants as controls because we assumed that they would behave as inert materials; in practice, however, a mild inflammation was produced by all these "controls." This was fortunate, because it provided us with a key for interpreting our results: the single experimental effect that our control and experimental cremasters shared was inflammation. This indicates that *the late phase of capillary leakage is not necessarily the effect of dead or*

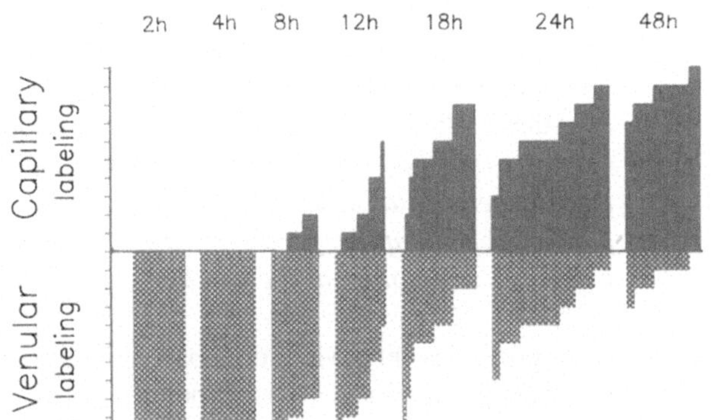

Figure 5. Graph combining the results obtained with "pseudoinfarcts" of all implants, illustrating the progressive shift from venular labeling to capillary labeling over a period of 48 hr.

dying tissue; it can be brought about by the inflammatory response itself. Glass or Teflon beads could certainly not release significant amounts of irritating materials into the tissues; however, because of their surface and mechanical properties, they did act as (mild) irritants, and inflammation ensued.

On the basis of our findings, we can conclude that *aseptic inflammation, at least in our model, is accompanied by an early phase of venular leakage, followed by a late phase of capillary leakage* (Fig. 5). The problem now shifts to finding the mechanism for the delayed capillary leakage.

Microvascular leakage could correspond to either of two basic changes: actual damage, or a physiologic response of intact cells. Damage can probably be ruled out; it would be difficult to conceive of a form of tissue damage that would hit exclusively the capillary wall; and furthermore, we have no electron microscopic evidence of cellular damage. As regards a physiologic response, we can probably rule out endothelial contraction, because there was no evidence of such a change on electron micrographs.

The choice is therefore narrowed down to a functional response of the capillary wall, unrelated to endothelial contraction. The likeliest answer is a diffuse form of angiogenesis, since it is well known that newly formed blood vessels leak[16] and that endothelial mitoses cause temporary leakage also in the arterial intima.[11] This form of angiogenesis could be a response to two types of stimulation, which are not mutually exclusive:

1. Angiogenesis due to the release of products by the inflammatory cells, possibly the macrophages. In favor of this interpretation is the fact that the phase of capillary leakage coincides with the predominance of macrophages in the inflammatory exudate; against it is the fact that some leaky capillaries appear to have no monocytes or macrophages in their vicinity.
2. Angiogenesis due to a *hemodynamic effect*. This is, in our opinion, the likeliest answer. We postulate that the persistent hyperemia of acute inflammation leads to a temporary, functional increase in the diameter of the vessels of the microcirculation, and that the dilated vessels adapt to this change: the stretched endothelial cells undergo mitosis, and the vessel becomes structurally enlarged. Endothelial mitoses were indeed observed. A mechanism of this kind is presumably involved in the development of a *collateral circulation*, whereby large arteries and veins develop from vessels of the microcirculation. If this is true, then we should expect a late (delayed) phase of capillary leakage in any form of long-lasting hyperemia, including of course inflammation. Further studies will have to show whether all forms of persistent inflammation do include such a phase; in any event, this hypothesis allows us to propose a new interpretation for the cases of "delayed capillary leakage" described in the past: *if inflammation is present in each one of those experimental situations—which appears very likely—then the progressive appearance of capillary leakage would have the same meaning as in our experiments: an adaptation of the capillary bed to increased blood flow.*

From the clinical point of view, our observations are relevant in that they point to a form of vascular leakage unrelated to endothelial contraction. Many of the drugs that are currently used to inhibit vascular leakage are also smooth muscle relaxants, presumably because—as relaxants—they also inhibit endothelial contraction in the venules.[15] Anti-inflammatory drugs of this kind should not be able to inhibit the phase of capillary leakage, because it is not the result of endothelial contraction.

To sum up, these studies have shown that capillaries—despite their known refractoriness to inflammatory mediators—do eventually become leaky in the course of a sterile inflammatory reaction. Further work is needed to find out whether the late phase of capillary leakage is due to flow conditions prevalent in this segment of the microcirculation or to mediators produced in the inflammatory focus.

ACKNOWLEDGMENTS. This work was supported in part by NIH Grants HL-25973 and HL-214291 and by a grant from Ciba-Geigy, Basel. We are indebted to Jane M. Manzi for preparation of the manuscript.

REFERENCES

1. Cotran, R. S., 1967, Studies on inflammation. Ultrastructure of the prolonged vascular response induced by *Clostridium oedematiens* toxin, *Lab. Invest.* **17**:39–60.
2. Cotran, R. S., and Majno, G., 1964, A light and electron microscopic analysis of vascular injury, *Ann. N.Y. Acad. Sci.* **116**:750–763.
3. Cotran, R. S., and Majno, G., 1964, The delayed and prolonged vascular leakage in inflammation. I. Topography of the leaking vessels after thermal injury. *Am. J. Pathol.* **45**:261–281.
4. Cuénoud, H. F., Joris, I., Langer, R. S., and Majno, G., 1987, Focal arteriolar insudation. A response of arterioles to chronic nonspecific irritation, *Am. J. Pathol.* **127**:592–604.
5. Heltianu, C., Simionescu, M., and Simionescu, N., 1982, Histamine receptors of the microvascular endothelium revealed in situ with a histamine–ferritin conjugate: Characteristic high-affinity binding sites in venules, *J. Cell Biol.* **93**:357–364.
6. Hurley, J. V., and Jago, M. V., 1976, Delayed and prolonged vascular leakage in inflammation: The effects of dehydromonocrotaline on blood vessels in the rat cremaster, *Pathology* **8**:7–20.
7. Joris, I., Majno, G., Corey, E. J., and Lewis, R. A., 1987, The mechanism of vascular leakage induced by leukotriene E_4. Endothelial contraction, *Am. J. Pathol.* **126**:19–24.
8. Joris, I., Cuénoud, H. F., Doern, G. V., Underwood, J. M., and Majno, G., 1990, Capillary leakage in inflammation: A study by vascular labeling, *Am. J. Pathol.* **137**:1353–1363.
9. Krizek, T. J., and Rabson, M. C., 1975, Evolution of quantitative bacteriology in wound management, *Am. J. Surg.* **130**:579–584.
10. Lenette, E. H., Balows, A., Hausler, W. J., and Shadomy, H. J. (eds.), 1985, *Manual of Clinical Microbiology, fourth edition*, American Society for Microbiology, Washington, D. C.
11. Lin, S.-J., Jan, K.-M., Weinbaum, S., and Chien, S., 1989, Transendothelial transport of low density lipoprotein in association with cell mitosis in rat aorta, *Arteriosclerosis* **9**:230–236.
12. Majno, G., and Palade, G. E., 1961, Studies on inflammation. I. The effect of histamine and serotonin on vascular permeability: An electron microscopic study, *J. Biophys. Biochem. Cytol.* **11**:571–605.
13. Majno, G., Palade, G. E., and Schoefl, G. I., 1961, Studies on inflammation. II. The site of action of histamine and serotonin along the vascular tree: A topographic study, *J. Biophys. Biochem. Cytol.* **11**: 607–626.
14. Majno, G., Shea, S. M., and Leventhal, M., 1969, Endothelial contraction induced by histamine-type mediators. An electron microscopic study, *J. Cell Biol.* **42**:647–672.
15. Persson, C. G. A., and Svensjo, E., 1985, Vascular responses and their suppression: Drugs interfering with venular permeability, in: *The Pharmacology of Inflammation* (I. L. Bonta, M. A. Bray, and M. J. Parnham, eds.), Elsevier, Amsterdam, pp. 71–82.
16. Schoefl, G. I., 1963, Studies on inflammation. III. Growing capillaries: Their structure and permeability, *Virchows Arch. Pathol. Anat.* **337**:97–141.
17. Simionescu, N., Simionescu, M., and Palade, G. E., 1978, Open junctions in the endothelium of the postcapillary venules of the diaphragm, *J. Cell Biol.* **79**:27–44.

18. Simionescu, N., Simionescu, M., and Palade, G. E., 1978, Structural basis of permeability in sequential segments of the microvasculature of the diaphragm. I. Bipolar microvascular fields, *Microvasc. Res.* **15:** 1–16.
19. Simionescu, N., Simionescu, M., and Palade, G. E., 1978, Structural basis of permeability in sequential segments of the microvasculature of the diaphragm. II. Pathways followed by microperoxidase across the endothelium, *Microvasc. Res.* **15:**17–36.
20. Wilhelm, D. L., 1973, Chemical mediators, in: *The Inflammatory Process*, Vol. II (B. W. Zweifach, L. Grant, and R. T. McCluskey, eds.), Academic Press, New York, pp. 251–301.
21. Willms-Kretschmer, K., Flax, M. H., and Cotran, R. S., 1967, The fine structure of the vascular response in hapten-specific delayed hypersensitivity and contact dermatitis, *Lab. Invest.* **17:**334–349.

III

Endothelial Cell in Viral Infections

15

Response of Human Vascular Cells to Viral Infection

Nicholas A. Kefalides

I. INTRODUCTION

Injury to blood vessel cells has been implicated in the development of a variety of vascular disorders including atherosclerosis,[2,15] disseminated intravascular coagulation,[9] and immune vasculitis.[56] Endothelial cell (EC) injury can arise from a variety of causes including hemodynamic stress,[64] mechanical trauma,[65] hypercholesterolemia,[57] infectious agents,[16,49,54] oxygen[33] and other chemical agents such as homocysteine.[74] Renewed interest in the relationship between vascular injury due to viral infection and atherosclerosis stems from recent observations which demonstrated that virus particles, virus antigens, and virus DNA can be detected in vascular lesions of chickens[11,49] and humans[4,23,48,76] with atherosclerosis. Viruses may produce vascular injury by mechanisms other than direct invasion of endothelium. Antigen-antibody complexes involving hepatitis B surface antigen have been isolated from the sera of some patients with periarteritis nodosa.[19] Tumor formation as a result of viral infection is another possible mechanisms by which viruses may induce vascular disease as is the case with cytomegalovirus and Kaposi's sarcoma.[22] At the time our laboratory began investigating the *in vitro* interaction between virus and EC,[8,17,45] few such studies had been reported.[1,20] Since the last review, Visser *et al.*[72] reported enhanced thrombin generation and platelet binding on herpes simplex virus (HSV)-infected endothelium and Dunn and Spear[80] demonstrated that cell surface heparan sulfate proteoglycan serves as the initial receptor for both HSV-1 and HSV-2 in HEp-2 cells. In this report I essentially summarize our published work as well as recently published work by other investigators. In our studies we attempted to (1) define further the viruses that are capable of infecting EC *in vitro*[18]; (2) examine the effect of viral infection on (a) adherence of granulocytes to EC,[45] (b) induction of Fc and C3b receptors,[8] and (c) matrix protein synthesis[32,35,44,77]; (3) determine the effect of viral infection on the composition of the

---

Nicholas A. Kefalides • Connective Tissue Research Institute and Department of Medicine, University of Pennsylvania, and University City Science Center, Philadelphia, Pennsylvania 19104.

Endothelial Cell Dysfunctions, edited by Nicolae Simionescu and Maya Simionescu. Plenum Press, New York, 1992.

subendothelial substratum and on the phenotypic expression of EC; and (4) demonstrate the inhibiting effect of lithium chloride on the ability of HSV to suppress host cell protein synthesis.[89]

II. EXPERIMENTAL PROCEDURES

A. Endothelial and Smooth Muscle Cell Cultures

Human EC were isolated from umbilical cord vein according to the method of Gimbrone *et al.*[21] Human EC were grown in tissue culture flasks that were coated with 1% gelatin in modified medium 199 supplemented with 90 μg/ml heparin[68] and 30 μg/ml EC growth factor.[46]

Bovine arterial and venous EC were isolated and grown as described by Macarak *et al.*[43] and Friedman *et al.*[18] Bovine smooth muscle cells (SMC) were obtained from the Institute for Medical Research, Camden, New Jersey and grown as recommended by the provider institute.

B. Viral Infection of Blood Vessel Cells

Virus pools for EC or SMC inoculation were prepared on human embryonic lung fibroblasts (MRC-5), rhesus monkey kidney or human epidermoid carcinoma cells (HEP-2) as described by Friedman *et al.*[17] Twenty-four hours before infection, the growth medium was removed and replaced with heparin-free medium and the concentration of the EC growth factor increased to 150 μg/ml. Confluent monolayers were infected with virus for 1 hr at 37°C at a multiplicity of infection (MOI) ranging from 0.003 to 50 depending on the nature of the experiment. After allowing the virus to adsorb to the EC or SMC, the medium and any unadsorbed virus were removed and the monolayer washed and refed with growth medium. Control cultures were either noninfected or mock-infected.

C. Indicators of Viral Infection

Viral infection of blood vessel cells was determined by immunofluorescence using specific antisera to detect viral antigens within EC or SMC cells and by observing monolayers for virus-induced cytopathology. To determine whether viral replication occurred, infected cultures were harvested for viral titrations 2 hr to 10 days after inoculation. Viral titers of infectious particles were expressed either as the highest dilution of fluid which produces cytopathology in 50% of the indicator cell cultures inoculated—50% tissue culture infectious dose endpoint ($TCID_{50}$)[17]—or as the number of infectious particles determined by a plaque assay.[60]

D. Assessment of Cell Response to Viral Infection

1. Granulocyte Adherence Assay

Granulocyte adherence to the endothelial monolayers grown in the 35-mm-diameter petri dishes was measured as described by MacGregor *et al.*[45] Medium 199 was decanted from the cultures and 1 ml of heparinized whole blood (5 U/ml) was added to triplicate plates, just covering the surface. Human blood was used in all experiments. The endothelial monolayer–blood overlay was incubated at 37°C and 100% humidity for 15 min without agitation and then the blood was aspirated. Comparison of the granulocyte counts before and after incubation permitted determination of the percentage of granulocytes to the endothelium. In some experiments, a pure suspension of granulocytes in Hanks' balanced salt solution (HBSS) was used rather than whole blood. Granulocytes were separated by Hypaque–Ficoll density gradient sedimentation, washed three times in modified Hanks' solution, and suspended in HBSS at a concentration of 5–10 × 10^6 cells/ml. Human umbilical vein EC or bovine aorta EC monolayers were infected with either adenovirus 7 (MOI = 0.3), polio 1 (MOI = 0.03), measles (MOI = 0.03), or herpes simplex 1 (MOI = 0.03) viruses.[45] EC cultures were harvested for viral titrations 2 hr after inoculation of cultures and then at daily intervals for up to 10 days.

2. Assay for Fc and C3 Receptors

To assay for Fc and C3 receptors, cells were grown to confluency in 24-well plates.[8] Half of the wells were infected with HSV-1 at an MOI of 0.5–1.0 (10^5 $TCID_{50}$/well). The remaining half of the wells were mock-infected to serve as controls. After a 1-hr adsorption at 37°C, the monolayers were washed and refed with 1 ml of medium. The cells were then incubated for 24 hr at 37°C, examined for cytopathic effect by phase-contrast microscopy, and used for the studies described below. Monolayers were also prepared from human mononuclear cells separated by Hypaque–Ficoll density centrifugation. These cells were utilized as positive controls for the binding of antibody or complement-coated erythrocytes. The ability of uninfected or virally infected EC to bind immunoglobulin or complement-coated particles was studied using fresh sheep erythrocytes radiolabeled with ^{51}Cr (Na_2CrO_4).[8]

3. Assessment of Matrix Protein Synthesis

Matrix protein synthesis was measured by metabolic labeling of the EC or SMC cultures as described by Kefalides and Ziaie[32] and Ziaie *et al.*[77] For metabolic studies, the cells were washed with PBS once and labeling medium was added at different time points postinfection. The labeling medium was Earle's minimal essential medium supplemented with 0.1% glucose, 1 mM glutamine, 50 μg/ml ascorbic acid, 50 μg/ml β-aminopropionitrile fumarate, and 15 mM *N*-2-hydroxyethylpiperazine-*N*-2-ethanesulfonic acid. For ^{14}C labeling, 6 μCi/ml [^{14}C]proline (250 mCi/mmole; Amersham Corp., Arlington Heights, Ill.) was added. For ^{35}S labeling, the minimal essential medium was replaced by Eagle's methionine-free medium supplemented with 50 μCi/ml [^{35}S]methionine (1400 Ci/mmole; Amersham). In the latter case, all cultures were starved with methionine-free medium for 1 hr prior to labeling. Following the labeling, at the specified times, the cell layer, medium, and matrix fractions were collected separately and analyzed.

For analysis, the medium was collected, protease inhibitors (*N*-ethylmaleimide (NEM) 10 mM and phenylmethanesulfonyl fluoride (PMSF) 1 mM] added, and the medium centrifuged at 500*g* for 20 min at 4°C to remove cell debris. Any virus that was released in the medium was either inactivated by the addition of Triton X-10 in 500 mM tris(hydroxymethyl)aminomethane (Tris), pH 7.2, to a final concentration of 1%or removed by centrifugation at 115,000*g* for 1 hr at 4°C. The cell layer was extracted by either of the following two methods: (1) 1% Triton X-100 in 50 mM Tris, pH 7.2, containing inhibitors (NEM and PMSF) was added and the lysate cleared by centrifugation at 18,000*g* for 40 min at 4°C or (2) extraction buffer containing 2 M urea, 50 mM NaCl, 5 mM NEM, 1 mM PMSF, 15 mM dithiothreitol, and 15 mM *N*-2-hydroxyethylpiperazine-*N*-2-ethanesulfonic acid was added to the cell layer, incubated at room temperature for 3 hr, and the layer removed by scraping. Complete virus particles and cell debris were removed by centrifugation at 115,000*g* for 1 hr at 4°C. The matrix was dissolved in 1% sodium dodecyl sulfate (SDS) in 50 mM Tris, pH 7.2, containing protease inhibitors (NEM and PMSF) and clarified by centrifugation at 18,000*g* for 40 min at 4°C. For analytical purposes, the samples were dialyzed against appropriate buffers.

Protein analysis and identification of specific matrix components was carried out by polyacrylamide gel electrophoresis (SDS–PAGE), electroimmunoblot, and enzyme-linked immunosorbent assay (ELISA). Proteins were analyzed by SDS–PAGE essentially according to Laemmli[34] as described by Kefalides and Ziaie.[32] After electrophoresis, the gels were stained with Coomassie brilliant blue. Fluorography was carried out according to Bonner and Laskey[5] using En^{3}Hance (New England Nuclear, Boston, Mass.). Dried gels were exposed to sensitized Kodak X-O-Mat XAR-5 film.[36]

For electroimmunoblot, proteins were separated by SDS–PAGE and transferred electrophoretically onto a nitrocellulose sheet (Bio-Rad Laboratories, Richmond, Calif.) according to Towbin *et al.*[69]

ELISA was performed essentially according to Engvall[10] as modified by Rennard *et al.*[53] To collect medium samples of control and infected cultures for competitive ELISA, growth medium was replaced by serum-free medium at 4 hr postinfection and the cultures were incubated for 13 hr (corresponding to 17 hr postinfection). Following this, the medium was collected as described previously and dialyzed against PBS–0.05% Tween and analyzed.

4. RNA Isolation

Total RNA was isolated from virus- and mock-infected human EC and SMC as described by London *et al.*[40] Cultures were harvested by trypsinization and the pelleted cells lysed and incubated at 42°C in 10 mM Tris-HCl (pH 7.5) containing 5 mM EDTA, 0.5% SDS, and proteinase K (125 μg/ml). RNA was extracted with phenol/chloroform, selectively precipitated overnight at −20°C with 3 M sodium acetate, 5 mM EDTA (pH 7.0), and washed with ethanol as described by Rowe *et al.*[58] RNA was quantitated by spectrophotometry at 260 nm. Twenty million cells yielded approximately 400 μg RNA.

5. RNA Blotting and Hybridization

Total cellular RNA, at several dilutions per sample, was heat denatured for 15 to 20 min in 4.6 M formaldehyde 7.5× SSC (1× SSC: 0.15 M NaCl, 0.015 M trisodium citrate) and applied to nitrocellulose filters using the Minifold II blot apparatus and recommended procedure of Wahl.[73] The filters were baked for 2 hr at 78°C under vacuum. cDNA probes were nick translated with [^{32}P]dCTP and [^{32}P]-TTP (Amersham, Arlington Heights, Ill.), without excision from the plasmid, to a specific gravity of 6–8 × 10^8cpm/μg. Hybridization was carried out as described by Lyons-Giordano *et al.*[41]

6. Proteoglycan Synthesis

To assess the effect of HSV infection on proteoglycan synthesis by EC, virus- and mock-infected cultures were treated in identical fashion as described by Kaner *et al.*[31] After 1 hr adsorption, the medium and any unadsorbed virus were removed, the monolayer was washed 3× with Dulbecco's phosphate saline (pH 7.3) (PBS), and refed with heparin-free growth medium. After an additional 6 hr, the cells were metabolically labeled with 100 μCi/ml $Na_2{}^{35}SO_4$ (43 Ci/mg S; ICN Radiochemicals, Irvine, Calif.). Radiolabeling was continued for a total of 48 hr, a period during which incorporation into proteoglycans synthesized by EC is linear. For analysis of newly synthesized proteoglycans, the medium was collected and made 4 M in guanidine·HCl in the presence of protease inhibitors, followed by centrifugation at 12,0000 rpm for 40 min at 4°C. The supernatant was collected and any residual virus was inactivated by the addition of Triton X-100 at a final concentration of 2%. The cell layer was extracted with 4 M guanidine·HCl, 2% Triton X-100, and protease inhibitors and scraped with a rubber policeman, followed by centrifugation. The unincorporated radioisotope was removed by gel filtration on disposable 10-ml Sephadex G-50 columns, equilibrated, and eluted with a 50 mM Tris, pH 7.5 buffer, containing 8 M urea, 25 mM EDTA, and 0.2% Triton X-100. Aliquots of the excluded volumes were counted for radioactivity. The excluded fractions were applied to a column of DEAE–Sephacel (0.9 × 15 cm) equilibrated and washed with the same buffer. The column was eluted with a continuous 0.1–0.7 M NaCl gradient in the same buffer. Fractions from the DEAE–Sephacel column were pooled and concentrated on a Centricon YM-10 membrane (Amicon Corp., Danvers, Mass.) or run directly on Sepharose CL-4B columns equilibrated and eluted with a 0.1 M Tris buffer, pH 7.0, containing 4 M guanidine·HCl, 0.1 M sodium sulfate, and 0.2% Triton X-100.

7. Exposure of Cells to Lithium Chloride

To study the effects of lithium chloride on HSV-infected EC, cultures of EC were infected with HSV-1 at a multiplicity of infection of 1 essentially as described by Ziaie *et al.*[77] Lithium chloride (20 or 30 mM) was added to EC cultures either with the virus or 1 hr after adsorption of the virus.[79] The concentration of LiCl was kept constant throughout the experiment. The cultures were labeled with 50 μCi/ml [^{35}S]methionine (ICN Radiochemicals) at 19 hr postinfection for 5 hr. For analysis the medium cell layer was separated and treated as described earlier,[77] before analysis by SDS–PAGE. Densitometric scannings of the fluorograms were performed by a laser scanner (Pharmacia LKB Biotechnology Inc., Piscataway, N. J.).

III. RESULTS

A. Replication of Virus

Results of viral infection of human and bovine EC are shown in Table 1 and Fig. 1. The data show that several human viruses can replicate in vascular cells. Although herpes simplex 1 and 2, adenovirus 7, measles, mumps, parainfluenza 3, polio 1, and echo 9 replicate in human venous EC, others, such as cytomegalovirus (CMV), influenza A/Victoria,

Table 1. Infection of Human Venous and Bovine Arterial Endothelium by Common Human Viruses[a]

Virus	Human endothelium				Bovine endothelium			
	Inoculum titer (MOI)	Viral replication	CPE	Positive IFA	Inoculum titer (MOI)	Viral replication	CPE	Positive IFA
Herpesviruses								
Herpes simplex virus	10^6 (6)	Yes	Yes	Yes	10^4 (0.03)	Yes	Yes	Yes
Cytomegalovirus	$10^{4.7}$ (0.3)	No	No	No	10^3 (0.003)	No	No	No
Respiratory viruses								
Adenovirus type 7	$10^{2.7}$ (0.003)	Yes	Yes	Yes	$10^{2.7}$ (0.001)	Yes	Yes	Yes
Measles	10^4 (0.05)	Yes	Yes	Yes	10^4 (0.03)	Yes	Minimal	Yes
Mumps	10^4 (0.05)	Yes	Yes	Yes	10^4 (0.03)	Yes	Minimal	ND
Parainfluenza type 3	10^7 (50)	Yes	Yes	Yes	10^7 (30)	Yes	Yes	Yes
Influenza A/Victoria	$10^{4.7}$ (0.3)	No	No	No	$10^{4.7}$ (0.1)	No	No	No
Respiratory syncytial	10^3 (0.0005)	No	No	No	10^3 (0.003)	No	No	No
Enteroviruses								
Poliovirus type I	$10^{2.7}$ (0.003)	Yes	Yes	ND	$10^{2.7}$ (0.001)	No	No	ND
Echovirus type 9	10^6 (5)	Yes	Yes	ND	10^6 (3)	Yes	No	ND
Coxsackievirus B4	10^5 (0.5)	No	No	ND	10^5 (0.3)	Yes	No	ND

[a]MOI (multiplicity of infection) = the ratio of the titer of infectious virus in the inocululm to the number of cells in the monolayer; IFA = immunofluorescence; ND = not done. Modified from Friedman *et al.*[17]

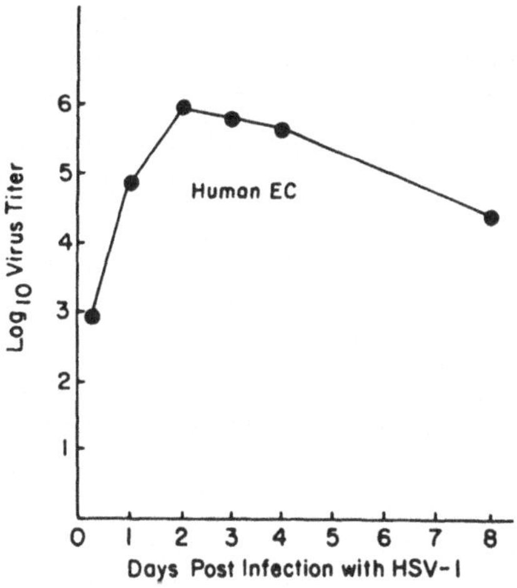

Figure 1. Growth curve of HSV-1 inoculated onto human EC cultures.

respiratory syncytial virus, and coxsackie B4, do not. The same viruses which grew in the human venous EC replicated in bovine arterial and venous EC, except that polio 1 did not grow whereas coxsackie B4 did grow in the latter. The failure of CMV to grow in either the human or bovine EC may be unique for these cells since replication of this virus has been reported in human arterial SMC.[71] These findings suggest that EC may provide an active barrier to some viruses, thus preventing certain viruses from infecting specific tissues supplied by a given vessel. Moreover, the development of significant cytopathology by some viruses implies that viral infection can potentially influence the functions of the host cell, as will become evident in later sections.

To evaluate the possibility that EC from different vessels vary in their susceptibility to viral infection, cultures of EC from different major vessels of a single species, i.e., bovine, were examined for their susceptibility to viral infection and replication.[18] Table 2 shows the

Table 2. Comparative Growth of Viruses in Bovine Pulmonary Artery, Thoracic Aorta, and Inferior Vena Cava Endothelial Cell Cultures

	Peak viral titers ($\log_{10}$ $TCID_{50}$/ml)[a]			
Virus	Pulmonary artery	Thoracic artery	Vena cava	Comments
Measles	5.2 (3.7)	2.7	2.7 (3.7)	Only minimal cytopathology was detectable
Mumps	(6.0)		1.0 (5.7)	Only minimal cytopathology was detectable
Herpes simplex I	5.7	5.7	5.0	Cytopathology characterized by rounding of cells and retraction of monolayer
Coxsackie B4	4.2	3.7	4.2	No cytopathology was detectable
Echo 9	3.7	4.4	4.2	No cytopathology was detectable

[a]Viral titers are calculated by determining the highest dilution of material harvested from infected endothelial cell cultures which produces viral cytopathic changes when inoculated onto indicator tissue culture cells which are known to be susceptible to the virus. Data from Friedman *et al.*[18]

results with five common viral pathogens of humans (herpes simplex 1, measles, mumps, echo 9, and coxsackie B4 viruses). These viruses were evaluated for growth in EC derived from bovine fetal pulmonary artery, thoracic aorta, and vena cava. All five species replicated in each type of EC. There were apparent differences in the quantities of measles and mumps viruses produced in pulmonary artery endothelium compared with thoracic aorta and vena cava when EC were obtained from different animals. However, when pulmonary artery EC cells were compared with vena cava cells from the same animal, growth of each virus was similar in the two cell types. Four of the viruses replicated in the various EC without producing appreciable changes in cell morphology. These results indicate that EC from different blood vessels are equally susceptible to the human viruses evaluated and that viral replication can occur without major alterations in cell morphology. In the latter case, EC could serve as permissive cells permitting viruses to leave the circulation, establish a persistent infection, and initiate infection in adjacent tissues, including the subendothelial SMC.

Studies in our laboratory have demonstrated that bovine aorta as well as human umbilical cord artery SMC will support growth of HSV-1.[35,40] Figure 2 shows a growth curve of HSV-1 in bovine SMC infected at an MOI of 5. Maximum titer was achieved at 24 hr postinfection after which it began to decrease slowly. Tumilowicz *et al.*[71] have shown that SMC from human umbilical cord artery will support replication of CMV. These authors found that the cytopathic effect of CMV in SMC was delayed when compared with infected IMR-90 fibroblasts, resulting in prolonged survival of a fraction of the starting cell population.

B. Effect of Viral Infection on Granulocyte Adherence

In collaboration with Dr. R. R. MacGregor, we examined the adherence of granulocytes on monolayers of human and bovine EC before and after viral infection with different viruses.[45] Adherence to uninfected human EC using 1.0 ml of whole blood averaged 17.9 ± 3.7% and to bovine EC was 20.3 ± 3.7%. The impact of infection of the endothelium by different viruses on subsequent granulocyte adherence was notable. Poliovirus produced an

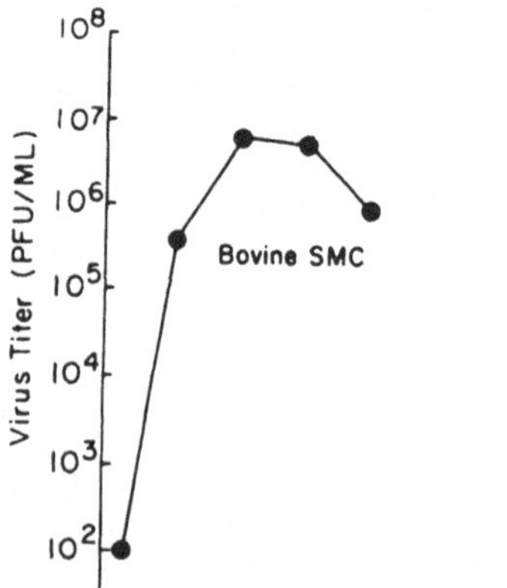

Figure 2. Growth curve of HSV-1 inoculated onto bovine SMC cultures.

acute lytic infection of human EC, with associated increased adherence to 185.4% of control 24 hr after inoculation. Significantly increased adherence was noted at 6 hr, before detectable cytopathic effect. HSV-1 caused a similar rapidly lytic infection of bovine endothelium in association with increased adherence to 213.7% of control 6 hr after inoculation. This augmented adherence could be demonstrated when granulocytes were suspended in physiologic saline solution, showing that antibody and complement need not be present. Trypsin treatment of infected monolayers did not prevent the augmentation, and supernate from infected monolayers increased the adherence of polymorphonuclear leukocytes to normal, uninfected monolayers. Chronic, slowly lytic infections, lasting 7 days or more, were induced with adenovirus in human endothelium and with measles virus in bovine cells. Adherence increased as virus was noted in the cell cultures on day 4, several days before cytotoxicity was seen. Thus, chronic viral infection of the endothelium appears possible and results in increased granulocyte adherence. The authors suggested that naturally occurring disease such as infection may act synergistically with adherent granulocytes to damage the endothelium and may represent an *in vitro* model of vasculitis.[45]

C. Induction of Fc and C3 Receptors

Studies by Westmorland and Watkins[75] demonstrated the induction of receptors for the Fc portion of IgG on fibroblasts infected with HSV. Fc and C3 receptors are sites for attachment of antigen–antibody complexes and can be demonstrated on the surfaces of monocytes, granulocytes, and lymphocytes. Although most reports indicate that Fc receptors are not readily detectable in EC from various tissues,[29,61] when a highly sensitive radioimmunoassay was used, Lyss *et al.*[42] showed binding of IgG to cultured human umbilical vein EC.

When human EC were infected with HSV-1 there was induction of receptors for the Fc portion of IgG and for C3.[8] In these studies, it was noted that when human EC were incubated with ^{51}Cr-labeled sheep erythrocytes sensitized with IgG, IgM, or IgM plus complement, no preferential binding of IgG- or complement-coated erythrocytes to EC monolayers was observed. In contrast, significant binding of erythrocytes coated with IgG or IgM plus complement was observed after viral infection. Phase-contrast and scanning electron microscopy demonstrated erythrocyte adherence around the infected EC in a rosette pattern. Binding of IgG-coated erythrocytes was fully inhibited by Fc (0.31 mg/ml) but not Fab′ fragments of nonimmune IgG. Binding of complement-coated cells was unaffected by the presence of IgG (1 mg/ml). With purified individual components, binding of complement-coated erythrocytes was proportional to the concentration of C3. Binding of IgG- or C3-coated cells was detected beginning 4 hr after infection.

These studies indicate that HSV-1 infection can induce IgG and C3 receptors on human EC thus promoting the deposition of immune complexes in vascular tissue. It is conceivable that other forms of injury or infection with other viruses may also induce similar receptors which may play a role in immunologically mediated vessel disease.

D. Suppression of Host-Cell Matrix Protein Synthesis

A number of viruses suppress and eventually shut off the synthesis of cellular proteins. The mechanism(s) is not clearly understood.[12,62] Competition for ribosomes by the large excess of viral mRNA has been offered as one possibility. Even when viral mRNA is not in excess, it may have a selective advantage in binding to ribosomes and initiating translation.[30] In cells infected with adenovirus, late shutoff of cellular protein may in part be due to inhibition of transport of cellular mRNA out of the nucleus.[3] Early studies in which permissive cells were infected with HSV-1 demonstrated a disruption in polysomes[67] which was followed by a rapid decline in host-cell protein synthesis.[14,55] HSV-1 also mediates the inhibition of cellular RNA[66] and DNA synthesis.[55] Studies by Fenwick and Clark[13] suggest that in Vero cells infected with a temperature-sensitive mutant of HSV-1 (tsB7) there is an early suppression of host-cell protein synthesis which is virion protein dependent and a delayed shutoff which requires virus protein synthesis. It has also been shown that several strains of HSV-2 cause a more rapid decrease in host-cell protein synthesis than certain HSV-1 strains[26,52] and that the shutoff by HSV-1 and HSV-2 is accomplished by different mechanisms.[26]

We chose to study the effect of viral infection on the synthesis of matrix proteins by vascular cells because EC and SMC rest on or are surrounded by a basement membrane matrix which contains several proteins having important functions in cell adhesion, cell proliferation, cell differentiation, and gene expression. We propose that *in vivo* cell injury, whether in the form of viral infection, radiation, or some immune process, can initiate a series of specific cellular responses that can result in the synthesis of altered extracellular matrix components and thus lead to long-term sequelae of chronic vascular disease.

Studies on the effects of HSV-1 infection (MOI = 1) on matrix protein synthesis by bovine EC were recently reported by our laboratory.[44] Protein synthesis was measured in infected and uninfected cultures by following the incorporation of [^{14}C]proline into nondialyzable protein at different times postinfection. Medium proteins were analyzed by polyacrylamide gel electrophoresis, gel filtration, and electroimmunoblotting. No change was detected in total ^{14}C incorporation into nondialyzable protein in infected compared to uninfected endothelial cells; however, there was a significant decrease in the synthesis of collagen and fibronectin in infected cultures. The hydroxy[^{14}C]proline content of fractionated medium proteins showed no significant differences between infected and uninfected cultures. It is suggested, therefore, that the decrease in collagen synthesis cannot be explained by increased collagen degradation.

The above studies examined the effects of prolonged viral infection of bovine EC on matrix protein synthesis. We thought it would be important to determine the early changes in human EC matrix protein synthesis following infection with HSV-1 or HSV-2.[32,77] Our results demonstrate that during infection, synthesis of host-specific proteins decreases with time, while an increase in the synthesis of viral proteins is observed. Figure 3 shows the electrophoretic analysis of samples from infected and mock-infected human EC cultures. In infected cultures, the synthesis of fibronectin (FN), type IV collagen, and thrombospondin (TSP) has decreased (Fig. 3, lanes 2 and 4 of the cell and medium fractions, respectively). However, the decrease in the level of TSP in the medium fraction is not as pronounced compared with FN and type IV collagen (Fig. 3, lanes 3 and 4). These data suggest that there is a suppression of host-cell protein synthesis and, specifically, of matrix proteins. The degree of this suppression appears to vary with the protein. Although the level of cellular proteins decreases during infection, new species of proteins appear in the infected cultures. These new proteins correspond to nearly synthesized HSV-1 polypeptides (compare lane 2 with lanes 7 and 8 in Fig. 3).[27,28]

In order to determine the time of onset of inhibition of cellular protein synthesis and

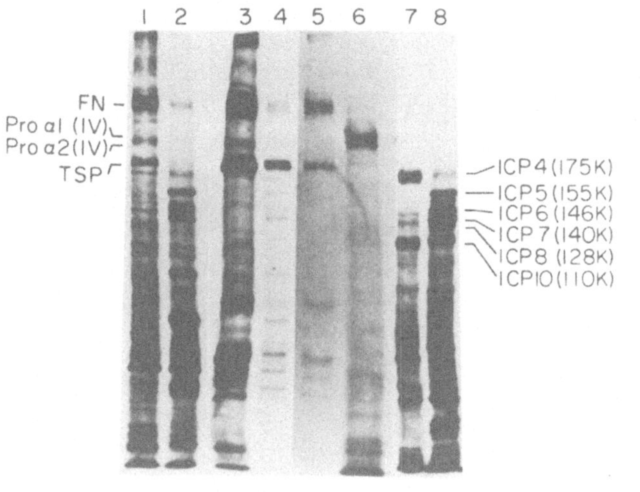

Figure 3. Pattern of protein synthesis by EC cultures infected with HSV-1. EC cultures were infected with HSV-1 at an MOI of 5 for 1 hr and then labeled with [^{14}C]proline at 5 hr postinfection for a period of 10 hr. The cell layer and medium fractions were collected and analyzed by SDS–PAGE. Cell layer: mock-infected (lane 1), infected (lane 2); medium: mock-infected (lane 3), infected (lane 4); FN and TSP standards (lane 5); procollagen type IV standard (lane 6); ICP = infected cell proteins, α- and β-polypeptides of HSV-1 (lanes 7 and 8, respectively). Note the substantial reduction in the level of FN, type IV collagen, and TSP in infected cultures (lanes 2 and 4) in comparison to controls (lanes 1 and 3). Data from Ziaie *et al.*[77]

the effect of the size of virus inoculum, human EC cultures were pulsed for 2 hr with [^{35}S]methionine at 2, 4, 6, 8, and 10 hr postinfection.[77] One set of experimental cultures was infected with HSV-1 at an MOI of 5 and another at an MOI of 20. For the cell layer fractions, the results are shown in Fig. 4. In cultures infected at an MOI of 20, inhibition of host-cell protein synthesis begins at 2 hr and becomes almost complete at 10 hr postinfection. However, at an MOI of 5, there appears to be some synthesis of matrix proteins, particularly of TSP even at 10 hr postinfection. Regardless of the size of inoculum, it appears that the degree of inhibition of synthesis after HSV-1 infection of human EC varies with each protein. At 10 hr postinfection, with an MOI of 5, TSP synthesis continued at a low level in the infected cells while FN and collagen type IV synthesis was not detectable under the same conditions. Densitometric analysis of SDS-PAGE fluorograms showed that at 6 and 10 hr postinfection, TSP decreases by 33.3 and 68.0% at a MOI of 5 and by 91.5 and 96.6% at a MOI of 20, respectively. The values obtained for FN for the same time points are decreased by 83.1 and 86.2% at a MOI of 5 and 96.6 and 99.4% at a MOI of 20, respectively. The value obtained for collagen type IV was 12.5% of control (87.5% decrease) at a MOI of 5 at 6 hr postinfection.

The specific matrix proteins were identified by electroimmunoblot using antibodies against type IV collagen, FN, and TSP.[32] Quantitative estimation of the decrease in FN and TSP synthesized during infection was accomplished in an ELISA system.[32,77] Although we have focused on the synthesis of three protein components of EC during viral infection, it is apparent that a variety of host proteins are affected by viral infection. The components migrating slightly slower than FN (Fig. 4) have tentatively been identified as von Wille-

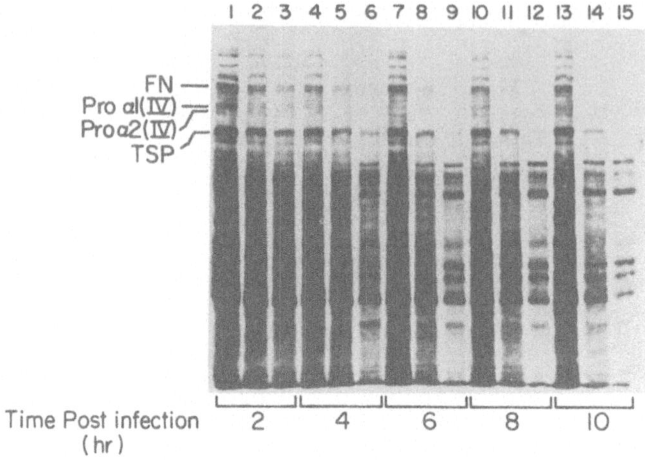

Figure 4. Effect of virus dose on the time and degree of inhibition of host protein synthesis. EC cultures were infected with HSV-1 at an MOI of 5 or 20 for 1 hr and then pulse-labeled with 50 μCi/ml [^{35}S]methionine for 2 hr at different times postinfection (2, 4, 6, 8, and 10 hr). A fluorogram of SDS–PAGE of the cell fraction is shown. Control cultures, lanes 1, 4, 7, 10, and 13; cultures infected with an MOI of 5, lanes 2, 5, 8, 11, and 14; cultures infected with an MOI of 20, lanes 3, 6, 9, 12, and 15. Positions of FN, type IV collagen, and TSP are indicated. Decline in the level of these proteins is observed at 2 hr postinfection with an MOI of 20 (lane 3) and continues throughout. The inhibition is much more effective with an MOI of 20 at any time point postinfection, but varies with the protein inoculum size. Data from Ziaie *et al.*[77]

brand factor, and as is evident, its synthesis is inhibited by the virus (lanes 2, 4, 6, 8, and 10). These results are consistent with the findings of Schek and Bachenheimer[59] who demonstrated that HSV-1 and HSV-2 infection of Vero cells decreased the cytoplasmic levels of β- and γ-actin, β-tubulin, and histone H3 and H4 mRNA, though not all at the same rate. Similarly, Mayman and Nishioka[47] observed that in Friend erythroleukemia cells infected with HSV-1, the ratio of the steady-state level of histone H3 mRNA to total RNA remained unchanged for the first 4 hr postinfection, while that of globin and actin mRNAs decreased progressively at early intervals postinfection. Unpublished studies in our laboratory demonstrate that the decrease in type IV collagen, FN, and von Willebrand synthesis correlates with decreases in the steady-state levels of mRNAs for these proteins.

Studies by Nishioka and Silverstein,[50] who used murine erythroid cells transformed by Friend leukemia virus, and Fenwick and Clark,[13] who infected Vero cells with HSV-1, suggest that early suppression of cellular protein following infection with these viruses is not dependent on protein synthesized after infection. Similar results were obtained by us[32] when human EC were infected with UV-inactivated HSV-1 and HSV-2. At 4 hr postinfection, control and infected cultures were pulsed for 1 hr with [^{35}S]methionine. Figure 5 shows the electrophoretic pattern of the cell-matrix fraction of cultures infected with either intact or UV-irradiated HSV-1 and HSV-2. It will be noted that, although in cultures infected with intact virus at MOIs of 5 or 20 (Fig. 5, HS1 5, HS2 5, HS1 20, HS2 20) there was suppression of matrix protein synthesis, viral protein synthesis was quite evident. UV irradiation of either HSV-1 or HSV-2 at an MOI of 20 caused suppression of matrix as well as virus protein synthesis (Fig. 5, HS1 irr 20, HS2 irr 20). In order to show that suppression of host-cell matrix protein synthesis with UV-irradiated virus was not occurring at low levels

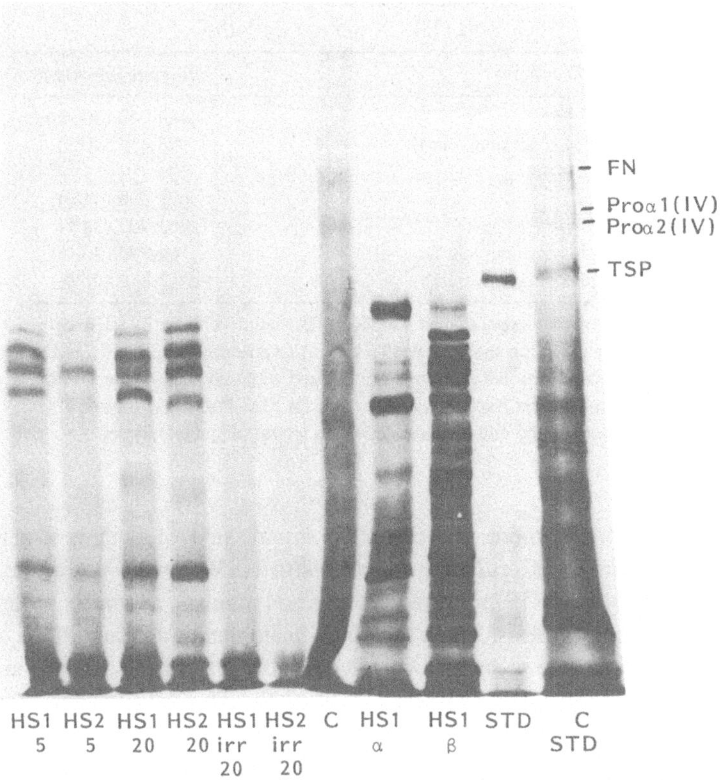

Figure 5. The effect on host-cell protein synthesis following infection of EC with intact HSV-1 (HS1) and HSV-2 (HS2) and UV-irradiated virus (HS1 irr and HS2 irr) is depicted in this fluorogram. Infection with HSV-1 or HSV-2 was carried out at an MOI of 5 or 20. Uninfected (C) and infected cultures were pulsed with [^{35}S]methionine for 1 hr at 4 hr postinfection and the cell-matrix fraction analyzed by SDS–PAGE. Standards: HS1 α, HS1 β = immediate early and early polypeptides, respectively, of HSV-1; STD = labeled standard of thrombospondin; C STD = matrix proteins in a cell fraction from a control uninfected EC culture that was labeled for 2 hr. Note the suppression of host-cell and viral proteins with the irradiated virus. Data from Kefalides and Ziaie.[32]

of viral DNA transcription, infection was carried out with intact virus in the presence and absence of actinomycin D, an inhibitor of mRNA synthesis.[32] Table 3 shows the total ^{35}S-labeled protein content of human EC cultures infected with HSV-1 or HSV-2 in the presence and absence of actinomycin D. It will be noted that protein-bound radioactivity decreased by only 60% in the presence of actinomycin in control uninfected cultures whereas in infected cultures actinomycin D resulted in further decrease, down to 2.8% with HSV-1 and 0.7% with HSV-2, respectively. These findings suggest that additional inhibition of host-protein synthesis occurred as a result of the viral infection but in the absence of new protein synthesis.

The electrophoretic pattern of the cell-matrix fraction of human EC cultures infected in the presence and absence of actinomycin D is shown in Fig. 6. In the presence of actinomycin D there was only partial shutoff of protein synthesis in control cultures (Fig. 6, C + ACT) due to continued translation of preexisting mRNA; however, in the presence of

Table 3. Effect of Actinomycin D on Protein Synthesis of HSV-1- and HSV-2-Infected Endothelial Cells

EC cultures	^{35}S (cpm/10^6 cells)[a]
Uninfected	2,488,740 (100%)
Uninfected plus actinomycin D	995,490 (40%)
Infected	
HSV-1	1,098,000 (44%)
HSV-2	962,000 (38%)
HSV-1 plus actinomycin D	69,390 (2.8%)
HSV-2 plus actinomycin D	17,880 (0.7%)

[a]Average of two experiments. Duplicate EC cultures were infected with HSV-1 or HSV-2 in the presence (2 μg/ml) or absence of actinomycin D. At 4 hr postinfection the cultures were pulsed with [^{35}S]methionine for 1 hr and the cell-matrix fraction was processed for SDS-PAGE. Actinomycin D was present in the culture throughout the length of the experiment.

virus plus actinomycin D, there was complete suppression of both host-cell and virus protein synthesis. Human EC cultures infected with HSV-1 or HSV-2 in the absence of actinomycin D (Fig. 6, HS1; HS2) showed suppression of host-cell matrix protein synthesis but active viral protein synthesis. These results indicate that under conditions of inhibition of synthesis of new mRNA, i.e., inhibition of viral protein synthesis, synthesis of host-cell proteins from preexisting host-cell mRNA is blocked. The data further suggest that in

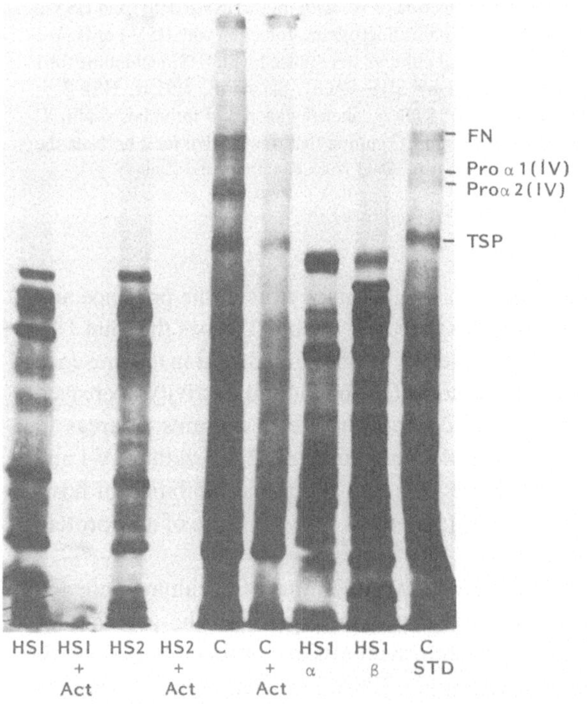

Figure 6. The effect of inhibition of mRNA synthesis on EC cultures infected with HSV-1 (HS1) and HSV-2 (HS2) is depicted in this fluorogram. Actinomycin D was added at the beginning of infection (HS1 + Act, HS2 + Act) and maintained until the end of the experiment. Cultures were pulsed with [^{35}S]methionine for 1 hr at 4 hr postinfection and the cell-matrix fraction analyzed by SDS–PAGE. C = control culture; C + Act = control + actinomycin D. Standards: HS1 α, HS1 β = immediate early and early polypeptides, respectively, of HSV-1; C STD = matrix proteins in a cell fraction from a control uninfected EC culture that was labeled for 2 hr. Note the suppression of host-cell and viral proteins in the cultures infected in the presence of actinomycin D (HS1 + Act; HS2 + Act). Data from Kefalides and Ziaie.[32]

human EC cultures infected with HSV-1 or HSV-2, inhibition of host-cell protein synthesis most likely occurs at the translational level and does not require prior synthesis of viral protein. The presence of virus proteins (virions) appears to be sufficient to shut off host-protein synthesis.

The effect of HSV-1 infection on matrix protein synthesis was also investigated in our laboratory using bovine aorta SMC.[35] Monolayers of SMC were infected at an MOI ranging from 0.1 to 20. Indicators of infection included cytopathology, viral titer, and immunofluorescence. Cytopathology varied with the time postinfection and was directly proportional to the MOI. Viral replication occurred at 0.1, 10, and 20 MOI and maximum titers were achieved by 24 hr postinfection. Infected and uninfected cultures of SMC were pulse-labeled with either [^{14}C]proline or [^{35}S]methionine at different hours postinfection. Incorporation of radioactivity into nondialyzable protein was determined in fluorograms following SDS–PAGE of the cell-matrix or medium fractions. The synthesis of FN and collagen types I and III was suppressed and the degree of suppression was dependent on the duration of infection and on the virus dose. These data suggest that SMC can support HSV-1 replication *in vitro* and that such infection can lead to altered matrix protein synthesis. Unpublished results from our laboratory also demonstrate that HSV-1 infection of SMC *in vitro* leads to qualitative alteration in the protein composition of the extracellular substratum. SDS–PAGE analysis of the extracellular substratum matrix after labeling infected SMC with [^{35}S]methionine for 2 hr and chasing for 4, 24, and 48 hr postlabeling revealed a decrease in the deposition of cell matrix proteins but considerable deposition of polypeptides corresponding to HSV-1 proteins. This is analogous to what we observed with HSV-1-infected human EC (see below).

London *et al.*,[40] in our laboratory, have shown that suppression of host protein synthesis in human SMC occurs early following infection with HSV-1 and this takes place before virus protein expression. To confirm that suppression of host protein synthesis occurs before virus protein expression, cultures were treated with actinomycin D (5 μg/ml) continuously from the beginning of infection. In the presence of actinomycin D, by 3.5 hr postinfection, virus-infected cultures showed much less incorporation of [^{35}S]methionine into SDS–PAGE-separated products than did identically treated mock-infected cultures. In the absence of actinomycin D, there was incorporation of radiolabel into viral proteins as early as 4.5 hr postinfection.

These data are consistent with the suggestion that suppression of host protein synthesis in SMC occurs before virus protein expression and that this is accomplished through a virion-competent factor.

E. Suppression of Proteoglycan Synthesis

The effects of HSV-1 infection on proteoglycan synthesis by human EC were studied in our laboratory by Kaner *et al.*[31] Confluent cultures of early passage EC from human umbilical vein were infected with HSV-1 at MOI from 0.001 to 1.0. HSV-1 infection produced a dose-dependent inhibition of total proteoglycan synthesis of up to 85%. Although there was a two- to threefold increase in the quantity of virus necessary to cause 50% inhibition of heparan sulfate compared to chondroitin/dermatan sulfate proteoglycan, the inhibition was relatively parallel, even up to high virus doses. There was no inhibition of an undersulfated heparan sulfate proteoglycan which contained glycosaminoglycan chains shorter than the predominant species. The decrease in the synthesis of heparan sulfate proteoglycans as well as chondroitin/dermatan sulfate proteoglycans relative to total proteoglycans appear in Table 4. The virus-induced changes in proteoglycan metabolism may influence cell–matrix interactions and lead to altered vessel wall function.

F. Steady-State Levels of mRNA for Host Proteins

To determine whether a correlation exists between suppression of synthesis of specific proteins and their mRNA levels, we measured the steady-state levels of mRNAs for several extracellular matrix proteins (Col IV, FN, and TSP) and intracellular proteins (actin and tublin) in human EC following HSV-1 infection. The results show that at 0–4 hr postinfection, when there is a rapid decrease in the synthesis of extracellular matrix proteins, the steady-state levels of the corresponding mRNAs remain high, but progressively decline to levels of 10% or less by 13 hr postinfection. These findings suggest that in the early hours postinfection there is an alteration in the translatability of the hybridizable message followed by degradation in the later hours. The data also indicated that not all host mRNA levels are affected equally. For example, the levels of hybridizable message for the pro α1 chain of type IV collagen and for actin decrease more slowly than do those for FN, TSP, and α-tubulin.[7]

Levels of mRNA for specific host proteins were also determined in HSV-1-infected human SMC by London *et al.*[40] in our laboratories. Hybridization of slot-blotted total RNA demonstrated that virus-infected cultures treated with and without actinomycin D showed a marked decrease in hybridizable RNA by 2 hr postinfection when compared with mock-infected controls. This decrease was seen with cDNA probes for the mRNA for secreted extracellular matrix macromolecules [FN, TSP, and collagen chains α2(I) α1(III)] as well as with cDNA probes for RNA of intracellular macromolecules actin and α-tubulin. *In vitro* translation of total RNA isolated from infected human SMC at 2 and 5 hr postinfection resulted in far less methionine incorporation into SDS–PAGE-separated products than would be predicted from the amount of hybridizable mRNA available. The results suggest that host message is rapidly rendered nonfunctional as well as degraded by means of a virion-competent mechanism(s).

Table 4. Quantitative Effect of HSV-1 Infection on Glycoproteins and Proteoglycans from Human Endothelial Cells

	% of uninfected controls	
	Medium	Cell layer
[^{35}S]methionine total proteins[a]	19%	48%
[^{35}]sulfate total proteoglycans[a]	6%	17%
Fibronectin[b]	5%	14%
Thrombospondin[b]	25%	32%
Collagen type IV[b]	0%	14%
Heparan sulfate proteoglycans[a]	5%	11%
Chondroitin/dermatan sulfate proteoglycan[a]	5%	8%

[a]From Kaner *et al.*[31]

[b]By densitometry of fluorograms, from SDS–PAGE analysis of cultures pulse-labeled at 10–12 hr postinfection. Modified from Ziaie *et al.*[77]

G. Qualitative Alteration in the Composition of the Extracellular Matrix

Studies from our laboratory[78] demonstrate that HSV-1 infection of human EC at MOI of 0.05, 0.1, and 1.0 followed by a 2-hr pulse with [^{35}S]methionine at 19 hr postinfection and a chase for 4, 24, and 48 hr resulted in a qualitatively altered extracellular matrix substratum. This was demonstrated by the presence of HSV-1 polypeptides in the matrix fraction upon SDS–PAGE. The matrix fraction of the EC cultures infected at either 0.05 or 0.1 MOI contained reduced amounts of cell-matrix proteins as well as proteins corresponding to HSV-1 polypeptides. The matrix fraction from the human EC cultures infected at an MOI of 1.0 contained labeled proteins corresponding to HSV-1 polypeptides only; host-cell proteins were almost completely absent.

When the substratum from infected human EC cultures, freed of their cell layer and virus inactivated, was subsequently used for the attachment of fresh, uninfected human EC, it was noted that attachment was not significantly different between "uninfected" and "infected" EC substratum. However, the amount of cell matrix proteins deposited by EC growing on the "infected" substratum was significantly diminished. These are exciting findings and suggest that viral infection can qualitatively alter the extracellular substratum which in turn could influence cellular gene expression.

H. Inhibition of Suppression of Host-Cell Matrix Protein Synthesis by Lithium Chloride

In a number of studies, remission of recurrent herpes infection was noted in patients under lithium therapy.[37,38,63] Trousdale *et al.*[70] and Patou *et al.*[51] have shown that lithium inhibits the replication of HSV-1 and HSV-2 in several cell culture systems. We decided, therefore, to examine the effect of lithium on host-cell protein synthesis in HSV-1-infected EC cultures. Protein synthesis was measured in HSV-1-infected EC in the presence and absence of 20 and 30 mM LiCl. Although LiCl restored synthesis of almost all host-cell proteins, [^{35}S]methionine incorporation was most pronounced in TSP and plasminogen activator inhibitor 1 and least in FN and type IV collagen (Fig. 7). LiCl was more effective at

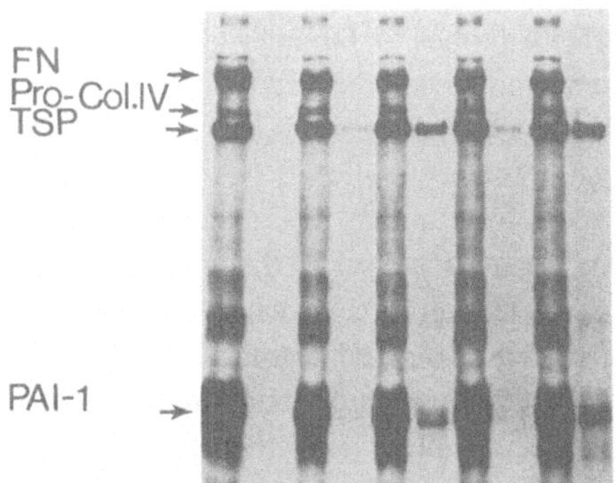

Figure 7. Pattern of protein synthesis by mock- and HSV-1-infected EC cultures in the presence and absence of lithium chloride. Cultures were labeled with [^{35}S]methionine at 19 hr postinfection for 5 hr. The fluorogram represents the medium fraction analyzed by SDS–PAGE. Lane 1: mock-infected. Lane 2: HSV-1-infected. Lanes 3–6: lithium added after virus adsorption: mock-infected, lithium 20 mM (lane 3) and 30 mM (lane 5); HSV-infected, lithium 20 mM (lane 4) and 30 mM (lane 6). Lanes 7–10: lithium added with the virus: mock-infected, lithium 20 mM (lane 7) and 30 mM (lane 9); HSV-1-infected, lithium 20 mM (lane 8) and 30 mM (lane 10). Positions of FN, procollagen type IV, TSP, and PAI-1 (plasminogen activator inhibitor-1) are shown. Data from Ziaie and Kefalides.[79]

the higher concentration (30 mM) and when it was added to the EC culture at the time of infection rather than after adsorption of HSV-1. Synthesis of virus proteins was markedly suppressed in LiCl-treated EC. The data suggest that LiCl interferes not only with virus replication, but may also, to some extent, interfere with a virion-associated inhibition of host protein synthesis. There was a direct correlation between the degree of increased host protein synthesis in lithium-treated HSV-infected EC and the mRNA levels for the specific protein (unpublished data).

IV. DISCUSSION

The way an individual's tissue responds to injury is a very important factor in determining whether chronic sequelae shall develop and tissue function shall be compromised. There are several types of vascular pathology, including vasculitis, thrombosis, atherosclerosis, aneurysmal dilation, capillary basement membrane thickening, vascular tumors, and others whose pathogenesis is incompletely understood. Little is known about the early, initiating events which lead to discernible pathologic and functional changes in blood vessels. We chose to study the *in vitro* infection of vascular cells with human viruses in order to learn more about the way EC and SMC respond to this particular form of injury. The parameters we have studied are affected in the early stages following viral infection *in vitro* and may reflect analogous responses of the same vessel cells in *in vivo* situations.

The studies reviewed here suggest that EC and SMC are susceptible to a number of common human viruses. Infection of EC or SMC leads to viral replication but not all viruses cause cytopathic changes. EC form our first line of defense in preventing the spread of a virus from the vascular lumen to the abluminal side of the EC and the subsequent spreading of the infection to tissues supplied by a given vessel.

Viral infection of EC leads to increased adherence of granulocytes which in turn may increase the changes of further injury to EC through immune mechanisms. Additional mechanisms of EC injury may be brought into play after viral infection as a result of induction of Fc and C3 receptors by HSV-1. Binding of immune complexes to such receptors may be critical in bringing about EC injury. Other early changes associated with viral infection of EC and SMC include the suppression of host-cell matrix protein synthesis and the incorporation of viral polypeptides into the extracellular matrix substratum. Using cDNA probes for human type IV procollagen, FN, TSP, and von Willebrand factor, we have determined that the steady-state levels of mRNA specific for these proteins are significantly decreased in the early hours postinfection. However, determination of steady-state levels for actin and β-tubulin mRNA showed a significant delay before there was a drop in mRNA levels to less than 10% of control cultures. Similarly, HSV-1 infection of SMC showed a moderate decrease in the mRNA levels for a variety of host proteins in the early hours postinfection (1–2 hr) which remained unchanged until about 9 hr postinfection after which there was a marked decrease to levels less than 10% of control cultures.[40] The mechanism(s) by which this comes about are not clearly understood. It is important to note, however, that new virus protein synthesis is not required for this effect. It has been suggested that disruption of polysomes and inactivation of mRNA are definite possibilities.

The demonstration of viral polypeptides in the matrix fraction of the EC cultures was an interesting and potentially significant finding. The altered substratum may cause detachment of newly deposited cells, may influence gene expression, and result in further alteration of the extracellular matrix composition or it may attract other cells which could release substances that stimulate or inhibit cell proliferation. Finally, the virus itself may persist in a latent state without destroying its host but it could perturb cell functions by disordering the synthesis of specific enzymes or hormones.

The ability of lithium chloride to prevent the suppression of host cell protein synthesis by HSV was a surprising finding since it was expected that by suppressing viral replication lithium would have allowed the virion-associated function to perform its task. We have evidence (unpublished observations) which shows that the suppression of host-cell protein synthesis that occurs in HSV-infected EC in the presence of actinomycin D is prevented by lithium. This suggests that lithium may interfere with the virion-associated suppression of host-cell protein synthesis. Although there is evidence to suggest that lithium salts interfere with viral DNA synthesis and viral replication, the mechanism by which this is accomplished is still unknown. It appears, therefore, that a clear understanding of the mechanisms that initiate the response of vascular cells to injury is an important requirement for understanding the pathogenesis of vascular disease.

ACKNOWLEDGMENTS. The author thanks Ms. Sheri Johnson and Ms. Margarette Wills for typing the manuscript. This work was supported by NIH grants HL-29492 and AM-20553 from the United States Public Health Service.

REFERENCES

1. Andrews, B. S., Theophilopoulos, A. N., Peters, C. J., Loskutoff, D. J., Brandt, W. E., and Dixon, F. J., 1978, Replication of dengue and junin viruses in cultured rabbit and human endothelial cells, *Infect. Immun.* **20:**776.
2. Ashford, T. P., and Freiman, D. G., 1968, Platelet aggregation at sites of minimal endothelial injury: An electron microscopic study, *Am. J. Pathol.* **53:**599.
3. Beltz, G. A., and Flint, S. J., 1979, Inhibition of HeLa cell protein synthesis during adenovirus infection: Restriction of cellular messenger RNA sequences to the nucleus, *J. Med. Biol.* **131:**353.
4. Benditt, E. P., Barret, T., and McDougall, J. K., 1983, Viruses in the etiology of atherosclerosis, *Proc. Natl. Acad. Sci. USA* **80:**6386.
5. Bonner, W. M., and Laskey, R. A., 1974, A film detection method for tritium-labeled proteins and nucleic acids in polyacrylamide gels, *Eur. J. Biochem.* **46:**83.
6. Brinker, J. M., Ziaie, Z., and Kefalides, N. A., 1987, Differential stability of host poly A^+ mRNAs in endothelial cells infected with herpes simplex type 1, *Fed. Proc.* **46:**1993.
7. Brinker, J. M., Ziaie, Z., and Kefalides, N. A., 1991, Differential suppression of host cell protein synthesis and mRNA levels in herpes simplex virus infected endothelial cells, *Virus Res.* **19:**209.
8. Cines, D. B., Lyss, A. P., Bina, M., Corkey, R., Kefalides, N. A., and Friedman, H. M., 1982, Fc and C3 receptors induced by herpes simplex virus on cultured human endothelial cells, *J. Clin. Invest.* **69:**123.
9. Colman, R. W., Robboy, S. J., and Minna, J. D., 1972, disseminated intra-vascular coagulation (DIC): An approach, *Am. J. Med.* **52:**679.
10. Engvall, E., 1980, Enzyme immunoassay ELISA and EMIT, *Methods Enzymol.* **70:**419.
11. Fabricant, C. G., Fabricant, J., Litrenta, M. M., and Minick, C. R., 1978, Virus-induced atherosclerosis, *J. Exp. Med.* **148:**335.
12. Fenwick, M. L., 1984, The effects of herpes viruses on cellular macromolecular synthesis, in: *Comprehensive Virology*, Vol. 19 (H. Fraenkel-Conrat and R. R. Wagner, eds.), Plenum Press, New York, pp. 359–390.
13. Fenwick, M. L., and Clark, J., 1982, Early and delayed shut-off of host protein synthesis in cells infected with herpes simplex virus, *J. Gen. Virol.* **61:**121.

14. Fenwick, M. L., and Walker, M. J., 1978, Suppression of the synthesis of cellular macromolecules by herpes simplex virus, *J. Gen. Virol.* **41**:37.
15. French, J. E., 1966, Atherosclerosis in relation to the structure and function of the arterial intima, with special reference to the endothelium, *Int. Rev. Exp. Pathol.* **5**:253.
16. Friedman, H. M., 1984, Viral infection of endothelium and the induction of Fc and C3 receptors, in: *Biology of Endothelial Cells* (E. A. Jaffe, ed.), Nijhoff, The Hague, pp. 268–276.
17. Friedman, H. M., Macarak, E. J., MacGregor, R. R., Wolfe, J., and Kefalides, N. A., 1981, Virus infection of endothelial cells, *J. Infect. Dis.* **143**:266.
18. Friedman, H. M., Wolfe, J., Kefalides, N. A., and Macarak, E. J., 1986, Susceptibility of endothelial cells derived from different blood vessels to common viruses, *In Vitro* **22**:397.
19. Fye, K. H., Becker, M. J., Theophilopoulos, A. N., Moutsopoulos, H., Feldman, J., and Talal, N., 1977, Immune complexes in hepatitis: β-antigen associated periarteritis nodosa, *Am. J. Med.* **62**:783.
20. Gimbrone, M. A., and Fareed, G. C., 1976, Transformation of cultured human vascular endothelium by SV40 DNA, *Cell* **9**:685.
21. Gimbrone, M. A., Cotran, R. S., and Folkman, J., 1974, Human vascular endothelial cells: Growth and DNA synthesis, *J. Cell Biol.* **60**:673.
22. Giraldo, G., Beth, E., and Huang, E. S., 1980, Kaposi's sarcoma and its relationship to cytomegalovirus (CMV), *Int. J. Cancer* **26**:23.
23. Gyorkey, F., Melnick, J. L., Guinn, G. A., Gyorkey, P., and DeBakey, M. E., 1984, Herpesviridae in the endothelial and smooth muscle cells of the proximal aorta in arteriosclerotic patients, *Exp. Mol. Pathol.* **40**:328.
24. Hajjar, D. P., Falcone, D. J., Fabricant, C. G., and Fabricant, J., 1985, Altered cholesteryl ester cycle is associated with lipid accumulation in herpes virus-infected arterial smooth muscle cells, *J. Biol. Chem.* **260**:6124.
25. Hajjar, D. P., Nicholson, A. C., Hajjar, K. A., Sando, G. N., and Summers, B. D., 1989, Decreased messenger RNA translation in herpes virus-infected arterial cells: Effects on cholesteryl ester hydrolase, *Proc. Natl. Acad. Sci. USA* **86**:3366.
26. Hill, T. M., Sinden, R. R., and Sadler, J. R., 1983, Herpes simplex virus types 1 and 2 induce shut-off of host protein synthesis by different mechanisms in Friend erythroleukemia cells, *J. Virol.* **45**:241.
27. Honess, R. W., and Roizman, B., 1974, Regulation of herpesvirus macromolecular synthesis. I. Cascade regulation of the synthesis of three groups of viral proteins, *J. Virol.* **14**:8.
28. Honess, R. W., and Roizman, B., 1975, Regulation of herpesvirus macromolecular synthesis. II. Sequential transition of polypeptide synthesis requires functional viral polypeptide, *Proc. Natl. Acad. Sci. USA* **72**:1276.
29. Johnson, P. M., Faulk, W. P., and Wang, A. C., 1976, Immunological studies of human placenta: Subclass fragment specificity of binding of aggregated IgG by placental endothelial cells, *Immunology* **31**:659.
30. Joklik, W. K., 1981, Structure and function of the reovirus genome, *Microbiol. Rev.* **45**:483.
31. Kaner, R. J., Iozzo, R. V., Ziaie, Z., and Kefalides, N. A., 1990, Inhibition of proteoglycan synthesis in human endothelial cells after infection with herpes simplex virus type 1 *in vitro*, *Am. J. Respir. Cell Mol. Biol.* **2**:423.
32. Kefalides, N. A., and Ziaie, Z., 1986, Herpes simplex virus suppression of human endothelial matrix protein synthesis is independent of viral protein synthesis, *Lab. Invest.* **55**:328.
33. Kistler, G. S., Caldwell, P. R. G., and Weibel, E. R., 1967, Development of fine structural damage to alveolar and capillary lining cells in oxygen poisoned rat lungs, *J. Cell Biol.* **33**:605.
34. Laemmli, U. K., 1970, Cleavage of structural proteins during the assembly of the head of bacteriophage T4, *Nature* **227**:680.
35. Lashgari, M., Friedman, H. M., and Kefalides, N. A., 1986, Herpes simplex (HSV-1) infection of bovine aorta smooth muscle cells (SMC) inhibits matrix protein synthesis, *Fed. Proc.* **45**:684.
36. Laskey, R. A., and Mills, A. D., 1976, Quantitative film detection of ^{3}H and ^{14}C polyacrylamide gels by fluorography, *Eur. J. Biochem.* **56**:335.
37. Lieb, J., 1979, Remission of recurrent herpes infection during therapy with lithium, *N. Engl. J. Med.* **301**:942.
38. Lieb, J., 1981, Immunopotentiation and inhibition of herpes virus activation during therapy with lithium carbonate, *Med. Hypothesis* **7**:885.
39. London, F. S., Ziaie, Z., Brinker, J. M., and Kefalides, N. A., 1988, Suppression of mRNA for host proteins in herpes simplex virus type I infected human smooth muscle cells, *J. Cell Biol.* **107**:135a.

40. London, F. S., Brinker, J. M., Ziaie, Z., and Kefalides, N. A., 1990, Suppression of host mRNA in human smooth muscle cells by a virion competent factor in herpes simplex virus type I *Lab. Invest.* **62:**189.
41. Lyons-Giordano, B., Brinker, J.M., and Kefalides, N. A., 1990, The effect of heparin on fibronectin and thrombospondin synthesis and mRNA levels in cultured human endothelial cells, *Exp. Cell Res.* **186:**39.
42. Lyss, A. P., Finko, R., Knight, K., Bina, M., Reeber, M., and Cines, D. B., 1982, Interaction of IgG with human endothelial cells, *Clin. Res.* **30:**323A.
43. Macarak, E. J., Howard, B. V., and Kefalides, N. A., 1977, Properties of calf endothelial cells in culture, *Lab. Invest.* **36:**62.
44. Macarak, E. J., Friedman, H. M., and Kefalides, N. A., 1985, Herpes simplex virus type 1 infection of endothelium reduces collagen and fibronectin synthesis, *Lab. Invest.* **53:**280.
45. MacGregor, R. R., Friedman, H. M., Macarak, E. J., and Kefalides, N. A., 1980, Virus infection of endothelial cells increases granulocyte adherence, *J. Clin. Invest.* **65;**1469.
46. Maciag, T., Cerundolo, J., Ilsley, S., Kelley, P. R., and Forand, R., 1979, An endothelial cell growth factor from bovine hypothalamus: Identification and partial characterization, *Proc. Natl. Acad. Sci. USA* **76:**5674.
47. Mayman, B. A., and Nishioka, Y., 1985, Differential stability of host mRNAs in Friend erythroleukemia cells infected with herpes simplex virus type 1, *J. Virol.* **54:**1.
48. Melnick, J. L., Petrie, B. L., Dreesman, G. R., Burek, J., McCullum, C. H., and DeBakey, M. E., 1983, Cytomegalovirus antigen within human arterial smooth muscle cells, *Lancet* **2:**644.
49. Minick, C. R., Fabricant, C. G., Fabricant, J., and Litrenta, M. M., 1979, Atheroarteriosclerosis induced by infection with a herpesvirus, *Am. J. Pathol.* **96:**673.
50. Nishioka, Y., and Silverstein, S., 1977, Degradation of cellular mRNA during infection by herpes simplex virus, *Proc. Natl. Acad. Sci. USA* **74:**2370.
51. Patou, G., Crow, T. J., and Taylor, G. R., 1986, The effects of psychotrophic drugs on synthesis of DNA and the infecting of herpes simplex virus, *Biol. Psychiatry* **21:**1221.
52. Pereira, L., Wolff, M. H., Fenwick, M. L., and Roizman, B., 1977, Regulation of herpesvirus macromolecular synthesis. V. Properties of polypeptides made in HSV-1 and HSV-2 infected cells, *Virology* **77:**733.
53. Rennard, S. I., Berg, R., Martin, G. R., Foidart, J. M., and Gehron-Robey, P., 1980, Enzyme-linked immunoassay (ELISA) for connective tissue components, *Anal. Biochem.* **104:**205.
54. Robbins, S. L., 1967, *Pathology*, Saunders, Philadelphia, pp. 293–294.
55. Roizman, B., Borgman, G. S., and Kamali-Rousta, M., 1967, Macromolecular synthesis in cells infected with herpes simplex virus, *Nature* **206:**1374.
56. Rosenberg, J. C., Hawkins, E., and Rector, F., 1971, Mechanisms of immunological injury during antibody-mediated hyperacute rejection of renal heterografts, *Transplantation* **11:**151.
57. Ross, R., and Harker, L., 1976, Hyperlipidemia and atherosclerosis: Chronic hyperlipidemia initiates and maintains lesions by endothelial cell desquamation and lipid accumulation, *Science* **193:**1094.
58. Rowe, D. W., Noen, R. C., Davidson, J. M., Byers, P., Bornstein, P., and Palmiter, R. D., 1978, Correlation of procollagen mRNA levels in normal and transformed chick embryo fibroblasts with different rates of procollagen synthesis, *Biochemistry* **17:**1581.
59. Schek, N., and Bachenheimer, S. L., 1985, Degradation of cellular mRNAs induced by a virion-associated factor during herpes simplex virus infection of Vero cells, *J. Virol.* **55:**601.
60. Schmidt, N. J., 1979, Laboratory diagnosis of viral infection, in: *Antiviral Agents and Viral Diseases* (G. J. Galasso, T. C. Merigan, and R. A. Buchanan, eds.), Raven Press, New York, pp. 209–252.
61. Shingu, M., Hashimoto, Y., Johnson, A. R., and Hurd, E. R., 1981, The search for Fc receptors on human tissues and human endothelial cells in culture, *Proc. Soc. Exp. Biol. Med.* **167:**147.
62. Shatkin, A. J., 1983, Molecular mechanisms of virus mediated cytopathology, *Philos. Trans. R. Soc. London Ser. B.* **303:**167.
63. Skinner, G. R. B., Hartley, C., Buchan, A., Harper, L., and Gallimore, P., 1980, The effect of lithium chloride on the replication of herpes simplex virus, *Med. Microbiol. Immunol.* **168:**139.
64. Stehbens, W. E., 1976, Hemodynamic production of lipid deposition, intimal tears, mural dissection and thrombosis of blood vessel walls, *Proc. R. Soc. London Ser. B.* **185:**357.
65. Stemerman, M., Spaet, T. H., Pitlick, F., Cintron, J., Lejnieks, I., and Tiell, M. L., 1977, Intimal healing: The pattern of re-endothelialization and intimal thickening, *Am. J. Pathol.* **87:**125.
66. Stenberg, R. M., and Pizer, L. I., 1982, Herpes simplex virus-induced changes in cellular and adenovirus RNA metabolism in adenovirus type 5-transformed human cell line, *J. Virol.* **42:**474.

67. Sydiskis, R. J., and Roizman, B., 1967, The disaggregation of host polysomes in productive and abortive infection with herpes simplex virus, *Virology* **32:**678.
68. Thornton, S. C., Mueller, S. N., and Levine, E. M., 1983, Human endothelial cells: Use of heparin in cloning and long-term serial cultivation, *Science* **222:**623.
69. Towbin, H., Staehelin, T. H., and Gordon, J., 1979, Electrophoretic transfer of proteins from polyacrylamide gels to nitrocellulose sheets: Procedure and some applications, *Proc. Natl. Acad. Sci. USA* **76:**4350.
70. Trousdale, M. D., Gordon, Y. J., Peters, A. C. B., Gropen, T. I., Nelson, E., and Nesburn, A. B., 1984, Evaluation of lithium as an inhibitory agent of herpes simplex virus in cell cultures and during reactivation of latent infection in rabbits, *Antimicrob. Agents Chemother.* **25:**522.
71. Tumilowicz, J. J., Gawlik, M. E., Powell, B. B., and Trentin, J. J., 1985, Replication of cytomegalovirus in human arterial smooth muscle cells, *J. Virol.* **56:**839.
72. Visser, M. R., Tracy, P. B., Vercelloti, G. M., Goodman, J. L., White, J. G., and Jacob, H. S., 1988, Enhanced thrombin generation and platelet binding on herpes simplex virus-infected endothelium, *Proc. Natl. Acad. Sci. USA* **85:**8227.
73. Wahl, G., 1981, Rapid detection of DNA and RNA using slot blotting. M. Schleicher and Schuell Sequences, Application update No. 371, Schleicher & Schuell, Inc., Keene, N.H.
74. Wall, R. T., Harlan, J. M., Harker, L. A., and Striker, G. E., 1980, Homocysteine-induced endothelial cell injury *in vitro*: A model for the study of vascular injury, *Thromb. Res.* **18:**113.
75. Westmorland, D., and Watkins, J. F., 1974, The IgG receptor induced by herpes simplex virus: Studies using radiolabeled IgG, *J. Gen. Virol.* **24:**167.
76. Yamashiroya, H. M., Ghosh, L., Yang, R., and Robertson, Jr., A. L., 1988, Herpesviridae in the coronary arteries and aorta of young trauma victims, *Am. J. Pathol.* **130:**71.
77. Ziaie, Z., Friedman, H. M., and Kefalides, N. A., 1986, Suppression of matrix protein synthesis by herpes simplex virus type 1 in human endothelial cells, *Collagen Relat. Res.* **6:**33.
78. Ziaie, Z., and Kefalides, N. A., 1987, Herpes simplex virus (HSV-1) incorporated in the extracellular matrix of HSV-1 infected endothelial cells (EC) influences behavior of freshly seeded EC, *Fed. Proc.* **46:**975.
79. Ziaie, Z., and Kefalides, N. A., 1989, Lithium chloride restores host protein synthesis in herpes simplex virus-infected endothelial cells, *Biochem. Biophys. Res. Commun.* **160:**1073.
80. Wu Dunn, D., and Spear, P. G., 1989, Initial interaction of herpes simplex virus with cells is binding to heparan sulfate, *J. Virol.* **63:**520.

IV

Endothelial Cell in Hypertension, Hyperlipidemia, and Atherosclerosis

16

Changes of Arterial Endothelial Cells in Chronic Hypertension with Special Reference to Regressive Changes and Replication

Fumitada Hazama and Masakiyo Sasahara

I. INTRODUCTION

Hypertension is a potent risk factor for atherosclerosis and causes hypertensive vascular changes by itself. On the other hand, injuries to endothelial cells have been shown to be a possible initiating event in atherogenesis[1,2] as well as in the development of hypertensive arterial changes.[3,4] Therefore, information about the morphological and functional changes of endothelial cells associated with chronic hypertension and the mechanisms by which hypertension injures endothelial cells is indispensable for a better understanding of the pathogenesis of atherosclerosis and hypertensive vascular changes. Recently it has been found that endothelial cells contain factors related to the contraction, dilation, migration, and proliferation of medial smooth muscle cells. Therefore, it is possible that endothelial cells have an important role in the regulation of the vascular functions. Information about the effects of chronic hypertension on such regulating mechanisms of endothelial cells is also very important.

The human cases of chronic hypertension are almost always complicated by atherosclerotic changes, and it is very difficult to differentiate pure hypertensive changes from atherosclerotic ones. An appropriate animal model of chronic hypertension is required to perform controlled studies. A strain of spontaneously hypertensive rats (SHR) was established by Okamoto and Aoki.[5] By further inbreeding of the animals in the SHR substrains which had a high incidence of developing cerebrovascular lesions,[6] a special strain was established, the animals of which showed a very high incidence of it. They were

Fumitada Hazama and Masakiyo Sasahara • Department of Pathology, Shiga University of Medical Science, Seta, Otsu 520-21, Japan.

Endothelial Cell Dysfunctions, edited by Nicolae Simionescu and Maya Simionescu. Plenum Press, New York, 1992.

named "stroke-prone spontaneously hypertensive rats (SHRSP)."[7] Because the animals in SHR or SHRSP show hypertensive vascular changes and sequelae in various organs,[3,6] the strains have been regarded as appropriate animal models for research on chronic hypertension.

In this chapter we review studies on the morphological and functional changes in endothelial cells in SHR, particularly in SHRSP with special reference to regressive changes and replication of the cells based mainly on our own experimental results. Emphasis will be placed on the changes of endothelial cells in cerebral arteries.

II. IMPORTANCE OF ENDOTHELIAL CHANGES IN THE DEVELOPMENT OF HYPERTENSIVE VASCULAR AND PARENCHYMAL CHANGES

A. Main Cerebral Lesions and Pathological Changes of Cerebral Arteries[3,6]

The main cerebral lesions observed in SHR (including SHRSP) are cerebral hemorrhages, and old and new necrosis with or without cyst formation. Edema is sometimes the sole finding observed in the markedly swollen brains in the animals showing neurological signs.

Various arterial changes are observed as the morphological basis of the cerebral lesions, namely, marked cellular hyperplasia of the pial arteries with narrowed or occluded lumina, hyaline degeneration, fibrinoid necrosis of the intracerebral arterioles accompanied by thickening of the wall, thrombosis, and microaneurysms.

B. Correlation between Endothelial Changes and the Hypertensive Vascular and Parenchymal Changes

The above-mentioned hypertensive vascular changes in the brains of SHR and SHRSP have been studied using various morphological methods, and our findings had suggested the importance of endothelial changes and hence induced increased vascular permeability in the development of the changes of vessel walls as well as brain parenchyma.[3,4]

Namely, (1) we had to assume that some growth factors are involved in cellular hyperplasia of the arterial walls in chronic hypertension. The proliferation of medial smooth muscle cells is often observed particularly beneath injured endothelial cells.[3] Insudation of plasma containing growth factors through injured endothelial cells seems to play a role in cellula hyperplasia. A platelet-derived growth factor may also take part in the cellular hyperplasia, because small clusters of platelets often adhered to the endothelial surface of the cerebral arteries in SHR.[4] In any case, endothelial injury seems to be an important condition for cellular hyperplasia. The endothelial cell-derived growth factors will be described later.

(2) Hyaline substances deposited in the arterial walls are electron microscopically thickened and confluent basement membranes having a lamellar pattern,[8] and biochemically are biosynthetic overproduced products of collagen by endothelial cells and medial smooth muscle cells.[9]

(3) Where there is fibrinoid necrosis, a deposition of fibrin materials and their precursors is observed in the arterial walls.[8] For the fibrin formation, the insudation of plasma containing fibrinogen through injured endothelial cells is a presupposed condition.

(4) Thrombosis is often observed in the cerebral arterial system in SHR.[3,6] Endothelial cells at sites where platelet adhesion is observed on the luminal surface of the cerebral arteries in SHR are in some cases entirely desquamated or degenerated.[3] The injury of endothelial cells plays an important role in the development of thrombosis.

(5)Edematous changes are a common finding in the SHR brain.[3,6] Such changes are the result of an increased vascular permeability due to functional and organic endothelial injury caused by chronic hypertension. Cerebrovascular permeability in SHRSP was investigated using horseradish peroxidase as a tracer, and an increased permeability was demonstrated in the animals particularly with extremely high blood pressure.[10] A leakage of plasma protein in SHRSP was also shown by immunohistochemical detection of extravasated fibrinogen.[11] The locations of the increased permeability proved to correspond well with the sites that tended to have cerebrovascular lesions.[10] We may, therefore, suppose that endothelial injury and thus induced increased cerebrovascular permeability due to hypertension leaves certain permanent cerebral lesions.

III. ULTRASTRUCTURAL CHANGES OF ENDOTHELIAL CELLS IN THE CEREBRAL ARTERIES IN SHR[3,4,8,12]

The findings of vascular changes in the brains of SHR and SHRSP suggest the importance of endothelial changes and hence induced vascular permeability in the development of the changes of vessel wall as well as brain parenchyma. Therefore, we tried to obtain further information concerning the ultrastructural changes of the cells using scanning[4,12] and transmission electron microscopes.[3,4,8]

In scanning electron microscopic studies, the endothelial cells reveal, on the one hand, progressive changes such as enlargement, increased plasmalemmal pits, numerous microvilli, and well-developed marginal folds. On the other hand, regressive changes such as balloonlike protrusions and craterlike concavities are noted with scanning electron microscopy and vacuoles, inclusion bodies, and coagulative necrosis with transmission electron microscopy.

In SHR, enlarged endothelial cells are observed not only at branching sites but also in other areas. Such enlarged endothelial cells contain large nuclei and are rich in organelles. They can be regarded as regenerated immature cells replacing injured cells, because the regenerated young endothelial cells are said to be characterized by a large cytoplasm, a well-developed Golgi apparatus, and rough-surfaced endoplasmic reticulum.[13] Some of such enlarged endothelial cells seem to participate in the production of hyaline substances, because in contact with thickened basement membranes enlarged endothelial cells with an increased number of free ribosomes and distended rough surfaced endoplasmic reticulums are observed.[3,8]

A large number of microvilli are observed on the surfaces of many endothelial cells in the SHR cerebral arteries. This abundance of microvilli also seems to be an expression of immaturity of the SHR endothelial cells, because the presence of profuse microvilli on the endothelial surface has been reported in fetal arteries.[14] The functions of the endothelial microvilli are unknown, although they are assumed to be related to augmented cellular activity.[15]

In highly magnified scanning electron micrographs, a large number of pinholes are observed in the luminal surface of the endothelial cells of cerebral arteries in SHR. Such minute holes correspond to the plasmalemmal pits demonstrated in transmission electron microscopic study.[16] In the latter study, the endothelial cells of the pial arteries as well as intracerebral arterioles have an increased number of pinocytotic vesicles in SHR as compared with control animals. Experiments on the permeability of the endothelial cells using horseradish peroxidase have shown that hypertension is a factor in the acceleration of vesicular transport in endothelial cells.[3] Degenerated cells with balloonlike protrusions or craterlike concavities have numerous plasmalemmal pits on their luminal surface (Fig. 1), a finding which suggests that the increased pinocytosis due to hypertension is one of the factors initiating endothelial degeneration in SHR.

Endothelial cells of the cerebral arteries in SHR display well-developed marginal folds. They are very irregular in arrangement and size. The junctions between the cells are irregular and partial detachment is seen. The various irregularities of the cells seem to be related to the rearrangement of regenerated cells replacing the injured cells.

On the luminal surface of endothelial cells of the cerebral arteries in SHR, severe regressive changes such as balloonlike protrusions, craterlike concavities, and transitional forms between these changes are often observed. Because some craters have plasmalemmal pits or microvilli on the surface of the concavity, the origin of the crater seems to be a collapsed balloon, as has been pointed out by Kawamura *et al.*[17] Transmission electron microscopic study shows vacuolar degeneration, the appearance of inclusion bodies and

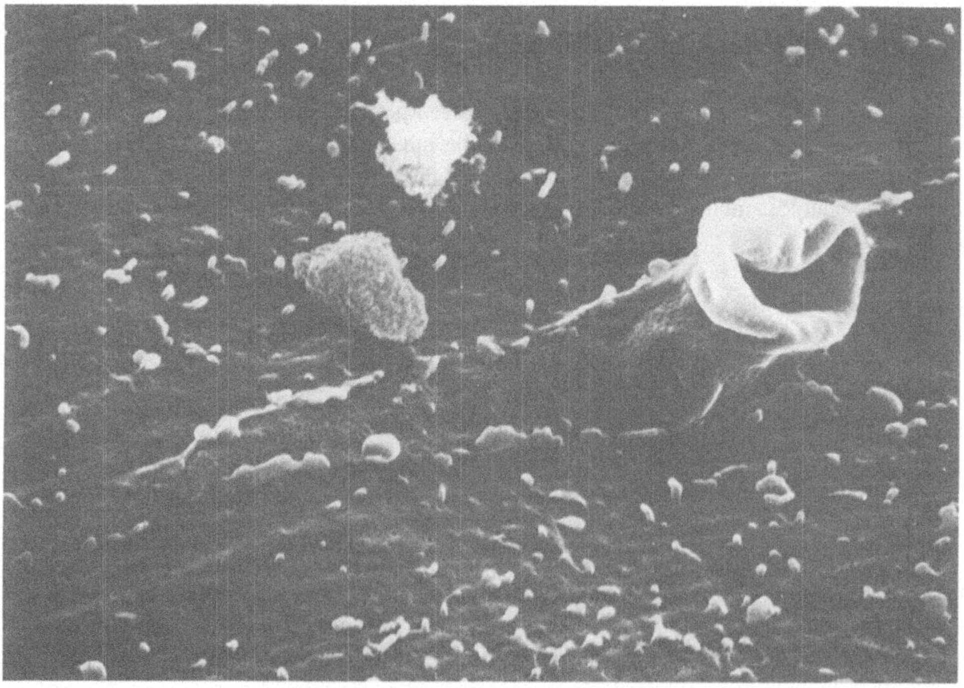

Figure 1. Scanning electron micrograph of the endothelial surface of the SHR cerebral artery. The endothelial cell with partial craterlike change shows more plasmalemmal pits than the surrounding cells. From Hazama *et al.*[12]

coagulative necrosis of the endothelial cells. The balloonlike protrusions also seen in them appear to correspond to the large vacuoles seen in transmission electron microscopic studies. These regressive changes of endothelial cells may contribute on the one hand to increased permeability of the arterial intima, and on the other hand to platelet adhesion, both of which may in turn play important roles in the development of hypertensive cerebrovascular lesions.

The fact that the endothelial cells of the cerebral arteries in SHR have both regressive and progressive changes may suggest an accelerated turnover of the cells.

IV. ROLE OF LYSOSOMAL ENZYMES IN ENDOTHELIAL INJURY IN CHRONIC HYPERTENSION

A. Lysosomal Enzyme Activities in the Arterial System in SHRSP

Information has been accumulating on the role of lysosomal enzymes in the pathogenesis of various connective tissue diseases[18,19] and vascular changes.[19–22]

In the histochemical study, there was a definite increase in acid phosphatase activity (Ac-Pase) in the cerebral arterial system of SHRSP as compared with that in control normotensive rats.[23] Wolinsky *et al.*[21] found an increased activity of aortic Ac-Pase and other lysosomal enzymes in a biochemical study on experimental renal hypertension. Our biochemical study on the Ac-Pase activity in the aorta of SHRSP showed that the enzyme activity in SHRSP is always higher than in age-matched control animals, and that antihypertensive drugs significantly suppressed the rise of blood pressure and decreased the aortic Ac-Pase activity.[24] Such findings indicate that hypertension is involved in the increase in the arterial lysosomal enzyme activity in SHR.

B. Lysosomal Enzymes in the Arterial Endothelial Cells in SHRSP

In order to obtain information about changes in lysosomal enzyme activities in the arterial endothelial cells under hypertensive conditions, semiquantitative histochemical investigations of Ac-Pase and *N*-acetyl-β-glucosaminidase (NAGase) activities in the aorta of SHRSP[25] and rats with renal hypertension (RHR)[26] were performed on *en face* endothelial monolayer preparations. The aortic endothelial cells in SHRSP and RHR showed increased Ac-Pase and NAGase activities as compared with those in control normotensive rats (Fig. 2), and the activities tended to increase with advancing age.

We performed a further biochemical study on five lysosomal enzymes (Ac-Pase, NAGase, cathepsin B, cathepsin D, and β-glucuronidase) in endothelial cells isolated by an enzymatic technique from the aorta of SHRSP and RHR.[27] There were increased activities of all enzymes examined in the aortic endothelial cells in the old SHRSP and RHR in comparison with those in age-matched control rats, and the enzyme activities tended to increase with aging.

The activities of lysosomal enzymes in the isolated microvessels from the cerebral cortex of SHRSP and RHR were also shown to be higher than those in normotensive rats.[28]

Cytochemically, Ac-Pase activity was demonstrated in the vacuolated secondary lysosomes often observed in the endothelial cells of intracerebral arterioles in SHR.[29] The

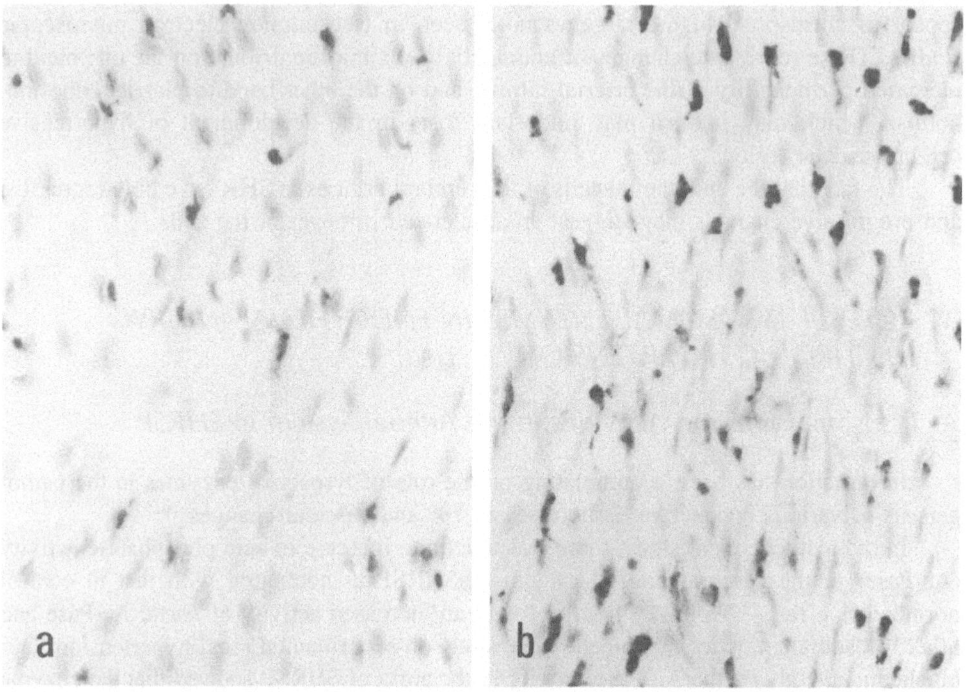

Figure 2. Histochemical reaction of Ac-Pase activity applied to *en face* endothelial monolayer preparations of the aorta from a control normotensive rat (a) and SHRSP (b). (a) Only a small number of endothelial cells show a positive reaction. (b) A large number of the cells show an intense reaction. From Yamada *et al.*[25]

cytoplasm surrounding ruptured lysosomes showed lytic changes suggesting that lysosomes are playing some role in the degeneration of endothelial cells (Fig. 3). Ac-Pase activity was also demonstrated in the intercellular spaces between and beneath endothelial cells adjacent to the intracytoplasmic vacuoles. These findings suggest that the lysosomal contents are discharged from the secondary lysosomes, which then become vacuoles, and that the discharged lysosomal lytic enzymes probably have an injurious effect on the vascular matrix and neighboring endothelial cells and medial smooth muscle cells.

Based on the above, we concluded that activated lysosomal enzymes seem to be involved in the mechanisms of arterial endothelial injury in chronic hypertension.

C. Mechanism of the Activation of Lysosomal Enzymes in the Endothelial Cells in Chronic Hypertension

It is highly probable that an enhanced autophagy of degenerated organelles due to mechanical injury caused by high blood pressure is involved in the activation of lysosomal enzymes. However, we are postulating an additional cause of lysosomal activation in chronic hypertension. Evidence has been accumulating that increased endocytosis activates lysosomal enzymes.[30,31] In our scanning electron microscopic study on the arteries of SHR brains, degenerated cells with balloon-like protrusions or crater-like concavities had a larger

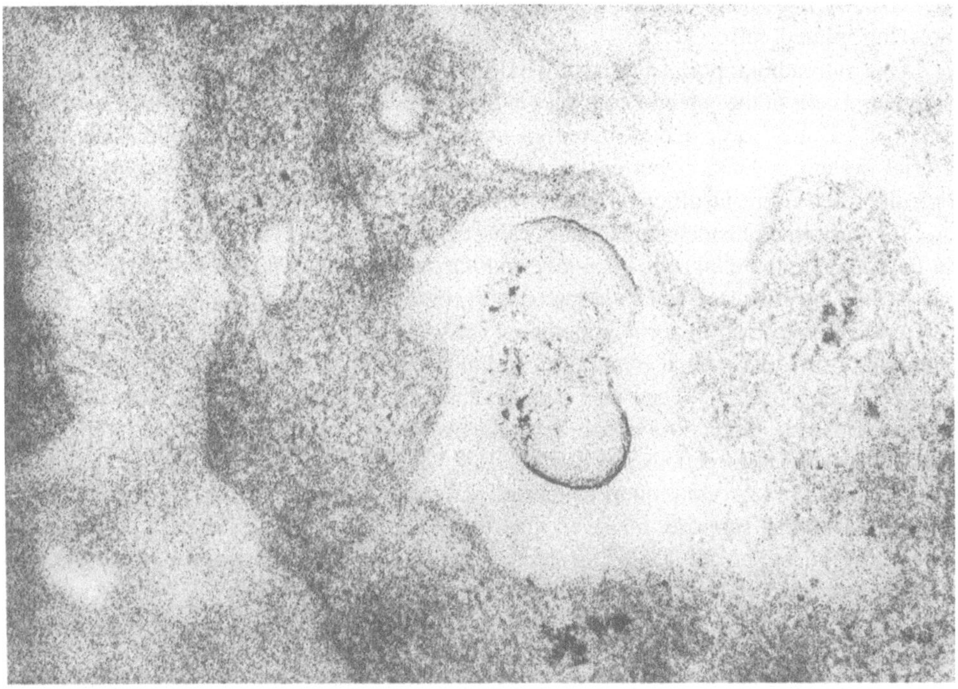

Figure 3. Cytochemical demonstration of Ac-Pase activity in endothelial cells of an intracerebral arteriole in SHRSP. The reaction products are observed on the inner side of the lysosomal membrane and around the ruptured portion. Cytoplasm around the lysosome shows lytic change.

number of plasmalemmal pits than neighboring endothelial cells did.[4,12] As with the case of acute hypertension[32–34] increased vesicular transport can be demonstrated in the endothelial cells of cerebral vasculatures in SHR.[3,11,35] It is difficult to differentiate transcytosis from pinocytosis in such morphological studies. If both phenomena take place in parallel, we may suppose that pinocytosis is activated in the SHR endothelial cells. Further, as will be shown later, the pinocytotic activity of cultured endothelial cells correlates very well with the proliferative activity of the cells.[36] Besides, the endothelial cells in the arterial system in SHRSP show an increased proliferative activity *in vivo*. Therefore, the pinocytosis of the endothelial cells in SHR must be activated in the arterial system. It can be concluded, then, that an increased pinocytosis due to hypertension causes the degeneration of endothelial cells through activated lysosomal enzymes.

V. REPLICATION OF ARTERIAL ENDOTHELIAL CELLS IN CHRONIC HYPERTENSION

A. Replication of Endothelial Cells in SHRSP

As mentioned in the previous section, some ultrastructural changes in the arterial endothelial cells observed in SHR can be regarded as the expressions of the immaturity of the cells. We may, therefore, assume such cells to be regenerated young endothelial cells replacing injured cells.

Our autoradiographic study with [^{3}H]thymidine revealed an increase in the labeled endothelial cells of the cerebral arterioles in SHRSP when compared with control animals.[37] Crane and Dutta[38] also demonstrated an increase in labeling of endothelial cells in the arterial system in DOC hypertensive rats. Because such studies were performed histologically, there were insufficient labeled endothelial cells to make a quantitative study.

In our immunohistochemical study using 5-bromo-2′-deoxyuridine (BrdU), the replication rate of endothelial cells in *en face* endothelial monolayer preparations from SHRSP was significantly elevated when compared with control animals (unpublished data) (Fig. 4). The replication rate of rat aortic endothelial cells has been reported to be increased in the early phase of experimental hypertension, but decreased to the level of normal animals during the late phase.[39,40] The cause of such an increased replication rate in the acute phase of hypertension is still controversial. Several possibilities have been proposed, such as that it is (1) a response to endothelial damage,[37,38] (2) a reaction to stretching of the lining endothelium,[41] or (3) a consequence of a humoral factor, in particular the renin–angiotensin system.[39] We must suppose, however, that there is an additional or different cause for an increased replication rate in SHRSP, because the increase continues even in the chronic phase.

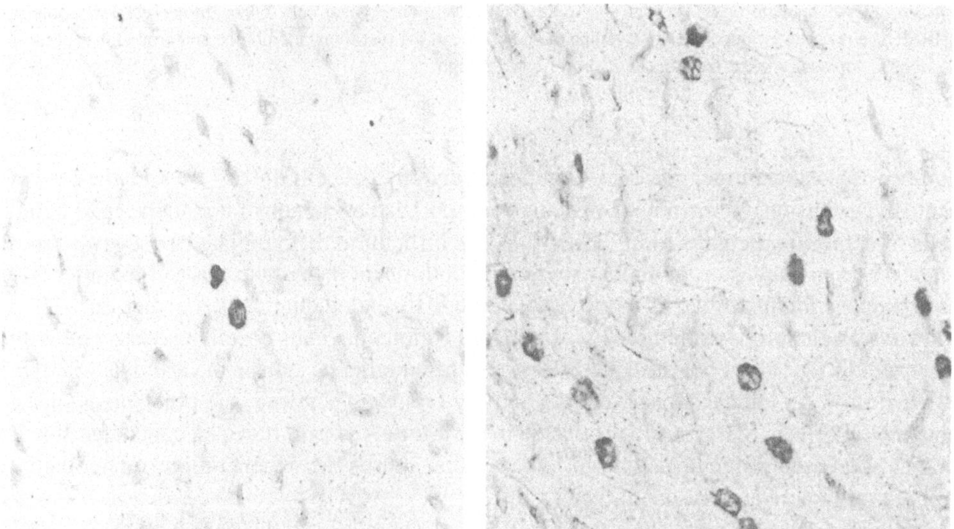

Figure 4. Immunohistochemical demonstration of BrdU in *en face* endothelial monolayer preparations of the aorta from a control rat (a) and SHRSP (b). More labeled cells are observed in SHRSP than in controls.

De Chastonay *et al.*[40] correlated such increased replication in acute hypertension with an increase of endothelial permeability to horseradish peroxidase based on their scanning electron microscopic findings that remodeling of endothelial cells develops discontinuous, tight junctions.

B. Proliferation of Cultured Aortic Endothelial Cells from Hypertensive Rats

Because there is little information about the proliferation of endothelial cells from hypertensive animals in a cultured state,[42] we studied the proliferative activity of cultured aortic endothelial cells from SHRSP, RHR, and control normotensive rats by counting cell numbers and [^{3}H]thymidine uptake.[36]

Cultured endothelial cells from SHRSP and RHR showed accelerated proliferation compared with those from control animals (Fig. 5). The proliferation rate of endothelial cells from SHRSP was very much higher than that of other groups at the beginning of growth, while the rate of endothelial cells from RHR was slightly higher than that of control. Because the direct effect of hypertension on the endothelial cells is excluded in the cultured state, we must regard the elevation of the replication as an attribute of the endothelial cells of hypertensive animals. Thus, the elevated replication of endothelial cells observed in hypertensive animals cannot be explained only as a regenerative reaction to injury of the cells due to hypertension. Moreover, because the accelerated proliferation of cultured endothelial cells was observed not only in the cells from SHRSP but also in those from RHR, it may be assumed that the tendency is acquired *in vivo* by an exposure of the cells to hypertension and persists in the *in vitro* condition. That is, the acquired functional

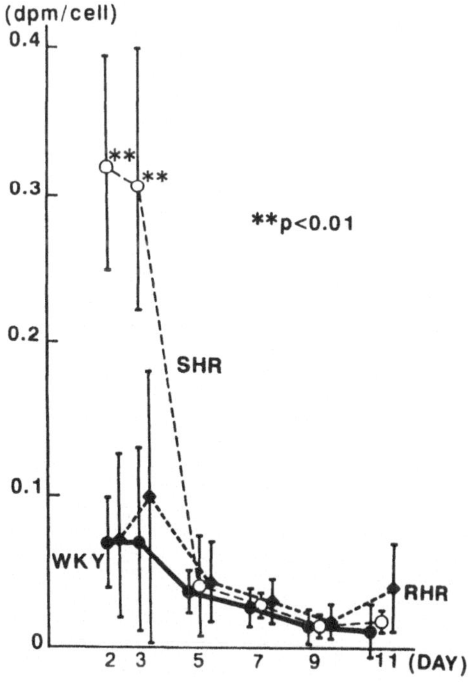

Figure 5. Changes in the [^{3}H]thymidine uptake of cultured aortic endothelial cells from SHRSP, RHR, and a control normotensive rat. From Hayase *et al.*[36]

changes of the endothelial cells chronically influenced by hypertension *in vivo* seem to persist even in the *in vitro* condition.

C. Pinocytosis of Cultured Aortic Endothelial Cells from Hypertensive Rats[36]

The pinocytotic activity of the above-mentioned cultured endothelial cells from SHRSP, RHR, and normotensive rats was studied by measuring [^{14}C[sucrose uptake and was correlated with the proliferative activity of the cells. The endothelial cells from both normotensive and hypertensive rats showed a high pinocytotic activity in the logarithmic phase and low pinocytosis in the postconfluent (Fig. 6). The rates of pinocytosis were demonstrated to be correlated closely with the replication rates as has also been reported by Davies *et al.*[43]

The basal rate of pinocytosis in nondividing cells in SHRSP was about half of that in control animals, and that in RHR was slightly smaller than in controls. This low rate of pinocytosis in cultured endothelial cells from hypertensive animals may also reflect one of the functional changes of endothelial cells which have been exposed chronically to hypertension *in vivo*.

D. Endothelial-Derived Growth Factors (EDGFs) in SHRSP

Vascular endothelial cells are known to produce several kinds of growth factors, which are thought to play an important role in the proliferation of vascular smooth muscle cells.[44,45]

From the fact that cultured aortic endothelial cells from hypertensive animals show an elevated replication, we may suppose the possibility of an increased production of EDGFs

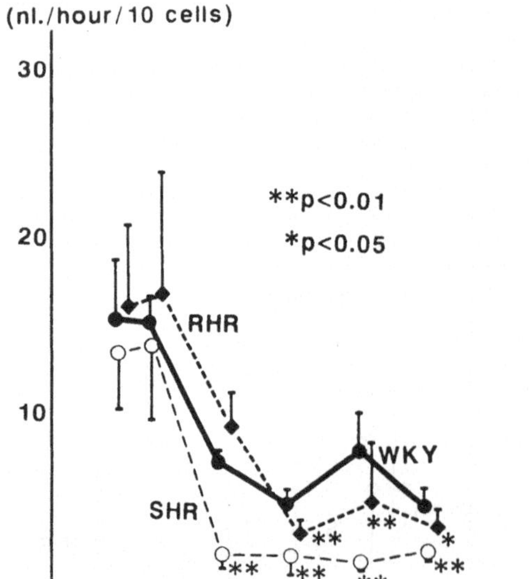

Figure 6. Changes in the pinocytosis of aortic endothelial cells from SHRSP, RHR, and a control rat. Pinocytosis was measured in a cultured state as cellular uptake of [^{14}C]sucrose from the tissue culture medium. From Hayase *et al.*[36]

in such animals. In order to test this hypothesis, we studied the activity of EDGFs in cultured aortic endothelial cells from SHRSP by bioassaying the proliferative effect of endothelial cell-conditioned medium on vascular smooth muscle cells, fibroblasts, and astrocytes, and obtained positive results (unpublished data). Such cells treated with endothelial cell-conditioned medium showed an accelerated proliferation compared with controls.

Furthermore, the release of EDGFs from cultured aortic endothelial cells was markedly increased by treatment with a tumor-promoting phorbol ester, phorbol-12,13-didecanoate, in a dose-dependent manner (unpublished data). On the other hand, 4-alpha-phorbol, which is a biologically inactive phorbol ester, did not increase the release of EDGFs. Neither did phorbol-12,13-didecanoate increase the release from aortic endothelial cells in which protein kinase C was down-regulated. These results indicate that protein kinase C plays a role in the regulation of EDGF release from the aortic endothelial cells.

Based on the above-mentioned findings, we may suppose that the increased growth factors in the arterial endothelial cells due to chronic hypertension may play a role in atherogenesis, and an excessive activation of protein kinase C, which releases the EDGFs, may trigger abnormal proliferation of medial smooth muscle cells leading to the formation of atheromatous lesions.

VI. CONCLUDING REMARKS

We have reviewed our studies on morphological and functional changes, particularly regressive changes and replication, of arterial endothelial cells in chronic hypertension.

Our pathological studies on hypertensive vascular changes in SHR (including SHRSP) suggested the importance of endothelial changes and hence induced increased vascular permeability in the development of the changes of vessel walls as well as brain parenchyma.

Our ultrastructural studies on arterial endothelial cells in SHR showed, on the one hand, progressive changes, some of which can be regarded as an expression of immaturity, and, on the other hand, regressive changes, suggesting an accelerated turnover of the cells in chronic hypertension.

Lysosomal enzyme activities in the arterial endothelial cells in SHR were demonstrated to be elevated and to play a role in the development of regressive endothelial changes. And we hypothesized that augmented lysosomal enzyme activities induced by increased pinocytosis and mechanically caused degeneration of cytoplasmic organelles are involved in the development of endothelial injury in chronic hypertension.

Our autoradiographic study with [^{3}H]thymidine and immunohistochemical studies with BrdU revealed an elevated replication of arterial endothelial cells in SHRSP. Cultured aortic endothelial cells from SHRSP and RHR also showed an accelerated proliferation compared with those from control animals. Thus, the increased replication of the cells in chronic hypertension cannot be explained only as a regenerative reaction to injury of the cells due to hypertension, but it must be assumed that endothelial cells have acquired an increased proliferative tendency by being chronically influenced by hypertension. In fact, the growth factors in endothelial cells in SHRSP proved to be increased.

All of the morphological and functional changes, including metabolic changes, of arterial endothelial cells associated with chronic hypertension can be presumed to be related to the further development of hypertensive and atherosclerotic vascular changes and also to the regulatory mechanisms to the vascular functions. The interrelationships among these various endothelial changes, such as that which has been found between pinocytosis and proliferation of the cells, must be studied further in order to clarify the developmental mechanisms of hypertensive and atherosclerotic vascular changes.

ACKNOWLEDGMENTS. This work was supported in part by a Grant-in-Aid for Scientific Research from the Ministry of Education, Science, and Culture of Japan (02670149).

The authors are grateful to Mr. P. Schneider, Shiga University of Medical Science, for assisting in the preparation of the manuscript.

Some figures are reproduced by permission of the American Heart Association, Inc., Elsevier Scientific Publisher Ireland Ltd., and Journal of Shiga University of Medical Science.

REFERENCES

1. Ross, R., and Glomset, J. A., 1976, The pathogenesis of atherosclerosis, *N. Engl. J. Med.* **295**:420–425.
2. Thorgeirsson, G., and Robertson, A. L., 1978, The vascular endothelium. Pathologic significance, *Am. J. Pathol.* **93**:803–848.
3. Hazama, F., Amano, S., Haebara, H., Yamori, Y., and Okamoto, K., 1976, Pathology and pathogenesis of cerebrovascular lesions in spontaneously hypertensive rats, in: *The Cerebral Vessel Wall* (J. Cervos-Navarro, E. Betz, F. Matakas, and R. Wüllenweber, eds.), Raven Press, New York, p. 245.
4. Hazama, F., and Amano, S., 1978, Pathological changes of cerebral vessel endothelial cells in spontaneously hypertensive rats, with special reference to the role of these cells in the development of hypertensive cerebrovascular lesions, *Adv. Neurol.* **20**:359–369.
5. Okamoto, K., and Aoki, K., 1963, Development of a strain of spontaneously hypertensive rats, *Jpn. Circ. J.* **16**:117–124.
6. Hazama, F., Ooshima, A., Tanaka, T., Tomimoto, K., and Okamoto, K., 1975, Vascular lesions in the various substrains of spontaneously hypertensive rats and the effects of chronic salt ingestion, *Jpn. Circ. J.* **39**:7–22.
7. Okamoto, K., Yamori, Y., and Nagaoka, A., 1974, Establishment of the stroke prone spontaneously hypertensive rat, *Cir. Res.* **34/35**(Suppl. I):143–145.
8. Amano, S., 1977, Vascular changes in the brain of spontaneously hypertensive rats: Hyaline and fibrinoid degeneration, *J. Pathol.* **121**:119–128.
9. Ooshima, A., Fuller, G., Cardinale, G,. Spector, S., and Udenfriend, S., 1975, Collagen biosynthesis in blood vessels of brain and other tissues of the hypertensive rats, *Science* **190**:898–900.
10. Hazama, F., Amano, S., Haebara, H., and Okamoto, K., 1975, Changes in vascular permeability in the brain of stroke-prone spontaneously hypertensive rats studied with peroxidase as a tracer, *Acta Pathol. Jpn.* **25**:565–574.
11. Fredriksson, K., Nordborg, C., Kalimo, H., Olsson, Y., and Johansson, B. B., 1988, Cerebral microangiopathy in stroke-prone spontaneously hypertensive rats. An immunohistochemical and ultrastructural study, *Acta Neuropathol.* **75**:241–252.
12. Hazama, F., Ozaki, T., and Amano, S., 1979, Scanning electron microscopic study of endothelial cells of cerebral arteries from spontaneously hypertensive rats, *Stroke* **10**:245–252.
13. Uehara, Y., Komuro, T., and Yamasaki, J., 1977, Structural problems of the blood vessel walls, in: *Basic Studies on Cardiovascular System*, Ministry of Education, Science and Culture of Japan, Tokyo, p. 74.
14. Takeshige, Y., Fujimoto, J., and Yamamoto, K., 1977, Structure and function of vascular endothelial cells, in: *Basic Studies on Cardiovascular System*, Ministry of Education, Science and Culture of Japan, Tokyo, p. 49.
15. Tokunaga, H., Osaka, M., and Fujita, T., 1973, Endothelial surface of rabbit aorta as observed by scanning electron microscope, *Arch. Histol. Jpn.* **36**:129–141.
16. Povlishock, J. T., Kantos, H. A., Rosenblum, W. I., Becker, D. P., Jenkins, L. W., and DeWitt, D. S., 1980,

Scanning electron-microscopic analysis of the intraparenchymal brain vasculature following experimental hypertension, *Acta Neuropathol.* **51**:203–212.

17. Kawamura, J., Gertz, S. D., Sunaga, T., Rennels, M. L., and Nelson, E., 1974, Scanning electron microscopic observations on the luminal surface of the rabbit common carotid artery subjected to ischemia by arterial occlusion, *Stroke* **5**:765–774.
18. Allison, A. C., Hariyton, J. S., and Birbeck, M., 1966, Examination of the cytotoxic effects of silica on macrophages, *J. Exp. Med.* **124**:141–154.
19. Weissmann, G., 1975, The molecular basis of acute gout, in: *Cell Membrane* (G. Weissmann, ed.), HP Publishing, New York, p. 257.
20. Wolinsky, H., Goldfisher, S., Schiller, B., and Kasak, L. E., 1973, Lysosomes in aortic smooth muscle cells. Effects of hypertension, *Am. J. Pathol.* **73**:727–734.
21. Wolinsky, H., Goldfisher, S., Schiller, B., and Kasak, L. E., 1974, Modification of the effects of hypertension on lysosomes and connective tissue in the rat aorta, *Circ. Res.* **34**:233–241.
22. Riede, U. N., and Staubesand, J. A., 1977, A unifying concept for the role of matrix vesicles and lysosomes in the formal pathogenesis of disease of connective tissue and blood vessels, *Beitr. Pathol.* **160**:3–37.
23. Yamada, E., Hazama, F., Amano, S., and Hanakita, J., 1979, Acid phosphatase activity of the brain in SHR. The first report of enzyme histochemical studies of spontaneously hypertensive rat brain, *Jpn. Circ. J.* **43**: 285–292.
24. Watanabe, M., Yamada, E., Hazama, F., and Nomura, T., 1981, Acid phosphatase activity in the aorta of spontaneously hypertensive rats and the effect of various antihypertensive drugs, *Atherosclerosis* **40**:167–174.
25. Yamada, E., Hazama, F., Amano, S., Sasahara, M., Kataoka, H., and Kayembe, K., 1983, Semiquantitative histochemical investigation of lysosomal enzyme activities in the aortic endothelial cells of spontaneously hypertensive rats, *Atherosclerosis* **47**:19–26.
26. Sasahara, M., Hazama, F., Yamada, E., Kataoka, H., Amano, S., and Kayembe, K., 1983, Semiquantitative histochemical investigation of lysosomal enzyme activities in the aortic endothelial cells of rats with renal hypertension, *Virchows Arch. B* **44**:267–274.
27. Sasahara, M., Hazama, F., Amano, S., Hayase, Y., Yukioka, N., Kawai, J., and Kataoka, H., 1988, Effects of hypertension on lysosomal enzyme activities in aortic endothelial cells, *Atherosclerosis* **70**:53–62.
28. Yukioka, N., Yamada, E., Sasahara, M., Kawai, J., Hayase, Y., Amano, S., Hazama, F., 1988, Lysosomal enzyme activities in the cerebral microvessels in spontaneously hypertensive rats, *Exp. Mol. Pathol.* **49**:111–117.
29. Yamada, E., Hazama, F., Amano, S., and Hanakita, J., 1980, Cytochemical investigation on acid phosphatase activity in cerebral arteries in spontaneously hypertensive rats, *Jpn. Circ. J.* **44**:467–475.
30. Dannenberg, A. N., Wabier, P. C., and Kapral, F. A., 1963, A histochemical study of phagocytic and enzymatic functions of rabbit mononuclear and polymorphonuclear exudate cells and alveolar macrophages, *J. Immunol.* **90**:448–465.
31. Stanton, D. A., and Cohn, Z. A., 1970, In vitro induction of lysosomal enzymes of phagocytosis, *J. Exp. Med.* **131**:1239–1260.
32. Hansson, H.-A., Johansonn, B. B., and Blomstand, C., 1975, Ultrastructural studies on cerebral permeability on acute hypertension, *Acta Neuropathol.* **32**:187–198.
33. Westergaard, E., van Deurs, B., and Bronsted, H. E., 1977, Increased vesicular transport of horseradish peroxidase across cerebral endothelium, evoked by acute hypertension, *Acta Neuropathol.* **37**:141–152.
34. Nag, S., Robertson, D., and Dinsdale, H. E., 1977, Cerebral cortical changes in acute experimental hypertension. An ultrastructural study, *Lab. Invest.* **36**:150–161.
35. Nag, S., Robertson, D. M., and Dinsdale, H. B., 1980, Morphological changes in spontaneously hypertensive rats, *Acta Neuropathol.* **52**:27–34.
36. Hayase, Y., Sasahara, M., Kawai, J., and Hazama, F., 1989, Proliferation and pinocytosis of cultured aortic endothelial cells from hypertensive rats, *J. Shiga Univ. Med. Sci.* **4**:25–36.
37. Hazama, F., Haebara, H., Amano, S., and Ozaki, T., 1977, Autoradiographic investigation of cell proliferation in the brian of spontaneously hypertensive rats, *Acta Neuropathol.* **37**:231–236.
38. Crane, W. A. J., and Dutta, L. P., 1963, the utilization of tritiated thymidine for deoxyribonucleic acid synthesis by the lesions of experimental hypertension in rats, *J. Pathol. Bacteriol.* **86**:83–97.
39. Daniel, R. E., Boitnott, J. K., Brown, G. D., and Heptinstall, R. H., 1982, Aortic endothelial cell activity in high renin and normal renin models of hypertension in the rat, *Lab. Invest.* **47**:451–458.

40. De Chastonay, C. G., Gabbiani, G., Elemerm, G., and Hüttner, I., 1983, Remodeling of the rat aortic endothelial layer during experimental hypertension. Changes in replication rate, cell density, and surface morphology, *Lab. Invest.* **48:**45–52.
41. Schwartz, S. M., and Benditt, E. P., 1975, Aortic endothelial cell replication. I. Effects of age and hypertension in the rats, *Circ. Res.* **41:**248–255.
42. Grünwald, J. M., Shaper, W., Shulte, W., and Hauss, W. H., 1984, About the effect of risk factor on the structure and the proliferation of cultivated aortic wall cells. Part 2: Changes in structure and proliferation of cultivated aortic endothelial cells from hypertensive minipigs, *Exp. Pathol.* **25:**163–168.
43. Davies, P. F., Selden, S. C., and Schwartz, S. M., 1980, Enhanced rate of fluid pinocytosis during exponential growth and monolayer regeneration by cultured arterial endothelial cells, *J. Cell. Physiol.* **102:**119–127.
44. Gajdusek, C., DiCorleto, P., Ross, R., and Schwartz, S. M., 1980, An endothelial cell-derived growth factor, *J. Cell Biol.* **85:**467–472.
45. DiCorleto, P. E., Gajdusek, C. M., Schwartz, S. M., and Ross, R., 1983, Biochemical properties of the endothelium-derived growth factor: Comparison to other growth factors, *J. Cell. Physiol.* **114:**339–345.

17

Specific Platelet Mediators and the Abrupt Progression of Coronary Artery Stenoses

Experimental Evidence and Potential Clinical Implications

James T. Willerson, Paolo Golino, and L. Maximilian Buja

I. INTRODUCTION

Acute coronary heart disease syndromes, including unstable angina and acute myocardial infarction, are usually caused by a primary decrease in coronary blood flow and myocardial oxygen delivery.[1–15] Unstable angina occurs in patients with endothelial injury or ulceration at the site of an atherosclerotic plaque and some patients have an intraluminal coronary artery thrombosis.[4–6,9–11] Falk[7] and Davies and Thomas[8] have shown that atherosclerotic plaque fissuring or rupture may lead to coronary arterial thrombosis and the development of unstable angina, acute myocardial infarction, or sudden death. Patients with Q-wave myocardial infarcts often have ulcerated or atherosclerotic plaques and the subsequent development of occlusive coronary artery thrombi.[2,3] Patients with non-Q-wave infarcts (usually subendocardial infarcts) much less commonly have coronary artery thrombi that are permanently occlusive, but instead these patients have transient reductions in coronary blood flow followed by reperfusion and sometimes partially occlusive thrombi.[3,16] We believe that transient reductions in coronary blood flow are most likely related to intermittent platelet aggregation and vasoconstriction contributed to by the release of selected

---

James T. Willerson • Department of Internal Medicine, University of Texas Medical School, and Cardiology Research, Texas Heart Institute, Houston, Texas 77225–0345. *Paolo Golino* • Division of Cardiology, University of Naples II School of Medicine, Naples 80131, Italy. *L. Maximilian Buja* • Department of Pathology, University of Texas Medical School, and Cardiology Research, Texas Heart Institute, Houston, Texas 77225–0345.

Endothelial Cell Dysfunctions, edited by Nicolae Simionescu and Maya Simionescu. Plenum Press, New York, 1992.

mediators from aggregating platelets at sites of coronary artery endothelial injury and stenosis.[4,5,17]

II. MECHANISMS POTENTIALLY RESPONSIBLE FOR THE DEVELOPMENT OR SUSTAINMENT OF UNSTABLE ANGINA

In 1979 and more recently, our group[4,5,10,17] hypothesized the following: (1) Increases in thromboxane and decreases in prostacyclin concentration at sites of endothelial injury and significant coronary artery stenosis cause platelet aggregation and dynamic coronary vasoconstriction. These, in turn, decrease regional coronary blood flow below a critical level and lead to the development or sustainment of unstable angina pectoris. (2) Mechanisms responsible for endothelial injury at sites of coronary artery stenosis vary but probably include (a) progression, fissuring, or rupture of atherosclerotic plaques[7,8]; (b) hemodynamic factors, including flow turbulence; (c) cigarette smoking; (d) systemic arterial hypertension; (e) infection or immune complex deposition; and, on occasion, (f) catheter-induced endothelial injury associated with coronary arteriography and/or angioplasty.

The hypothesis that unstable angina is associated with increases in transmyocardial thromboxane A_2 concentrations in patients was tested from 1979 to 1980.[10] In 60 patients undergoing cardiac catheterization, blood samples were obtained from the coronary sinus and ascending aorta, and prostaglandins were measured by radioimmunoassay.[10] By history, noninvasive evaluation, and the results of cardiac catheterization, patients were separated into five groups. Group A patients ($n = 6$) had congenital and acquired noncoronary cardiac lesions, but they were without significant coronary stenoses; group B patients ($n = 14$) complained of chest pain, but they had no significant coronary artery stenoses or provocable coronary artery spasm; group C patients ($n = 18$) had significant coronary artery disease, and their most recent episodes of chest pain occurred more than 96 hr before study; group D patients ($n = 15$) had chest pain that occurred 24–96 hr before study; and group E patients ($n = 7$) had unstable angina with chest pain within 24 hr of study.

Thromboxane B_2 (the inactive metabolite of thromboxane A_2) concentrations were elevated across the coronary bed in group E patients, i.e., those with continuing rest angina (Fig. 1), and values in this group were significantly higher than those of patients in groups A, B, and C. Group D patients had a bimodal distribution; 12 patients had low concentrations, and 3 had markedly elevated values. It is not clear why 3 patients in group D had elevated transcardiac concentrations of thromboxane B_2. However, two possible explanations include (1) persistent increases in transcardiac thromboxane concentration after the relief of unstable angina and (2) continuing unstable angina not detected clinically or identified by the patient, i.e., "silent myocardial ischemia." We were unable to show differences in transmyocardial prostacyclin concentrations among these patients.[10]

These data showed a temporal relation between continuing unstable angina and increases in transcardiac thromboxane concentrations, and they were consistent with the hypothesis that thromboxane A_2 accumulation and consequent platelet aggregation and coronary vasoconstriction are important factors in the pathogenesis of unstable angina in patients. However, these data did not prove a causal relation.

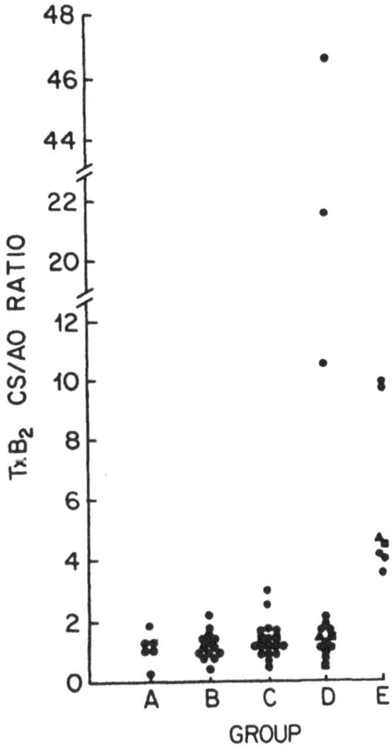

Figure 1. Plot of ratios of thromboxane B_2 (TxB_2) in the coronary sinus (CS) and ascending aorta (AO) in the five groups of patients ($n = 60$) studied. Each point represents the data from one patient. ■, patients who received a cyclooxygenase inhibitor within 5 days of study; ▼, patients with coronary artery spasm. In group A (valvular and congenital nonischemic heart disease), B (chest pain syndrome without ischemic heart disease), and C (ischemic heart disease without chest pain for at least 96 hr), the patients had thromboxane B_2 coronary sinus:aorta ratios of 3.1 or less. Group D patients (unstable angina pectoris with chest pain 24–96 hr before study) had a bimodal distribution: 12 patients had low ratios, whereas 3 had very high ratios. Group E patients (unstable angina pectoris with chest pain within 24 hr before study) had ratios that were higher than those in patients in groups A, B, and C ($p < 0.05$). Reprinted from Hirsh *et al.*[10] by permission.

Subsequently, Fitzgerald *et al.*[11] at Vanderbilt and Hamm *et al.*[12] have shown that many patients with unstable angina have elevated urinary concentrations of a major thromboxane metabolite, i.e, 2-3 dinor-thromboxane B_2, and that in some patients, further increases occur in the thromboxane urinary metabolite with new episodes of chest pain.

More recently, subsets of patients with limiting angina and significant coronary heart disease have been shown to have increased transcardiac serotonin concentrations (Fig. 2).[13] Subsets of patients with limiting angina have a vasoconstrictor substance in their coronary sinus effluents whose vasoconstrictor effect is blocked by a serotonin receptor antagonist.[14] Thus, both thromboxane A_2 and serotonin have been shown to increase across the coronary arterial bed in patients with limiting angina.

III. EVALUATION OF THE PHYSIOLOGIC IMPORTANCE OF THROMBOXANE A_2 ACCUMULATION IN LEADING TO INTRACORONARY PLATELET AGGREGATION, THROMBUS DEVELOPMENT, AND REDUCTIONS IN CORONARY BLOOD FLOW

In the canine heart, a severe proximal coronary artery stenosis with associated endothelial injury causes cyclic coronary blood flow reductions as originally described by Folts *et al.* (Fig. 3).[18] In this model, endothelial injury is caused by the application of an external constrictor and by gentle stroking of the left anterior descending coronary artery

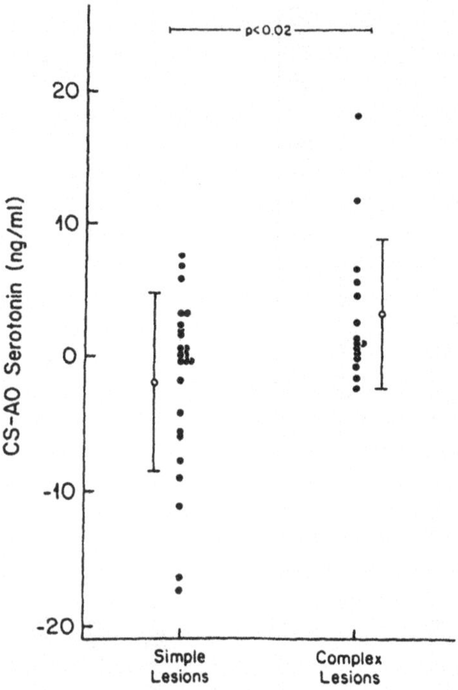

Figure 2. Transcardiac alterations in serotonin production. Approximately one-third to one-half of patients with limiting angina and coronary heart disease produce serotonin across the coronary bed during the period in which they have limiting angina. Patients with "complex" coronary artery lesions (irregular borders, eccentric lesions, and sometimes intraluminal filling defects) often produce serotonin across their coronary artery bed, whereas patients with "simple" atherosclerotic lesions have a more heterogeneous pattern and some are extracting serotonin across their coronary artery system. Reproduced from Van den Berg *et al.*[13] by permission.

with cloth-covered forceps. The cyclic coronary flow alterations are related to platelet aggregation and leukocyte and red blood cell accumulation at the stenotic site (Figs. 3 and 4). Dazoxiben (UK-37-248, Pfizer Pharmaceuticals, New York), a thromboxane synthetase inhibitor,[19] or SQ29,548 (Squibb Pharmaceuticals, Princeton, N.J.), a thromboxane receptor antagonist,[20] abolishes or significantly attenuates the frequency of cyclic flow reductions in approximately 70% of treated animals (Fig. 3). Thromboxane B_2 concentrations in the distal portion of the narrowed coronary artery and at the site of the coronary stenosis and endothelial injury are increased during cyclic coronary flow reductions and are reduced to control values after administration of dazoxiben.[19,21] In contrast, 6-keto-prostaglandin $F_{1\alpha}$ (the inactive metabolite of prostacyclin) generation is reduced at the site of coronary artery constriction compared with generation in the nonconstricted, noninjured coronary artery.[21] The 6-keto-prostaglandin $F_{1\alpha}$ levels distal to the narrowed coronary artery increase significantly during cyclic flow reductions and remain elevated after the administration of dazoxiben.[19] In the canine model, systemically administered dazoxiben (2.5 mg/kg body wt i.v.) suppresses arachidonic acid-induced thromboxane A_2 (but not prostaglandin E_2) production in platelets, and it does not significantly influence prostacyclin synthesis by coronary artery rings.[19] Furthermore, the topical application or intraatrial administration of a thromboxane mimetic (U46619) generally restores cyclic coronary flows in this model after they are abolished by a thromboxane synthesis inhibitor.[19] Folts *et al.*[18] have shown that aspirin often eliminates cyclic coronary artery flow reductions in the same experimental model.

Therefore, thromboxane concentration increases at the site of a coronary artery stenosis and endothelial injury in the canine model, and the narrowed, injured artery makes

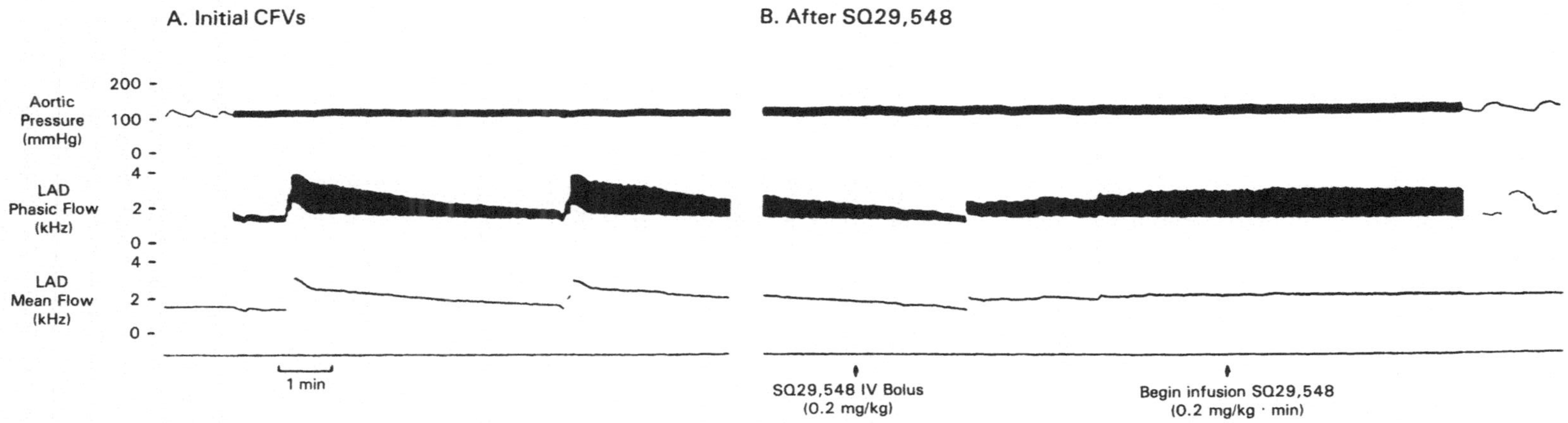

Figure 3. Representative recording from a dog with cyclic coronary flow variations (CFVs) and abolition of the cyclic flows with SQ29,548, a thromboxane receptor antagonist. (A) Initial CFVs. Note the reoccurring pattern of decline and restoration of mean and phasic left anterior descending coronary artery (LAD) flow. (B) Normal pattern of flow restored minutes after administration of SQ29,548. CFVs occur in the canine model with a tight proximal coronary artery stenosis and endothelial injury as previously reported (see Refs. 18–27). Reproduced from Ashton *et al.*[20] by permission.

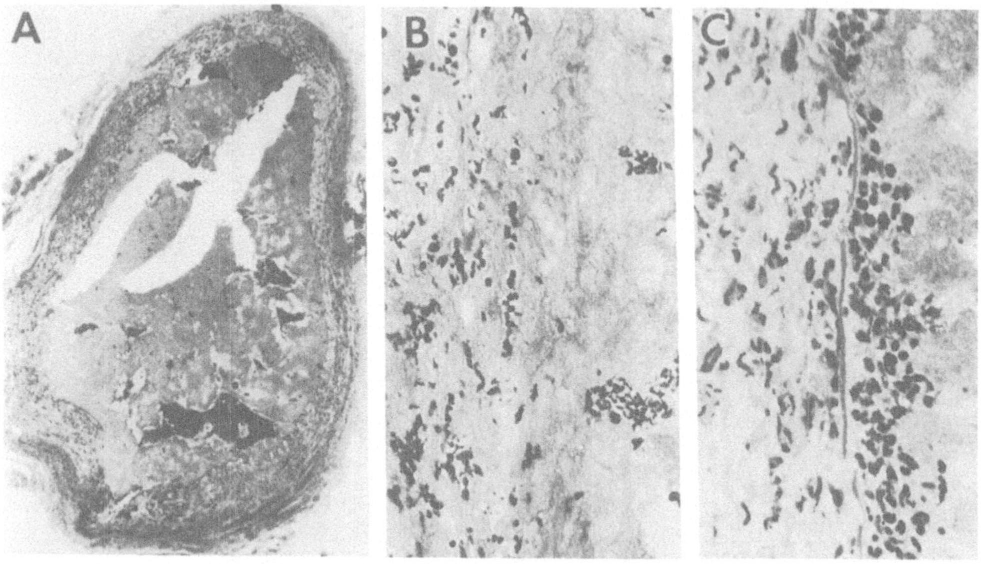

Figure 4. Photomicrographs of periconstrictor region of a left anterior descending coronary artery of a dog sacrificed during a cyclic coronary artery flow reduction in the absence of pharmacologic therapy. (A) Lumen of the artery is practically occluded by a large thrombus. Original magnification, ×50. (B) Thrombus is composed primarily of platelets (small granular bodies) and also contains erythrocytes and leukocytes (dark structures). Original magnification, ×317. (C) Vessel wall is devoid of endothelium and has damaged smooth muscle adjacent to the internal elastic lamella (IEL). Note the large numbers of platelets and neutrophils attached to the denuded IEL and the neutrophils that have entered the damaged media. Original magnification, ×390.

substantially more thromboxane when arachidonic acid is added *in vitro*[19,21] The administration of a thromboxane synthetase inhibitor or receptor antagonist usually abolishes cyclic flow alterations in this model.[18–20] More recently, we have shown that dynamic coronary vasoconstriction occurs adjacent to the site of stenosis in this canine model and that thromboxane and serotonin receptor antagonists prevent the dynamic vasoconstriction.[22] Thromboxane synthesis inhibitors and receptor antagonists also reduce *in situ* platelet aggregation[23] in this experimental model.

IV. SEROTONIN'S ROLE IN MEDIATING PLATELET AGGREGATION, CORONARY THROMBUS FORMATION, AND DYNAMIC CHANGES IN CORONARY ARTERY TONE

Subsequent studies in our laboratory have shown that serotonin concentration increases by at least 18-fold at the site of a coronary artery stenosis and endothelial injury when cyclic coronary flow reductions occur.[24] Ketanserin, a serotonin receptor antagonist, usually abolishes cyclic flow reductions[24,25] and works in essentially every instance in which a thromboxane synthetase inhibitor or receptor antagonist, or both, have failed (Fig. 5).[26] Other serotonin receptor antagonists without α-adrenergic antagonist properties at the concentrations at which they were used also abolish cyclic coronary flow alterations in

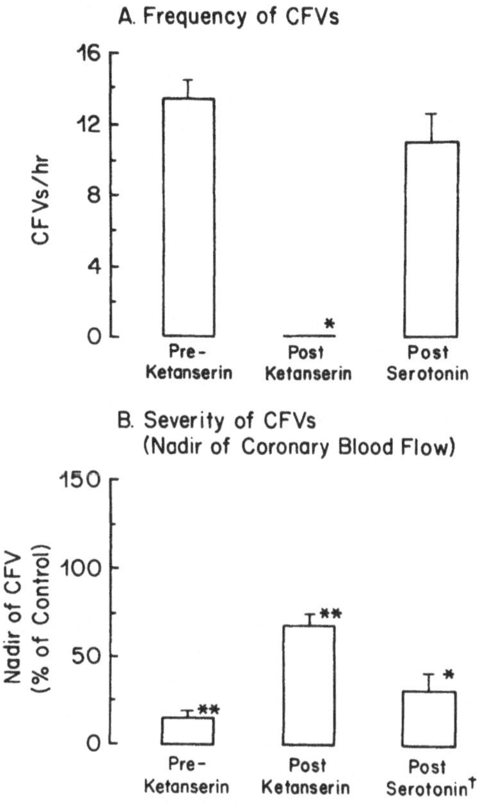

Figure 5. (A) Bar graph of number of cyclic coronary artery flow variations (CFVs) per hour in dogs during a control period, after abolition of cyclic coronary flows by ketanserin (a serotonin receptor antagonist), and after reestablishment of cyclic coronary artery flows by the infusion of serotonin (1–3 μg/min) directly into the left atrium. (B) Bar graph of severity of CFVs, obtained by averaging the two lowest mean flow velocity values (nadirs) as a percentage of controlled unconstricted blood flow velocity throughout a 30-min interval, is shown during the control period of cyclic flows and after the serotonin receptor antagonist, ketanserin, had abolished cyclic alterations in coronary blood flow. The severity of cyclic flows after their restoration by serotonin administration is shown in the far right. The mean flow velocity nadir in serotonin-restored CFVs was taken as a percentage of mean flow velocity after infusion of serotonin but before the restoration of CFVs. $^{**}p < 0.05$, significantly different from control, unconstricted coronary blood flow velocity; $^{*}p < 0.05$, significantly different from coronary blood flow velocity during infusion of serotonin and before CFVs were restored. Reprinted from Ashton *et al.*[24] by permission.

approximately 90% of animals and are effective when thromboxane synthetase inhibitors or receptor antagonists have failed.[26] After cyclic flows are abolished by serotonin receptor antagonists, they may be restored by the intraatrial administration of serotonin.[24] There is also an amplification of thromboxane's and serotonin's effects on cyclic coronary flow variations, such that interfering with the effect of either is generally protective in eliminating cyclic flow alterations.[26] Thus, both thromboxane and serotonin are important in initiating or sustaining cyclic flow reductions in this experimental model.[26] If the contribution from both thromboxane and serotonin receptor stimulation is eliminated, cyclic coronary flow reductions are nearly always abolished in this experimental model[26] and are very difficult to restore even if systemic epinephrine concentrations are markedly increased.[27]

V. CORONARY ARTERY STENOSIS AND ENDOTHELIAL INJURY IN CLOSED-CHEST, AWAKE, UNSEDATED DOGS

Chronically instrumented, awake, unsedated dogs develop cyclic coronary flow variations or a persistently reduced coronary blood flow 2–7 days after their original instrumentation with an external coronary artery constrictor and after endothelial injury.[28]

Cyclic coronary flow variations and the low-flow state are often reversed to normal coronary blood flow patterns by thromboxane and serotonin receptor antagonists in this experimental model,[28] just as they are in the open-chest, anesthetized animals.[19–27] In the awake, closed-chest dog with coronary artery stenosis and endothelial injury, there is a continuum from cyclic coronary flow alterations to a low -flow state and finally an occlusive coronary thrombus.[28]

Interestingly, the awake, unsedated, closed-chest dog also develops cyclic coronary flow alterations or a severe and sustained reduction in coronary blood flow when excited by the presence of a person, by food, or by low-level exercise (Fig. 6).[29] This suggests that excitement and low-level effort may lead to platelet aggregation with consequent mechanical obstruction and dynamic reductions in coronary blood flow when coronary artery stenosis and endothelial injury coexist. Thromboxane and serotonin receptor antagonists may prevent the cyclic coronary flow alterations caused by low-level exercise in this model.[29]

One to three weeks after the application of the coronary artery constrictor and endothelial injury and after frequent cyclic coronary flow alterations, chronically instrumented dogs develop marked intimal proliferation that converts a moderately severe coronary artery stenosis to a critical one. Similar intimal proliferation has been described previously as occurring in endothelially injured carotid arteries in the rat[30] and in the patient who develops restenosis after coronary artery angioplasty.[31] We have suggested that in the stenosed coronary artery with endothelial injury and chronic cyclic coronary flow alterations in the awake, unsedated dog and in patients with restenosis after coronary artery angioplasty, the marked intimal proliferative responses found may be caused by the accumulation of selected growth factors from aggregating platelets, infiltrating white blood cells, endothelial or smooth muscle cells, and/or activated T cells, including one or more of the following: platelet-derived growth factors, transforming growth factors, epidermal growth factor, fibroblast growth factor, serotonin, endothelin, or a heparin-associated growth factor.[32–38] Further studies are needed to test this suggestion and to identify which growth factor(s) and which cell population(s) contribute to the fibroproliferative reaction.

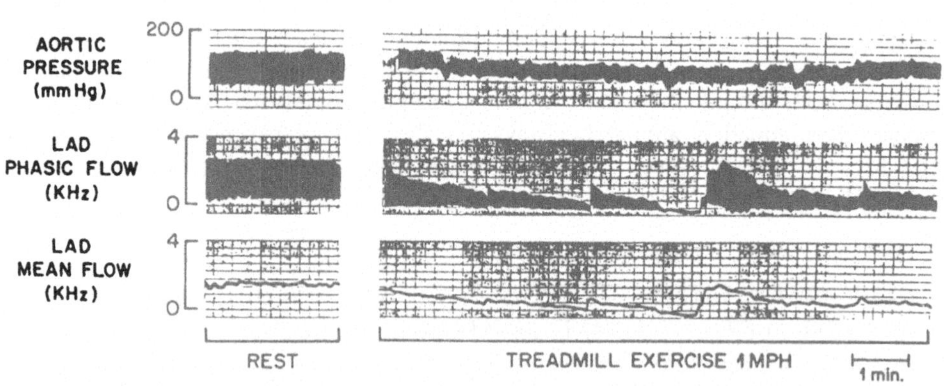

Figure 6. Recordings of cyclic alterations in coronary blood flow that occur in a chronically instrumented, awake dog with a coronary artery stenosis and endothelial injury when the animal exercises at a low level of effort or becomes excited.

VI. CLINICAL RELEVANCE

If platelet aggregation and the consequent release of thromboxane and serotonin are important in the initiation or sustaining of selected acute coronary heart disease syndromes, including unstable angina, an intervention that interferes with platelet aggregation and the subsequent release of thromboxane or serotonin should reduce morbidity and mortality in patients with unstable angina. Clinical studies have shown that aspirin, which is an inhibitor of cyclooxygenase, thromboxane synthesis, and platelet aggregation, reduces the risk of subsequent nonfatal myocardial infarction and death in patients with unstable angina.[39–41] Specifically, a multicenter Veterans Administration Medical Center study established that one tablet of aspirin daily (325 mg/day) reduces mortality and the risk of nonfatal myocardial infarction in the weeks after the onset of unstable angina in men.[39] A Canadian study showed that the equivalent of four aspirin tablets per day reduces mortality and the risk of subsequent nonfatal myocardial infarction in the ensuing 2 years in men and women with unstable angina.[40] These data are consistent with the hypothesis that platelet aggregation and subsequent increases in local thromboxane concentrations may affect the abrupt conversion from chronic to acute coronary heart disease syndromes in humans through thromboxane and serotonin accumulation leading to further platelet aggregation, coronary vasoconstriction, and, sometimes, the development of occlusive coronary thrombi.

VII. CONCLUSIONS

We have speculated previously that the abrupt conversion from chronic stable to unstable angina and the continuum to acute myocardial infarction may result from myocardial ischemia caused by progressive platelet aggregation and dynamic vasoconstriction themselves caused by local increases in thromboxane and serotonin at sites of coronary artery stenosis and endothelial injury (Fig. 7).[3,4,17] Platelet aggregation and dynamic coronary artery vasoconstriction probably result from the local accumulation of thromboxane and serotonin and possible other mediators as well as relative decreases in the local concentrations of endothelially derived vasodilators and inhibitors of platelet aggregation, such as endothelium-derived relaxing factors (EDRF) and prostacyclin (Fig. 7).[3,4,17,42,43] With severe reductions in coronary blood flow caused by these mechanisms, platelet aggregates may increase, and an occlusive thrombus composed of platelets and white and red blood cells in a fibrin mesh may develop. When coronary arteries are occluded or narrowed for a sufficient period of time by these mechanisms, myocardial necrosis, electrical instability, or sudden death may occur. We believe that unstable angina and acute myocardial infarction are a continuum in relationship to the process of coronary artery thrombosis and vasoconstriction. When the period of platelet aggregation or dynamic vasoconstriction at sites of endothelial injury and coronary artery stenosis is brief, unstable angina or non-Q-wave infarction may occur. However, when the coronary artery obstruction by these mechanisms is prolonged for several hours, Q-wave myocardial infarction results. Chronic endothelial injury and coronary artery stenosis are probably associated with the accumulation of platelets, white and red blood cells, and a fibrin mesh at the site of stenosis and endothelial injury. This endothelial injury may lead to the accumulation of other mediators that contribute to the development of dynamic vasoconstriction and platelet

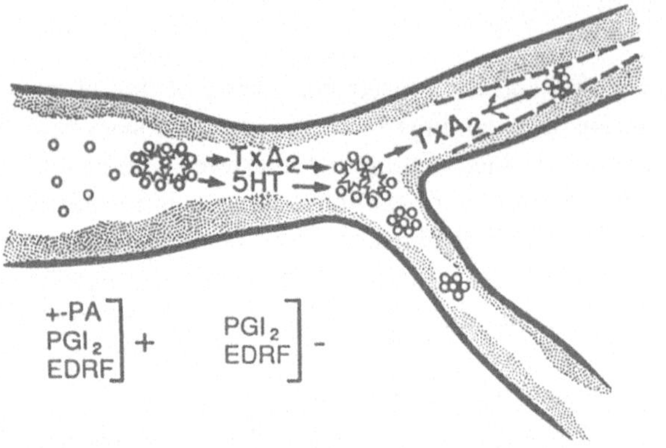

Figure 7. Schematic diagram indicating the possible mechanisms by which thromboxane A_2 (TxA_2) and serotonin (5HT) promote platelet aggregation and decrease coronary blood flow in the patient with unstable angina. Aggregating platelets (stars) release thromboxane A_2 and serotonin at sites of coronary artery stenosis and endothelial injury, which cause further platelet aggregation at that site and downstream, dynamic coronary vasoconstriction, and partial or total coronary artery thrombosis. Absence or reductions in endothelially derived relaxing factors (EDRF) and prostacyclin (PGI_2) at vascular sites with endothelial injury probably contribute to the development of vasoconstriction and thrombosis. See text for details. Modified and reproduced from Willerson *et al.*[17] by permission.

aggregation, including histamine, platelet activating factor, selected leukotrienes (including LTC_4 and LTD_4), prostaglandin D_2, and possibly endothelin. Future clinical studies should test these hypotheses and determine whether selected synthesis inhibitors and/or receptor antagonists provide clinical protection in patients.

ACKNOWLEDGMENTS. We express our appreciation to Angie Ruiz for expert secretarial assistance in the preparation of the manuscript, and to Ms. Janice McNatt for outstanding technical assistance in the performance of these studies. Supported in part by the National Heart, Lung, and Blood Institute Ischemic SCOR HL-17669.

REFERENCES

1. Maseri, A., Pesola, A., Marzilli, M., Severi, S., Parodi, O., L'Abbate, A., Ballestra, A. M., Maltinti, G., DeNes, D. M., and Biagini, A., 1977, Coronary vasospasm in angina pectoris, *Lancet* **1:**713–717.
2. DeWood, M. A., Spores, J., Notske, R., Mouser, L. T., Burroughs, R., Golden, M. S., and Lang, H. T., 1980, Prevalence of total coronary occlusion during the early hours of transmural myocardial infarction, *N. Engl. J. Med.* **303:**897–902.
3. Buja, L. M., and Willerson, J. T., 1981, Clinicopathologic correlates of acute ischemic heart disease syndromes, *Am. J. Cardiol.* **47:**343–356.
4. Willerson, J. T., Campbell, W. B., Winniford, M. D., Schmitz, J., Apprill, P., Firth, B. G., Ashton, J., Smitherman, T., Bush, L., and Buja, L. M., 1984, Conversion from chronic to acute coronary artery disease: Speculation regarding mechanisms, *Am. J. Cardiol.* **54:**1349–1354.
5. Willerson, J. T., Hillis, L. D., Winniford, M. D., and Buja, L. M., 1986, Speculations regarding mechanisms responsible for acute ischemic heart disease syndromes, *J. Am. Coll. Cardiol.* **8:**245–250.
6. Sherman, C. T., Litvak, F., Grundfest, W., Lee, M., Hickey, A., Chaux, A., Kass, R., Blanche, C., Matloff,

J., and Morgenstern, L., 1986, Coronary angioscopy in patients with unstable angina pectoris, *N. Engl. J. Med.* **315:**913–919.

7. Falk, E., 1983, Plaque rupture with severe pre-existing stenosis precipitating coronary thrombosis: Characteristics of coronary atherosclerotic plaques, *Br. Heart J.* **50:**127–134.
8. Davies, M. J., and Thomas, A. C., 1985, Plaque fissuring—The cause of acute myocardial infarction, sudden ischemic death, and crescendo angina, *Br. Heart J.* **53:**363–373.
9. Gotoh, K., Minamino, T., Katoh, O., Hamano, Y., Fukui, S., Hori, M., Kusuoka, H., Mishima, M., Inoue, M., and Kamada, T., 1988, The role of intracoronary thrombus in unstable angina: Angiographic assessment and thrombolytic therapy during ongoing anginal attacks, *Circulation* **77:**526–534.
10. Hirsh, P. D., Hillis, L. D., Campbell, W. B., Firth, B. G., and Willerson, J. T., 1981, Release of prostaglandins and thromboxane into the coronary circulation in patients with ischemic heart disease, *N. Engl. J. Med.* **304:** 685–691.
11. Fitzgerald, D. J., Roy, L., Catella, F., and Fitzgerald, G. A., 1986, Platelet activation in unstable coronary disease, *N. Engl. J. Med.* **315:**983–989.
12. Hamm, C. W., Lorenz, R., Bleifeld, W., Kupper, W., Wober, W., and Weber, P. C., 1987, Biochemical evidence of platelet activation in patients with persistent unstable angina, *J. Am. Coll. Cardiol.* **9:**998–1004.
13. Van den Berg, E. K., Schmitz, J. M., Benedict, C. R., Malloy, C. R., Willerson, J. T., and Dehmer, G. J., 1989, Transcardiac serotonin concentration is increased in selected patients with limiting angina and complex coronary lesion morphology, *Circulation* **79:**116–124.
14. Rubany, G. M., Frye, R. L., Holmes, D. R., and Vanhoutte, P. M., 1987, Vasoconstrictor activity of coronary sinus plasma from patients with coronary artery disease, *J. Am. Coll. Cardiol.* **9:**1243–1249.
15. Ambrose, J. A., Winters, S. L., Arora, R. R., Eng, A., Riccio, A., Gorlin, R., and Fuster, V., 1986, Angiographic evolution of coronary artery morphology in unstable angina, *J. Am. Coll. Cardiol.* **7:**472–478.
16. Gibson, R. S., Beller, G. A., Gheorghiade, M., Nygaard, T. W., Watson, D. D., Huey, B. L., Sayre, S. L., and Kaiser, D. L., 1986, The prevalence and clinical significance of residual myocardial ischemia 2 weeks after uncomplicated non-Q wave infarction: A prospective natural history study, *Circulation* **73:**1186–1198.
17. Willerson, J. T., Golino, P., Eidt, J., Campbell, W. B., and Buja, L. M., 1989, Specific platelet mediators and unstable coronary artery lesions: Experimental evidence and potential clinical implications, *Circulation* **80:** 198–205.
18. Folts, J. D., Crowell, E. B., Jr., and Rowe, C. G., 1976, Platelet aggregation in partially obstructed vessels and its elimination by aspirin, *Circulation* **54:**365–370.
19. Bush, L. R., Campbell, W. B., Tilton, G. D., Buja, L. M., and Willerson, J. T., 1984, Effects of the selective thromboxane synthetase inhibitor, dazoxiben, on variations in cyclic blood flow in stenosed canine coronary arteries, *Circulation* **69:**1161–1170.
20. Ashton, J. H., Schmitz, J. M., Campbell, W. B., Ogletree, M. L., Raheja, S., Taylor, A. L., Fitzgerald, C., Buja, L. M., and Willerson, J. T., 1986, Inhibition of cyclic flow variations in stenosed canine coronary arteries by thromboxane A_2/prostaglandin H_2 receptor antagonists, *Circ. Res.* **59:**568–578.
21. Schmitz, J., Apprill, P., Buja, L. M., Willerson, J. T., and Campbell, W. B., 1985, Vascular prostaglandin and thromboxane production in a canine model of myocardial ischemia, *Circ. Res.* **57:**223–231.
22. Golino, P., Ashton, J. H., Buja, L. M., Rosolowsky, M., Taylor, A. L., McNatt, J., Campbell, W. B., and Willerson, J. T., 1989, Local platelet activation causes vasoconstriction of large epicardial canine coronary arteries in vivo: Thromboxane A_2 and serotonin are possible mediators, *Circulation* **79:**154–166.
23. Golino, P., Buja, L. M., Ashton, J. H., Kulkarni, P., Taylor, A., and Willerson, J. T., 1988, Effect of thromboxane and serotonin receptor antagonists on intracoronary platelet deposition in dogs with experimentally stenosed coronary arteries, *Circulation* **78:**701–711.
24. Ashton, J. H., Benedict, C. R., Fitzgerald, C., Raheja, S., Taylor, A., Campbell, W. B., Buja, L. M., and Willerson, J. T., 1986, Serotonin is a mediator of cyclic flow variations in stenosed canine coronary arteries, *Circulation* **73:**572–578.
25. Bush, L. R., Campbell, W. B., Kern, K., Tilton, G. D., Apprill, P., Ashton, J., Schmitz, J., Buja, L. M., and Willerson, J. T., 1984, The effects of alpha$_2$-adrenergic and serotonergic receptor antagonists on cyclic blood flow alterations in stenosed canine coronary arteries, *Circ. Res.* **55:**642–652.
26. Ashton, J. H., Ogletree, M. L., Michel, I. M., Golino, P., McNatt, J. M., Taylor, A. L., Raheja, S., Schmitz, J., Buja, L. M., Campbell, W. B., and Willerson, J. T., 1987, Cooperative mediation by serotonin S_2 and thromboxane A_2/prostaglandin H_2 receptor activation of cyclic flow variations in dogs with severe coronary artery stenoses, *Circulation* **76:**952–959.

27. Ashton, J. H., Golino, P., McNatt, J. M., Buja, L. M., and Willerson, J. T., 1989, Serotonin S_2 receptor blockade (LY53857) combined with thromboxane A_2/prostaglandin H_2 receptor blockade (SQ29,548) provide protection against epinephrine-induced intermittent thrombosis in severely narrowed canine coronary arteries, *J. Am. Coll. Cardiol.* **13**:755–763.
28. Eidt, J. F., Ashton, J., Golino, P., McNatt, J., Buja, L. M., and Willerson, J. T., 1989, Thromboxane A_2 and serotonin mediate coronary blood flow reductions in unsedated dogs, *Am. J. Physiol.* **26**:H873–H882.
29. Eidt, J. F., Ashton, J., McNatt, J., Buja, L. M., and Willerson, J. T., 1989, Treadmill exercise promotes cyclical alterations in coronary blood flow in dogs with coronary artery stenosis and endothelial injury, *J. Clin. Invest.* **84**:517–527.
30. Fishman, J. A., Ryan, G. B., and Karnovsky, M. J., 1975, Endothelial regeneration in the rat carotid artery and the significance of endothelial denudation in the pathogenesis of myointimal thickening, *Lab. Invest.* **32**: 339–351.
31. Austin, G. E., Ratliff, N. G., Hollman, J., Tabei, S., and Phillips, D. F., 1985, Intimal proliferation of smooth muscle cells as an explanation for recurrent coronary artery stenosis after percutaneous transluminal coronary angioplasty, *J. Am. Coll. Cardiol.* **6**:369–375.
32. Itoh, Y., Yanagisawa, M., Ohkubo, S., Kimura, C., Kosaka, T., Inoue, A., Ishida, N., Mitsui, Y., Onda, H., and Fujino, M., 1988, Cloning and sequence analysis of cDNA encoding the precursor of a human endothelium-derived vasoconstrictor peptide, endothelin: Identity of human and porcine endothelin, *FEBS Lett.* **231**:440–444.
33. Jonasson, L., Holm, J., and Hansson, G. K., 1988, Cyclosporin A inhibits smooth muscle proliferation in the vascular response to injury, *Proc. Natl. Acad. Sci. USA* **85**:2302–2306.
34. Berk, B. C., Alexander, R. W., Brock, T. A., Gimbrone, M. A., Jr., and Webb, R. C., 1986, Vasoconstriction: A new activity for platelet-derived growth factor, *Science* **232**:87.
35. Waterfield, M. D., Scrace, G. T., Whittle, N., Stroobant, P., Johnsson, A., Wasteson, A., Westermark, B., Heldin, C. H., Huang, J. S., and Deuel, T. F., 1983, Platelet-derived growth factor is structurally related to the putative transforming protein $p28^{cis}$ of simian sarcoma virus, *Nature* **304**:35–39.
36. Williams, L. T., Tremble, P. M., Lavin, M. R., and Sunday, M. E., 1984, Platelet-derived growth factor receptors form a high affinity state in membrane preparations: Kinetics and affinity cross-linking studies, *J. Biol. Chem.* **259**:5287–5294.
37. Roberts, A. B., Sporn, M. B., Assoian, R. K., Smith, J. M., Roche, N. S., Wakefield, L. M., Heine, U. I., Liotta, L. A., Falanga, V., and Kehrl, J. H., 1986, Transforming growth factor type beta: Rapid induction of fibrosis and angiogenesis in vivo and stimulation of collagen formation in vitro, *Proc. Natl. Acad. Sci. USA* **83**:4167–4171.
38. Akhurst, R. J., Fee, F., and Balmain, A., 1988, Localized production of TGF-beta mRNA in tumour promoter-stimulated mouse epidermis, *Nature* **331**:363–364.
39. Lewis, H. D., Jr., Davis, J. W., Archibald, D. G., Steinke, W. E., Smitherman, T. C., Doherty, J. E., III, Schnaper, H. W., LeWinter, M. M., Linares, E., Pouget, J. M., Sabharwal, S. C., Chesler, E., and DeMots, H., 1983, Protective effects of aspirin against acute myocardial infarction and death in men with unstable angina, *N. Engl. J. Med.* **309**:396–403.
40. Cairns, J. A., Gent, M., Singer, J., Finnie, K. J., Froggatt, G. M., Holder, D. A., Jablonsky, G., Kostuk, W. J., Melendez, L. J., and Myers, M. G., 1985, Aspirin, sulfinpyrazone, or both in unstable angina, *N. Engl. J. Med.* **313**:1369–1375.
41. Hirsh, P. D., Campbell, W. B., Willerson, J. T., and Hillis, L. D., 1981, Prostaglandins and ischemic heart disease, *Am. J. Med.* **71**:1009–1026.
42. Ludmer, P. L., Selwyn, A. P., Shook, T. L., Wayne, R. R., Mudge, G. H., Alexander, R. W., and Ganz, P., 1986, Paradoxical vasoconstriction induced by acetylcholine in atherosclerotic coronary arteries, *N. Engl. J. Med.* **315**:1046–1051.
43. Nabel, E. G., Ganz, P., Gordon, J. B., Alexander, R. W., and Selwyn, A. P., 1988, Dilation of normal and constriction of atherosclerotic coronary arteries caused by the cold pressor test, *Circulation* **77**:43–52.

18

Endothelial Dysfunction and Atherosclerosis

Russell Ross

I. INTRODUCTION

The concept of endothelial "injury," or dysfunction, has been central to the "response to injury" hypothesis of atherosclerosis, which was formulated in 1973[1] and has been modified and retested for the past 16 years.[2–4] This hypothesis proposes that many forms of functional alteration may occur in the endothelium, some of which may have morphologic manifestations, whereas others may induce no visible change. These changes may lead to a series of cellular interactions at focal sites within the artery wall. They involve white cells of the blood (T lymphocytes and monocyte/macrophages) and cells of the artery wall (endothelium and smooth muscle) in a series of complex interactions that lead to the development of fatty streaks, fibrofatty (or intermediate) lesions, and ultimately to fibrous plaques, which may become the advanced, complicated, ulcerated, or calcified occlusive lesions of atherosclerosis.

Endothelial cells have numerous functions (described in other chapters in this book). These include serving as a lining for the blood container, a permeability barrier, and a source of connective tissue matrix, as well as providing a nonthrombogenic surface, vasoactive substances, and growth factors. Alterations in these and other functional activities of the endothelium may lead to adherence of T lymphocytes and monocytes followed by their entry between endothelial cells into the artery wall, culminating in the formation of the first and ubiquitous lesion of atherosclerosis, the fatty streak.[5–8] The fatty streak continues to expand as long as the elements that provide the original "injury" are present. Fatty streaks may occur at sites of preexisting collections of intimal smooth muscle cells, sometimes referred to as "intimal cushions." Many of the macrophages in the fatty streak may release growth-regulatory molecules that lead to proliferation of these smooth muscle cells. At other anatomic sites, smooth muscle cells may be stimulated to migrate from the underlying media into the intima to accomplish the same result. The lesions may

Russell Ross • Department of Pathology, University of Washington, Seattle, Washington 98195.

Endothelial Cell Dysfunctions, edited by Nicolae Simionescu and Maya Simionescu. Plenum Press, New York, 1992.

then expand due to the further formation of growth-regulatory substances (both stimulators and inhibitors), the net effect of which will determine whether the lesion expands, remains stable, or regresses. This chapter will review some of these events in greater detail.

II. THE PROCESS OF ATHEROGENESIS

Atherogenesis has been recognized for over 100 years. However, it is only in the past 25 years that smooth muscle proliferation has been recognized as the key event involved in lesions that may occlude the arterial lumen and thus induce clinical sequelae. Although lipid accumulation is one of the earliest events in hyperlipidemic individuals, if lipid accumulation occurred in the absence of smooth muscle proliferation, there would be little to no clinical sequelae. It is the change from a flat, lipid-rich fatty streak to a thickened, acentric, proliferative smooth muscle lesion that may precede alterations in the surface endothelium, which may result in thrombosis and ultimately may lead to occlusion of the artery with induction of a myocardial or cerebral infarct. Thus, the process of atherogenesis must be considered a dynamic one in which lesions may be constantly in the process of forming and/or regressing, and, of course, until the lumen of the artery is sufficiently occluded, the entire process remains clinically silent.

All of the cells involved in the process of atherogenesis, at all stages of the process, have the capacity to make both growth stimulators and growth inhibitors. T lymphocytes can synthesize interferon-gamma (IFNγ), lymphotoxin [tumor necrosis factor-beta (TNFβ)], interleukin 2 (IL-2), and colony stimulating factor 1 (CSF-1).[9] Macrophages can provide an even larger number of growth-regulatory molecules, including both chains of platelet-derived growth factor (PDGF),[10] CSF-1,[11] interleukin 1 (IL-1),[12] transforming growth factor-alpha (TGFα),[13] transforming growth factor-beta (TGFβ),[14] and tumor necrosis factor-alpha (TNFα).[11] Smooth muscle cells and endothelium can secrete CSF-1,[15] IL-1,[16] and TGFβ.[17,18] Thus, all of the cells involved in the process of atherogenesis have the capacity to provide growth-regulatory molecules that can affect one or more of the numerous cells with which they may interact within the arterial intima. If platelet thrombi form, as they may at branches and bifurcations, at sites of endothelial alteration, or at sites of occlusive thrombosis, they can release a host of growth factors locally, including PDGF, TGFβ, epithelial growth factor (EGF/TGFα), platelet-derived ECGF (PD-ECGF), and others.[19,20]

It follows that whether lesions progress from fatty streaks to complicated advanced lesions will depend upon numerous factors that act upon the local microenvironment. These include the rheologic properties of the vessel at particular anatomic sites, as well as injurious agents that may form locally as a result of risk factors such as hypercholesterolemia, cigarette smoking, diabetes mellitus, hypertension, or less frequent conditions such as homocystinuria. Moreover, all of these factors and conditions occur within the background of the genetic constitution of the individual.

The lesions of atherosclerosis that form may be full of lipid, or they may contain relatively little lipid and be quite fibrotic. Again, the ultimate nature of the lesion depends upon the factors noted above. Consequently, the lesions of atherosclerosis are not uniformly alike and may have diverse manifestations, depending upon the risk factors, the anatomic sites, and the genetic makeup of the individual at risk.

III. ALTERATIONS IN ENDOTHELIAL FUNCTION

As noted by the Simionescus and in Chapter 1, alterations in endothelial permeability may be among the first changes that occur after exposure to risk factors such as hyperlipidemia. In individuals who are hypercholesterolemic, as has been demonstrated in rabbits[21] and monkeys,[7,8] one of the first changes to occur results in increased transport of lipid and lipoprotein particles into the subendothelial intimal space. These particles vary in appearance. Some look like liposomes, whereas others resemble lipoprotein particles, many of which may be altered due to the capacity of endothelium to modify (oxidize) the lipoproteins as they are taken up and transported through the endothelium. In some sites, the endothelium may be sufficiently altered that intercellular junctions become leaky, allowing some plasma constituents to actually pass between the junctions of the endothelial cells. Most data suggest that the entrance of lipids into the subendothelium occurs via transport mechanisms within the endothelial cells themselves.[21] Numerous pharmacologic agents, as well as hyperlipidemia, can lead to increased transport and thus increased permeability of the endothelium not only to lipids, but to other solutes present within the plasma. If the endothelium oxidizes the lipoproteins, as discussed in Chapter 19, these oxidized lipoproteins may serve as chemotactic substances for monocytes after the altered lipoproteins enter the subendothelial space.[22] This in turn may help bring leukocytes into the artery wall; they could then become activated and provide additional monocyte chemotactic agents as well as substances that could be injurious to the cells, including the endothelial cells in the local microenvironment.

IV. ADHERENCE OF LEUKOCYTES TO ENDOTHELIUM

Following the increased permeation of lipid into the artery wall in hyperlipidemic individuals, the first clearly measurable alteration in all of the experimental animals that have been carefully studied appears to be the increased adherence of monocytes and T lymphocytes at particular anatomic sites in the artery wall.[5,7,23] These include branches and bifurcations and sites where the flow characteristics of the blood are altered so that flow is decreased and there are back currents and eddy currents. These regions of the artery do not appear to be regions of high shear stress. The increased adherence of the leukocytes to the endothelium appears to be due to the formation of a series of cell-surface glycoproteins on the endothelium (ELAMs) (discussed in Chapter 9), as well as to formation of a series of cell-surface glycoproteins on the leukocytes themselves, including the CD11/CD18 complex and other cell-surface glycoproteins (discussed in Chapter 9). Figure 1, taken from the work of Masuda and Ross,[7] demonstrates the typical appearance of adherent leukocytes at a branch point in the artery wall of a nonhuman primate that had been hyperlipidemic with cholesterol levels of approximately 300 mg/dl for 6 months. A transmission electron micrograph of such an adherent monocyte, taken from the work of Faggiotto *et al.*[5] (Fig. 2), demonstrates the intimate nature of the adherence, as evidenced by the observation that the monocytes send processes into infoldings of the endothelial surface much like fingers into a glove. After these leukocytes adhere, which they appear to do in increasing numbers as long as the animal remains hypercholesterolemic, they continue to move upon the surface of the endothelium.

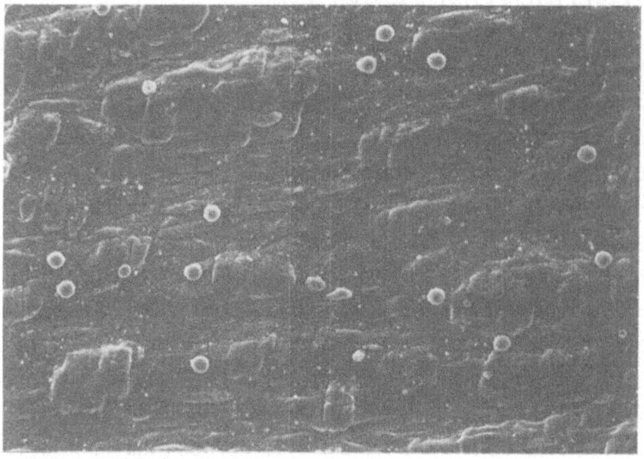

Figure 1. Scanning electron micrograph from a nonbranching region of the thoracic aorta in a monkey (*Macaca nemestrina*) fed a low-level hypercholesterolemic diet for 6 months. Many small irregularly shaped elevated lesions (fatty streaks) and scattered adherent leukocytes are seen on the intact endothelial surface. × 260. (Reproduced by permission of the American Heart Association from Masuda and Ross.[7])

The formation of chemotactic factors, by either the endothelium and/or the monocytes and T lymphocytes that have entered between junctions of the endothelial cells, could cause additional chemotaxis of leukocytes to occur. Like their predecessors, these leukocytes migrate between the endothelial cells and localize in the subendothelial intimal space (Fig. 3). This increased adherence of leukocytes and their migration into the intima leads to the development of the first and ubiquitous lesion of atherosclerosis, the fatty streak. The monocytes that localize in the intima become converted to macrophages and act as scavenger cells to take up the lipid, particularly the modified lipoproteins, and through this process, they develop into foam cells (Fig. 4). As long as the factors, such as hyper-

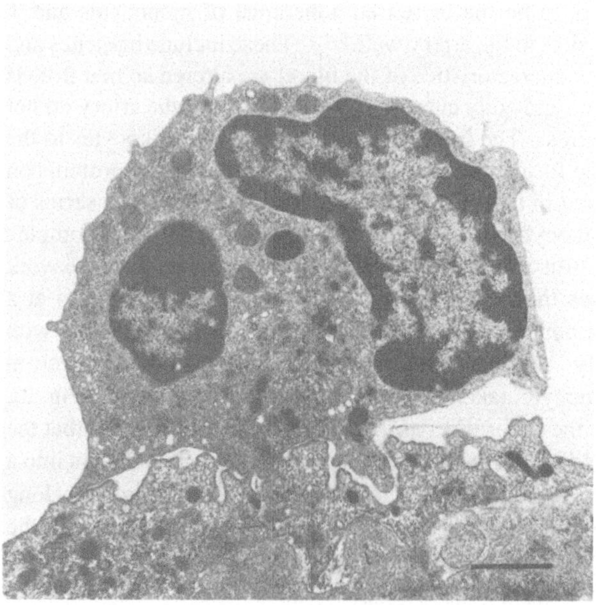

Figure 2. Electron micrograph demonstrating a monocyte adherent to the endothelium of a monkey (*Macaca nemestrina*) after 12 days on a high-level hypercholesterolemic diet. Bar = 1 μm. (Reproduced by permission of the American Heart Association from Faggiotto *et al.*[6])

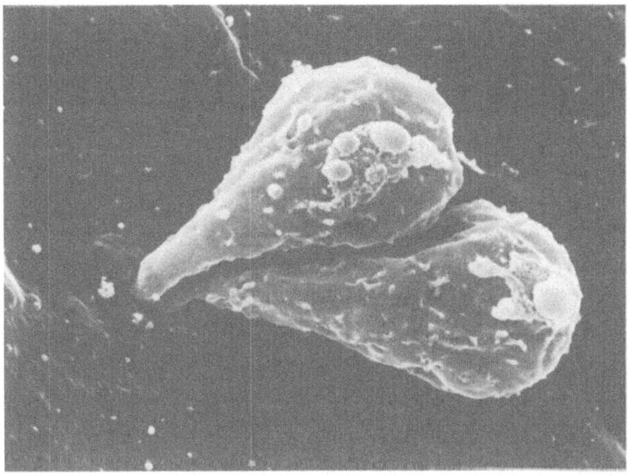

Figure 3. Scanning electron micrograph demonstrating leukocytes entering the artery wall between endothelial junctions of thoracic aorta of a monkey (*Macaca nemestrina*) after 6 months on a low-level hypercholesterolemic diet. × 4000. (Reproduced by permission of the American Heart Association from Masuda and Ross.[7])

lipidemia, that are associated with increased adherence and subendothelial migration of leukocytes remain present, the lesions of atherosclerosis will continue to expand by continuous recruitment of these cells at focal sites within the arterial tree.

V. GROWTH-REGULATORY MOLECULES

As indicated above, all of the cells involved in the lesions of atherosclerosis have the capacity to make growth-regulatory molecules that are both inhibitory and stimulatory to each of the cells localized in the lesions.

A. TGFβ

Of these growth factor molecules, a number of them can be mitogenic and/or inhibitory to the endothelial cells. For example, TGFβ, a potent inhibitor of endothelial cell proliferation, can be formed and secreted by activated macrophages, T lymphocytes, smooth muscle cells, and endothelial cells themselves.[24] When platelet interactions occur during the process of thrombosis, for example, one of the major growth-regulatory molecules released from the storage granules present in platelets is TGFβ.[24] TGFβ has been shown to be a potent inhibitor of both endothelial cell migration and proliferation.[25] Furthermore, it is a potent inducer of connective tissue synthesis by cells such as fibroblasts and smooth muscle cells.[26] It is not yet clear whether TGFβ is active in terms of inducing synthesis by the endothelium of connective tissue elements such as basement membrane proteins and fibronectin.

Although there are increasing data to suggest that the principal activities of TGFβ are related to its capacity either to induce connective-tissue formation by cells or to push cells into a differentiative pathway, and thus decrease their proliferative response to mitogens, there are also accumulating data that TGFβ induces many of its changes in cells by causing them to express a second set of genes. Examples of this include the capacity of TGFβ to induce PDGF gene expression in smooth muscle cells and fibroblasts.[27] This will undoubtedly be the case for some of the other growth-regulatory molecules discussed below.

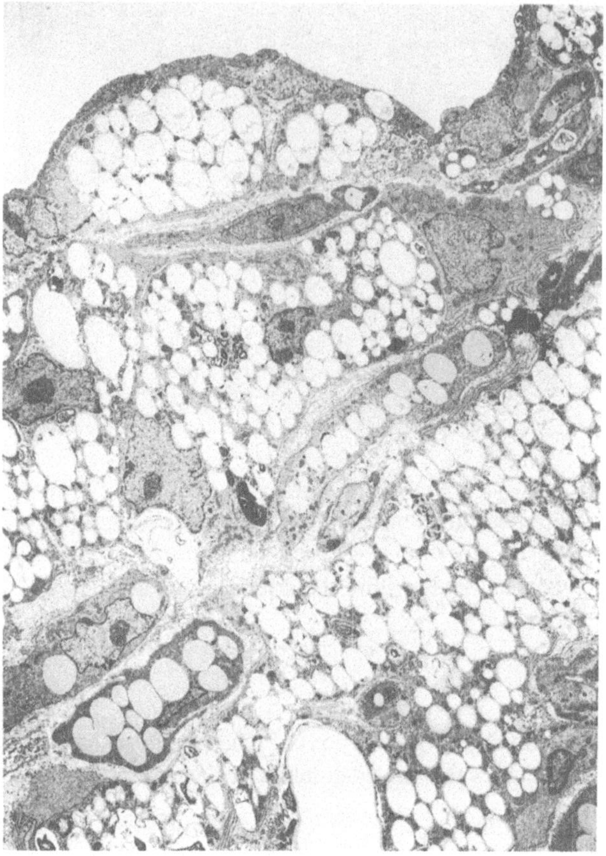

Figure 4. A transmission electron micrograph of an advanced fatty streak from the thoracic aorta of a monkey (*Macaca nemestrina*) fed a low-level hypercholesterolemic diet for 6 months. This fatty streak contains multiple layers of subendothelial macrophage-derived foam cells with intervening lipid-laden smooth muscle cells. Note the difference in density of lipid droplets in the macrophages and the smooth muscle cells. They appear to be less extracted in the smooth muscle cells. × 2200. (Reproduced by permission of the American Heart Association from Masuda and Ross.[7])

B. PD-ECGF

Platelets also contain a potent endothelial mitogen, PD-ECGF. PD-ECGF is the only endothelial mitogen in platelets, and has a molecular weight of 45,000. Unlike other angiogenic factors, such as FGF (see below), it does not bind to heparin and does not stimulate fibroblasts or smooth muscle.[28] The PD-ECGF molecule can act both as an angiogenic agent on microvascular endothelium *in vivo*, and as an agent capable of inducing endothelial migration and proliferation *in vitro*.[29] Release of PD-ECGF by platelets at sites of endothelial injury could lead to migration and regeneration of endothelium over such sites.

C. FGF

Another endothelial mitogen that is present in the cells involved in atherogenesis is fibroblast growth factor (FGF). Both acidic and basic FGF are potent endothelial mitogens and have been observed in macrophages and can also be formed by the endothelial cells themselves.[30] Since FGF has no signal sequence, it is not clear whether it is secreted from cells or, if so, how this occurs. It has been found in the extracellular environment, in association with matrix proteins within the basement membrane of endothelium, and in other basement-membrane-like structures such as Descemet's membrane in the eye. It has been suggested that FGF may be secreted as part of a complex with heparan sulfate or with other matrix proteins. Thus far, however, it has only been found intracellularly by inducing lysis of cells such as macrophages and/or endothelium, both of which have the capacity to make this potent angiogenic and mitogenic molecule.

Substances such as FGF can have a striking effect upon endothelium. After endothelial denudation by use of an intraarterial balloon catheter, Reidy and colleagues (unpublished data) have demonstrated that a bolus intraarterial injection of basic FGF can enhance endothelial migration and proliferation so that regeneration and coverage of the entire denuded surface can occur. Under normal circumstances, in the absence of any FGF, balloon catheter denudation of the common carotid artery in the rat leads to regeneration of approximately only 50% of the surface of the denuded artery. In the presence of administered FGF, Reidy and colleagues have been able to obtain coverage of 100% of the surface, not only after continuous infusion of FGF for prolonged periods (6 hr), but even after infusion of a bolus amount of FGF. Thus, the presence of mitogens and inducers of migration and proliferation such as FGF could be important in the maintenance of endothelial integrity over prolonged periods of time at sites where endothelial injury and denudation may possibly occur.

D. TNFα

TNFα has been found to be angiogenic in some cases and inhibitory to endothelium in other cases. This depends upon the type of endothelium involved. Since angiogenesis occurs in relation to capillary endothelium, TNFα appears to be stimulatory whereas arterial endothelium appears to be inhibited by TNFα. TNFα is formed by activated macrophages, and represents one of the principal cytokines that these cells release.[11] A cytokine with similar activity, lymphotoxin or TNFβ, can also be released by activated T lymphocytes.[9] Thus, in tissues at sites of chronic inflammation where these cytokines may be present, the capacity of endothelium to respond may be markedly modified by these paradoxically angiogenic and inhibitory agents.

Similar to TGFβ, TNFα can also have effects on cells by inducing secondary gene expression. An example of this interaction is the effect of TNFα upon arterial endothelial cells. TNFα induces PDGF B-chain gene expression in arterial endothelial cells, and release of this growth factor, which can have a profound paracrine effect on the surrounding smooth muscle cells by stimulating them to proliferate.[31]

E. IL-1

IL-1 is one of the principal cytokines synthesized and secreted by activated macrophages. Since activated macrophages are ubiquitous in all lesions of atherosclerosis, IL-1 may be involved in the process of atherogenesis. IL-1 may play a regulatory role in terms of the interactions between the T lymphocytes, the macrophages, and the smooth muscle cells in the lesions. In addition, similar to activities discussed above for TNFα and TGFβ, IL-1 has also been described as a proliferative agent for fibroblasts and smooth muscle cells. In fact, when added to smooth muscle cells or fibroblasts, IL-1 induces PDGF A-chain gene expression followed by synthesis and secretion of PDGF-AA. Thus, IL-1 can stimulate proliferation by autocrine stimulation of either the fibroblasts or the smooth muscle cells by the PDGF-AA that they have formed.[32] As a consequence, IL-1 is itself not mitogenic for smooth muscle cells and fibroblasts, as originally believed, but rather induces their mitogenesis by PDGF A-chain autocrine stimulation.

F. PDGF

PDGF is one of the mitogens principally responsible for the stimulation of connective tissue cells, such as smooth muscle cells and fibroblasts, to migrate chemotactically and to proliferate. This molecule was observed to be the principal source of mitogenic activity in whole blood serum, and consists of three dimeric forms derived from the two separate chains of PDGF, A and B. Each of these peptides is coded for on separate chromosomes, and each in turn has specific affinities for one of the two receptor subunits for PDGF, termed alpha and beta, respectively. Like the two chains of PDGF that can form the three PDGF dimers (AA, AB, BB), the two receptor subunits can also form three dimeric forms ($\alpha\alpha$, $\alpha\beta$, $\beta\beta$). The A chain can bind to only the α subunit, whereas the B chain can bind to either the α or the β subunit. Thus, each of the three dimeric forms of PDGF has specific affinities for the various combinations of receptor subunits on the cell surface. Each cell type contains differing numbers of α and β receptor subunits. Thus, the relative mitogenicity or chemotactic capacity of PDGF depends both on the dimeric form presented to the cell and on the number of α and β subunits that can form mature dimeric forms of the receptor on the cell, as well.

Endothelial cells can express the genes for both PDGF-A and PDGF-B,[33] and do so when the cells are appropriately stimulated. Thus, in tissue responses as diverse as wound healing, atherosclerosis, and various tumors, endothelial cells could potentially serve as an important source of this potent connective tissue cell mitogen. By releasing PDGF, endothelial cells could act in a paracrine fashion to stimulate neighboring cells such as fibroblasts or smooth muscle cells to proliferate. In the case of atherogenesis, activation of endothelium to provide PDGF could represent an important response of the endothelium to atherogenic stimuli. Since arterial endothelial cells have no PDGF receptors, they themselves cannot respond to this mitogen.

VI. CELLULAR INTERACTIONS AMONG ENDOTHELIUM, SMOOTH MUSCLE, PLATELETS, AND LEUKOCYTES IN ATHEROGENESIS

The earliest grossly observable lesion of atherosclerosis, the fatty streak, consists of principally monocyte-derived macrophages and CD8- and CD4-positive T lymphocytes. The initial cellular interactions that occur in the development of these lesions in experimental animals take place among each of these cells as they interact with the endothelium. The initial events of atherogenesis involve increased adhesion of monocytes and lymphocytes to the endothelium,[5–8] which can provide CSF. The lymphocytes and monocytes can also provide factors that both stimulate and inhibit one another. Further macrophage activation

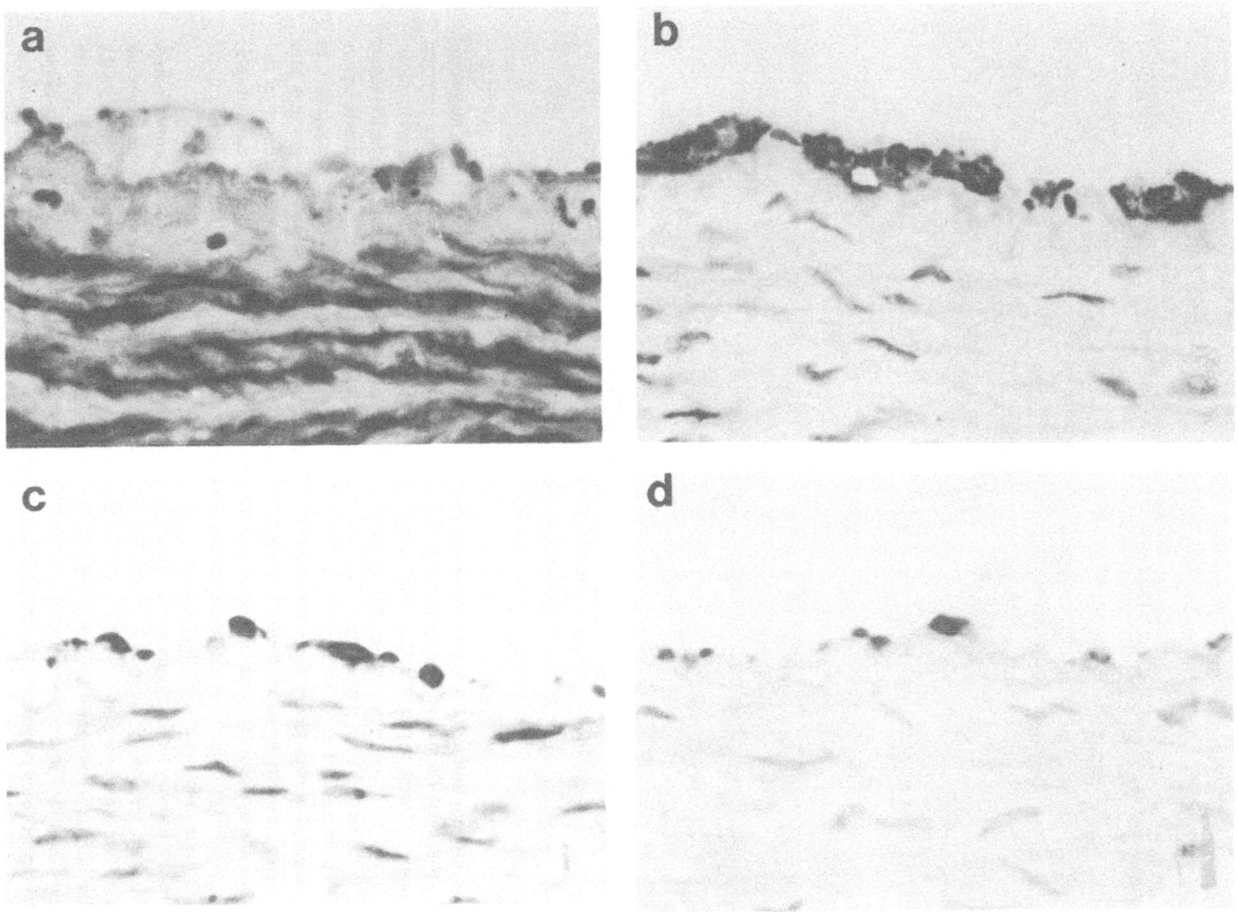

Figure 5. Immunoperoxidase staining of an early fatty streak, serial frozen sections from thoracic aorta of a monkey (*Macaca nemestrina*) after 6 months on a low-level hypercholesterolemic diet. (a) Antimuscle antibody, HHF35. (b) Antimacrophage antibody, HAM56. (c) Anti-CD8 antibody, G10.1. (d) Anti-CD4 antibody, OKT4a. ABC method with OsO_4 intensification, counterstained with nuclear fast red. × 150. (Reproduced by permission of the American Heart Association from Masuda and Ross.[7])

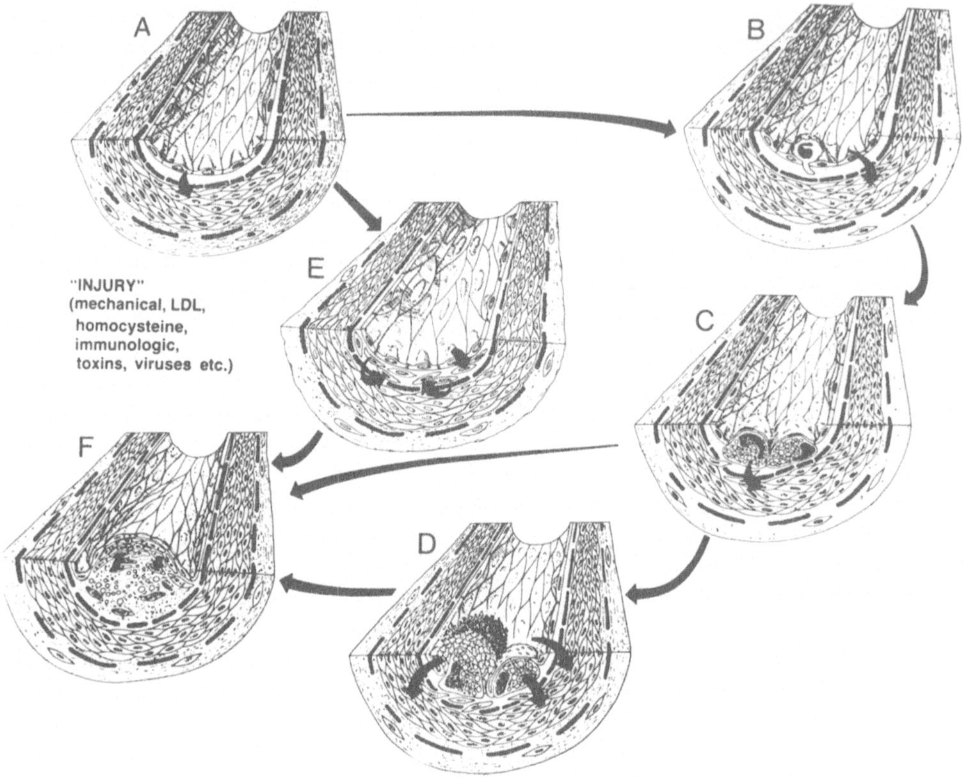

Figure 6. The response-to-injury hypothesis. Advanced intimal proliferative lesions of atherosclerosis may occur by at least two pathways. The pathway demonstrated by the clockwise (long) arrows to the right has been observed in experimentally induced hypercholesterolemia. Injury to the endothelium (A) may induce growth factor secretion (short arrow). Monocytes attach to endothelium (B), which may continue to secrete growth factors (short arrow). Subendothelial migration of monocytes (C) may lead to fatty-streak formation and release of growth factors such as PDGF (short arrow). Fatty streaks may become directly converted to fibrous plaques (long arrow from C to F) through release of growth factors from macrophages or endothelial cells or both. Macrophages may also stimulate or injure the overlying endothelium. In some cases, macrophages may lose their endothelial cover and platelet attachment may occur (D), providing three possible sources of growth factors—platelets, macrophages, and endothelium (short arrows). Some of the smooth muscle cells in the proliferative lesion itself (F) may form and secrete growth factors such as PDGF (short arrows).

An alternative pathway for development of advanced lesions of atherosclerosis is shown by the arrows from A to E to F. In this case, the endothelium may be injured but remain intact. Increased endothelial turnover may result in growth-factor formation by endothelial cells (A). This may stimulate migration of smooth muscle cells from the media into the intima, accompanied by endogenous production of PDGF by smooth muscle as well as growth factor secretion from the "injured" endothelial cells (E). These interactions could then lead to fibrous-plaque formation and further lesion progression (F). (Reproduced by permission of *The New England Journal of Medicine* from Ross.[4])

may occur through IFNγ secretion by the lymphocytes, as well as through substances transported by the endothelium such as modified lipoproteins.[22] Each of these cells can present antigen and MHC complexes.[34–36] The presence of the T cells suggests that some lesions of atherosclerosis may represent an immune or perhaps even an autoimmune response (Fig. 5).

When smooth muscle cells are present at preexisting sites in the intima, or when they migrate into the intima from the media, they can respond to factors such as PDGF and TGFβ, which can induce proliferation and connective tissue elaboration, respectively. The PDGF and TGFβ may be derived from activated macrophages and/or from activated endothelial cells. Thus, in the mature lesion of atherosclerosis, one could anticipate the possibilities of endothelial–smooth muscle interactions, endothelial-macrophage interactions, endothelial–lymphocyte interactions, lymphocyte–macrophage interactions, macrophage–smooth muscle interactions, and smooth muscle–smooth muscle interactions. Thus, gene expression for growth-regulatory molecules in each of these individual cells, the controls of such gene expression, and the different secretory products that can activate or inhibit each of the individual cells within the microenvironment of the developing lesion of atherosclerosis will ultimately determine whether the lesion progresses, remains stable, or regresses. The formation of the different growth-regulatory molecules may be closely related to the levels of potentially injurious agents that may be present within the plasma or within the interstitial fluid, depending upon whether the patient is hypercholesterolemic, hypertensive, a cigarette smoker, a diabetic, or some combination of these. Each of these components will be superimposed upon the specific genetic background of the individual and will ultimately determine whether or not the patient will develop proliferative smooth-muscle lesions (Fig. 6).

We now know the nature of the cells as well as the different macromolecules they may potentially form during the cellular interactions in the different lesions of atherosclerosis. Although the possible combinations of cells and factors are numerous, one should not be deterred by the seeming complexity of the process. The tools are at hand to determine which cells are activated, and markers are available to determine their levels of activation. Which growth-regulatory molecules and other substances may be formed, and how these interactions may contribute to the formation of occlusive proliferative lesions under different circumstances, need to be determined. These approaches herald the day when new diagnostic tools and therapeutic approaches may permit early diagnosis, treatment, and prevention.

ACKNOWLEDGMENT. This work was supported in part by a grant from NIH (HL-18645).

REFERENCES

1. Ross, R., and Glomset, J., 1973, Atherosclerosis and the arterial smooth muscle cell, *Science* **180**:1332–1339.
2. Ross, R., and Glomset, J. A., 1976, The pathogenesis of atherosclerosis, *N. Engl. J. Med.* **295**:369–377, 420–425.
3. Ross, R., 1981, Atherosclerosis—A problem of the biology of arterial wall cells and their interaction with blood components, *Arteriosclerosis* **1**:293–311.
4. Ross, R., 1986, The pathogenesis of atherosclerosis—An update, *N. Engl. J. Med.* **314**:488–500.

5. Faggiotto, A., Ross, R., and Harker, L., 1984, Studies of hypercholesterolemia in the nonhuman primate. I. Changes that lead to fatty streak formation, *Arteriosclerosis* **4:**323–340.
6. Faggiotto, A., and Ross, R., 1984, Studies of hypercholesterolemia in the nonhuman primate. II. Fatty streak conversion to fibrous plaque, *Arteriosclerosis* **4:**341–356.
7. Masuda, J., and Ross, R., 1990, Low-level hypercholesterolemia in the nonhuman primate. I. Fatty streak formation, *Arteriosclerosis* **10:**164–177.
8. Masuda, J., and Ross, R., 1990, Atherogenesis during low-level hypercholesterolemia in the nonhuman primate. II. Fatty streak conversion to fibrous plaque, *Arteriosclerosis* **10:**178–187.
9. Vilcek, J., Gray, P. W., Rinderknecht, E., and Sevastopoulos, 1985, Interferon-γ: A lymphokine for all seasons, *Lymphokines* **11:**1–22.
10. Shimokado, K., Raines, E. W., Madtes, D. K., Barrett, T. B., Benditt, E. P., and Ross, R., 1985, A significant part of macrophage-derived growth factor consists of at least two forms of PDGF, *Cell* **43:**277–286.
11. Nathan, C. F., 1987, Secretory products of macrophages, *J. Clin. Invest.* **79:**319–326.
12. Dinarello, C. A., 1984, Interleukin-1, *Rev. Infect. Dis.* **6:**51–95.
13. Madtes, D. K., Raines, E. W., Sakariassen, K. S., Assoian, R. K., Sporn, M. B., Bell, G. I., and Ross, R., 1988, Induction of transforming growth factor-α in activated human alveolar macrophages, *Cell* **53:**285–293.
14. Assoian, R. K., Fleudelys, B. E., Stevenson, H. C., Miller, P. J., Madtes, D. K., Raines, E. W., Ross, R., and Sporn, M. B., 1987, Expression and secretion of type β transforming growth factor by activated human macrophages, *Proc. Natl. Acad. Sci. USA* **84:**6020–6024.
15. Ralph, P., 1989, Colony stimulating factors, in: *Human Monocytes* (M. Zembala and G. Asherson, eds.), Academic Press, New York, pp. 228–246.
16. Dinarello, C. A., 1988, Biology of interleukin 1, *FASEB J.* **2:**108–115.
17. Antonelli-Orlidge, A., Saunders, K. B., Smith, S. R., and D'Amore, P. A., 1989, An activated form of transforming growth factor β is produced by cocultures of endothelial cells and pericytes, *Proc. Natl. Acad. Sci. USA* **86:**4544–4548.
18. Sato, Y., and Rifkin, D. B., 1989, Inhibition of endothelial cell movement of pericytes and smooth muscle cells: Activation of a latent transforming growth factor-β1-like molecule by plasmin during co-culture, *J. Cell Biol.* **109:**309–315.
19. Raines, E. W., Bowen-Pope, D. F., and Ross, R., 1990, Platelet-derived growth factor, in: *Handbook of Experimental Pharmacology: Peptide Growth Factors and Their Receptors*, Vol. 95/I (M. B. Sporn and A. B. Roberts, eds.), Springer-Verlag, Berlin, pp. 173–262.
20. Ross, R., Raines, E. W., and Bowen-Pope, D. F., 1986, The biology of platelet-derived growth factor, *Cell* **46:**155–169.
21. Simionescu N., Vasile, E., Lupu, F., Popescu, G., and Simionescu, M., 1986, Prelesional events in atherogenesis. Accumulation of extracellular cholesterol-rich liposomes in the arterial intima and cardiac valves of the hyperlipidemic rabbit, *Am. J. Pathol.* **123:**109–125.
22. Parthasarathy, S., Quinn, M. T., Schwenke, D. C., Carew, T. E., and Steinberg, D., 1989, Oxidative modification of beta-very low density lipoprotein. Potential role in monocyte recruitment and foam cell formation, *Arteriosclerosis* **9:**398–404.
23. Gerrity, R. G., 1981, The role of the monocyte in atherogenesis. I. Transition of blood-borne monocytes into foam cells in fatty lesions, *Am. J. Pathol.* **103:**181–190.
24. Sporn, M. B., Roberts, A. B., Wakefield, L. M., and de Crombrugghe, B., 1987, Some recent advances in the chemistry and biology of transforming growth factor-beta, *J. Cell Biol.* **105:**1039–1045.
25. Heimark, R. L., Twardzik, D., and Schwartz, S. M., 1986, Inhibition of endothelial regeneration by type-beta transforming growth factor from platelets, *Science* **233:**1078–1080.
26. Ignotz, R. A., and Massague, J., 1986, Transforming growth factor-β stimulates the expression of fibronectin and collagen and their incorporation into the extracellular matrix, *J. Biol. Chem.* **261:**4337–4345.
27. Leof, E. B., Proper, J. A., Goustin, A. S., Shipley, G. D., DiCorleto, P. E., and Moses, H. L., 1986, Induction of c-sis mRNA and activity similar to platelet-derived growth factor by transforming growth factor beta: A proposed model for indirect mitogenesis involving autocrine activity, *Proc. Natl. Acad. Sci. USA* **83:**2453–2457.
28. Miyazono, K., Okabe, T., Urabe, A., Takaku, F., and Heldin, C.-H., 1987, Purification and properties of an endothelial cell growth factor from human platelets, *J. Biol. Chem.* **262:**4098–4103.
29. Ishikawa, F., Miyazono, K., Hellman, U., Drexler, H., Wernstedt, C., Hagiwara, K., Usuki, K., Fumimaro,

T., Risau, W., and Heldin, C.-H., 1989, Identification of angiogenic activity and the cloning and expression of platelet-derived endothelial cell growth factor, *Nature* **338:**557–562.

30. Baird, A., Mormede, P., and Bohlen, P., 1985, Immunoreactive fibroblast growth factor in cells of peritoneal exudate suggests its identity with macrophage-derived growth factor, *Biochem. Biophys. Res. Commun.* **126:** 358–364.
31. Hajjar, K. A., Hajjar, D. P., Silverstein, R. L., and Nachman, R. L., 1987, Tumor necrosis factor-mediated release of platelet-derived growth factor from cultured endothelial cells, *J. Exp. Med.* **166:**235–245.
32. Raines, E. W., Dower, S. K., and Ross, R., 1989, IL-1 mitogenic activity for fibroblasts and smooth muscle cells is due to PDGF-AA, *Science* **243:**393–396.
33. DiCorleto, P. E., and Bowen-Pope, D. F., 1983, Cultured endothelial cells produce a platelet-derived growth factor-like protein, *Proc. Natl. Acad. Sci. USA* **80:**1919–1923.
34. Emeson, E. E., and Robertson, A. L., Jr., 1988, T lymphocytes in aortic and coronary intimas. Their potential role in atherogenesis, *Am. J. Pathol.* **130:**369–376.
35. Hansson, G. K., Jonasson, L., Holm, J., and Claesson-Welsh, L., 1986, Class II MHC antigen expression in the atherosclerotic plaque: Smooth muscle cells express HLA-DR, HLA-DQ and the invariant gamma chain, *Clin. Exp. Immunol.* **64:**261–268.
36. Jonasson, L., Holm, J., and Hansson, G. K., 1988, Smooth muscle cells express Ia antigens during arterial response to injury, *Lab. Invest.* **58:**310–315.

19

Artery Wall Interactions in Early Atherogenesis

Judith A. Berliner, Tripathi B. Rajavashisth, Mahamad Navab, Ali Andalibi, Susan Imes, Joy S. Frank, Mary C. Territo, Aldons J. Lusis, and Alan M. Fogelman

I. INTRODUCTION

The early events in the development of the atherosclerotic lesion include lipoprotein entry, retention and modification in the artery wall, monocyte and lymphocyte recruitment into the subendothelial space, conversion of the monocytes into macrophage foam cells, and the migration of smooth muscle cells from the media into the subendothelial space where some become foam cells.[1] In our laboratory we have considered these events to result from interactions between circulating lipoproteins and the cells and matrix of the artery wall. We have also considered the possibility that cell–cell interactions may have an important role in these events. In this chapter we will address lipoprotein transport, lipoprotein retention, monocyte recruitment, and conversion of monocytes into macrophages in the subendothelial space.

II. LIPOPROTEIN TRANSPORT

The general consensus is that lipoprotein transport into the artery wall is among the earliest if not the earliest event in atherogenesis. There is, however, some disagreement as to

Judith A. Berliner • Department of Pathology, UCLA School of Medicine, Los Angeles, California 90024–1732. *Tripathi B. Rajavashisth* • Division of Medical Genetics, Harbor UCLA Medical Center, Torrance, California 90502–2064. *Mahamad Navab, Susan Imes, Mary C. Territo, and Alan M. Fogelman* • Department of Medicine, UCLA School of Medicine, Los Angeles, California 90024–1679. *Ali Andalibi* • Department of Microbiology, UCLA School of Medicine, Los Angeles, California 90024–1679. *Joy S. Frank* • Departments of Medicine and Physiology, UCLA School of Medicine, Los Angeles, California 90024–1679. *Aldons J. Lusis* • Departments of Medicine and Microbiology, UCLA School of Medicine, Los Angeles, California 90024–1679.

Endothelial Cell Dysfunctions, edited by Nicolae Simionescu and Maya Simionescu. Plenum Press, New York, 1992.

whether transport is increased or is maintained at normal levels. Chobanian and colleagues[2] reported that after 16 weeks on a 1.5% cholesterol diet there was a marked increase in arterial permeability to large molecules such as albumin but not to small molecules such as sugar and water. Navab *et al.*[3] found that the transport of macromolecules such as albumin and low-density lipoproteins (LDL) was increased across rabbit aortic endothelial cell monolayers *in vitro* after preincubation with rabbit beta migrating very-low-density lipoprotein (β-VLDL) but that the transport of small molecules such as sodium ions was not altered. Based on a kinetic analysis, Schwenke and Carew[4,5] have concluded that the rate of LDL transport was not increased into the artery wall of cholesterol-fed rabbits 16 days after cholesterol feeding in areas of lesion predilection compared to areas that were relatively resistant to lesion formation.

The mechanism of LDL transport into the artery wall appears to be independent of the LDL receptor. *In vivo* studies by Wiklund *et al.*[6] revealed that LDL penetration into the aorta of rabbits was not affected by methylation, a modification which prevents its recognition by the LDL receptor. Navab *et al.*[3] subsequently demonstrated *in vitro* that monolayers of aortic endothelial cells from rabbits genetically deficient in the LDL receptor (WHHL rabbits) transported LDL at the same rate as did monolayers from normal rabbits. Additionally, Navab *et al.*[3] demonstrated that *in vitro* approximately half of the LDL transported by rabbit aortic endothelial monolayers was transported by an energy-dependent process (e.g., pinocytosis) while the remainder appeared to be transported by energy-independent processes.

In other studies, Navab *et al.*[7] showed that *in vitro* lipoprotein complexed to bacterial lipopolysaccharide was transported across aortic endothelial monolayers at the same rate as native lipoprotein.

III. LIPOPROTEIN RETENTION

Schwenke and Carew[4,5] concluded from their kinetic analyses that the increased concentration of LDL in the artery wall was due to increased retention of LDL at sites of predilection as compared to retention at relatively resistant sites in the aorta. It has long been suspected that lipoprotein lipid retention in the artery wall is mediated by interactions with the extracellular components. Nearly three decades ago it was observed that there is a close spatial relationship between metachromatic regions of the artery wall (presumably mucopolysaccharide-rich regions) and those regions that stained for lipids.[8,9] It has also been known for some time that glycosaminoglycans can form complexes with lipoproteins[10,11] and that VLDL and LDL can bind to arterial glycosaminoglycans at physiological pH and ionic strength.[12] Lipoprotein–glycosaminoglycan complexes have been isolated from human plaques. A working hypothesis is that proteoglycans or glycosaminoglycan within the matrix of the subendothelial space trap lipid and lipoproteins. Based on thin sections stained with ruthenium red, Wight and Ross[13] and Wight[14] suggested that the arterial wall contained a three-dimensional network of proteoglycan and interconnecting collagen fibrils. Frank and Fogelman[15] examined the aortas of normal, WHHL, and cholesterol-fed rabbits by ultrarapid freezing without chemical fixation followed by rotary shadow freeze-etching. The extracellular matrix in areas devoid of cells was seen in extraordinary detail. It consisted of a reticulum of thick filaments, finer branching filaments, collagen fibrils, and granules of varying sizes. No lipid deposits were seen in normal aortic intima. In contrast, the subendothelial space of the WHHL rabbit aorta was filled with collagen fibrils surrounding and entwined between clusters of discrete lipid vesicles that ranged in size from 23 to 169 nm. Approximately 80% of the lipid vesicles in the WHHL rabbit intima measured between 70 and 169 nm. The lipid particles in the WHHL intima always appeared in clusters, many of which appeared to be fusing into larger size vesicles. These aggregates were clearly linked to the matrix filaments. A similar deposition of lipid particles was seen in the extracellular matrix of cholesterol-fed rabbits, but in contrast to the particle size distribution of the WHHL intima, more than 75% of the particles in the cholesterol-fed intima had a diameter between 23 and 68 nm and 51% were between 23 and 45 nm. The differences in particle size in the WHHL intima as compared to the cholesterol-fed intima may in part have been due to the difference in the duration of the hyperlipidemia, 10 days versus 2 to 8 months for the cholesterol-fed and WHHL animals, respectively. These results confirmed and extended those of Simionescu *et al.*[16] After only 2 weeks of cholesterol feeding, these investigators identified unesterified cholesterol-rich lipid particles in the intima.[16] Furthermore, they reported that apolipoprotein B was associated with the outer aspect of these vesicles after only 1 week of a high-fat diet.[17] A striking feature of the findings by Frank and Fogelman[15] in both the WHHL and the cholesterol-fed rabbits was the extensive cagelike structure of the matrix filaments and the extent of their interactions with collagen. These images supported the possibility that complex macromolecules such as lipoproteins could have multiple interactions if caught in a three-dimensional meshwork. Following the demonstration by Goldstein *et al.*[18] that acetyl-LDL could cause massive cholesteryl ester accumulation in macrophages, Fogelman *et al.*[19] demonstrated that LDL modified *in vitro* by malondialdehyde produced a similar accumulation of cholesteryl esters in macrophages. Fogelman *et al.*[19] postulated that malondialdehyde produced by lipid peroxidation might modify LDL in the artery wall. Subsequently, Haberland *et al.*[20] demonstrated that malondialdehyde modification of LDL bound to glycosaminoglycan led to release of the lipoprotein and a change in recognition from the LDL receptor to the scavenger (acetyl-LDL) receptor. Havel *et al.*[21] showed that in addition to accumulating LDL, WHHL rabbits also accumulated triglyceride-rich lipoproteins such as VLDL. Mazzone and colleagues[22] have demonstrated that modified VLDL can also be recognized by the scavenger receptor *in vitro*. All of these *in vitro* studies required very high concentrations of reactants to modify the lipoproteins in solution. Until Haberland and colleagues[23] directly demonstrated the presence of malondialdehyde–protein conjugates in the artery wall of WHHL rabbits, it was disputed as to whether such events occurred *in vivo*. Subsequently, these results were confirmed and extended by Steinberg and colleagues[24] and Chait and colleagues.[25] Presumably the three-dimensional extracellular matrix traps the lipoproteins *in vivo* creating a microenvironment that permits high levels of reactants sufficient to modify the lipoproteins. Fogelman *et al.*[19] demonstrated in 1980 that cross-linking LDL molecules with glutaraldehyde produced macromolecular aggregates of LDL that were taken up by human monocyte-macrophages producing massive cholesteryl ester accumulation. Subsequently, Hoff *et al.*[26] demonstrated that 4-hydroxynonenal, an aldehyde which is produced as a result of lipid peroxidation and which has been shown to be present in lesions,[24] can cross-link LDL and produce aggregates that cause massive cholesteryl ester accumulation upon being ingested by macrophages. Suits *et al.*[27] showed that LDL aggregates formed *in vitro* can bind to the surface of macrophages via the LDL receptor. This was followed by phagocytosis of the aggregates resulting in massive cholesteryl ester formation. Khoo *et al.*[28] also demonstrated enhanced macrophage uptake of LDL aggregates that were formed by simply vortexing LDL. Wolfbauer *et al.*[29] demonstrated, *in vitro*, that cholesteryl ester-rich lipid droplets from macrophages could be taken up by smooth muscle cells if the lipid droplets were allowed to physically come in contact with the smooth muscle cells. The studies of the Simionescus[16,17] and those of Frank and Fogelman[15] are consistent with the formation of aggregates in the artery walls of both cholesterol-fed and WHHL rabbits.

IV. MONOCYTE RECRUITMENT

The earliest cellular event thus far recognized in the development of an atherosclerotic lesion is the adherence of monocytes to the endothelium. Neutrophils have not been observed to participate in the early events in the artery wall following cholesterol feeding or in neonatal WHHL rabbits with endogenous hyperlipidemia.

Territo *et al.*[30] reported that treatment of rabbit aortic endothelial cells or human umbilical vein endothelial cells for as little as 1 day with rabbit β-VLDL caused an increased binding of human peripheral blood monocytes to the endothelium. Neutrophil binding to the endothelial cells was not significantly increased by exposure of the endothelial cells to β-VLDL. In contrast, exposure of the endothelial cells to bacterial lipopolysaccharide (endotoxin; LPS) resulted in increased binding of both neutrophils and monocytes to the endothelial cells. Thus, exposure of endothelial cells to β-VLDL, the lipoprotein that accumulates after cholesterol-feeding, altered the endothelium in such a way that only monocyte binding was enhanced.

As noted above, a number of laboratories have studied LDL which has been modified to the extent that it is no longer recognized by the LDL receptor but is recognized by the scavenger (acetyl-LDL) receptor. Such LDL can be prepared by chemical modification[18,19] or by incubation with iron or copper in the absence of serum and this modification is enhanced by incubation with cells in iron- or copper-containing media if the media are devoid of serum.[31–35] Usually, these modifications of LDL have been extensive and the resulting LDL is highly oxidized containing 10 to 50 nM thiobarbaturic acid reactive substances (TBARS) as malondialdehyde equivalents per mg of LDL cholesterol. One of us (J.A.B.) reasoned that such an extensive modification of LDL was less likely at the very beginning of lesion formation because at this time there would be relatively few cells in the region of the subendothelial space. Therefore, we undertook extensive investigations of minimally modified LDL (MM-LDL) which was only minimally oxidized. Freshly isolated LDL (i.e., stored in the presence of EDTA for less than 3 weeks at 4°C) contained zero to 2 nM TBARS/mg LDL cholesterol. MM-LDL whether prepared by simply storing the LDL under the same conditions for longer periods (i.e., 3–6 months) or prepared by mild iron oxidation[36] contained 2 to 5 nM TBARS/mg LDL cholesterol. In addition to the 1.5-fold increase in TBARS in MM-LDL compared to freshly isolated LDL, there was also a 1.5-fold increase in hydroperoxides and a 4-fold increase in cholesterol epoxide.[37] MM-LDL obtained by either storage or mild iron oxidation demonstrated some fragmentation of apolipoprotein B (Fig. 1). Nonetheless, both preparations of MM-LDL (stored and that prepared by mild iron oxidation) were taken up by cells at the same rate as freshly isolated

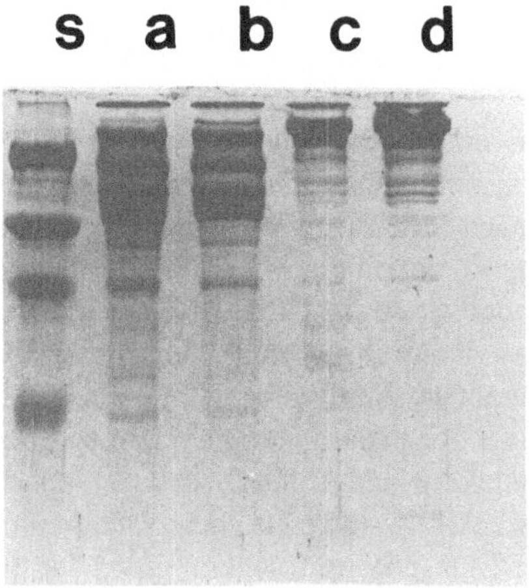

Figure 1. SDS–polyacrylamide gel electrophoresis of freshly isolated and minimally modified (MM) LDL. Samples (10 μg) were run on 5% polyacrylamide gels and stained with Coomassie blue. Lane s: Molecular weight markers. The first three darkly stained bands from top to bottom are 220, 116, and 92 kDa, respectively. Lane A: MM-LDL prepared from the LDL of donor #1 by storage at 4°C for 3 months in the presence of 0.01% EDTA. Lane b: MM-LDL produced from the LDL of donor #2 by mild iron oxidation. This preparation contained 3 nM TBARS/mg cholesterol. Lane c: Freshly isolated LDL from donor #1. Lane d: Freshly isolated LDL from donor #2.

LDL and the uptake of radiolabeled MM-LDL preparations was prevented by the addition of excess unlabeled freshly isolated LDL but not by malondialdehyde-altered LDL which was recognized by the scavenger (acetyl-LDL) receptor.[37] Additionally, addition of excess unlabeled MM-LDL prevented the uptake of radiolabeled freshly isolated LDL but not radiolabeled malondialdehyde–LDL.[37] Thus, despite the fragmentation of some of the apolipoprotein B, the LDL receptor could not distinguish between MM-LDL and freshly isolated LDL.

Preexposure of endothelial cells to as little as 0.12 μg/ml MM-LDL caused the endothelial cells to bind significantly more monocytes. In these assays the MM-LDL was removed before the introduction of the monocytes and therefore the MM-LDL was acting on the endothelial cells to induce increased monocyte binding.[37] Neutrophil binding did not increase after exposure to MM-LDL but neutrophil and monocyte binding increased after exposure of the endothelial cells to LPS.[37] The induction of increased monocyte binding by exposure of the endothelial cells to MM-LDL, was maximal at a concentration of 1 μg/ml MM-LDL and resulted in a sevenfold increase in monocyte binding when the endothelial cells were derived from rabbit aorta.[37] Monocyte binding increased significantly after only 2 hr of preexposure of the endothelial cells to the MM-LDL and was maximal after 4 hr and remained at near-maximal levels for 48 hr.[37] Neutrophil adhesion was not stimulated by MM-LDL even after 24 hr of exposure to the endothelial cells but neutrophil binding was enhanced after only 4 hr of exposure to bacterial LPS.[37] Western blotting techniques which detected as little as 10 ng of LPS did not detect any LPS in 200 μg of MM-LDL. Moreover, endothelial cells had to be exposed to 500 pg/ml of free LPS or 20 μg/ml of LPS complexed to LDL before the endothelial cells demonstrated increased leukocyte binding. As noted above, exposure of endothelial cells to as little as 0.12 μg/ml of MM-LDL (which could not contain more than 3 pg/ml of LPS) significantly increased monocyte binding to the endothelial cells.[37] Thus, the increased binding of monocytes following preexposure of the endothelial cells to MM-LDL was not due to LPS contamination. The increased binding of monocytes following exposure of the endothelial cells to MM-LDL was markedly reduced by treating the endothelial cells with 1 μg/ml of cycloheximide for 2 hr prior to the addition of the MM-LDL.[37] This suggests that the MM-LDL induced the synthesis of a monocyte binding protein on the endothelial cells.

Lipid extracts of MM-LDL, particularly a fraction containing charged lipids, induced the endothelial cells to bind monocytes.[37] The extracted polar lipids of MM-LDL at a concentration of 10 μg/ml increased monocyte binding to the same extent as did 3 μg/ml of intact MM-LDL.[37] The protein component(s) of MM-LDL did not induce increased monocyte binding.[37] Cholesterol epoxides at concentrations tenfold greater than that contained in MM-LDL did not induce monocyte binding by the endothelial cells.[37]

Exposure of endothelial cells to MM-LDL not only resulted in the induction of monocyte binding protein(s) on the endothelial cells but also resulted in a sevenfold induction in the amount of monocyte chemotactic factor produced by the endothelial cells.[37] Steinberg and colleagues have reported that the highly oxidized LDL that they have extensively studied is chemotactic for monocytes.[38] Unlike the highly oxidized LDL, MM-LDL is not itself chemotactic for monocytes.[37] Freshly isolated LDL was also not chemotactic for monocytes and it did not stimulate endothelial cells to produce monocyte chemotactic factor.[37] As little as 0.05 μg/ml of MM-LDL induced a significant increase in the endothelial cell production of monocyte chemotactic factor.[37] However, even at high concentrations MM-LDL did not induce the endothelial cells to produce chemotactic factor for neutrophils.[37]

Metcalf and colleagues[39] have reported that transgenic mice expressing the gene for the colony-stimulating factor GM-CSF accumulated macrophages in those tissues expressing this gene and these accumulations of macrophages were associated with tissue damage. Since it is known that endothelial cells produce colony-stimulating factors (CSFs)[40–47] and it is also known that CSFs can induce monocytes to migrate[48] and endothelial cells to migrate and proliferate,[49] we studied the effect of MM-LDL on CSF production by endothelial cells. As shown in the Northern blot in Fig. 2, exposure of endothelial cells to MM-LDL resulted in the induction of mRNA for both granulocyte–macrophage (GM-CSF) and macrophage (M-CSF) colony-stimulating factors but there was no induction of mRNA for interleukin-1-β (IL-1-β) or transforming growth factor-β (TGF-β). This was true whether the MM-LDL was prepared by prolonged storage or by mild iron oxidation.[50] The time courses for the induction by MM-LDL of M-CSF and GM-CSF were similar peaking at 4 hr after exposure and decreasing thereafter.[50] This time course is similar to that observed using other inducers of CSFs in endothelial cells including IL-1, tumor necrosis factor (TNF-α), and LPS. Thus, it is possible that MM-LDL and the other inducers share a common signaling pathway for CSF induction. The induction of mRNA for the CSFs was paralleled by the induction of bioactivity.[50]

As shown in Fig. 3, exposure of human aortic endothelial cells to MM-LDL induced a marked increase in the mRNA for GM-CSF after only 4 hr. In contrast, exposure of human umbilical vein endothelial cells to MM-LDL for periods as long as 24 hr did not induce mRNA for GM-CSF (Fig. 3). This is one of the few examples of a different biologic response between aortic endothelial cells and umbilical vein endothelial cells. Freshly isolated LDL without evidence of oxidation did not induce mRNA or bioactivity for CSFs.[50] High levels of M-CSF mRNA were detected in atherosclerotic lesions of WHHL and cholesterol-fed rabbits compared to control rabbits.[50] The levels of GM-CSF were below the level of detection and hence no conclusion regarding this CSF *in vivo* can be drawn.[50] The

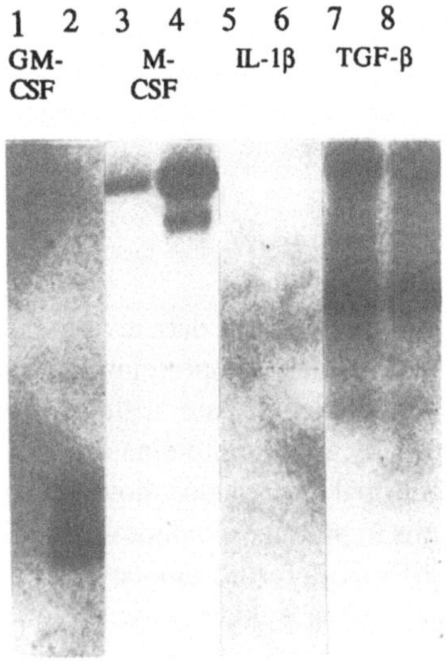

Figure 2. MM-LDL induces mRNA for colony-stimulating factors in endothelial cells. Total RNA was extracted from rabbit aortic endothelial cells after 4 hr of culture in the absence (lanes 1, 3, 5, 7) or presence (lanes 2, 4, 6, 8) of MM-LDL (2 μg/ml) and examined by Northern analysis for mRNA for GM-CSF, M-CSF, interleukin-1-β (IL-1-β), or transforming growth factor-β (TGF-β).

presence of M-CSF in the artery wall with an increase in lesions, however, suggests that CSFs may be produced *in vivo* and account in part for the accumulation of monocyte-macrophages in the subendothelial space of the developing atherosclerotic lesion. In support of this notion are the findings of Averill *et al.*[51] who presented evidence suggesting that M-CSF was induced in cholesterol-fed swine.

Under normal conditions there are relatively few cells in the subendothelial space of the artery wall and the smooth muscle cells are largely separated from the endothelial cells by the internal elastic lamina. As the lesion develops, however, smooth muscle cells migrate into the subendothelial space in close apposition to the endothelial cells. Coculture of human adult aortic endothelial cells together with human adult aortic smooth muscle cells resulted in cell–cell interactions that greatly enhanced the migration of human monocytes

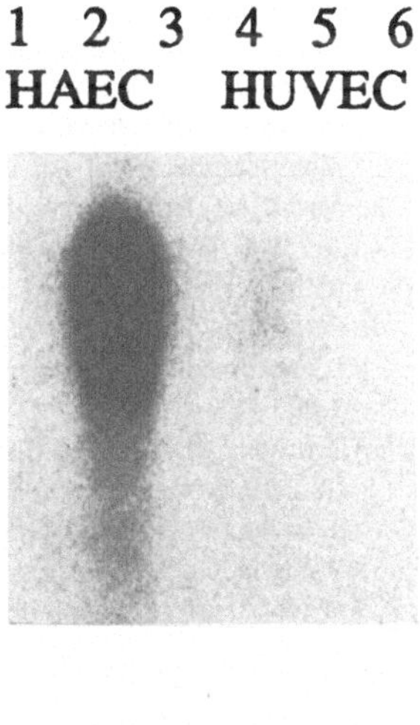

Figure 3. MM-LDL induces mRNA for GM-CSF in aortic endothelial cells but not in umbilical vein endothelial cells. Human aortic endothelial cells (HAEC) and human umbilical vein endothelial cells (HUVEC) were exposed to MM-LDL (20 μg/ml) for 0 hr (lanes 1, 3, 5), 4 hr (lanes 2, 4), or 24 hr (lane 6). Total RNA was then extracted and examined by Northern analysis for mRNA for GM-CSF.

into the subendothelial space of the coculture.[52] In this coculture system as *in vivo*, the overlying endothelium remained intact and largely maintained its permeability characteristics despite a massive monocyte migration into the subendothelial space.[52]

V. SUMMARY

Based on the data reviewed above, we hypothesize that LDL and other apolipoprotein B-containing lipoproteins are retained in the three-dimensional matrix network of the subendothelial space at sites of predilection. Since there are relatively few cells in this area initially, the lipoproteins are only minimally modified. The resulting MM-LDL and other modified lipoproteins, however, are biologically active and induce the overlying endothelial cells to produce a monocyte-binding protein as well as monocyte chemotactic factor(s) and CSFs. As a result, monocytes are recruited to the subendothelial space where they become macrophages and release products which further modify the lipoproteins so that they are now recognized by the scavenger (acetyl-LDL) receptor or form aggregates (lipid vesicles). The modified lipoproteins are then taken up by the monocyte-macrophages producing foam cells. Early in the development of the lesion, smooth muscle cells migrate into the subendothelial space and interact with the endothelial cells inducing further monocyte recruitment. As the monocyte foam cells die, their lipid droplets are taken up by the smooth muscle cells and they too are converted to foam cells. The elements of this hypothesis are amenable to testing and the validity or lack of validity of these speculations will soon become known.

ACKNOWLEDGMENTS. This work was supported in part by United States Public Health Service grants HL-30568, T32 HL-07412, RR-865, the Laubisch Fund, and the M.K. Grey Fund. A.J.L. is an Established Investigator of the American Heart Association. We thank Susan Lee for the experiment described in Fig. 1.

REFERENCES

1. Ross, R., 1986, The pathogenesis of atherosclerosis—An update, *N. Engl. J. Med.* **314**:488–500.
2. Chobanian, A. V., Menzoian, J. O., Shipman, J., Heath, K., and Haudenschild, C. C., 1983, Effects of endothelial denudation and cholesterol feeding on in vivo transport of albumin, glucose, and water across rabbit carotid artery, *Circ. Res.* **53**:805–814.
3. Navab, M., Hough, G. P., Berliner, J. A., Frank, J. A., Fogelman, A. M., Haberland, M. E., and Edwards, P. A., 1986, Rabbit beta-migrating very low density lipoprotein increases endothelial macromolecular transport without altering electrical resistance, *J. Clin. Invest.* **78**:389–397.
4. Schwenke, D. C., and Carew, T. E., 1989, Initiation of atherosclerotic lesions in cholesterol-fed rabbits. I. Focal increases in arterial LDL concentration precede development of fatty streak lesions, *Arteriosclerosis* **9**:895–907.
5. Schwenke, D. C., and Carew, T. E., 1989, Initiation of atherosclerotic lesions in cholesterol-fed rabbits. II. Selective retention of LDL vs. selective increases in LDL permeability in susceptible sites of arteries, *Arteriosclerosis* **9**:908–918.
6. Wiklund, O., Carew, T. E., and Steinberg, D., 1984, Role of the low density lipoprotein receptor in penetration of low density lipoprotein into rabbit aortic wall, *Arteriosclerosis* **5**:135–141.

7. Navab, M., Hough, G. P., Van Lenten, B. J., Berliner, J. A., and Fogelman, A. M., 1988, Low density lipoproteins transfer bacterial lipopolysaccharides across endothelial monolayers in a biologically active form, *J. Clin. Invest.* **81**;601–605.
8. Curran, R. G., and Crane, A. J., 1962, Mucopolysaccharides in the atheromatous aorta, *J. Pathol. Bacteriol.* **84**:405–412.
9. Hollander, W., 1976, Unified concept on the role of acid mucopolysaccharides and connective tissue proteins in the accumulation of lipids, lipoproteins and calcium in the atherosclerotic plaque, *Exp. Mol. Pathol.* **25**:106–120.
10. Iverius, P. H., 1972, The interaction between human plasma lipoprotein and connective tissue glycosaminoglycans, *J. Biol. Chem.* **247**:2607–2613.
11. Srinivasan, S.R ., Dolan, P., Radhakrishnamurthy, B., and Berenson, G. S., 1972, Isolation of lipoprotein–acid mucopolysaccharide complexes from fatty streaks of human aortas, *Atherosclerosis* **16**:95–105.
12. Srinivasan, S. R., Radhakrishnamurthy, B., and Berenson, G. S., 1975, Studies on the interaction of heparin with serum lipoproteins in the presence of Ca^{2+}, Mg^{2+}, and Mn^{2+}, *Arch. Biochem. Biophys.* **170**:334–342.
13. Wight, T. N., and Ross, R., 1975, Proteoglycan in primate arteries, I. Ultrastructural localization and distribution in the intima, *J. Cell Biol.* **67**:660–674.
14. Wight, T. N., 1980, Vessel proteoglycans and thrombogenesis, in: *Progress in Hemostasis and Thrombosis* (T. H., Spaet, ed.), Grune & Stratton, New York, pp. 1–39.
15. Frank, J. S., and Fogelman, A. M., 1989, Ultrastructure of the intima in WHHL and cholesterol-fed rabbit aortas prepared by ultra-rapid freezing and freeze-etching, *J. Lipid Res.* **30**:967–978.
16. Simionescu, N., Vasile, E., Lupu, F., Popescu, G., and Simionescu, M., 1986, Prelesional events in atherogenesis. Accumulation of extracellular cholesterol-rich liposomes in the arterial intima and cardiac valves of the hyperlipidemic rabbit, *Am. J. Pathol.* **123**:109–125.
17. Mora, R., Lupu, F., and Simionescu, N., 1987, Prelesional events in atherogenesis. Colocalization of apolipoprotein B unesterified cholesterol and extracellular phospholipid liposomes in the aorta of hyperlipidemic rabbits, *Arteriosclerosis* **67**:143–154.
18. Goldstein, J. L., Ho, Y. K., Basu, S. K., and Brown, M. S., 1979, Binding site on macrophages that mediates uptake and degradation of acetylated low density lipoprotein producing massive cholesterol deposition, *Proc. Natl. Acad. Sci. USA* **76**:333–337.
19. Fogelman, A. M., Shechter, I., Seager, J., Hokom, M., Child, J. S., and Edwards, P. A., 1980, Malondialdehyde alteration of low-density lipoprotein leads to cholesteryl-ester accumulation in human monocyte macrophages, *Proc. Natl. Acad. Sci. USA* **77**:2214–2218.
20. Haberland, M. E., Fogelman, A. M., and Edwards, P. A., 1982, Specificity of receptor-mediated recognition of malondialdehyde-modified low density lipoproteins, *Proc. Natl. Acad. Sci. USA* **79**:1712–1716.
21. Havel, R. J., Kita, J., Kotite, L., Kane, J. P., Hamilton, R. L., Goldstein, J. L., and Brown, M. S., 1982, Concentration and composition of lipoproteins in blood plasma of the WHHL rabbit. An animal model of familial hypercholesterolemia, *Arteriosclerosis* **2**:467–474.
22. Mazzone, T., Lopez, C., and Bergstrasser, L., 1988, Modification of very low density lipoproteins leads to macrophage scavenger receptor uptake and cholesteryl ester deposition, *Arteriosclerosis* **7**:191–196.
23. Haberland, M. E., Fong, D., and Cheng, L., 1988, Malondialdehyde-altered protein occurs in atheroma of Watanabe Heritable Hyperlipidemic rabbits, *Science* **241**:215–218.
24. Palinski, W., Rosenfeld, M. E., Yla-Herttuala, S., Gurtner, G. C., Socher, S. S., Butler, S. W., Parthasarathy, S., Carew, T. E., Steinberg, D., and Witztum, J. L., 1989, Low density lipoprotein undergoes oxidative modification in vivo, *Proc. Natl. Acad. Sci. USA* **86**:1372–1376.
25. Boyd, H. C., Gown, A. M., Wolfbauer, G., and Chait, A., 1989, Direct evidence for a protein recognized by a monoclonal antibody against oxidatively modified LDL in atherosclerotic lesions from a Watanabe Heritable Hyperlipidemic Rabbit, *Am. J. Pathol.* **135**:815–826.
26. Hoff, H. F., Morel, D. W., Jurgens, G., Esterbauer, H., and Chisolm, G. M. III, 1987, Modification of LDL by 4-hydroxynonenal induces enhanced uptake by macrophages and cytotoxicity to proliferating fibroblasts, *Arteriosclerosis* **7**:523a.
27. Suits, A. G., Chait, A., Aviram, M., and Heinecke, J. W., 1989, Phagocytosis of aggregated lipoprotein by macrophages: Low density lipoprotein receptor-dependent foam-cell formation, *Proc. Natl. Acad. Sci. USA* **86**:2713–2717.

28. Khoo, J. C., Miller, E., McLoughlin, P., and Steinberg, D., 1988, Enhanced macrophage uptake of low density lipoprotein after self aggregation, *Arteriosclerosis* **8**:348–358.
29. Wolfbauer, G., Glick, J. M., Minor, L. K., and Rothblat, G. H., 1986, Development of smooth muscle foam cell: Uptake of macrophage lipid inclusions, *Proc. Natl. Acad. Sci. USA* **83**:7760–7764.
30. Territo, M. C., Berliner, J. A., Almada, L., Ramirez, R., and Fogelman, A. M., 1989, β-very low density lipoprotein pretreatment of endothelial monolayers increases monocyte adhesion, *Arteriosclerosis* **9**:824–828.
31. Morel, D. W., DiCorleto, P. E., and Chisolm, G. M., 1984, Endothelial and smooth muscle cells alter low density lipoprotein in vitro by free radical oxidation, *Arteriosclerosis* **4**:357–364.
32. Steinbrecher, U. S., Parthasarathy, S. P., Leake, D. S., Witztum, J. L., and Steinberg, D., 1984, Modification of low density lipoprotein by endothelial cells involves lipid peroxidation and degradation of low density lipoprotein phospholipids, *Proc. Natl. Acad. Sci. USA* **81**:3883–3887.
33. Heinecke, J. W., Rosen, H., and Chait, A., 1984, Iron and copper promote modification of LDL by human arterial smooth muscle cells, *J. Clin. Invest.* **74**:1890–1894.
34. Cathcart, M. K., Morel, D. W., and Chisolm, G. M., 1985, Monocytes and neutrophils oxidize low density lipoprotein making it cytotoxic, *J. Leukocyte Biol.* **38**:341–350.
35. Steinberg, D., Parthasarathy, S. P., Carew, T. E., Khoo, J. C., and Witztum, J. L., 1989, Modifications of LDL that increase its atherogenicity, *N. Engl. J. Med.* **320**:916–924.
36. Kosugi, K., Morel, D. W., DiCorleto, P. E., and Chisolm, G. M., 1987, Toxicity of oxidized LDL to cultured fibroblasts is selective for S phase of the cell cycle, *J. Cell. Physiol.* **130**:311–320.
37. Berliner, J. A., Territo, M. C., Sevanian, A., Ramin, S., Kim, J. A., Bamshad, B., Esterson, M., and Fogelman, A. M., 1990, Minimally modified LDL stimulates monocyte endothelial interactions, *J. Clin. Invest.* **85**:1260–1266.
38. Quinn, M. T., Parthasarathy, S., Fong, L. G., and Steinberg, D., 1987, Oxidatively modified low density lipoproteins: A potential role in recruitment and retention of monocyte/macrophages during atherogenesis, *Proc. Natl. Acad. Sci. USA* **84**:2995–2998.
39. Lang, R. A., Metcalf, D., Cuthbertson, R. A., Lyons, I., Stanley, E., Kelso, A., Kannourakis, G., Williamson, D. J., Klintworth, G. K., Gonda, T. J., and Dunn, A. R., 1987, Transgenic mice expressing a hemopoietic growth factor gene (GM-CSF) develop accumulations of macrophages, blindness, and a fatal syndrome of tissue damage, *Cell* **51**:675–686.
40. Sieff, C. A., Niemeyer, C. M., and Faller, D. V., 1987, The production of hematopoietic growth factors by endothelial accessory cells, *Blood Cells* **13**:65–74.
41. Segal, G. M., McCall, E., Stueve, T., and Bagby, G. C., Jr., 1987, Interleukin 1 stimulates endothelial cells to release multilineage human colony-stimulating activity, *J. Immunol.* **138**:1722–1728.
42. Broudy, V. C., Kanshansky, K., Harlan, J. M., and Adamson, J. W., 1987, Interleukin 1 stimulates endothelial cells to produce granulocyte–macrophage colony-stimulating factor and granulocyte colony-stimulating factor, *J. Immunol.* **139**:464–468.
43. Seelentag, W. K., Mermod, J. J., Montesano, R., and Vassalli, P., 1987, Additive effects of interleukin 1 and tumor necrosis factor-alpha on the accumulation of the three granulocyte and macrophage colony stimulating factor mRNAs in human endothelial cells, *EMBO J.* **6**:2261–2265.
44. Griffin, J. D., Rambaldi, A., Vellenga, E., Young, D. C., Osapovicz, D., and Canistra, S. A., 1987, Secretion of interleukin-1 by myeloblastic leukemia cells in vitro induces endothelial cells to secrete colony stimulating factors, *Blood* **70**:1218–1221.
45. Zsebom, M., Yuschenkoff, U. N., Schiffer, S., Chang, D., McCall, E., Dinarello, C. A., Brown, M. A., Altrock, B., and Bagby, G. C., Jr., 1988, Vascular endothelial cells and granulopoiesis: Interleukin-1 stimulates release of G-CSF and GM-CSF, *Blood* **71**:99–103.
46. Malone, D. G., Pierce, J. H., Falko, J. P., and Metcalfe, D. D., 1988, Production of granulocyte–macrophage colony-stimulating factor by primary cultures of unstimulated rat microvascular endothelial cells, *Blood* **71**:684–689.
47. Dainiak, N., Warren, H. B., Kreczko, Riordan, M. A., Feldman, L, Lawler, J., Cohen, A. M., and Davies, P. F., 1988, Acetylated lipoproteins impair erythroid growth factor release from endothelial cells, *J. Clin. Invest.* **81**:834–843.
48. Wang, J. M., Griffin, J. D., Rambaldi, A., Chen, Z. G., and Mantovani, A., 1988, Induction of monocyte migration by recombinant macrophage colony-stimulating factor, *J. Immunol.* **141**:575–579.
49. Bussolino, F., Wang, J. M., Defilippi, P., Turrini, F., Sanavio, F., Edgell, C.-J. S., Aglietta, M., Arese, P., and

Mantovani, A., 1989, Granulocyte- and granulocyte–macrophage-colony stimulating factors induce human endothelial cells to migrate and proliferate, *Nature* **337**:471–473.

50. Rajavashisth, T. B., Andalibi, A., Territo, M. C., Berliner, J. A., Navab, M., Fogelman, A. M., and Lusis, A. J., 1990, Modified low density lipoproteins induce endothelial cell expression of granulocyte and macrophage colony stimulating factors, *Nature* **344**:254–257.
51. Averill, L. E., Meagher, R. C., and Gerrity, R. G., 1989, Enhanced monocyte progenitor cell proliferation in bone marrow of hyperlipidemic swine, *Am. J. Pathol.* **135**:369–377.
52. Navab, M., Hough, G. P., Stevenson, L. W., Drinkwater, D. C., Laks, H., and Fogelman, A. M., 1988, Monocyte migration into the subendothelial space of a co-culture of adult human aortic endothelial and smooth muscle cells, *J. Clin. Invest.* **82**:1853–1863.

20

Endothelial Cell Dysfunction and Injury in Experimental Hyperlipidemic Atherogenesis

Nicolae Simionescu

I. INTRODUCTION

There is substantial evidence that the onset of atherosclerosis requires neither physical damage to endothelium nor platelet involvement.[18,33,34,44,58,65–67] Endothelial denudation followed by platelet adhesion to the subendothelial structure appears to be a late event occurring in advance atherosclerotic lesions with marked tendency to thrombosis and subsequent lumen obstruction.[64,66] It is known that vascular endothelium participates in all major physiopathologic processes involving the vessel wall, and that the earliest cellular event so far reported has been the focal adherence of monocytes to arterial endothelium. Bearing on this, we tried to identify and characterize the subtle biochemical and ultrastructural changes which appear in the lesion-prone areas of the artery wall, at the inception of atherogenesis before monocyte recruitment, during the so-called "prelesional" stage.[65,66] As experimental model, we chose the diet-induced hypercholesterolemia, a condition reputedly known to be alone capable of inducing both in humans and in experimental animals lipid accumulation in the intima leading to fatty streak formation that can evolve into a fibrous plaque.[14,15,35,43,54,62,64,74]

II. EXPERIMENTS

A. Experimental Design

In cholesterol-fed rabbits[65] and hamsters[54] we examined the lesion-prone areas of the aortic arch and the aortic valve, preferentially during the first 2 weeks on the diet. The

Nicolae Simionescu • Institute of Cellular Biology and Pathology, Bucharest 79691, Romania.

Endothelial Cell Dysfunctions, edited by Nicolae Simionescu and Maya Simionescu. Plenum Press, New York, 1992.

investigations were focused on the chemistry, electrochemistry, and properties of the endothelial cell (EC) surface and the transport and accumulation of lipoproteins. The findings were correlated with the bona fide atherosclerotic lesions developed at later stages in such regions of the vessel wall.

B. Animal Models

The experiments were conducted on 499 Chinchilla male adult rabbits and 214 controls, and on 1609 golden male adult hamsters and 770 controls. The animals were fed a lipid-rich diet (0.5–2% cholesterol and 5% butter) for 1–2 weeks (as standard time for the prelesional stage), for up to 16 weeks (for early lesional stages), and for up to 1 year (for the advanced lesions).

C. Plasma Lipid Analysis

Hyperlipidemia was assessed by measuring plasma total lipids, cholesterol, triglycerides, β-lipoproteins, and lipid peroxides as detailed in Refs. 47–49, 65.

D. Tissue Sampling

In our experimental conditions, we found that in more than 95% of cases, the lesion begins on two locations: (1) on the arterial aspect of the aortic valve, and (2) on the inner lesser curvature of the aortic arch.[66] Such lesion-prone areas (previously determined by backwards screening of lesional stages) were especially examined at the prelesional stage, generally at 1 week on the diet for the hamster aortic valve, and at 2 weeks for the rabbit aortic arch (Fig. 1). In some experiments, as shown hereafter, additional investigations were carried out on *in situ* or *in vitro* systems.

III. INVESTIGATIONS

The materials collected were analyzed by several sets of investigations which according to the goal and nature of the inquiry were performed by either *in vivo*, *in situ*, or *in vitro* experiments (Fig. 2).

A. EC Surface Charge and Chemistry

The following moieties of the luminal endothelial glycocalyx were detected with the cognate markers such as:

- Cationic sites of high pK_a values were revealed using hemeundecapeptide or anionic hemeundecapeptide of pI 4.8 and 4.3, respectively.
- Anionic groups of strongly charged residues of low pK_a values were visualized with cationized ferritin, pI 8.4; as additional probes, cationic hemeundecapeptide and safranine O were also used.

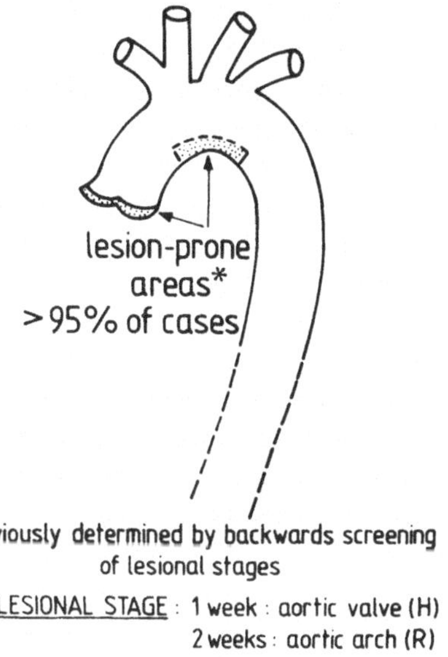

Figure 1. Diagram of tissue sampling from the lesion-prone areas of the rabbit aortic arch and aortic valve. The determination of these regions in which in more than 95% of cases the inception of atherogenesis occurred was achieved by backwards screening of the lesional stages. For the evaluation of the prelesional stage, hamster aortic valve at 1 week, and rabbit aortic arch at 2 weeks of fat-rich diet were commonly used.

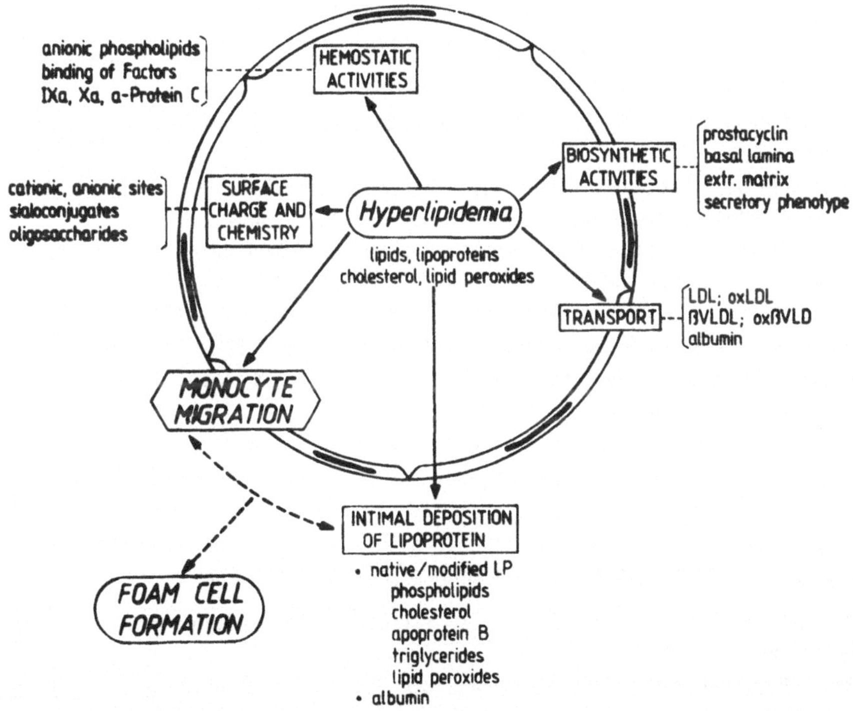

Figure 2. Diagrammatic representation of the set of investigations carried out in the experiments addressing the early modifications in the arterial and valvular intima during the diet-induced hyperlipidemic atherogenesis. According to the nature of the inquiry, the experiments were carried out *in vivo*, *in situ*, or *in vitro*.

- Sialyl residues non-*O*-acetylated at C-8 or C-9 were labeled with ferritin hydrazide on specimens previously oxidized by Na periodate.
- *N*-Acetylneuraminic acid and *N*-acetylglucosamine were detected with wheat germ agglutinin (WGA) followed by a mucin–gold conjugate.
- Terminal galactosyl groups were labeled with peanut agglutinin in sequence with lactosaminated bovine serum albumin–gold conjugate.
- Mannosyl and glucosyl residues were visualized with Con A followed by HRP–gold complex.
- Terminal galactose and *N*-acetylgalactosamine were marked with the sequence galactose oxidase–ferritin hydrazide.
- Subterminal galactosyl groups were reacted with *Ricinus communis* agglutinin followed by lactosaminated bovine serum albumin–gold complex.

The specimens were prepared and examined by electron microscopy and the micrographs evaluated by morphometric and statistical analyses as described in Ref. 18.

B. Hemostatic Activities of Endothelial Cells

In experiments carried out in our laboratory by F. Lupu *et al.* either *in situ*, *ex vivo*, or *in vitro*, we examined primarily the effect of β-VLDL on:

- The procoagulant activity by determining
  - The expression of anionic phospholipids on the surface of the EC membrane (using [^{125}I]polymyxin B and merocyanine 540)
  - The binding of some coagulation factors, namely the radioiodinated factor IXa and factor Xa
- The anticoagulant activity determined by the protein C activation (using chromogenic S-2366) and binding of the radioiodinated activated protein C
- The fibrinolytic and antifibrinolytic activities evaluated through assays for plasminogen activators and for plasminogen activator inhibitors: for the latter, using [^{125}I]fibrin and [^{125}I]plasminogen activators (chromogenic S-2251)
- Microfluidity of endothelial plasma membrane employing trimethylammonium-diphenylhexatriene (TMA-DPH) as fluorescent probe work by M. G. Badea *et al.* in our laboratory.

C. Biosynthetic Activities

Among the numerous secretory functions of EC, we looked at:

- The prostacyclin and thromboxane A_2 production [by the appropriate RIA determined by S. Tasca]
- The phenotypic shift to a secretory type such as that reported for "activated endothelium"[9,10]
- The *in vivo* production of basal lamina and extracellular matrix material (examined by electron microscopy and cytochemistry with safranine O and ruthenium red)

D. Transport of Lipoproteins

Two groups of experiments were conducted:

1. *Experiments with exogenous probes* consisted of either adding radiolabeled lipoproteins to the food (i.e., ^{3}H]cholesterol), perfused *in vivo* or *in situ*, or put in the incubation medium of cultured endothelial cells from bovine aorta. Low-density lipoprotein (LDL) or β-very-low-density lipoprotein (β-VLDL) was used either in native form or chemically modified by methylation, glycation, or oxidation. The probe was coupled with either the fluorescent DiI (1,1′-dioctadecyl-3,3,3′,3′-tetramethylindocarbocyanine perchlorate) for examination by fluorescent microscopy, radioiodinated for radioassay and autoradiography, or tagged with 5-nm gold particles for visualization in electron microscopy and morphometric analysis.[65,71–73]

2. *Experiments with endogenous molecules* consisted of the *in situ* immunochemical detection either of apo B (β-lipoproteins) or of the unesterified cholesterol by incubating the specimens with filipin or tomatine (reportedly known to bind specifically to 3-β-hydroxysterols) followed by electron microscopic examination.[65–67,71–73]

E. In Situ Detection of Intimal Lipid Deposits

According to the nature of the lipids detected *in situ* (phospholipids, unesterified cholesterol and cholesteryl esters, apolipoprotein B, triglycerides, lipid peroxides), we have used fluorescence microscopy, cytochemistry, immunocytochemistry, freeze-fracture electron microscopy.[41,42,47–49,54,65–67]

F. Isolation and Characterization of Intimal Lipid Deposits

Recently we have been able to isolate, purify, and partially characterize the lipid deposits from the rabbit aorta at the prelesional or lesional stage of atherogenesis.[49] Briefly, saline extracts of aortic intima collected from rabbits on 2 or 16 weeks of cholesterol-rich diet were sequentially subjected to gel filtration chromatography and density gradient ultracentrifugation: a crude lipoprotein fraction was obtained consisting of complex extracellular liposomes,[15,47–49,54,65–67] particulate β-lipoproteins, and albumin. This fraction was passed through two successive affinity chromatography columns using as immunoadsorbent anti-apo B (which retained β-lipoprotein particles and the apo B fragments), then antialbumin (to remove albumin from the fraction). Finally, a purified extracellular liposome (EL) fraction consisting exclusively of EL vesicles was obtained. Along the procedure, each step was monitored by negative staining and assays for lipids, protein, lipid peroxides, and a special group of analyses to determine the location of albumin which, surprisingly, was not removed by affinity chromatography.[49]

All the data emerging from these lines of investigations were correlated with the onset of monocyte recruitment and migration, the physical state of the endothelium, and the formation of foam cells in the subendothelial compartment.

IV. RESULTS AND DISCUSSION

A. Plasma Cholesterol

Although most animals developed rather rapidly hypercholesterolemia, relatively large individual variations were noted. In rabbits, the serum cholesterol increased from 40 mg/ml (control) to 250–500 mg/ml (2nd week of diet) and to 2000 mg/ml (within 8 weeks of diet). In hamsters, plasma cholesterol level doubled within 3 weeks, increased about 40-fold in the 4th week, and progressively amounted to 15-fold values at 10–12 months on diet.

B. EC Surface Charge and Chemistry

In the lesion-prone areas examined, both in the prelesional stage and in the early phases of plaque formation, endothelium was morphologically intact and no platelet adhesion could be detected.[35,36,44,65]

The resistance of the endothelial luminal glycocalyx to very high levels of plasma cholesterol was remarkable (Figs. 3–5). At these early stages and later during the development of fatty streaks, no relevant perturbations were noticeable in the density and distribution of the anionic, cationic, sialyl, and oligosaccharide residues. In more advanced stages, particularly when EC become loaded with lipid deposits,[42] a decrease in anionic sites and sialoconjugates was observed[18] (Figs. 3–6). Such a change may contribute to the augmented transport of plasma LDL.[21] Circulating monocytes showed modulations in the surface charge and chemistry which paralleled those of EC.[18] These observations revealed that during the initiation and early progression of hypercholesterolemic atherogenesis, the early intimal lesions are not preceded by significant general alterations in the charge and the oligosaccharide moieties of the cell surface of the endothelium and monocytes. Even at that stage of knowledge it became conceivable that more specific molecules are involved in monocyte adhesion to endothelium, such as we know today.[9,19,20,52] Under the intact endothelium, however, consistently, the lesion-prone regions became more rapidly the sites of intimal accumulation of modified β-lipoproteins reassembled as vesiclelike structures named "extracellular liposomes" (EL).[65]

C. Endothelial Hemostatic Functions

The *ex vivo* or in culture exposure of EC to high concentrations of β-VLDL resulted in a decreased protein C activation and a significant switch to procoagulant activities. The latter was expressed by the surface exposure of anionic phospholipids, enhanced biding of factors IXa and Xa, and a pronounced prothrombinase activity. Concomitantly, an increase in the t-PA activity, paralleled by a slight diminution of the PAI-1 expression could be noted. These preliminary observations, in partial agreement with some findings reported in Refs. 25 and 59, are under further investigations.

D. Biosynthesis of Prostacyclin and Thromboxane A_2

Cultured bovine aortic EC incubated either with native or acetylated β-VLDL, or with hydroxynonenal showed a differentiated increased production of prostacyclin. While

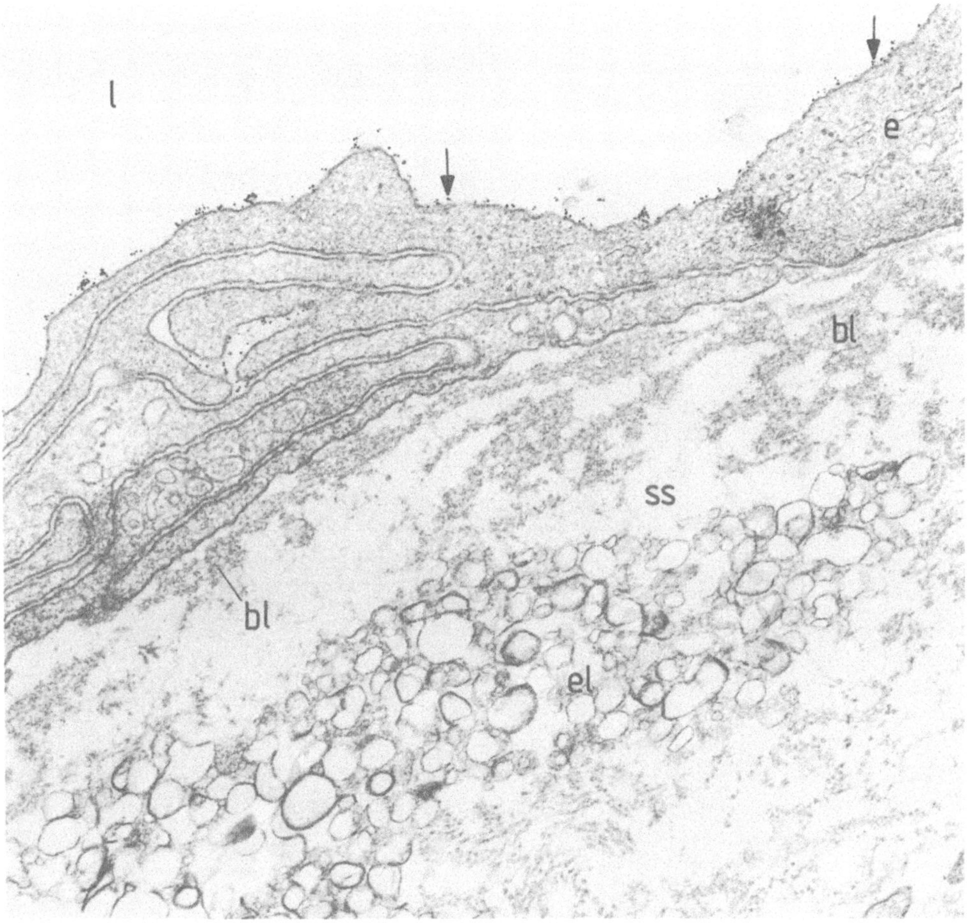

Figure 3. Aortic intima of a rabbit at 2 weeks of hyperlipidemic diet. The subendothelial space (ss) contains polymorphic extracellular liposomes (el) and a hyperplastic but disrupted basal lamina (bl). Endothelial cell surface was decorate by Con A/horseradish peroxidase–gold complex, the probe used for the indirect visualization of mannosyl residues (arrows). 1, lumen. × 40,000.

prostacyclin synthesis is stimulated by the native β-VLDL, it is inhibited by both acetyl-β-VLDL and hydroxynonenal, presumably due to their lipid peroxides (Fig. 7).

The marked augmentation in the thromboxane A_2 production induced by the native β-VLDL may be part of the procoagulant activation generated by hypercholesterolemia.

E. Production of Basal Lamina and Extracellular Matrix

While the lesion-prone regions examined were still devoid of adhering monocytes or intimal macrophages, a remarkable overproduction of successive layers of basal lamina material was occurring under the endothelium. Frequently these layers were interspersed or

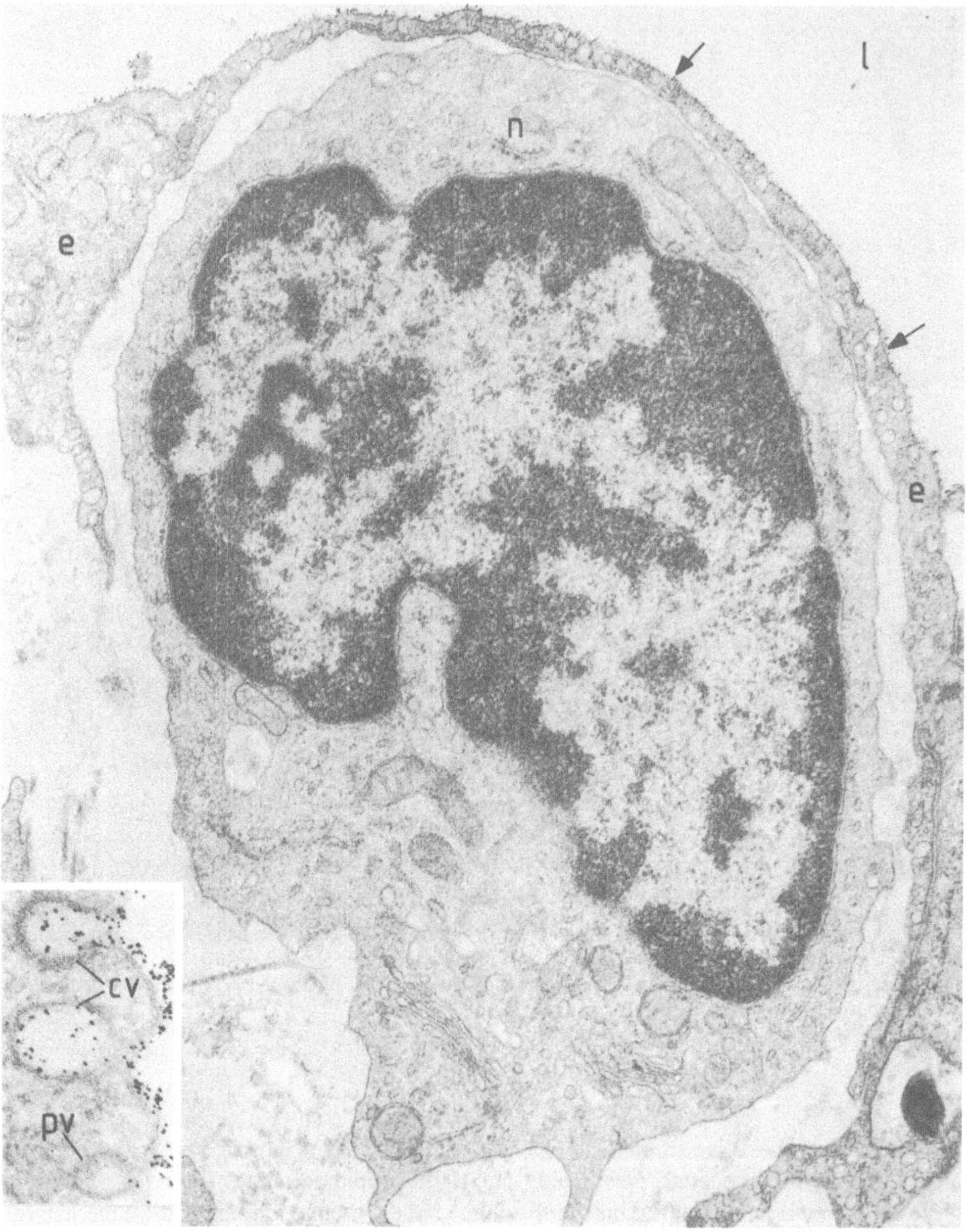

Figure 4. Rabbit aorta at 4 weeks of hyperlipidemic diet. A monocyte has migrated under a morphologically intact endothelium (e). The latter shows a normal distribution of terminal and/or subterminal galactosyl moieties indirectly visualized with *Ricinus communis* agglutinin–lactosaminated BSA–gold complex (arrows). 1, lumen. × 18,600. Inset: The tracer labeled coated pits and coated vesicles (cv) but failed to decorate most plasmalemmal vesicles (pv). × 120,000.

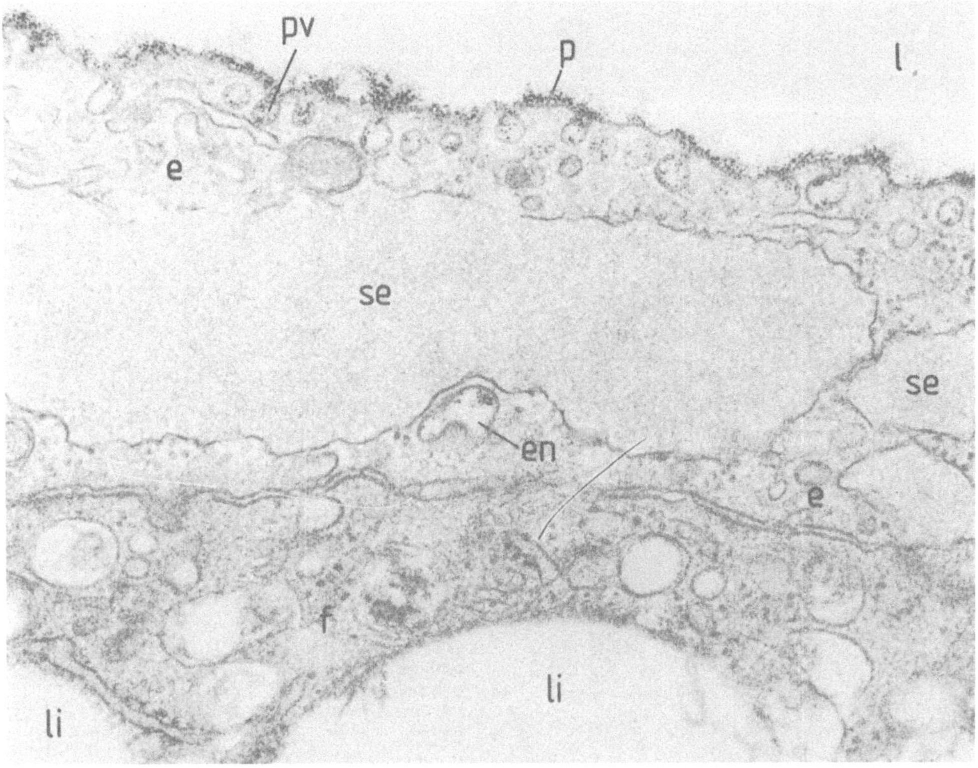

Figure 5. Segment of a rabbit aortic endothelial cell (8 weeks of diet) (e) displaying a markedly dilated smooth endoplasmic reticulum (se). Underneath the endothelial cell is a macrophage-derived foam cell (f). On the endothelial cell, cationized ferritin decorates rather uniformly the plasmalemma proper (p) and most of the plasmalemmal vesicles (pv) open to the luminal front. In en, presumably an endosome labeled by the tracer. l, lumen; li, lipid inclusion. × 62,000.

alternating with rows of numerous extracellular liposomes.[15,54,65] In many areas, the basal lamina proper appeared focally detached from the abluminal front of the EC. Concomitantly there was a marked proliferation of the apparently amorphous extracellular matrix, collagen fibers, and small bundles of elastin. The latter was difficult to define as a result either of a hyperplastic process or of the elastolysis of the lamina elastica interna.

In these regions displaying various degrees of EL accumulation, the overlying EC were often characterized by a relatively thick (1–2 μm) cell body with elaborate smooth and rough endoplasmic reticulum, extensive perinuclear Golgi complex, and sometimes intracytoplasmic membrane-bound vacuoles or granules. This apparent shift to a secretory phenotype was reminiscent of an activated EC as described in inflammation and immune reactions.[9,10]

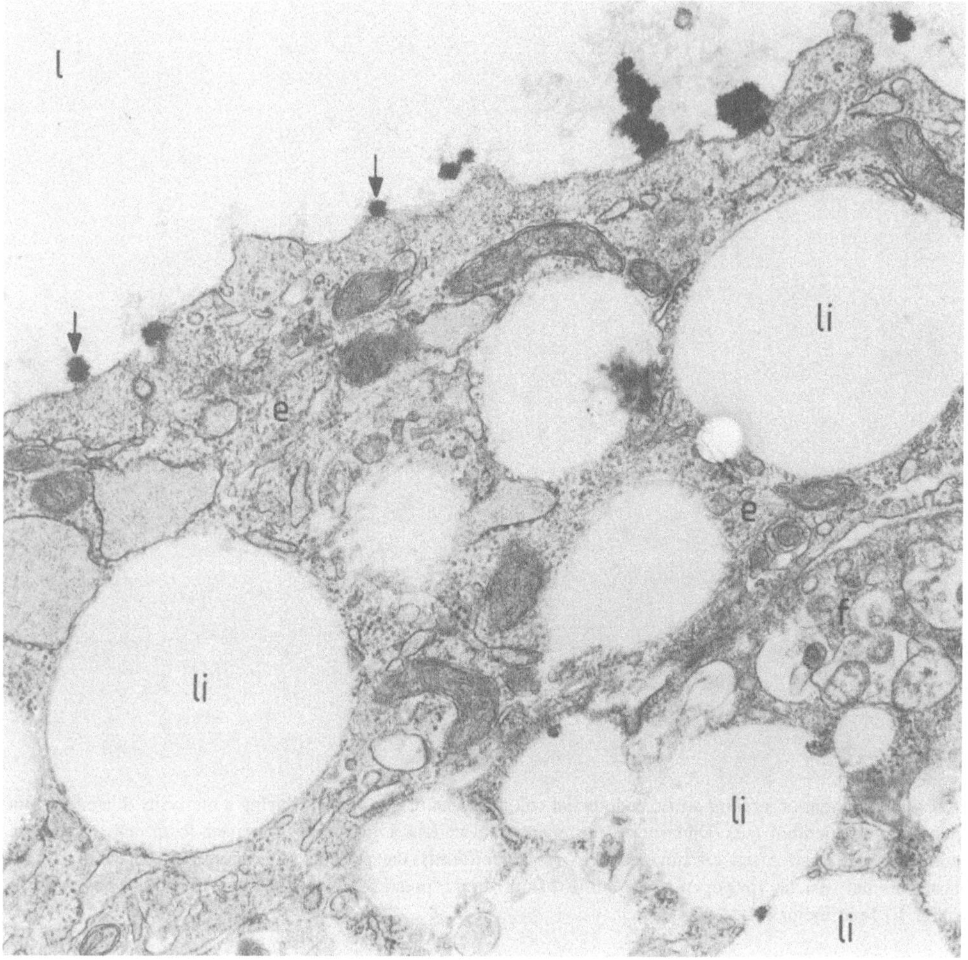

Figure 6. Segment of the rabbit aortic intima (16 weeks of diet). When an endothelial cell (e) becomes loaded with lipid inclusions (li), the cell surface labeling with cationized ferritin (CF) is drastically reduced and occurs mostly in patches of CF clusters (arrows). f, macrophage-derived foam cell; 1, lumen. × 30,000.

F. Transport of Lipoproteins

1. Incorporation of Dietary [³H]-Cholesterol into the Aortic Wall

In these experiments (Fig. 8), rabbits which were fed for 10 or 25 days the fat-rich diet received in the last 46 hr 1–5 mCi of [^{3}H]cholesterol added to 100 g food. After 40 hr, plasma aliquots and specimens from the aortic intima (separately the aortic arch) were collected. In animals on the diet for 25 days, serum radioactivity was recovered mainly in the β-VLDL fraction; within the vessel wall about 60% of the radioactivity was found in the intima at values almost twice those measured in the control animals. After density gradient

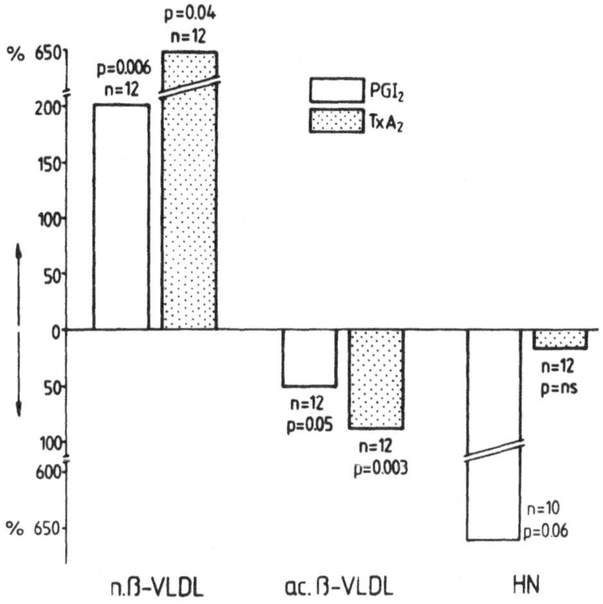

Figure 7. Production of prostacyclin and thromboxane A_2 by cultured bovine aortic endothelial cells incubated with native β-VLDL, acetyl-β-VLDL, or 4-hydroxynonenal (RIA assays by S. Tasca). (From Simionescu *et al.*, with permission.)

ultracentrifugation, the material extracted from intima was recovered as a band in the range densities of VLDL and LDL.[65,66]

2. *Experiments with [¹²⁵I]β-VLDL Injected in Vivo*

Rabbits fed the fat-rich diet received weekly an i.v. injection of 0.3–0.5 mCi of radioiodinated β-VLDL (collected from other animals at the same stage of diet). Twenty-four hours after the injection, blood samples and aorta were collected and analyzed by radioassay and autoradiography. By comparison with the lipids accumulated in the vessel wall, the protein deposition was much lower, presumably due to the steep increase in the cholesterol content of VLDL particles. It was found that while in the first week of diet the cholesterol/protein ration was 1.5:1, it amounted to 3.8:1 in the 5th week and about 12.0:1 in the 8th week of diet. In the first 2 weeks of diet, the [^{125}I]β-VLDL uptake by the aortic wall was 30–40% higher in the hypercholesterolemic than in normal animals. The same difference was recorded at 8 weeks of diet (15 ± 2 ng protein [^{125}I]β-VLDL/mg wet tissue in hyperlipidemic animals versus 10 ± 2 ng protein [^{125}I]β-VLDL in normal aortas). At more advanced stages, when large portions of arterial intima contained fatty streaks, the decay curve of radioiodinated β-VLDL in normal plasma was almost double that in hypercholesterolemic animals. At 22 weeks of diet, the uptake of [^{125}I]β-VLDL by the aortic wall increased to 57 ± 6 ng radioiodinated β-VLDL protein/mg wet tissue as compared to 12 ± 2 ng in controls.

Light microscopic autoradiography revealed a progressive intimal accumulation of autoradiographic grains (ARG) that paralleled the development of foam cells. Electron microcopy showed that most ARG were localized to plasmalemma vesicles, some vacuoles, and extracellular matrix (Fig. 9).

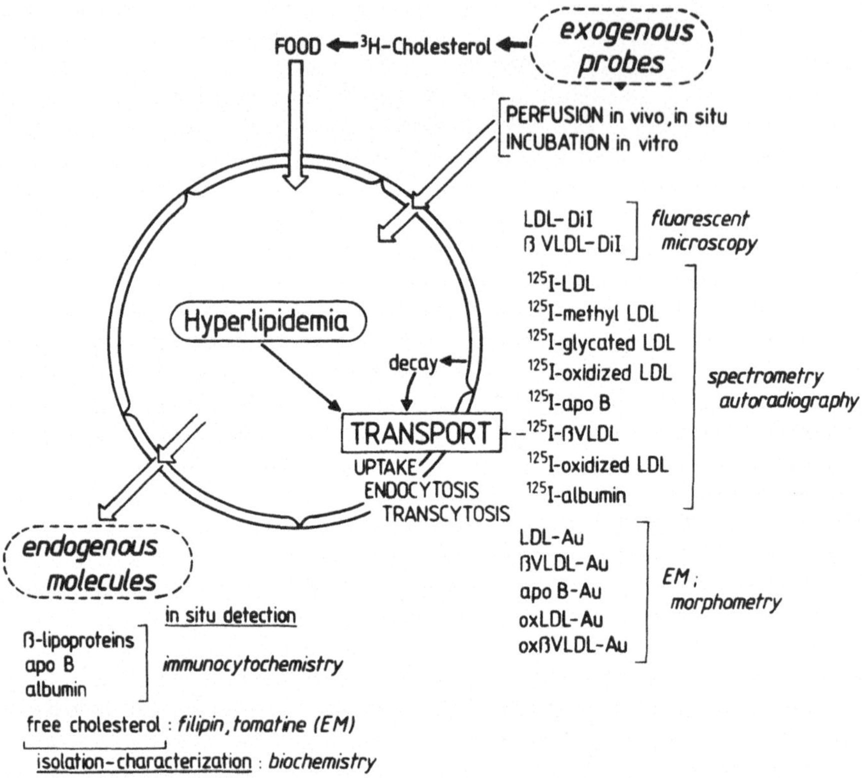

Figure 8. Schematic representation of the experimental design and the probes used to assess the transport of lipoproteins and albumin by the arterial endothelium in normal and hyperlipidemic rabbits. (From Simionescu *et al.*, with permission.)

3. Experiments with [^{125}I]β-VLDL-DiI Perfused in Situ

This probe was perfused *in situ* over a total period of 1–2 hr and aortic segments were excised and examined by radioassay and fluorescence microscopy. Both methods revealed an augmented deposition of β-VLDL in the arterial intima in hypercholesterolemic animals as compared with controls,[65,66] results similar to those reported with different methods by others.[11]

4. Experiments with β-VLDL–Gold Complex Perfused in Situ

The tracer was perfused *in situ* for up to 15 min and specimens from different segments of the aorta (especially the aortic arch) were collected, prepared, and examined by electron microscopy. The probe was visualized mostly in plasmalemma vesicles in all locations (luminal, inside the cytoplasm, and abluminal), in the subendothelial space, and was absent from the intercellular junctions. To a lesser extent, the complex ligand labeled endothelial structures normally involved in endocytosis (coated pits, coated vesicles, endosomes, multivesicular bodies, and what structurally have been identified as lysosomes). Morpho-

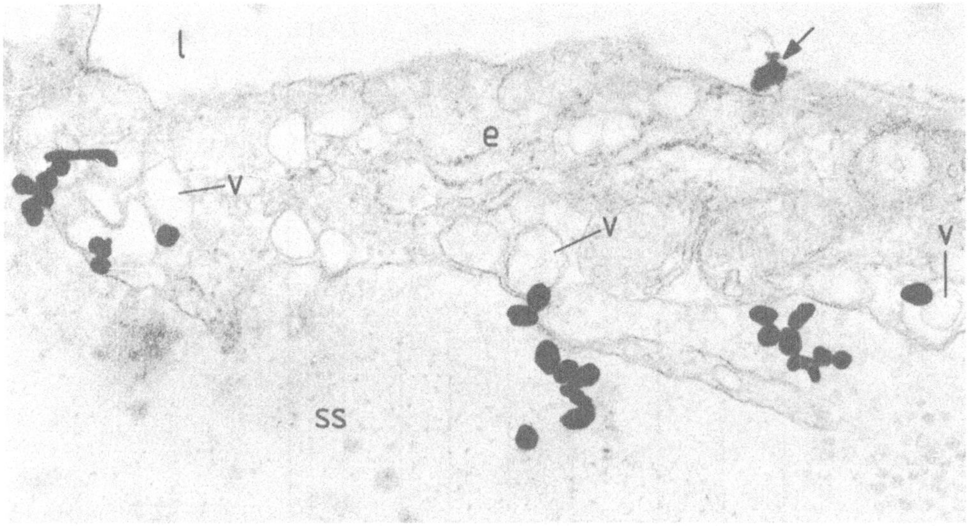

Figure 9. Electron microscopic autoradiography displaying the transendothelial transport of radioiodinated β-VLDL in the rabbit aorta. Silver grains appear associated with luminal plasmalemma (arrow), with plasmalemmal vesicles in various positions (v) and within the subendothelial space (ss). e, endothelium; 1, lumen. × 60,000. (From Vasile *et al.*, with permission.)

metric and statistical analysis showed that transcytosis of β-VLDL–Au particles was the prevalent process, being significantly pronounced by hypercholesterolemia[65,73] (Fig. 10).

During early stages, both apo B and apo E were immunocytochemically detected (especially in the aortic arch) predominantly in the extracellular matrix. Frequently they delineated the contour of EL.

5. Experiments with Oxidized LDL and β-VLDL

Grown in Millicell chambers of a bicompartmental system, bovine aortic EC (BAEC) were incubated with native or copper-oxidized human LDL (oLDL) or rabbit β-VLDL (o-β-VLDL). The same probes were also perfused *in situ* in normal or hypercholesterolemic rabbits whose aortas were examined by radioassay, autoradiography, or were used to prepare saline extracts further subjected to ultracentrifugation and assayed for lipid peroxides and lysophosphatidylcholine. The findings revealed that cultured BAEC degraded and transcytosed oLP at values double those of native LP. Similar differences were recorded *in situ*. Unlike controls, aortic intima of the hypercholesterolemic rabbits contained detectable amounts of lipid peroxides.

These observations indicate that transcytosis is a major mechanism for monitoring the excessive LP accumulation in the plasma. Both in normal and hyperlipidemic animals, transcytosis occurs via plasmalemma vesicles. Hypercholesterolemia induces a significant augmentation of LP transport across EC as well as transport of other plasma proteins.[5,53] Opening of endothelial junctions may occur only in advanced lesions, adding a convective pathway to the LP transport. Other cellular and especially extracellular alterations may

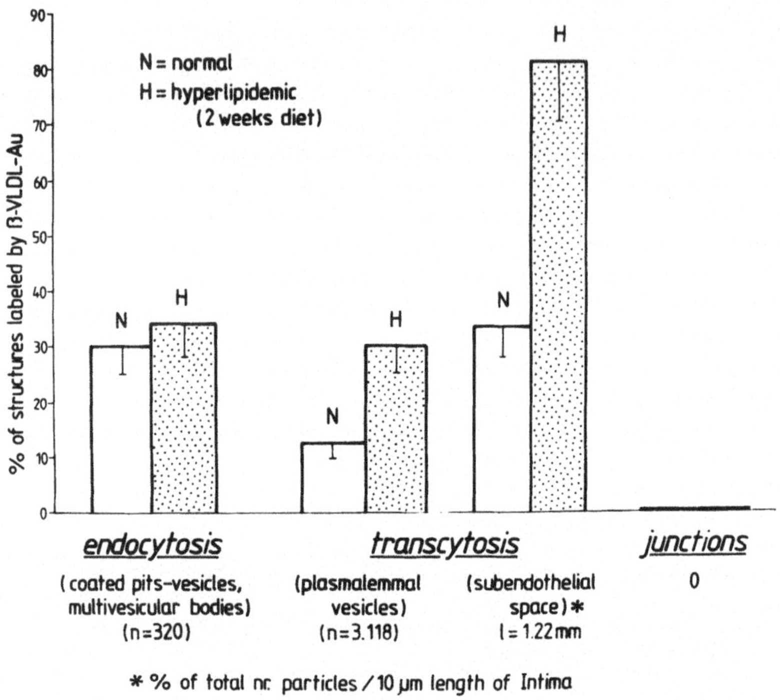

Figure 10. Morphometric analysis of the β-VLDL–Au complex distribution in various structures of aortic endothelium of normal and hyperlipidemic rabbits, 15 min after tracer perfusion *in situ*. (Data from Vasile *et al.*)

impede the efflux of the LP-derived material, this promoting plaque progression and complication.

6. Intimal Accumulation of LP-Derived Components

During the first 2 weeks of fat-rich diet, in the lesion-prone areas of the aortic arch and aortic valve, no grossly visible modifications of the vessel wall were seen. At the ultrastructural level, however, a progressive deposition within the subendothelial matrix predominantly of small (100–300 nm in diameter) vesicles (or lipid compact particles) formed by uni- or multilamellar phospholipid lamellae was noticed. These extracellular liposomes (Figs. 3 and 11) were found to contain in their lamellae unesterified cholesterol (UC) (visualized with filipin),[65,66] and appeared to be surrounded by small deposits of apo B (detected immunocytochemically).[47–49] Apo B and A were localized immunocytochemically in human aortic fibrinoid lesions,[6] and apo B fragments have been shown to mediate their LDL interactions with arterial proteoglycans.[7] By reacting the same specimen with an anti-apo B–HRP conjugate and with tomatine (reportedly specific for UC) we could demonstrate that apo B and UC colocalize in close association with EL.[47–49] EL occurred in a cell-free subendothelial compartment that commonly displayed a proliferation of the extracellular matrix: successive layers of basal laminae, proteoglycans (visualized with safranine O or ruthenium red), microfibrils, bundles of banded collagens, and elastic tissue.

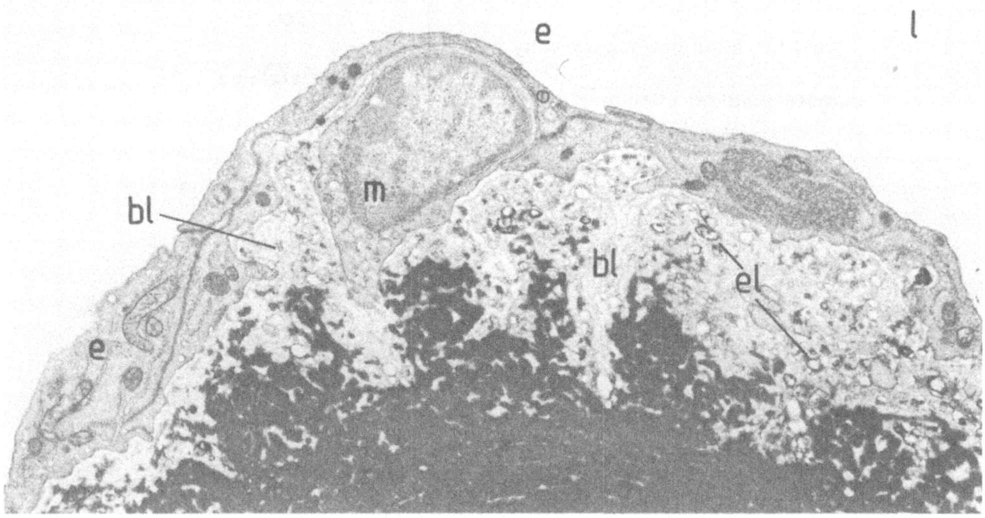

Figure 11. Segment of the aortic arch of a rabbit at 2 weeks of hyperlipidemic diet. A monocyte (m) has migrated into the subendothelial space of an intimal region containing extracellular liposomes (el). bl, hyperplastic basal lamina; e, endothelium; 1, lumen, × 30,000. (From Simionescu *et al.*, with permission.)

Same areas also displayed various amounts of apo B-reactive particles of LDL or VLDL size. In early atherosclerosis, vesiclelike features similar to EL have previously been reported,[65,66] and more recently such structures have been in more detail described in the WHHL or cholesterol-fed rabbit aortas.[8,17,60] Such material was identified either by standard electron microscopy[60,8] or by rapid freezing and freeze-etching.[3,17] Similar lipid deposits were also found in the atherosclerotic human aortas.[8,22,23] The EL deposits may represent the ultrastructural equivalent of the filipin-positive particles visualized by other investigators by fluorescence microscopy.[39] A preliminary observation which deserves a more detailed analysis is that in regions with sero-fibrinous exudate only, tomatine forms spicules suggestive of the presence of UC carried by either lipoprotein particles or/and albumin, the latter being immunocytochemically detectable in such areas.

G. Isolation and Characterization of Intimal Extracellular Liposomes

Several laboratories have already reported the isolation and partial characterization of lipid or lipoprotein phases of advanced animal or human atherosclerotic lesions.[28–30,37,51,61,63,68,75–77] The studies were mainly focused on defining the chemical alteration of the extracted aortic β-LP. Reports on the extraction and characterization of the lipid deposits from early stages of the experimental atherosclerosis are very scarce and were based on gel filtration and ultracentrifugation.[8]

Recently we have developed a four-step procedure for isolation and purification of the EL from rabbit aortas at the prelesional (2 weeks) or lesional (16 weeks of fat-rich diet) stages of atherogenesis (Fig. 12). Briefly, isotonic saline extracts of aortic intima after gel

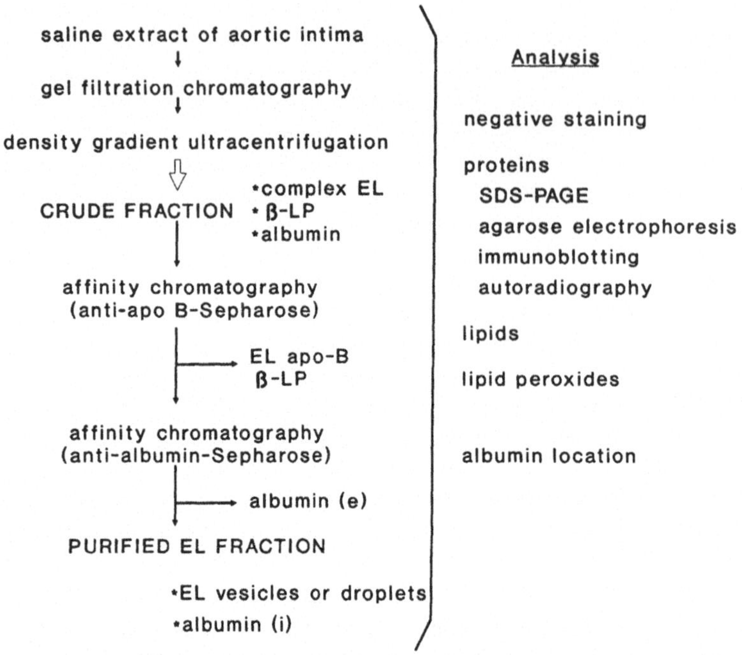

Figure 12. Schematic flow sheet of the procedure used for the isolation and purification of the extracellular liposome fraction from saline extracts of rabbit aortic intima. (Data from Mora *et al.*)

filtration yielded a crude EL fraction which my density gradient ultracentrifugation floated in a relatively broad density range (1.02–1.08 g/ml). In addition to lipids, the fraction contained apo B and albumin which were removed by affinity chromatography on anti-apo B and antialbumin–Sepharose columns, to obtain a purified EL fraction. Each step and especially the EL fractions were examined by negative staining (Fig. 13), SDS–PAGE, immunoelectroblotting, autoradiography, uronic acid and analyzed for their content in total cholesterol (TC), UC, cholesterol esters (CE), phospholipids (PL), triglycerides (TG), and lipid peroxides (LP) determined as malondialdehyde adducts (MDA[24]). By comparison with serum LDL and β-VLDL, the crude EL fraction was characterized by a higher relative percentage mass of UC, PL, apo B, and albumin. These differences became more prominent in the lesional stage. The purified EL fraction contained almost exclusively UC, PL, and (surprisingly) albumin (Fig. 14). Delipidation or pronase digestion of the purified EL, followed by dot-blot immunolabeling and autoradiography revealed that albumin was located within the core of EL (Figs. 15 and 16). Lipid peroxides were found in increased amounts only in the crude EL fraction (Table 1), and they eluted mostly with immunoadsorbed β-LP.

These findings suggested that aortic EL represent chemically modified lipoproteins which are reassembled extracellularly trapping albumin (a natural antioxidant[26,57] presumably upon interaction with lysophosphatidylcholine produced by lipid peroxidation.[40]

In our experimental conditions, monocytes are recruited mostly in regions which contain intimal EL. Ultrastructural images suggest that EL are phagocytosed by macrophages and later on also by smooth muscle cells committed to form foam cells.

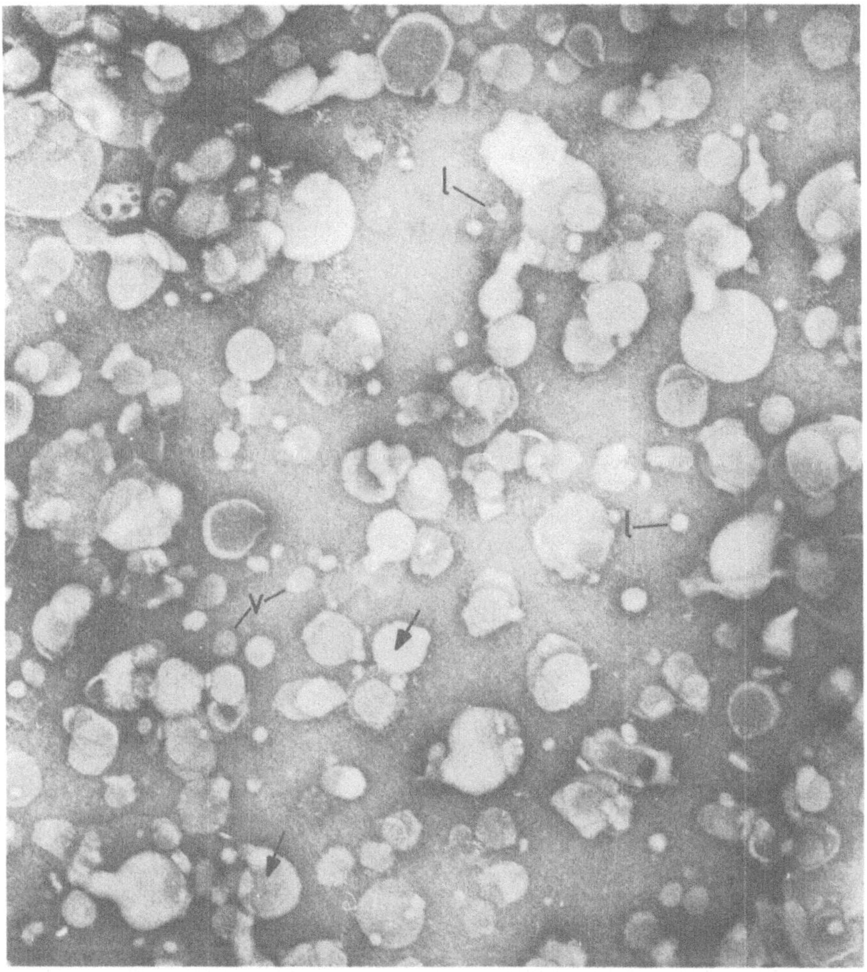

Figure 13. Negative staining with phosphotungstic acid of an aliquot of the crude fraction of extracellular liposomes obtained after subjecting a saline extract from hyperlipidemic rabbit aorta to gel filtration chromatography and density gradient ultracentrifugation. The material consists of particulate lipoproteins in the size range of LDL (l) and β-VLDL (v), and a relatively large proportion of polymorphic extracellular liposomes (arrows). × 96,000. (From Simionescu *et al.*, with permission.)

V. TENTATIVE INTEGRATION OF FINDINGS—WORKING HYPOTHESIS

Since the findings described in this review article emerged from separate, although correlated, lines of investigation, their integration into a global picture remains an unattainable aim. The following tentative integration of these observations and other data produced by various laboratories is far from being conclusive, and is intended as a working hypothesis.

In our experiments, in spite of a very pronounced hypercholesterolemia, arterial

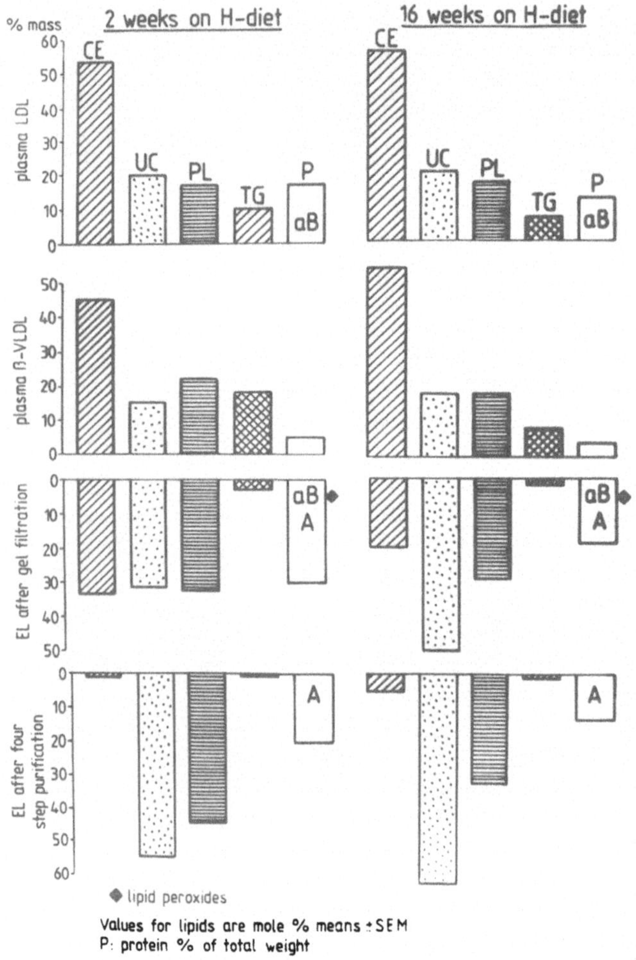

Figure 14. Histogram of the chemical composition determined at 2 weeks (prelesional stage) and 16 weeks (lesional stage) of hypercholesterolemic (H) diet of rabbit plasma LDL and β-VLDL as compared with the crude and affinity-purified fraction of extracellular liposomes (EL). CE, cholesteryl esters; UC, unesterified cholesterol; PL, phospholipids; TG, triglycerides; P, proteins; A, albumin; aB, apoprotein B. (From Simionescu *et al.*, with permission.)

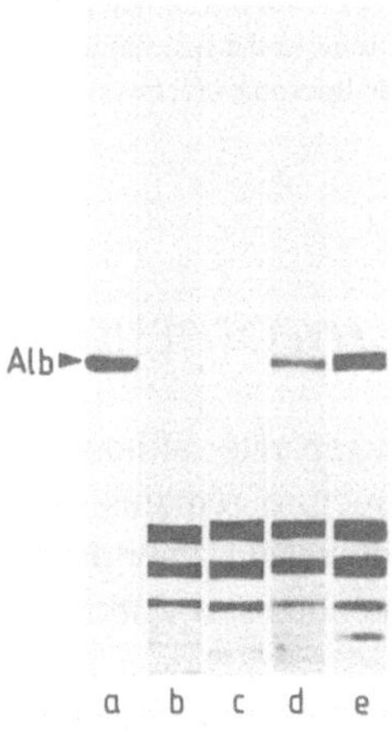

Figure 15. SDS–PAGE analysis of purified EL after treatment with 100 μg pronase E for 7 hr at 37°C (a) Rabbit serum albumin (RSA). (b) Pronase E. (c) RSA after proteolysis with pronase E; note that the protein is completely digested by the enzymatic proteolysis, the only protein bands visualized being those of the pronase E used. (d) EL fraction obtained at prelesional (2 weeks) and (e) lesional stages (16 weeks) of hyperlipidemia, after proteolytic digestion with pronase E; note that albumin associated with EL fractions (lanes d and 3) was not affected by the proteolytic enzyme. Alb, albumin. (From Mora *et al.*, with permission.)

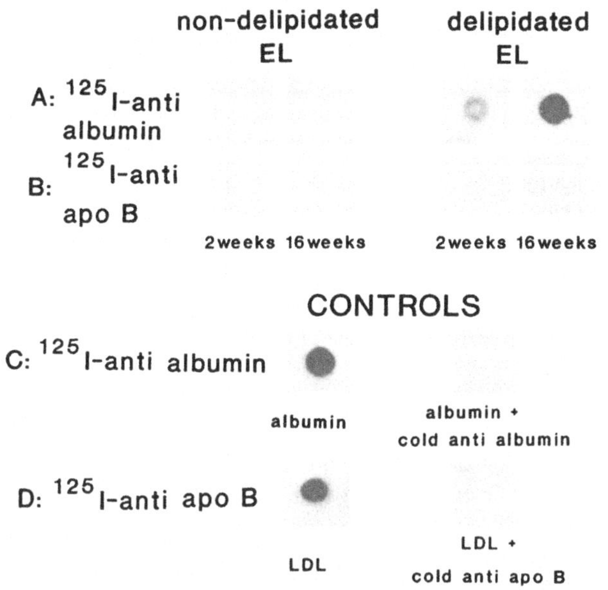

Figure 16. Autoradiography of the dot-blot immunoanalysis of purified extracellular liposomes (EL) isolated from extracts of aortic intima at 2 weeks (prelesional stage) and 16 weeks (lesional stage) of diet-induced hyperlipidemia, in rabbits. (A) Upon incubation with radioiodinated anti-rabbit serum albumin there is a lack of immunoreactivity in the dots of the nondelipidated EL and a positive reaction of the EL blots after delipidation with chloroform–methanol. (B) Incubation with radioiodinated anti-apo B does not real any immunoreactivity of either the nondelipidated or delipidated EL. (C, D) Control for the specificity of the immunoreaction, in which dots of the albumin alone, or LDL alone, or each of them overlayed with the corresponding cold antibody, were incubated with the specific radioiodinated antibody. Note the negative results obtained when the ligand was previously reacted with its nonlabeled antibody. (From Mora *et al.*, with permission.)

endothelium is morphologically intact over an extended period which covers the prelesional and early lesional stages. Neither endothelial damage nor platelet involvement was observed until advanced stages of fatty streak development, when EC themselves showed signs of marked physical injury such as their transformation into veritable foam cells.

During the prelesional stage (1–2 weeks of diet) in the lesion–prone areas examined (aortic arch and aortic valve), one could distinguish *modulations* of some constitutive functions of the EC such as an enhanced transcytosis of β-VLDL (rabbits) or LDL (hamsters), presumably generated by a complex of factors among which is the plasma–interstitial fluid concentration gradient of LP.

Upon interaction with EC, LP undergo chemical modifications, including oxidation,[16,46,50,56,69] the sites of which are difficult to detect at present. Potential candidates for the location of this process are the EC surface, the plasmalemma vesicles, or the subendothelial compartment where LP can possibly undergo autoxidation[13] (Fig. 17).

The transcytosed LP occur either as particulate native LDL and β-VLDL or as degraded LP which upon interactions with the extracellular matrix components, are progressively reassembled as EL. Areas rich in EL also contain albumin and other plasma proteins. Self-aggregation of LP[31,32,38,70] may also play a role in EL formation.

Table 1. TBA Reactants (TBAR) in the Extracellular Liposome Fraction (EL), as Compared to the Aortic and Serum β-Lipoproteins in the Lesional Stage of Hyperlipidemia

Sample	TBAR (nmoles malondialdehyde/mole lipid) × 10^4
Serum β-VLDL	1.0 ± 0.1
Serum LDL	0.9 ± 0.2
EL, gel filtered	3.1 ± 0.4
Aortic β-LP[a]	14.3 ± 1.7
EL, purified	0.9 ± 0.1

[a]Retained by the anti-apo B immunoadsorbent.

EL appear to be a key element in the prelesional modifications of the intima; presumably they can result from modified and aggregated LP whose phospholipids are organized as uni-, then multilamellar vesicles or droplets. The lamellae entrap UC that may represent a local protective reaction against the sclerogenic effect of cholesterol.[1,2] The EL core retains albumin, probably locally segregated, as natural antioxidant.[26,57] Closely associated with the EL contour appears immunoreactive apo B (or fragments) which contains lipid peroxides.

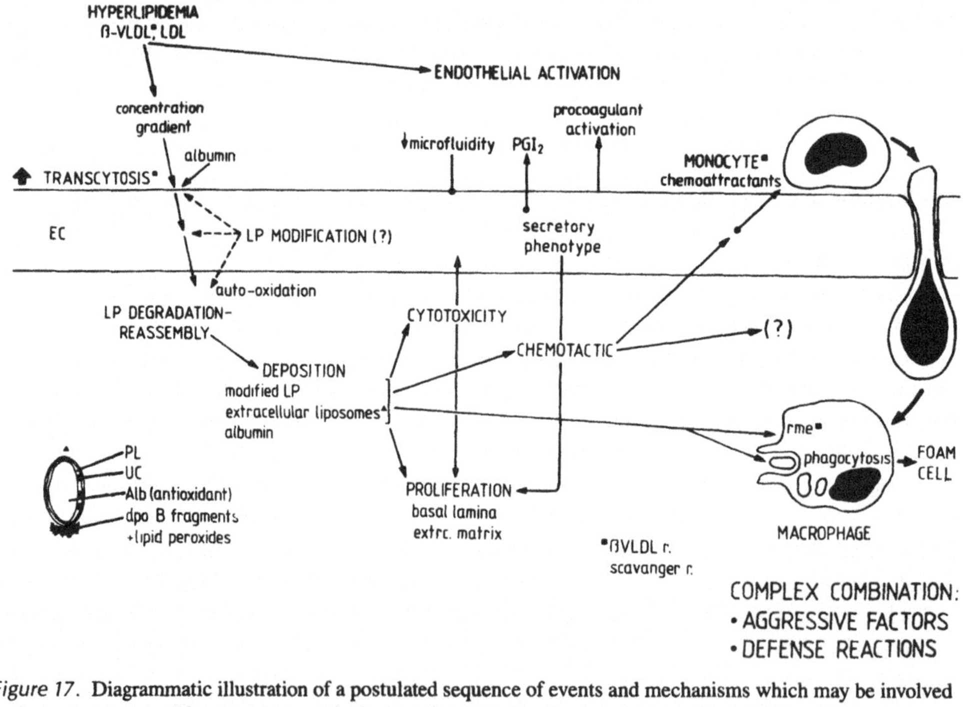

Figure 17. Diagrammatic illustration of a postulated sequence of events and mechanisms which may be involved in the prelesional modifications detected in the vessel wall during the early stages of hyperlipidemic atherogenesis (working hypothesis). (From Simionescu *et al.*, with permission.)

Yet not fully proved, there is suggestive evidence that altered LP and their derived EL have cytotoxic effects on the overlying EC, and induce proliferation of basal lamina and extracellular matrix especially proteoglycans which tightly interact with these LP thus contributing to their retention in the intima.

At this early stage of atherogenesis, it seems that EC are affected from both sides: luminally by hypercholesterolemia and abluminally by the subendothelially accumulated altered LP and foam cells.[12,14,16,24,27,33,55,62,65–67,69] As a result, EC undergo activation and dysfunction expressed by a shift to a secretory phenotype, augmented production of basal lamina and extracellular matrix, prostacyclin and thromboxane A_2, enhanced expression of procoagulant activity, chemoattractants[4,45] and adhesion molecules for monocytes.[9,10,19,20] The latter migrate within the subendothelial space, are activated as macrophages, and by taking up altered LP including EL (via scavenger receptor, β-VLDL receptor, and phagocytosis) become foam cells marking the inception of the lesional stage.

At more advanced stages, EC become thinner, frequently loaded with lipid deposits (endothelial-derived foam cells), and show signs of physical damage and *injury* that may lead to their desquamation, exposure of the subendothelial matrix constituents, platelet adhesion, and thrombus formation.

Thus, along those prelesional cellular and molecular events underlining the atherogenetic process, EC go successively through *modulation*, *dysfunction*, and (only in the advanced plaque) *injury*. The prelesional changes constitute a very complex and dynamic combination and interplay between *injurious factors* and *defense reactions*. Since very often it is difficult to discern between the two, much caution has to e exercised when considering a biochemical or structural change as a *cause* of their disease, for it may be just a *consequence* of a protective mechanism.

ACKNOWLEDGMENTS. This review article is largely based on work supported by the Ministry of Education, Romania, and by the National Institutes of Health (USA), Grant HL-26343.

REFERENCES

1. Adams, C. W. M., Bayliss, P. B., and Ibrahim, M. Z. M., 1963, The distribution of lipids and enzymes in the aortic wall in dietary rabbit atheroma and human atherosclerosis, *J. Pathol. Bacteriol.* **86**:421–430.
2. Adams, C. W. M., and Morgan, R. S., 1967, The effect of saturated and polyunsaturated lecithins on the resorption of 4-^{14}C-cholesterol from subcutaneous implants, *J. Pathol. Bacteriol.* **94**:73–76.
3. Amanuma, K., Kanaseki, T., Ikeuchi, Y., Ohkuma, S., and Takano, T., 1986, Studies on fine structure and location of lipids in quick-freeze replicas of atherosclerotic aorta of WHHL rabbits, *Virchows Arch. A* **410**:231–238.
4. Berliner, J. A., Territo, M., Almada, L., Carter, A., Shafonsky, E., and Fogelman, A. M., 1986, Monocyte chemotactic factor produced by large vessel endothelial cells in vitro, *Arteriosclerosis* **6**:254–258.
5. Bini, a., Femeglio, J. J., Jr., Mesa-Tejada, R., Kudryk, B., and Kaplan, K. L., 1989, Identification and distribution of fibrinogen, fibrin and fibrin(ogen) degradation products in atherosclerosis. Use of monoclonal antibodies, *Arteriosclerosis* **9**:109–121.
6. Bocan, T. M. A., Brown, S. A., and Guyton, J. R., 1988, Human aortic fibrolipid lesions. Immunochemical localization of apolipoprotein B and apolipoprotein A, *Arteriosclerosis* **8**:499–508.
7. Camejo, G., Olofsson, S.-O., Lopez, F., Carlsson, P., and Bondjers, G., 1988, Identification of apo B-100 segments mediating the interaction of low density lipoproteins with arterial proteoglycans, *Arteriosclerosis* **8**:368–377.

8. Chao, F. F., Amende, L. M., Blanchette-Mackie, E. J., Skarlatos, S. I., Gamble, W., Resan, J. H., Merguer, W. T., and Kruth, H. S., 1988, Unesterified cholesterol-rich lipid particles in atherosclerotic lesions of human and rabbit aortas, *Am. J. Pathol.* **131:**73–86.
9. Cotran, R. S., 1987, New role for the endothelium in inflammation and immunity, *Am. J. Pathol.* **129:** 407–413.
10. Cotran, R. S., and Pober, J.S ., 1988, Endothelial activation: Its role in inflammatory and immune reactions, in: *Endothelial Cell Biology in Health and Disease* (N. Simionescu and M. Simionescu, eds.), Plenum Press, New York, pp. 335–347.
11. Daugherty, A., Lange, L. G., Sobel, B. E., and Schonfeld, G., 1985, Aortic accumulation and plasma clearance of beta-VLDL and HDL: Effects of diet-induced hypercholesterolemia in rabbits, *J. Lipid Res.* **26:**955–963.
12. DiCorleto, P. E., and Chisolm, G. M., III, 1986, Participation of the endothelium in the development of the atherosclerotic plaque, *Prog. Lipid Res.* **25:**365–374.
13. Esterbauer, H., Jurgens, G., Quchenberger, O., and Koller, E., 1987, Autoxidation of human low density lipoprotein: Loss of polyunsaturated fatty acids and vitamin E, and generation of aldehydes, *J. Lipid Res.* **28:** 495–509.
14. Fagiotto, A., and Ross, R., 1984, Studies on the hypercholesterolemia in the human primate. II. Fatty streak conversion to fibrous plaque, *Arteriosclerosis* **4:**341–356.
15. Filip, D. A., Nistor, A., Bulla, A., Radu, A., Lupu, F., and Simionescu, M., 1987, Cellular events in the development of valvular atherosclerotic lesions induced by experimental hypercholesterolemia, *Atherosclerosis* **67:**199–214.
16. Fogelman, .A M., Shechter, I., Seager, J., Hokom, M., Child, J. S., and Edwards, P. A., 1980, Malondialdehyde alteration of low density lipoprotein leads to cholesteryl ester accumulation in human monocyte-macrophages, *Proc. Natl. Acad. Sci. USA* **77:**2214–2218.
17. Frank, J. S., and Fogelman, A. M., 1989, Ultrastructure of intima of WHHL and cholesterol fed rabbit aorta prepared by ultrarapid freezing and freeze etching, *J. Lipid Res.* **30:**967–978.
18. Ghinea, N., Leabu, M., Hasu, M., Muresan, V., Colceag, J., and Simionescu, N., 1987, Prelesional events in atherogenesis: Changes induced by hypercholesterolemia in the cell surface chemistry of arterial endothelium and blood monocytes in rabbit, *J. Submicrosc. Cytol.* **19:**209–227.
19. Gimbrone, M. A., Jr., 1986, Endothelial dysfunction and the pathogenesis of atherosclerosis, in: *Atherosclerosis VII* (N. H. Fidge and P. J. Nestel, eds.), Elsevier, Amsterdam, pp. 367–381.
20. Gimbrone, M. A., Jr., 1989, Endothelial dysfunction and atherosclerosis, *J. Cardiac. Surg.* **4:**180–183.
21. Gorog, G., and Born, G. V. R., 1982, Increased uptake of circulating low density lipoproteins and fibrinogen by arterial wall after removal of sialic acid from their endothelial surface, *Br. J. Exp. Pathol.* **63:**447–451.
22. Guyton, J. R., and Klemp, K. F., 1988, Ultrastructural discrimination of lipid droplets and vesicles in atherosclerosis: Value of osmium–thiocarbohydrazide–osmium and tannic acid–paraphenylendiamine technique, *J. Histochem. Cytochem.* **36:**1319–1328.
23. Guyton, J. R., and Klemp, K. F., 1989, The lipid-rich core region of human atherosclerotic fibrous plaques. Prevalence of small lipid droplets and vesicles by electron microscopy, *Am. J. Pathol.* **134:**705–717.
24. Haberland, M. E., Fong, D., and Cheng, L., 1988, Malondialdehyde-altered protein occurs in atheroma of Watanabe heritable hyperlipidemic rabbits, *Science* **241:**215–218.
25. Hajjar, K. A., Gavish, D., Breslow, J. L., and Nachman, R. L., 1989, Lipoprotein (a) modulation of endothelial cell surface fibrinolysis and its potential role in atherogenesis, *Nature* **339:**303–305.
26. Halliwell, B., 1988, Albumin—An important extracellular antioxidant? *Biochem. Pharmacol.* **37:**569–571.
27. Hansson, G. K., and Bondjers, G., 1987, Endothelial dysfunction and injury in atherosclerosis, *Acta Med. Scand. Suppl.* **715:**11–17.
28. Hoff, H. F., and Gaubatz, J. W., 1982, Isolation, purification and characterization of a lipoprotein containing apo B from the human aorta, *Atherosclerosis* **42:**273–291.
29. Hoff, H. F., and Morton, R. E., 1985, Lipoproteins containing apo B extracted from human aortas: Structure and function, *Ann. N.Y. Acad. Sci.* **454:**183-‹94.
30. Hollander, W., Paddock, J., and Colombo, M., 1979, Lipoproteins in human atherosclerotic vessels. Part I. Biochemical properties of arterial low density lipoproteins, very low density lipoproteins and high density lipoproteins, *Exp. Mol. Pathol.* **30:**144–158.
31. Hoover, G. A., McCormick, S., and Kalant, N., 1988, Interaction of native and cell-modified low density lipoprotein with collagen gel, *Arteriosclerosis* **8:**525–534.

32. Hunter, J. A., Shahrokh, Z., Forte, T. M., and Nichols, A. V., 1982, Aggregation of low density lipoproteins with unilamellar phosphatidylcholine vesicles, *Biochem. Biophys. Res. Commun.* **105:**828–834.
33. Jialal, I., and Chait, A., 1989, Differences in the metabolism of oxidatively modified low density lipoprotein and acetylated low density lipoprotein by human endothelial cells: Inhibition of cholesterol esterification by oxidatively modified low density lipoprotein, *J. Lipid Res.* **30:**1561–1568.
34. Joris, I., and Majno, G., 1977, Atherosclerosis and inflammation, *Adv. Exp. Med. Biol.* **104:**227–238.
35. Joris, I., Zand, T., Nunnary, J. L., Krolikowski, F. J., and Majno, G., 1983, Studies on the pathogenesis of atherosclerosis. I. Adhesion and emigration of mononuclear cells in the aorta of hypercholesterolemic rats, *Am. J. Pathol.* **113:**341–358.
36. Joris, I., Billingham, M.-E., and Majno, G., 1984, Human coronary arteries: An ultrastructural search for the early changes of atherosclerosis, *Fed. Proc.* **43:**710.
37. Katz, S. S., and Small, D. M., 1980, Isolation and partial characterization of the lipid phases of human atherosclerotic plaques, *J. Biol. Chem.* **255:**9573–9758.
38. Khoo, J. C., Miller, E., McLoughlin, P., and Steinberg, D., 1988, Enhanced macrophage uptake of low density lipoprotein after self-aggregation, *Arteriosclerosis* **8:**348–358.
39. Kruth, H. S., 1984, Filipin-positive, oil red O negative particles in atherosclerotic lesions induced by cholesterol feeding, *Lab. Invest.* **50:**87–98.
40. Kumar, V. V., Malewicz, B., and Baumann, W. J., 1989, Lysophosphatidylcholine stabilizes small unilamellar phosphatidylcholine vesicles. Phosphorus-31 NMR evidence for the "wedge" effect, *Biophys. J.* **55:**789–792.
41. Lupu, F., Danaricu, I., and Simionescu, N., 1986, Endothelial cell-derived foam cells in experimental atherosclerosis: A physical, cytochemical, and ultrastructural study, *4th Int. Symp. Biol. Vasc. Endoth. Cell*, Noordwijkerhout, Holland, Abstr. Vol., p. 124.
42. Lupu, F., Danaricu, I., and Simionescu, N., 1987, The development of intracellular lipid deposits in the lipid-laden cells of the atherosclerotic lesions: A cytochemical and ultrastructural study, *Atherosclerosis* **67:** 127–142.
43. Mahley, R. W., 1982, Atherogenic hyperlipoproteinemia. The cellular and molecular biology of plasma lipoproteins altered by dietary fat and cholesterol, *Med. Clin. North Amr.* **66:**375–402.
44. Majno, G., 1985, Atherosclerosis: New Horizons, *Hum. Pathol.* **16:**3–5.
45. Mazzone, T., Jensen, M., and Chait, A., 1983, Human arterial wall cells secrete factors that are chemotactic for monocytes, *Proc. Natl. Acad. Sci. USA* **80:**5094–5097.
46. Molenaar, R., Visser, W. J., Verkerk, A., Koster, J. F., and Jongkind, J. F., 1989, Peroxidative stress and in vitro ageing of endothelial cells increases the monocyte–endothelial cell adherence in a human in vitro system, *Atherosclerosis* **76:**193–202.
47. Mora, R., Lupu, F., and Simionescu, N., 1987, Prelesional events in atherogenesis. Colocalization of apolipoprotein B, unesterified cholesterol and extracellular phospholipid liposomes in the aorta of hyperlipidemic rabbit, *Atherosclerosis* **67:**143–154.
48. Mora, R., Lupu, F., and Simionescu, N., 1989, Cytochemical localization of beta-lipoproteins and their components in successive stages of hyperlipidemic atherogenesis of rabbit aorta, *Atherosclerosis* **79:**183–195.
49. Mora, R., Simionescu, M., and Simionescu, N., 1990, Purification and partial characterization of extracellular liposomes isolated from the hyperlipidemic rabbit aorta, *J. Lipid Res.* **31:**1793–1807.
50. Morin, R. J., and Peng, S.-K., 1989, The role of cholesterol oxidation products in the pathogenesis of atherosclerosis, *Ann. Clin. Lab. Sci.* **19:**225–237.
51. Morton, R. E., West, G. A., and Hoff, H. F., 1986, A low density-lipoprotein-sized particle isolated from human atherosclerotic lesions is internalized by macrophages via a non-scavenger receptor mechanism, *J. Lipid Res.* **27:**1124–1134.
52. Munro, J. M., and Cotran, R. S., 1988, The pathogenesis of atherosclerosis: Atherogenesis and inflammation, *Lab. Invest.* **53:**249–261.
53. Navab, M., Hough, G. P., Berliner, J. A., Frank, J. A., Fogelman, A. M., Haberland, M. E., and Edwards, P. A., 1986, Rabbit beta-migrating very low density lipoprotein increases endothelial macromolecular transport without altering electrical resistance, *J. Clin. Invest.* **78:**389–397.
54. Nistor, A., Bulla, A., Filip, D. A., and Radu, A., 1987, The hyperlipidemic hamster as a model of experimental atherosclerosis, *Atherosclerosis* **68:**159–173.
55. Palinski, W., Rosenfeld, M. E., Yla-Herttuala, S., Gurtner, G. C., Socher, S. S., Butler, S. W., Parthasarathy, S., Carew, T. E., Steinberg, D., and Witzum, J. L., 1989, Low density lipoprotein undergoes oxidative modification in vivo, *Proc. Natl. Acad. Sci. USA* **86:**1372–1376.

56. Parthasarathy, S., Quinn, M. T., Schwenke, D. C., Carew, T. E., and Steinberg, D., 1989, Oxidative modification of beta-very low density lipoprotein. Potential role in monocyte recruitment and foam cell formation, *Arteriosclerosis* **9**:398–404.
57. Pirisino, R., DiSimplicio, P., Ignesti, G., Bianchi, G., and Barbera, P., 1988, Sulfhydryl groups and peroxidase-like activity of albumin as scavenger of organic peroxides, *Pharmacol. Res. Commun.* **20**: 545–552.
58. Reidy, M. A., 1985, A reassessment of endothelial injury and arterial lesion formation, *Lab. Invest.* **53**: 513–520.
59. Rodgers, G. M., Kane, W. H., and Pitas, R. E., 1988, Formation of factor Va by atherosclerotic rabbit aorta mediates factor Xa-catalyzed prothrombin activation, *J. Clin. Invest.* **81**:1911–1919.
60. Rosenfeld, M. E., Tsukada, T., Gown, A. M., and Ross, R., 1987, Fatty streak initiation in Watanabe Heritable Hyperlipemic and comparably hypercholesterolemic fat-fed rabbits, *Arteriosclerosis* **1**:9–23.
61. Rosenfeld, M. E., Chait, A., Bierman, E. L., King, W., Goodwin, P., Walden, C. E., and Ross, R., 1988, Lipid composition of aorta of Watanabe Heritable Hyperlipemic and comparably hypercholesterolemic fat-fed rabbits. Plasma lipid composition determines aortic lipid composition of hypercholesterolemic rabbits, *Arteriosclerosis* **8**:338–347.
62. Ross, R., 1986, The pathogenesis of atherosclerosis: An update, *N. Engl. J. Med.* **314**:488–500.
63. Shaikh, M., Martini, S., Quiney, J. R., Baskerville, P., La Ville, A. E., Browse, N. L., Duffield, R., Turner, P. R., and Lewis, B., 1988, Modified plasma-derived lipoproteins in human atherosclerotic plaques, *Atherosclerosis* **69**:165–172.
64. Sima, A., Bulla, A., and Simionescu, N., 1990, Experimental obstructive coronary atherosclerosis in the hyperlipidemic hamster, *J. Submicrosc. Cytol. Pathol.* **22**:1–16.
65. Simionescu, N., Vasile, E., Lupu, F., Popescu, G., and Simionescu, M., 1986, Prelesional events in atherogenesis. Accumulation of extracellular cholesterol-rich liposomes in the arterial intima and cardiac valves of the hyperlipidemic rabbit, *Am. J. Pathol.* **123**:109–125.
66. Simionescu, N., 1988, Prelesional changes of arterial endothelium in hyperlipoproteinemic atherogenesis, in: *Endothelial Cell Biology in Health and Disease* (N. Simionescu and M. Simionescu, eds.), Plenum Press, New York, pp. 385–429.
67. Simionescu, N., Mora, R., Vasile, E., Lupu, F., Filip, D. A., and Simionescu, M., 1990, Prelesional modifications of the vessel wall in the hyperlipidemic atherogenesis: Extracellular accumulation of modified and reassembled lipoproteins, *Ann. N.Y. Acad. Sci.* **598**:1–16.
68. Smith, E. B., and Staples, E. M., 1980, Distribution of plasma proteins across the human aortic wall. Barrier functions of endothelium and internal elastic lamina, *Atherosclerosis* **37**:579–592.
69. Steinberg, D., Parthasarathy, S., Carew, T. E., Khoo, J. C., and Witztum, J. L., 1989, Beyond cholesterol. Modifications of low density lipoprotein that increase its atherogenicity, *N. Engl. J. Med.* **320**:915–924.
70. Suits, A. G., Chait, A., Aviram, M., and Heinecke, J. W., 1989, Phagocytosis of aggregated lipoprotein by macrophages: Low density lipoprotein receptor-dependent foam cell formation, *Proc. Natl. Acad. Sci. USA* **86**:2713–2717.
71. Vasile, E., Simionescu, M., and Simionescu, N., 1983, Visualization of the binding, endocytosis and transcytosis of low density lipoproteins in the arterial endothelium in situ, *J. Cell Biol.* **96**:1677–1689.
72. Vasile, E., and Simionescu, N., 1985, Transcytosis of low density lipoprotein through vascular endothelium, in: *Glomerular Dysfunction and Biopathology of the Vascular Wall* (E. Seno, A. L. Copley, M. A. Venkatachalam, Y. Yamashida, and T. Tsujii, eds.), Academic Press, New York, pp. 87–102.
73. Vasile, E., Antohe, F., Simionescu, M., and Simionescu, N., 1989, Transport pathways of beta-VLDL by aortic endothelium of normal and hypercholesterolemic rabbits, *Atherosclerosis* **75**:77–109.
74. Yla-Herttuala, S., Solakivi, T., Hirvonen, J., Laaskonen, H., Mottonen, M., Pesonen, E., Raekallio, J., Akerblom, H. K., and Nikkari, T., 1987, Glycosaminoglycans and apolipoproteins B and A-I in human aortas. Chemical and immunological analysis of lesion-free aortas from children and adults, *Arteriosclerosis* **7**:333–340.
75. Yla-Herttuala, S., Jaakkola, O., Enholm, C., Tikkanen, M. J., Solaviki, T., Sarkioja, T., and Nikkari, T., 1988, Characterization of two lipoproteins containing apolipoproteins B and E from lesion-free human aortic intima, *J. Lipid Res.* **29**:563–572.
76. Yla-Herttuala, S., Palinski, W., Rosenfeld, M. E., Parthasarathy, S., Carew, T. S., Butler, S., Witztum, J. L., and Steinberg, D., 1989, Evidence for the presence of oxidatively modified low density lipoprotein in atherosclerotic lesions of rabbit and man, *J. Clin. Invest.* **84**:1086–1095.

V

Endothelial Cell in Diabetes

21

Brain Capillary Endothelial Transport of Insulin

William M. Pardridge

I. INTRODUCTION

The most immediate short-term action of insulin on cells is the increase in plasma membrane glucose transporter activity in such tissues as fat, skeletal muscle, and heart.[1] Although insulin does not cause a short-term increase in glucose transport into brain,[2,3] it does have important long-term actions on brain at the physiologic concentrations occurring in plasma, i.e., 1–10 ng/ml (25–250 μU/ml). For example, physiologic concentrations of insulin increase the glutamate-induced firing rate of cholinergic retinal neurons in tissue culture,[4] and increase the coupling across nascent synapses of superior cervical ganglion cells or sympathetic neurons in primary tissue culture.[5,6] *In vivo* studies have shown that the hypothalamic injection of picomolar quantities of insulin increases the firing rate of rat sympathetic nerves innervating brown fat,[7] and that the chronic intraventricular infusion of insulin in baboons enhances the suppressive effect of intravenous cholecystokinin-8 on meal size.[8]

Given the above findings that physiologic concentrations of insulin have profound effects on the central nervous system (CNS) and the finding that insulin is present in both brain and cerebrospinal fluid (CSF) at concentrations in excess of those expected for nonspecific diffusion into brain from blood, an important question relates to the origin of intracerebral insulin. This origin could either be (1) *de novo* synthesis in brain or (2) receptor-mediated transport of insulin into brain from blood. Northern blot studies have shown that insulin mRNA transcripts are not detectable in brain.[9] Therefore, the major origin of intracerebral insulin is transport from blood. This chapter will review the data available in the literature describing the transport mechanisms mediating the brain uptake of insulin from blood. Prior to this discussion, the chapter will provide brief overviews of the dual barrier systems in brain, of receptor-mediated transport of peptides through the blood–

William M. Pardridge • Department of Medicine, Division of Endocrinology, and Brain Research Institute, UCLA School of Medicine, Los Angeles, California 90024-1682.

Endothelial Cell Dysfunctions, edited by Nicolae Simionescu and Maya Simionescu. Plenum Press, New York, 1992.

brain barrier, and of the model that two different populations of insulin receptors exist in brain.

II. TWO BARRIER SYSTEMS IN BRAIN: THE BLOOD–BRAIN BARRIER AND BLOOD–CSF BARRIER

There are two barriers in brain that separate brain extracellular fluids from circulating blood (Fig. 1). The blood–brain barrier (BBB) is formed by the brain capillary wall and arises from epithelial-like tight junctions which cement cerebral microvascular endothelia together. The blood–CSF barrier is formed by tight junctions within the ventricular ependymal epithelium at circumventricular organs (CVOs) of brain such as the choroid plexus and a half-dozen other tiny areas of brain surrounding the ventricles.[10] The CVOs are perfused by capillaries that are highly porous and lack a BBB (Fig. 1). Whereas the BBB separates brain parenchymal interstitial fluid from blood, the blood–CSF barrier separates CSF from blood.[11] A general misconception is that although a substance in blood may not be transported across the brain capillary barrier, this substance may enter brain parenchymal interstitial fluid by transport into CSF followed by free diffusion from the ventricular surface. Although there is no anatomical barrier separating the interstitial fluid deep within brain parenchyma from the surface of the brain, there is a functional barrier. This functional

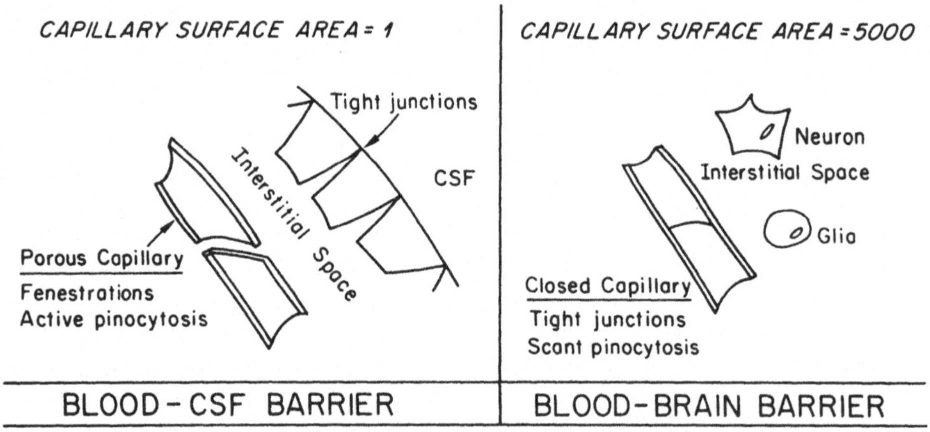

Figure 1. The blood–brain barrier (BBB) and the blood–CSF barrier are shown on the right and the left, respectively, and comprise the two major membrane systems separating brain extracellular fluid from the circulating plasma. The blood–CSF barrier is found at the circumventricular organs (CVOs),[10] wherein capillaries are porous and allow for rapid distribution of circulating peptides such as insulin and other small molecules into the interstitial space. The presence of low-resistance tight junctions on the ventricular side of the ependymal cells impedes distribution of circulating substances into the CSF in the CVOs.[11] Conversely, capillaries perfusing the vast area of brain are not porous and are characterized by high-resistance epithelial-like tight junctions and scant pinocytosis.[11] Since the surface area of the BBB is 5000-fold greater than the surface area of the blood–CSF barrier,[12] the rapid distribution of circulating peptides into brain interstitial space is dependent upon the transport properties at the brain capillary barrier and not at the choroid plexus or other CVOs. In turn, the uptake of insulin into the CSF compartment is regulated by insulin transport systems within the ependymal epithelia at the blood–CSF barrier.[13]

barrier arises from the fact that the total CSF volume in man of 140 ml is turned over with a half-time of approximately 3 hr.[14] In rats, the CSF volume of approximately 100 μl is turned over with a half-time of approximately 30 min.[15] As CSF is constantly produced at the choroid plexus epithelium, it is drained out of brain into the systemic circulation via absorption through the arachnoid villi at the superior sagittal sinus. These bulk flow pathways are manyfold faster than the rates of solute or hormone diffusion into brain parenchyma.[16] Thus, while cells at the surface of the brain may be exposed to solute within the ventricular compartment, cells deep within the brain parenchyma will be exposed to solute concentrations log orders lower than those found at the surface of the brain. The finding that drug, solute, or hormone is transported into CSF from blood means that this compound is transportable across the blood–CSF barrier but not necessarily transportable across the BBB. For example, active thymidine transport systems are present in the choroid plexus and allow for thymidine transport into CSF from blood,[17] but a thymidine transport system is absent in the BBB.[18,19]

The studies of Woods, Porte, and their co-workers[20,21] have shown that insulin transport systems are present in the blood–CSF barrier and that this transport system is saturable. Following initial studies showing that CSF and blood insulin concentrations correlate in the dog,[20] it has subsequently been shown that insulin is transportable into the CSF in man.[21] For example, under basal conditions, the CSF insulin content is 7.5% of the plasma insulin concentration in humans.[21] This ratio is severalfold higher than the CSF/plasma ratio expected if insulin was available for transport to the CSF only via nonspecific diffusion pathways.[13] Moreover, the concentration of insulin in CSF rises more than 3-fold with a 20-fold increase in blood insulin concentration.[21] Under these conditions, however, the CSF/plasma ratio is only 1.0%.[21] The inverse relationship between the CSF/plasma ratio and the plasma insulin content indicates that the blood–CSF barrier insulin transport system is saturable. Similarly, the transport of insulin into the CSF compartment of rats is saturable. When lean and obese Zucker rats are compared, the plasma level is increased nearly 10-fold in the obese animals but only 5-fold in the CSF compartment.[22] Thus, the CSF/plasma ratio in the obese animals of 0.9% is one-half that of the lean littermates at 1.8%. The CSF/plasma ration in lean Wistar rats is 1.2%. The finding that the basal CSF/plasma ratio in rats is one-sixth the value found in humans suggests either that the activity of the blood–CSF barrier insulin transport system in rats is much less than that found in humans, or that the rate of bulk flow of CSF out of the rat brain is much faster than that of the human brain. The presence of insulin transport systems at the blood–CSF barrier allows for increased insulin delivery to cells near the ventricular surface. For example, immunoreactive cells have been localized near the CSF surface in rat brain.[23]

As reviewed in subsequent sections of this chapter, insulin transport systems in brain are believed to be identical to the insulin receptor in peripheral tissues. Support for this model is the finding of high concentrations of insulin receptor both in the choroid plexus and in brain microvessels relative to insulin receptor content in neurons and glia.[24–26] In addition to being a repository for receptor transport systems for peptides such as insulin or insulinlike growth factors (IGFs), the choroid plexus ependyma have also been shown to be the major, if not the only, site of synthesis of IGF-II in rat brain.[27] The concentration of IGF-II mRNA transcripts in brain is higher than in any other tissue in the adult rat, but *in situ* hybridization studies have shown that IGF-II is made exclusively in the choroid plexus.[27] This is analogous to choroid plexus synthesis of transport proteins in high concentrations such as prealbumin or transferrin.[28] Thus, the high concentrations in CSF of proteins such as transferrin, prealbumin, or IGF-II may reflect synthesis and local choroid plexus secretion of these hormones and/or specific transport across the blood–CSF barrier.

III. OVERVIEW OF RECEPTOR-MEDIATED TRANSPORT OF PEPTIDES THROUGH THE BBB

Water-soluble molecules such as insulin are not transportable through the BBB by free diffusion owing to the absence of pores in this microvascular barrier.[11] However, some peptides such as insulin are capable of traversing the BBB via receptor-mediated transcytosis mechanisms.[29–31] This process is believed to be composed of three components: (1) receptor-mediated endocytosis of peptide at the blood or luminal border of the BBB, (2) diffusion through the 0.3 μm of endothelial cytoplasm, and (3) receptor-mediated exocytosis of the peptide into brain interstitial fluid.[30] In an analogous manner, some polycationic proteins such as cationized albumin, cationized immunoglobulin G, or histone are capable of transport through the BBB via absorption-mediated transcytosis mechanisms.[32–34] The latter processes represent interactions between the protein and membrane components of an electrostatic nature and, like receptor-mediated transcytosis, are highly specific, saturable, and subject to competitive inhibition. The known peptides or plasma proteins that have receptors on brain capillaries are listed in Table 1. Not all of the receptors shown in Table 1 may be transport systems. However, morphologic or physiologic evidence has been previously reported that indicates that the group of peptides capable of receptor- or absorptive-mediated transport through the BBB includes at least insulin,[31] transferrin,[35] cationized albumin,[32] cationized immunoglobulin G,[33] and histone.[34]

The model of receptor- or absorption-mediated transcytosis of peptides or plasma proteins across the brain capillary barrier is analogous to processes that have been shown to take place in peripheral tissues. For example, blood-borne molecules such as insulin or albumin undergo receptor-mediated transcytosis across myocardial capillaries,[41,42] prolactin undergoes receptor-mediated transcytosis across choroid plexus epithelium,[43] and

Table 1. Brain Capillary Peptide or Plasma Protein Binding Sites

Brain capillary species	Ligand	Ref.	K_D (nM)	B_{max} (pmoles/mg_p)
Human	Transferrin	36	5.6 ± 1.4	0.10 ± 0.02
	Insulin	29	1.2 ± 0.5	0.17 ± 0.08
	IGF-I	37	2.1 ± 0.4	0.17 ± 0.02
	IGF-II	37	1.1 ± 0.1	0.21 ± 0.01
Bovine	Atriopeptin	38	0.11	0.058
	Insulin	39	0.44	0.18
	IGF-II	40	1.8	1.0
	IGF-I	40	2.0	1.7
	Cationized albumin	32	81 ± 12	79 ± 10
	Cationized IgG	33	900 ± 370	1400 ± 400
	Histone	34	15,200 ± 2800	7700 ± 1000

insulin undergoes receptor-mediated transport across monolayers of cultured bovine aortic endothelium.[44] The presence of insulin receptor transport systems in the microcirculation of various tissues was anticipated by the early studies of Rasio,[45] which showed that insulin is sequestered in a reversible manner by microvascular endothelium and that the capillary wall acts as barrier to insulin transport into muscle but not into organs, such as liver, that have highly porous microvessels.

The study of BBB peptide receptors and receptor transport systems has been performed by two fundamentally different methodologic approaches, one *in vitro* and one *in vivo*. The characterization of a receptor and its saturable binding sites in the BBB can be demonstrated *in vitro* using isolated brain microvessels as an *in vitro* model system of the brain capillary (Fig. 2).[13,30] As shown in Fig. 3, [^{125}I]insulin is bound by isolated human brain microvessels in a manner that is saturable by high concentrations of unlabeled insulin or proinsulin, but not by unrelated peptides.[29] These saturation data can be analyzed with

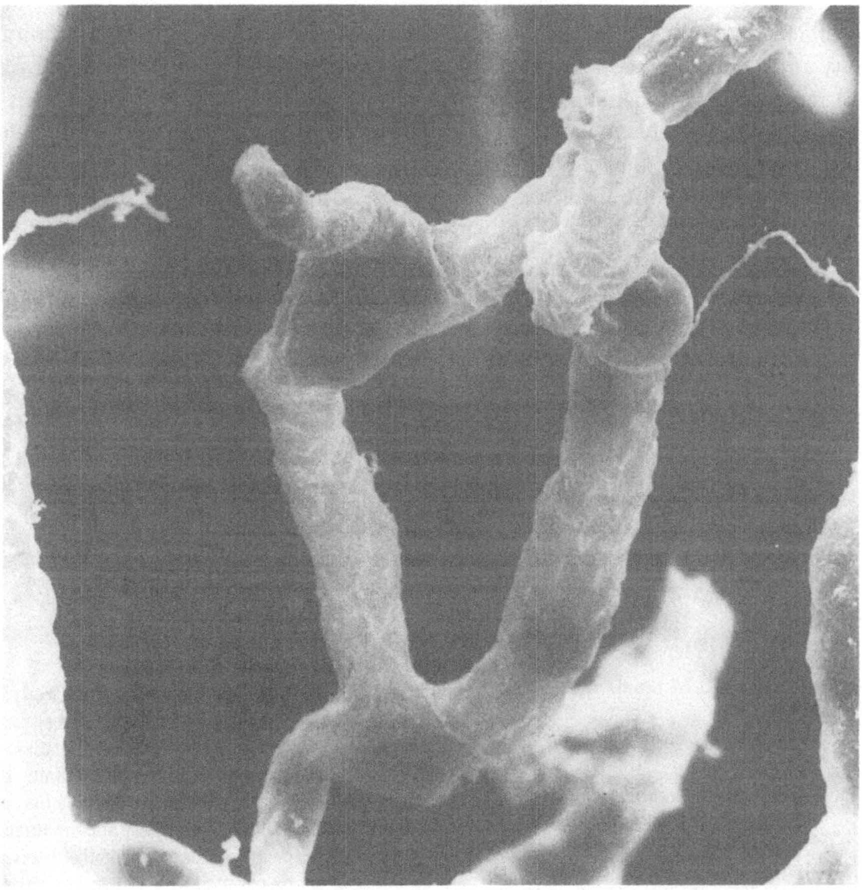

Figure 2. Scanning electron micrograph of isolated bovine brain capillaries shows the capillaries to be relatively free of brain tissue but to be intact and encased by a basement membrane.

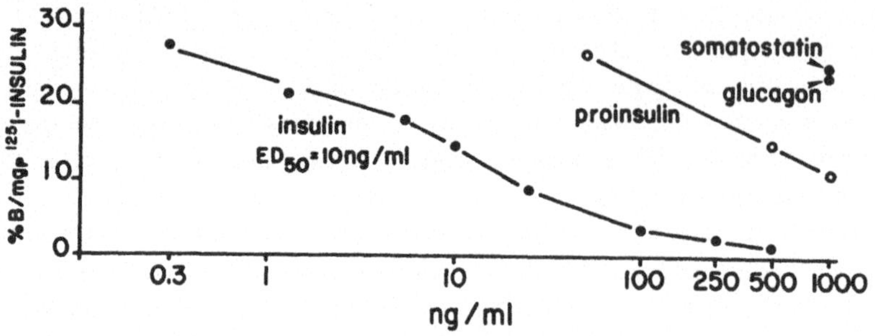

Figure 3. [^{125}I]Insulin binding to human brain capillaries is measured in the presence of various concentrations of unlabeled bovine insulin, porcine proinsulin, somatostatin, or glucagon.[29] The concentration (ED_{50}) of unlabeled insulin that results in a 50% decrease in specific insulin binding is 10 ng/ml.

classical Scatchard plots, and these findings are consistent with a single high-affinity, low-capacity and a single low-affinity, high-capacity binding site (Fig. 4). The first arm of the overall receptor-mediated transcytosis process, i.e, receptor-mediated endocytosis, may also be studied *in vitro* using isolated brain capillaries.[29] A mild acid, salt wash technique has been employed to demonstrate endocytosis of ligands and, as shown in Fig. 5, approximately 75% of the insulin bound by isolated human brain microvessels at 37°C represents internalized insulin. The acid wash technique has been shown to release molecules bound to the exterior of cells but not the release of intracellular molecules, even those as small as 3-*O*-methylglucose.[46] However, the overall transcytosis process for insulin movement through the BBB cannot be demonstrated with isolated capillaries *in vitro*,[40] and must be demonstrated *in vivo* with either physiologic or morphologic techniques (see below).

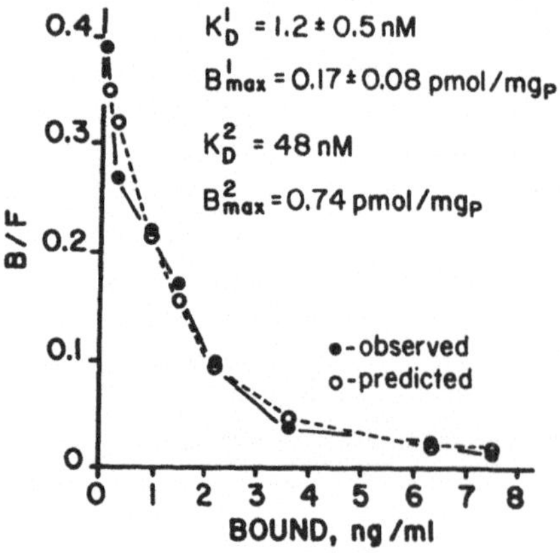

Figure 4. The insulin saturation data from Fig. 3 were analyzed with a two-site Scatchard analysis using a nonlinear regression least-squares fitting procedure.[29] The dissociation constant and binding capacity of the high-affinity site are noted as K_D^1 and B_{max}^1, respectively. The dissociation constant and binding capacity of the low-affinity site are represented by K_D^2 and B_{max}^2, respectively. Experimentally observed data shown in Fig. 3 are represented by the closed circles and the open circles are predicted by fitting the data to the two-site model.

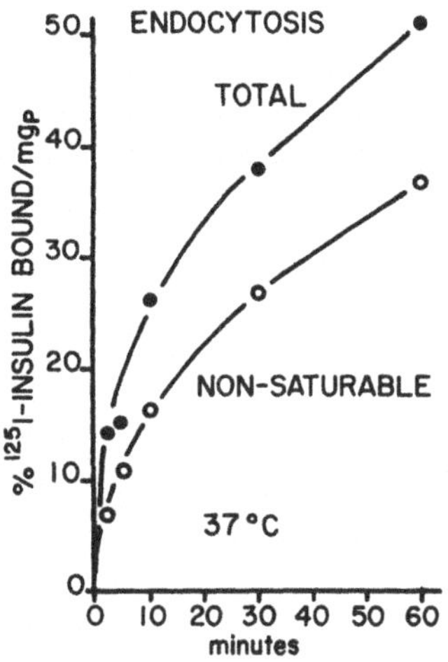

Figure 5. The total and nonsaturable binding of [^{125}I]insulin to human brain capillaries at 37°C rises with time for at least 60 min of incubation.[29] The nonsaturable binding represents internalized insulin and is approximately 75% of the total binding at the physiologic temperature. The nonsaturably bound [^{125}I]insulin represents internalized insulin since previous studies showed that the portion of insulin bound to isolated human brain microvessels that was resistant to a mild acid, salt wash was identical to the nonsaturable component of total insulin binding.[29]

IV. TWO INSULIN RECEPTORS IN BRAIN: CAPILLARY ENDOTHELIAL VERSUS BRAIN CELL RECEPTORS

The insulin receptor is a tetrameric structure of approximately 440,000 molecular weight comprised of two α subunits of approximately 130,000 molecular weight and two β subunits of approximately 90,000 molecular weight.[47] The α subunit represents the insulin binding site and the β subunit represents the tyrosine kinase activity of the insulin receptor.[48] The molecular weight of the α subunit may be measured using affinity cross-linking reagents, [^{125}I]insulin, and SDS–PAGE. As shown in Fig. 6, the molecular weight of the α subunit of either the human BBB insulin receptor or IGF receptor has been measured using disuccinimidyl suberate as the affinity cross-linking reagent.[29,37] These studies show that the molecular weight of the α subunit of the BBB insulin receptor is approximately 135,000, which is equivalent to 130,000 when the weight of the insulin peptide is subtracted from the α subunit-insulin complex. The BBB insulin receptor, like insulin receptors in peripheral tissues, is bound by a wheat germ agglutinin (WGA) affinity column and is eluted with high concentrations of *N*-acetylglucosamine (NAGA), as shown in Fig. 7.

Earlier studies of the insulin receptor in brain led to conflicting results in regard to whether the brain insulin receptor binds to a WGA column and whether the molecular weight of the brain insulin receptor is on the order of 130,000 to 135,000, which is found in peripheral tissues, or is a smaller molecule with a truncated carbohydrate moiety with a molecular weight of 115,000–120,000. These discrepancies were subsequently resolved in 1985 when two different groups independently provided evidence that there are two populations of insulin receptors in brain.[29,51] One population at the BBB is typical of insulin receptors in fat or liver with a molecular weight of the α subunit of 130,000–135,000 and high affinity for WGA affinity columns.[29,51,55] The second receptor with a lower molecular weight of the α subunit and poor binding to WGA columns, owing to a truncated carbohydrate moiety, is found in neurons and glia.[49–55] One study suggests that differences

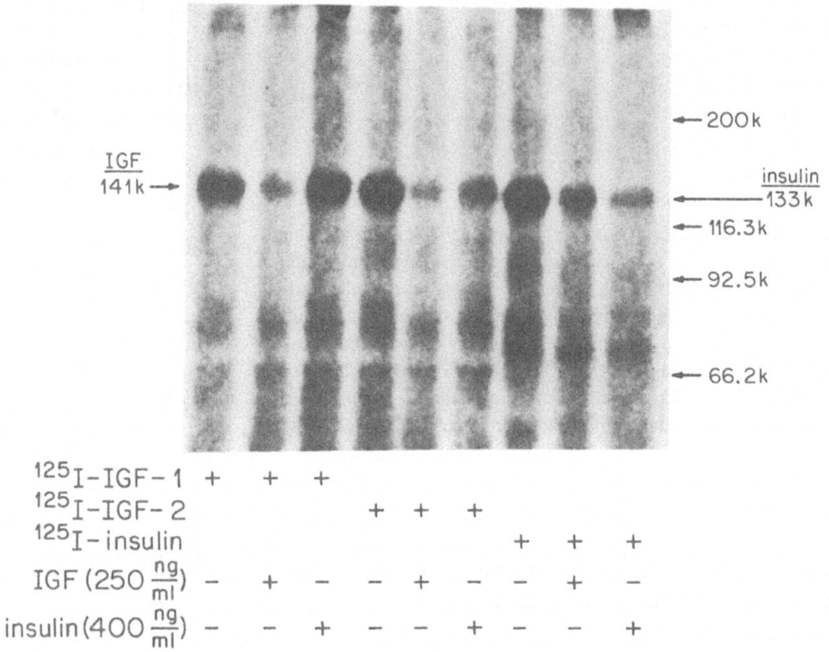

Figure 6. Affinity cross-linking of [^{125}I]insulin, [^{125}I]insulinlike growth factor (IGF)-I, and [^{125}I]-IGF-II to isolated human brain capillaries using disuccinimidyl suberate (DSS).[37] The capillaries were incubated with labeled hormone in the presence of either no additive, 250 ng/ml IGF, or 400 ng/ml porcine insulin. The migration of molecular weight markers is shown on the right-hand side of the figure, together with the molecular weight of the insulin binding site (133k). The molecular weight of the IGF binding site (141k) is shown on the left-hand side of the figure. The solubilized affinity cross-linked capillary proteins were separated on a 7.5% SDS–polyacrylamide gel before slab drying and autoradiography. The binding of [^{125}I]-IGF-I was nearly completely displaced by 250 ng/ml IGF, but there was no visible displacement by 400 ng/ml insulin. The binding of [^{125}I]-IGF-II was nearly completely displaced by 250 ng/ml IGF, and was partially displaced by 400 ng/ml insulin. The binding of [^{125}I]insulin was nearly completely displaced by 400 ng/ml insulin and approximately 50% displaced by 250 ng/ml IGF. The IGF mixture was equally comprised of unlabeled IGF-I and IGF-II.[37]

in lots or commercial sources of WGA–Sepharose explain the conflicting reports regarding binding of the brain insulin receptor to the affinity resin.[56] However, this factor would appear to be of less significance. The data in Fig. 7 show the marked differences in affinity of the BBB versus the brain cell insulin receptor using the same batch of affinity resin.

Depending upon the method used to prepare brain membranes, one or both of the brain insulin receptors will be detected. For example, if a homogenate of brain is made and centrifuged at low speed to remove the nuclear pellet and subsequent membranes are obtained from the low-speed supernatant, then the brain microvascular insulin receptor will not be detected since microvessels sediment with nuclei in the initial low-speed pellet.[29] These methodologic differences explain conflicting reports. For example, the molecular weight of the α subunit of the human brain insulin receptor reported in one study is 118,000,[57] whereas in another study that described the purification of the human brain insulin receptor with an initial WGA affinity column, the molecular weight of the α subunit

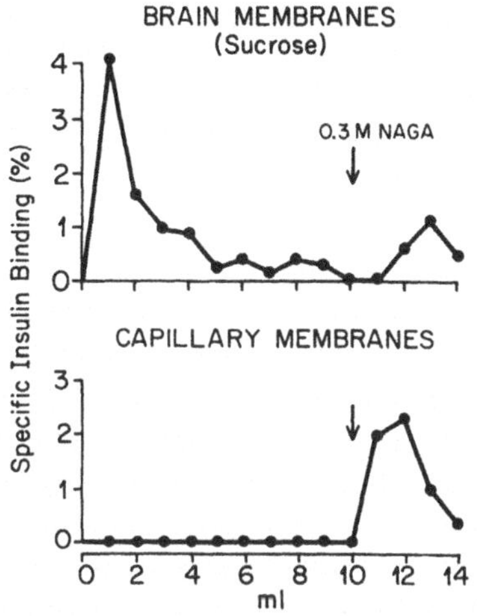

Figure 7. Binding of [^{125}I]insulin to bovine capillary-depleted brain synaptosomal membranes (top) or bovine brain capillary plasma membranes (bottom) following elution from an affinity column using wheat germ agglutinin (WGA)–Sepharose 6MB (Sigma Chemical Co., St. Louis, Mo.). To prepare the synaptosomal membranes, bovine brain was homogenized in isotonic sucrose and capillary-depleted synaptosomal membranes were prepared as described previously.[68] Capillary plasma membranes were isolated from bovine brain capillaries using methods described previously.[29] Triton-solubilized membranes were recycled three times through a WGA–Sepharose column and the column was then eluted sequentially with 10 × 1 ml of HEPES-buffered saline (0.1% Triton X-100) and 4 × 1 ml of 0.3 M *N*-acetylglucosamine (NAGA) as described previously.[29] The individual fractions were dialyzed and divided into two equal aliquots. [^{125}I]-Insulin binding to each aliquot was measured in the absence or presence of 10 μg/ml bovine insulin and the difference provided a measure of specific insulin biding for each fraction off the column. The results show that all of the brain capillary or BBB insulin receptor specifically elutes from the WGA column in the presence of NAGA, whereas virtually all of the synaptosomal insulin receptor does not specifically bind to the WGA column.

is 135,000.[58] Thus, the former study selectively measured the neuronal–glial insulin receptor, whereas the latter study selectively purified the microvascular receptor. The difference between the microvascular and brain cell insulin receptors probably resides in the carbohydrate moiety and not in the polypeptide portion. Posttranslational modifications in the carbohydrate portion of the insulin receptor likely account for the differences in molecular weight of the brain cell versus the microvascular receptor.[53] Interestingly, human glial tumors have an insulin receptor that is comparable in size to the microvascular receptor, suggesting altered posttranslational modification of the glial insulin receptor associated with malignancy.[57]

The structural differences between the brain microvascular versus the brain cell insulin receptor also hold for the type I IGF receptor. The molecular weight of the bovine BBB IGF-I receptor α subunit is 132,000.[59] Conversely, using the capillary-depleted brain cell membranes described in Fig. 7, studies have shown that the molecular weight of the brain cell IGF-I receptor α subunit is 114,000.[59] Whether the brain microvascular and brain cell IGF-I receptors show differential binding to WGA is not known at present. These structural differences between the microvascular and brain cell insulin or IGF-I receptors may have functional consequences. For example, the microvascular receptors are involved in transcytosis processes, whereas the brain cell receptors may only endocytose extracellular peptide.

V. BBB INSULIN TRANSPORT

Following the initial reports that intracarotid insulin administration could selectively affect brain function and bind saturably to brain blood vessels,[60,61] isolated brain capillaries were used in 1981 to characterize the BBB insulin receptor.[39] These studies showed for the first time that brain capillaries possessed a classical insulin receptor that demonstrated the same binding kinetics as insulin receptors in peripheral tissues. These initial studies using bovine brain capillaries were subsequently confirmed in a number of different species,[55,62,63] and in 1985, studies of binding and endocytosis with isolated human brain capillaries led to the initial proposal that the BBB insulin receptor was in fact a transport system that mediated the transcytosis of insulin through the brain capillary.[29]

The hypothesis that circulating insulin was transportable through the BBB required *in vivo* confirmation since previous *in vivo* studies in the rat led to the conclusion that insulin was not transported into brain from blood.[64] These conclusions were based on the finding that the uptake of [^{125}I]insulin by rat brain was less than 5% of the corresponding uptake by heart.[64] However, the blood volume of myocardium is more than tenfold greater than the blood volume in brain.[34] Therefore, conclusions regarding the uptake of insulin by brain following peripheral administration cannot be made by comparing brain uptake to that by myocardium. Rather, the uptake of [^{125}I]insulin by brain should be compared to the uptake of a corresponding plasma volume marker such as [^{3}H]albumin. When this is done,[31] it can be shown that brain uptake of [^{125}I]insulin, relative to [^{3}H]albumin, is measurable and that this uptake is saturable by the inclusion of high concentrations of unlabeled insulin in the intracarotid infusate (Fig. 8).

The experiments in Fig. 8 do support the conclusion that blood-borne insulin is transported into brain. However, this conclusion cannot be made with certainty without

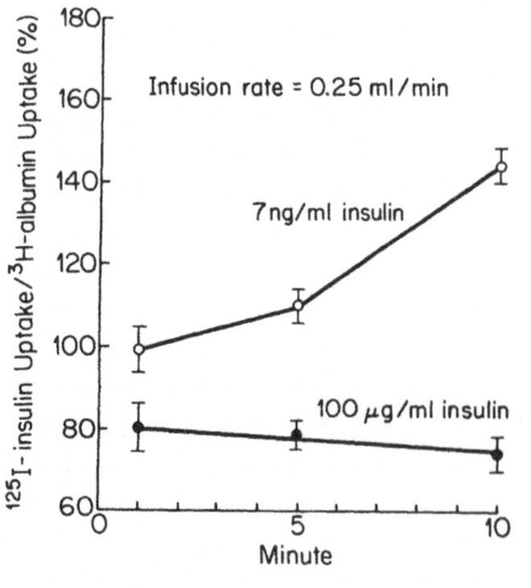

Figure 8. Rabbit brain uptake of [^{125}I]insulin relative to [^{3}H]albumin as a function of duration of common carotid artery infusion of the two labeled proteins.[31] The open circles represent data from rabbits infused with only 7 ng/ml of radiolabeled insulin while the closed circles represent data from animals that were infused with an additional 100 μg/ml of unlabeled insulin. Each point represents the mean and standard error for seven animals.

ancillary biochemical and morphologic studies to address two principal caveats. First, it is possible that blood-borne [^{125}I]insulin is rapidly metabolized to [^{125}I]tyrosine. Since tyrosine is readily transported through the BBB via a neutral amino acid transport system within the barrier,[13] the systemic administration of [^{125}I]insulin could lead to artifactual uptake of ^{125}I radioactivity. However, this possibility was addressed by reverse-phase high-performance liquid chromatography (HPLC) studies showing that there was minimal if any conversion of [^{125}I]insulin to [^{125}I]tyrosine during the 10-min carotid artery infusion in the 1-month-old rabbits used in the studies shown in Fig. 8. Suckling animals were used as model system since it is known that serum proteases which convert peptides such as insulin to iodotyrosine are poorly expressed in developing animals.[31] The second caveat interpreting data such as those shown in Fig. 8 is that the brain uptake of radiolabeled peptide may simply represent sequestration by the vasculature and not actual transcytosis into brain parenchyma.

The differentiation between vascular sequestration of blood-borne peptide and actual transcytosis into the brain interstitial fluid is the crucial element in studies of peptide transport into brain from blood. The differentiation between these two processes can be made either by morphologic techniques or by cell isolation techniques wherein homogenates of brain are depleted of brain capillaries.[65] Thaw-mount autoradiography was chosen as the initial morphologic technique to demonstrate insulin transcytosis into brain,[31] and Fig. 9 shows the darkfield illumination of developing rabbit brain obtained 10 min after an intracarotid infusion of [^{125}I]insulin corresponding to the study shown in Fig. 8. Similarly, thaw-mount autoradiography has been used to demonstrate the transcytosis of blood-borne cationized albumin,[32] or cationized immunoglobulin G,[33] into brain parenchyma following carotid artery infusion of labeled protein. The disadvantage of autoradiography is that it often takes several weeks or months to develop the autoradiograms to allow for a complete assessment of the amount of labeled peptide or protein that enters brain parenchyma as opposed to vascular sequestration. On this basis, an internal carotid artery perfusion/capillary depletion method has recently been developed for quantifying the transcytosis of blood-borne peptides or plasma proteins by rat brain and to allow for quantitative differentiation of this process from simple vascular sequestration of blood-borne labeled peptide or protein.[34,65]

An indirect determination of insulin transport into brain involves the measurement of the brain to plasma insulin ratio. Since insulin is not synthesized in brain,[9] and is derived via the circulation, the brain to plasma insulin ratio or cerebral insulin volume of distribution should parallel BBB transport activity at the brain capillary. As shown in Table 2, the brain to plasma insulin ratio in newborn rabbits is threefold higher than that in adult rabbits.[67] This suggests that the rate of insulin transport across the BBB in newborn animals is severalfold faster than that at the adult rabbit BBB. This was confirmed by measurements of insulin receptor activity in capillaries isolated from either newborn or adult rabbit brain. Moreover, the sequestration of insulin by newborn rabbit brain following intracarotid injection is much greater than that found in adult rabbit brain.[67] A second example is the case of brain insulin content in streptozotocin experimental diabetes mellitus. The brain to plasma insulin ratio is decreased approximately threefold in the diabetic rat brain (Table 2). This decreased volume of distribution suggests that the BBB insulin receptor transport activity is decreased in the diabetic condition. This was confirmed in studies employing capillaries isolated from control and diabetic rat brain showing a

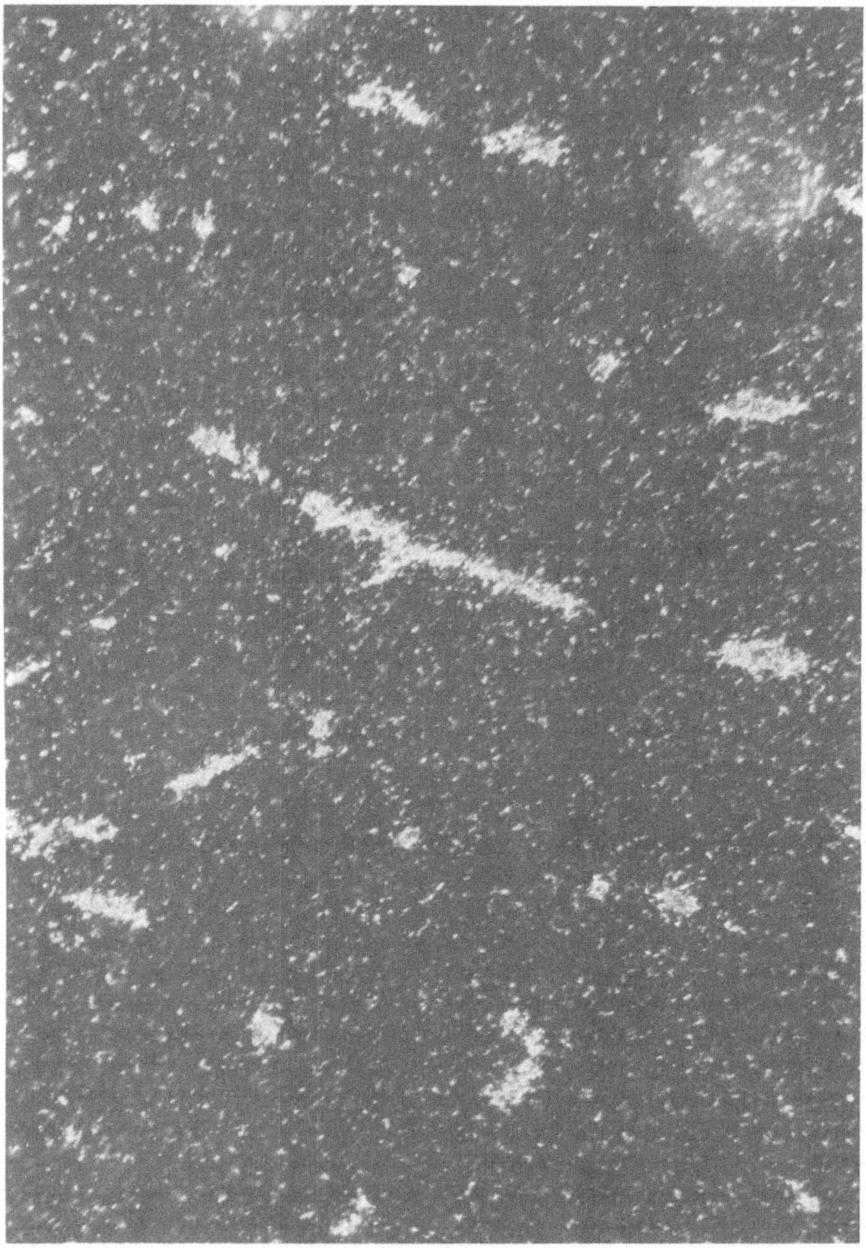

Figure 9. Darkfield illumination of thaw-mount autoradiography of rabbit brain after a 10-min common carotid artery infusion of [^{125}I]insulin shows extensive distribution of silver grains well within the brain parenchyma.[31] This photograph corresponds to the 10-min uptake of the data shown in Fig. 8. 70×.

Table 2. Brain Insulin Concentrations[66,67]

Species	Condition	Plasma insulin (μU/ml)	Brain insulin (μU/g)	Brain/plasma ratio (ml/g)
Rabbit	Newborn	26.2 ± 3.5	17.5 ± 0.9	0.67 ± 0.09
	Adult	42.5 ± 7.1	9.6 ± 3.4	0.22 ± 0.08
Rat	Control	33.4 ± 3.5[a]	2.6 ± 0.4	0.092 ± 0.024
	Diabetic	21.9 ± 2.1[a]	0.32 ± 0.30	0.029 ± 0.011

[a]Serum.

decreased number of insulin transporter receptor sites in capillaries isolated from diabetic rat brain.[66] The brain insulin concentrations reported in Table 2 are comparable to those measured by Baskin *et al.*,[69] but are about 50-fold less than rat brain insulin concentrations originally reported by Havrankova *et al.*[70] Owing to differences in extraction methods,[66,69,70] the measurements of Havrankova *et al.*[70] appear to be confounded by interfering substances, and this may explain their failure to observe decreased brain insulin concentrations in experimental diabetes.[70]

VI. CONCLUSIONS

Based on the observation that insulin mRNA transcripts are not detectable in brain,[9] yet insulin is found in cerebral tissues at relatively high volumes of distribution (Table 2), it is clear that the major, if not entire, origin of brain insulin is via the circulation. The mechanism by which insulin gains access to brain tissue is receptor-mediated transport of the molecule across the brain capillary endothelium which makes up the BBB *in vivo*. Autoradiography studies at the light microscopic level are consistent with the hypothesis that the mechanism by which insulin traverses the BBB is receptor-mediated transcytosis through the brain capillary endothelium.[31] However, this mechanism is provisional at the present time since ultrastructural studies have not been performed to date in regard to insulin transport. However, unlike capillaries in other organs, the microvessels in brain have no pores.[11] Therefore, it is improbable that insulin traverses the BBB via intercellular routes. Further support that peptides traverse the BBB via receptor-mediated transcytosis mechanisms comes from recent ultrastructural studies showing that WGA–horseradish peroxidase conjugates traverse the BBB directly through the endothelial cytoplasm and into brain interstitial fluid via absorption-mediated transcytosis mechanisms.[71] The actual intraendothelial transport organelles involved in the overall endothelial transcytosis process and the intracellular route through which these organelles traverse are unknown at present. Hopefully these issues will be clarified by future ultrastructural studies involving the receptor-mediated transcytosis of insulin and other peptides or plasma proteins through the BBB.

ACKNOWLEDGMENTS. Dawn Brown skillfully prepared the manuscript. Work in the author's laboratory is supported by NIH grant AI-28760 and by the U.S. Army Medical Research and Development Command under Contract No. DAMD17-87-C-7137. Opinions, interpretations, conclusions, and recommendations are those of the author's and are not necessarily endorsed by the U.S. Army.

REFERENCES

1. Pilkis, S. J., and Park, C. R., 1974, Mechanism of action of insulin, *Annu. Rev. Pharmacol.* **14**:365-388.
2. Betz, A. L., Gilboe, D. D., Yudilevich, D. L., and Drewes, L. R., 1973, Kinetics of unidirectional glucose transport into the isolated dog brain, *Am. J. Physiol.* **225**:586–592.
3. Daniel, P. M., Love, E. R., and Pratt, O. E., 1975, Insulin and the way the brain handles glucose, *J. Neurochem.* **25**:471–476.
4. Puro, D. G., and Agardh, E., 1984, Insulin-mediated regulation of neuronal maturation, *Science* **225**:1170–1172.
5. Kessler, J. A., Spray, D. C., Saez, J. C., and Bennett, M. V. L., 1984, Determination of synaptic phenotype: Insulin and cAMP independently initiate development of electrotonic coupling between cultured sympathetic neurons, *Proc. Natl. Acad. Sci. USA* **81**:6235–6239.
6. Wolinsky, E. J., Patterson, P. H., and Willard, A. L., 1985, Insulin promotes electrical coupling between cultured sympathetic neurons, *J. Neurosci.* **5**:1675–1679.
7. Sakaguchi, T., and Bray, G. A., 1987, Intrahypothalamic injection of insulin decreases firing rate of sympathetic nerves, *Proc. Natl. Acad. Sci. USA* **84**:2012–2014.
8. Figlewicz, D. P., Stein, L. J., West, D, Porte, Jr., D., and Woods, S. C., 1986, Intracisternal insulin alters sensitivity to CCK-induced meal suppression in baboons, *Am. J. Physiol.* **250**:R856–R860.
9. Giddings, S. J., Chirgwin, J., and Permutt, M. A., 1985, Evaluation of rat insulin messenger RNA in pancreatic and extrapancreatic tissues, *Diabetologia* **28**:343–347.
10. Weindl, A., 1973, Neuroendocrine aspects of circumventricular organs, in: *Frontiers in Neuroendocrinology* (W. F. Ganong and L. Martini, eds.), Oxford University Press, London, p. 3.
11. Brightman, M. W., 1977, Morphology of blood–brain interfaces, *Exp. Eye Res.* **25**(Suppl.):1–25.
12. Crone, C., 1971, The blood–brain barrier—Facts and questions, in: *Ion Homeostasis of the Brain* (B. K., Siesjo and S. C. Sorenson, eds.), Copenhagen, Munksgaard, p. 52.
13. Pardridge, W. M., 1983, Neuropeptides and the blood–brain barrier, *Annu. Rev. Physiol.* **45**:73–82.
14. Pardridge, W. M., 1989, Strategies for drug delivery through the blood–brain barrier, *Neurobiol. Aging* **10**:636–637.
15. Pardridge, W. M., and Frank, H. J. L., 1983, Mechanisms of peptide transport from blood to brain, in: *Neuroendocrine Perspectives II* (E. E. Muller and R. M. Macleod, eds.), Elsevier/North-Holland, Amsterdam, p. 107.
16. Covell, D. G., Narang, P. K., and Poplack, D. G., 1985, Kinetic model for disposition of 6-mercaptopurine in monkey plasma and cerebrospinal fluid, *Am. J. Physiol.* **248**:R147–R156.
17. Spector, R., and Berlinger, W. H., 1982, Localization and mechanism of thymidine transport in the central nervous system, *J. Neuorchem.* **39**:837–841.
18. Cornford, E. M., and Oldendorf, W. H., 1975, Independent blood–brain barrier transport systems for nucleic acid precursor, *Biochim. Biophys. Acta* **394**:211–219.
19. Terasaki, T., and Pardridge, W. M., 1988, Restricted transport of AZT and dideoxynucleosides through the blood–brain barrier, *J. Infect. Dis.* **158**:630–632.
20. Woods, S. C., and Porte, D., 1977, Relationship between plasma and cerebrospinal fluid insulin levels of dogs, *Am. J. Physiol.* **233**:E331–E334.
21. Wallum, B. J., Taborsky, G. J., Porte, Jr., D., Figelwicz, D. P., Jacobson, L., Beard, J. C., Ward, W. K., and Dorsa, D., 1987, Cerebrospinal fluid insulin levels increase during intravenous insulin infusions in man, *J. Clin. Endocrinol. Metab.* **64**:190–194.
22. Stein, L. J., Dorsa, D. M., Baskin, D. G., Figlewicz, D. P., Ikeda, H., Frankmann, S. P., Greenwood, M. R. C., Porte, Jr., D., and Woods, S. C., 1983, Immunoreactive insulin levels are elevated in the cerebrospinal fluid of genetically obese Zucker rats, *Endocrinology* **113**:2299–2301.
23. Baskin, D. G., Woods, S. C., West, D. B., van Houten, M., Posner, B. I., Dorsa, D. M., and Porte, Jr., D., 1983, Immunocytochemical detection of insulin in rat hypothalamus and its possible uptake from cerebrospinal fluid, *Endocrinology* **113**:1818–1825.

24. Young, W. S., Kuhar, M. J., Roth, J., and Brownstein, M. J., 1980, Radiohistochemical localization of insulin receptors in the adult and developing rat brain, *Neuropeptides* **1**:15–22.
25. Baskin, D. G., Brewitt, B., Davidson, D. A., Corp, E., Paquetta, T., Figlewicz, D. P., Lewellen, T. K., Graham, M. K., Woods, S. G., and Dorsa, D. M., 1986, Quantitative autoradiographic evidence for insulin receptors in the choroid plexus of the rat brain, *Diabetes* **35**:246–249.
26. Werther, G. A., Hogg, A., Oldfield, B. J., McKinley, M. J., Figdor, R., Allen, A. M., and Mendelsohn, F. A. O., 1987, Localization and characterization of insulin receptors in rat brain and pituitary gland using in vitro autoradiography and computerized densitometry, *Endocrinology* **21**:1562–1570.
27. Hynes, M. A., Brooks, P. J., Van Wyk, J. J., and Lund, P. K., 1988, Insulin-like growth factor II messenger ribonucleic acids are synthesized in the choroid plexus of the rat brain, *Mol. Endocrinol.* **2**:47–54.
28. Aldred, A. R., Dickson, P. W., Marley, P. D., and Schreiber, G., 1987, Distribution of transferrin synthesis in brain and other tissues in the rat, *J. Biol. Chem.* **262**:5293–5297.
29. Pardridge, W. M., Eisenberg, J., and Yang, J., 1985, Human blood–brain barrier insulin receptor, *J. Neurochem.* **44**:1771–1778.
30. Pardridge, W. M., 1986, Receptor-mediated peptide transport through the blood–brain barrier, *Endocrine Rev.* **7**:314–330.
31. Duffy, K. R., and Pardridge, W. M., 1987, Blood–brain barrier transcytosis of insulin in developing rabbits, *Brain Res.* **420**:32–38.
32. Kumagai, A. K., Eisenberg, J., and Pardridge, W. M., 1987, Absorptive-mediated endocytosis of cationized albumin and a β-endorphin-cationized albumin chimeric peptide by isolated brain capillaries. Model system of blood–brain barrier transport, *J. Biol. Chem.* **262**:15214–15219.
33. Triguero, D., Buciak, J. B., Yang, J., and Pardridge, W. M., 1989, Blood–brain barrier transport of cationized immunoglobulin G. Enhanced delivery compared to native protein, *Proc. Natl. Acad. Sci. USA* **86**:4761–4765.
34. Pardridge, W. M., Triguero, D., and Buciak, J. B., 1989, Transport of histone through the blood–brain barrier, *J. Pharmacol. Exp. Ther.* **251**:821–826.
35. Fishman, J. B., Rubin, J. B., Handrahan, J. V., Connor, J. R., and Fine, R. E., 1987, Receptor-mediated transcytosis of transferrin across the blood–brain barrier, *J. Neurosci. Res.* **18**:299–304.
36. Pardridge, W. M., Eisenberg, J., and Yang, J., 1987, Human blood–brain barrier transferrin receptor, *Metabolism* **36**:892–895.
37. Duffy, K. R., Pardridge, W. M., and Rosenfeld, R. G., 1988, Human blood–brain barrier insulin-like growth factor receptor, *Metabolism* **37**:136–140.
38. Chabrier, P. E., Boubert, P., and Braquet, P., 1987, Specific binding of atrial natriuretic factor in brain microvessels, *Proc. Natl. Acad. Sci. USA* **84**:2078–2081.
39. Frank, H. J. L., and Pardridge, W. M., 1981, A direct in vitro demonstration of insulin binding to isolated brain microvessels, *Diabetes* **30**:757–761.
40. Frank, H. J. L., Pardridge, W. M., Morris, W. L., Rosenfeld, R. G., and Choi, T. B., 1986, Binding and internalization of insulin and insulin-like growth factors by isolated brain microvessels, *Diabetes* **35**:654–661.
41. Bar, R. S., Boes, M., and Sandra, A., 1988, Vascular transport of insulin to rat cardiac muscle. Central role of the capillary endothelium, *J. Clin. Invest.* **81**:1225–1231.
42. Milici, A. J., Watrous, N. E., Stukenbrok, H., and Palade, G. E., 1987, Transcytosis of albumin in capillary endothelium, *J. Cell Biol.* **105**:2603–2609.
43. Walsh, R. J., Slaby, F. J., and Posner, B. I., 1987, A receptor-mediated mechanism for the transport of prolactin from blood to cerebrospinal fluid, *Endocrinology* **120**:1846–1851.
44. King, G. L., and Johnson, S. M., 1985, Receptor-mediated transport of insulin across endothelial cells, *Science* **227**:1583–1586.
45. Rasio, E. A., 1969, Preliminary Communications: The displacement of insulin from blood capillaries, *Diabetologia* **5**:416–419.
46. Hinkle, P. M., and Kinsella, P. A., 1982, Rapid temperature-dependent transformation of the thyrotropin-releasing hormone–receptor complex in rat pituitary tumor cells, *J. Biol. Chem.* **257**:5462–5470.
47. Boni-Schnetzler, M., Scott, W., Waugh, S. M., DiBella, E., and Pilch, P. F., 1987, The insulin receptor. Structural basis for high affinity ligand binding, *J. Biol. Chem.* **262**:8395–8401.
48. White, M. F., Haring, H.-U., Kasuga, M., and Kahn, C. R., 1984, Kinetic properties and sites of autophosphorylation of the partially purified insulin receptor from hepatoma cells, *J. Biol. Chem.* **259**: 255–264.

49. Lowe, W., and LeRoith, D., 1986, Insulin receptors from guinea pig liver and brain: Structural and functional studies, *Endocrinology* **118:**1669–1677.
50. Heidenreich, K.A., Zahniser, N. R., Berhanu, P., Brandenburg, D., and Olefsky, J. M., 1983, Structural differences between insulin receptors in the brain and peripheral target tissues, *J. Biol. Chem.* **258:**8527–8530.
51. Haskell, J. F., Meezan, E., and Pillion, D. J., 1985, Identification of the insulin receptor of cerebral microvessels, *Am. J. Physiol.* **248:**E115–E125.
52. Lowe, W. L., Boyd, F. T., Clarke, D. W., Raizada, M. K., Hart, C., and LeRoith, D., 1986, Development of brain insulin receptors: Structural and functional studies of insulin receptors from whole brain and primary cell cultures, *Endocrinology* **119:**25–35.
53. Heidenreich, K. A., Brandenburg, D., 1986, Oligosaccharide heterogeneity of insulin receptors. Comparison of N-linked glycosylation of insulin receptors in adipocytes and brain, *Endocrinology* **118:**1835–1842.
54. Gammeltoft, S., Kowalski, A., Fehlmann, M., and van Obberghen, 1984, Insulin receptors in rat brain: Insulin stimulates phosphorylation of its receptor β-subunit, *FEBS Lett.* **172:**87–90.
55. Haskell, J. F., Meezan, E., and Pillion, D. J., 1984, Identification and characterization of the insulin receptor of bovine retinal microvessels, *Endocrinology* **115:**698–704.
56. Shemer, J., and LeRoith, D., 1987, The interaction of brain insulin receptors with wheat germ agglutinin, *Neuropeptides* **9:**1–8.
57. Grunberger, G., Lowe, W. L., McElduff, A., and Glick, R. P., 1986, Insulin receptor of human cerebral gliomas. Structure and function, *J. Clin. Invest.* **77:**997–1005.
58. Roth, R. A., Morgan, D. O., and Beaudoin, J., 1986, Purification and characterization of the human brain insulin receptor, *J. Biol. Chem.* **261:**3753–3757.
59. Rosenfeld, R. G., Pham, H., Keller, B. T., Borchardt, R. T., and Pardridge, W. M., 1987, Demonstration and structural comparison of receptors for insulin-like growth factor-I and -II (IGF-I and -II) in brain and blood–brain barrier, *Biochem. Biophys. Res. Commun.* **149:**159–166.
60. Szabo, A. J., and Szabo, O., 1975, Influence of the insulin-sensitive central nervous system glucoregulator receptor on hepatic glucose metabolism, *J. Physiol. (London)* **253:**121–133.
61. Van Houten, M., and Posner, B. I., 1979, Insulin binds to brain blood vessels in vivo, *Nature* **282:**623–625.
62. Pillion, D. J., Haskell, J. F., and Meezan, E., 1982, Cerebral cortical microvessels: An insulin-sensitive tissue, *Biochem. Biophys. Res. Commun.* **104:**686–692.
63. Albrecht, J., Wroblewska, B., Mossakowski, M. J., 1982, The binding of insulin to cerebral capillaries and astrocytes of the rat, *Neurochem. Res.* **7:**489–494.
64. Goodner, C. J., and Berrie, M. A., 1977, The failure of rat hypothalamic tissues to take up labeled insulin in vivo or to respond to insulin in vitro, *Endocrinology* **101:**605–612.
65. Triguero, D., Buciak, J. B., and Pardridge, W. M., 1990, Capillary depletion method for quantifying blood–brain barrier transcytosis of circulating peptides and plasma proteins, *J. Neurochem.* **54:**1882–1888.
66. Frank, H. J. L., Pardridge, W. M., and Jankovic-Vokes, T., 1986, Insulin binding to the blood–brain barrier in the streptozotocin diabetic rat, *J. Neurochem.* **47:**405–411.
67. Frank, H. J. L., Jankovic-Vokes, T., and Pardridge, W. M., 1985, Enhanced insulin binding to blood–brain barrier in vivo and to brain microvessels in vitro in newborn rabbits, *Diabetes* **34:**728–733.
68. Pardridge, W. M., Yang, J., and Eisenberg, J., 1985, Blood–brain barrier protein phosphorylation and dephosphorylation, *J. Neurochem.* **45:**1141–1147.
69. Baskin, D. G., Porte, D., Guest, K., and Dorsa, D. M., 1983, Regional concentrations of insulin in the rat brain, *Endocrinology* **112:**898–903.
70. Havrankova, J., Roth, J., and Brownstein, M. J., 1979, Concentrations of insulin and of insulin receptors in the brain are independent of peripheral insulin levels. Studies of obese and streptozotocin-treated rodents, *J. Clin. Invest.* **64:**636–642.
71. Broadwell, R. D., Balin, B. J., and Salcman, M., 1988, Transcytotic pathway for blood-borne protein through the blood–brain barrier, *Proc. Natl. Acad. Sci. USA* **85:**632–635.

22

Vascular Endothelium and Diabetes Mellitus

Robert S. Bar

I. INTRODUCTION

Diabetes mellitus is a disease complex defined by abnormalities of glucose homeostasis and characterized by a group of chronic complications that affect the function of the eyes (retinopathy), kidneys (nephropathy), peripheral nerves (neuropathy), large blood vessels (atherosclerosis), and microvessels (microangiopathy). Indeed, the vascular dysfunction(s) in diabetes may underlie all of the chronic complications of the disease. These complications account for the vast majority of morbidity and mortality that accompanies diabetes, with atherosclerosis underlying the susceptibility to myocardial infarction and stroke, and the microvascular disease underlying retinopathy, nephropathy, and, perhaps, neuropathy. Diabetes mellitus is divided into two general categories, type I [insulin-dependent diabetes mellitus (IDDM)] and type II [non-insulin-dependent diabetes mellitus (NIDDM)]. The general characteristics of type I and type II diabetes are outlined in Table 1. Both type I and type II patients have similar complications, although the frequency of each complication differs between the two groups, with type I patients having a much greater incidence of microvascular complications (nephropathy and retinopathy) and type II patients having a somewhat higher frequency of macrovascular complications that are associated with atherosclerosis. Although the precise etiologies of type I and II diabetes, as well as the causes of the diabetic complications, are not known, abnormalities of the vascular endothelium have been shown to occur early in the disease and have been implicated in the pathogenesis of chronic complications in this complex illness. In this chapter, I will first review information pertaining to endothelial dysfunction in human and animal models of diabetes, as well as the effects of the diabetic "milieu" on endothelium of isolated vasculature and culture endothelial cells. I will then review a group of recent studies that suggest a fundamental role of endothelium in processes that are critical to glucose homeostasis and vascular integrity, but whose direct involvement in diabetes mellitus has not yet been evaluated.

Robert S. Bar • Diabetes and Endocrinology Research Center, University of Iowa, Iowa City, Iowa 52246.

Endothelial Cell Dysfunctions, edited by Nicolae Simionescu and Maya Simionescu. Plenum Press, New York, 1992.

Table 1. Features of Type I and Type II Diabetes Mellitus

Type I (insulin-dependent)	Type II (non-insulin-dependent)
Onset usually prior to 20–40 years of age	Onset usually after age 40
Ketosis-prone	Not ketosis-prone
Insulin-sensitive	Insulin-insensitive
Usually thin	Usually obese
Low or absent plasma insulin levels	Plasma insulin levels often normal
Unstable	Stable
Autoimmune features	No autoimmune features
Primary treatment = insulin	Primary treatment = diet
Similar complications	
Mortality = renal and cardiovascular	Mortality = cardiovascular, primarily

II. VASCULAR SENSITIVITY TO VASODILATING AGENTS

Because of the overwhelming vascular pathology associated with diabetes, it has been suggested for some time that the vasculopathies are caused or propagated by altered sensitivity/reactivity to agents inducing vasoconstriction or vasodilation of vascular smooth muscle.[1,2] Extensive studies pertaining to this concept have been reported since the early investigations of Brody *et al.*[3] in the 1960s, with most reports assessing responses of vascular tissue from animal models of diabetes and, in limited study, human diabetes.[3–19] The findings related to vasoconstricting agents have been conflicting. Vasodilatory responses, however, tend to be more consistently impaired, with many laboratories now reporting a defect in endothelium-dependent dilation of several blood vessels (aorta, femoral artery, mesenteric arterioles, cerebral arterioles, and arterioles of skeletal muscle) from streptozotocin-treated animals,* BB/w diabetic rats, and, in limited study, diabetic dogs.[13]

Furchgott and others[20–25] first demonstrated that the vascular endothelium plays a vital role in the vasodilatory response of the aorta to certain vasodilating agents, such as acetylcholine. Subsequently, additional vasodilating substances (e.g., the calcium ionophore A23187, substance P, thrombin, ADP, ATP, and histamine)[18,24] have been demonstrated to indirectly cause smooth muscle relaxation via primary stimulation of the release of an endothelium-derived relaxation factor (EDRF). In contrast, other vasodilating agents (e.g., nitroprusside, adenosine, atrial natriuretic factor, forskolin, and papaverine) act directly on smooth muscle, causing vasodilation in the absence of endothelium. EDRF, once released from the endothelial cells, stimulates smooth muscle guanylate cyclase, with the resultant elevated smooth muscle cGMP level causing smooth muscle relaxation.[25,26] Recently, it has been shown that EDRF is nitric oxide or a compound which releases nitric oxide. Nitric oxide is produced by cultured endothelial cells, is a potent vasodilator that acts

*The two animal models most commonly studied are streptozotocin (STZ)-induced diabetic rats[23] and Bio-Breeding Wistar (BB/w)[24] rats. Streptozotocin is a glucose-containing nitrosourea with selective toxicity for the pancreatic beta cell. Depending on the dose, frequency, and timing of injections, STZ-induced diabetes can have characteristics of type I or type II diabetes. The BB/w rat is a model of spontaneous diabetes with many immunologic and biochemical properties of type I diabetes.

directly on smooth muscle and is indistinguishable from EDRF in biological activity, stability, and susceptibility to inhibiting and potentiating agents.[27,28] In diabetic animals and in isolated vessels from diabetic animals, the response to EDRF is attenuated. It is not yet clear whether the impaired response is secondary to decreased production of EDRF by endothelium or whether the smooth muscle response to EDRF is deficient. Insulin treatment of the diabetic animals rapidly reversed the impaired vasodilatory response to the endothelium-dependent agents and, in two studies, direct application of insulin to isolated vessels improved or reversed the defect *in vitro*, suggesting that insulin deficiency *per se* may have a direct role in this vascular defect.[14] Most recently, Kamata *et al.*[9] have demonstrated that, in addition to the previously described defects, aortic strips from diabetic animals have decreased basal and acetylcholine-induced production of cGMP, consistent with a defect in EDRF production or EDRF responsivity in diabetes. In contrast to the findings with endothelium-dependent vasodilators, the vasodilatory response to endothelium-independent agents is not affected by diabetes.

A similar vascular defect has also been demonstrated in human diabetes in a study of diabetic men with impotence,[17] an unfortunately common problem in diabetes that is associated with abnormalities of both the vasculature and the autonomic nervous system. Since relaxation of smooth muscle of the penile corpus cavernosum is required for erection, de Tejada and colleagues[17] investigated smooth muscle relaxation on corpus cavernosum tissue obtained from impotent diabetic and nondiabetic men at the time of surgery for penile implants. In addition to having more pronounced defects in muscle relaxation by autonomic nervous innervation, the tissue from diabetics also had significantly greater impairment of endothelium-dependent relaxation (acetylcholine response). Endothelium-independent relaxation (nitroprusside and papaverine stimulation) was similar in tissue with diabetic and nondiabetic men. Finally, it is of interest to note that the endothelium-dependent relaxation defect described in human and animal diabetes may not be specific for diabetes mellitus. Two conditions commonly accompanying diabetes, atherosclerosis[29] and hypertension,[30] have also been reported to have defective endothelium-mediated relaxation of vascular smooth muscle.

Thus, altered endothelium-dependent responses are present in the vasculature of humans and animals with diabetes mellitus. Such changes have been observed in several large arteries and in arteriolar beds from muscle and brain tissue suggesting a generalized impairment of vasodilation in diabetic animals and, perhaps, diabetic humans. These changes appear to occur early in the course of the disease and are likely to contribute to the altered blood flow and, perhaps, subsequent vascular changes that characterize diabetes. With the recent advances in the understanding of the basic mediators of endothelium-dependent smooth muscle relaxation, it is likely that the precise mechanisms of the defects in vasodilation in diabetes can be established and the relevance of such defects to the vasculopathies of diabetes determined.

III. VON WILLEBRAND FACTOR (vWF), PROSTACYCLIN (PGI_2), AND PLASMINOGEN ACTIVATOR

vWF, PGI_2, and plasminogen activator are all produced by vascular endothelial cells. Their production in diabetes mellitus, and their potential roles in diabetic vasculopathy have been extensively reviewed in the past few years.[33,40,47–49] These areas will therefore only be briefly summarized in the following section.

A. vWF

Factor VIII circulates as a complex composed of a smaller factor VIII coagulant (factor VIII:Ag), produced by parenchymal cells in the liver and by macrophages in the liver,[31,32] and a larger oligomeric glycoprotein (monomer 270,000 daltons), vWF,[33] produced primarily by endothelium[34,35] and megakaryocytes.[36] vWF is required for platelet adherence to subendothelium at relatively high shear rates[37,38] and is also involved in thrombus formation.[39] As summarized by Colwell *et al.*,[40,41] in the vast majority of 13 separate reports, the plasma levels of vWF and factor VIII:Ag were increased in children and adults with diabetes. This finding occurred in patients with type I and type II diabetes, and in individuals with and without vascular disease. Patients with microangiopathy tended to have even higher levels of vWF antigen.[42] Similar elevations in plasma levels of vWF and factor VIII:Ag have been observed in diabetic animal models and the chronology of the onset of the elevated plasma levels has been defined. In rats made diabetic with streptozotocin, elevations of both factor VIII:Ag and vWF were observed and occurred at ≤ 4 weeks of diabetes, i.e., prior to the onset of macroscopic vessel disease in the animals.[44] Because these abnormalities precede clinical or histologic evidence of vascular disease, they have been implicated in the pathogenesis of the vascular changes and are also thought to reflect chronic stimulation of the endothelium throughout the course of the disease. Despite the finding of elevated plasma vWF in early diabetes, higher plasma levels of vWF are usually observed in patients with clinically detectable vascular disease,[42] the magnitude of the elevations generally correlating with the duration of the disease.[40] Short-term improvement of glycemic control with insulin, resulted in improvement, but not normalization, of elevated plasma vWF levels.[43] Whether the changes in vWF levels reflect endothelial reparative mechanisms, responses to structural changes in the vessel wall, or adaptations to a changing metabolic milieu has not yet been determined.

B. PGI_2

PGI_2 is a metabolite of arachidonic acid that is produced by many cells, including endothelial cells. It is a potent vasodilator and inhibits platelet adhesion and aggregation. Its production by endothelium is thought to be relevant in the control of vascular resistance and maintenance of the nonthrombogenic surface of the vascular endothelium.[40,41] PGI_2 production by vascular tissue from diabetic humans and animals has generally been shown to be reduced by the diabetic condition (see Ref. 10 for summary of literature). Culture of human umbilical vein endothelial cells with human diabetic sera has caused inhibition of PGI_2 synthesis by the endothelial cells, with the inhibition not due to hyperglycemia, hypoinsulinemia, or hyperlipidemia.[45] Long-term treatment of diabetic animals with insulin or treatment with vitamin E reversed the lowered synthetic rates of PGI_2. However, neither the precise biochemical defect nor the etiologic agent(s) causing the defect in PGI_2 synthesis has been established, although recent studies suggest that diabetic serum may contain a specific inhibitor of PGI_2 production by endothelium.[46]

C. Fibrinolytic Activity (Plasminogen Activator)

Diabetes mellitus is frequently associated with thromboembolic events, hence the particular interest in both the thrombotic process as well as fibrinolytic pathways in diabetes. Plasminogen activator, produced by endothelial cells, is an enzyme which converts plasminogen to plasmin—a key step in the fibrinolytic process.[40] As tabulated by Colwell *et al.*,[40] several studies have evaluated plasminogen activator activity in human and animal diabetes.[40] Approximately 50% of the studies showed decreased plasminogen activator activity and diminished fibrinolytic activity, with the remainder showing no change or, occasionally, increased activity. Furthermore, it was not clear whether fibrinolytic activity correlates with the presence/absence or degree of vascular disease, glycemic/metabolic control, duration of disease, or form of therapy instituted for control of blood sugar. The complicated nature of the fibrinolytic cascade and the various methods used for its assessment have been reviewed by Colwell and others,[33,40] and may account, in part, for the disparate findings in diabetes mellitus.

In summary, diabetes is generally associated with vascular responses resulting in increased production of endothelium-derived vWF and plasminogen activator, and decreased production of PGI_2 from endothelium. It remains to be determined how each of these abnormalities is involved in the pathogenesis of the vascular diseases of diabetes.

IV. GLUCOSE AND ENDOTHELIUM

Hyperglycemia (elevated blood sugar) is the common finding in all forms of diabetes mellitus and, in fact, defines the condition. Since the vascular endothelium is directly bathed by the hyperglycemic circulation, any direct effect of glucose on endothelial cell function would be of obvious importance. The effects of ambient glucose on endothelial function have been evaluated primarily in studies of cultured endothelial cells. Elevated concentrations of glucose and diabetic serum have been reported to: (1) decrease, increase, or have no effect on thymidine incorporation into DNA and cell proliferation[50–55]; (2) increase the production and release of factor VIII-related antigen[56]; (3) increase the formation of glycosylated albumin[57] and glycosylated LDL[58] with a resultant increase in the endothelial uptake of the albumin and a decrease in the uptake and degradation of the glycosylated LDL (as compared to similar studies with native albumin and native LDL); and (4) in certain endothelial cells, such as aortic endothelial cells,[55] to increase the formation of the intracellular polyol, sorbitol, and decrease cellular levels of myoinositol and phosphoinositol. The latter observation may be of particular relevance to vascular complications of diabetes. The polyol/myoinositol hypothesis has been invoked to explain several of the complications of diabetes, particularly neuropathy.[60–62] When blood glucose levels rise, several cells of the body, especially those associated with diabetic complications, accumulate glucose intracellularly and convert the glucose to the polyol, sorbitol, via the enzyme aldose reductase. The increased concentration of intracellular sorbitol, which is not freely permeable, is associated with reductions of intracellular myoinositol, decreases in membrane phosphoinositol, and decreased membrane Na/K-ATPase activity. When sorbitol formation is prevented by inhibitors of aldose reductase activity, the decrease in myoinositol and phosphoinositol, as well as the decrease in Na/K-ATPase activity is not observed. It is apparent that certain endothelial cells, *in vivo* and *in vitro*, have aldose reductase activity,[55,63–66] whereas other endothelial cells may not possess this enzyme activity.[59] Studies of polyol/myoinositol metabolism in endothelium are still at an early stage and the precise role of the polyol/myoinositol pathways in endothelial and pericyte[59,67] function and their relationship to the vascular complications of diabetes remain to be determined.

V. VASCULAR PERMEABILITY IN DIABETES MELLITUS

Defects in vascular permeability are present in the early stages of diabetes. Clinically, they are most prominently manifested by the leakage of fluorescein dyes from retinal capillaries during fluorescein angiography and, at the kidney level, by loss of albumin through glomerular capillaries. In addition, the intravenous administration of insulin in certain diabetic patients causes a reduction in both plasma volume and the intravascular pool of albumin, implying a direct effect of insulin on capillary permeability.[68] Several mechanisms have been invoked to explain the altered vascular permeability, including a thickened but more porous vascular basement membrane and direct effects of the diabetic milieu on the barrier function of the capillary endothelium.

A. Vascular Basement Membrane

In human and animal diabetes, all vascular basement membranes become thickened and yet, at the same time, become more porous to substances normally excluded from passage across the basement membrane.[69–71] Both the thickness and the porosity of the vascular basement membrane increase with duration of the diabetes. Specific changes in the composition of diabetic basement membranes have been suspected, but not consistently observed.[72–74] The role of the endothelium in the basement membrane changes has not been directly studied. *In vitro*, endothelial cells have been shown to produce most constituents of basement membranes including type IV collagen,[75] heparan sulfate and other glycosaminoglycan-containing proteoglycans,[76,77] and fibronectin.[78] The contribution of endothelial cells to basement membrane formation in health and diabetes is not clear, but certainly is an area deserving further study.

B. Diabetic Milieu

Both hypoinsulinemia and hyperglycemia have been implicated in the etiology of the vascular permeability defects of diabetes. Diabetic animals have greatly reduced inflammatory responses and respond poorly to permeability factors[79] such as histamine and 5-hydroxytryptamine[80]; both of these microvessel permeability defects are either reversed by insulin or prevented by prior insulin treatment. Osterby *et al.*[81] have reported that insulin treatment of diabetic rats results in an increase in the number of micropinocytotic vesicles of skeletal muscle capillary endothelium. This endothelial vesicular network has been shown by many other studies to likely play a major role in the transendothelial transport of plasma macromolecules.

Many studies indicate that the endothelial metabolism of histamine, a vasodilatory substance that increases endothelial permeability to small and large molecules,[82–84] is affected by diabetes. Streptozotocin-induced diabetes is associated with increased accumulation of histamine, and increased vascular permeability. Both the permeability defect and histamine accumulation are prevented by insulin treatment of the animal, with concomitant normalization of the enzymatic activity of histamine decarboxylase and histaminase.

Several indirect observations have also led to the postulate that the high ambient glucose levels of diabetes directly cause the morphological (thickened basement membrane, pericyte dropout, endothelial cell proliferation) and functional (increased vascular permeability) alterations observed in capillaries from diabetic animals and humans. For example, several studies of the pathogenesis of diabetic vascular disease have implicated hyperglycemia in (1) glycation of endothelial membrane proteins, circulating proteins, lipoproteins, and basement membrane components[58,85–89] and (2) the direct alteration of carbohydrate metabolism in endothelial cells. In one particularly relevant report, Bendayan and Rasio[90] took advantage of several naturally occurring phenomena of the eel. This teleost adapts to cold water by increasing its blood sugar concentrations sixfold, from 100 to 600 mg/dl. The eel also has a particularly well-characterized capillary network since its swim bladder rete is easily isolated in large quantities free of parenchymal cells. Bendayan and Rasio studied the microvascular of the eel in the normoglycemic and hyperglycemic state. They clearly demonstrated that the hyperglycemia induced microvascular changes that characterize human microangiopathy, including basement membrane thickening and increased capillary permeability.

VI. INSULIN, INSULINLIKE GROWTH FACTORS, AND VASCULAR ENDOTHELIUM

Despite some inconsistency, the previous information (Sections II–V) strongly indicates that dysfunction of the vascular endothelium is an intrinsic component of diabetes mellitus. The next section of this chapter will review recent studies which demonstrate complex interactions of the vascular endothelium with two hormones central to the development of diabetes and its complications, insulin and the insulinlike growth factors, IGF-I and IGF-II. These studies demonstrate a critical role for the vascular endothelium in the normal metabolism and transport of insulin and the IGFs. By extension, they suggest that even minimal endothelial dysfunction, as undoubtedly occurs in the early stages of diabetes mellitus, will result in significant alterations in the actions of insulin and the IGFs, which will likely contribute to the metabolic and anatomic abnormalities which characterize the disease.

This section will discuss five related areas: (1) receptors for insulin and IGFs in endothelial cells, (2) functions of insulin and IGFs in endothelial cells, (3) processing of insulin and IGFs by cultured endothelial cells, (4) transport of insulin and IGFs by vascular endothelium, and (5) IGF-binding proteins produced by endothelial cells. Several of these areas have been reviewed in detail in the past 2–3 years,[91–95] and therefore will only be summarized.

A. Receptors for Insulin and IGFs

Specific receptors for insulin, IGF-I, and IGF-II have been described on endothelial cells from micro- and macrovessels,[96–103] isolated microvessels,[104,105] endothelium of perfused organs,[106–109] and endothelium of intact animals.[110] The characteristics of the insulin receptors of endothelial cells are similar to receptors described in other cell types and are summarized in Table 2. Endothelial receptors for IGF-I and IGF-II are also similar to type I and type II IGF receptors that have been characterized in other cells and tissues. The IGF-I receptor has high affinity for IGF-I and II and lower affinity for insulin. The holoreceptor is ~ 350,000 Da, consisting of a heterotetramer ($\alpha_2\beta_2$) formed by two identical extracellular α subunits containing the IGF-1 binding domain and two identical transmembrane β subunits, containing the kinase domain which causes autophosphorylation when IGF-I binds to the α subunit. The subunit structure, i.e., $\alpha_2\beta_2$, is similar for the insulin and IGF-I receptors. The IGF-II receptor has high affinity for IGF-II and IGF-I but does not recognize insulin. The receptor consists of a monomer of ~ 250,000 Da, which contains an extracellular, transmembrane, and intracellular domain. The IGF-II receptor of endothelial cells is likely to be identical to the mannose-6-phosphate receptor.

Three additional points are worthy of note. (1) The concentration of IGF (IGF-1 and -II) receptors is substantially higher than insulin receptors in the cultured endothelial cells. Tracer binding of [^{125}I]-IGF-I and -II is 6–12 times greater than insulin binding, regardless of the source of the cultured endothelial cells. (2) Both insulin and IGF-I have cross-reactivity with each other's receptors. It has recently been suggested by Siddle, Jonas, Harrison, Jacobs, and Pessin[111–113] that this apparent cross-reactivity may reflect a minor population of hybrid receptors having one heterodimer ($\alpha\beta$) from the insulin receptor and the other heterodimer from the IGF-I receptor. Of interest, endothelial cells have been shown to have a subpopulation of "atypical" receptors[114] possessing equal affinity for insulin and IGF-I. Whether these "atypical" endothelial receptors represent such hybrid receptors and the physiologic role of these hybrid or "atypical" receptors are presently unknown. (3) Endothelial cells make large quantities of IGF-binding proteins which specifically recognize IGF-I and -II and have no cross-reactivity with insulin. Some of these binding proteins adhere to the surface of the endothelial cells and as much as 20–30% of total IGF binding to cultured cells may reflect the association of IGF with cell binding

Table 2. Characteristics of Insulin Binding (Insulin Receptor) in Endothelial Cells

High affinity for insulin, low affinity for IGF-I and -II
Rapid, saturable
pH, temperature optima
Curvilinear Scatchard plot
Negative cooperativity
High affinity ($ID_{50} = 10^{-10}$ M)[a]
Homologous down-regulation (sensitivity $\leq 10^{-11}$ M insulin)
Binding α subunit 130,000–140,000 Da, β subunit ~ 90,000 Da, and phosphorylated by insulin holoreceptor $\alpha_2\beta_2$
Specificity—insulin analogue, antireceptor antibody specificity
Concentration differences among vascular sources of endothelial cells

[a]ID_{50} = concentration of unlabeled insulin causing a 50% decrease in tracer binding.

protein rather than true IGF membrane receptors.[115,116] The significance of the IGF bound to the cell-associated binding proteins, *in vitro* and *in vivo*, is currently being investigated in several laboratories.

B. Functions of Insulin and IGFs in Endothelial Cells

Insulin and the IGFs have both overlapping and distinct biological effects in cultured endothelial cells. Insulin and the IGFs stimulated glucose transport, amino acid transport, glucose oxidation, and thymidine incorporation into DNA in cultured microvessel endothelial cells derived from bovine retina, bovine fat, and rabbit cardiac muscle.[102,117–121] Insulin and the IGFs have also demonstrated several of these effects in isolated intact microvessels. Despite the overlapping nature of these effects, each hormone appears to mediate its function through its homologous receptor, i.e., insulin effects at physiologic concentrations of insulin are mediated through the insulin receptor, and not via insulin's cross-reactivity with the IGF-I receptor.[121] Similar biologic effects have not been observed in endothelial cells cultured from a variety of large-diameter vessels such as bovine aorta, pulmonary artery, pulmonary vein, and human umbilical vein. The IGFs, but not insulin, stimulate proteoglycan synthesis in both microvessel and macrovessel endothelial cells.[122,123]

As previously mentioned, both insulin and IGFs are potent stimulators of glucose uptake in cultured microvessel endothelial cells. Of particular interest to glucose metabolism in diabetes mellitus is the additional finding that the transport of glucose in certain cultured microvessel cells is also stimulated by several other circulating and regional factors relevant to diabetes, including basic fibroblast growth factor (bFGF),[124] platelet-derived growth factor (PDGF),[125] transforming growth factor-β (TGF-β)[124] and IGF-binding proteins[115] which are all produced by endothelial cells *in vitro*[126–128] and perhaps *in vivo*. The precise mechanism of the stimulation of glucose uptake by any of these factors has not been established, nor is it known whether each acts through similar or different pathways. However, should microvascular endothelium *in vivo* be involved in either plasma glucose disposal or the transport of plasma glucose to subendothelial tissues, the control of glucose uptake by capillary/microvascular endothelium becomes especially relevant in diabetes, a disease characterized by disordered glucose metabolism.

C. Processing of Receptor-Bound Insulin, IGF-I, and IGF-II

Endothelial cells from large-diameter blood vessels and microvessels, process insulin and each IGF by distinct and somewhat unusual pathways[129–132] (Fig. 1). Receptor-bound insulin is internalized and then rapidly released from the cell, predominantly as intact insulin. Receptor-bound IGFs are also rapidly internalized but are retained within the cell for longer time periods than is insulin. Intact IGF-I is slowly released from the cell, while IGF-II is released as both intact and degraded IGF-II.

D. Transport of Insulin and IGFs by Vascular Endothelium

Common to the processing of receptor-bound insulin and IGFs by cultured cells is the subsequent release of at least a portion of the hormone in the form of undegraded, biologically active hormone. Should similar hormonal processing occur *in vivo* and have

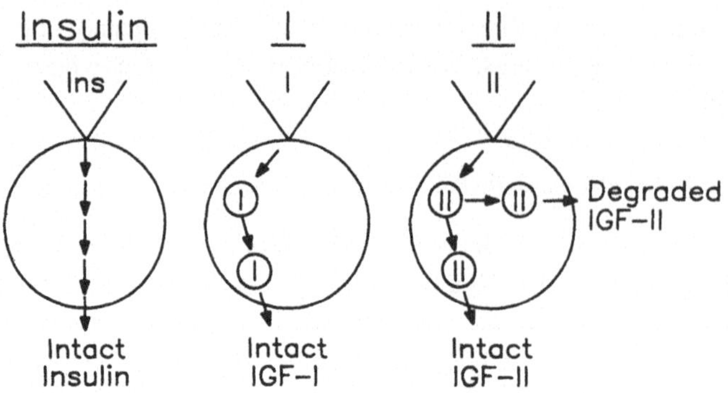

Figure 1. Processing of receptor-bound insulin (Ins), IGF-I, and IGF-II by endothelial cells. Reprinted with permission from Bar *et al.*[93]

appropriate vectoral polarity, the endothelium would have the potential of transporting intact hormone from the circulation to subendothelial tissues. That such transport occurs *in vivo* has been conclusively demonstrated for insulin[133] and has been suggested for IGF-I.[134] Studies in which insulin is perfused in the isolated, beating heart show that 70–75% of the insulin appearing in cardiac muscle must first bind to capillary endothelial cell surface receptors and is then transported across the capillary en route to cardiac muscle.[133] Of the several steps involved in this transport process, the initial binding of insulin to the endothelial cell surface receptor for insulin, and not other intraendothelial or subendothelial processing, is the rate-limiting step in the transport of insulin to cardiac muscle.[133]

Diabetes is characterized by inadequate effects of insulin. In patients with type I diabetes, the insufficient effects of insulin are primarily due to reduced amounts of circulating insulin, whereas in type II patients there is also a component of tissue insensitivity to insulin. Given the role of endothelial cells in the transport of insulin, we must now also consider the possibility of impairment of endothelial transport of insulin to relevant target tissue in the genesis of impaired insulin effect in the type II patients and in the type I patients who are treated with exogenous insulin. The potential of having different transport characteristics in regional capillary beds, as well as the likelihood of selective tissue abnormalities in diabetes may also contribute to the tissue-specific differences in insulin action that are frequently observed in diabetes. Obviously, we will need to develop totally new technology in order to assess the role of endothelial transport of plasma insulin in the intact animal before the precise role of this endothelial transport process can be determined.

E. IGF-Binding Proteins

The IGFs, or somatomedins, mediate the growth-promoting effects of growth hormone on cartilage and long bone and also have many overlapping functions with insulin.[135–137] The IGFs circulate in plasma in high concentrations (near microgram levels) attached to specific serum-binding proteins.[138] Alterations in both IGF levels and changes in serum-binding proteins occur in diabetes, with the alterations of IGF-binding proteins implicated in several abnormalities observed in diabetes, such as the disordered angiogenesis accompanying proliferative retinopathy, the impaired skeletal growth, the diminished cellular repair, and the poor wound healing. In the past 5 years, a series of important observations have indicated that both IGF-I/II and multiple IGF-binding proteins are produced by tissues throughout the body.[139–144] These findings have led to the hypothesis that the IGFs have a diverse group of autocrine, paracrine, and endocrine functions with a specific effect of IGF determined by both the regional concentration of the IGF and the nature of the binding protein with which the IGF is associated.

Endothelial cells produce large quantities of IGF-binding proteins.[115,116] These binding proteins have several features of particular importance to the understanding of IGF physiology, in general, and IGF biology in diabetes, in particular. First, in contrast to other cellular and circulating IGF-binding proteins, some of the endothelium derived IGF-binding proteins have intrinsic biologic activity,[116,145] being able to stimulate acute metabolic processes in microvessel endothelial cells. Second, the endothelial binding proteins, when perfused through isolated beating hearts, can cross the capillary boundary and distribute in subendothelial tissues.[146] Other serum binding proteins can also cross the capillary boundary; the subendothelial anatomic regions of distribution of endothelial binding proteins and the lower-molecular weight IGF serum binding protein (IGF BP-1) are entirely different, with endothelial binding proteins having much higher affinity for connective tissue elements (basement membranes and endomysia), whereas the serum IGF BP-1 primarily associates with cardiac myocytes.[146] Third, and of most immediate relevance to diabetes, the rates of transcapillary transport of endothelial binding proteins and IGF BP-1 in the perfused heart are both significantly influenced by insulin, with insulin increasing the transport of BP-1 but decreasing transcapillary permeability of endothelial cell binding proteins.[146,147] In recent studies in humans, it has been demonstrated by several groups that any rise in serum insulin levels is followed within 30–60 min by a decrease in serum levels of IGF BP-1.[148,149] When considered in light of the transcapillary permeability studies in isolated hearts, these clinical data would suggest that insulin causes an acute efflux of IGF BP-1 from the circulation. Such transcapillary movement of serum IGF-binding proteins would provide a means of rapidly altering the biologic effects of IGFs in diabetes, with a specific effect dependent on the association of IGF with the IGF BP-1, the subendothelial tissue distribution of IGF BP-1/[BP-1–IGF-1], and the presence of IGFs or other binding proteins in the subendothelial tissue. In fact, given the nearly ubiquitous presence of a diverse group of IGF-binding proteins, the binding proteins could orchestrate a complex movement of IGFs from cell to cell, tissue to tissue, and between tissues and the vascular compartment. This would provide a highly controlled system with multiple levels for regulation of endocrine, paracrine, and autocrine effects of the IGFs. The vascular endothelium would likely have a major role in such a system, related to both its production of specialized IGF-binding proteins and its regional control of vascular permeability of selected binding proteins.[146] The effects of insulin (and perhaps other components of the diabetic milieu) on the vascular permeability of IGF-binding proteins also suggest a critical interaction of endothelium with diabetes and IGF function.

VII. OTHER AREAS

This review has highlighted selected areas which strongly indicate an involvement of the vascular endothelium with diabetes mellitus. Several additional relevant topics that were only alluded to or were not covered in this chapter include: (1) increased adherence of platelets and circulating cells to vascular endothelium in diabetes,[150–156] (2) lipid metabolism by endothelium in diabetes,[157–159] (3) morphologic changes of microvessel and macrovessel endothelium in diabetes,[160–172] and (4) alteration of endothelial enzyme activities induced by diabetes.[173–175]

VIII. CONCLUSION

In this chapter I have reviewed studies assessing the involvement of the vascular endothelium with the disease, diabetes mellitus. Taken together, the findings indicate that endothelial cell abnormalities occur early in the course of the disease and influence endothelial function at multiple levels. However, what is missing from all of these reports is a clear definition of the precise role of any of the endothelial abnormalities in the etiology of the vascular disease that characterize diabetes. In the next decade, it is hoped that the relevance of specific endothelial defects to the pathogenesis of diabetic complications can be determined and the means of halting or reversing the endothelial abnormalities established. It is felt that such goals and expectations are now feasible, given the rapidly advancing technologies for the study of vascular endothelium *in vitro* and in the intact host, and the increasing application of such technology to the complex metabolic disorder, diabetes mellitus.

Acknowledgments. This work was supported by NIH grants DK25295 (Diabetes and Endocrinology Research Center), DK25421, and VA research funds.

REFERENCES

1. Christrieb, A. R., 1973, Diabetes and hypertensive vascular disease. Mechanism and treatment, *Am. J. Cardiol.* **32:**592–606.
2. Weidmann, P., Berreta-Piccoli, C., Keusch, G., Glück, Z., Mujagic, M., Grimm, M., Meier, A., and Ziegler, W. H., 1979, Sodium-volume factor, cardiovascular reactivity and hypotensive mechanism of diuretic therapy in mild hypertension associated with diabetes mellitus, *Am. J. Med.* **67:**779–784.
3. Brody, M. J., and Dixon, R. L., 1964, Vascular reactivity in experimental diabetes mellitus, *Circ. Res.* **14:** 494–501.
4. Pfaffman, M. A., Ball, C. R., Darby, A., and Hilman, R., 1982, Insulin reversal of diabetes-induced inhibition of vascular contractility in the rat, *Am. J. Physiol.* **242:**H490–H495.
5. Scarborough, N. L., and Carrier, G. O., 1984, Nifedipine and alpha$_2$-adrenoceptors in rat aorta. II. Role of extracellular calcium in enhanced alpha$_2$-adrenoceptor-mediated contraction in diabetes, *J. Pharmacol. Exp. Ther.* **236:**603–609.
6. Agrawal, D. K., and McNeill, J. H., 1987, Vascular responses to agonists in rat mesenteric artery from diabetic rats, *Can. J. Physiol. Pharmacol.* **65:**1484–1490.
7. Head, R. J., Longhurst, P. A., Panek, R. L., and Stitzel, R. E., 1987, A contrasting effect of the diabetic state upon the contractile responses of aortic preparation from rat and rabbit, *Br. J. Pharmacol.* **91:**275–286.
8. Kamata, K., Miyata, N., and Kasuya, Y., 1988, Mechanisms of increased responses of the aorta to α-adrenoceptor agonists in streptozotocin-induced diabetic rats, *J. Pharmacobio-Dyn.* **11:**707–713.
9. Kamata, K., Miyata, N., and Kasuya, Y., 1989, Impairment of endothelium-dependent relaxation and changes in levels of cyclic GMP in aorta from streptozotocin-induced diabetic rats, *Br. J. Pharmacol.* **97:**614–618.

10. Kamata, K., Miyata, N., and Kasuya, Y., 1989, Involvement of endothelial cells in relaxation and contraction responses of the aorta to isoproterenol in naive and streptozotocin-induced diabetic rats, *J. Pharmacol. Exp. Ther.* **249:**890–894.
11. Pieper, G. M., and Gross, G. J., 1988, oxygen free radicals abolish endothelium-dependent relaxation in diabetic rat aorta, *Am. J. Physiol.* **225:**H825–H833.
12. Mayhan, W. G., 1989, Impairment of endothelium-dependent dilatation of cerebral arterioles during diabetes mellitus, *Am. J. Physiol.* **256:**H621–H625.
13. Gebremedhin, D., Koltai, M. Z., Pogatsa, G., Magyar, K., and Hadházy, P., 1987, Differential contractile responsiveness of femoral arteries from healthy and diabetic dogs: Role of endothelium, *Arch. Int. Pharmacodyn.* **288:**100–108.
14. Fortes, Z. B., Leme, J. G., and Scivoletto, R., 1984, Vascular reactivity in diabetes mellitus: Possible role of insulin on the endothelial cell, *Br. J. Pharmacol.* **83:**635–643.
15. Durante, W., Sen, A. K., and Sunahara, F. A., 1988, Impairment of endothelium-dependent relaxation in aortae from spontaneously diabetic rats, *Br. J. Pharmacol.* **94:**463–468.
16. Gebremedhin, D., Koltai, M. Z., Pogatsa, G., Magyar, K., and Hadházy, P., 1988, Influence of experimental diabetes on the mechanical responses of canine coronary arteries: Role of endothelium, *Cardiovasc. Res.* **22:**537–544.
17. de Tejada, I. S., Goldstein, I., Azadzoi, K., Krane, R. J., and Cohen, R. A., 1989, Impaired neurogenic and endothelium-mediated relaxation of penile smooth muscle from diabetic men with impotence, *N. Engl. J. Med.* **320:**1025–1030.
18. Tanz, R. D., Chang, K. S. K., and Weller, T. S., 1989, Histamine relaxation of aortic rings from diabetic rats, *Agents Actions* **28:**1–8.
19. Van de Voorde, J., and Leusen, I., 1982, Vascular endothelium and the relaxation effect of histamine on aorta of the rat, *Arch. Int. Pharmacodyn.* **256:**329–330.
20. Furchgott, R. F., and Zawadzki, J. V., 1980, The obligatory role of endothelial cells in the relaxation of arterial smooth muscle by acetylcholine, *Nature* **288:**373–376.
21. Furchgott, R. F., Zawadzki, J. V., and Cherry, P. D., 1981, Role of endothelium in the vasodilator response to acetylcholine, in: *Vasodilation* (P. M. Vanhoutte, and I. Leusen, eds.), Raven Press, New York, pp. 49–66.
22. Cherry, P. D., Furchgott, R. F., Zawadzki, J. V., and Jothianandan, D., 1982, Role of endothelial cells in relaxation of isolated arteries by bradykinin, *Proc. Natl. Acad. Sci. USA* **79:**2106–2110.
23. Chand, N., and Altura, B. M., 1981, Acetylcholine and bradykinin relax intrapulmonary arteries by acting on endothelial cells: Role in lung vascular diseases, *Science* **213:**1376–1379.
24. Furchgott, R. F., 1983, Role of endothelium in responses of vascular smooth muscle, *Circ. Res.* **53:**557–573.
25. Holzmann, S., 1982, Endothelium-induced relaxation by acetylcholine associated with larger rises in cyclic GMP in coronary arterial strips, *J. Cyclic Nucleotide Res.* **8:**409–419.
26. Rapoport, R. M., and Murad, F., 1983, Agonist-induced endothelium-dependent relaxation in rat thoracic aorta may be mediated through cGMP, *Circ. Res.* **52:**352–357.
27. Khan, M. T., and Furchgott, R. F., 1987, Similarities of behavior of nitric oxide (NO) and endothelium-derived relaxing factor in a perfusion cascade bioassay system, *Fed. Proc.* **46:**385.
28. Palmer, R. M. J., Ferridge, A. G., and Moncada, S., 1987, Nitric oxide release accounts for the biological activity of endothelium derived relaxing factor, *Nature* **327:**524–526.
29. Jayakody, L., Kappagoda, T., Senaratne, P. J., and Thomson, A. B., 1988, Impairment of endothelium-dependent relaxation: An early marker for atherosclerosis in the rabbit, *Br. J. Pharmacol.* **94:**335–346.
30. Sim, M. K., and Singh, M., 1987, Decreased responsiveness of the aortae of hypertensive rats to acetylcholine, histamine and noradrenaline, *Br. J. Pharmacol.* **90:**147–150.
31. Zelechowska, M. G., von Mourik, J. A., and Brodniewicz-Proba, T., 1985, Ultrastructural localization of factor VIII procoagulant antigen in human liver hepatocytes, *Nature* **317:**729–738.
32. Wion, K. L., Kelly, D., Summerfield, J. A., Tuddenham, E. G., and Lawn, R. M., 1985, Distribution of factor VIII mRNA and antigen in human liver and other tissues, *Nature* **317:**726–729.
33. Banga, J. D., and Sixma, J. J., 1986, Diabetes mellitus, vascular disease and thrombosis, *Clin. Haematol.* **15:** 465–492.
34. Jaffe, E. A., Hoyer, L. W., and Nachman, R. L., 1974, Synthesis of von Willebrand factor by cultured human endothelial cells, *Proc. Natl. Acad. Sci. USA* **71:**1906–1909.
35. Bloom, A. L., 1979, The synthesis of factor VIII, *Clin. Haematol.* **8:**53–76.

36. Nachman, R., Levine, R., and Jaffe, E. A., 1977, Synthesis of factor VIII antigen by cultured guinea pig megakaryocytes, *J. Clin. Invest.* **60:**914–921.
37. Tschopp, T. P., Weiss, H. J., and Baumgartner, H. R., 1974, Decreased adhesion of platelets to subendothelium in von Willebrand's disease, *J. Lab. Clin. Med.* **83:**296–300.
38. Sakariassen, K. S., Bolhuis, P. A., and Sixma, J. J., 1979, Human blood platelet adhesion to artery subendothelium is mediated by factor VIII–von Willebrand factor bound to the subendothelium, *Nature* **279:**636–638.
39. Turitto, V. T., Weiss, H. J., and Baumgartner, H.R ., 1984, Platelet interaction with rabbit subendothelium in von Willebrand's disease: Altered thrombus formation distinct from platelet adhesion, *J. Clin. Invest.* **74:** 1730–1741.
40. Colwell, J. A., Lopes-Virella, M., Winocour, P. D., and Halushka, P. V., 1988, New concepts about the pathogenesis of atherosclerosis in diabetes mellitus, in: *The Diabetic Foot* (M. E. Levin and L. W. O'Neal, eds.), Mosby, St. Louis, pp. 51–70.
41. Colwell, J. A., Lopes-Virella, M., and Halushka, P. V., 1981, Pathogenesis of atherosclerosis in diabetes mellitus, *Diabetes Care* **4:**121–133.
42. Porta, M., Ricchetii, I., la Selva, M., Bertagna, A., and Molinatti, G. M., 1984, Quantitative and qualitative assessment of plasma von Willebrand factor variations, as induced by forearm venous stasis in patients with diabetic microangiopathy, *Diabetes Res.* **1:**219–221.
43. Gonzalez, J., Colwell, J. A., Sarji, K. E., and Nair, R. M., 1980, Effect of metabolic control with insulin on plasma von Willebrand factor activity (VIIIR:WF) in diabetes mellitus, *Thromb. Res.* **17:**261–266.
44. Winocour, P. D., Lopes-Virella, M., Laimins, M., and Colwell, J. A., 1983, Time course of changes in *in vitro* platelet function and plasma von Willebrand factor activity (VIIIR:WF) and factor VIII-related antigen (VIIIR:Ag) in the diabetic rat, *J. Lab. Clin. Med.* **102:**795–804.
45. Aanderud, S., Krane, H., and Nordoy, A., 1985, Influence of glucose, insulin and sera from diabetic patients on the prostacyclin synthesis *in vitro* in cultured human endothelial cells, *Diabetologia* **28:**641–644.
46. Umeda, F., Inoguchi, T., and Nawata, H., 1989, Reduced stimulatory activity on prostacyclin production by cultured endothelial cells in serum from aged and diabetic patients, *Atherosclerosis* **75:**61–66.
47. Colwell, J. A., and Lopes-Virella, M. F., 1988, A review of the development of large-vessel disease in diabetes mellitus, *Am. J. Med.* **85**(Suppl. 5A)**:**113–118.
48. Colwell, J. A., Winocour, P. D., Lopes-Virella, M., and Halushka, P. V., 1983, New concepts about the pathogenesis of atherosclerosis in diabetes mellitus, *Am. J. Med.* **75:**67–80.
49. Porta, M., la Selva, M., Molinatti, P., and Molinatti, G. M., 1987, Endothelial cell function in diabetic microangiopathy, *Diabetologia* **30:**601–609.
50. Stout, R. W., 1982, Glucose inhibits replication of cultured human endothelial cells, *Diabetologia* **23:**436–439.
51. Lorenzi, M., Cagliero, E., Toledo, S., 1985, Glucose Toxicity for human endothelial cells in culture: Delayed replication, disturbed cell cycle, and accelerated death, *Diabetes* **34:**621–627.
52. Weimann, B. J., Lorch, E., and Baumgartner, H. R., 1984, High glucose concentrations do not influence replication and prostacyclin release of human endothelial cells, *Diabetologia* **27:**62–63.
53. Lorenzi, M., Cagliero, E., and Toledo, S., 1985, Glucose toxicity for human endothelial cells in culture: Delayed replication, disturbed cell cycle, and accelerated death, *Diabetes* **34:**621–627.
54. Gillion, K. R. W., and Hawthorne, J. N., 1983, Transport of myo-inositol in endoneurial preparations of sciatic nerve from normal and streptozotocin-diabetic rats, *Biochemistry* **210:**775–781.
55. Yorek, M. A., and Dunlap, J. A., 1989, The effect of elevated glucose levels on myo-inositol metabolism in cultured bovine aortic endothelial cells, *Metabolism* **38:**16–22.
56. Mordes, D. B., Lazarchick, J., Colwell, J. A., and Sens, D. A., 1983, Elevated glucose concentrations increase factor VIIIR:Ag levels in human umbilical vein endothelial cells, *Diabetes* **32:**876–878.
57. Williams, S. K., Devenny, J. J., and Bitensky, M. W., 1981, Micropinocytic ingestion of glycosylated albumin by isolated microvessels: Possible role in pathogenesis of diabetic microangiopathy, *Proc. Natl. Acad. Sci. USA* **78:**2393–2397.
58. Lorenzi, M., Cagliero, E., Markey, B., Henriksen, T., Witztum, J. L., and Sampietro, T., 1984, Interaction of human endothelial cells with elevated glucose concentrations and native and glycosylated low density lipoproteins, *Diabetologia* **26:**218–222.
59. Sussman, I., Carson, M. P., Schultz, V., Wu, X. P., McCall, A. L., Ruderman, N. B., and Thornheim, K., 1988, Chronic exposure to high glucose decreases myo-inositol in cultured cerebral microvascular pericytes but not in endothelium, *Diabetologia* **31:**771–775.

60. Winegrad, A. I., 1987, Does a common mechanism induce the diverse complications of diabetes? *Diabetes* **36**:396–406.
61. Greene, D. A., Lattimer, S. A., and Sima, A. A. F., 1987, Sorbitol, phosphoinositides, and sodium-potassium-ATPase in the pathogenesis of diabetic complications, *N. Engl. J. Med.* **316**:599–606.
62. Gabbay, K. H., 1973, The sorbitol pathway and the complications of diabetes, *N. Engl. J. Med.* **288**: 831–836.
63. Kennedy, A., Frank, R. N., and Varma, S. D., 1983, Aldose reductase activity in retinal and cerebral microvessels and cultured vascular cells, *Invest. Ophthalmol. Vis. Sci.* **24**:1250–1258.
64. Ludvigson, M. A., and Sorenson, R. L., 1980, Immunohistochemical localization of aldose reductase. II. Rat eye and kidney, *Diabetes* **29**:450–459.
65. Lorenzi, M., Toledo, S., Boss, G. R., Lane, M. J., and Montisano, D. F., 1987, The polyol pathway and glucose 6-phosphate in human endothelial cells cultured in high glucose concentrations, *Diabetologia* **30**:222–227.
66. Williamson, J. R., Chang, K., Rowold, E., Marvel, J., Tomlinson, M., Sherman, W.R., Ackerman, K.E., and Kilo, C., 1985, Sorbinil prevents diabetes-induced increases in vascular permeability but does not alter collagen cross-linking, *Diabetes* **34**:703–705.
67. Li, W., Khatami, M., and Rockey, J. H., 1985, The effects of glucose and an aldose reductase inhibitor on the sorbitol and collagen synthesis of bovine retinal capillary pericytes in culture, *Exp. Eye Res.* **40**: 439–444.
68. Gundersen, H. J. G., and Christensen, N. J., 1977, Intravenous insulin causing loss of intravascular water and albumin and increased adrenergic nervous activity in diabetics, *Diabetes* **26**:551–557.
69. Siperstein, M. D., Unger, R. H., and Madison, L. L., 1968, Studies of capillary basement membranes in normal subjects, diabetic, and prediabetic patients, *J. Clin. Invest.* **47**:1973–1999.
70. Vracko, R., 1970, Skeletal muscle capillaries in diabetics: A quantitative analysis, *Circulation* **41**:271–284.
71. Bloodworth, J. M. B., Jr., Engerman, R. L., and Powers, K. L., 1969, Experimental diabetic microangiopathy. I. Basement membrane statistics in the dog, *Diabetes* **18**:455–458.
72. Osterby Hansen, R., and Lundback, K., 1970, The basement membrane morphology in diabetes mellitus, in: *Diabetes Mellitus: Theory and Practice* (M. Ellenbert and H. Rifkin, eds.), McGraw–Hill, New York, pp. 178–209.
73. Williamson, J.R ., and Kilo, C., 1976, Basement-membrane thickening and diabetic microangiopathy, *Diabetes* **25**(Suppl. 2):925–927.
74. Beisswenger, P. J., and Spiro, R. G., 1970, Human glomerular basement membrane: Chemical alteration in diabetes mellitus, *Science* **168**:596–598.
75. Kramer, R. H., Fuh, G. M., and Bensch, K. G., 1985, Synthesis of extracellular matrix glycoproteins by cultured endothelial cells, *J. Cell. Physiol.* **123**:1–9.
76. Bar, R. S., Dake, B. L., and Spanheimer, R. G., 1985, Sulfated glycosaminoglycans in cultured endothelial cells from capillaries and large vessels of human and bovine origin, *Atherosclerosis* **56**:11–26.
77. Bar, R. S., Dake, B. L., and Stueck, S., 1987, Stimulation of proteoglycans by IGF-I and IGF-II in microvessel and large vessel endothelial cells, *Am. J. Physiol.* **253**:E21–E27.
78. McAuslan, B. R., Hannan, G. N., Reilly, W., and Stewart, F. H., 1980, Variant endothelial cells. Fibronectin as a transducer of signals for migration and neovascularization, *J. Cell. Physiol.* **104**:177–186.
79. Garcia, L. J., Hamamura, L., Migliorini, R. H., and Leite, M. P., 1973, Influence of diabetes upon the inflammatory response of the rat. A pharmacological analysis, *Eur. J. Pharmacol.* **23**:74–81.
80. Llorach, M. A. S., Bohm, G. M., and Garcia, L. J., 1976, Decreased vascular reaction to permeability factors in experimental diabetes, *Br. J. Exp. Pathol.* **57**:747–754.
81. Osterby, R., Gundersen, H. J., and Christensen, N. J., 1978, The acute effect of insulin on capillary endothelial cells, *Diabetes* **27**:745–749.
82. Fortes, Z. B., Leme, J. G., and Scivoletto, R., 1983, Vascular reactivity in diabetes mellitus: Role of the endothelial cell, *Br. J. Pharmacol.* **79**:771–781.
83. Orlidge, A., and Hollis, T. M., 1982, Aortic endothelial and smooth muscle histamine metabolism in experimental diabetes, *Arteriosclerosis* **2**:142–150.
84. Hollis, T. M., Gallik, S. G., Olidge, A., et al., 1983, Aortic endothelial and smooth muscle histamine metabolism, *Arteriosclerosis* **3**:599–606.
85. Stevens, V. J., Rouzer, C. A., Monnier, V. M., and Cerami, A., 1978, Diabetic cataract formation: Potential role of glycosylation of lens crystallins, *Proc. Natl. Acad. Sci. USA* **75**:2918–2922.

86. Day, J. F., Ingebretsen, C. G., Ingebretsen, W. R., Jr., Baynes, J. W., and Thorpe, S. R., 1980, Nonenzymatic glycosylation of serum proteins and hemoglobin: Response to changes in blood glucose levels in diabetic rats, *Diabetes* **29:**524–527.
87. Brownlee, M., and Cerami, A., 1981, The biochemistry of the complications of diabetes mellitus, *Annu. Rev. Biochem.* **50:**385–432.
88. Witztum, J. L., Steinbrecher, U. P., Fisher, M., and Kesaniemi, A., 1983, Nonenzymatic glucosylation of homologous low density lipoprotein and albumin renders them immunogenic in the guinea pig, *Proc. Natl. Acad. Sci. USA* **80:**2757–2761.
89. Brownlee, M., Cerami, A., and Vlassara, H., 1988, Advanced products of nonenzymatic glycation and the pathogenesis of diabetic vascular disease, *Diabetes Metab. Rev.* **4:**437–451.
90. Bendayan, H., and Rasio, E. A., 1981, Hyperglycemia and microangiopathy in the eel, *Diabetes* **30:**317–325.
91. Bar, R. S., Lowe, W. L., and Spanheimer, R. G., 1989, Interactions of insulin and IGFs with cellular components of the arterial wall: potential impact on atherosclerosis, in: *Molecular and Cellular Biology of Insulin-like Growth Factors and Their Receptors*, (D. LeRoith and M. K. Raizada, eds.), Plenum Press, New York, pp. 473–484.
92. Bar, R. S., Boes, M., Dake, B. L., Henley, S. A., and Booth, B. A., 1990, Receptor-mediated transport of circulating insulin and IGF by vascular endothelium, in: *Endothelial Cell Function in Diabetic Microangiopathy: Problems in Methodology and Clinical Aspects. Frontiers in Diabetes*, Vol. 9 (G. M. Molinatti, R. S. Bar, F. Belfiore, and M. Porta, eds.), Karger, Basel, pp. 97–107.
93. Bar, R. S., Boes, M., Dake, B., Booth, B., Henley, S. A., and Sandra, A., 1988, Insulin, IGFs and vascular endothelium, *Am. J. Med.* **85**(5A):59–70.
94. Bar, R. S., 1984, Receptors for insulin and multiplication stimulating activity (MSA) in endothelial cells, in: *Biology of Endothelial Cells* (E. A. Jaffe, ed.), Nijhoff, The Hague.
95. Bar, R. S., Boes, M., Dake, B. L., Henley, S., and Sandra, A., 1988, Vascular endothelium and the IGFs, in: *Diabetes 1988* (R. G. Larkins, P. Z. Zimmet, and D. J. Chisholm, eds.), Elsevier Science Publishers, B.V., pp. 151–154.
96. Bar, R. S., Hoak, J. C., and Peacock, M. L., 1978, Insulin receptors in human endothelial cells: Identification and characterization, *J. Clin Endocrinol. Metab.* **47:**699–702.
97. Bar, R. S., Peacock, M. L., Spanheimer, R. G., Veenstra, R., and Hoak, J. C., 1980, Differential binding of insulin to human arterial and venous endothelial cells in primary culture, *Diabetes* **29:**991–995.
98. Peacock, M. L., Bar, R. S., and Goldsmith, J. C., 1982, Interactions of insulin with bovine endothelium, *Metabolism* **31:**52–56.
99. Bar, R. S., and Boes, M., 1984, Distinct receptors for IGF-I, IGF-II, and insulin are present on bovine capillary endothelial cells and large vessel endothelial cells, *Biochem. Biophys. Res. Commun.* **124:** 203–209.
100. King, G. L., Buzney, S., Kahn, C. R., Hetu, N., Buchwald, S., Macdonald, S.G., and Rand, L. I., 1983, Differential responsiveness to insulin in endothelial and support cells from micro-and macrovessels, *J. Clin. Invest.* **71:**974–979.
101. Bar, R. S., Goldsmith, J.C , and Owen, W. G., 1982, Properties of endothelial cells in culture: Recent studies, *Lymphokines* **7:**249–271.
102. Bar, R. S., Dolash, S., Dake, B. L., and Boes, M., 1986, Cultured capillary endothelial cells from bovine adipose tissue: A model for insulin binding and action in microvascular endothelium, *Metabolism* **35:** 317–322.
103. King, G. L., Goodman, A. D., Buzney, S., Moses, A., and Kahn, C. R., 1985, Receptors and growth-promoting effects of insulin and insulin-like growth factors on cells from bovine and retinal capillaries and aorta, *J. Clin. Invest.* **75:**1028–1036.
104. Frank, H. J. L., and Partridge, W. M., 1981, A direct *in vitro* demonstration of insulin biding to isolated brain microvessels, *Diabetes* **30:**757–761.
105. Pillion, D. J., Haskell, J. F., and Meezan, E, 1982, Cerebral cortical microvessels: An insulin sensitive tissue, *Biochem. Biophys. Res. Commun.* **104:**686–692.
106. Bar, R. S., DeRose, A., Sandra, A., Peacock, M. L., and Owen, W. G., 1983, Insulin binding to microvascular endothelium of intact heart: A kinetic and morphometric analysis, *Am. J. Physiol.* **244:**E447–E452.
107. Bar, R. S., Boes, M., and Sandra, A., 1985, Receptors of insulin-like growth factor I (IGF-I) in myocardial capillary endothelium of the intact perfused heart, *Biochem. Biophys. Res. Commun.* **133:**724–730.

108. Bar, R. S., Boes, M., and Sandra, A., 1988, IGF receptors in myocardial capillary endothelium: Potential regulation of IGF-I transport to cardiac muscle, *Biochem. Biophys. Res. Commun.* **152:**93–98.

109. Bar, R. S., Boes, M., and Sandra, A., 1988, Vascular transport of insulin to rat cardiac muscle: Central role of the capillary endothelium, *J. Clin. Invest.* **81:**1225–1233.

110. van Houten, M., and Posner, B. I., 1979, Insulin binds to brain blood vessels *in vivo*, *Nature* **282:**623–625.

111. Jonas, H. A., Newman, J. D., and Harrison, L. C., 1986, An atypical insulin receptor with high affinity for insulin-like growth factors copurified with placental insulin receptor, *Proc. Natl. Acad. Sci. USA* **83:**4124–4128.

112. Moxham, K., Duronio, V., and Jacobs, S., 1989, Insulin-like growth factor I receptor β subunit heterogeneity, *J. Biol. Chem.* **264:**13238–13244.

113. Treadway, J., Morrison, B. D., Goldpine, I., and Pessin, J. E., 1989, Assembly of insulin/insulin-like growth factor I hybrid receptors *in vitro*, *J. Biol. Chem.* **264:**21450–21453.

114. Cox, A. J., Bar, R. S., Jonas, H. A., and Newman, J. D., 1987, Atypical receptors for insulin-like peptides in endothelial cells, *Diabetes* **36:**54A.

115. Bar, R. S., Harrison, L. C., Baxter, R. C., Boes, M., Dake, B.L., Booth, B., and Cox, A., 1987, Production of IGF-binding proteins by vascular endothelial cells, *Biochem. Biophys. Res. Commun.* **148:**734–739.

116. Bar, R. S., Booth, B. A., Boes, M., and Dake, B. L., 1989, Insulin-like growth factor-binding proteins from vascular endothelial cells: Purification, characterization and intrinsic biological activities, *Endocrinology* **125:**1910–1920.

117. King, G. L., Buzney, S., Kahn, C. R., Hetu, N., Buchwald, S., Macdonald, S. G., and Rand, L. I., 1983, Differential binding of insulin to human arterial and venous endothelial cells in primary culture, *J. Clin. Invest.* **71:**974–979.

118. King, G. L., Goodman, A. D., Buzney, S., Moses, A., and Kahn, C. R., 1985, Receptors and growth-promoting effects of insulin and insulin-like growth factors on cells from bovine retinal capillaries and aorta, *J. Clin. Invest.* **75:**1028–1036.

119. Gerritsen, M. E., and Burke, T. M., 1985, Insulin binding and effects of insulin on glucose uptake and metabolism in cultured rabbit coronary microvessel endothelium, *Proc. Soc. Exp. Biol. Med.* **180:**17–23.

120. Vinters, H. V., Berliner, J. A., Beck, D. W., Maxwell, K., Bready, J. V., and Cancilla, P., 1985, Insulin stimulates DNA synthesis in cerebral microvessel endothelium and smooth muscle, *Diabetes* **34:**964–969.

121. Bar, R. S., Siddle, K., Dolash, S., Boes, M., and Dake, B., 1988, Actions of insulin and insulin-like growth factors I and II in cultured capillary endothelial cells from bovine adipose tissue, *Metabolism* **37:**714–720.

122. Bar, R. S., Stueck, S., Dake, B. L., and Spanheimer, R. G., 1984, Multiplication stimulating activity (MSA) specifically augments proteoglycan synthesis in cultured endothelial cells, *Endocrinology* **115:**2487–2489.

123. Bar, R. S., Dake, B. L., and Stueck, S., 1987, Stimulation of proteoglycans by IGF-I and IGF-II in microvessel and large vessel endothelial cells, *Am. J. Physiol.* **253:**E21–E27.

124. Boes, M., Dake, B. L., and Bar, R. S., 1991, Interactions of cultured endothelial cells with TGF-β, bFGF, PDGF and IGF-I, *Life Sciences* **48:**811–821.

125. Bar, R. S., Boes, M., Booth, B. A., Dake, B. L., Henley S., and Hart, M. N., 1989, The effects of platelet-derived growth factor (PDGF) in cultured microvessel endothelial cells, *Endocrinology* **124:**1841–1848.

126. Vlodavsky, I., Fridman, R., Sullivan, R., Sasse, J., and Klagsbrun, M., 1987, Aortic endothelial cells synthesize basic fibroblast growth factor which remains cell associated and platelet-derived growth factor-like protein which is secreted, *J. Cell. Physiol.* **131:**402–408.

127. Hannan, R. L., Kourembanas, S., Flanders, K., Rogelj, S. J., Roberts, A. B., Faller, D. V., and Klagsbrun, M., 1988, Endothelial cells synthesize basic fibroblast growth factor and transforming growth factor beta, *Growth Factors* **1:**7–17.

128. Kourembanas S., and Faller, D. V., 1989, Platelet-derived growth factor production by human umbilical vein endothelial cells is regulated by basic fibroblast growth factor, *J. Biol. Chem.* **264:**4456–4459.

129. Dernovsek, K. D., Bar, R. S., Ginsberg, B. H., and Lioubin, M. N., 1984, Rapid transport of biologically intact insulin through cultured endothelial cells, *J. Clin. Endocrinol. Metab.* **58:**761–763.

130. Dernovsek, K. D., and Bar, R. S., 1985, Processing of cell-bound insulin by capillary and macrovascular endothelial cells in culture, *Am. J. Physiol.* **248:**E244–E251.

131. Jialal, I., King, G. L., and Buchwald, S., 1984, Processing of insulin by bovine endothelial cells in culture: Internalization without degradation, *Diabetes* **33:**794–800.

132. Bar, R. S., Boes, M., and Yorek, M., 1986, Processing of insulin-like growth factors I and II by capillary and large vessel endothelial cells, *Endocrinology* **118:**1072–1080.

133. Bar, R. S., Boes, M., and Sandra, A., 1988, Vascular transport of insulin to rat cardiac muscle: Central role of the capillary endothelium, *J. Clin. Invest.* **81**:1225–1233.
134. Bar, R. S., Boes, M., and Sandra, A., 1988, IGF receptors in myocardial capillary endothelium: Potential regulation of IGF-I transport to cardiac muscle, *Biochem. Biophys. Res. Commun.* **152**:93–98.
135. Van Wyk, J. J., Underwood, L. E., Hintz, R. L., Clemmons, D. R., Viona, S J., and Weaver, R. P., 1974, The somatomedins: A family of insulin-like hormones under growth hormone control, *Recent Prog. Horm. Res.* **30**:259–318.
136. Froesch, E. R., Schmid, D., Schwander, J., and Zapf, J., 1985, Actions of insulin-like growth factors, *Annu. Rev. Physiol.* **47**:443–467.
137. Zapf, J., and Foresch, E. R., 1986, Insulin-like growth factors/somatomedins: Structure, secretion, biological actions and physiological role, *Horm. Res.* **24**:121–130.
138. Baxter, R. C., and Martin, J. L., 1989, Binding proteins for the insulin-like growth factors: Structure, regulation and function, *Prog. Growth Factor Res.* **1**:49–68.
139. D'Ercole, A. J., Applewhite, G. T., and Underwood, L. E., 1980, Evidence that somatomedin is synthesized by multiple tissues in the fetus, *Dev. Biol.* **73**:315–328.
140. Clemmons, D. R., Underwood, L. E., and Van Wyk, J. J., 1981, Hormonal control of immunoreactive somatomedin production by cultured human fibroblasts, *J. Clin. Invest.* **67**:10–19.
141. Moses, A. C., Frienkel, A. J., Knowles, B. B., and Aden, D. P., 1983, Demonstration that a human hepatoma cell line produces a specific insulin-like growth factor carrier protein, *J. Clin. Endocrinol. Metab.* **56**:1003–1008.
142. Romanus, J. A., Yang, Y. W.-H., Nissley, S. P., and Rechler, M. M., 1987, Biosynthesis of the low molecular weight carrier protein for the insulin-like growth factor in rat liver and fibroblasts, *Endocrinology* **121**:1041–1050.
143. McCusker, R. H., and Clemmons, D. R., 1988, Insulin-like growth factor binding protein and secretion by muscle cells: Effects of cellular differentiation and proliferation, *J. Cell. Physiol.* **137**:505–512.
144. D'Ercole, A. J., Stiles, A. D., and Underwood, L. E., 1984, Tissue concentrations of somatomedin C: Further evidence for multiple sites of synthesis and paracrine or autocrine mechanisms of action, *Proc. Natl. Acad. Sci. USA* **81**:935–939.
145. Booth, B. A., Bar, R. S., Boes, M., Dake, B. L., Bayne, M., and Cascieri, M., 1990, Intrinsic bioactivity of IGF-binding proteins from vascular endothelial cells, *Endocrinology* **127**:2630–2638.
146. Bar, R. S., Clemmons, D. R., Boes, M., Busby, W. H., Booth, B. A., Dake, B. L., and Sandra, A., 1990, Transcapillary permeability and subendothelial distribution of endothelial and amniotic fluid insulin-like growth factor binding proteins in the rat heart, *Endocrinology* **127**:1078–1086.
147. Bar, R. S., Boes, M., Dake, B., Booth, B., Busby, W. H., and Sandra, A., 1990, Transcapillary permeability of IGF-binding proteins (abstract 286), p. 96, in: Endocrine Society 72nd Annual Meeting (June 20–23, 1990) Programme and Abstracts.
148. Suikkari, A.-M., Koivisto, V. A., Rutanen, E.-M., Yki-Järvinen, H., Karonen, S. L., and Seppälä, M., 1988, Insulin regulates the serum levels of low molecular weight insulin-like growth factor-binding protein, *J. Clin. Endocrinol. Metab.* **66**:266–272.
149. Suikkari, A.-M., Koivisto, V. A., Koistinen, R., Seppala, M., and Yki-Järvinen, H., 1989, Dose–response characteristics for suppression of low molecular weight plasma insulin-like growth factor-binding protein by insulin, *J. Clin. Endocrinol. Metab.* **68**:135–140.
150. Wautier, J. L., Paton, R. C., Wautier, M. P., Pintigny, D., Abadie, E., Passa, P., and Caen, J. P., 1981, Increased adhesion of erythrocytes of endothelial cells in diabetes mellitus and its relation to vascular complications, *N. Engl. J. Med.* **305**:237–242.
151. Schmid-Schonbein, H., and Volger, E., 1976, Red-cell aggregation and red-cell deformability in diabetes, *Diabetes* **25**(Suppl. 2):897–902.
152. Barnes, A. J., Locke, P., Scudder, P. R., Dormandy, T. L., Dormandy, J. A., and Slack, J., 1977, Is hyperviscosity a treatable component of diabetic microcirculatory disease? *Lancet* **2**:789–791.
153. McMillan, D. E., Utterback, N. G., and La Puma, J., 1978, Reduced erythrocyte deformability in diabetes, *Diabetes* **27**:895–901.
154. Wali, R. K., Jaffe, S., Kumar, D., and Kalra, V. K., 1988, Alterations in organization of phospholipids in erythrocytes as factors in adherence to endothelial cells in diabetes mellitus, *Diabetes* **37**:104–111.
155. Wautier, J. L., Wautier, M. P., Pintigny, D., Galacteros, F., Courillon, A., Passa, P., and Caen, J. P., 1983, Factors involved in cell adhesion to vascular endothelium, *Blood Cells* **9**:221–234.

156. Colwell, J. A., Winocour, P. D., and Halushka, P. V., 1983, Do platelets have anything to do with diabetic microvascular disease? *Diabetes* **32**(Suppl. 2):14–19.
157. Berliner, J. A., Frank, H. J. L., Karasic, D., and Capdeville, M., 1984, Lipoprotein-induced insulin resistance in aortic endothelium, *Diabetes* **33**:1039–1044.
158. Fielding, C. J., 1981, The endothelium, triglyceride-rich lipoproteins, and atherosclerosis: Insights from cell biology and lipid metabolism, *Diabetes* **30**(Suppl. 2):19–23.
159. Ramirez, I., and Severson, D. L., 1986, Effect of diabetes on acid and neutral triacylglycerol lipase and on lipoprotein lipase activities in isolated myocardial cells from rat heart, *Biochem. J.,* **238**:233–238.
160. Yee, R. W., Matsuda, M., Kern, T. S., Engerman, R. L., and Edelhauser, H. F., 1985, Corneal endothelial changes in diabetic dogs, *Curr. Eye Res.* **4**:759–766.
161. Sharma, N. K., Gardiner, T. A., and Archer, D. B., 1985, A morphologic and autoradiographic study of cell death and regeneration in the retinal microvasculature of normal and diabetic rats, *Am. J. Ophthalmol.* **100:** 51–60.
162. Itoi, M., Nakamura, T., Mizobe, K., Kodama, Y., Nakagawa, N., and Itoi, M., 1989, Specular microscopic studies of the corneal endothelia of Japanese diabetics, *Cornea* **8**:2–6.
163. Busted, N., Olsen, T., and Schmitz, O., 1981, Clinical observations on the corneal thickness and the corneal endothelium in diabetes mellitus, *Br. J. Ophthalmol.* **65**:637–690.
164. Schultz, R. O., Matsuda, M., and Yee, R. W., 1984, Corneal endothelial changes in type I and type II diabetes mellitus, *Am. J. Ophthalmol.* **98**:401–410.
165. Moore, S. A., Bohlen, H. G., Miller, B. G., and Evans, A. P., 1985, Cellular and vessel wall morphology of cerebral cortical arterioles after short-term diabetes in adult rats, *Blood Vessels* **22**:265–277.
166. Wu, J. X., Maes, L., Andries, R., Warson, F., Gepts,W., and Bourgain, R. H., 1985, Early morphologic changes in coronary arteries of experimental diabetic rats, *Acta Diabetol. Lat.* **22**:317–326.
167. Dolgov, V. V., Zaikina, O. E., Bondarenko, M. F., and Repin, V. S., 1982, Aortic endothelium of alloxan diabetic rabbits: A quantitative study using scanning electron microscopy, *Diabetologia* **22**:338–343.
168. Malik, R. A., Newrick, P. G., Sharma, A. K., Jennings, A., Ah-See, A. K., Mayhew, T. M., Jakubowski, J., Boulton, A. J., and Ward, J. D., 1989, Microangiopathy in human diabetic neuropathy: Relationship between capillary abnormalities and the severity of neuropathy, *Diabetologia* **32**:92–102.
169. Powell, H. C., and Myers, R. R., 1984, Axonopathy and microangiopathy in chronic alloxan diabetes, *Acta Neuropathol.* **65**:128–137.
170. Dyck, P. J., Hansen, S., Karnes, J., O'Brien, P., Yasuda, H., Windeband, A., and Zimmerman, B., 1985, Capillary number and percentage closed in human diabetic sural nerve, *Proc. Natl. Acad. Sci. USA* **82**:2513–2517.
171. Powell, H. C., Rosoff, J., and Myers, R. R., 1985, Microangiopathy in human diabetic neuropathy, *Acta Neuropathol.* **68**:295–305.
172. Vracko, R., 1982, A comparison of the microvascular lesions in diabetes mellitus with those in normal aging, *Am. Geriatr. Soc.* **30**:201–205.
173. Agren, A., Rehn, G., and Naeser, P., 1979, Morphology and enzyme activities of the retinal capillaries in streptozotocin-diabetic mice, *Acta Ophthalmol.* **57**:1065–1069.
174. Naeser, P., and Agren, A., 1978, Morphology and enzyme activities of the retinal capillaries in mice with the obese-hyperglycemic syndrome (gene symbol OB), *Acta Ophthalmol.* **56**:607–616.
175. Toop, M. J., Dallinger, K. J. C., Jennings, P. E., and Barnett, A. H., 1986, Angiotensin-converting enzyme (ACE): Relationship to insulin-dependent diabetes and microangiopathy, *Diabetic Med.* **3**:455–457.

23

Cellular and Molecular Mechanism of Glucose-Induced Diabetic Microangiopathy

Stuart K. Williams

I. INTRODUCTION

The cellular and biochemical changes which occur as a result of diabetes mellitus all can be traced, albeit both directly and indirectly, to elevated blood glucose concentrations. The eventual cure of this disease will result from our understanding of insulin synthesis by the pancreatic beta cell. Until this time, treatments for the pathologic sequelae of diabetic hyperglycemia will continue to be a major effort in diabetes research.

Altered function of vascular endothelial cells in diabetes mellitus has seen increased study during the last 10 years due primarily to our understanding of the pivotal role of endothelium in organ physiology, and the well-documented micro- and macroangiopathies observed in diabetes. In fact, the major overt pathologies observed in diabetes, namely blindness, renal failure, and vascular disease, are now believed to occur as a direct result of endothelial cells responding to biochemical changes pursuant to hyperglycemia. To fully understand the cellular and molecular dysfunctions of endothelium in diabetes, the biochemical response of cells to increased glucose concentrations has received increased attention.

II. BIOCHEMISTRY OF GLUCOSE TOXICITY

The response of endothelium to elevated plasma glucose concentrations can be generally categorized into three major areas. The initial effect of elevated plasma glucose is hyperosmotic alterations in endothelial integrity. This represents an immediate effect dependent directly on glucose concentrations. Chronic elevation of glucose concentration

---

Stuart K. Williams • Department of Surgery, University of Arizona Health Sciences Center, Tucson, Arizona 85724.

Endothelial Cell Dysfunctions, edited by Nicolae Simionescu and Maya Simionescu. Plenum Press, New York, 1992.

results in the acceleration of both enzymatic and nonenzymatic reactions. The enzymatic processing of glucose involves the sorbitol (polyol) pathway which converts glucose to fructose via an intermediate molecule known as sorbitol.[1] Alternatively, glucose may also be converted to glucose-6-*P* due to the action of hexokinase (Fig. 1). Due to the action of these enzyme systems, the endothelial cell has the capacity to increase the cytoplasmic concentration of a number of molecular species. With increased glucose concentrations inherent in diabetes, many of the angiopathies associated with diabetes have been linked to increased concentrations of the metabolic products sorbitol, fructose, and glucose-6-*P*.[2] For example, elevated levels of sorbitol in cells as a result of the initial metabolism of glucose by the enzyme aldose reductase have been implicated in both diabetic angiopathy and neuropathy.[3] Specific inhibitors of enzymes involved in sorbitol production now have been synthesized.[4] Initial results with these compounds indicate that several pathologies associated with increased glucose concentrations can be reduced in severity.[3] Owing to the long onset of these complications in diabetes and the biochemical complexities of glucose enzymatic metabolism, the exact mechanism linking glucose metabolism and specific pathologies has not been established.

III. NONENZYMATIC GLYCATION

Glucose has also been shown to have a direct effect on molecules without the need for enzymatic metabolism to an alternate reactive species, or without the need of a specific enzyme to catalyze the reaction of glucose with molecules. This reaction is known as nonenzymatic glycosylation (NEG) and is defined as nonenzymatic glucosylation when glucose is the monosaccharide involved in the modification[5] (Fig. 2). This reaction has received increased attention by investigators studying diabetes due to the observation that the products of this modification are elevated in diabetics and especially in those who exhibit poor control of their serum glucose levels.[6,7] Since this is a nonenzymatic process, the major factor limiting the reaction rate is the external concentration of glucose.[8] At higher glucose concentrations, as observed in a diabetic exhibiting poor metabolic control, the rate of glucosylation is enhanced and results in enhanced production of glucosylated proteins.[9,10]

The initial reaction of glucose with amino acid residues on proteins is now known to be just the initial step in a series of reactions leading to the production of a large number of

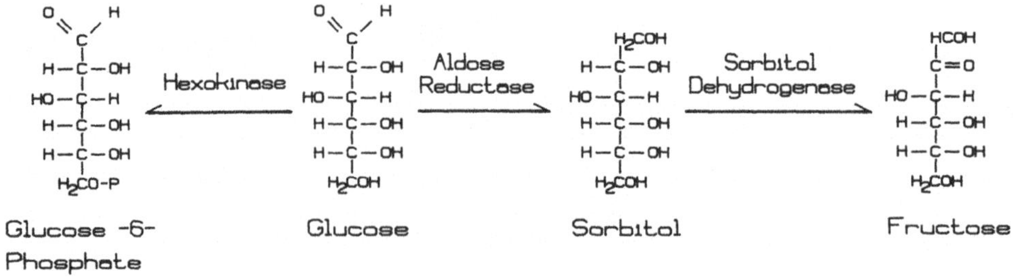

Figure 1. Initial reactions in the enzymatic metabolism of glucose.

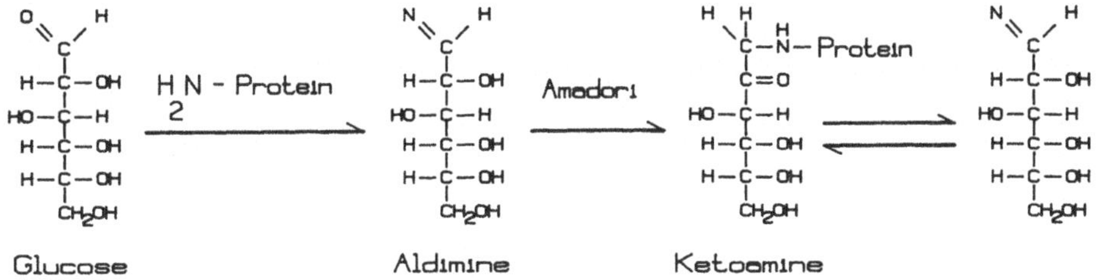

Figure 2. Initial reactions in the nonenzymatic reactions of glucose with proteins.

secondary reaction products.[5,11,12] These molecules are known as advanced glycosylation end-products (AGE).[13] This process has been studied in the past predominantly by food biochemists. In diabetics it is now recognized that AGE exhibit brown or fluorescent properties. One example of these complex AGE identified by Pongor *et al.*[14] is 2-(2-furoyl)-4(5)-2(furanoyl)-1*H* imidazole, the chemical structure of which is shown in Fig. 3. Thus, as observed with enzymatic glucose metabolism, the nonenzymatic glycosylation reactions also exhibit a great deal of complexity requiring careful analysis of the putative effects of nonenzymatic glycosylation reactions on cellular and molecular function.

IV. RELATIONSHIP BETWEEN NONENZYMATIC GLYCOSYLATION AND DIABETIC MICROANGIOPATHIES

Our laboratory's initial interest in nonenzymatic glycosylation stemmed from studies of microvascular endothelial cell permeability. Since serum albumin is present in the blood at high concentrations, and albumin plays a critical role in maintaining a hydrostatic pressure gradient across the endothelial barrier, the effects of possible changes in albumin function due to NEG were of interest. Furthermore, serum albumin is unique among almost all serum proteins in that it is not enzymatically glycosylated during its synthesis by hepatocytes.[15] The nonenzymatic addition of glucose to serum albumin subsequently creates a glycoprotein from a protein normally devoid of carbohydrates. Employing a system of isolated microvascular endothelial cells, we observed that the endocytosis of glucosylated serum albumin (gA) was much greater than native albumin and speculated that the increased endocytosis of gA by endothelium may result in the observed leakiness of the microvasculature in diabetes mellitus.[16] Enhanced endocytosis was also observed for glucosylated myoglobin and ovalbumin.[17] These initial findings have prompted additional

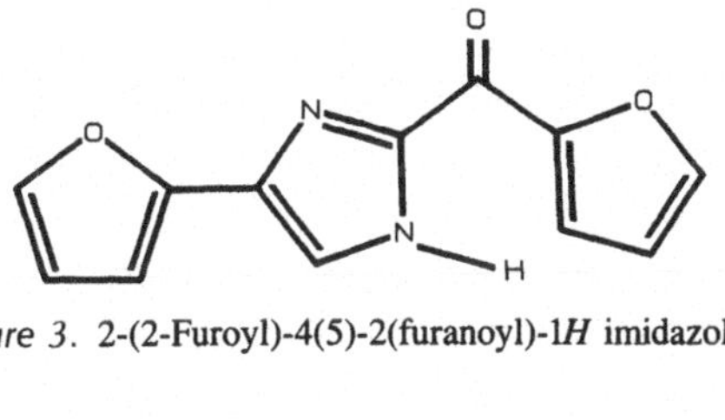

Figure 3. 2-(2-Furoyl)-4(5)-2(furanoyl)-1*H* imidazole.

studies of albumin structural modification and the interactions of glycosylated proteins with the microvasculature.[18–23] Prior to discussion of those studies, a description of the specific structure of serum albumin is important to a complete understanding of albumin interaction with endothelium.

V. SERUM ALBUMIN STRUCTURE

Serum albumin is a 69-kDa protein which is anionic in neutral buffers with a pI of approximately 4.6.[15] The structure of albumin permits it to interact with both hydrophilic as well as hydrophobic molecules. For this reason, numerous pharmacologic agents, toxins, cellular metabolites bind to serum albumin and are carried throughout the vascular system as a complex.[24,25] The tertiary structure of albumin is highly sensitive to environmental factors presumably due to albumin's function as a carrier protein. For investigators studying the function of albumin, this sensitivity is extremely important. Even the method of albumin isolation from blood and subsequent storage will affect albumin structure and function. For example, the heat pasteurization used by many manufacturers in the processing of serum for subsequent protein isolations results in albumin unfolding and rearrangements.[15]

The nonenzymatic reaction of glucose (as well as other monosaccharides) with native serum albumin is highly specific and involves lysine residue No. 525.[26,27] Therefore, the *in vivo* NEG of serum albumin results in a single glucose molecule per molecule of albumin. Molar ratios of glucose to albumin greater than 1 are presumably due to changes in the tertiary structure of albumin which uncovers additional reactive amino acid residues for glucose attack. For our *in vitro* studies we have used serum albumin modified under conditions resulting in a final glucose-to-albumin molar ratio of 1. AGE products of albumin were also formed during this modification; however, neither the extent nor the implications of AGE-albumin products on permeability functions have been elucidated.

On the basis of *in vitro* studies with isolated microvessels[16] and *in vivo* studies,[21,18,23] it is apparent that vascular endothelium recognizes glucosylated serum albumin differentially from native albumin. Based on *in vitro* studies, we hypothesized that if gA is more actively ingested by vascular endothelium as compared to native albumin, gA should exhibit preferential uptake by vascular endothelial cells *in vivo* as well. As seen in Fig. 4, when the sequestration of glucosylated (Fig. 1) and native albumin following injection into the bloodstream is compared, there is a preferential deposition of gA into all organs except the spleen and liver. Moreover, the deposition of these proteins in the kidney shows an overwhelming preference for gA. Several conclusions can be made. First, the addition of a single molecule of glucose to serum albumin dramatically alters albumin's recognition by different organs. This altered recognition is not unique to the continuous microvascular endothelium observed in adipose tissue and muscle, as it is also observed in fenestrated microcirculations. A question remains: is the enhanced deposition a result of increased binding of albumin to the vascular wall without subsequent transendothelial transport, or does gA preferentially gain access to the extravascular space?

The preferential transendothelial transport of gA has now been confirmed in a number of systems.[18–23] Our initial *in vivo* original studies of transendothelial gA flux focused on the kidney since proteins transported across the glomerular endothelial cell barrier can

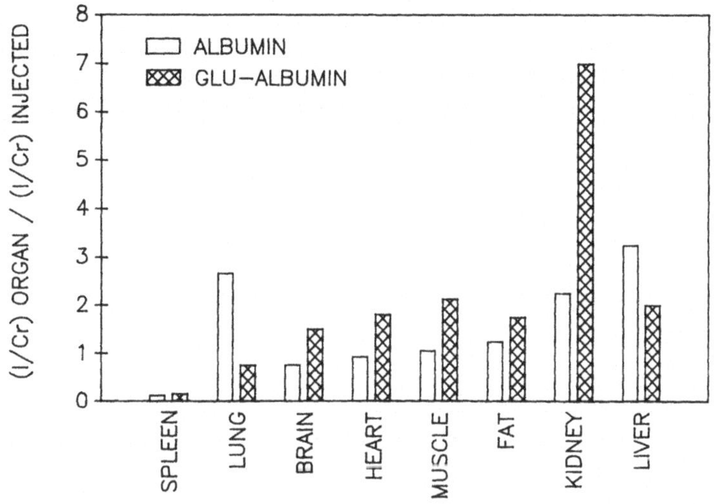

Figure 4. The accumulation of ^{125}I-labeled albumin and glucosylated albumin is different in each of these organs. ^{51}Cr-labeled red cells were injected with radiolabeled albumins and served as an intravascular control.

easily be collected as urine. Our examination of the kidney first was also prompted by our clearance data which indicated the kidney was a major site of preferential gA deposition. As seen in Fig. 5, the relative concentration of gA is much higher in urine as compared to plasma. Higher concentrations of gA were found in urine from both healthy and diabetic animals, indicating that even in the nondiabetic population preferential transport of gA occurs. Ultrastructural studies with NEG ferritin confirmed the enhanced transport of

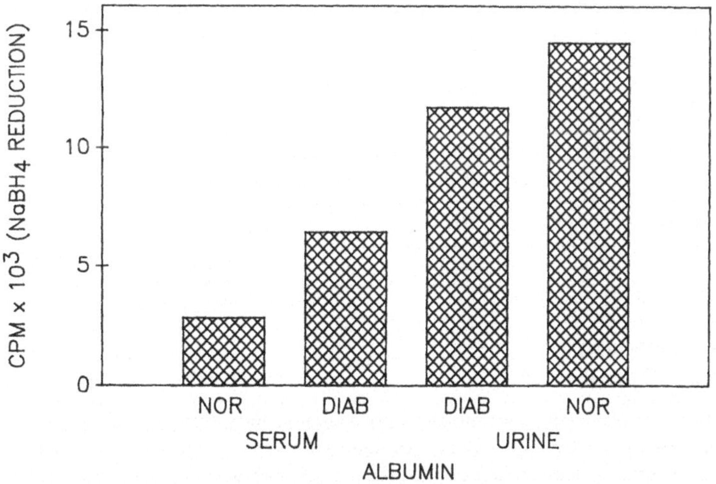

Figure 5. Protein samples were prepared from plasma and urine of normal and diabetic rats. Proteins were reduced with sodium borotritide and the quantity of radiolabel associated with serum albumin was measured. Increased radiolabeling is indicative of increased glucosylation.

glucosylated proteins. Therefore, it appears across the glomerulus that the preferential recognition of gA by the microvascular[28] endothelium is not limited to the hyperglycemic state. We suggest that the normal loss of albumin into the urine may be due, in part, to the NEG of albumin in all individuals and the preferential transport of gA by the glomerulus. The increased flux of albumin observed during the early stages of diabetes may be simply the result of increased concentrations of gA in plasma rather than a direct alteration of the glomerular function.

VI. SEARCH FOR THE MECHANISM OF PREFERENTIAL gA PERMEABILITY

Studies of nonenzymatically glucosylated protein recognition by both continuous and fenestrated microvascular endothelial cells have led several authors to suggest molecular mechanisms.[21,29] Our initial hypothesis was that the addition of glucose to albumin was recognized by endothelium via a specific binding site on the endothelial cell surface. With the current interest in specific receptors on endothelial cells, this hypothesis seemed logical. However, a series of experiments have led us to search for alternate hypotheses. First, as shown in Table 1 the endocytosis of gA by isolated microvessels is not inhibited by free glucose. If gA recognition by endothelium occurs via glucose recognition, free glucose should competitively reduce gA endocytosis. While gA does reduce the endocytosis of labeled gA, native albumin has the same effect on gA endocytosis. Further evidence against a glucose-specific binding site was our observation that albumin modified with a number of other monosaccharides including mannose, fructose, and galactose also exhibited preferential endocytosis over native albumin. Finally, if a specific finding site for gA mediates transendothelial transport of this molecule, then a mechanism for gA release from the abluminal endothelial surface must also be found.

Further evidence that binding sites are not necessary for the preferential transendothelial transport of solutes came from studies of glomerular permeability. Transport of glucosylated proteins across the glomerulus is regulated by the filtration characteristics of

Table 1. Endocytosis[a] of Glycosylated Albumin in the Presence of Putative Competitive Inhibitors

Additions[b]	Glu-Alb endocytosed[c] (pmole)
—	0.389 ± 0.05
Glucose (5 mM)	0.356 ± 0.06
Fructose (5 mM)	0.364 ± 0.03
Albumin (250 μM)	0.184 ± 0.03
Glu-Alb (250 μM)	0.189 ± 0.01

[a]Endocytosis of glucosylated albumin was measured by incubating fluorescein isothiocyanate-labeled glycosylated albumin with isolated rete fat microvessel endothelial cells. The amount of Glu-Alb endocytosed was measured spectrofluorometrically after 30 min at 37°C.
[b]Cells were preincubated with substances for 10 min prior to Glu-Alb addition.
[c]Endocytosis is expressed as pmole Glu-Alb endocytosed per 30 min per μg DNA.

the glomerular basement membrane. While glomerular endothelial cells may exhibit preferential uptake of glucosylated proteins via membrane association, the bulk of proteins directly associated with the glomerular basement membrane, passing essentially unobstructed through endothelial cells via fenestrations. If glucosylated proteins exhibited preferential binding to the glomerular basement membrane, we would expect these binding sites to rapidly be saturated without the ability to clear the bound material. For this reason we have begun to look for alternate explanations for the preferential transport of glucosylated proteins which do not necessitate the involvement of binding sites.

Physical-chemical studies comparing the structure of native albumin with nonenzymatically glucosylated albumin have been performed.[26] As previously mentioned, the extent of glucose addition is minimal with the addition of a single 120-Da glucose molecule to the 69-kDa albumin protein. The net effect on total charge is also minimal due to the coverage of only a single epsilon amino lysine group.[26,27] However, even this single modification results in a slightly more negative molecule. One may speculate that localized charge effects may have profound effects on albumin functions. While glucose modification does not significantly alter charge, one dramatic effect of glucose modification of albumin has been the observed change in the solubility of gA in aqueous solution. We have observed increased solubility of gA as compared to native albumin using an assay based on solubility in increasing concentrations of polyethylene glycol.[30] Thus, changes in solubility may provide an alternate explanation for enhanced gA transendothelial transport. In addition, the hypothesis that endothelial cell recognition of albumin is solubility dependant is compatible with the observation of increased glucosylated protein transport across the kidney glomerular basement membrane.

VII. CONCLUSIONS

Elevated plasma glucose concentration as a result of diabetes mellitus is now understood to cause a number of cellular biochemical changes. As discussed, the toxicity of glucose may be a result of both enzymatic and nonenzymatic processes. While the mechanism for the reported increase in the permeability of the microcirculation to glucosylated proteins is not clear, these studies have added an additional factor which must be considered in any description of general microvascular permeability. In addition to the well-documented effects of molecular size, shape, and net charge, the presence of carbohydrate groups on molecules also influences permeability. Diabetes mellitus represents the ultimate example of posttranslational protein modification by the process of nonenzymatic glucosylation. Regardless of the final mechanism(s) elucidated to explain the diabetic microangiopathies, the interaction of solutes with vascular endothelium provides a rich area to study cellular communication at the molecular level. While the complete elucidation of these processes will be complex, it is interesting to speculate that all of the miscommunication observed in diabetes mellitus may result from a single glucose modification.

ACKNOWLEDGMENTS. I acknowledge the technical assistance of Debbie Rose, Nina Solenski, Sandy McKenney, and David Carter, and thank Cathy Sawyer for her secretarial help.

REFERENCES

1. Sakamoto, N., Kinoshita, J. H., Kador, P. F., and Hotta, N. (eds.), 1988, Polyol pathway and its role in diabetic complications, in: *Excerpta Medica International Congress Series* **760**, Elsevier, Amsterdam.
2. Engerman, R., Bloodworth, J. M. B., Jr., and Nelson, S., 1977, Relationship of microvascular diabetes to metabolic control, *Diabetes* **26**:760–769.
3. Kador, P. F., 1988, The role of aldose reductase in the development of ocular diabetic complications, in: *Complications of Diabetes Mellitus*, Vol. 3 (B. Draznin, S. Melmed, and D. LeRoith, eds.), Liss, New York, pp. 103–114.
4. Kador, P. F., Robison, W. G., Jr., and Kinoshita, J. H., 1985, Pharmacology of aldose reductase inhibitors, *Annu. Rev. Pharmacol. Toxicol.* **25**:691–714.
5. Brownlee, M., Vlassara, H., and Cerami, A., 1984, Nonenzymatic glycosylation and the pathogenesis of diabetic complications, *Ann. Intern. Med.* **101**:527–537.
6. Yue, D. K., Morris, K., McLennas, S., and Turtle, J. R., 1980, Glycosylation of plasma protein and its relation to glycosylated hemoglobin in diabetes, *Diabetes* **29**:296–300.
7. Dolhoffer, R., and Wieland, O. H., 1979, Glycosylation of serum albumin: Elevated glycosyl-albumin in diabetic patients, *FEBS Lett.* **103**:282–286.
8. Eble, A. S., Thorpe, S. R., and Baynes, J. W., 1983, Nonenzymatic glucosylation and glucose-dependent cross-linking of protein, *J. Biol. Chem.* **258**:9406–9412.
9. Day, J. F., Thornburg, R. W., Thorpe, S. R., and Baynes, J. W., 1979, Nonenzymatic glucosylation of rat albumin, *J. Biol. Chem.* **254**:9394–9400.
10. Olufemi, S., Talwar, D., and Robb, D. A., 1987, The relative extent of glycation of haemoglobin and albumin, *Clin. Chim. Acta* **163**:125–136.
11. Monnier, V. M., and Cerami, A., 1981, Nonenzymatic browning in vivo: Possible process for aging of long-lived proteins, *Science* **211**:491–493.
12. Armbruster, D. A., 1987, Fructosamine: Structure, analysis, and clinical usefulness, *Clin. Chem.* **33**:2153–2163.
13. Brownlee, M., 1988, The role of nonenzymatic glycosylation in the pathogenesis of diabetic angiopathy, in: *Complications of Diabetes Mellitus*, Vol. 3 (B. Draznin, S. Melmed, and D. LeRoith, eds.), Liss, New York, pp. 9–17.
14. Pongor, S., Ulrich, P. C., Bencsath, F. A., and Cerami, A., 1984, Aging of proteins: Isolation and identification of a fluorescent chromophore from the reaction of polypeptides with glucose, *Proc. Natl. Acad. Sci. USA* **81**:2684–2688.
15. Rothschild, M. A., Oratz, M., and Schreiber, S. S., 1988, Serum albumin, *Hepatology* **8**:385–401.
16. Williams, S. K., Devenny, J. J., and Bitensky, M. W., 1981, Micropinocytic ingestion of glycosylated albumin by isolated microvessels: Possible role in pathogenesis of diabetic microangiopathy, *Proc. Natl. Acad. Sci. USA* **78**:2393–2397.
17. Williams, S. K., and Solenski, N. J., 1984, Enhanced vesicular ingestion of nonenzymatically glucosylated proteins by capillary endothelium, *Microvasc. Res.* **28**:311–321.
18. Ghiggeri, G. M., Candiano, G., Delfino, G., and Queirolo, C., 1984, Glycosyl albumin and diabetic microalbuminuria: Demonstration of an altered renal handling, *Kidney Int.* **25**:565–570.
19. McVerry, B. A., Hopp, A., Fisher, C., and Huehns, E. R., 1980, Production of pseudodiabetic renal glomerular changes in mice after repeated injections of glucosylated proteins, *Lancet* **1**:738–740.
20. Kowluru, A., Kowluru, R., Bitensky, M. W., Corwin, E.-J., Solomon, S. S., and Johnson, J. D., 1987, Suggested mechanism for the selective excretion of glucosylated albumin. The effects of diabetes mellitus and aging on this process and the origins of diabetic microalbuminuria, *J. Exp. Med.* **166**:1259–1279.
21. Villaschi, S., Johns, L., Cirigliano, M., and Pietra, G., 1986, Binding and uptake of native and glycosylated albumin–gold complexes in perfused rat lungs, *Microvasc. Res.* **32**:190–199.
22. Layton, G. J., and Jerums, G., 1988, Effect of glycation of albumin on its renal clearance in normal and diabetic rats, *Kidney Int.* **33**:673–676.
23. Sampietro, T., Colantuoni, a., Lenzi, S., Berguglia, S., Bionda, A., and Donato, L., 1987, Increased permeability of hamster microcirculation to glycosylated albumin, *Lancet* **2**:994–996.
24. Mereish, K. A., Rosenberg, H., and Cobby, J., 1982, Glycosylated albumin and its influence on salicylate binding, *J. Pharm. Sci.* **71**:235–238.

25. Tsuchiya, S., Sakurai, T., and Sekiguchi, S., 1984, Nonenzymatic glucosylation of human serum albumin and its influence on binding capacity of sulfonylureas, *Biochem. Pharmacol.* **33**:2967–2971.
26. Garlick, R. L., and Mazer, J. S., 1983, The principal site of nonenzymatic glycosylation of human serum albumin in vivo, *J. Biol. Chem.* **258**:6142–6146.
27. Bohney, J. P., Fonda, M. L., and Feldhoff, R. C., 1989, Nonenzymatic glycosylation and pyridozylation of human serum albumin: Site-specificity of Schiff base and ketoamine formation, *FASEB J.* **3**:A926.
28. Williams, S. K., and Siegal, R. K., 1985, Preferential transport of nonenzymatically glucosylated ferritin across the kidney glomerulus, *Kidney Int.* **28**:146–152.
29. Predescu, D., Simionescu, M., Simionescu, N., and Palade, G. E., 1988, Binding and transcytosis of glycoalbumin by the microvascular endothelium of the murine myocardium: Evidence that glycoalbumin behaves as a bifunctional ligand, *J. Cell Biol.* **107**:1729–1738.
30. Williams, S. K., Carter, D., McKenney, S. L., and Rose, D. G., 1988, Solubility dependent endocytosis of proteins by vascular endothelium, *FASEB J.* **2**:A945.

VI

Endothelial Cell in Neoplasia and Metastasis

24

Role of the Vascular Endothelium in Cancer Metastasis

Paula N. Belloni and Garth L. Nicolson

I. INTRODUCTION

The overwhelming majority of patients who die from cancer do so because of metastatic disease. The prevention of cancer deaths is therefore dependent on understanding the mechanism of malignant tumor cell spread from primary tumor sites to distant organ sites. Cancer metastases form by way of a complex series of sequential steps involving a variety of tumor cell and host properties. These steps typically involve growth and invasion of malignant cells at primary sties, followed by their penetration of the lymphatics, the circulatory system, and body cavities. Once in this compartments, the malignant cells can detach and be transported to distant sites. There they can implant and invade surrounding tissues and establish suitable microenvironments for survival and growth.

Although the location of most regional metastases can be explained on the basis of anatomical and mechanical factors,[1] clinical and experimental observations indicate that many malignant tumors preferentially metastasize to particular distant organ sites.[2–6] Historically, two arguments have been advanced to explain the organ preference of distant metastases. Ewing argued that tumor cells mechanically lodge in the first capillary bed encountered during their blood-borne transit and that the anatomic structure and hemodynamics of the vascular system determine the secondary locations of tumor colonization.[7] In contrast, the "seed and soil" hypothesis of Paget[8] proposed that distant metastasis was a consequence of the successful interaction of the tumor cells or "seeds" with the microenvironment of a particular target organ or "soil," which allowed for specific implantation, survival, and growth of the tumor cells. These two hypotheses are not mutually exclusive: organ colonization is probably determined both by hemodynamic factors and by tumor cell and host characteristics unique to particular tumor cells and host environments.[2,9–14]

During the transit of tumor cells or normal leukocytes, the vascular endothelium is the

---

Paula N. Belloni and Garth L. Nicolson • Department of Tumor Biology, The University of Texas M. D. Anderson Cancer Center, Houston, Texas 77030.

Endothelial Cell Dysfunctions, edited by Nicolae Simionescu and Maya Simionescu. Plenum Press, New York, 1992.

primary barrier between blood or lymph and the extravascular tissues. Specific interactions of circulating cells with components of the vessel lumen and underlying basement membrane are thought to be important for cell targeting to different tissues and their exit from the circulation. For example, lymphocytes that circulate continuously between the bloodstream and lymphatics cross the vascular barrier at specialized regions known as high endothelial venules (HEV).[15–18] These specialized endothelial cells are morphologically and antigenically distinct from other endothelial cells,[19] and HEVs from different lymphoid tissues express antigenic differences that can be recognized by circulating lymphocytes.[20–22] Recruitment of blood leukocytes into tissues containing foreign antigens requires interaction with and invasion of capillary endothelium.[16,23] Similarly, malignant tumor cells that preferentially colonize certain organ sites must adhere to and invade through endothelial cells and basement membranes during blood-borne metastasis.[5]

The adhesion of malignant cells to microvascular endothelial cells at specific organ sites appears to be involved, in part, in determining the distribution and organ preference of certain metastases.[2,5,6,14,25,26] The differential expression of organ-associated molecules or domains on the endothelial cell surfaces in capillaries is thought to play a role in preferential tissue localization of blood-borne tumor cells.[2,6,14,27–29] The dynamic expression of common and organ-specific properties by normal and activated endothelial cells at diverse tissue sites forms the basis of tremendous heterogeneity among microvascular tissues. This review will focus on both common and organ-specific properties of microvascular endothelial cells and their role in organ-associated metastasis of tumor cells.

II. STRUCTURAL HETEROGENEITY OF ENDOTHELIAL CELLS AT VARIOUS SITES

Each segment of the microvascular bed has its own characteristic structure, and the length, diameter, wall thickness and ultrastructure vary considerably among endothelial cells in different tissues and organs.[30,31] Continuous endothelium is termed thick or thin depending on the thickness of the endothelial cells. Continuous thick capillaries (cell layer > 2 μm) are found in skeletal tissue, cardiac smooth muscle, testes, and ovarian tissues,[31] while continuous thin capillaries (cell layer < 1 μm) are typical in tissues of the central nervous system, lung, dermis, vasa rectum of the kidney, spleen, thymus, bone marrow, and bone. These endothelial cells are generally flat and hexagonal and contain the normal complement of cytoplasmic organelles. Both the apical and basolateral endothelial cell surfaces are rich in cytoplasmic and pinocytotic vesicles (~ 600–700 Å in diameter), most of which lie just below the plasma membrane. Weibel–Palade bodies can be found in the arterial capillaries, but these are usually rare.[32] The continuity and integrity of the endothelial monolayer is maintained by gap junctions and desmosomes, except in the central nervous system where tight junctions predominate. The cells also have multiple points of attachment to a basement membrane containing laminin, various proteoglycans (particularly heparan sulfate proteoglycan), fibronectin, and collagen types III, IV, and V. The composition of the basement membrane is also a source of vascular heterogeneity since variations exist not only in the distribution of classical matrix proteins,[33–37] but also in the types of regulatory proteins involved in growth control and hemostasis that are stored in the extracellular matrix.[38–41] Although endothelial cells are morphologically similar, there are distinctive regional differences in the size and thickness of cells (e.g., large vessel endothelium is thicker than microvascular endothelium) as well as variations in the numbers of specific organelles, types of junctions, and cell surface receptors. Specialized structures such as fenestrations, transendothelial channels, and gap junctions that participate in the transfer of circulating molecules and cells across the endothelium in other tissues are absent in brain capillaries but are common in visceral fenestrated capillaries.

A. Differential Lectin Binding to Vascular Endothelial Cells

Lectins—carbohydrate-binding proteins that can be used to detect cell surface determinants *in vivo* and *in vitro*—and cationic proteins have been used to detect differences in endothelial cells in different organs. Using cationic ferritin and lectin–peroxidase conjugates to probe the fenestrated surfaces of murine pancreas and intestinal mucosa, Simionescu and others[42–48] have demonstrated that the microvascular system is a complex network of differentiated cell surface microdomains, identifiable by differences in negative charge and glycoconjugate composition. Differences in net negative surface charge have been attributed to changes in endothelial cell-surface sialic acid content,[49] presumably on sialoglycoproteins and sialoglycolipids. The existence of endothelial cell surface microdomains has also been reported for sinusoidal endothelium.[50,51]

Lectin–colloidal gold complexes have been used to demonstrate differential expression of glycoconjugates by ultrastructural localization of receptors on the luminal and abluminal surface of brain microvessels.[52] Lectins that recognize mannose and *N*-acetylgalactosamine-containing glycoconjugates selectively labeled the abluminal surfaces and basement membranes. Some differences were also noted in the distribution of lectin receptors on the three types of brain microvessels (capillaries, arterioles, and venules) associated with the maintenance of the blood–brain barrier.[53] Similar techniques were used by Ponder and Wilkinson,[54] who investigated the *Dolichos biflorus* agglutinin-binding sites on glycoconjugates expressed at endothelial cell surfaces in different mouse organs. Differential binding was noted not only on endothelial cell surfaces in different organs, but also on different vessels within the same organ. Relatively uniform lectin binding to the endothelium was found in several tissues. For example, in lung a high number of binding sites were found on pulmonary arteries, fewer on pulmonary veins, a barely detectable binding to pulmonary microvessels, and none on the endothelia in lymph nodes, adrenal gland, or brain cortex, demonstrating the variable distribution of *N*-acetylgalactosamine-containing glycoconjugates on endothelia of different tissues.

Biochemical techniques have also been used to demonstrate differences in glycoconjugate expression on endothelial cells in various organs. *In situ* cell-surface labeling techniques combined with Western enzyme-linked lectin analysis have been used to demonstrate that microvessel endothelia *in vivo* and isolated microvessel endothelial cells *in vitro* express organ differences in cell surface glycoproteins.[55] In this study, endothelial cell surface proteins were selectively labeled with ^{125}I *in situ* in perfused animals, and the cell surface glycoproteins were solubilized from isolated microvessels and analyzed using a panel of lectin–horseradish peroxidase (HRP) conjugates. Certain lectins, such as concanavalin A (from *Canavalia ensiformis*), reacted with the microvascular glycoproteins from all organs, whereas other lectins reacted selectively to the microvascular glycoproteins from specific organs. For example, wheat germ agglutinin (WGA) was highly reactive with brain microvessel glycoproteins, and *Ricinus communis* agglutinin (RCA)-binding protein bound preferentially to lung microvessel glycoproteins. Some of the glycoproteins appeared to be selectively expressed in a particular organ, such as the WGA-binding brain glycoproteins gp135 and gp90, while others, such as RCA-binding gp120, appeared to be expressed in several organ-derived microvessels (Table 1).[55] The results demonstrate that microvessel endothelial cells express cell surface glycoprotein profiles that are unique to their organ environments and suggest that specific glycosyl transferase activities may be expressed differentially or have different activities in certain organs. In addition, some organ-associated endothelial cell glycoproteins appeared to be conserved across species barriers,[55–57] while others showed variations between species.[58–60] These studies strengthen the conclusion that endothelial cells from different tissues express heterogeneous cell-surface glycoconjugates.

B. Antigenic Heterogeneity of Vascular Endothelium

The heterogeneity of murine vascular endothelium was first demonstrated by Pressman,[61,62] who used antibodies generated against vascular trees. Differences in the antigenic

Table 1. Differentially Expressed Organ-Derived Microvascular Glycoproteins[55]

	Differentially expressed glycoproteins ($M_r \times 10^3$)				
Lectins	Brain	Heart	Lung	Liver	Kidney
WGA	220		160		150
acid-sensitive	78	—	—	—	140
	135		180	180	
	90		160	110	
WGA	75	85	120		
acid-resistant	50				65
	25			35	40
			140		170
		120	120		120
RCA_I		110	110	90	
	84		80		66
	60		55		50
	47				47
		180	180	165	
	110	120	125	140	
Con A	105	85	85	85	
		45	48	43	18
				33	16
			160		
PNA	31		90		
	25		55	48	
			45	35	
				35	
	31	29	32, 30, 28	32, 30, 28	
UEA_I	30	28		25	
		19		23	
	16, 14, 12	16, 14, 12	16, 14, 12	16, 14, 12	
LTA	—	—	—	—	

composition of various organ microvascular endothelia were demonstrated after intravenous injection of labeled polyclonal antisera raised against isolated organ-derived vessels. The labeled antibodies localized to the organ used as a source of immunogen, demonstrating organ-specific expression of certain determinants on endothelial cell surfaces. More recently, experiments with panels of monoclonal antibodies produced by immunization with isolated brain microvascular cells or heterogeneous populations of lung cells indicated that the luminal surfaces of certain capillary endothelia may be organ associated.[63]

Several endothelial surface antigens have been identified using polyclonal and monoclonal antibodies (MAb) generated against the blood–brain barrier. Many of these antibodies express distinctive tissue distributions, but few have been found to be specific to brain endothelium. For example, Ghandour *et al.*[64] generated a monoclonal antibody (MESA-1) against mouse cerebellum that was localized to endothelial cells of capillaries, veins, and arteries of all tissues with the exception of liver sinusoids. Zanetta *et al.*[65] have produced antibodies against a lectin-like molecule of $M_r \sim 130{,}000$ that is specific to brain and concentrated in endothelial cells and neurons. Michalak *et al.*[66] generated a panel of MAbs against isolated rat brain microvessels and found that only 1 (EBM) out of 156 was specific to brain endothelia. The antigen-reactive epitope was also expressed by the brush border of kidney proximal tubules and liver bile canaliculi. The authors suggested that the epitope may be associated with the specialized transport properties of these three tissues. The expression of two protein antigens was reported to be restricted to brain endothelial cells, a $M_r \sim 46{,}000$ glycoprotein associated with the lateral membrane of brain capillaries[67] and a $M_r \sim 76{,}000$ protein associated with the acquisition of barrier function in brain microvessels.[68]

Many of the endothelial–monocyte antigens defined by organ transplantation studies are expressed in an organ-restricted manner. For example, MAb UCHM 1 produced intense staining of liver and lymphoid endothelia but not lung, brain, skin, or kidney endothelia.[69] In contrast, MAb Mac-3 recognizes a $M_r \sim 110{,}000$ protein, a β-integrin subunit, commonly expressed on macrophages and all vascular endothelia.[70] Several other transplantation-related antigens have been identified on endothelial cells.[71,72]

Evidence for endothelial cell-associated expression of organ-specific antigens has also been reported for the lung. Two MAbs generated against lung homogenates reacted with a protein (p112) that has expression restricted to lung capillary endothelium.[73] Administration of these MAbs into mice increased lung colonization of B16–F10 melanoma cells.[74] In other studies, MAbs generated against rat lung angiotensin-converting enzyme (ACE) preferentially reacted with lung relative to kidney ACE on frozen tissue sections.[75] Another antigen commonly used to identify the endothelial nature of cells, factor VIII-related antigen (FVIII-RAg), is present in most organ endothelia except liver sinusoids,[76] renal glomerular capillaries,[77] or lymphatic endothelium.[78]

C. Homing Receptor Heterogeneity of Vascular Endothelium

Further evidence supporting the tissue specificity of endothelia has evolved from investigations on the molecular basis of lymphocyte homing and identification of vascular addressins involved in homing.[20] MAbs have been generated against tissue-specific endothelial cell surface antigens involved in lymphocyte homing to mucosal lymphoid tissue[21] or peripheral lymphoid tissue.[22] Other endothelial cell molecules involved in lymphocyte adhesion, intercellular adhesion molecule-1 (ICAM-1),[79] and endothelial–leukocyte adhesion molecule (ELAM-1),[80] are inducible in most tissues and contribute to transient heterogeneity at sites of inflammation. Although many other organ-specific endothelial adhesion molecules or vascular addressins may exist, none have yet been identified.

III. FUNCTIONAL DIVERSITY OF THE VASCULAR ENDOTHELIUM

Endothelial cells in various tissues are metabolically distinct as well as structurally dissimilar. They can participate to various degrees in: (1) the maintenance of a nonthrombogenic luminal surface via production of prostacyclin; (2) the regulation of the coagulation pathway (clot formation and fibrinolysis); (3) the presentation of antigens to T lymphocytes; (4) the secretion of chemotactic monokines; (5) leukocyte adhesion and extravasation; (6) secretion of angiogenic factors and other growth factors that coordinately stimulate the growth of new blood vessels, smooth muscle cells, and pericytes; (7) the clearance and metabolism of vasoactive substances such as angiotensin I, bradykinin, ADP, and thrombin; (8) the process of vasomotion; (9) the regulation of perfusion and permeability at the microvascular level and responses to and secretion of autocoids such as PGI_2, epinephrine, and vasopressin; and (10) the regulation of macromolecular transport between the plasma and interstitial fluids via large pores, vesicular transport, and specific cell surface receptors. Although the highly regulated endothelial functions associated with hemostasis and metabolism of vasoactive substances undoubtedly contribute to the metastatic process of certain tumors, details on the heterogeneity of endothelial cell metabolism are beyond the scope of this review. The following discussion on the functional diversity of endothelia will be limited to metabolic properties commonly used for the characterization of endothelial cultures and model systems of the vascular endothelium. The reader is referred to other reviews for more information.[14,24,81,82]

A. Selective Barrier Functions of the Vascular Endothelium

Transcapillary movement of solutes, such as hormones, drugs, sugars, amino acids, lipids, and electrolytes, is regulated by endothelial cells. The regulatory mechanisms include: (1) cell surface microdomains that attract or exclude molecules based on their size and charge: (2) receptor-specific membrane transport; (3) ectoenzymes that chemically alter solutes transferred into, out of, or between endothelial cells; and (4) synthesis and release of solutes by endothelial cells resulting in concentration gradients between plasma and the interstitial tissues. Classical theories and contemporary models of the microvascular membrane describe it as a static pore system, in which solute and fluid flux are regulated by changes in flow, pressure, and surface area. This may be a suitable description of fluxes under basal conditions; however, it cannot explain the increase in protein efflux that occurs during inflammatory reactions. The concept of a variable mediator-stimulated pathway in which microvascular membrane permeability is physiologically controlled has evolved from a static pore system in which fluid and solute flux are regulated by changes in blood flow to a highly regulated macromolecular transport system. Such a variable macromolecular transport pathway has been shown to be operable in most vascular beds. For example, lung microvessels are not responsive to most classical inflammatory mediators, and certain mediators show tissue-selective effects.[83] The regulation of microvascular permeability *in vivo* has been reviewed extensively.[84–88]

Endocytosis, transendothelial cell channels, vasoactive stimuli, and endothelial desquamation all contribute to the regulation of the transport of components through the vascular endothelium. Regional and local tissue variability also exist in this complicated regulatory process. Cultured endothelial cell monolayers derived from a number of different vascular sites and species have been used extensively to study transendothelial cell permeability.[89] Barrier function is partially attributable to components in the subendothelial basement membrane as well as to the presence of tight junctional complexes between endothelial cells.[90]

Localized tumors are known to dramatically alter the permeability and transport properties of blood vessels. In brain tumors a common problem is vasogenic edema that can cause subsequent neurological deficits.[91] Ultrastructural studies have demonstrated structural alterations in tumor-induced blood vessels[92,93] and peritumoral brain microvessels,[94] such as increased numbers of pinocytotic vesicles, large vacuoles, and open junctions, any of which may cause increased permeability. Other changes that may contribute to increased cerebrovascular permeability include destruction of brain capillaries,[91,94] increased hydrostatic pressure,[92] and biochemical mediators.[95,96]

B. Synthetic and Metabolic Functions of the Vascular Endothelium

The tissue factor thromboplastin is a glycoprotein produced by endothelial cells ($M_r \sim$ 43,000) that forms an enzymatically active complex at the cell surface with factor VII to activate the extrinsic coagulation system via activation of factor X.[97] Antibody staining indicates that it is not uniformly expressed on endothelial cell surfaces; it is particularly abundant in brain, lung, and placenta,[98] whereas other tissues show little or no reactivity to antithromboplastin.[81] *In vitro* studies have linked disseminated intravascular coagulation, which is often associated with acute bacterial infections, with increased expression of tissue factor on endothelial cells exposed to endotoxin.[99] Tissue factor expression by human umbilical cord endothelial cells is increased fivefold by the immune modulator interleukin 1 (IL-1), and a similar response is induced by tumor necrosis factor, suggesting that a more general link exists between inflammation and coagulation.[100,101]

Another high-molecular-weight glycoprotein complex ($M_r \sim 1 \times 10^6$) that is important in hemostasis and blood coagulation is factor VIII. It consists of two glycoproteins that have distinct functional, biochemical, and immunological properties and are under separate genetic regulation.[101,103] One component of the complex, factor VIII:C, is a glycoprotein of $M_r \sim$ 250,000–300,000 that has antihemophilic procoagulant activity. MAbs against VII:C/Ag have been used to determine that liver sinusoidal endothelial cells are the primary source of the glycoprotein, as well as mononuclear cells in the lymph nodes and alveolar septa of lung.[104] However, reactivity was not found in the vascular beds of other organs.

Factor VIII-related protein or von Willebrand factor (vWF) is the major component (> 90% by weight) of plasma factor VIII. vWF is composed of several subunits of $M_r \sim$ 200,000–225,000. Plasma vWF binds to the subendothelial matrix and interacts with the platelet membrane receptor GpIIbIIIa to mediate platelet adhesion. After thrombin stimulation, platelet adhesion to vWF occurs in an RGD-dependent manner via the platelet integrin GpIIb/IIIa.[105,106] Initial immunofluorescent studies have identified vWF in the endothelial cells of arteries, veins, arterioles, capillaries, throughout the body, as well as in megakaryocytes and platelets.[107] Factor VIIIR:Ag has also been detected in the perinuclear region of cultured endothelial cells, in small rodlike structures in the cytoplasm, and on filaments in the extracellular matrix.[108] Although factor VIIIR:Ag has been identified in cultured endothelial cells from various species and types of vessels,[109] its expression on endothelial cells during long-term culture is often reduced or undetectable, depending on the age and vessel source.[110] Recent studies by Dejana *et al.*[111] indicate that vWF also supports the adhesion and spreading of cultured endothelial cells via an RGD-dependent mechanism.

Several naturally occurring vasoactive substances have been shown to induce vascular and endothelial cell prostaglandin PGI_2 synthesis (ADP, angiotensin II, platelet-derived growth factor, thrombin, and histamine). The response appears to be both species- and site-specific. For example, Hong[112] has demonstrated thrombin-inducible PGI_2 synthesis in human umbilical cord endothelial cells, but this response was not detected in calf or pig aortic endothelial cells. Increased synthesis of PGI_2 after thrombin stimulation was also reported for sheep umbilical cord vein EC, but no effect was detected in sheep aortic segments.[113] Prostacyclin metabolism also varies in different endothelial cell cultures. Cultured rat aortic endothelial cells degrade PGI_2 into keto-acids by the enzyme 15-hydroxyprostaglandin dehydrogenase,[114] but degradation was not measurable in canine pulmonary microvascular endothelial cells.[115] In addition, differences have been detected in PGI_2 synthetic capacities expressed by human pulmonary arteries and veins.[116]

C. Subendothelial Matrix-Associated Molecules

Endothelial cells synthesize and secrete various molecules associated with the subendothelium of blood vessels, such as collagens, laminin, fibronectin, and glycosaminoglycans, as well as factor VIII and thrombospondin.[117,118] There is increasing evidence that the secretion and deposition of extracellular matrix components differs in various types of endothelial cells. Differences in the types of collagen produced by endothelial cells and their distribution have been reviewed elsewhere.[119] The major collagens produced by endothelial cells from various species, including human, appear to be types IV and V.[120]

D. Other Properties of Cultured Microvessel Endothelial Cells

Our analyses of brain-, lung-, and liver-derived microvascular endothelial cell lines isolated from various species suggest that many of the cultures differentially express components that are consistent with their *in vivo* properties.[121] Endothelial cells were characterized on the basis of cell morphology, contact inhibition of growth, nonthrombogenic apical surfaces, presence of factor VIII/vWF antigen, binding of acetylated low-density lipoprotein (Ac-LDL), and synthesis of basement membrane-associated components. The endothelial cell lines that were established from microvessel trees expressed most but not all of these characteristics (Table 2). For example, factor VIII/vWF antigen was detected in all of the cell lines, except in the liver-derived endothelial cells. The intensity of antibody reactivity to VIII/vWF antigen, however, varied for different endothelial cells within a particular culture. Furthermore, as Maciag *et al.*[122] and Hormonia[123] reported, antibody reactivity decreased with culture passage. The finding that liver-derived sinu-

Table 2. Characteristics of Organ-Derived Endothelial Cell Cultures

Cell line[a]	Cobblestone morphology[b]	Thrombogenic surface[c]	FVIII/vWF Ag[d]	UEA[e]	Dil-Ac-LDL[g]
BAE	++[g]	−	++	+	+
BCE	++	−	+	−	±
RLE-8	++	−	+	++	+
C57-LE-1	++	−	+	±	±
C57-LE-2	++	−	++	±	±
C57-HSE	++	−	±	−	++
C57-HSE-4.5	++	−	−	−	++
C57-MBE-12	++	−	++	−	−
HBE-1	+	−	++	++	−
HBE-2	+	−	++	++	−
HBE-2.11	+	−	++	++	−
HME-1	±	−	+	++	nd
HHB-1.6	++	−	++	++	−
HHB-1.8	±	−	+	++	nd
HF	−	++	−	−	−

[a]Low-passage cultures of endothelial cells were examined for properties associated with cultured endothelia: BAE (bovine aortic); BCE (bovine corneal); RLE (rat lung); HBE (human brain); HME (human menangioma); HHB (human hemangioblastoma); C57-LE (mouse lung); C57-HSE (mouse hepatic sinusoids); and C57-MBE-12 (mouse brain). Human forskin fibroblasts (HF) are used as a control.[121]

[b]Contact inhibition and cobblestone monolayer morphology determined by phase-contrast microscopy.

[c]Thrombogenic surface determined by platelet-binding assay as described in methods.

[d]Factor VIII/von Willebrand factor antigen (vWF:Ag) determined by indirect immunofluorescence using rabbit anti-human vWF:Ag antibody and FITC-conjugated goat anti-rabbit Ig.

[e]*Ulex europaeus* lectin (UEA) staining determined by TRITC-conjugated lectin.

[f]Acetylated-LDL uptake determined by staining with fluorescent probe (Dil-Ac-LDL).

[g]−, negative; ±, weakly positive (≥ 25% cells); +, positive (≥ 50% cells), ++, strongly positive (≥ 80% cells).

soidal endothelial (HSE) cells were negative for factor VIII/vWF antigen was consistent with our inability to stain mouse liver sinusoids with this antibody and with previous reports on the distribution of this antigen *in vivo*.[76] HSE cells selectively internalized modified LDL, Dil-Ac-LDL, as did bovine aortic, bovine pulmonary, and rat lung endothelial cells; however, both mouse and human brain-derived microvascular endothelial cells lacked this property. A similar *in vivo* distribution of receptors for Ac-LDL in rat endothelia has been reported.[124] Rat brain microvascular endothelial cell cultures also appeared to lack this property[125] and variably expressed other endothelial cell markers, suggesting that the brain microvasculature is a mosaic or that morphologically distinct brain cultures may originate from different microanatomical sties. Differences between organ-derived endothelial cell cultures were consistent with phenotypic variations that have been observed *in vivo* in different organs.

Cultures of organ-derived endothelial cells possessed distinct patterns of growth responsiveness to purified growth factors (Fig. 1). Microvascular endothelial cells established from each organ tested responded to endothelial cell growth factor (ECGF). While maximum growth stimulation of lung endothelial cells (LE) was achieved by thrombin alone, brain endothelial cells (MBE) required combinations of thrombin and ECGF, or thrombin and subendothelial extracellular matrix. Liver sinusoidal endothelial cells (HSE)

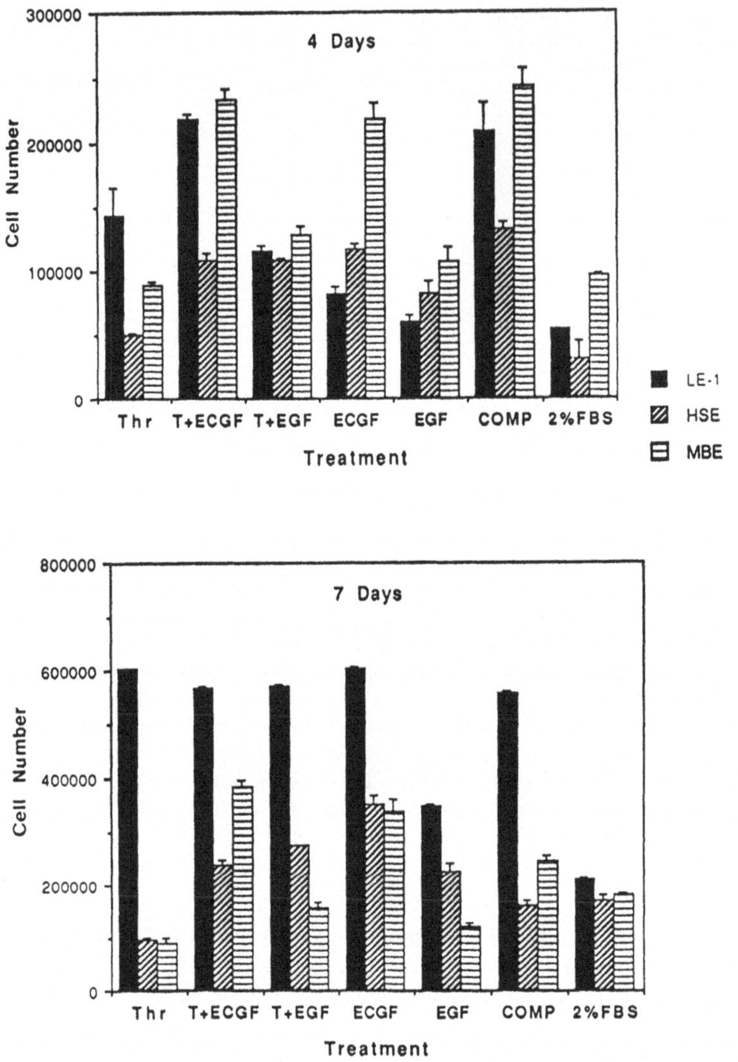

Figure 1. Effects of various growth factors on proliferation of microvascular endothelial cells. Endothelial cells were plated at an initial density of 10,000 cells/well in media containing 2% FBS and various combinations of growth factors: 3 U/ml thrombin (Thr); 3 U/ml thrombin plus 50 μg/ml ECGF (T + ECGF); 3 U/ml thrombin plus 100 μg/ml EGF (T + EGF); 50 μg/ml ECGF (ECGF); 100 μg/ml EGF (EGF); no additions (2% FBS); complete medium (COMP) containing 5% FBS, 2% PPHS, 100 μg/ml ECGF, and 10 U/ml heparin. Cell counts were determined at days 4 and 7 in triplicate. Mouse lung (LE-1); mouse sinusoids (HSE); mouse brain (MBE-12).

were relatively unresponsive to thrombin, but were the most responsive cells to EGF. High-affinity thrombin binding sites were detected that were differentially expressed on the organ-derived endothelial cell cultures (LE > MBE > HSE > BAE), suggesting that the growth responsiveness of these cells correlates with their number of high-affinity receptors.[121]

IV. INTERACTIONS OF METASTATIC TUMOR CELLS WITH THE VASCULAR ENDOTHELIUM

Numerous clinical and experimental studies indicate that malignant tumors tend to metastasize to particular secondary tissues.[1,3,5,126,127] For example, human mammary carcinomas often metastasize to bone and brain,[128] small cell carcinomas of the lung form secondary colonies in brain,[129] and bone is a common metastatic site for clear cell carcinomas of the kidney[130] and prostatic carcinomas.[131] Model metastasis systems also demonstrate preferential organ localization.[5] The murine B16 melanoma tumor system has been selected *in vivo* for organ colonization to yield tumor cell lines with enhanced ability to colonize lung, ovary, or brain.[132–134] Kinetic distribution analyses of B16 melanoma cells injected into the circulation via tail vein or left ventricle of the heart indicated that lung was preferentially colonized at equivalent efficiencies by lung-selected B16–F10 cells, after either route of injection.[132] Since tumor cells that have entered the bloodstream from the left ventricle must travel through the extrapulmonary circulation before reaching the lung microvasculature, arrest in the lungs appeared to involve specific recognition rather than random trapping of the B16 cells. Since tumor cell arrest and penetration occur most frequently in the capillaries or postcapillary venules, adhesion or recognition molecules expressed on microvascular endothelial cells and in the underlying matrix are probably important in the initial implantation of tumor cells.

A. Experimental Systems for Analyzing Tumor Cell–Endothelial Cell Interactions

The significance of tumor cell–host interactions in preferential tumor colonization has been demonstrated in several different experimental systems. Greene and Harvey[26] suggested that the organ distribution patterns of metastatic foci were dependent on the formation of sufficient adhesive bonds between arrested tumor cells and endothelial cells, and they hypothesized that the interactions were similar to lymphocyte/endothelial cell interactions at sites of inflammation. Furthermore, they suggested that endothelial cells of various organs expressed differential receptivity for certain tumor cells. Evidence for specific organ cell recognition and adhesion in preferential organ metastasis was demonstrated by testing the ability of lung-colonizing B16–F melanoma cell lines to aggregate single cell suspensions derived from the lung, liver, spleen, or kidney.[27] The highly metastatic B16–F13 cells aggregated lung cells into a single clump, while B16–F1 cells caused only slight aggregation. Aggregation of cells derived from nontarget organs with B16–F cells was not significant, suggesting that tumor colonization may occur through specific adhesive interactions via organ-specific recognition determinants that are shared by the parenchyma and endothelial cells. Similar correlations between organ-specific adhesion and metastatic properties in other tumor systems exist, such as in a spleen-colonizing leukemia[135] and two liver-colonizing murine lymphomas, RAW117[136] and Eb/Esb.[137] In other *in vitro* experiments, suspensions of organ tissues, such as lung, ovary, and heart, were used to demonstrate the preferential adhesion and invasion of target tissues by B16[138] and RAW117[139] sublines of differing metastatic colonizing potentials.

Tissue-specific tumor cell adhesion has also been demonstrated using cryostat tissue sections. Netland and Zetter[142] noted that metastatic tumor cells preferentially adhered to cryostat sections of the same organs the tumor cells selectively colonize *in vivo*. When B16-F1 melanoma cells were selected for adhesion to lung tissue sections, the selected cells produced a sixfold increase in lung metastases *in vivo*. In contrast, tumor cells selected for adhesion to brain tissue *in vitro* did not show altered metastatic properties compared with the unselected cells.

Cultured endothelial cell monolayers derived from bovine aorta (BAE) or human umbilical vein (HUVE) have been used to investigate the adhesive properties of various tumor cells *in vitro*. With the exception of platelets,[143] significant differences in adhesion to endothelial cell monolayer surfaces were not found using large vessel endothelial cell cultures. Metastatic cells stimulated endothelial cell retraction at cell junctions adjacent to the adherent malignant cells and were invasive. This behavior was uncharacteristic of most noninvasive, nonneoplastic cells. After the metastatic cells stimulated endothelial cell retraction, they generally moved toward and adhered to the exposed subendothelial matrix. This was followed by tumor cell spreading on the subendothelial matrix, cell movement, dissolution of the matrix, and penetration of this structure (Fig. 2). This same sequence is known to occur *in vivo*.[140,141]

The kinetics of tumor cell adhesion to endothelial cell monolayers differs among tumor subpopulations and often correlates with their observed metastatic properties.[139,144–148] For example, in an analysis of the adhesive characteristics of 11 cloned cell lines derived from a rat rhabdomyosarcoma, four of five clones displaying low lung-colonizing capacity showed low attachment rates to endothelial cell monolayers, while all six of the high lung-colonizing lines studied had high attachment rates.[148] Although the rates of attachment to extracellular matrix were greater than to the cell monolayers, differences in adhesiveness observed between high and low metastatic lines were not evident using extracellular matrix. In general, the subendothelial matrix provides a better adhesive substrate for most tumor cells than endothelial cell surfaces, but differences in metastatic potentials may or may not be reflected in differential rates of tumor cell adhesion to subendothelial matrix.[149–152] Correlations between metastatic potentials of various tumor cells and their adhesion to particular matrix components such as laminin, fibronectin,[153–157] and type IV collagen,[157,158] have been reported, but will not be reviewed here (see Refs. 5, 159).

B. Tumor Cell–Endothelial Cell Adhesion and Metastasis

Organ-derived microvascular endothelial cell cultures have recently been developed for examining the basis for preferential metastatic cell localization. Nicolson[28] demonstrated that brain-colonizing B16 melanoma cells adhered to brain-derived endothelial cells faster than to endothelial cells of unrelated origin. Auerbach[6,160] examined the adhesive interactions of tumor cells with microvascular endothelial cells and correlated the rates of adhesion with their tissue origin or metastatic organ preference. For example, hepatoma and glioma cell lines preferentially adhered to the endothelial cells derived from the organ of tumor origin, whereas metastatic teratoma cells preferentially attached to endothelial cells derived from ovaries, the target organ for metastasis of these cells. Adhesion preference was not observed when large vessel endothelial cell monolayers were substituted for microvessel endothelial cells. Roos *et al.*[13,147] have described high-affinity binding between the murine MB6A lymphosarcoma that normally colonizes mouse liver and freshly isolated rat liver endothelial cells or hepatocytes. The binding was inhibited by antibodies specific for

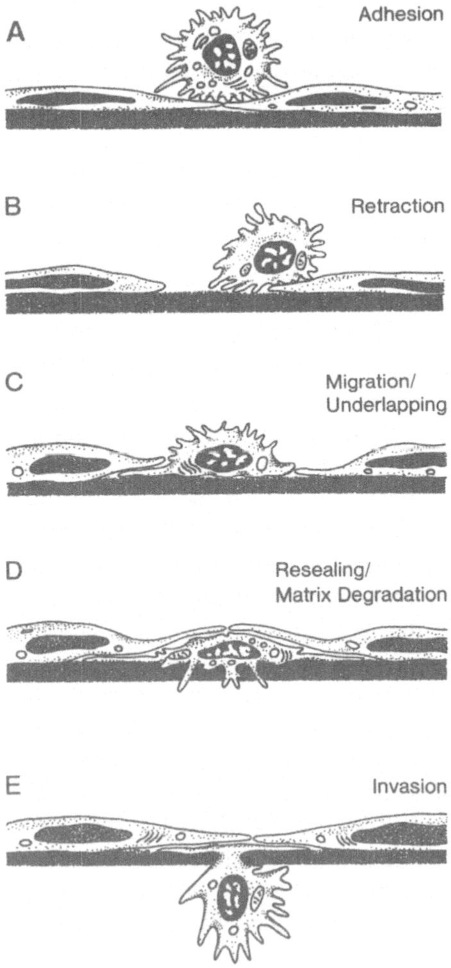

Figure 2. Metastatic tumor cell attachment and invasion of vascular endothelial cells and their underlying basal lamina extracellular matrix. (A) Tumor cell attachment to endothelial cells; (B) endothelial cell retraction and exposure of basal lamina; (C) tumor cell invasion and underlapping of adjacent endothelial cells; (D) tumor cell-mediated destruction of basal lamina and re-formation of endothelial cell interactions; (E) penetration of the sub-endothelial basal lamina and surrounding tissue.

the tumor cell plasma membrane. These adhesion properties were different from those of another liver-colonizing tumor, the TA3/HA mammary adenocarcinoma, which adhered to hepatocytes, but not to sinusoidal endothelial cells. The above experiments suggest that certain tissue-specific adhesion molecules are shared by endothelia and organ parenchyma, while others may be cell type-specific, and that different tumor populations may use one or more adhesion mechanisms in attaching to a particular tissue.

Our results using three different metastatic systems also suggest that specific tumor cell/endothelial cell interactions contribute to organ-specific metastasis. For example, B16 melanoma sublines that colonize lung at high efficiencies (B16–F10) attached to lung endothelial cell monolayers at rates significantly higher than other B16 sublines, whereas brain-colonizing B16–B15b cells showed higher rates of adhesion to brain endothelial cell monolayers.[5] In similar studies on the *in vitro* adhesion properties of human A375 melanoma sublines selected for lung (A375-MM) or liver (A375-L) colonization, preferential adhesion to lung or liver endothelial cell monolayers was found, consistent with metastatic properties *in vivo* (Fig. 3). In comparison, in the murine RAW117 large cell lymphoma system, high liver-colonizing RAW117-H10 and -L17 sublines adhered at higher rates to murine liver HSE cells, while only the lung-colonizing RAW117-L17 subline bound at high rates to lung LE-1 endothelial cell monolayers (Fig. 4). Preferential adhesion to target organ microvessel endothelia was not found at the level of the subendothelial matrix in this system, and the rates of adhesion of all RAW117 sublines to the subendothelial matrix were lower than to the endothelial cell monolayers; there was no organ preference of adhesion to the subendothelial matrix.[5,161] In contrast, in a rat mammary adenocarcinoma metastatic system the rate of adhesion of metastatic sublines to lung and other subendothelial matrix was related to the spontaneous metastatic properties of the sublines.[152] Thus, different metastatic systems can show variations in endothelial and subendothelial matrix adhesive properties, but in general these have been related to the metastatic properties of cell systems.

The organ-associated endothelial cell surface determinants involved in binding metastatic tumor cells have been examined using affinity procedures. To identify organ-associated recognition molecules expressed on endothelial cell surfaces, ^{125}I-labeled lysates of LE-1, HSE, and BAE cells were adsorbed onto glutaraldehyde-fixed tumor cells, solubilized in SDS sample buffer, and analyzed by SDS–polyacrylamide gel electrophoresis and autoradiography. The majority of the adsorbed radiolabeled proteins migrated in the molecular weight region known to contain the superfamily of integrin adhesion subunits ($M_r \sim$ 90,000–180,000); moreover, the individual protein profiles were unique for each endothelial cell line (Fig. 5). The tumor-adsorbed HSE lysates contained at least three proteins of $M_r \sim$ 120,000, ~ 90,000, and ~ 40,000, that were not adsorbed from LE-1 or BAE lysates, whereas protein bands of $M_r \sim$ 180,000, ~ 80,000, and ~ 25,000 were preferentially adsorbed from LE-1 lysates. The most prominent components adsorbed from BAE were ones of $M_r \sim$ 130,000 and ~ 45,000. These results suggest that tumor cell adhesion may be directed to specific "addresses" by selective expression of specific integrin-like molecules by the microvascular endothelium in different environments, consistent with the recent studies on lymphocyte trafficking.[162,163]

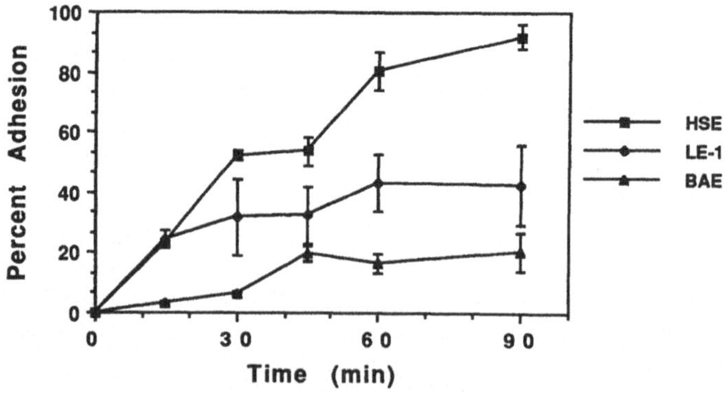

Figure 3. Adhesion potentials of human melanoma cell variant selected for liver (A375-L) on endothelial cell monolayers. ^{35}S-labeled tumor cells (1 × 10^5/well) were incubated with endothelial cell monolayers, adherent cells were harvested at various times, and the number of adherent cells determined by scintillation counting.

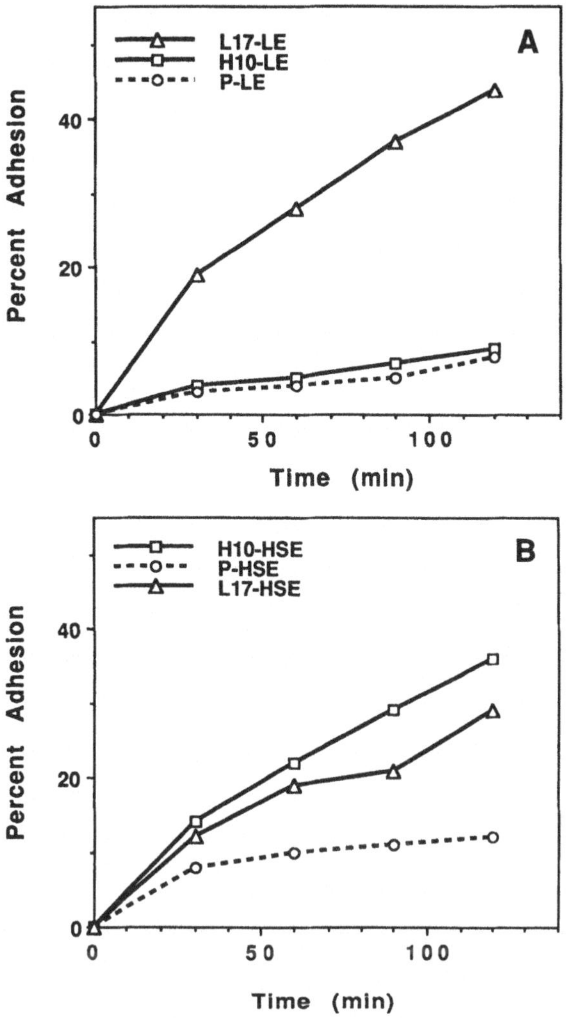

Figure 4. Adhesion potentials of RAW117 lymphoma variants on lung- or liver-derived endothelial cell monolayers. ^{35}S-labeled tumor cells (1×10^5/well) were incubated with endothelial cell monolayers and adherent cells harvested at various times. (A) Adhesion to mouse lung endothelial cells (LE-1). (B) Adhesion to mouse hepatic sinusoidal endothelial cells (HSE).[139]

Other microvascular endothelial cell surface components have been identified as important determinants in mediating tumor cell–endothelial cell adhesion. The cell-binding components were identified by adding viable biotinylated or radiolabeled tumor cells to Western transfers containing detergent-solubilized endothelial cell surface membrane proteins.[156] Five endothelial cell-derived glycoproteins (M_r ~ 48,000, ~ 32,000, ~ 30,000, ~ 25,000, and ~ 18,000) were identified using biotinylated RAW117-H10 cells, and the RAW117 cells bound to the Western transfers on the basis of their metastatic potentials (H10 >> P) and on the basis of organ preference of metastasis (HSE > LE-1 >> BAE). Liver-colonizing RAW117 tumor cell binding also occurred to lung endothelial cell components of similar M_r, but to a lesser extent, suggesting that differential adhesion to organ-derived endothelia may be due, in part, to quantitative, not qualitative, differences in the expression of endothelial cell surface components. This finding is consistent with the

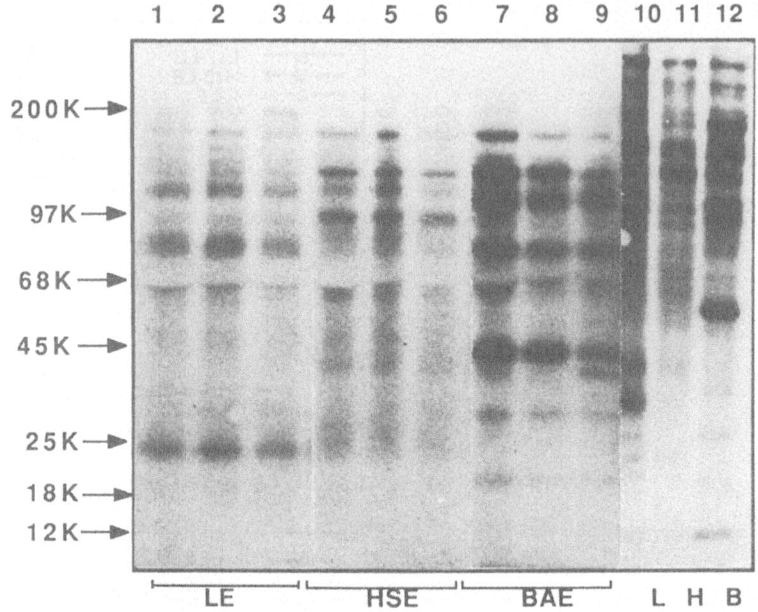

Figure 5. Tumor affinity-adsorption of ^{125}I-labeled endothelial cell glycoproteins. ^{125}I-labeled endothelial cell lysates (60 mg protein) were incubated with 2×10^6 glutaraldehyde-fixed RAW117 tumor cells for 1 hr, washed extensively with buffer containing 0.05% NP-40, and the bound glycoproteins analyzed by SDS–PAGE and lectin staining.

observation that quantitative, not qualitative, differences occur in the rates of adhesion of organ-specific tumor cells to target organ endothelial cells. Alternatively, specificity may be conferred by other surface molecules, such as those identified in the tumor adsorption experiments. Interestingly, we found in the Western adhesion experiment that normal leukocytes and human A375 melanoma cells also bound to the same five endothelial cell surface glycoproteins, suggesting that they are also involved in normal cell–cell interactions in the microcirculation. These endothelial cell-binding proteins have been designated endothelial common adhesion molecules (ECAM 1–4); although they are probably involved, it is unlikely that these molecules alone can confer adhesive specificity.

Endothelial cell-binding components that are responsible, in part, for tumor cell adherence have been identified as glycoproteins by their ability to bind to certain lectin-affinity columns.[161] It is not clear whether the glycoproteins are five distinctive adhesive receptors, or fewer proteins with alternate forms of glycosylation and sulfation. Polyclonal antisera have been raised against three of the individual endothelial cell-binding glycoproteins (gp32, gp25, and gp18) and their normal tissue distributions were determined.[164] We found that gp32 and gp25 had similar tissue distributions; the highest concentrations were found on the apical surfaces of kidney microvessels, glomeruli, and liver sinusoids; less was found on lung and brain microvessels. In contrast, gp18 was localized in the subendothelial spaces of blood vessels, particularly in lung. Additionally, at high dilutions, anti-gp18 selectively stained subendothelial components in lung, anti-gp25 and specifically stained the cell surfaces of hepatic sinusoids and bile ducts, and kidney glomeruli were selectively stained by anti-gp32. However, at lower dilutions the antibodies stained microvessels in most tissues. For example, anti-gp25 stained microvessels in lung, liver, and kidney, although the intensity was greatest in the liver. These studies indicate that these endothelial cell adhesion-associated glycoproteins are differentially distributed in various organs. Adhesion of biotinylated RAW117-H10 cells was only partially inhibited (~ 40%) by anti-gp25 plus anti-gp32, indicating that multiple components mediate the adhesive interactions at microvascular surfaces. At this time we do not know the relationship of these glycoproteins to other reported adhesion molecules; however, tumor cell binding to these components is not affected by treatment with RGD-containing peptides, indicating that they are not integrin-related components. Additionally, the basal expression of these glycoproteins precludes any relationship to ELAM expressed by immune-activated endothelial cells.[80,165] Wood *et al.*[166] have recently described a set of endothelial–monocyte antigens expressed by immune-activated human endothelial cells that appear to be similar to the glycoproteins described by us. Butcher has proposed that lymphocyte adhesive factors vary with cell differentiation and antigenic stimulation, while the endothelial recognition/attachment components vary with anatomic site.[167] Undoubtedly, multiple types of adhesion molecules selectively expressed by various organ endothelia, including lectins, proteoglycans, and integrins, are also involved in site-specific adhesion and migration of metastatic tumor cells (Table 3).

C. Integrin-Mediated Tumor Cell Adhesion Processes

Tumor cell adhesion to endothelial cells in target organs involves multiple molecular interactions, in a process dictated by cell adhesion molecules whose expression is dependent on the position of the tissue or address of a particular organ.[5,6,161,168] This cell homing is also important in normal biological processes such as neural development, trafficking of

Table 3. Cell Adhesion to Specific Vascular Endothelial Cell Molecules

Function	Molecule	Size (K)	Distribution	Ref.
Leukocyte adhesion	ELAM-1	115–97	Inflammatory lesions Cytokine-activated EC	165
	ICAM-1	90	Wide distributed Increased expression upon activation	171
	ICAM-2	46	Constitutively expressed Distribution unknown	200
Lymphocyte adhesion	MECA-79	92	Murine PNHEV MN HEV	21
	MECA-367	58–66	Murine PPHEV MN HEV	22
Platelet adhesion	GMP-140	140	WP bodies rapidly mobilized to plasma membrane upon stimulation	21
Tumor cell adhesion	ECAM-1	48	Constitutively expressed by microvascular EC	201
	ECAM-2	32–30		
	ECAM-3	25		
	ECAM-4	18		
	Hepatic EC specific	120–100	Unknown	29, 161
	Brain EC specific	25	Unknown	6

lymphocytes, and embryogenesis.[20,163,169] The families of cell adhesion molecules involved in these processes include cadherins,[170] intercellular adhesion molecule (ICAM-1),[79,171] endothelial–leukocyte adhesion molecule-1 (ELAM-1),[80,165] and integrins.[172] Many among the integrin superfamily of adhesion receptors use the tripeptide Arg-Gly-Asp (RGD) as the primary recognition site for binding to adhesive proteins such as fibronectin, vitronectin, laminin, von Willebrand factor, or ICAM-1[172,173] and are divalent cation-dependent in their binding activity.[174] They are integral membrane glycoproteins composed of two noncovalently associated α and β subunits having molecular weight ranges of 105,000 to 210,000 and 95,000 to 205,000, respectively. Integrin and integrinlike adhesion components are expressed on a variety of cell types, including lymphoid[175,176] and endothelial cells.[177,178] Studies on tumor cell adhesion to components of the extracellular matrix (ECM) demonstrate that these interactions are an important mechanism by which tumor cells can adhere to ECM when they metastasize.[179]

Evidence for the role of integrins in metastasis comes primarily from experiments where *in vitro* invasion or *in vivo* dissemination of tumor cells could be inhibited by fibronectin, laminin, RGD-containing peptides, or proteins.[180,181] Welch *et al.*[182] reported on the inhibition of experimental metastasis of murine B16 melanoma cells by pretreatment of the tumor cells with soluble laminin or fibronectin. Studies by Welch *et al.*[182] have demonstrated the ability of RGD-containing peptides to inhibit the invasion of human A375 melanoma cells in a membrane invasion culture system (MICS) assay. To determine whether integrins play a role in RAW117 murine lymphoma–endothelial cell adhesion, we used RGD-containing peptides and assessed their effects on tumor cell–endothelial cell adhesion *in vitro*. Initial studies using the tetrapeptide RGDS at concentrations as high as 1 mg/ml as well as antibodies directed against various known lymphoid cell integrins (LFA-1, Mac-1, gp150/95 β_1 subunit) did not inhibit RAW117 cell–endothelial cell adhesion. However, RGD polymers containing specific pentameric combinations of GRGDS $(GRGDS)_4$ or D_4 and/or GRGES $(GRGES)_4$ or E_4 were able to inhibit RAW117-H10 adhesion to murine HSE cell monolayers at polymer concentrations $\leq$ 10 μg/ml,[183] indicating that peptide secondary and tertiary structure are important in conferring inhibitory potential and specificity. For example, the homopolymers D_4 and E_4 and heteropolymer DE_2D significantly inhibited the binding of H10 cells to HSE monolayers while heteropolymers E_2D_2 and ED_2E had no effect (Fig. 6). Other studies have also suggested that the secondary structure of RGD-containing peptides is important for integrin-inhibitory activities.[173,184–186] Pierschbacher and Ruoslahti[184] cyclized peptides to restrict their secondary structure and found that attachment to vitronectin could be inhibited using tenfold lower concentrations than the uncyclized peptide. Recently, Danilov and Juliano[185] reported that the valency and distribution of RGD sequences on a protein play a role in determining the efficiency of integrin-mediated cell–substratum adhesion and spreading. The fact that the ED_2E and E_2D_2 polymers had no effect on adhesion even though they contained two RGD sequences suggests that simple avidity alone cannot explain the inhibitory effects of the D_4, DE_2D, and E_4 polymers. Thus, the secondary structure and spatial relationships of the RGD- and RGE-containing polymers may be important for the polymers' inhibitory activity in RAW117 tumor cell adhesion to different endothelial cell monolayers. We used Garnier secondary structure analysis[186] of the peptides to demonstrate that in the case of RAW117-H10/HSE adhesion, the polymers with greater structural homology to the D_4 polymer had greater inhibitory effect.

We failed to confirm the usefulness of the YIGSR sequence, derived from the β1 chain

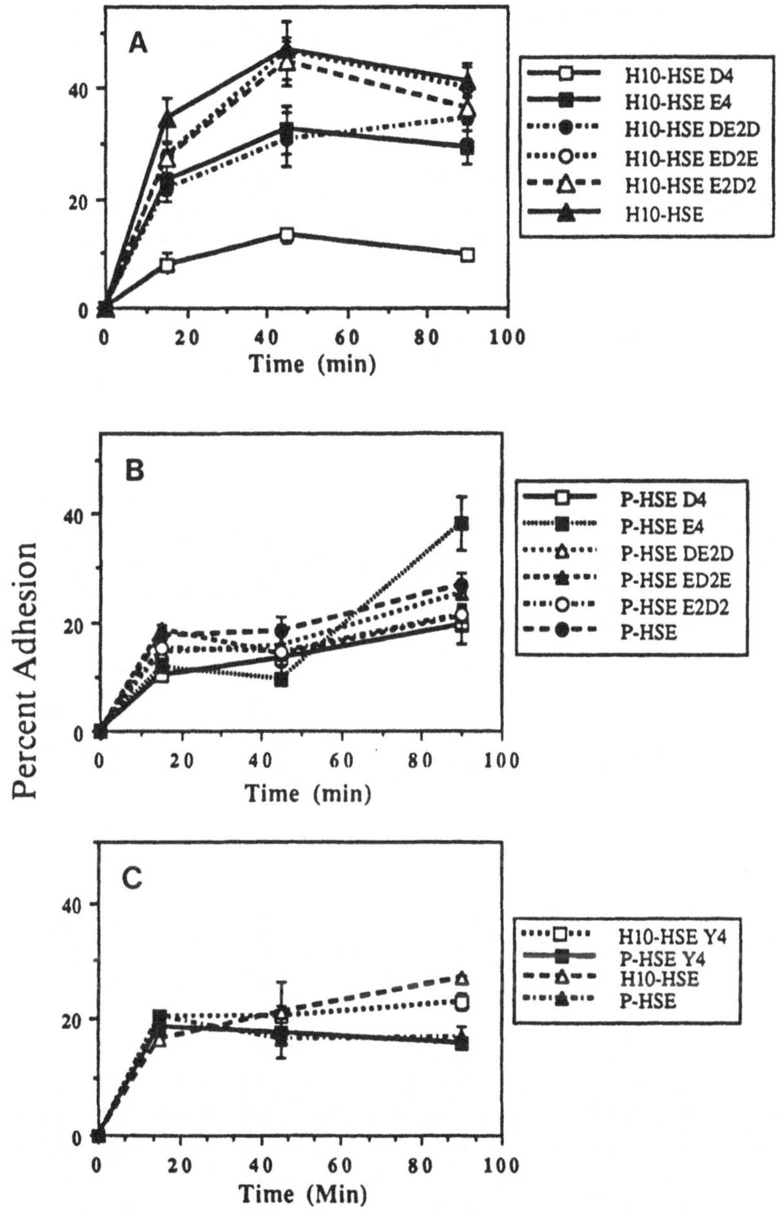

Figure 6. Kinetics of adhesion of parental (P) and high liver-colonizing (H10) murine large-cell lymphoma cells to murine hepatic sinusoidal endothelial (HSE) cell monolayers in the presence of various synthetic peptide polymers. Values are expressed as percent of total cells added per well of a 24-well tissue culture plate (1×10^5 cells added/well). (A) Adhesion of H10 cells in the presence or absence of various RGD-containing polymers to HSE monolayers. (B) Adhesion of P cells in the presence or absence of various RGD-containing polymers to HSE monolayers. (C) Adhesion of P and H10 cells in the presence or absence of Y4 polymers to HSE monolayers.[183]

of laminin, in inhibiting RAW117 cell adhesion. Although Iwamoto *et al.*[187] have shown that a peptide containing the YIGSR sequence can partially inhibit B16–F10 melanoma invasion *in vitro* and lung colony formation *in vivo*, we could not demonstrate that YIGSR or its homopolymer (Y_4) had any effect on the endothelial cell adhesive properties of RAW117 cells. H10 cells can adhere poorly to immobilized laminin, but there is no correlation of their *in vitro* laminin adhesion with their metastatic behavior *in vivo*.[139] The adhesion of RAW117 cells to intact extracellular matrix synthesized by various endothelial cells was lower than their adhesion to endothelial cell monolayers and did not correlate with metastatic potential.[139]

Differences in the abilities of different RGD-containing polymers to inhibit RAW117 cell adhesion to HSE and BAE cells suggest that the organ specificity of adhesion seen in this tumor system may be determined, in part, by integrin-mediated adhesion mechanisms. We found that the D_4 polymer could preferentially decrease RAW117-H10/HSE adhesion to the level of RAW117-P/HSE adhesion, while having no further inhibitory effect on RAW117-P/HSE adhesion (Fig. 6A–C), suggesting that integrin-mediated adhesion may be responsible for the quantitative difference in adhesion of RAW117-H10 versus -P cells to HSE monolayers. H10 cells may express greater amounts of integrin-type molecules or RGD-containing molecules on their surface than P cells. Alternatively, differences in accessibility of an RGD epitope on H10 and P cells due to differences in glycosylation, stearic hindrance, or other factors may account for the quantitative difference in adhesion to HSE monolayers. The inability of the polymers to significantly affect the adhesion of the low metastatic P cell line or to completely inhibit H10 adhesion suggests that these tumor cells use both integrin-dependent and -independent mechanisms to adhere to endothelial cells.[29]

D. Endothelial Cell Secretion of Metastatic Cell Growth Factors

The proliferation of malignant cells at particular host sites could depend on their response to different concentrations of hormones in individual organs, differentially expressed local factors, and paracrine growth factors. Horak *et al.*[188] and Naito *et al.*[189] found that lung-conditioned medium differentially stimulated the growth *in vitro* of lung-colonizing metastatic cells. Using adenocarcinoma cells of low and high metastatic potential for mouse colon, Tsuruo *et al.*[190] found that growth of both cell lines was stimulated by a lung extract; however, the more metastatic cell line was differentially stimulated at lower extract concentrations. Using lung- and ovary-colonizing B16 melanoma sublines, we found that cell growth in serum-limited medium was differentially stimulated by nondialyzable factors from target organ tissue. High lung-colonizing B16–F10 cells were growth stimulated by lung-conditioned medium and high ovary-colonizing B16–O10 cells were growth stimulated by ovary-conditioned medium significantly more than the other organ-conditioned media tested.[191] The only B16 subline that did not fit the overall pattern of growth stimulation by target organ-conditioned medium was the brain-colonizing B16–B15b subline. Exclusively lung-metastasizing rat 13762NF mammary adenocarcinoma cells were growth stimulated by both lung- and liver-conditioned medium. The highly metastatic MTLn3 cells, however, were growth stimulated at significantly higher rates than the low metastatic MTC cells.[192]

Some of the tissue-conditioned media were growth inhibitory. Liver tissue factors and, at high concentrations, kidney tissue factors were inhibitory to all of the B16 sublines, correlating with the low potential of the B16 sublines to colonize liver and kidney.[191] However, when high liver-colonizing RAW117-H10 cells were tested for their growth in serum-limited, liver-conditioned medium, they were significantly more growth stimulated than the parental RAW117-P cells.[193] Lung-selected RAW117-L17 cells that are capable of colonizing lung as well as liver were significantly more growth stimulated by lung- and liver-conditioned media than parental cells. Interestingly, high liver-colonizing RAW117-H10 cells that do not form significant numbers of lung tumors were also growth stimulated by the lung growth factors. Kidney-conditioned medium was only slightly stimulatory to the lung-selected RAW117-L17 cells, but this subline does not colonize kidney.[193]

Using organ-derived microvessel endothelial cells we examined the effects of endothelial cell-derived growth factors on the proliferation of RAW117 cells. The patterns of stimulation/inhibition were almost exactly the same as found with organ-conditioned media. That is, high liver-colonizing RAW117-H10 cells were significantly more growth stimulated than the parental RAW117-P cells by conditioned medium from HSE cells. Lung-selected RAW117-L17 cells that are capable of colonizing lung as well as liver were significantly more growth stimulated by LE-1-conditioned medium than parental cells (Fig. 7). Thus, the organ growth properties of metastatic cells may be determined by the release of paracrine growth factors and inhibitors from organ microvessel endothelial cells.

Few organ-derived paracrine growth factors have been identified and isolated. Using BSC-1 kidney cells, Holley *et al.*[194] isolated a growth inhibitor of $M_r \sim 25{,}000$ that is released into kidney cell-conditioned medium. This growth inhibitor was later identified as TGF-β2.[195] A hepatic growth factor of $M_r \sim 26{,}000$ has been purified from rat liver extracts. This inhibitor reversibly blocks the proliferation of untransformed rat liver cells but has no apparent effect on hepatoma cells.[196] Szaniawska *et al.*[197] recently reported that lung tissue-conditioned medium contains both cell growth-stimulating molecules of $M_r \sim$ 50,000–70,000 and cell growth-inhibiting molecules of $M_r \sim$ 12,000–20,000 and ~ 3000–5000. One of the most potent metastatic cell growth-stimulating factors has now been purified to homogeneity from lung tissue- and LE-1-conditioned medium in our laboratory using a five-step purification procedure.[198] This $M_r \sim 66{,}000$ glycoprotein appears to be related to transferrin based on its sequence and other data.[199] A different lung paracrine growth factor has been isolated and partially purified by Yamori *et al.*[200] They found that murine lung tissue homogenates stimulated the growth of four of four highly metastatic cell clones of murine colon 26 carcinoma cells, but only one of four poorly metastatic cell clones. The activity was trypsin- and heat-labile and was partially purified by gel chromatography. The active component was eluted in the M_r range of 90,000 to 120,000.[200] Parenchymal cell- and endothelial cell-derived paracine growth factors might be involved in normal local tissue regeneration and inflammation processes. When tissues are locally damaged by invading tumor cells, such paracrine growth factors may be released by parenchymal and endothelial cells to stimulate normal organ tissue repair.

V. CONCLUSIONS

In general, the seed and soil hypothesis explains the clinical and experimental observations of organ-specific metastasis. Tumor cell and host organ properties, such as

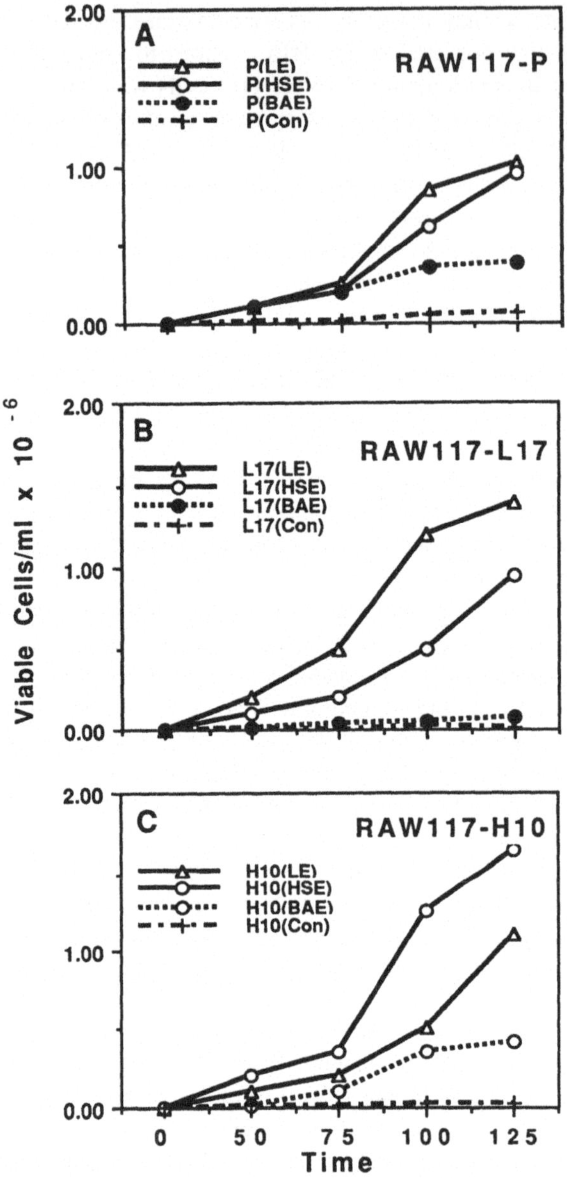

Figure 7. Effects of endothelial cell-derived growth factors on the proliferation of RAW117 cell variants. (A) RAW117-P; (B) RAW117-L17; (C) RAW117-H10.

differential tumor cell adhesion to organ-derived microvessel endothelial cells, differential invasion of organ basement membranes and tissues, and differential tumor cell responses to organ-derived growth-stimulatory and inhibitory factors, all appear to be important determinants in explaining the organ preference of metastasis. Both common and organ-specific endothelial cell determinants contribute to tumor cell/endothelial cell recognition and adhesive interactions. Circumstantial evidence suggests that multiple adhesion mechanisms in tumor cell/endothelial cell interactions work in parallel, and thus intervention of the metastatic process at this stage will require multiple modalities.

The following statements summarize the conceptual role of the vascular endothelium in cancer metastasis as discussed in this text.

1. The adhesion of malignant cells to microvascular endothelial cells at specific organ sites is a significant factor in determining the distribution and organ preference of certain metastatic cells.
2. The dynamic expression of common and organ-specific properties by normal and activated endothelial cells at diverse tissue sites forms the basis of tremendous heterogeneity among microvascular tissues.
3. The kinetics of tumor cell adhesion to endothelial cell monolayers differs among tumor subpopulations and often correlates with their observed metastatic properties.
4. The differential expression of organ-associated molecules or domains on microvessel endothelial cell surfaces is thought to serve as tissue "addressins" in preferential tissue localization of blood-borne tumor cells.
5. Tumor cell adhesion to endothelial cells in target organs involves multiple molecular interactions, including integrin complexes.
6. The growth of metastatic cells at secondary sites is also a determinant in the organ preference of metastasis. In a given organ microenvironment, the regulation of tumor cell growth is determined by the collective actions of cell–cell, cell–matrix, and cell–soluble molecule interactions.

ACKNOWLEDGMENTS. The authors' studies were supported by grants from the National Institutes of Health, National Cancer Institute (R35-CA44352), National Foundation for Cancer Research, and Alkermes, Inc. (to G. L. N.).

REFERENCES

1. Sugarbaker, E. V., 1981, Patterns of metastasis, *Cancer Biol. Rev.* **2**:235–278.
2. Nicolson, G. L., 1982, Cancer metastasis: Organ colonization and the cell-surface properties of malignant cells, *Biochim. Biophys. Acta* **695**:113–176.
3. Naito, S., von Eschenbach, A. C., and Fidler, I. J., 1987, Different growth pattern and biologic behavior of human renal cell carcinoma implanted into different organs of nude mice, *J. Natl. Cancer Inst.* **78**:377–385.
4. Murphy, P., Alexander, P., Senior, P. V., Flemming, J., Kirkham, N., and Taylor, I., 1988, Mechanisms of organ selective tumour growth by blood-borne cancer cells, *Br. J. Cancer* **57**:19–31.
5. Nicolson, G. L., 1988, Organ specificity of tumor metastasis: Role of preferential adhesion, invasion and growth of malignant cells at specific secondary sites, *Cancer Metastasis Rev.* **7**:143–188.
6. Auerbach, R., Lu, W., Pardon, E., Gukowski, F., Kaminska, G., and Kaminska, M., 1987, Specificity of adhesion between murine tumor cells and capillary endothelium: An *in vitro* correlate of preferential metastasis *in vivo*, *Cancer Res.* **47**:1492–1496.
7. Ewing, J., 1928, A treatise on tumors, in: *Neoplastic Diseases*, 3rd ed., Saunders, Philadelphia.
8. Paget, S., 1889, The distribution of secondary growths in cancer of the breast, *Lancet* **1**:571–573.
9. Weiss, L., 1983, Random and nonrandom processes in metastasis, and metastatic efficiency, *Invasion Metastasis* **3**:193–208.

10. Irimura, T., and Nicolson, G. L., 1984, Carbohydrate chain analysis by lectin binding to mixtures of glycoproteins separated by polyacrylamide slab-gel electrophoresis with *in situ* chemical modification, *Carbohydr. Res.* **115**:209–220.
11. Hart, I. R., 1982, The role of animal models in the study of experimental metastasis, in: *Tumor Invasion and Metastasis* (L. Liotta and I. R. Hart, eds.), Nijhoff, The Hague, pp. 1–14.
12. Hart, I. R., and Fidler, I. J., 1980, The role of organ selectivity in the determination of metastatic patterns of B16 melanoma, *Cancer Res.* **40**:2281–2287.
13. Roos, E., Middlekoop, O. P., and Van de Pavert, I. V., 1984, Adhesion of tumor cells to hepatocytes: Different mechanisms for mammary carcinoma compared with lymphosarcoma cells, *J. Natl. Cancer Inst.* **73**:963–969.
14. Belloni, P. N., and Tressler, R.J., 1990, Microvascular heterogeneity: Interactions with leukocytes and tumor cells, *Cancer Metastasis Rev.* **8**:353–389.
15. Ford, W. L., 1973, The cellular basis for immune responses, in: *Defense and Recognition* (R. Porter, ed.), Butterworths, London, pp. 65–102.
16. Ford, W. D., Sedgely, M., and Smith, M. E., 1976, The migration of lymphocytes across specialized endothelium, *Cell Tissue Kinet.* **9**:351–361.
17. Gowans, J. L., and Knight, E. J., 1964, The route of recirculation of lymphocytes in the rat, *Proc. R. Soc. London* **159**:275–282.
18. Stamper, H. B., and Woodruff, J. J., 1976, Lymphocyte homing into lymph nodes: *In vitro* demonstration of the selective affinity of recirculating lymphocytes for HEV, *J. Exp. Med.* **144**:828–841.
19. Dujvestijn, A. M., Kerkhove, M., Bargatze, R. F., and Butcher, E. C., 1987, Lymphoid tissue and inflammation specific endothelial cell differentiation defined by monoclonal antibodies, *J. Immunol.* **138**: 713–719.
20. Gallatin, W. M., Weissman, I. L., and Butcher, E. C., 1983, A cell surface molecule involved in organ-specific homing of lymphocytes, *Nature* **304**:30–34.
21. Streeter, P. R. Lakey-Berg, E., Rouse, B., Bargatze, R., and Butcher, E. C., 1988, A tissue specific endothelial cell molecule involved in lymphocyte homing, *Nature* **331**:41–46.
22. Streeter, P. R., Rouse, B. T., and Butcher, E. C., 1988, Immunohistologic and functional characterization of vascular addressin involved in lymphocyte homing into peripheral lymph nodes, *J. Cell Biol.* **107**:1853–1862.
23. deBono, D., 1977, Lymphocytes and the microcirculation, *Adv. Microcirc.* **7**:68–95.
24. Weiss, L., Orr, F. W., and Honn, K. V., 1988, Interaction of cancer cells with the microvasculature during metastasis, *FASEB J.* **2**:12–21.
25. Tressler, R. J., Belloni, P. N., and Nicolson, G. L., 1989, Correlation of inhibition of tumor cell–endothelial cell adhesion by RGD-containing peptide polymers and metastatic potential: Role of integrin-dependent and -independent adhesion mechanisms, *Cancer Commun.* **1**:55–63.
26. Greene, H. S., and Harvey, E. K., 1964, The relationship between the dissemination of tumor cells and the distribution of metastases, *Cancer Res.* **24**:799–811.
27. Nicolson, G. L., and Winkelhake, J. L., 1975, Organ specificity of blood-borne tumour metastasis determined by cell adhesion? *Nature* **255**:230–234.
28. Nicolson, G. L., 1982, Metastatic tumor cell attachment and invasion assay utilizing vascular endothelial cell monolayers, *J. Histochem. Cytochem.* **30**:214–220.
29. Tressler, R. J., and Nicolson, G. L., 1991, Butanol-extractable and detergent-solubilized cell surface components from murine large cell lymphoma cells associated with adhesion to organ microvessel endothelial cells, *J. Cell. Biochem.* in press.
30. Rhoden, L., 1980, Structure and metabolism of connective tissue proteoglycans, in: *The Biochemistry of Proteoglycans* (W. J. Lennarz, ed.), Plenum Press, New York, pp. 180–221.
31. Majno, G., 1965, Ultrastructure of the vascular membrane, in: *Handbook of Physiology: Circulation*, Vol. III, pp. 2293–2375.
32. Stemmerman, M. B., 1980, General properties of blood vessels: Vascular endothelia, in: *Blood Vessels and Lymphatics in Organ Systems* (D. Abramson and P. Dobrin, eds.), Academic Press, New York, pp. 25–31.
33. Timpl, R., and Martin, G. R., 1982, Components of basement membranes, in *Immunochemistry of the Extracellular Matrix* (H. Furthmayer, ed.), CRC Press, Boca Raton, pp. 119–150.
34. Carley, W. W., Milici, A. J., and Madri, J. A., 1988, Extracellular matrix specificity for the differentiation of capillary endothelial cells, *Exp. Cell Res.* **178**:426–434.

35. Kefalides, N. A., 1975, Basement membranes: Structural and biosynthetic considerations, *J. Invest. Dermatol.* **65:**85–92.
36. Madri, J. A., and Pratt, B. M., 1986, Endothelial cell–matrix interactions: *In vitro* models of angiogenesis, *J. Histochem. Cytochem.* **34:**85–91.
37. Wang, Z. W., Irimura, T., Nakajima, M., Belloni, P. N., and Nicolson, G. L., 1985, Characterization of the ECM associated GAG produced by untransformed and transformed bovine corneal endothelial cells in culture, *Eur. J. Biochem.* **153:**125–130.
38. Gospodarowicz, D., Gonzalez, R., and Fujii, D., 1983, Are factors originating from serum, plasma, or cultured cells involved in the growth promoting effect of the ECM produced by cultured bovine corneal endothelial cells? *J. Cell. Physiol.* **114**:191–199.
39. Vlodavsky, I., Folkman, J., and Klagsburn, M., 1987, Heparin-binding endothelial cell growth factors are sequestered by the extracellular matrix, in: *Angiogenesis: Mechanisms and Pathology* (D. B. Rifkin and M. Klagsburn, eds.), Cold Spring Harbor Laboratories, Cold Spring Harbor, N.Y., pp. 58–64.
40. Liotta, L. A., Rao, C. N., and Barsky, S. H., 1983, Tumor invasion and the extracellular matrix, *Lab. Invest.* **49:**636–649.
41. Nakajima, M., Irimura, T., and Nicolson, G. L., 1987, Basement membranes degradative enzymes and tumor metastasis, *Cancer Bull.* **39:**142–149.
42. Simionescu, M., Simionescu, N., and Palade, G., 1981, Differentiated microdomains on the luminal surfaces of capillary endothelium, *J. Cell Biol.* **90:**605–613.
43. Simionescu, M., Simionescu, N., Silbert, E., and Palade, G., 1981, Differential microdomains on the luminal surfaces of capillary endothelium. II. Partial characterization of their anionic sites, *J. Cell Biol.* **90:**614–621.
44. Simionescu, M., and Simionescu, N., 1980, Hydrophobic pathways of capillary endothelium, a dynamic system, in: *Alfred Benson Symposium 15* (H. H. Ussing, N. B. Bindilev, and O. Sten-Knudsen, eds.), Munksgaard, Copenhagen, pp. 1–21.
45. Simionescu, M., Simionescu, N., and Palade, G., 1982, Differential microdomains on the luminal surface of capillary endothelium: Distribution of lectin receptors, *J. Cell Biol.* **94:**406–413.
46. Simionescu, M., Simionescu, N., and Palade, G., 1982, Preferential distribution of anionic sites on the basement membrane and abluminal aspect in fenestrated endothelium, *J. Cell Biol.* **95:**425–434.
47. Simionescu, M., Simionescu, N., Santoro, F., and Palade, G., 1985, Differentiated microdomains of the luminal plasmalemma of murine muscle capillaries: Segmental variations in young and old, *J. Cell Biol.* **100:** 1296–1307.
48. Bankston, P. W., and Milici, A. J., 1983, A survey of the binding of polycationic ferritin in several fenestrated capillary beds. Indication of heterogeneity in the luminal glycocalyx of fenestrated diaphragms, *Microvasc. Res.* **26:**36–48.
49. Born, G. V. R., and Palinski, D., 1985, Unusually high concentrations of sialic acid on the surface of vascular endothelia, *Br. J. Exp. Pathol.* **66:**543–549.
50. Soda, R., and Tavassolli, M., 1983, Mapping of the bone marrow sinus endothelium with lectins and glycosylated ferritins: Identification of different microdomains and their functional significance, *J. Ultrastruct. Res.* **84:**299–310.
51. Pino, R. M., 1984, Ultrastructural localization of lectin receptors on the bone marrow sinusoidal endothelium of the rat, *Am. J. Anat.* **169:**259–272.
52. Vorbrodt, A. W., Dobrogowska, D. H., Lossinsky, A. S., and Wisniewski, H. M., 1986, Ultrastructural localization of lectin receptors on the luminal and abluminal aspects of brain micro-blood vessels, *J. Histochem. Cytochem.* **34:**251–261.
53. Vorbrodt, A. W., 1986, Changes in the distribution of endothelial surface glycoconjugates associated with altered permeability of brain blood vessels, *Acta Neuropathol.* **70:**103–111.
54. Ponder, B. A., and Wilkinson, M. M., 1983, Organ related differences of *Dolichos biflorus* agglutinin to vascular endothelium, *Dev. Biol.* **96:**535–544.
55. Belloni, P. N., and Nicolson, G. L., 1988, Differential expression of cell surface glycoproteins on various organ-derived microvascular endothelia and endothelial cell cultures, *J. Cell. Physiol.* **136:**398–410.
56. Fatehi, M. I., Gerhurt, D. Z., Myers, T. G., and Drews, L. R., 1987, Characterization of the blood–brain barrier: Glycoconjugate receptors of 14 lectins in canine brain, cultured endothelial cells and blotted membrane proteins, *Brain Res.* **415:**30–39.

57. Belloni, P. N., and Nicolson, G. L., 1988, Biochemical and immunological detection of organ specific glycoproteins associated with the blood brain barrier, Biology of the Vascular Endothelial Cell Vth Symp. (Abstract) VI-3.
58. Holthoffer, H., Virtanen, I., Kariniemi, A. L., Hormonia, M., Linder, E., and Miettinen, A., 1982, *Ulex europeus* I lectin as a marker for vascular endothelium in human tissue, *Lab. Invest.* **47:**60–66.
59. Alroy, J., Goyal, V., and Skutelsky, E., 1987, Lectin histochemistry of mammalian endothelium, *Histochemistry* **86:**603–607.
60. Laitinen, L., 1987, *Griffonia simplicifolia* lectins bind specifically to endothelial cells and some epithelial cells in mouse tissues, *Histochem. J.* **19:**225–234.
61. Pressman, D., 1955, Tissue localizing antibodies, *Ann. N.Y. Acad. Sci.* **59:**376–379.
62. Pressman, D., 1964, Certain aspects of tissue specific antigens, *Can. Cancer Conf.* **5:**363–376.
63. Auerbach, R., Alby, L., Morissey, L., Tu, M., and Joseph, J., 1985, Expression of organ-specific antigens on capillary endothelial cells, *Microvasc. Res.* **29:**401–406.
64. Ghandour, M. S., Langley, O. K., Gombos, G., Hirn, M., Hirsch, M. R., and Goridis, C. A., 1982, A surface marker for murine vascular endothelial cells defined by monoclonal antibody, *J. Histochem. Cytochem.* **30:**165–170.
65. Zanetta, J. P., Dontenwill, M., Meyer, A., and Roussel, G., 1985, Isolation and immunohistochemical localization of a lectin-like molecule from rat cerebellum, *Dev. Brain Res.* **17:**233–243.
66. Michalak, T., White, F. P., Gard, A. L., and Dutton, G. R., 1986, A monoclonal antibody to the endothelium of rat brain microvessel, *Brain Res.* **379:**320–328.
67. Partridge, W. M., Yang, J., Eisenberg, J., and Mietus, L. J., 1986, Antibodies to blood–brain barrier bind selectively to brain capillary endothelial lateral membranes and to a 46K protein, *J. Cereb. Blood Flow Metab.* **6:**203–211.
68. Risau, W., Hallman, R. Albrecht, U., and Henke-Fahle, S., 1986, Brain induces the expression of an early cell surface marker for blood–brain barrier-specific endothelium, *EMBO J.* **5:**3179–3183.
69. Hogg, N., MacDonald, S., Slusarenky, M., and Beverly, P. C., 1984, Monoclonal antibody specific for human monocytes, granulocytes and endothelium, *Immunology* **53:**753–767.
70. Flotte, T. J., Springer, T. A., and Thorbecke, G. J., 1983, Dendritic cell and macrophage staining by monoclonal antibodies in tissue sections and epidermal sheets, *Am. J. Pathol.* **111:**112–119.
71. Paul, L.C., Baldwin, W. M., and van Es, L. S., 1985, Vascular endothelial alloantigens in renal transplantation, *Transplantation* **40:**117–123.
72. Blankert, J. J., Paul, L. C., and van Es, L. A., 1985, Immunogenicity of various organs and genetic control of the response to a non-MHC endothelial antigen in rat, *Transplant. Proc.* **17:**785–791.
73. Kennel, S. J., 1987, Rat monoclonal antibodies to mouse lung components for analysis of fibrosis, *Exp. Mol. Pathol.* **47:**110–124.
74. Kennel, S. J., Lankford, T. K., Ullrich, R. L., and Jamashbi, R. J., 1988, Enhancement of lung tumor colony formation by treatment of mice with monoclonal antibodies to pulmonary capillary endothelial cells, *Cancer Res.* **48:**4964–4968.
75. Moore, M. G., Chrzanowski, R. R., McCormick, J. R., and Schwink, A., 1984, Production of monoclonal antibodies to rat lung angiotensin-converting enzyme, *Clin. Immunol. Immunopathol.* **33:**301–305.
76. Irving, M. G., Roll, F. J., Huang, S., and Montgomery-Bissel, D., 1984, Characterization and culture of sinusoidal endothelium from normal rat liver; lipoprotein uptake and collagen phenotype, *Gastroenterology* **82:**1203–1217.
77. Mukai, K., Rosai, J., and Burgdorf, W. H. C., 1980, Localizing of factor VIII related antigen in vascular endothelial cells using immunoperoxidase method, *Am. J. Surg. Pathol.* **4:**273–280.
78. Harrach, H. R., Jasani, B., and Williams, E. D., 1983, Factor VIII as a marker of endothelial cells in follicular carcinoma of the thyroid, *J. Clin. Pathol.* **36:**1050–1058.
79. Dustin, M. L., Rothlein, R., Bhan, A. K., Dinarello, C. A., and Springer, T. A., 1986, Induction by IL-1 and interferon-γ tissue distribution, biochemistry, and function of a natural adherence molecule (ICAM-1), *J. Immunol.* **37:**245–254.
80. Bevilacqua, M. P., Pober, J. S., Mendrick, D. L., Cotran, R. S., and Gimbrone, M. A., 1987, Identification of an inducible endothelia–leukocyte adhesion molecule, *Proc. Natl. Acad. Sci. USA* **84:**9238–9242.
81. Shepro, D., and Dunham, B., 1986, Endothelial cell metabolism of biogenic amines, *Annu. Rev. Physiol.* **48:** 335–368.

82. Zetter, B. R., 1988, Endothelial heterogeneity: Influence of vessel size, organ location, and species specificity, in: *Biology of Vascular Endothelial Cells* (U. Ryan, ed.), CRC Press, Boca Raton, pp. 63–80.
83. Grega, G. J., Svensjo, E., and Haddy, F., 1981, Macromolecular permeability of the microvascular membrane: Physiological and pharmacological regulation, *Microcirculation* **1**:325–341.
84. Schneeberger, E. E., 1982, Circulating proteins and macromolecular transport across continuous nonfenestrated endothelium, *Ann. N.Y. Acad. Sci.* **401**:25–37.
85. Svensjo, E., and Roempke, K., 1984, Microvascular aspects of edema formation and its inhibition by β-receptor stimulants and some other anti-inflammatory drugs, in: *Progress in Microcirculation II* (F. C. Courtice, D. G., Garlick, and M. A. Perry, eds.), University of New South Wales, pp. 447–463.
86. Robertson, A. L., and Rosen, L. A., 1977, The arterial endothelium: Characteristics and function of the endothelial lining of large arteries, in: *Microcirculation* (G. Kaley and B. M. Altura, eds.), University Park Press, Baltimore, pp. 145–165.
87. Huttner, I., Boufet, M., and More, R. H., 1973, Studies on protein passage through arterial endothelium. II. Regional differences in permeability to fine structural protein tracers in arterial endothelium of normotensive rats, *Lab. Invest.* **28**:678–685.
88. Thorgeirsson, G., and Robertson, A. L., 1978, The vascular endothelium: Pathobiologic significance, *Am. J. Pathol.* **93**:802–848.
89. Albelda, S. M., Sampson, P. M., Hastelton, F. R., McNiff, J. M., Mueller, S. N., Williams, S. K., Fishman, A. P., and Levine, E.M., 1988, Permeability characteristics of cultured endothelial cell monolayers, *J. Appl. Physiol.* **64**:308–322.
90. DelVecchio, P. J., Siflinger-Birnboim, A., Shepard, J. M., Bizios, R., Cooper, J. A., and Malik, A. B., 1987, Endothelial monolayer permeability to macromolecules, *Fed. Proc.* **46**:2511–2515.
91. Klatzo, I., 1967, Neuropathological aspects of brain edema, *J. Neuropathol. Exp. Neurol.* **26**:1–14.
92. Groothuis, D. R., and Vick, N. A., 1982, Brain tumors and the blood–brain barrier, *Trends Neurosci.* **5**: 232–235.
93. Long, D. M., 1979, Capillary ultrastructure and the blood–brain barrier in human malignant tumors, *J. Neurosurg.* **51**:53–58.
94. Roy, S., and Chitra, S., 1989, Ultrastructural study of microblood vessels in human brain tumors and peritumora tissue, *J. NeuroOncol.* **7**:283–294.
95. Criscuolo, G. R., Merrill, M. J., and Oldfield, E. H., 1988, Further characterization of malignant glioma-derived vascular permeability factor, *J. Neurosurg.* **69**:254–262.
96. Belloni, P. N., Ohigashi, H., and Nicolson, G. L., 1990, Human melanoma-derived factors increase brain microvascular permeability, in preparation.
97. Nemerson, Y., 1966, The reaction between brain tissue factor and factor VII and X, *Biochemistry* **5**:601–606.
98. Gonmori, H., and Takeda, Y., 1976, Properties of human tissue thromboplastin from brain, lung and placenta, *Thromb. Haemostas.* **36**:90–103.
99. Colucci, M., Balconi, R., Lorenzel, A., Pietra, D., and Semarara, N., 1983, Cultured endothelial cells generate tissue factor in response to endotoxin, *J. Clin. Invest.* **71**:1893–1901.
100. Bevilacqua, M. P., Pober, J. S., Wheeler, M. E., Cotran, R. S., and Gimbrone, M. A., 1985, Interleukin-1 activation of vascular endothelium. Effects on procoagulant activity, *Am. J. Pathol.* **121**:393–403.
101. Bevilacqua, M. P., Pober, J. S., Majeua, G. R., Fiers, W., Cotran, R. S., and Gimbrone, M. A., 1986, Recombinant TNF induce procoagulant activity in cultured human endothelia; Characterization and comparison with the actions of IL-1, *Proc. Natl. Acad. Sci. USA* **83**:4533–4541.
102. Hoyer, L. W., Santos, R., and Hoyer, J. R., 1973, Antihemophilic factor antigen. Localization in endothelial cells by immunofluorescent microscopy, *J. Clin. Invest.* **52**:2737–2744.
103. Jaffe, E., 1984, Synthesis of factor VIII by endothelial cells, in: *Biology of Endothelial Cells*, Nijhoff, The Hague, pp. 209–214.
104. Kwast, T. H., Stel, H. V., Cristen, E., Bertina, R. M., and Veerman, E. C., 1986, Localization of FVIII-procoagulant antigen, *Blood* **67**:222–227.
105. Ruggeri, Z. M., DeMarco, L., Gatti, L., Bader, R., and Montgomery, R. R., 1983, Platelets have more than one binding site for VonWillebrand factor, *J. Clin. Invest.* **72**:1–12.
106. Plow, E. F., Piersbacher, M. D., Ruoslahti, E., Marguerie, G. A., and Ginsberg, M. H., 1985, The effect of Arg-Gly-Asp-containing peptides on fibrinogen and Von Willebrand factor binding to platelets, *Proc. Natl. Acad. Sci. USA* **82**:8057–8062.

107. Hoyer, L. W., 1981, The factor VIII complex: Structure and function, *Blood* **58:**1–13.
108. Hormonia, M., Lehto, V. P., and Virtanen, I., 1984, Intracellular localization of factor VIII-RAg and fibronectin in cultured human endothelial cells, *Eur. J. Cell Biol.* **33:**217–228.
109. Zetter, B. R., 1981, The endothelial cells of large and small blood vessels, *Diabetes* **30:**244–248.
110. Hormonia, M., 1982, Expression of factor VIII-related antigen and UEA lectin binding sites in endothelial cells during long-term culture, *Cell Biol. Int. Rep.* **6:**1123–1134.
111. Dejana, E., Lampugnani, M. G., Giorgio, M., Gaboli, M., Federici, A. B., Ruggeri, Z. M., and Marchisio, P. C., 1989, VonWillebrand factor promotes endothelial cell adhesion via an Arg-Gly-Asp-dependent mechanism, *J. Cell Biol.* **109:**367–375.
112. Hong, S. L., 1980, Effect of bradykinin and thrombin on PGI synthesis in endothelial cells from calf and pig aorta and human umbilical cord vein, *Thromb. Res.* **18:**787–795.
113. Goldsmith, J. C., and Kisker, C. T., 1982, Thrombin–endothelial cell interactions: Critical importance of vessel origin, *Thromb. Res.* **25:**131–136.
114. Sun, F. F., and Taylor, B. M., 1978, Metabolism of prostacyclin in rat, *Biochemistry* **17:**4996–5000.
115. Dusting, G. J., Moncada, S., and Vane, J. R., 1978, Recirculation of prostacyclin (PGI_2) in the lung, *Br. J. Pharmacol.* **64:**315–319.
116. Johnson, A. R., 1980, Human pulmonary endothelial cells in culture: Activities of cells from arteries and veins, *J. Clin. Invest.* **65:**841–849.
117. Mosher, D. F., Doyle, M. J., and Jaffe, E. A., 1982, Synthesis and secretion of thrombospondin by cultured human endothelia cells, *J. Cell Biol.* **93:**343–348.
118. Kefalides, N. A., 1980, Chemistry of basement membranes: Structure and biosynthesis, in: *Vascular Endothelium and Basement Membranes* (B. M. Alttura, E. Davis, and H. Harder, eds.), Karger, Basel, pp. 295–322.
119. Sage, H., 1984, Collagen synthesis by endothelial cells in culture, in: *Biology of Endothelial Cells* (E. Jaffe, ed.), Nijhoff, The Hague, pp. 161–177.
120. Madri, J. A., Dreyer, B., Pitlick, A., and Furthmeyer, H., 1980, The collagenous components of the subendothelium—Correlation of structure and function, *Lab. Invest.* **43:**303–315.
121. Belloni, P. N., Carney, D. H., and Nicolson, G. L., 1991, Organ-derived endothelial cells express differential responsiveness to thrombin and other growth factors, *Microvasc. Res.* in press.
122. Maciag, T., Hoover, G. A., and Stemmerman, M. B., 1981, Serial propagation of human endothelial cells *in vitro*, *J. Cell Biol.* **91:**420–426.
123. Hormonia, M., 1982, Expression of factor VIII-related antigen and UEA lectin binding sites in endothelial cells during long-term culture, *Cell Biol. Int. Rep.* **6:**1123–1134.
124. Pitas, R. E., Boyles, J., Mahley, R. W., and Montgomery-Bissell, D., 1985, Uptake of chemically modified low density lipoproteins *in vivo* is mediated by specific endothelial cells, *J. Cell Biol.* **100:**103–117.
125. Rupnick, M. A., Carey, A., and Williams, S. K., 1988, Phenotypic diversity in cultured cerebral microvascular endothelial cells, *In Vitro* **24:**435–444.
126. Del Regato, D., 1977, Pathways of metastatic spread of malignant tumors, *Semin. Oncol.* **4:**33–38.
127. Fidler, I. J., and Nicolson, G. L., 1981, Immunobiology of experimental metastatic melanoma, *Cancer Biol. Rev.* **2:**171–234.
128. Viadana, E., Bross, I. D. J., and Pickren, J. E., 1973, An autopsy study of some routes of dissemination, *Br. J. Cancer* **27:**336–340.
129. Hansen, H. H., and Muggia, F. M., 1972, Staging of inoperable patients with bronchogenic carcinoma with special reference to bone marrow examination, *Cancer* **30:**1395–1401.
130. Elliott, R. H. E., and Frantz, V. K., 1960, Metastatic carcinoma masquerading as primary thyroid cancer: A report of authors' 14 cases, *Ann. Surg.* **151:**551–561.
131. Prout, G. R., 1973, Prostate gland, in: *Cancer Medicine* (J. F. Holland and E. Frei, eds.), Lea & Febiger, Philadelphia, pp. 1680–1694.
132. Fidler, I. J., and Nicholson, G. L., 1976, Organ selectivity for implantation, survival, and growth of B16 melanoma variant tumor lines, *J. Natl. Cancer Inst.* **57:**1199–1202.
133. Brunson, K. W., Beatty, G., and Nicolson, G. L., 1978, Selection and altered tumor cell properties of brain colonizing metastatic melanoma, *Nature* **272:**543–545.
134. Nicolson, G. L., Miner, K. M., and Reading, C. L., 1981, Tumor cell heterogeneity and blood-borne metastasis, in: *Fundamental Mechanisms in Human Cancer Immunology* (J. Saunders, J. Daniels, B. Serrous, C. Rosenfeld, and C. Denny, eds.), Elsevier, Amsterdam, pp. 31–39.

135. Fidler, I. J., 1973, Selection of successive tumor lines for metastasis, *Nature New Biol.* **242**:148–149.
136. Nicolson, G. L., Muscali, J. J., and McGuire, E. J., 1983, Metastatic RAW117 lymphosarcoma as a model for malignant–normal cell interaction. Possible roles for cell surface antigens in determining the quantity and localization of secondary tumors, *Oncodev. BioMed.* **4**:149–152.
137. Schirrmacher, V., Chiensong Popov, R., and Arnheiter, H., 1980, Hepatocyte tumor cell interactions *in vitro*, *J. Exp. Med.* **151**:984–989.
138. Nicolson, G. L., Dulski, K. M., Basson, C., and Welch, D. R., 1985, Preferential organ attachment and invasion *in vitro* by B16 melanoma cells selected for different metastatic colonization and invasive properties, *Invasion Metastasis* **5**:144–158.
139. Nicolson, G. L., Belloni, P. N., Tressler, R. J., Dulski, K., Inoue, T., and Cavanaugh, P. G., 1989, Adhesive, invasive and growth properties of selected metastatic variants of murine large cell lymphoma, *Invasion Metastasis* **9**:102–116.
140. Kawaguchi, T., Kawaguchi, M., Miner, K. M., Lembo, T. M., and Nicolson, G. L., 1983, Brain meninges tumor formation by *in vitro*-selected metastatic B16 melanoma variants in mice, *Clin. Exp. Metastasis* **3**: 247–259.
141. Lapis, K., Pakus, S., and Liotta, L. A., 1988, Endothelialization of embolized tumor cells during metastasis formation, *Clin. Exp. Metastasis* **6**:73–89.
142. Netland, P. A., and Zetter, B. R., 1985, Metastatic potential of B16 melanoma cells after *in vitro* selection for organ-specific adherence, *J. Cell Biol.* **101**:720–724.
143. Menter, D. G., Hatfield, J. S., Harkins, C., Sloane, B. F., Taylor, J. D., Crissman, J. D., and Honn, K. V., 1987, Tumor cell–platelet interactions *in vitro* and their relationship to *in vivo* arrest of hematogenously circulating tumor cells, *Clin. Exp. Metastasis* **5**:65–78.
144. Kramer, R. H., and Nicolson, G. L., 1981, Invasion of vascular endothelial cells and underlying extracellular matrix by metastatic human cancer cells, in: *International Cell Biology* (S. Schweiger, ed.), Springer-Verlag, Berlin, pp. 794–799.
145. Varani, J., Orr, W., and Ward, P. A., 1980, Adhesion characteristics of tumor cell variants of high and low tumorigenic potential, *J. Natl. Cancer Inst.* **64**:1173–1178.
146. Varani, J., Grimstad, I. A., Knibbs, R. N., Hovig, T., and McCoy, J. P., 1985, Attachment spreading, and growth *in vitro* of highly malignant and low malignant murine fibrosarcoma cells, *Clin. Exp. Metastasis* **3**:45–59.
147. Roos, E., Tulp, A., Middlekoop, O. P., and Pavert, I. V., 1985, Interaction between lymphoid tumor cells and isolated liver endothelial cells, *J. Natl. Cancer Inst.* **72**:1173–1180.
148. Korach, S., Poupon, M. F., DuVillard, J. A., and Becker, M., 1986, Differential adhesiveness of rhabdomyosarcoma-derived cloned metastatic cell lines to vascular endothelial monolayers, *Cancer Res.* **46**: 3624–3629.
149. Kramer, R. H., Gonzalez, R., and Nicolson, G. L., 1980, Metastatic tumor cells adhere preferentially to the extracellular matrix underlying vascular endothelial cells, *Int. J. Cancer* **26**:639–645.
150. Vlodavsky, I., Schirrmacher, V., Ariav, Y., and Fuks, Z., 1983, Lymphoma cell interaction with cultured vascular endothelial cells and with the subendothelial basal lamina: Attachment, invasion and morphological appearance, *Invasion Metastasis* **3**:81–97.
151. Nicolson, G. L., Irimura, T., Gonzales, R., and Ruoslahti, E., 1981, The role of fibronectin in adhesion of metastatic melanoma cells to endothelial cells and their basal lamina, *Exp. Cell Res.* **135**:461–465.
152. Lichtner, R. B., Belloni, P. N., and Nicolson, G. L., 1989, Differential adhesion of metastatic rat mammary carcinoma cells to organ derived microvascular endothelial cells and subendothelial matrix, *Exp. Cell Biol.* **57**:146–152.
153. Netland, P. A., and Zetter, B. R., 1986, Melanoma cell adhesion to defined extracellular matrix components, *Biochem. Biophys. Res. Commun.* **139**:515–522.
154. Murray, C. J., Liotta, L. A., Rennard, S. I., and Martin, G. R., 1980, Adhesion characteristics of murine metastatic and nonmetastatic tumor cells *in vitro*, *Cancer Res.* **40**:347–351.
155. Terranova, V. P., Williams, J. E., Liotta, L. A., and Martin, G. R., 1984, Modulation of metastatic activity of melanoma cells by laminin and fibronectin, *Science* **226**:982–985.
156. Chung, D. C., Zetter, B. R., and Brodt, N. P., 1988, Lewis lung carcinoma variants with different metastatic specificities adhere preferentially to differentially defined extracellular matrix molecules, *Invasion Metastasis* **8**:103–117.
157. Liotta, L. A., Rao, C. N., and Barsky, S. H., 1983, Tumor invasion and the extracellular matrix, *Lab. Invest.* **49**:636–649.

158. Irimura, T., Hester, J. E., Yamori, T., Belloni, P. N., Ota, D., and Nicolson, G. L., 1991, Preferential adhesion of metastatic human colon carcinoma cells and extracellular matrix, submitted for publication.
159. McCarthy, J. B., Basara, M. L., Palm, S. L., Sas, D. F., and Furcht, L. T., 1985, The role of cell adhesion proteins laminin and fibronectin in the movement of malignant and metastatic cells, *Cancer Metastasis Rev.* **4**:125–152.
160. Alby, L., and Auerbach, R., 1984, Differential adhesion of tumor cells to capillary endothelial cells *in vitro*, *Proc. Natl. Acad. Sci. USA* **81**:5739–5743.
161. Belloni, P. N., Tressler, R. J., and Nicolson, G. L., 1991, Adhesion of metastatic tumor cells to organ-derived microvascular endothelial cells: Identification of non-integrin endothelial cell adhesion molecules, submitted for publication.
162. Yednock, T. A., and Rosen, S. D., 1989, Lymphocyte homing, *Adv. Immunol.* **44**:313–378.
163. Berg, E. L., Goldstein, L. A., Jutila, M. A., Nakache, M., Picker, L. J., Streeter, P. R., Wu, N. W., Zhou, D., and Butcher, E. C., 1989, Homing receptors and vascular addressins: Cell adhesion molecules that direct lymphocyte traffic, *Immunol. Rev.* **108**:5–18.
164. Belloni, P. N., and Nicolson, G. L., 1989, Identification and partial characterization of endothelial cell surface glycoproteins associated with adhesion of murine RAW117 large cell lymphoma cells and leukocytes. M. D. Anderson Symposium: Critical determinants in cancer progression and metastasis.
165. Bevilacqua, M. P., Stengelin, S., Gimbrone, M. A., and Seed, B., 1989, Endothelial leukocyte adhesion molecule 1: Receptor for neutrophils related to complement regulatory proteins and lectins, *Science* **243**: 1160–1165.
166. Wood, N. L., Schook, L. B., Stoder, E. J., and Mohanakumar, T., 1988, Biochemical characterization of human vascular endothelial cell monolayer antigens defined by monoclonal antibodies, *Transplantation* **45**: 787–792.
167. Butcher, E. C., and Weissman, I., 1984, Lymphocytes, tissues and organs, in: *Fundamental Immunology* (W. E. Paul, ed.), Raven Press, New York, pp. 109–127.
168. Bargatze, R. F., Wu, N. W., Weissman, I. L., and Butcher, E. C., 1987, High endothelial venule binding as a predictor of the dissemination of passaged murine lymphomas, *J. Exp. Med.* **166**:1125–1131.
169. Hatta, K. S., Takagi, S., Fujisawa, H., and Takeichi, M., 1987, Spatial and temporal expression pattern of N-cadherin cell adhesion molecules correlated with morphogenetic processes of chicken embryos, *Dev. Biol.* **120**:215–227.
170. Crossin, K. L., Chuong, C. M., and Edelman, G.M., 1985, Expression sequences of cell adhesion molecules, *Proc. Natl. Acad. Sci. USA* **82**:6942–6946.
171. Dustin, M. L., and Springer, T. A., 1988, Lymphocyte function associated antigen-1 (LFA-1) interaction with intercellular adhesion molecule-1 (ICAM-1) is one of at least three mechanisms for lymphocyte adhesion to cultured endothelial cells, *J. Cell Biol.* **107**:321–331.
172. Hynes, R. O., 1987, Integrins: A family of cell surface receptors, *Cell* **48**:549–554.
173. Ruoslahti, E., and Pierschbacher, M. D., 1987, New perspectives in cell adhesion, *Science* **238**:491–497.
174. Cheresh, D. A., Pytela, R., Pierschbacher, M. D., Klier, F. G., Ruoslahti, E., and Reisfeld, R. A., 1987, An Arg-Gly-Asp-directed receptor on the surface of human melanoma cells exists in a divalent cation-dependent functional complex with the disialoganglioside GD2, *J. Cell Biol.* **105**:1163–1173.
175. Miller, L. J., Schwarting, R., and Springer, T. A., 1986, Regulated expression of the Mac-1, LFA-1, p150,95 glycoprotein family during leucocyte differentiation, *J. Immunol.* **137**:2891–2900.
176. Springer, T. A., Dustin, M. L., Kishimoto, T. K., and Marlin, S. D., 1987, The lymphocyte function-associated LFA-1, CD2, and LFA-3 molecules: Cell adhesion receptors of the immune system, *Annu. Rev. Immunol.* **5**:223–252.
177. Fitzgerald, L. A., Charo, I. F., and Phillips, D. R., 1985, Human and bovine endothelial cells synthesize membrane proteins similar to human platelet glycoproteins IIb and IIIa, *J. Biol. Chem.* **260**:10893–10896.
178. Fujimoto, T., and Singer, S. J., 1988, Immunocytochemical studies of endothelial cells *in vivo*. II. Chicken aortic and capillary endothelial cells exhibit different cell surface distributions of the integrin complex, *J. Histochem. Cytochem.* **36**:1309–1317.
179. Knudsen, K. A., Smith, L., Smith, S., Karczewski, J., and Tuszynski, G. P., 1988, Role of IIb-IIIa-like glycoproteins in cell–substratum adhesion of human melanoma cells, *J. Cell. Physiol.* **136**:471–478.
180. McCarthy, J. B., Chelberg, M. K., Mickelson, D. J., and Furcht, L. T., 1988, Localization and chemical synthesis of fibronectin peptides with melanoma adhesion and heparin binding activities, *Biochemistry* **27**: 1380–1388.

181. Humphries, M. J., Yamada, K. M., and Olden, K., 1988, Investigation of the biological effects of anti-cell adhesive synthetic peptides that inhibit experimental metastasis of B16-F10 murine melanoma cells, *J. Clin. Invest.* **81:**782–790.
182. Welch, D. R., Lobl, T. J., Seftor, E. A., Wack, P. I. J., Aeed, P. A., Rohem, K. H., Seftor, R. E. B., and Hendrix, M. J. C., 1989, Use of the membrane invasion culture system (MICS) as a screen for anti-invasive agents, *Int. J. Cancer* **43:**449–457.
183. Tressler, R. J., Belloni, P. N., and Nicolson, G. L., 1989, Correlation of inhibition of adhesion of large cell lymphoma and hepatic sinusoidal endothelial cells by RGD-containing peptide polymers with metastatic potential: Role of integrin-dependent and -independent adhesion mechanisms, *Cancer Commun.* **1:**55–63.
184. Pierschbacher, M. D., and Ruoslahti, E., 1987, Influence of stereochemistry of the sequence Arg-Gly-Asp-Xaa on binding specificity of cell adhesion, *J. Biol. Chem.* **262:**17294–17298.
185. Danilov, Y. N., and Juliano, R. L., 1989, $(Arg\text{-}Gly\text{-}Asp)_n$–albumin conjugates as a model substratum for integrin-mediated cell adhesion, *Exp. Cell Res.* **182:**186–196.
186. Levin, J. M., and Garnier, J., 1988, Improvements in a secondary structure prediction method based on a search for local sequence homologies and its use as a model building tool, *Biochim. Biophys. Acta* **955:** 283–295.
187. Iwamoto, F. A., Robey, J., Graf, M., Sasaki, S., Kleinman, H. K., Yamada, Y., and Martin, G. R., 1987, YIGSR, a synthetic laminin pentapeptide, inhibits experimental metastasis formation, *Science* **238:**1132–1134.
188. Horak, E., Darling, D. L., and Tarin, D., 1986, Analysis of organ-specific effects on metastatic tumor formation by studies *in vitro*, *J. Natl. Cancer Inst.* **75:**913–922.
189. Naito, S., Giavazzi, R., and Fidler, I. J., 1987, Correlation between the *in vitro* interaction of tumor cells with an organ environmental and metastatic properties *in vivo*, *Invasion Metastasis* **7:**126–129.
190. Tsuruo, T., Kawabata, H., Iida, H., and Yamori, T., 1986, Tumor-induced platelet aggregation and growth promoting factors as determinants for successful tumor metastasis, *Clin. Exp. Metastasis* **4:**25–33.
191. Nicolson, G. L., and Dulski, K. M., 1986, Organ specificity of metastatic tumor colonization is related to organ-selective growth properties of malignant cells, *Int. J. Cancer* **38:**389–394.
192. Nicolson, G. L., 1988, Differential organ tissue adhesion, invasion, and growth properties of metastatic rat mammary adenocarcinoma cells, *Breast Cancer Res. Treat.* **12:**167–176.
193. Nicolson, G. L., 1987, Differential growth properties of metastatic large cell lymphoma cells in target organ conditioned medium, *Exp. Cell Res.* **168:**572–577.
194. Holley, R. W., Bohlen, P., Fava, R., Baldwin, J. H., Kleeman, G., and Armour, R., 1980, Purification of kidney epithelial cell growth inhibitors, *Proc. Natl. Acad. Sci. USA* **77:**5989–5992.
195. Tucker, R. F., Shipley, G. D., Moses, H. L., and Holley, R. W., 1984, Growth inhibitor from BSC-1 cells closely related to platelet type b transforming growth factor, *Science* **226:**705–707.
196. McMahon, J. B., Farelly, J. G., and Iype, P. T., 1982, Purification and properties of rat liver protein that specifically inhibits the proliferation of non-malignant epithelial cells from rat liver, *Proc. Natl. Acad. Sci. USA* **79:**456–460.
197. Szaniawska, B., Majewski, S., Kaminski, M. J., Noremberg, K., Swierz, M., and Janik, P., 1985, Stimulatory and inhibitory activities of lung-conditioned medium on the growth of normal and neoplastic cells *in vitro*, *J. Natl. Cancer Inst.* **75:**303–306.
198. Cavanaugh, P. G., and Nicolson, G. L., 1989, Purification and some properties of a lung-derived growth factor that differentially stimulates the growth of tumor cells metastatic to the lung, *Cancer Res.* **49:**3928–3933.
199. Cavanaugh, P. G., and Nicolson, G. L., 1991, Lung-derived growth factor that stimulates the growth of lung-metastasizing tumor cells: identification as transferrin. *J. Cell. Biochem.* in press.
200. Yamori, T., Iida, H., Tsukagoshi, S., and Tsuruo, T., 1988, Growth stimulatory activity of lung extract on lung colonizing colon 26 clones and its partial characterization, *Clin. Exp. Metastasis* **6:**131–139.
201. Staunton, D. E., Dustin, M. L., and Springer, T. A., 1989, Functional cloning of ICAM-2, a cell adhesion ligand for LFA-1 homologous to ICAM-1, *Nature* **339:**61–64.

25

Endothelial Cell Heterogeneity

Its Role as a Determinant of Selective Metastasis

Robert Auerbach

I. INTRODUCTION

The organ-selective pattern by which tumor cells establish metastatic foci has long been recognized, but it is only within the last decade that serious attention has been given to the possible role played by vascular endothelial cells in this selective process.[1–5] In large part this has been due to the fact that the heterogeneity of endothelial cells, especially of microvascular endothelial cells, was not appreciated.[6] This lack of recognition was the result of several factors: (1) the inability to grow microvascular endothelial cells *in vitro*; (2) the central importance of large vessel endothelial cells in well-studied disease states such as atherosclerosis; and (3) the absence of a theoretical framework to justify the concept that microvascular endothelial cells from different organs are not basically similar to each other. But over the last 10 years or so we have learned how to isolate and grow endothelial cells *in vitro*, we have gained at least some rudimentary understanding of the basic developmental processes underlying microvascular endothelial cell differentiation, and we have begun to recognize the importance of the microvascular endothelium in disease processes ranging from microangiopathies to tumor cell metastasis.[7–9]

The microvascular endothelium represents the first barrier to neoplastic cells that detach from the primary tumor mass to enter the circulation.[10] The microvascular endothelial cell also is the first target of tumor cells in the bloodstream that seek a new site for extravasation and the subsequent establishment of metastases.[1,5,11] We know little about the role of endothelial cells in permitting intravasation, what events occur at the abluminal surface of endothelial cells, or whether endothelial cells from different organ sites differ with respect to their response to tumor cells from different primary tumors. On the other hand, we have learned a considerable amount about microvascular endothelial cell heterogeneity and how this heterogeneity may influence tumor cell arrest and extravasation.[1,2,12–15] This chapter focuses on studies documenting organ-specific differences among

Robert Auerbach • Center for Developmental Biology, University of Wisconsin, Madison, Wisconsin 53706.

Endothelial Cell Dysfunctions, edited by Nicolae Simionescu and Maya Simionescu. Plenum Press, New York, 1992.

microvascular endothelial cells: (1) expression of organ-specific cell surface antigens; (2) heterogeneity in cell surface ligands as evidenced by terminal sugar moieties; (3) organ-selective patterns of tumor cell adhesion; and (4) consideration of the developmental processes that underlie the establishment of organ-specific differences during embryonic development.

II. EXPRESSION OF ORGAN-SPECIFIC ANTIGENS AT THE ENDOTHELIAL CELL SURFACE

In order to demonstrate directly the heterogeneity of microvascular endothelial cells we designed experiments to document the presence of unique organ-specific cell surface antigens on these cells.[17,18] These experiments relied on three observations:

1. Cell recognition processes such as those that permit aggregation and sorting out of embryonic cells in histotypical manner are not species specific. For example, Moscona and his colleagues showed that liver cells and cartilage cells would sort out from each other in mixed aggregates, but that chicken and mouse liver cells, or chicken and mouse cartilage cells, would form chimeric aggregates.[19]
2. Adhesion molecules expressed at the cell surface are principally responsible for the aggregation and sorting out of these cells.[20]
3. Key molecules involved in cell adhesion are highly conserved among species as diverse as invertebrates and vertebrates.[20]

Thus, we argued, we should in principle be able to use monoclonal antibody technology to generate serological reagents that would be directed at organ-specific antigens that might play a role in selective adhesion of tumor cells. Because of our interest in specific tumor types, gliomas, lung tumors, and teratomas, we elected to generate antibodies against brain, lung, and ovary antigens. By varying the species used for providing the antigen from the species used to screen the antibodies, the selection of hybridomas was aimed at obtaining cross-species reactive antibodies important in cell recognition processes.[17]

By immunizing host animals with cells derived from selected organs, it was possible to sensitize these animals against multiple antigens, including—but not limited to—those that might be organ-restricted. Hybrids generated by fusion of splenocytes from these animals with B-lymphoma fusion partners were then screened using protocols designed to select only those hybrids producing organ-specific antibodies. The manner in which this was achieved can be exemplified by the following protocol (Fig. 1): Mice were immunized with rat ovary cells. Hybrids were screened using Chinese *hamster ovary*-derived *fibroblasts*, and those identified by the first screening were then tested for their ability to bind to various *mouse endothelial cells*. When antibodies were identified as positive for both these target cells, they were then also tested on other murine endothelial cells and on murine fibroblasts. In this manner, hybrids were identified that produced antibodies that were organ specific, i.e., reactive against ovary fibroblasts and ovary endothelial cells, but nonreactive against nonovary fibroblasts and endothelial cells (Fig. 1A).

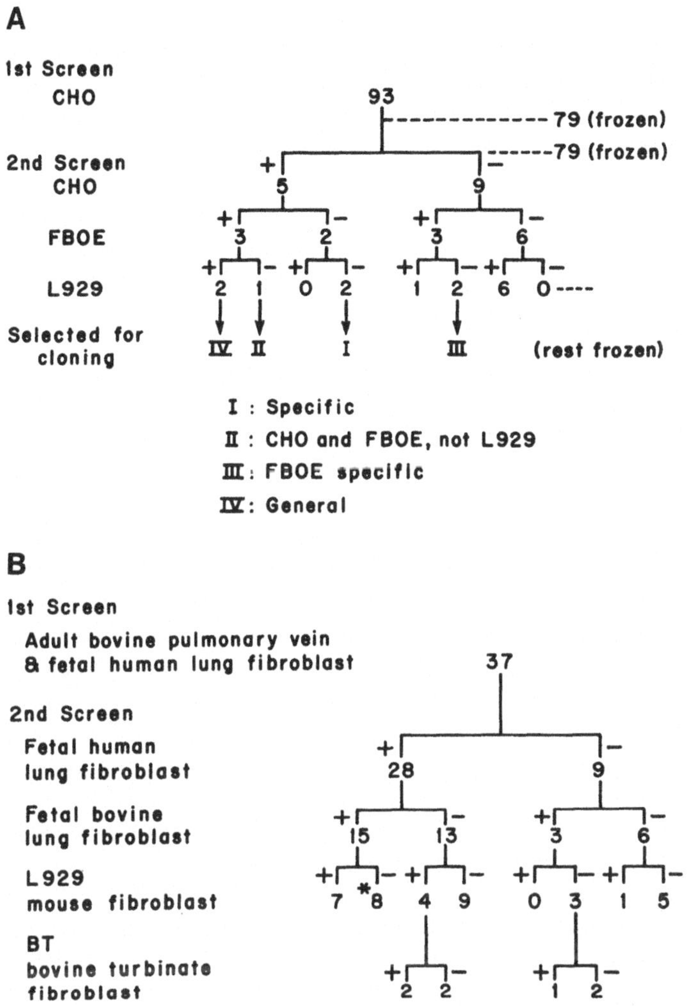

Figure 1. Generation of organ-specific antibodies. (A) Antiovary protocols and results. (B) Antilung protocols and results. (Data from Ref. 17).

Similar protocols were used to generate lung-specific antibodies, by using rat minced lung tissue as immunogen and human lung fibroblasts for primary screening (Fig. 1B). For brain, we discovered that an organ-specific antigen already existed: Thy 1, a cell surface antigen known to be involved in cell–cell interactions, had already been shown to be primarily expressed on thymocytes and most mouse brain cells. Indeed, the early polyclonal anti-Thy 1 antibodies were generated in rabbits by immunizing them against rat brain cells. (In retrospect, we recognize that these experiments were an early demonstration of the validity of the concept of using cross-species immunization protocols to obtain organ-specific antibodies.) Among our panel of endothelial cells, only cerebrum-derived microvascular endothelial cells expressed Thy 1, thus providing evidence that this antigen was expressed in an organ-specific manner.[17]

We also generated organ-specific antibrain antibodies in a different manner: Mouse brain-derived microvascular endothelial cells were used to immunize rats, a rat/mouse fusion protocol was used to obtain rat anti-mouse brain endothelial cell antibodies, and the antibodies were then screened against nonbrain endothelial cells and various fibroblasts. In this manner, other monoclonal antibodies with brain-specific distribution were isolated.[17]

III. LECTIN-BINDING SITES ON ENDOTHELIAL CELLS

Cell-surface interactions almost universally involve glycoproteins, and much of the specificity is associated with changes in the terminal sugar residues. Thus, it seemed reasonable to examine the thesis that organ-specific differences among endothelial cells may be reflected in their ability to bind lectins specific for different sugar moieties. The availability of a large panel of fluorochrome- or biotin-coupled plant lectins permitted us to use flow cytometry to characterize microvascular endothelial cell surface phenotypes on the basis of their lectin-binding properties.[12] What has emerged from studies in our laboratory and others[1,21] is the realization that the pattern of expression of receptors for plant lectins is distinct for endothelial cells isolated from different organ beds. In some instances, moreover, heterogeneity was found even among endothelial cells isolated from a single microvascular bed, suggesting that microheterogeneities may further define endothelial cells subsets.

As shown in Table 1, some lectins bind to virtually all endothelial cells, while others are selective. We have now extended the analysis to include 21 different lectins and each additional lectin augments our ability to identify any given endothelial cell.

It should be emphasized that microvascular endothelial cells may show strain and species differences which may be unrelated to organ specificity. For example, Ulex I lectin is an excellent marker for bovine and human endothelial cells but is absent or only weakly expressed on murine endothelial cells. Even intraspecies differences exist, as illustrated by the finding that Dolichos lectin binds selectively to endothelial cells from the RIII strain of mice but not to endothelial cells from most other mouse strains.[22]

Two cautionary notes need to be inserted: (1) Microvascular endothelial cells will not maintain all their surface characteristics indefinitely *in vitro*. In general, it is wise to restrict analysis to primary cultures or cells in early passage. (2) Reagents may be different when purchased from different vendors. Even different lots of Ulex I lectin from a single vendor were found to show marked differences in efficacy. For example, murine endothelioma cells failed to stain with one lot of Ulex I (see Table 1), stained strongly (2 log over background) with a second lot, and showed only marginal labeling with a third lot.

Studies involving two- and three-color analysis using different lectins are now under way, in order to develop a labeling pattern matrix that can be used to define microvascular endothelial cells with even more precision. In combination with the serological reagents already described, they promise to provide unique phenotypes of microvascular endothelial cells of different origins.

Table 1. Lectin-Binding Properties of Different Endothelial Cells

Cell	PBS[a]	ConA	DBA	PNA	RCAI	SBA	UEAI	WGA
Lung								
%[b]		94.6	1.6	16.3	97.0	21.1	0	97.7
Mean[c]	58.6	120.7	57.2	70.6	129.9	75.2	53.6	130.2
MBE								
%		98.7	1.1	98.6	90.7	35.6	0	98.5
Mean	53.5	124.4	54.5	122.3	91.3	70.3	50.9	131.9
MLyE								
%		97.2	0.3	37.7	98.0	77.5	0.4	97.7
Mean	58.5	125.2	62.3	77.5	121.9	86.7	60.1	137.5
ABAE								
%		97.7	0	6.3	98.3	82.5	0	97.6
Mean	60.1	118.1	56.8	67.7	128.3	88.5	52.4	141.3
Endothelioma								
%		97.6	0	0.5	98.2	64.1	0	95.9
Mean	62.6	115.6	59.9	66.4	137.6	89.2	55.0	146.6
Liver								
%		98.4	3.7	92.6	97.8	31.3	0.5	98.4
Mean	56.9	124.9	58.3	89.9	139.0	71.1	56.2	132.6
Mixed liver								
%		95.0	11.5	29.7	92.8	17.0	3.5	94.9
Mean	61.3	109.8	63.6	70.6	123.8	65.1	58.0	128.8
Fibroblasts								
%		98.4	7.8	18.5	94.1	73.5	0	98.8
Mean	55.1	114.5	57.7	62.0	127.1	86.8	51.1	146.1

[a]PBS, phosphate-buffered saline; Con A, concanavalin A; DBA, Dolichos biflorus agglutinin; PNA, peanut agglutinin; RCAI, *Ricinus communis* agglutinin; SBA, soybean agglutinin; UEAI, *Ulex europaeus* agglutinin; WGA, wheat germ agglutinin; MBE, mouse brain endothelium; MLyE, mouse lymphatic endothelium; ABAE, adult bovine aortic endothelium.

[b]Percentage corrected for PBS control values (PBS = 0%).

[c]Mean channel value. Log amplification: 36 channels = 1 decade increase in fluorescence.

IV. ORGAN SPECIFICITY OF TUMOR CELL ADHESION/ENDOTHELIAL CELL INTERACTIONS

The most important interaction between tumor cells in the circulation and microvascular endothelial cells is that which leads to the adhesion and thus retention of tumor cells in a specific target organ. Without such adhesion the subsequent process of extravasation cannot occur. Such interactions depend on properties expressed on the endothelial cell surface and on the surface of metastasizing tumor cells. Heterogeneity of both the microvascular endothelial cells and the tumor cells may be presumed to control the process of organ-specific adhesion.

A. Selective Adhesion

To test the thesis that organ-specific differences among tumor cells are associated with organ-selective metastasis, studies both in other laboratories and in our own were carried out in which labeled tumor cells were permitted to adhere *in vitro* to monolayers of cultured microvascular endothelial cells.[1,2,13,14,23,24]

Our own studies initially were carried out using a murine glioma cell line and a teratoma cell line which showed spontaneous metastasis to the ovary.[13] To test organ-specific adhesion to endothelium, we established murine brain and ovary endothelial cell cultures, expanded them through several passages *in vitro*, and then prepared monolayers for adhesion assays. Glioma and teratoma cells were labeled overnight with [^{3}H]-thymidine, added to wells containing endothelial cells, and permitted to adhere for 0–45 min under constant agitation to reduce nonspecific adhesion. At various time points, nonadherent cells were removed and the retained cells enumerated by measuring radioactivity. The results of these initial studies are shown in Fig. 2. Glioma cells adhered preferentially to brain microvascular endothelial cells while the teratoma cells adhered preferentially to ovary-derived endothelial cell monolayers.

Extending these studies, we found that different tumors had different adhesive propensities.[14] In general, these followed predictions: hepatoma cells adhered preferentially to liver-derived endothelium, glioma cells to brain-derived endothelium, teratomas to endothelial cells obtained from the ovary, and mammary tumor cells to lymphatic endothelium. Some tumors such as the S180 sarcoma showed no selectivity, while yet others such as a murine endothelioma cell line gave unexpected specificity patterns, adhering best to lung fibroblasts (Fig. 3).

Several other laboratories have made similar observations using other tumors and other endothelial cell targets.[1,2,23,24] Most striking have been other studies using hepatomas which almost universally adhere preferentially to hepatic endothelium, and various melanoma sublines preselected for various metastatic patterns, which then manifest some selectivity in adhesion.[23] In a recent study of lymphoma variants, a lung-selective variant adhered preferentially to pulmonary vascular endothelial cells while another variant with preferential metastasis to the liver adhered preferentially to liver-derived microvascular endothelium.[25] Mizrachi *et al.* observed specific adhesion between pheochromocytoma cells and adrenal medullary endothelial cells in coculture.[26]

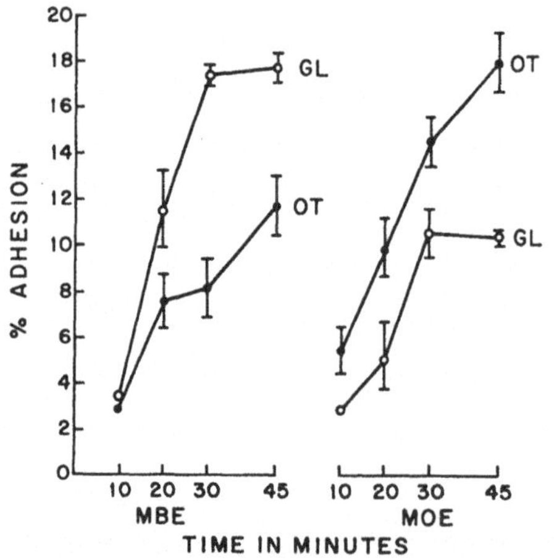

Figure 2. Adhesion of GL26 glioma and OTT6050 teratoma cells to murine brain- and ovary-derived microvascular endothelial cells. Radiolabeled tumor cells were added to confluent monolayer cultures of endothelial cells, agitated at room temperature at a rotational speed of 100 rpm. Replicate cultures were gently washed at different time points and the amount of radioactivity retained on the monolayer was used to calculate the percent adhesion of tumor cells. (From Ref. 14.)

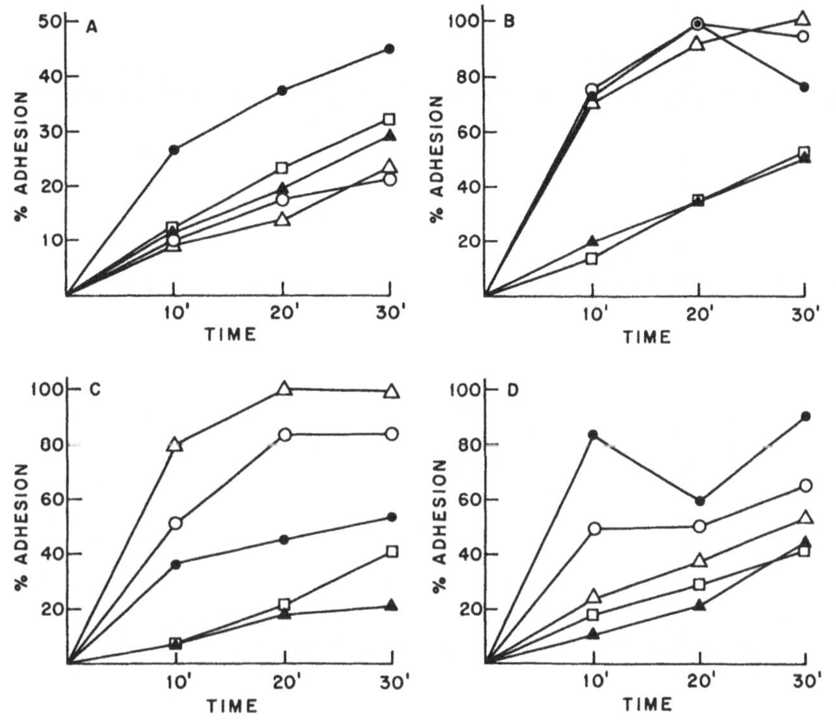

Figure 3. Adhesion of different mouse tumors (A, endothelioma; B, Sarcoma 180; C, OTT6050 teratoma; D, MBT-2 bladder tumor) to different murine confluent culture monolayers; Δ, lung endothelium; ○, 3T3 fibroblasts; ●, liver endothelium; □, thoracic duct endothelium; ▲, brain endothelium. (From Ref. 13.)

B. Role of the Extracellular Matrix

Discussions of the role of the extracellular matrix may be found in other chapters in this volume (see in particular the chapter by Belloni and Nicolson). Reid and her colleagues have carried out an extensive analysis of the role of extracellular matrix components in the metastatic process.[16] An intriguing finding was made recently by Pauli and Lee who observed that large vessel endothelial cells can acquire organ-specific adhesive properties by being cultured on cell matrix components of different endothelial cell origins.[21] Their laboratory is now enlarging these studies by also using lectins to help define these differences. While matrix components are of primary importance in cell adhesion, there are cell–cell interactions both superimposed and partially independent of those molecules.[27]

C. Change in Tumor Cell Phenotype following Selection for Specific Organ Affinity

Several tumor types have been subjected to selection for preferential metastasis to different organ sites. Among these have been a teratoma cell line selected for seeding to the spleen,[28] melanoma cell lines seeding to lung, brain, and liver,[29–31] selection of a mammary tumor cell line which grows preferentially in the lung and liver,[31] and a variety of less well studied tumors isolated from mice and rats, each with distinct and characteristic site preferences. Several human tumors have also been selected for preferential growth in specific organs of athymic (*nu/nu*) mice.[33] In our own laboratory, we decided to use the protocol developed by Fidler[29] to select a glioma subline which would preferentially settle in the lung. Our purpose was to determine whether the selection of a lung-selective variant would be accompanied by specific changes in cell surface lectin-binding sites and lectin receptors of the tumor subline. In turn, we would then determine whether these changes correlated with the expression of lectin-binding sites and lectin receptors on pulmonary microvascular endothelial cells.

The first phase of this study has now been completed. To select for lung-seeding tumor variants, GL26 tumor cells were injected intravenously, an occasional tumor found in the lung was isolated, cultivated *in vitro*, passaged once to obtain more cells, and then reinjected into the tail vein of syngeneic mice. By the time the fourth cycle was concluded ($GL26F_4$), a tenfold enrichment in lung-seeding capacity was achieved. At this time the 21-lectin panel was used to compare the initial GL26 and $GL26F_1$ and $GL26F_4$. Of the 21 lectins, one had lost efficacy of binding and one had gained in labeling the tumor subline. It remains to be determined whether these two changes are independent or correlated, and whether a *de novo* selection protocol would result once again in the same shift in cell surface-associated sugar residues.

V. DEVELOPMENTAL PROCESSES CONTROLLING MICROVASCULAR ENDOTHELIAL CELL DIFFERENTIATION

A major unresolved question in developmental biology is that dealing with the nature of endothelial cell differentiation.[18,34] Initial formation of endothelial cells in the blood islands and other undifferentiated mesenchymal beds prior to the establishment of true vessels appears to occur in a diffuse manner, with individual angioblasts becoming apparent as the mesenchymal beds undergo condensation and initial morphogenesis. Subsequently, vasculogenesis appears to involve both a coalescence of previously formed endothelial cells within this local environment as well as an ingrowth of vessel-associated endothelial cells from previously established major vessels of the embryo. Whether the initial local vasculature is even retained once ingrowth of extrinsic vessels occurs has never been determined. It is quite conceivable, for example, that the initial endothelial cell population is replaced by a second population just as the embryonic erythrocytic lineage is sequentially replaced by intermediate cells from the fetal liver and ultimately by cells derived from the bone marrow; or much as during the development of the renal system the pronephros is replaced by the mesonephros and then by the metanephros with no continuity between each of these rather similar excretory organs.[35]

The fact that this question has not been resolved is unfortunate with respect to the issue of organ specificity of the microvasculature, for the rules that govern *in situ* differentiation of endothelial cell precursors may not be the same as those that regulate the differentiation of already formed endothelial cells entering the organ rudiment from a previously established vessel. On the other hand, in either case it is clear that both the morphology and the biochemical attributes of the endothelial cells ultimately found in a given organ reflect the properties of that organ. Thus, endothelial cells entering the brain anlage develop the attributes of the blood–brain barrier both structurally and functionally,[36] cells entering the developing glomerulus become fenestrated,[37] and cells invading the developing limb bud develop the typical tight junctions found in the capillaries of these organs.[35]

There are, however, several points which seem firmly established: (1) Endothelial cells form first in the embryonic yolk sac, where the blood islands give rise to individual endothelial-like cells which coalesce to become small microvessels. (2) Endothelial cells form separately in the heart, from which they migrate, forming the primary large vessels. (3) Individual endothelial cell precursors may migrate from the yolk sac throughout the body to differentiate *in situ* in different organ environments. (4) Major endothelial cell movement, however, occurs in sheets, with endothelial cells from preexisting vessels entering various organ rudiments as these differentiate and enlarge.

Whether it is preendothelial cells which migrate into specific organ sites or already functional endothelial cells from extrinsic vessels does not appear to influence the nature of endothelial cell differentiation: In each documented instance, endothelial cells *acquire* organ specificity after entering the developing rudiment, differentiating in accordance with the local microenvironment. This appears to be true postnatally as well, both in late-developing organs such as the brain, and in transplants which become vascularized by endothelial cells of the host. A particularly intriguing question is what developmental changes are essential during the process of angiogenesis: When tumors evoke new blood vessel formation, do they induce dedifferentiation in the responding endothelial cells? If so, this would seem particularly important if one wishes to approach specific tumor therapy by interfering with the process of angiogenesis.

VI. CONCLUDING COMMENTS

We are only now beginning to appreciate the vast complexity of the vascular system. Heterogeneity seems to describe almost all biological systems, but what at one time was thought to be true only for the immune system and the nervous system more and more seems to be true for other systems as well. Not only are all endothelial cells not alike, but they may differ even within a given organ, they may change in response to microenvironmental cues, they may vary in the products they secrete or the substances to which they respond, they may change shape, they may modulate their extracellular matrix, they may secrete numerous cytokines, they can interact directly with tumor cells, muscle cells, astrocytes, and pericytes. In all likelihood, heterogeneity results from specific tissue interactions between endothelial cells and the cells and tissues in which they differentiate.

To understand selective metastasis we must understand the basis for organ-specific differentiation of endothelial cells. Almost daily, new adhesion-mediating molecules are described, comprising several molecular families, each one of which will almost certainly include a large, possibly near-infinite number of variants. Studies with lectins and antibodies may provide a means of defining this complex differentiation process. At the same time, they may serve as a handle for developing tumor therapy strategies. Clearly, much work needs to be done.

REFERENCES

1. Nicolson, G. L., 1988, Organ specificity of tumor metastasis role of preferential adhesion, invasion and growth of malignant cells at specific secondary sites, *Cancer Metastasis Rev* **7**:143–188.
2. Nicolson, G. L., 1988, Cancer metastasis: Tumor cell and host organ properties important in metastasis to specific secondary sites, *Biochim. Biophys. Acta* **948**:175–224.
3. Zetter, B. R., 1990, The cellular basis of site-specific tumor metastasis, *N. Engl. J. Med.* **332**:605–612.
4. Auerbach, R., 1988, Patterns of tumor metastasis: Organ selectivity in the spread of cancer cells, *Lab. Invest.* **58**:361–364.
5. Weiss, L., Orr, F. W., and Honn, K. V., 1989, Interactions between cancer cells and the microvasculature: A rate-regulator for metastasis, *Clin. Exp. Metastasis* **7**:127–167.
6. Jaffe, E. A., 1983, *The Biology of Endothelial Cells*, Nijhoff, The Hague.
7. Ryan, U. (ed.), 1989, *Endothelial Cells*, CRC Press, Boca Raton.
8. Simionescu, N., and Simionescu, M., 1988, *Endothelial Cell Biology in Health and Disease*, Plenum Press, New York.
9. Simionescu, N., and Simionescu, M., 1991, *Endothelial Cell Dysfunction*, Plenum Press, New York.
10. Shepro, D., 1988, Endothelial cells, inflammatory edema and the microvascular barrier: Comments by a "free radical," *Microvasc. Res.* **35**:247.
11. Fidler, I. J., 1989, Origin and biology of cancer metastasis, *Cytometry* **10**:673–680.
12. Gumkowski, F., Kaminska, G., Kaminski, M., Morrissey, L. W., and Auerbach, R., 1987, Heterogeneity of mouse vascular endothelium. In vitro studies of lymphatic, large blood vessel and microvascular endothelial cells, *Blood Vessels* **24**:11–23.
13. Auerbach, R., Lu, W. C., Pardon, E., Gumkowski, F., Kaminska, G., and Kaminski, M., 1987, Specificity of adhesion between murine tumor cells and capillary endothelium: An in vitro correlate of preferential metastasis in vivo, *Cancer Res.* **47**:1492–1496.
14. Alby, L., and Auerbach, R., 1984, Differential adhesion of tumor cells to capillary endothelial cells in vitro, *Proc. Natl. Acad. Sci. USA* **81**:5739–5743.
15. Zetter, B. R., 1989, Endothelial heterogeneity: Influence of vessel size, organ localization and species specificity on the properties of cultured endothelial cells, in: *Endothelial Cells*, Vol. 2 (U.S. Ryan, ed.), CRC Press, Boca Raton, pp. 63–79.
16. Doerr, R., Zvibel, I., Chiuten, D., D'Olimpio, J., and Reid, L. M., 1989, Clonal growth of tumors on tissue-specific biomatrices and correlation with organ site specificity of metastases, *Cancer Res* **49**:384–392.
17. Auerbach, R., Alby, L., Morrissey, L. W., Tu, M., and Joseph, J., 1985, Expression of organ-specific antigens on capillary endothelial cells, *Microvasc. Res.* **29**:401–411.
18. Auerbach, R., and Joseph, J., 1983, Cell surface markers on endothelial cells: A developmental perspective, in: *The Biology of Endothelial Cells* (E. A. Jaffe, ed.), Nijhoff, The Hague, pp. 393–400.
19. Moscona, A., 1957, The development of chimeric aggregates of dissociated embryonic chick and mouse cells, *Proc. Natl. Acad. Sci. USA* **43**:184–194.
20. Rutishauser, U., and Jessell, T. M., 1988, Cell adhesion molecules in vertebrate neural development, *Physiol. Rev.* **68**:819–857.
21. Pauli, B. M., and Lee, C.-L., 1988, Organ preference of metastasis: The role of organ-specific modulated endothelial cells, *Lab. Invest.* **58**:379–387.
22. Ponder, A. J., and Wilkinson, M. M., 1983, Organ-related differences in binding of Dolichos biflorus agglutinin to vascular endothelium, *Dev. Biol.* **96**:535–541.
23. Schirrmacher, V., 1985, Cancer metastasis, *Adv. Cancer Res.* **43**:1–73.
24. Netland, P. A., and Zetter, B. R., 1984, Organ-specific adhesion of metastatic tumor cells in vitro, *Science* **224**:1113–1115.
25. Brodt, P., 1989, Tumor cell adhesion to frozen lymph node sections—An in vitro correlate of lymphatic metastasis, *Clin. Exp. Metastasis* **7**:135–139.
26. Mizrachi, Y, Lelkes, P. I., Ornberg, R. L., Goping, G., and Pollard, H. B., 1989, Specific adhesion between pheochromocytoma (PC12) cells and adrenal medullary endothelial cells in co-culture, *Cell Tissue Res.* **256**: 365–372.
27. Gilbert, S., 1991, *Developmental Biology*, 3rd Edition, Sinauer Associates, Stamford, Conn.
28. Auerbach, R., 1971, The use of tumors in the analysis of inductive tissue interactions, *Dev. Biol.* **28**:304–309.

29. Fidler, I. J., 1973, Selection of successive tumor lines for metastasis. *Nature* **242**:148–149.
30. Nicolson, G. L., and Winkelhake, J. L., 1975, Organ specificity of blood-borne tumour metastasis determined by cell adhesion? *Nature* **255**:230–232.
31. Killion, J. J., and Fidler, I. J., 1989, The biology of tumor metastasis, *Semin. Oncol.* **16**:106–115.
32. Nicolson, G. L., Belloni, P. N., Tressler, R. J., Dulski, K., Inoue, T., and Cavanaugh, P. G., 1989, Adhesive, invasive and growth properties of selected metastatic variants of a murine large-cell lymphoma, *Invasion Metastasis* **9**:102–116.
33. Price, J. E., Daniels, L. M., Campbell, D. E., and Giavazzi, R., 1989, Organ distribution of experimental metastases of a human colorectal carcinoma injected in nude mice, *Clin. Exp. Metastasis* **7**:55–68.
34. Auerbach, R., 1991, Interactions between cancer cells and the endothelium, in: *The Microcirculation in Cancer Metastasis* (F. W. Orr, M. Buchanan, and L. Weiss, eds.), CRC Monograph, Boca Raton, pp. 169–181.
35. Feinberg, R., Sherer, G., and Auerbach, R., 1991, *Development of the Vascular System*, Karger, Basel.
36. Risau, W., and Wolburg, H., 1990, Development of the blood–brain barrier, *Trends Neurosci.* **13**:174–178.
37. Sariola, H., Ekblom, P., Lehtonen, E., and Saxen, L., 1983, Differentiation and vascularization of the metanephric kidney grafted on the chorioallantoic membrane, *Dev. Biol.* **96**:427–435.

26

Organotypic Endothelial Cell Surface Molecules Mediate Organ Preference of Metastasis

Bendicht U. Pauli, Robert C. Johnson, and Marwan E. El-Sabban

I. INTRODUCTION

Metastases are tumor colonies that develop in distant, often multiple organ sites by dissemination from a malignant primary tumor. Although the most frequent organ sites of distant metastases are the first organs encountered by blood-borne tumor cells, many cancers metastasize to sites which are unrelated to the initial organ entered by circulating tumor cells and, thus, display unique organ colonization patterns that do not fit simple, anatomical–mechanical trapping theories of tumor cell dissemination.[1–7] For example, breast adenocarcinomas metastasize frequently to liver, bone, brain, and adrenals, in addition to the expected, high incidence of metastasis to the lungs, and carcinomas of the prostate metastasize most often to bone. This nonrandom pattern of metastasis was first described by Paget,[8] who championed the "seed and soil" hypothesis of the metastatic spread. He postulated that gross tumor development was a consequence of the provision of a fertile environment (the soil) in which compatible tumor cells (the seeds) could proliferate.[4] Numerous studies on the mechanisms of preferential metastasis have since shown that tumor cell implantation, invasion, survival, and growth at secondary organ sites depend upon a number of tumor cell and host characteristics that provide the proper cellular and stromal environment for metastasis to develop.[9–16]

Abbreviations used in this chapter: BAEC, bovine aortic endothelial cells; TAMs, organ-specific, extracellular matrix-associated tumor attachment modulators; L-TAMs, TAMs derived from lung; H-TAMs, TAMs derived from liver; K-TAMs, TAMs derived from kidney; BAEC-(L-TAMs); BAEC grown on L-TAMs; BAEC-(H-TAMs), BAEC grown on H-TAMs; BAEC-(K-TAMs), BAEC grown on K-TAMs.

Bendicht U. Pauli, Robert C. Johnson, and Marwan E. El-Sabban • Cancer Biology Laboratories, Department of Pathology, College of Veterinary Medicine, Cornell University, Ithaca, New York 14853.

Endothelial Cell Dysfunctions, edited by Nicolae Simionescu and Maya Simionescu. Plenum Press, New York, 1992.

Recent studies indicate that the initial colonization of a secondary organ occurs at the time of attachment of blood-borne cancer cells to the microvascular endothelium of that organ.[17–19] This process is thought to be mediated by organotypic adhesion molecules expressed on the surface of endothelial cells. Attempts to characterize these molecules have been frustrated by complex isolation and culture procedures of organ-specific microvascular endothelial cells,[20–22] contamination and subsequent overgrowth of endothelial cell cultures with smooth muscle cells and/or fibroblasts,[20–22] and loss of the organ-specific microvascular phenotype after prolonged culture periods on plastic,[23–28] typically consumed during cloning of pure cell populations. By taking advantage of recent observations to the effect that components of the extracellular matrix can mediate phenotypic expression, growth, and differentiation of cells,[29–35] we have bypassed the problems of culturing microvascular endothelial cells and grown large vessel endothelial cells [e.g., bovine aorta (BAEC)] that can be easily isolated and established as pure, early passage cells in culture, on organ-derived biomatrix to induce organ-specific endothelial cell modulation.[36] Using monolayers of modulated BAEC, we show here that tumor cells which metastasize to a given organ, preferentially bind to BAEC grown on biomatrix of that organ.

II. MATERIALS AND METHODS

A. Cell Culture

Endothelial cells were routinely isolated and grown from aortas of 18-month-old bovines using the procedure of Jaffe.[37] Endothelial cells were grown in DME medium supplemented with 2 mM L-glutamine and 10% bovine calf serum, frozen at passage 2, and used at passage 2 to 8.

The following tumor cell systems were used in this study: (1) murine B16 melanoma cells[38] obtained from Dr. I. J. Fidler, University of Texas System Cancer Center, M.D. Anderson Hospital and Tumor Institute, Houston): B16–F1 (parental cell line: low lung colonization) and B16–F10 (selected *in vivo* for high lung colonization); (2) rat R3230AC mammary adenocarcinoma cells[39] obtained from Dr. J. A. Kellen, Sunnybrook Medical Center, University of Toronto, Toronto, Canada: R3230AC (parental cell line: intermediate lung colonization), R3230AC-MET (selected *in vivo* for high lung colonization), and R3230AC-LR [nonmetastatic, lectin (WGA and Con A)-resistant variant of R3230AC]; (3) M5076 reticulum sarcoma cells[40] obtained from Dr. J. E. Talmadge, NCI–Frederick Cancer Research Facility, Frederick, Md.: high liver, intermediate ovary and kidney, and low spleen colonization; (4) RAW117 Abelson–Moloney transformed Balb/c splenocytes[41] obtained from Dr. K. W. Brunson, Pfizer Central Research, Groton, Conn.: RAW117-P (parental cell line: forms low numbers of metastases in lungs, liver, and spleen) and RAW117-H10 (selected *in vivo* for high liver colonization); (5) human urinary bladder carcinoma cells[42] obtained from J. F. Kieler, The Fibiger Institute, Copenhagen, Denmark: 647V (tumorigenic, nonmetastatic in nude mice); and (6) rat urinary bladder carcinoma cells derived from chemical carcinogen *N*-[4-(5-nitro-2-furyl)-2-thiazolyl]formamide-induced bladder tumors[43]: well-differentiated, nontumorigenic, nonmetastatic RBTCC-1 tumor cell line.

B. Preparation and Extraction of Organ-Specific Biomatrix

Lung, liver, and kidney biomatrices, used throughout this study, were of bovine origin and were isolated by a modification[36] of the procedure of Rojkind *et al.*[44] Biomatrices devoid of nucleic acids and lipids were sterilized by gamma irradiation and stored at −40°C until used. Acetic acid extraction (10 ml of 0.5 N acetic acid/g biomatrix) was carried out overnight of 4°C.[29] Extracts were centrifuged for 30 min at 23,000*g*. The resultant pellet was resuspended in 5 volumes of 0.5 N acetic acid and further extracted for 18 hr. The supernatants of the 2-day extracts were pooled, neutralized with 1 N NaOH, and the protein concentration estimated by the method of Lowry.[45] The protein extract was then aliquoted, lyophilized, and stored at −40°C. These extracts contained the molecules that were responsible for modulating large vessel endothelial cells to assume phenotypic traits of organ-specific, microvascular endothelium. These as yet undefined molecules were termed "tumor attachment modulators" (TAMs).[36]

C. The Tumor Cell Attachment Assay

The tumor cell/endothelial cell adherence assay was carried out in 24-well plates coated with organ-specific TAMs. Wells were filled with 50 μg TAMs/0.5 ml distilled water, incubated overnight at 4°C, washed thoroughly with serum-free DMEM, and then seeded with endothelial cells (5×10^4 cells/well in DMEM + 10% bovine calf serum). Wells containing confluent endothelial cell monolayers were washed with DMEM supplemented with 0.4% bovine serum albumin (BSA, Sigma Fraction V, specially treated with periodic acid to remove glycoprotein impurities[46]) (DMEM–BSA). Tumor cells to be analyzed for attachment to vascular endothelial cells were labeled with [^{3}H]thymidine (2 Ci/mmole; 200 nCi/ml) for 24 hr at 37° C, then either directly used in the attachment assay (RAW117-P, RAW117-H10) or after removal from the growth surface by brief trypsinization (all other cell lines). Trypsinized tumor cells were allowed to recover from the enzyme treatment in their respective growth medium + 10% fetal bovine serum for 30 min at 37° C. After this recovery time, tumor cells were washed twice in DMEM–BSA, plated onto endothelial cell monolayers at a concentration of 1×10^5 tumor cells/0.5 ml DMEM–BSA, and incubated for 30 min at 37°C (if not otherwise stated). Wells were then washed three times with DMEM–BSA to remove unattached tumor cells. Endothelial cell monolayers with adherent tumor cells were solubilized with 1% SDS in 0.5 ml distilled water. Counts were determined in a Beckman 7000 scintillation counter and transformed into actual cell numbers by comparing with standards.

D. Extraction and Two-Dimensional Gel Electrophoresis of Endothelial Cell Surface Components

Endothelial cell monolayers (approximately 2×10^7 cells) grown in uncoated 100-mm dishes or dishes coated with organ-specific TAMs were extracted with 1 M urea in PBS, pH 7.4, for 30 min at room temperature.[47] Extracts were centrifuged at 10,000*g* for 30 min, and the supernatant dialyzed against distilled water at 4°C overnight and lyophilized. Aliquots of 80 μg endothelial cell surface proteins were dissolved in 20 μl of 10 mM Tris–HCl buffer, pH 7.4, containing 5 mM $MgCl_2$, 9 M urea, 1% NP-40, 1% Ampholine (pH 6–8), and 2.5% mercaptoethanol, and subjected to two-dimensional gel electrophoresis essentially as described by O'Farrell.[48] In the first dimension, samples were loaded onto tubes (135 mm × 1.8 mm i.d.) of 4% polyacrylamide and proteins separated by isoelectric focusing (IEF). In the second dimension, IEF tube gels were electrophoresed on 20 × 20-cm slab gels of an 8–18% polyacrylamide linear gradient. Gels were fixed, stained with silver, then analyzed with a BioImage Visage 110 computerized imaging system.

E. Preparation of Endothelial Cell Surface Vesicles in Vivo

Closed perfusion systems of rat lungs and legs were established by cannulating the pulmonary artery and vein through the left and right ventricles of the heart and by cannulating the left femoral artery and vein, respectively. Using a peristaltic pump, lungs and legs were slowly perfused with a solution of 100 mM formaldehyde and 2 mM DTT in PBS (containing 1 mM Ca^{+} and Mg^{2+}) for 3 hr at 37°C. Outside-out membrane vesicles released from the luminal membranes of rat lung and leg vascular endothelia were harvested by centrifugation (30,000*g*, 30 min, 4°C). Similar vesicles were prepared from BAEC monolayers, using the method of Scott.[49] Membrane vesicles iodinated as described by Soule *et al.*[50] were then used in a tumor cell attachment assay. In brief, tumor cells were grown in 96-well plates to subconfluency, then washed with several changes of RPMI 1640–BSA. Endothelial membrane vesicles were added in 40 μl PBS (approximately 1 × 10^6 cpm/well). Vesicles were spun onto the anchored tumor cells at 1000 rpm for 5 min and incubated for 30 min at 37°C. Unattached vesicles were removed by washing wells three times with RPMI. After solubilization of tumor cells and adherent vesicles with 1% SDS, wells were counted with a gamma counter and data expressed as percent of total counts added. Attachment assays were also performed with vesicles that had been incubated with 0.5 U/ml neuraminidase for 30 min at 37°C.

III. RESULTS

A. Preferential Adhesion of Tumor Cells to Endothelial Cells Grown on the Biomatrix from Various Organs

Tumor cell lines which were selected *in vivo* to metastasize to a specific organ site, displayed a significantly higher binding affinity for BAEC monolayers grown on TAMs of the preferred, metastasized organ than for BAEC monolayers grown on TAMs of any other organ which was not colonized by the tumor cells (Fig. 1). For example, RAW117–H10 lymphoma cells selected *in vivo* for high liver colonization, and M5076 reticulum sarcoma cells which spontaneously metastasized to the liver, adhered preferentially to BAEC-(H-TAMs) (Fig. 1A). The parental RAW117–P cell line, which upon tail vein injection generated low numbers of both lung and liver metastases, exhibited correspondingly low attachment preferences for BAEC-(L-TAMs) and BAEC-(H-TAMs). Murine B16–F10 melanoma cells, and rat R3230AC-MET mammary adenocarcinoma cells, which were selected for high lung colonization, attached insignificantly higher numbers to BAEC-(L-TAMs) than to BAEC-(H-TAMs) or BAEC-(K-TAMs) (Fig. 1B). The attachment of these

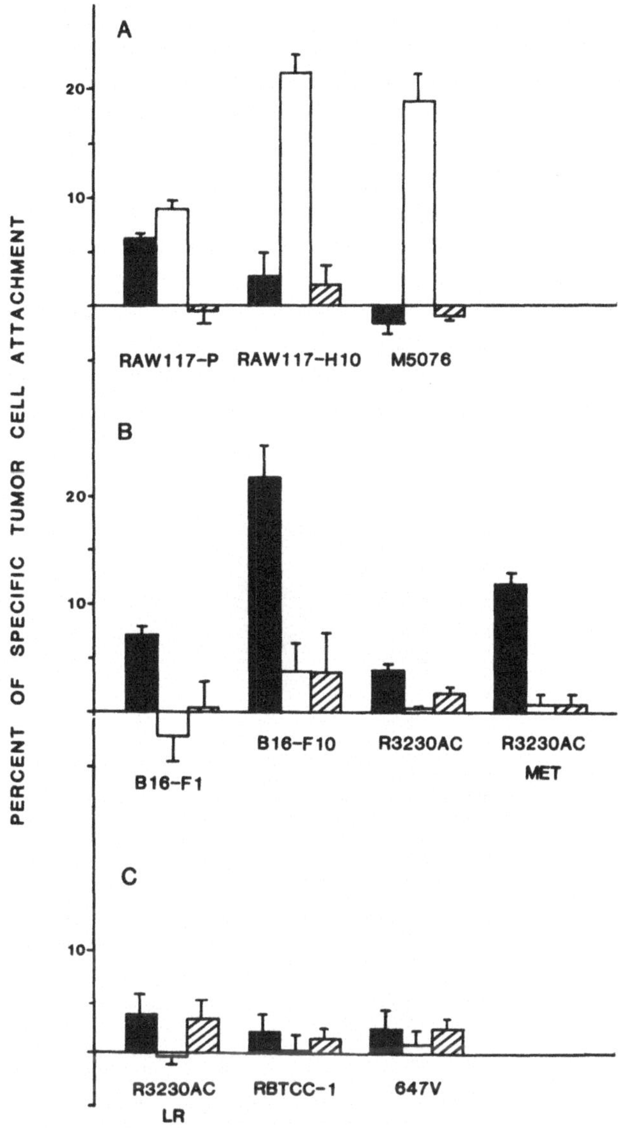

Figure 1. Preferential attachment of liver-metastatic (A), lung-metastatic (B), and nonmetastatic (C) tumor cells to monolayers of BAEC grown on organ-specific TAMs: percent specific attachment to BAEC-(L-TAMs) (solid bars), BAEC-(H-TAMs) (open bars), and BAEC-(K-TAMs) (hatched bars). Mean and standard deviation are from triplicate experiments. (From Ref. 36, with permission.)

tumor cells to BAEC-(L-TAMs) was also significantly higher than that of their parental cell lines, B16-F1 and R3230AC, respectively. R3230AC-LR cells, a low tumorigenic, *non*-metastasizing, Con A- and WGA-resistant cell subpopulation isolated from the malignant R3230AC rat mammary carcinoma cell line, showed weak, nonpreferential binding to BAEC-(L-TAMs), BAEC-(H-TAMs), and BAEC-(K-TAMs) that was statistically not different from the attachment of these cells to BAEC monolayers grown on *non*coated tissue culture plastic (Fig. 1C). No preferential attachment was noticed for any of the metastatic cell lines, when BAEC were substituted with skin fibroblasts (bovine, rat) in our tumor cell attachment assay.

Metastatic tumor cells predominantly attached to TAM-modulated endothelial cell monolayers at the apposition zone of neighboring endothelial cells (Fig. 2). After anchoring with foot processes to both endothelial cells, liver-metastatic RAW117–H10 tumor cells penetrated the monolayer within an 8-hr time period. This was achieved by "wedging" and was aided by the active participation of endothelial cells (Fig. 3). Similar to the reendothelialization of microthrombi or latex spheres, endothelial cells crawled over the tumor cells with a leading lamella, actively removing the tumor cell from the monolayer surface to the subendothelial space. At no time did endothelial cells expose unoccupied subendothelial matrix, but instead hugged the tumor cells closely during the entire extravasation process.

In the subendothelial space, tumor cells also expressed preferential attachment to biomatrix extracts derived from the metastasized organ. However, attachment to these extracts was significantly lower than that to correspondingly modulated BAEC. For example, total attachment of R3230AC-MET adenocarcinoma cells to BAEC-(L-TAMs)

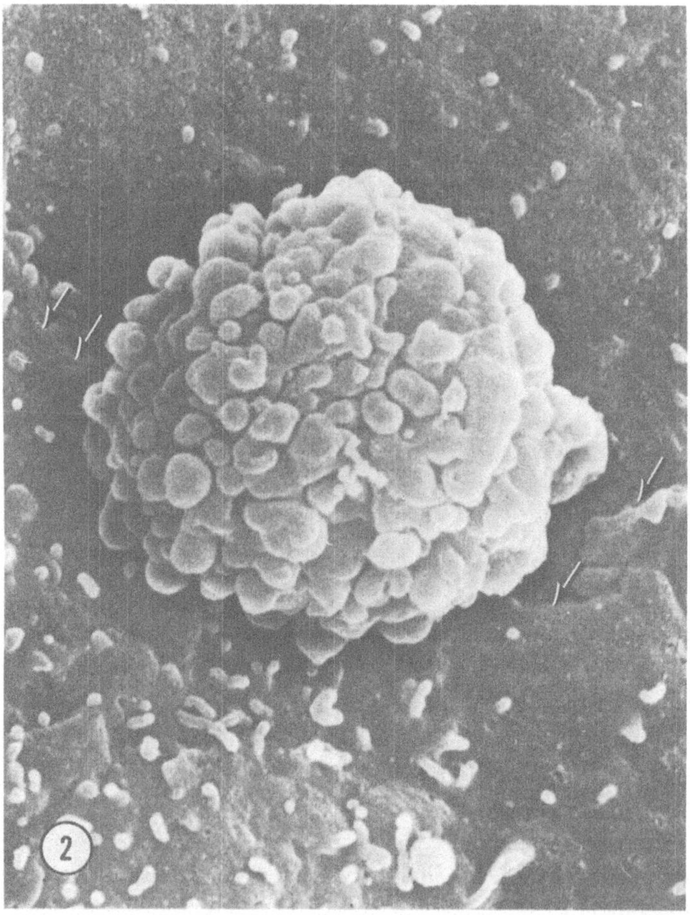

Figure 2. RAW117–H10 cell adheres to the border zone of two adjacent endothelial cells. Endothelial cells interdigitate by delicate, lamellar marginal folds (arrows). Incubation period: 30 min; SEM × 8500.

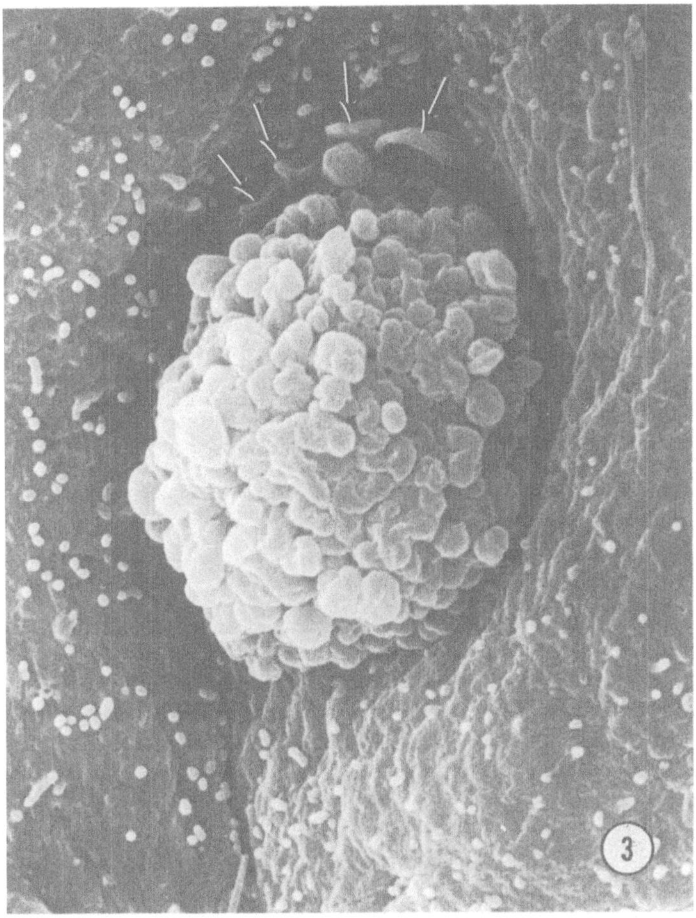

Figure 3. Granular RAW117–H10 cell has halfway penetrated the endothelial cell monolayer between two adjacent cells. Endothelial cells and their lifted marginal folds (arrows) tightly hug the penetrating tumor cell, exposing no free extracellular matrix. Incubation period: 4 hr; SEM × 6400. (From Ref. 36, with permission.)

was 70% compared to only 20% total attachment to L-TAMs after a 30-min incubation period (Fig. 4). Similar observations were made for the liver-metastatic cell line RAW117–H10, with 70% adhering to BAEC-(H-TAMs) and only 25% to H-TAM-containing biomatrix. Biomatrix extracts prepared from nonmetastasized organs showed similar adherence affinities as plastic surfaces coated with BSA.

B. Effects of Continuous Exposure of BAEC to Organ-Specific TAMs on the Preferential Attachment of Metastatic Tumor Cells

Preferential attachment of both lung- and liver-metastatic tumor cells to BAEC monolayers was significantly increased, when BAEC were grown continuously for ten

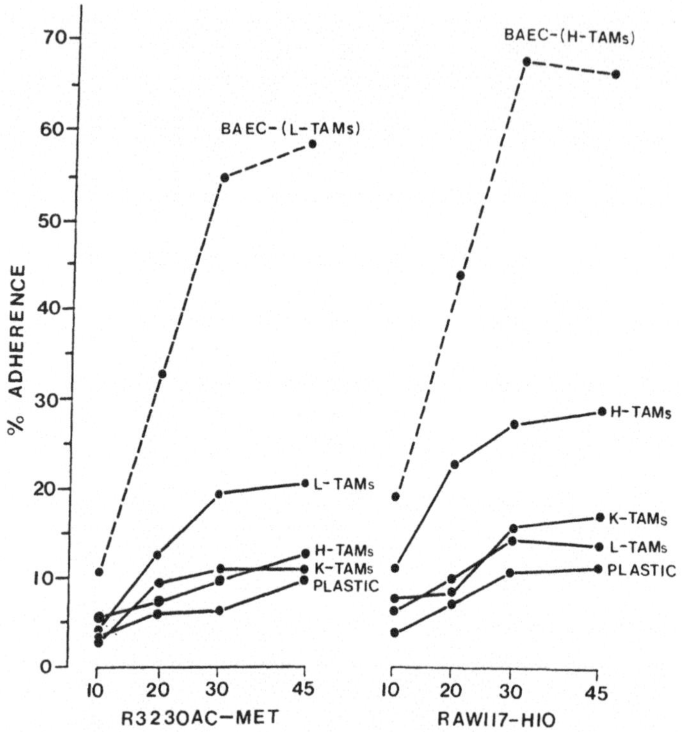

Figure 4. Preferential attachment of lung-metastatic R3230AC-MET and liver-metastatic RAW117–H10 to BAEC grown on L-TAMs or H-TAMs, respectively, is compared with their adhesion to organ-specific TAMs (without endothelial cells). Both tumor cells adhere in significantly higher numbers to BAEC grown on TAMs of the metastasized organ than they adhere to corresponding TAMs alone. However, binding to TAMs of the metastasized organ is still significantly higher than that to TAMs of the nonmetastasized organs or plastic.

generations (three passages) in dishes coated with TAMs prepared from the metastasized organ (adapted BAEC) before performing an attachment assay. The highest increases were observed for R3230AC-MET, M5076, and RAW117–H10 cells ($p < 0.001$; relative to the attachment of tumor cells to nonadapted BAEC grown on either L-TAMs or H-TAMs) (Fig. 5). In contrast, attachment of B16–F10 melanoma cells to adapted monolayers was only slightly increased relative to the attachment to nonadapted BAEC monolayers ($p = 0.1$). If adapted BAEC were grown on uncoated plastic surfaces, the specific adherence of tumor cells decreased, but, except for B16–F10 melanoma cells, it remained significantly higher than that to unadapted BAEC grown in dishes coated with L-TAMs or H-TAMs ($p < 0.01$) (Fig. 5). BAEC adapted to TAMs other than those of the metastasized organ exhibited the same nonspecific tumor cell attachment as unadapted BAEC grown on polystyrene.

C. Preferential Tumor Cell Attachment to Extracts from BAEC-(TAMs)

Tumor cell attachment to detergent-solubilized surface proteins from TAM-modulated BAEC was nonspecific, while urea extraction preserved the BAEC components that were responsible for preferential tumor cell attachment. Lung-metastatic tumor cells (B16-F10,

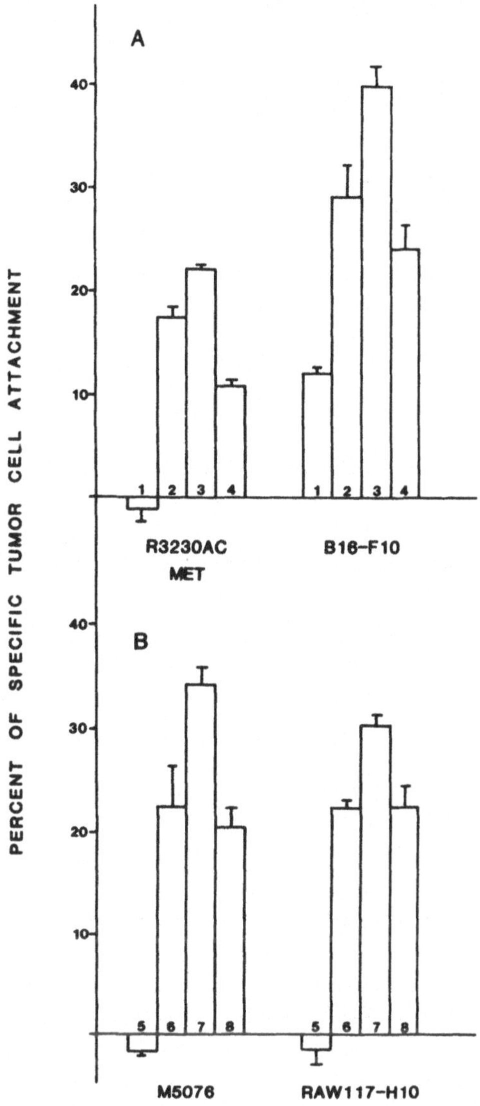

Figure 5. Effect of continuous exposure of BAEC to organ-specific TAMs on the preferential attachment of lung-metastatic (R3230AC-MET, B16–F10) (A) and liver-metastatic (RAW117–H10, M5076) (B) tumor cells. BAEC are grown for ten generations on either L-TAMs (A) or H-TAMs (B) (adapted BAEC). Specific tumor cell attachment to adapted BAEC grown for ten generations on plastic (1, 5), adapted BAEC grown on plastic (2, 6), adapted BAEC grown on L-TAMs (3) or H-TAMs (7), and unadapted BAEC grown on L-TAMs (4) or H-TAMs (8) is depicted. Mean and standard deviation are from triplicate analyses. (From Ref. 36, with permission.)

R3230AC-MET) attach preferentially to urea extracts from BAEC-(L-TAMS), while liver metastatic tumor cells (RAW117–H10, M5076) exhibited preferential attachment to extracts from BAEC-(H-TAMs). Attachment to extracts from BAEC modulated with TAMs derived from nonmetastasized organs, were statistically not different from tumor cell attachment to extracts obtained from BAEC grown on plastic. This observation together with the findings that BAEC could be adapted to express increased, preferential tumor cell attachment, and that fibroblasts expressed no specific tumor cell attachment, excluded the possibility that preferential tumor cell attachment was mediated directly by TAMs which might have been accessible through small gaps in the BAEC monolayers.

D. Two-Dimensional Gel Electrophoresis of Endothelial Cell Surface Extracts

Silver-stained two-dimensional gels from surface extracts of BAEC-(L-TAMs) and BAEC-(K-TAMs) containing an average of 400 protein spots, were scanned, matched, normalized, and then compared, using a BioImage Visage 110 image analysis system. While the bulk of these spots were matched in both gels (approximately 320 spots), there were a number of qualitative and/or quantitative differences that were partly identified in Fig. 6A,B. Different spots were characterized by their isoelectric point, apparent molecular weight, and ratio of spot area in gels of BAEC-(L-TAMs) and BAEC-(K-TAMs). Quantitative spot differences greatly exceeded the few qualitative spot differences.

E. Attachment of Endothelial Cell Membrane Vesicles to Tumor Cells

Membrane vesicles prepared from the luminal surface of lung and leg endothelial *in vivo* and BAEC grown on plastic were tested for their ability to bind to B16–F1 and B16–F10 melanoma cells. B16–F10 cells bind significantly more lung endothelial cell membrane vesicles than B16–F1 (100% for B16–F10 versus 53% for B16–F1) (Table 1). However, both B16–F1 and B16–F10 bound similar numbers of leg endothelial cell vesicles and vesicles from cultured BAEC. These numbers were also comparable with the number of lung endothelial vesicles binding to B16–F1. Preferential binding of lung endothelial vesicles to B16–F10 was abolished by prior incubation of vesicles with neuraminidase.

IV. DISCUSSION

Paget's "seed-and-soil" hypothesis of metastatic spread states that tumor cell implantation, invasion, survival, and growth at secondary organ sites depend upon a number of tumor cell and host characteristics that provide the proper stromal and cellular environment for metastasis formation.[9–16] In this study, we have shown that the initial organ-selection for metastasis occurs at the time of attachment of blood-borne tumor cells to the organ-specific, microvascular endothelium and that this process is conceptually similar to that of lymphocytes "homing" to high endothelial venules of specific lymphoid tissues.[51–55] While "homing" studies with lymphocytes were facilitated by the availability of a reproducible and sensitive *in vitro* assay system, in which lymphocytes could be tested for specific adherence to vascular endothelium in frozen sections from various lymphoid tissues,[56] preliminary studies with tumor cells in a similar assay system failed, possibly due to the enormous complexity and variability of tumor cell surface components, and the concurrent adhesion of tumor cells to many structural components of a frozen tissue section (e.g.,

→

Figure 6. Two-dimensional gel electrophoresis of urea extracts from BAEC grown on L-TAMs (A) or K-TAMs (B). Silver-stained gels are scanned, normalized, and matched using a BioImage Visage 110 image analysis system. The following quantitative and qualitative differences are illustrated: Spot #1, pI 5.40, 86 kDa, 1:19.9 (spot ratio A:B); spot #2, pI 6.78, 77 kDa, 1:1.7; spot #3, pI 4.75, 38 kDa, 1:1.9; spot #4, pI 4.95, 59 kDa, 1:2.9; spot #5, pI 10.30, 47 kDa, nondetectable in A; spot #6, pI 9.10, 77 kDa, 2.9:1.

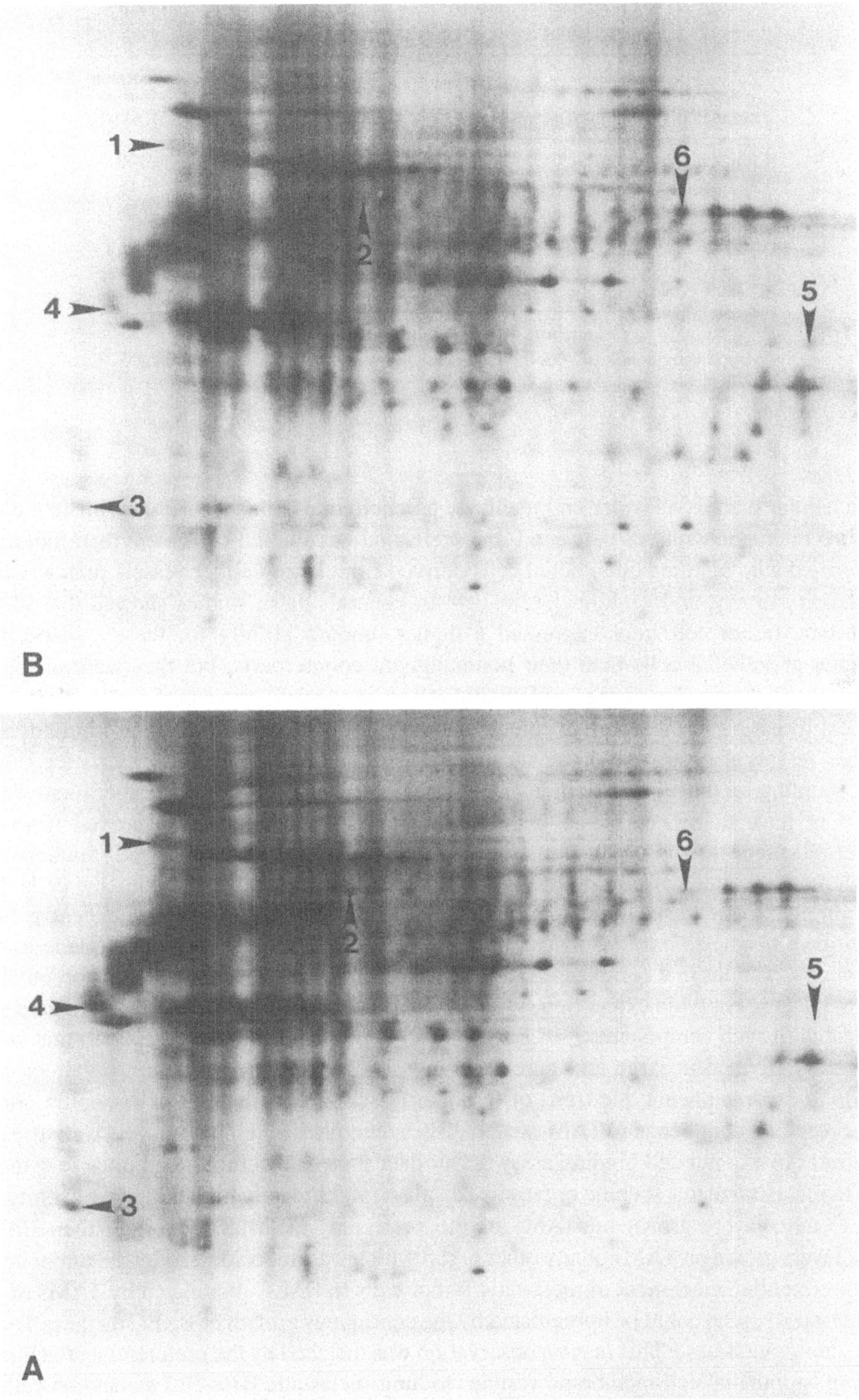
1
6
2
4
5
3
B
1
6
2
5
4
3
A

Table 1. Preferential Attachment of Endothelial Cell Membrane Vesicles to B16 Melanoma Cells

Endothelial cell membrane vesicles	Vesicle treatment	Relative % adhesion B16–F1	Relative % adhesion B16–F10
Lung endothelium	—	53[a]	100
Leg endothelium	—	42	46
BAEC monolayer	—	46	46
Lung endothelium	Neuraminidase[b]	46	45
Leg endothelium	Neuraminidase	45	51
BAEC monolayer	Neuraminadase	41	42

[a]Percent of lung endothelium vesicle binding to B16–F10.
[b]Vesicles are incubated with 0.5 U/ml of neuraminidase in PBS for 30 min at 37°C prior to performing an adhesion assay.

extracellular matrix, vascular endothelium, parenchymal cells) (Pauli, unpublished data). *In vitro* interaction studies between tumor cells and endothelial cells were therefore often performed with cultured endothelial cells derived from large-caliber vessels such as aorta, pulmonary artery, and umbilical vein.[57–59] In general, these studies showed that highly metastatic tumor cell lines expressed a higher binding affinity for these "unspecific" vascular endothelial cells than their nonmalignant counterparts, but they were unable to shed light on the involvement of endothelial cells in metastatic organ preference. This result is not surprising since it is well documented that endothelial cells originating from different caliber vessels and different organ sites vary greatly in their surface characteristics.[28,60–65] In accounting for the organ specificity of endothelial cells, an appropriate experiment would therefore have to be conducted with microvascular endothelial cells derived from the preferred, metastasized organ site. However, these studies have been greatly hindered by problems associated with the culturing and maintenance of capillary endothelial cells.[20–22] An alternative *in vitro* tumor cell/endothelial cell adhesion assay has recently been developed in our laboratory to bypass these problems.[36] This assay system made use of recent observations that components of the extracellular matrix might act as historegulators and mediate phenotypic expression, growth, and differentiation of cells which are grown in contact with such components.[29–34] Large vessel endothelial cells (e.g., BAEC) that could be readily isolated in large and pure quantities, were therefore grown on organ-specific TAMs to express phenotypic traits of the microvascular endothelium in a specific organ. After reaching confluence in TAM-coated dishes, endothelial cell monolayers were used as substrates in a tumor cell binding assay.[36] Our data showed that tumor cell lines selected *in vivo* to metastasize to a specific organ site, display a significantly higher binding affinity for BAEC monolayers grown on TAMs of the preferred, metastasized organ than BAEC monolayers grown on TAMs of any other organ which was not colonized by the tumor cells. The preferential attachment of metastatic tumor cells to BAEC modulated by TAMs of the metastasized organ could be upregulated by the continuous growth of BAEC on these TAMs (e.g., ten generations). This *in vitro* observation was matched by the preferential attachment of lung endothelial cell membrane vesicles to lung-metastatic B16–F10 melanoma cells.[66]

The preferential attachment of tumor cells to BAEC grown on TAMs of the metastasized organ was maintained across species barriers, and was not appreciably stronger for homologous endothelial cells and TAMs.[36] These data support previous findings that lectin-binding sites on microvascular endothelial cells are quantitatively similar from one species to another, yet vary greatly from organ to organ.[61,64,65] Although the biochemical nature of tumor cell/endothelial cell interadhesion is as yet unknown, recent studies by Rice and Bevilacqua[67] indicate that tumor cell attachment to endothelial cells can be mediated by cytokine [e.g., tumor necrosis factor (TNF)] inducible endothelial cell adhesion molecules which, at least in some instances, were identical to known endothelial cell leukocyte adhesion molecules (e.g., ELAM-1,[68] ICAM-1[69]). However, the cytokine concentrations that triggered the expression of these adhesion molecules greatly exceeded the synthetic capacity of a single tumor cell that usually gives rise to a metastatic colony. Even if such cytokine concentrations could be achieved by a single tumor cell, they would be counterproductive to tumor survival as a concomitantly enhanced host immune surveillance would result from such activity. This development is also at variance with the finding that tumor cells adherent to endothelial cells become refractory to cytolysis by natural killer cells,[70] a cell type that is usually activated by cytokines such as TNF. Furthermore, the tissue distribution of a melanoma cell adhesion molecule expressed on cytokine-stimulated human umbilical vein endothelial cells, is so ubiquitous that it could not account for preferential organ metastasis. Nevertheless, these and similar molecules might account for the observation that blood-borne tumor cells are often trapped and promoted to grow in inflammatory tissues,[71,72] possibly due to an upregulation of leukocyte adhesion molecule and an associated increase in stickiness of tumor cells.[73] Thus, cytokine-induced expression of adhesion molecules, while of significant importance in inflammatory processes, may have no role in tumor cell spread to specific organ sites, and may merely increase unspecific stickiness of tumor cells to sites of inflammation. Alternatively, our data[36] and those of Auerbach[18,63] and Nicolson[28] indicate that adhesion molecules involved in metastatic organ colonization are sets of molecules that are organotypically expressed and restricted to defined segments of the vasculature. These molecules are lost as endothelial cells are removed from their microenvironment and are cultured on artificial substrates (e.g., plastic).[36,63] It seems that tumors or clones thereof acquire ligands for such organotypic adhesion molecules as one of the many mechanisms that tumor cells use for survival in a hostile host environment.

The present test system is thought to be well suited for the study of the molecular mechanisms that govern the adherence of blood-borne tumor cells to vascular endothelium at the site of preferential metastasis formation and for shedding new light on the structural and functional complexity of endothelial cell adhesion molecules.[54] By determining the molecular mechanisms of tumor cell adhesion to endothelial cells, it may be possible to eventually develop new strategies for interfering with this process by specifically targeting agents to organ sites at predictable risk for metastatic involvement.[12]

ACKNOWLEDGMENTS. The authors thank Ms. Cho-Lea Lee, Ms. Ursula Nanna, and Dr. Hellmut Augustin-Voss for their expert technical assistance. This study was supported by Public Health Service Grant CA-47668 from the National Cancer Institute and a grant from the Cornell Biotechnology Program.

REFERENCES

1. Kinsey, D. L., 1960, An experimental study of preferential metastasis, *Cancer* **13**:674–676.
2. Patel, J. K., Didolkar, M. S., Pickren, J. W., and Moore, R. H., 1978, Metastatic pattern of malignant melanoma: A study of 216 autopsy cases, *Am. J. Surg.* **135**:807–810.
3. Sugarbaker, E. V., 1981, Patterns of metastasis in human malignancies, *Cancer Biol. Rev.* **2**:235–303.
4. Hart, I. R., 1982, "Seed and soil" revisited: Mechanisms of site-specific metastasis, *Cancer Metastasis Rev.* **1**:5–16.
5. Tarin, D., 1985, Clinical and experimental studies on the biology of metastasis, *Biochim. Biophys. Acta* **780**: 227–235.
6. Graf, A. H., Buchberger, W., Langmayr, H., and Schmid, K. W., 1988, Site preference of metastatic tumors of the brain, *Virchows Arch. A* **412**:493–498.
7. Nicolson, G. L., 1988, Organ specificity of tumor metastasis: Role of preferential adhesion, invasion and growth of malignant cells at specific secondary sites, *Cancer Metastasis Rev.* **7**:143–188.
8. Paget, S., 1889, The distribution of secondary growths in cancer of the breast, *Lancet* **1**:571–573.
9. Netland, P. A., and Zetter, B. R., 1984, Organ-specific adhesion of metastatic tumor cells *in vitro*, *Science* **224**:1113–1115.
10. Schirrmacher, V., 1985, Cancer metastasis: Experimental approaches, theoretical concepts and impacts for treatment strategies, *Adv. Cancer Res.* **43**:1–73.
11. Weiss, L., 1985, *Principles of Metastasis*, Academic Press, New York.
12. Nicolson, G. L., 1986, Organ preference of metastasis, *Prog. Clin. Biol. Res.* **212**:25–43.
13. Auerbach, R., 1988, Patterns of tumor metastasis: Organ selectivity in the spread of cancer cells, *Lab. Invest.* **58**:361–364.
14. Nicolson, G. L., 1988, Cancer metastasis: Tumor cell and host organ properties important in metastasis to specific secondary sites, *Biochim. Biophys. Acta* **948**:175–224.
15. Sher, B. T., Bargatze, R., Holzmann, B., Gallatin, W. M., Mathews, D., Wu, N., Picker, L., Butcher, E. C., and Weissman, I. L., 1988, Homing receptors and metastasis, *Adv. Cancer Res.* **51**:361–390.
16. Zetter, B. R., 1990, The cellular basis of site-specific tumor metastasis, *N. Engl. J. Med.* **322**:605–612.
17. Alby, L., and Auerbach, R., 1984, Differential adhesion of tumor cells to capillary endothelial cells in vitro, *Proc. Natl. Acad. Sci. USA* **81**:5739–5743.
18. Auerbach, R., Lu, W. C., Pardon, E., Gumkowski, F., Kaminska, G., and Kaminski, M., 1987, Specificity of adhesion between murine tumor cells and capillary endothelium: An *in vitro* correlate of preferential metastasis *in vivo*, *Cancer Res.* **47**:1492–1496.
19. Roos, E., Tulp, A., Middelkoop, O. P., and van de Pavert, I. V., 1984, Interactions between lymphoid tumor cells and isolated liver endothelial cells, *J. Natl. Cancer Inst.* **72**:1173–1180.
20. Bowman, P. D., Betz, A. L., Ar, D., Wolinsky, J. S., Penney, J. B., Shivers, R. R., and Goldstein, G. W., 1981, Primary culture of capillary endothelium from rat brain, *In Vitro* **17**:353–362.
21. Carson, M. P., and Haudenschild, C. C., 1986, Microvascular endothelium and pericytes: High yield, low passage cultures, *In Vitro Cell Dev. Biol.* **22**:344–354.
22. Folkman, J., Haudenschild, C. C., and Zetter, B. R., 1979, Long-term culture of capillary endothelial cells, *Proc. Natl. Acad. Sci. USA* **76**:5217–5221.
23. Borsum, T., Hagen, T., Henrickson, T., Hagen, I., Henriksen, T., and Carlander, B., 1982, Alterations in the protein composition and surface structure of human endothelial cells during growth in primary culture, *Atherosclerosis* **44**:367–378.
24. McAuslan, B. R., Hannan, G. N., and Reilly, W., 1982, Signals causing change in morphological phenotype, growth mode, and gene expression of vascular endothelial cells, *J. Cell Physiol.* **112**:96–106.
25. Rogers, K. A., and Kalnmus, V. I., 1983, Comparison of the cytoskeleton in aortic endothelial cells *in situ* and *in vitro*, *Lab. Invest.* **49**:650–654.
26. deBono, D. P., and Green, C., 1984, The adhesion of different cell types to vascular endothelium: Effects of culture density and age, *Br. J. Exp. Pathol.* **65**:145–154.
27. Bartlet, C. P., Heale, G., and Levene, C. I., 1985, Some factors regulating collagen polymorphism in cultured porcine and bovine aortic endothelium, *Atherosclerosis* **54**:301–309.
28. Belloni, P. N., and Nicolson, G. L., 1988, Differential expression of cell surface glycoproteins on various organ-derived microvascular endothelia and endothelial cell cultures, *J. Cell. Physiol.* **136**:398–410.

29. Reid, L., Morrow, B., Jubinsky, P., Schwartz, E., and Gatmaitan, Z., 1981, Regulation of growth and differentiation of epithelial cells by hormones, growth factors, and substrates of extracellular matrix, *Ann. N.Y. Acad. Sci.* **372**:354–370.
30. Wicha, M. S., Lowrie, G., Kohn, E., Bagavandass, P., and Mahn, T., 1982, Extracellular matrix promotes mammary epithelial growth and differentiation *in vitro*, *Proc. Natl. Acad. Sci. USA* **79**:3213–3217.
31. Gatmaitan, Z., Jefferson, D. M., Ruiz-Opazo, N., Biempica, L., Arias, I. M., Dudas, G., Leinwand, L. A., and Reid, L. M., 1983, Regulation of growth and differentiation of a rat hepatoma cell line by the synergistic interaction of hormones and collagenous substrate, *J. Cell Biol.* **97**:1179–1190.
32. Madri, J. A., and Williams, S. K., 1983, Capillary endothelial cell cultures: Phenotypic modulation by matrix components, *J. Cell Biol.* **97**:153–165.
33. Edelman, G. M., 1985, Cell adhesion and the molecular process of morphogenesis, *Annu. Rev. Biochem.* **54**: 135–169.
34. Montesano, R., 1986, Cell–extracellular matrix interactions in morphogenesis: An in vitro approach, *Experientia* **42**:977–985.
35. Carley, W. W., Milici, A. J., and Madri, J. A., 1988, Extracellular matrix specificity for the differentiation of capillary endothelial cells, *Exp. Cell Res.* **178**:426–434.
36. Pauli, B. U., and Lee, C. L., 1988, Organ preference of metastasis: The role of organ-specifically modulated endothelial cells, *Lab. Invest.* **58**:379–387.
37. Jaffe, E. A., Nachman, R. L., Becker, C. G., and Minick, C. R., 1973, Culture of human endothelial cells derived from umbilical veins: Identification of morphological and immunological criteria, *J. Clin. Invest.* **52**:2745–2758.
38. Fidler, I. J., 1973, Selection of successive tumor lines for metastasis, *Nature* **242**:148–149.
39. Pauli, B. U., Kellen, J. A., and Ng, R., 1987, Correlation of fibrinolytic activity with invasion and metastasis of R3230AC rat mammary carcinoma cell lines, *Invasion Metastasis* **7**:158–171.
40. Hart, I. R., Talmadge, J. E., and Fidler, I. J., 1981, Metastatic behavior of a murine reticulum cell sarcoma exhibiting organ-specific growth, *Cancer Res.* **41**:1281–1287.
41. Brunson, K. W., and Nicolson, G. L., 1978, Selection and biological properties of malignant variants of a murine lymphosarcoma, *J. Natl. Cancer Inst.* **61**:1499–1502.
42. Kieler, J. F., 1984, Invasiveness of transformed bladder epithelial cells, *Cancer Metastasis Rev.* **3**:265–296.
43. Pauli, B. U., Anderson, S. N., Memoli, V. A., and Kuettner, K. E., 1980, Development of an *in vitro* and *in vivo* epithelial tumor model for the study of invasion, *Cancer Res.* **40**:4571–4580.
44. Rojkind, M., Gatmaitan, Z., Mackensen, S., Gimbrone, M. A., Ponce, P., and Reid, L. M., 1980, Connective tissue biomatrix: Its isolation and utilization for long-term cultures of normal rat hepatocytes, *J. Cell Biol.* **87**:255–263.
45. Lowry, O. H., Rosebrough, N. J., Farr, A. L., and Randall, R. J., 1951, Protein measurement with the folin phenol reagent, *J. Biol. Chem.* **193**:265–275.
46. Glass, W. F., Briggs, R. C., and Hnilica, L. S., 1981, Use of lectins for detection of electrophoretically separated glycoproteins transferred onto nitrocellulose sheets, *Anal. Biochem.* **115**:219–224.
47. Heimark, R. L., and Schwartz, S. M., 1985, The role of membrane–membrane interactions in the regulation of endothelial cell growth, *J. Cell Biol.* **100**:1934–1940.
48. O'Farrell, P. H., 1975, High resolution two-dimensional electrophoresis of proteins, *J. Biol. Chem.* **250**: 4007–4021.
49. Scott, R. E., 1976, Plasma membrane vesiculation: A new technique for isolation of plasma membranes, *Science* **194**:743–745.
50. Soule, H. R., Lanford, R. E., and Butel, J. S., 1982, Detection of simian virus 40 surface-associated large tumor antigen by enzyme-catalyzed radioiodination, *Int. J. Cancer* **29**:337–344.
51. Gallatin, W. M., Weissman, I. L., and Butcher, E. C., 1983, A cell surface molecule involved in organ specific homing of lymphocytes, *Nature* **304**:30–34.
52. Lasky, L. A., Singer, M. S., Yednock, T. A., Dowbenko, D., Fennie, C., Rodriguez, H., Nguyen, T., Stachel, S., and Roscu, S. D., 1989, Cloning of a lymphocyte homing receptor reveals a lectin domain, *Cell* **56**:1045–1055.
53. Siegelman, M. H., van de Rijn, M., and Weissman, I. L., 1989, Mouse lymph node homing receptor cDNA clone encodes a glycoprotein revealing tandem interaction domains, *Science* **243**:1165–1172.
54. Stoolman, L. M., 1989, Adhesion molecules controlling lymphocyte migration, *Cell* **56**:907–910.
55. Butcher, E. C., 1990, Cellular and molecular mechanisms that direct leukocyte traffic, *Am. J. Pathol.* **136**:3–11.

56. Stamper, H. B., Jr., and Woodruff, J. J., 1976, Lymphocyte homing into lymph nodes: *In vitro* demonstration of the selective affinity of recirculating lymphocytes to high-endothelial venules, *J. Exp. Med.* **144**:828–833.
57. Korach, S., Poupon, M. F., Du Villard, J. A., and Becker, M., 1986, Differential adhesiveness of rhabdomyosarcoma-derived cloned metastatic cell lines to vascular endothelial monolayers, *Cancer Res.* **46**:3624–3629.
58. Kramer, R. H., and Nicolson, G. L., 1979, Interaction of tumor cells with vascular endothelial monolayers: A model for metastatic invasion, *Proc. Natl. Acad. Sci. USA* **76**:5704–5708.
59. Nicolson, G. L., Irimura, T., Nakajima, M., Irimura, T., Nakajima, M., and Estrada, J., 1984, Metastatic cell attachment to and invasion of vascular endothelium and its underlying basal lamina using endothelial cell monolayers, in: *Cancer Invasion and Metastasis: Biological and Therapeutic Aspects* (G. L. Nicolson and L. Milas, eds.), Raven Press, New York, p. 145.
60. Pressman, D., and Yagi, Y., 1964, Chemical differences in vascular beds, in: *Small Blood Vessel Involvement in Diabetes Mellitus* (A. R. Cowell and K. Meyer, eds.), Am. Inst. Biol. Sci., Washington, D.C., p. 177.
61. Simionescu, M., Simionescu, N., and Palade, G. E., 1982, Differentiated microdomains on the luminal surface of the capillary endothelium: Distribution of lectin receptors, *J. Cell Biol.* **94**:406–413.
62. Joseph, J., Miao, T., Alby, L., Grieves, J., Hauser, B., Kubai, L, Morrisey, L., Sidky, Y. A., Watt, S. A., and Auerbach, R., 1983, Immunological probes for the study of endothelial cell diversity, in: *The Endothelial Cell—A Pluripotent Control Cell of the Vessel Wall* (D. G. S. Thio-Korner and R. I. Fredshney, eds.), Karger, Basel, p. 55.
63. Auerbach, R., Alby, L., Morrisey, L. W., Tu, M., and Joseph, J., 1985, Expression of organ specific antigens on capillary endothelial cells, *Microvasc. Res.* **29**:401–411.
64. Mills, A. N., and Haworth, S. G., 1986, Changes in lectin binding patterns in the developing pulmonary vasculature of the pig, *J. Pathol.* **149**:191–199.
65. Irie, S., and Tavassoli, M., 1986, Mapping of the rat liver endothelial membrane with lectins and glycosylated ferritins, *Am. J. Anat.* **177**:403–413.
66. Johnson, R., Augustin-Voss, H., Zhu, D., and Pauli, B. U., 1991, Endothelial cell membrane vesicles in the study of organ preference of metastasis, *Cancer Res.* **51**:394–399.
67. Rice, G. E., and Bevilacqua, M. P., 1989, An inducible endothelial cell surface glycoprotein mediates melanoma adhesion, *Science* **246**:1303–1306.
68. Bevilacqua, M. P., Pober, J. S., Mendrick, D. L., Cotran, R. S., and Gimbrone, M. A., Jr., 1987, Identification of an inducible endothelial–leukocyte adhesion molecule, *Proc. Natl. Acad. Sci. USA* **84**:9238–9242.
69. Dustin, M. L., and Springer, T. A., 1988, Lymphocyte function-associated antigen-1 (LFA-1) interaction with intercellular adhesion molecule-1 (ICAM-1) is one of at least three mechanisms for lymphocyte adhesion to cultured endothelial cells, *J. Cell Biol.* **107**:321–331.
70. Kaminski, M., and Auerbach, R., 1988, Tumor cells are protected from NK-cell mediated lysis by adhesion to endothelial cells, *Int. J. Cancer* **41**:847–849.
71. Orr, F. W., Adamson, I. Y., and Young, L., 1985, Pulmonary inflammation generates chemotactic activity for tumor cells and promotes lung metastasis, *Am. Rev. Respir. Dis.* **131**:607–611.
72. Van den Brenk, H. A. S., Stone, M., Kelly, H., Orton, C., and Sharpington, C., 1974, Promotion of growth of tumor cells in acutely inflamed tissues, *Br. J. Cancer* **30**:246–260.
73. Weiss, L., Orr, F. W., and Honn, K. V., 1989, Interactions between cancer cells and the microvasculature: A rate-regulator for metastasis, *Clin. Exp. Metastasis* **7**:127–167.

27

The Microvascular Phases of Metastasis

Leonard Weiss and F. William Orr

I. INTRODUCTION

Metastases are cancers which are not in contiguity with the malignant lesions generating them, and in one way or another, constitute the major clinical problem in the majority of patients with cancer.

The presence and site of metastases are important factors in the diagnosis, staging, treatment, and prognosis of cancer.

Cancer cells disseminate from their parent tumors by a number of anatomic routes, the most common of which are the bloodstream, the lymphatic system, and transcoelomic routes; we focus on the bloodstream and hematogenous metastasis.

Within the context of this review, the sequential reactions between cancer cells and the microvasculature which are involved in metastasis are:

1. The intravasation of cancer cells
2. The arrest of cancer cells in the microvasculature
3. Interactions resulting in cancer cell trauma
4. Interactions resulting in microvascular trauma
5. The inflammatory response
6. The coagulatory response
7. Neovascularization and the subsequent development of metastases

II. INTRAVASATION OF CANCER CELLS

By means of invasive processes, cancer cells enter the bloodstream via small veins or the neovasculature of their parent tumors. Less commonly, entry may be into vascular clefts in some sarcomas, or rarely via arteries as a near-terminal event.

Leonard Weiss • Department of Experimental Pathology, Roswell Park Memorial Institute, Buffalo, New York 14263. *F. William Orr* • Department of Pathology, McMaster University, Hamilton, Ontario, L8N 3Z5, Canada.

Endothelial Cell Dysfunctions, edited by Nicolae Simionescu and Maya Simionescu. Plenum Press, New York, 1992.

Much recent emphasis has been placed on the role of enzymes released directly from cancer cells in the degradation of perivascular basement membranes, including plasminogen activators which serve to plasmin-activate type IV collagenase, cathepsin B, and heparin sulfate endoglycosidase. Exclusive focus on enzymes released by cancer cells as the major cause of tissue degradation is misleading, since it overlooks noncancer cell sources of these and other histolytic enzymes, including leukocytes,[1,2] fibroblasts,[3,4] and endothelial cells[5]; the activities or diminished activities of host tissue-produced enzyme inhibitors, including tissue inhibitors of metalloproteases (TIMP); altered synthesis of basement membrane, which covers the spectrum ranging from tumor encapsulation, to repair of local histolysis, or impaired synthesis; and finally, the role of active movements of cancer cells and their modulation, the invasive process.[6]

Recent studies of the interaction of the cytomatrix with cell membranes and the extracellular matrix through a variety of signals and transduction mechanisms linking cell motility with other changes, serve to emphasize the complexity of these aspects of the invasive process.

Although cancer cells may enter blood vessels by a variety of mechanisms, depending on the amount of entry trauma and consequent intravascular coagulation, entry may not necessarily result in the circulation of tumorigenic cells in the bloodstream. At present, we cannot assess the contributions to the overall inefficiency of the metastatic process, of these aborted attempts at intravasation. However, a variety of observations on humans[7] and animals[8] indicate that many cancer cells do in fact gain access to the circulation during the natural history of even small cancers, and that intravasation failure is therefore unlikely to be a major rate regulator for metastasis. In contrast, *initiation* of invasion, as judged by the decades taken for the progression of cancers from the in *situ* to the invasive state, as a major cause of (temporal) metastatic inefficiency.[9]

III. CANCER CELL ARREST

Only single cancer cells can enter and be arrested in the capillaries, and multicellular emboli are arrested in larger vessels. It is of interest that on a per cell basis, multicellular emboli are more efficient at generating metastases than single cells. This could be due to the protection of cancer cells within a cluster by those surrounding them; alternatively, it could be due to the capillaries constituting a more hostile environment than the larger vessels.

In the case of single cancer cells, entry into capillaries results in their deformation. Passage through the capillaries is expected to be facilitated by the presence of a boundary layer of plasma, which acts as a lubricant, between the cancer cell surface and the vessel wall. As they are driven along the capillaries by blood-pressure differentials, the viscoelastic cancer cells tend to recover from the entry deformation, and to recoil toward the vessel walls, thereby driving out the interposed plasma. Direct adhesions due to short-range forces between the opposing surfaces of sufficient magnitude to cause cancer cell arrest, can only occur when the distance between them is reduced to less than 2 nm, or by mechanical trapping. Attempts have been made to calculate the recoil times of cancer cells, to determine whether close enough approach to form direct adhesive bonds could be achieved during the time taken for cancer cells to pass along capillary segments.[10] The results indicated that the recoil times required for adhesion via short-range forces, were considerably longer than transit times. However, if long-range attractive forces[11] which act over distances of tens of nanometers, are of sufficient magnitude to resist the shear forces generated by blood flow, then cells could be arrested in the presence of interposed plasma films within their capillary transit times. In this case, adhesion macromolecules could stabilize the positions of previously arrested cancer cells as a secondary rather than a primary event.

The fact that arrest of cancer cells within the microvasculature does occur, probably by mechanisms not requiring direct contact between the cells and capillary walls, could also be taken to indicate that the primary arrest mechanism involves adhesive macromolecules acting as bridges between the two adherents, when they are separated by distances of approximately 20 nm; this distance is of the same order as the lengths of molecules or molecular assemblies, of some members of the immunoglobulin supergene family.

If the partial inhibition of metastasis by antibodies or peptides against adhesion molecules and discussed below, is in fact due to their antiarrest activities, the evidence does not permit discrimination between inhibition of primary or secondary (stabilization) adhesive processes.

A. Site Specificity of Arrest

Specificity in site of cancer cell arrest has been identified as one factor contributing to the organ-specific metastatic patterns observed in many cancers,[12] and is generally thought to involve specific interactions between the surfaces of circulating cancer cells and the endothelium and/or subendothelium. Early support for this hypothesis came from the work of Pressman and Yagi[13] and Auerbach,[14] which showed antigenic differences between different parts of vascular beds in different organs. Several types of experiments have been used in attempts to demonstrate the role of adhesive, or arrest-specificity in the generation of metastatic patterns. These include the selective involvement of organs in parabionts,[15] or areas within organs by some cancers[16,17] and in ectopic transplants of organs[15,18]; and also by the demonstration of organ-specific adhesion *in vitro*[19–21]; and the selection of specific organ-"seeking" variants of cancer cells by repetitive cycles of growth and injection of cells in the organs in question.[22] Although these various observations are *consistent* with the concept of specific arrest sites, technical problems associated with at least some of these experiments make proof of the concept elusive. Thus, the detection of metastases in some organs but not others, may indicate that the detection systems were relatively insensitive. In the end, it is unlikely that organ-specific adhesion alone can account for site-specific metastasis. Organ specificity may relate both to specific adhesive interactions with target organ endothelial cells and to localized growth factors[23,24] supporting Paget's original seed and soil hypothesis.[25]

B. Adhesive Interactions and Adhesion Molecules

Cell–substrate adhesion has been thought to promote metastasis by providing an anchor for migrating cells in extravascular tissue or by anchoring intravascular cells to the vessel wall at arrest sites. However, recent experiments indicate that a *loss* of adhesion may also correlate with metastasis.[26] For example, loss of intercellular adhesion molecule (uvomorulin) expression or the use of antibodies to block this molecule in transformed cells has been associated with acquisition of the invasive phenotype *in vitro*.[27] Indeed, in some models, cellular adhesiveness seems to have little influence on metastatic potential.[28]

1. Cancer Cell Adhesion Molecules and Metastasis

Carbohydrates on the surface of cancer cells are involved in metastatic interactions with other cancer cells,[29] host cells, including platelets,[30] and the extracellular matrix. Attempts to identify metastasis-specific molecules have been made by many investigators. Many of these molecules have counterparts in untransformed cells. The expression of others seems to correlate (at least in part) with acquisition of the malignant phenotype. Notable examples of these molecules include: (1) intercellular adhesion molecules (CAMs),[31] the expression of which may account for aggregation and epithelioid arrangements in some cancers[32]; (2) lectins and lectin binding proteins[33,34]; (3) lymphocyte function-associated antigens (LFAs) which promote intercellular adhesion of lymphocytes and leukocytes[35]; anti-LFA-1 antibody blocked invasion of hepatocyte cultures and mutant cells defective in LFA-1 were defective for invasion; (4) Galα1-3Gal residues, which correlate with the metastatic potential of some murine cell lines, have been found on malignant human breast tumors but not their benign counterparts[36]; (5) complex-type asparagine-linked oligosaccharides. P2B, an example of the latter and a major PHA-L-binding protein, was isolated from metastatic lymphoma cells and shown to bind collagen, fibronectin, and laminin with increased affinity after removal of *N*-linked sialic acid.[37]

It seems likely that knowledge of the structure of these molecules will resolve much of the current confusion in this area. For example, the amino acid sequence of an 89-kDa molecule, p3.58 on melanoma cells was deduced from the cDNA nucleotide sequence pertaining to this metastasis-associated molecule. p3.58 was found to be identical to the leukocyte intercellular adhesion molecule ICAM-1,[38,39] emphasizing that there are points of resemblance between mechanisms of metastasis, inflammation, and embryogenesis. The potential relevance of such homeostatic homing molecules to metastasis is also illustrated by experiments in which fusion of nonmetastatic cells with normal lymphocytes resulted in acquisition of a metastatic phenotype and by the observation that virtually all lymph node metastases of murine lymphomas contained a high proportion of cells with the homing receptor MEL-14.[40]

2. Endothelial Adhesion Molecules

The initial attachments of cancer cells to vessel walls occur on endothelial surfaces at junctions between adjacent endothelial cells. This is followed by endothelial retraction and by adhesion of cancer cells to the underlying basement membrane.[41–43] Evidence for the existence of endothelial cell adhesion molecules includes correlations between the rates of attachment of cancer cells to endothelial monolayers *in vitro* and their malignancy *in vivo*,[42,44] the identification of organ-specific endothelial cell glycoproteins,[45] organ-specific endothelial cell antigens,[14] and tissue-specific molecules involved in lymphocyte homing.[46,47]

Endothelial cells bind fibrinogen, vitronectin, von Willebrand factor, and fibronectin, inferring that receptors for these molecules exist on endothelial cells. In most cases, the specificity of this binding has been defined by the ability of monoclonal antibodies to inhibit cellular attachment or to detach endothelial cells from these extracellular substrates. Most of the endothelial cell receptors for extracellular matrix components are members of the integrin class of adhesion molecules. These proteins are transmembrane glycoproteins comprised of noncovalent heterodimers which participate in cell–matrix and cell–cell adhesion, dependent upon the presence of an Arg-Gly-Asp (RGD) sequence within the ligand. These proteins are present on many cell types and have been purified by affinity chromatography on a matrix linked to RGD-containing peptides.[48–50] Fibronectin and vitronectin are each recognized by a distinct receptor on the same human umbilical vein endothelial cell. These receptors cluster, cause microfilament rearrangement, and promote adhesion only in response to their particular ligand.[51] Some endothelial surface receptors appear to be constitutively expressed with a role in the attachment of endothelial cells to the adjacent matrix. The importance of these receptors for cancer cell attachment to the apical surface of endothelial cells is not known since many of the identified receptors are actually localized at cell *substratum* contact points.[48] Thus, the identification of endothelial adhesion molecules isolated from cells in suspension may be difficult to relate to metastasis since the basal and apical surfaces of attached endothelial cells exhibit polarization in the distribution of their integral membrane proteins.[52]

Endothelial cells carry binding sites for activators such as interleukin-1, tumor necrosis factor, interferon, and thrombin. Perturbation by these agents results in synthesis and/or expression/activation of a second group of adhesion molecules which are either not present, or not activated on the unstimulated endothelium.[53] The expression of this group of inducible molecules is associated with enhanced attachment of neutrophils, monocytes, and lymphocytes to the activated endothelial cell surface, for example, at sites of inflammation.[54–56] The study of endothelial adhesion molecules is an area of increasing activity. It is likely that several different types of receptors are involved in any one interaction. For example, the adhesion of lymphocytes to endothelium involves at least three different interactions: (1) an LFA-1- and ICAM-1-dependent interaction which is induced by cytokines but is independent of RGD; (2) an LFA-1-dependent ICAM-1-independent interaction which is not affected by cytokine treatment, and (3) an LFA-1- and ICAM-1-independent interaction which is regulated by cytokines.[57]

Such endothelial adhesion molecules may be important in metastasis and could provide an explanation for preferential metastasis at sites of inflammation. Independent studies have demonstrated enhanced attachment of cancer cells to human endothelial monolayers *in vitro* when the endothelial cells were pretreated with an appropriate activator.[58–60] The effects of stimulation were time-dependent and blocked by treatment of the endothelial cells with inhibitors of protein synthesis, suggesting *de novo* synthesis of adhesion molecules. In one study, enhanced adhesion of cancer cells to activated endothelium was blocked when the assays were conducted in the presence of the peptide GRGDS.[61] Antibody-blocking studies suggest the adhesive molecule is the vitronectin receptor.[62] In nude mice, injection of 10 μg IL-1 α before i.v. injection of tumor cells also enhanced the retention of tumor cells in the lung 24 hr later.[63]

3. Subendothelial Attachment Sites and Adhesion Molecules

Cancer cells adhere more readily to the basement membrane than to the endothelium.[64,65] Many basement membrane components have the ability to act directly as cellular binding sites, e.g., type IV collagen,[66] laminin,[67] vitronectin,[51] fibrinogen,[68] fibronectin[69] and heparan sulfate proteoglycan.[70,71] Cancer cell surface receptors have been identified for fibronectin, type I collagen, vitronectin, and laminin.[72–77] In the case of laminin and fibronectin, there are functional domains which have differing effects on cell attachment and spreading, migration, growth, or metastasis.[78–80] Metastatic cells have greater numbers of receptors for laminin than nonmetastatic cells with expression of a second population of binding sites.[81]

As noted above, the amino acid sequence, arginine-glycine-aspartic acid (RGD), is common to many adhesion molecules, having been identified in fibronectin, vitronectin, type I collagen, fibrinogen, von Willebrand factor, and osteopontin.[50,51,68,69,82] The importance of the RGD sequence for cell attachment has ben demonstrated by using RGD-containing peptides to competitively block cell adhesion or cause cellular detachment.[83] Several groups have reported that injection of the RGD peptide blocked the development of pulmonary metastasis in mice injected simultaneously with intravenous cancer cells.[84–86] In a model of spontaneous metastasis, multiple i.v. administrations of a polymerized peptide containing repeating RGD sequences also caused significant reduction of metastatic lung colonies but had no effect on the size of the primary tumor.[87] In contrast, peptides with amino acid inversions or substitutions and unpolymerized RGD peptides had a lesser effect. There were also correlations between the relative inhibitory effects of these peptides *in vivo* and their abilities to disrupt cell adhesion *in vitro*. RGD-containing peptides have also been reported to inhibit the penetration of cancer cells through human amniotic basement membrane,[88] an assay used to study tissue invasion by cancer cells. These experiments infer that receptor–ligand interactions between cancer cells and the vessel wall can promote metastasis but they do not indicate which cell type carries the receptor and which the ligand.

Other evidence for interaction between cancer cells and the integrin like molecules in the basement membrane comes from studies in which polyclonal antibodies prepared against integrin-related complexes have been used to inhibit attachment of (B16 melanoma) cancer cells to complex basement membrane matrices.[89] In these experiments, integrinlike receptors were found on the surfaces of the cells by immunoprecipitation with these antibodies, which (incidentally) failed to react with laminin-binding protein. Thus, a variety of adhesion molecules are likely to be involved in the penetration of tissue barriers by metastasizing cells.

C. Therapeutic Implications

It remains to be seen whether the use of antibodies or peptides can provide the basis for effective therapy in man.[90,91] One potential problem relates to the ability of RGD peptides to detach endothelial cells from their basement membranes (see Section V). Another difficulty is raised by the multiplicity of potential adhesion molecules and to the variation in adhesion molecules or receptor expression between different cancers. Many of the potential targets are involved in regulation of normal homeostatic mechanisms. The cost of producing antibodies in sufficient quantity to be of long-term therapeutic value poses still another problem. Attempts have also been made to inhibit metastasis by pharmacologic modification of cancer cell surface glycoproteins involved in adhesion.[92–94] However, in view of recent studies, it seems more likely that the preferential patterns observed in many cancers are due not only to preferential arrest but also to preferential *growth* of cancer cells in the different sites.[95–98]

IV. CANCER CELL TRAUMA

Many observations extending back to the beginning of the century, indicate that the vast majority of cancer cells do not form metastases.[9] Bioassays of the lungs,[8] myocardium,[99] and skeletal muscle[100] of mice, initiated immediately after intravascular injections, reveal that at least 85% of the injected cancer cells delivered to these sites were lethally damaged in less than 5 min. Longer-term studies reveal that all but a small fraction of 1% of the surviving cells are killed over the next few hours or days. In addition, studies of cancer cell traffic indicate that a substantial amount of cancer cell killing occurs in the microvasculature of the lungs[101] and liver.[102]

It should be emphasized that although more than 99.9% of cancer cells delivered to the microvasculature of all organs examined are killed by fast and slow processes, unless therapeutic steps are taken, the surviving fraction will kill the patient. The effect of this marked inefficiency in the intravascular phase of metastasis is to retard, and in some cases to abort the subsequent steps in the metastatic process. However, even an inefficient process will succeed, if repeated often enough!

A. Rapid Lethal Trauma

As this is thought to occur almost instantaneously, it appears likely that the underlying mechanisms are due to direct interactions between cancer cells and the microvessel walls, including:

1. Mechanical Trauma

When cancer cells enter capillaries having smaller diameters than themselves, in those cases where the cancer cells are more deformable than the capillaries, the cells will be deformed from spherical configurations into sausage-shape. At constant cell volume, any change in shape from spherical configuration mandates an increase in surface area. It has been suggested that these increases are brought about by two sequential mechanisms.[103,104] The first is an *apparent* increase in surface area, accomplished by nonlethal, reversible unfolding of the normally convoluted cancer cell surface. Although in the present context the end point is surface membrane-unfolding, the underlying cell deformation depends upon the rheologic properties of the whole cell. If the geometric demands for increased surface area can be met by unfolding alone, then cancer cells are expected to survive the shape transitions. However, if the demands for increased surface area are not met by the apparent increases associated with unfolding or other forms of membrane recruitment, a second mechanism, namely *true* increase in surface area, is invoked. When the true surface area of most if not all biologic membranes is increased by more than 3 or 4%, the accompanying increases in membrane tension result in lethal rupture and cell death. This shape-transition hypothesis is supported by experiments in which demands for increased surface area are created by hypotonic cell swelling, when membrane unfolding is followed by membrane rupture.[99,105,106]

Rapid, lethal, mechanical intravascular trauma depends not only on the mechanical properties of the cancer cells, but also on the relative mechanical properties of the vessels and tissues surrounding them. In the myocardium[99] and skeletal muscle,[100] contraction has been associated with the rapid death of cancer cells delivered via the arterial route; this could at least partially account for the relatively low incidence of metastases in these sites.

It may also be mentioned here that shape transitions analogous to those encountered in the microvasculature, can induce a state of dormancy in surviving cancer cells.[107] The potential role of mechanically induced shape changes and compression on cancer and endothelial cells, and the surrounding tissue matrices is speculative. However, it is possible that these interactions can act as biologic rate regulators by means of mechanochemical transduction across surface membranes.[108]

2. *Trauma Induced by Oxygen Radicals*

Another possible nonexclusive mechanism of rapid intravascular trauma is the production of oxygen radicals by endothelial cells[109]; the property of radical production appears to be shared by all types of reticuloendothelial cells. The arrest of cancer cells, in common with other microembolic events, generates an inflammatory response, resulting in the release of oxygen radicals. The initial oxygen product is the superoxide anion (O_2^-), which is converted to H_2O_2, hydroxyl radical ($HO^\cdot$), and possibly singlet oxygen ($^1O_2$), which have tumoricidal activity. An indication that some rapid lethal intravascular trauma is associated with release of oxygen radicals, comes from experiments showing that cancer cell death following arrest in the lungs, is reduced in mice given prior injections of superoxide dismutase (SOD).[110] As the only known activity of SOD is to scavenge superoxide anion, and as it has a very short lifetime *in vivo*, it is distinctly possible that its action under the conditions of Glaves's experiments was mainly on endothelial cell products. Although leukocytes and NK cells are involved in some tumoricidal interactions involving oxygen radicals, their role in *rapid* intravascular trauma is difficult to assess. This is partly because the ratios of these cells to cancer cells required for rapid killing *in vivo* are not known, and partly because adherent cancer cells block blood flow in involved capillaries, and therefore restrict rapid access of leukocytes.

B. *Slow Intravascular Trauma*

This involves the generally accepted, interacting humoral and cellular defense mechanisms, including immune reactions, inflammation, and coagulation. A number of these reactions have recently been reviewed in the present context,[6] but limitations in space prevent their meaningful review here.

V. MICROVASCULAR TRAUMA

Cytotoxic agents (including antineoplastic drugs),[111–113] high concentrations of oxygen,[114] or X irradiation[115,116] can promote experimental or spontaneous metastasis, possibly by exposing the more adhesive basement membrane through denudation of the endothelium or endothelial retraction.[117] Some experiments have demonstrated a direct relationship between microvascular damage and the effect on metastasis with a loss of metastatic enhancement after endothelial repair.[118] Oxygen-derived free radicals are generated by most of these agents and could be a common effector pathway for endothelial injury.[119,120] The effects of endothelial injury on metastasis could also result from the release of interleukin-1 from perturbed endothelium (acting as a cancer cell chemoattractant)[121] enhanced expression of endothelial surface adhesion molecules,[58,60] or activation of endothelial collagenases (to initiate basement membrane degradation).

To leave the circulation, blood-borne cancer cells must also penetrate the vascular basement membrane. Morphologic studies on this step of metastasis have demonstrated dissolution of the basement membrane,[43] and the ability of cancer cells to degrade basement membrane constituents *in vitro* has often been correlated with their malignant potential.[122–128] Nakajima *et al.* have also shown that collagenolysis could be attributed to the action of more than one proteinase, and that much of the type IV collagenolytic activity was secreted in a latent form, requiring activation by trypsinlike enzymes or the presence of serum for its expression.[124] Cancer cells,[129] leukocytes,[130] platelets,[131,132] or endothelial cells[109] are all potential sources of free radicals. It is possible that oxygen-derived free radicals from these cells could activate such latent tumor cell proteases[133,134] or inactivate protease inhibitors[135–137] as they do in the case of leukocyte-mediated basement membrane damage. The relevance of enzyme production by cancer cells to the degradation of vascular basement membrane *in vivo* requires further analysis since, in at least one group of experiments, infusion of biologically effective doses of inhibitors of these enzymes failed to block experimental metastasis of the B16 murine melanoma.[138] However, if the relevant enzymes were membrane-bound, or active at the intravasation step, humoral inhibitors might not have an effect on metastasis.

VI. EFFECTS OF INFLAMMATORY RESPONSES

Leukocytes account for a significant proportion of circulating host cells and are able to interact with intravenous cancer cells, forming heterotypic aggregates.[139–141] The effects of such interactions on metastasis have not been widely studied and are currently controversial.[6] On the one hand, it appears that leukocytes can promote metastasis. The early studies by Wood suggested that rabbit VX2 carcinoma cells leave the microvasculature through defects caused by leukocytes.[142] In man, leukocytosis has been associated with poor prognosis for some solid tumors with increased metastatic burden.[143] In experimental systems, intravenous injection of cancer cells into animals with granulocytosis,[144] injection of preformed leukocyte–tumor aggregates,[140] infusion of neutrophils with tumor cells,[145] or injection of cancer cells into animals with localized inflammatory processes[146,147] has resulted in greater retention of cancer cells in target organs and/or more metastatic tumors. These results have been explained on the basis of several mechanisms including embolic localization of heterotypic aggregates,[140] elevated levels of heparanase and type IV collagenolytic activity in the leukocytes of rats with highly metastatic tumors,[145] the generation of localized chemoattractants for cancer cells,[146] induction of microvascular damage via leukocyte-derived oxygen free radicals,[139,148] and release of neutrophil-derived growth factors.[147] Contradictory results have been obtained by others, in particular Glaves who has given evidence that neutrophil-derived superoxide anion can *release* trapped intravascular tumor cells from arrest sites in target organs[110,149] and Jones *et al.*[150] who have identified heterogeneous effects of leukocytes, possibly related to the activation state of leukocytes and/or the differential sensitivity of different cancer cells to the cytolytic effects of activated neutrophils.

VII. COAGULATION

Wood's[142] direct cinematographic observations on the *in vivo* arrest of cancer cells in rabbit ear preparations, served to focus attention on the involvement of platelets and fibrin in metastasis. This interest was reinforced by the observations of Gasic *et al.*[151] that induction of thrombocytopenia in mice, prior to tail vein injections of cancer cells, reduced the number of subsequent lung colonies. Numerous experiments have shown that cancer cells induce platelet aggregation *in vitro* and, in some cases, the amount of aggregation apparently correlated with the metastatic capacity of the cancer cells. However, a recent review of the literature[6] reveals that such correlations are by no means general. Regardless of uncertainties, the roles of platelets and thrombogenesis in intravascular events associated with metastasis require clarification, because many potent agents are currently available which affect platelet interactions and coagulation, and which could theoretically at least form the basis for antimetastatic therapy.

The metastatic potential of some cancer cell lines has been correlated with their ability to activate platelets, i.e., the induction of thrombocytopenia *in vivo* and/or platelet aggregation *in vitro*.[152,153] Other evidence for platelet involvement in metastasis is that thrombocytopenia or antiplatelet agents (antibodies or pharmacologic inhibitors) can retard the metastasis of some cancer cells. Tumor cells activate platelets by (1) generating thrombin through activation of the coagulation system (thought to be of major importance[153–155]); (2) releasing ADP[156] or cathepsin B[154,156]; (3) activating platelet arachidonic acid metabolism[156]; (4) interacting with platelets via membrane proteins having the Arg-Gly-Asp amino acid sequence[157,158]; (5) a combination of these mechanisms. In turn, activated platelets are thought to enhance metastasis by facilitating the attachment of tumor cells to blood vessel walls, by inducing vasopermeability, by sequestering cancer cells from host immune surveillance (reviewed in Ref. 159), and by generating platelet-derived growth factors, i.e., PDGF and TGF-β.[160]

In a series of studies made on rats given intravascular injections of Walker 256 cancer cells, Wallace *et al.*[161] observed that within 2 min after injection, cancer cells were "loosely" adherent to the pulmonary vascular endothelium, and associated with platelets and occasionally leukocytes. Over the next 3 hr, the platelets showed progressive degeneration and fibrin deposition first increased and then decreased in amount. By 12 hr after injection, both fibrin and platelets disappeared. Many of the apparently contradictory observations on the presence or absence of fibrin around arrested cancer cells are reconciled by taking into account the times after arrest at which the observations were made. The transient appearance of platelets associated with cancer emboli has been confirmed and extended by Crissman *et al.*[162] and others. In these studies, platelets were not observed lying between cancer and endothelial cells, and early associations between cancer cells and platelets were topographically limited. These results, together with others,[163] indicate that cancer cell-induced platelet aggregation *in vivo*, occurs after cancer cell arrest and attachment to the vessel wall. Early postarrest interactions, during which both activated and unactivated platelets are seen adherent to the luminal surfaces of cancer cells, are associated with processes extended by the cancer cells exclusively in regions in contact with activated platelets. Also associated in some way with the progressive increases in platelet interaction, is the retraction of endothelial cells which exposes the subendothelium which in turn is the site of additional interactions with cancer cells and platelets.[162,164]

Studies on endothelial cells in monolayers *in vitro*[42,165,166] indicate that cancer cell adhesion and endothelial cell retraction can result from direct interaction of these two cell types alone, albeit under different hydrodynamic conditions than experienced *in vivo*. However, these studies to not exclude a role for platelets, because *in vitro* experiments also reveal that thrombin reversibly induces endothelial cell retraction produced by a 12-lipoxygenase metabolite of arachidonic acid, 12(*S*)-hydroxyeicosatetraenoic acid [12(*S*)-HETE], which is generated during cancer cell–platelet–endothelial cell interactions. The 12(*S*)-HETE also increases expression of the GpIIb/IIIa receptor at the cancer cell surfaces, thereby promoting their adhesion to the subendothelial matrix,[167] and it has been shown that platelets enhance cancer cell adhesion to both endothelial cells and subendothelial matrix.[41,168] The underlying mechanisms for platelet-enhanced adhesion include elements of their surface membranes (GpIIb/IIIa) which are classed with the integrins, and intact microfilaments.[6,169] The existence of some type of eicosanoid-driven balance in the subendothelial adhesion of cancer cells, is suggested by the observation that endothelial cell-derived prostacyclin inhibits cancer cell–platelet interactions.[169] Nonetheless, the significance of platelets in the initial arrest phase of cancer cells remains open to question, since short-term arrest studies *in vivo*, on thrombocytopenic mice revealed no changes when compared with controls, in arrest in either the pulmonary[170] or the hepatic[171] microvasculature. In addition, various anticoagulants did not affect the short-term arrest of cancer cells in the pulmonary microcirculation of mice.[172] Therefore, either enough platelets were present in thrombocytopenic animals to permit a platelet-dependent arrest process or their putative roles in metastasis occur in the later phases, including the stabilization of initial adhesions and extravasation.

The role of intravascular coagulation in metastasis is perplexing! The concept that a fibrin cocoon protects arrested cancer cells from various arms of host defense, has also prompted numerous attempts to use anticoagulants as antimetastatic agents. Even when this approach has had limited success with metastases in a single site, it is doubtful whether the primary mode of action of these agents was through anticoagulation.[173] In addition, a recent study of the effects of a number of different potent antithrombotic agents on the spontaneous metastasis of Lewis lung carcinoma in mice, showed that although some of these agents indeed inhibited pulmonary (i.e., first organ) metastasis, other organs were subsequently involved as a result of treatment and there were no significant increases in either the number of survivors or the median survival times.[174] This treatment-induced change in metastatic pattern is analogous to that observed in heparinized mice given B16 melanoma cells.[175] It remains to be seen whether anticoagulant therapy will prove successful in other tumor systems.

VIII. ANGIOGENESIS

Arrested cancer cells extravasate between endothelial cells by means of active movement[41,162,176]; alternatively, and more commonly, they "burst" out of blood vessels by enzymatic degradation involving cancer and noncancer cells,[6] with or without intravascular proliferation of the arrested cancer cells.[41,161,177–179]

Extravascular foci of cancer cells are without a viable blood supply, and their growth is limited by diffusion-dependent mechanisms. In this situation, metastases cannot grow to diameters greater than approximately 0.5 mm.[180,181]

Stimuli from a variety of pathologic foci, including cancer cells, act on nearby venules causing focal fragmentation of their basement membranes and activation of their endothelium. The products of matrix degradation act as a trigger for subsequent phases of neovascularization.

Endothelial cells next migrate toward the micrometastases in the form of solid capillary "sprouts" which initially do not require endothelial cell proliferation. Gross *et al.*[182] showed that capillary endothelial cells can produce both plasminogen activator and inactive or latent collagenase, and endothelial migration may be associated with a transient increase in urokinase-type plasminogen activator[183] in common with many other cell migration processes. It has been suggested that angiogenesis factors act on the migratory phase of angiogenesis by causing the endothelial cells to secrete increased levels of these enzymes. The plasminogen activator converts plasminogen to plasmin and, in turn, degradation produced by these enzymes enhances the invasion processes required to proximate the neovasculature and the tumor.

Later, endothelial cell proliferation occurs under the action of a number of mitogens, and lumina are formed; adjacent capillary "sprouts" fuse to form capillary loops, around which new basement membranes are generated.[180,184–186] Although it is generally accepted that the neovasculature invades tumors, some consider that the tumor infiltrates the neovascular net.[187,188]

Ischemia-associated tissue injury and death provide a common stimulus for angiogenesis, due to released mitogens which are taken up by endothelial cells and/or the extracellular matrix. In addition to cancer cells, cells involved in the inflammatory response also generate angiogenic stimuli; and in the absence of inflammation, the endothelial cells themselves may exert autocrine control over neovascularization.[189]

Angiogenic factors, including acidic and basic fibroblast growth factors and transforming growth factor-α and -β, appear to act directly on endothelial cells. However, other factors act indirectly, via macrophages and the extracellular matrix for example, causing them to release agents acting on endothelial cells.[189] In relating angiogenesis to other processes, it is of interest that PGE_1 and PGE_2 are angiogenic, whereas those of the A and F series are not. Mast cells and mast cell-derived heparin, which binds to both acidic and basic fibroblast growth factors, stimulate endothelial cell migration and augment the angiogenic activity of acidic FGF, but in the absence of some corticosteroids, heparin fragments lacking anticoagulant activity may actually inhibit angiogenesis.[190] It is of potential therapeutic interest that in some model systems, angiogenic activity may be greatly inhibited by a synthetic heparin substitute administered with hydrocortisone.[191] However, as angiogenesis is such a common "physiologic" process, it remains to be seen whether or not it can be used as a major strategy for antimetastasis therapy.[192] Although the various mechanisms involved in angiogenesis are extremely complex and not completely understood, the importance of the process is indisputable. Without it, tumors would tend to be arrested at the level of micrometastases, which are nonlethal; in its presence, they progress and ultimately kill the patient. In addition to this role, the neovasculature provides a ready route of intravasation for cancer cells from tumors, not only in the case of metastasis from primary lesions, but also during metastasis of metastases.[12]

IX. CONCLUSIONS

Metastasis is one of the most feared and frequent complications of cancer; affecting diagnosis, treatment, and prognosis. There is good evidence that pivotal phases of metastasis involve interactions of cancer cells, the microvasculature and other noncancer cells. These interactions act as major rate regulators for the metastatic process.

REFERENCES

1. Woolley, D. E., 1984, Collagenolytic mechanisms in tumor cell invasion, *Cancer Metastasis Rev.* **3**:361–272.
2. Welch, D. R., Schissel, D. J., Howrey, R. P., and Aeed, P. A., 1989, Tumor-elected polymorphonuclear cells, in contrast to "normal" circulating polymorphonuclear cells, stimulate invasive and metastatic potentials of rat mammary adenocarcinoma cells, *Proc. Natl. Acad. Sci. USA* **86**:5859–5863.
3. Biswas, C., 1982, Host–tumor cell interactions and collagenase activity, in: *Tumor Invasion and Metastasis* (L. A. Liotta, and I. R. Hart, eds.), Nijhoff, The Hague, p. 405.
4. Murphy, G., Hembry, R. M., McGarrety, A. M., and Reynolds, J. J., 1989, Gelatinase (type IV collagenase) immunolocalization in cells and tissues: Use of an antiserum to rabbit bone gelatinase that identifies high and low Mr forms, *J. Cell Sci.* **92**:487–495.
5. Franks, A. J., and Ellis, E., 1989, Immunohistochemical localization of tissue plasminogen activator in human brain tumors, *Br. J. Cancer* **59**:462–466.
6. Weiss, L., Orr, F. W., and Honn, K. V., 1989, Interactions between cancer cells and the microvasculature: A rate-regulator for metastasis, *Clin. Exp. Metastasis* **7**:127–167.
7. Glaves, D., Huber, R. P., and Weiss, L., 1988, Hematogenous dissemination of cells from human renal adenocarcinomas, *Br. J. Cancer* **57**:32–35.
8. Glaves, D., 1986, Detection of circulating metastatic cells in cancer metastasis, in: *Cancer Metastasis: Experimental and Clinical Strategies* (D. R. Welch, B. K. Bhuyan, and L. A. Liotta, eds.), Liss, New York, p. 151.
9. Weiss, L., 1990, Metastatic inefficiency, *Adv. Cancer Res.* **54**:159–211.
10. Weiss, L., Dimitrov, D. S., and Angelova, M., 1985, The hemodynamic destruction of intravascular cancer cells in relation to myocardial metastasis, *Proc. Natl. Acad. Sci. USA* **82**:5737–5741.
11. Israelachvili, J. N., and McGuiggau, P. M., 1988, Forces between surfaces in liquids, *Science* **241**:795–798.
12. Weiss, L., 1985, *Principles of Metastasis*, Academic Press, New York, p. 1.
13. Pressman, D., and Yagi, Y., 1964, Chemical differences in vascular beds, in: *Small Blood Vessel Involvement in Diabetes Mellitus* (A. R. Colwell and K. Meyer, eds.), AIBS, Washington, D.C., p. 177.
14. Auerbach, R., Alby, L., Morrissey, W., Tu, M., and Joseph, J., 1985, Expression of organ-specific antigens on capillary endothelial cells, *Microvasc. Res.* **29**:401–411.
15. Hart, I. R., and Fidler, I. J., 1980, Role of organ selectivity in the determination of metastatic patterns of B16 melanoma, *Cancer Res.* **40**:2281–2287.
16. Azzarelli, B., Easterling, K., and Norton, J. A., 1989, Leukemic cell–endothelial cell interactions in leukemic cell dissemination, *Lab. Invest.* **60**:45–64.
17. Barbera-Gulliem, E., Alonso-Varoona, A., and Vidal-Vanaclocha, F., 1989, Selective implantation and growth in rats and mice of experimental liver metastasis in acinar zone one, *Cancer Res.* **49**(14):4003–4010.

18. Kinsey, D. L., 1960, An experimental study of preferential metastasis, *Cancer* **13**:674–676.
19. Brodt, P., 1989, Tumor cell adhesion to frozen lymph node sections—An in vitro correlate of lymphatic metastasis, *Clin. Exp. Metastasis* **7**:343–352.
20. Nicolson, G. L., Belloni, P. N., Tressler, R. J., Dulski, K., Inoue, T., and Cavanaugh, P. G., 1989, Adhesive, invasive, and growth properties of selected metastatic variants of a murine large-cell lymphoma, *Invasion Metastasis* **9**:102–116.
21. Netland, P. A., and Zetter, B. R., 1984, Organ-specific adhesion of metastatic tumor cells in vitro, *Science* **224**:1113–1115.
22. Nicolson, G. L., 1982, Cancer metastasis. Organ colonization and the cell-surface properties of malignant cells, *Biochim. Biophys. Acta* **695**:113–176.
23. Nicolson, G. L., 1988, Differential organ tissue adhesion, invasion, and growth properties of metastatic rat mammary adenocarcinoma cells, *Breast Cancer Res. Treat.* **12**:167–176.
24. Doerr, R. , Zvibel, I., Chiuten, D., D'Olimpio, J., and Reid, L. M., 1989, Clonal growth of tumors on tissue-specific biomatrices and correlation with organ site specificity of metastases, *Cancer Res.* **49**:384–392.
25. Paget, S., 1889, The distribution of secondary growths in cancer of the breast, *Lancet* **1**:571–573.
26. Leung-Tack, J., Capo, C., De-Lapeyriere, O., Benoleil, A.-M., Arnaud, D., and Bongrand, P., 1988, Relationship between cellular adhesiveness and metastatic activity in polyomavirus-transformed FR3T3 rat cell lines, *Int. J. Cancer* **42**:946–951.
27. Behrens, J., Mareel, M. M., Van-Roy, F. M., and Birchmeier, W., 1989, Dissecting tumor cell invasion: Epithelial cells acquire invasive properties after the loss of uvomorulin-mediated cell–cell adhesion, *J. Cell Biol.* **108**:2435–2447.
28. Zoller, M., and Matzku, S., 1989, Changes in adhesive properties of tumor cells do not necessarily influence metastasizing capacity, *Clin. Exp. Metastasis* **7**:227–242.
29. Meromsky, L., Lotan, R., and Raz, A., 1986, Implications of endogenous tumor cell surface lectins as mediators of cellular interactions and lung colonization, *Cancer Res.* **46**:5270–5275.
30. Bastida, E., Almiral, L., Jamieson, G. A., and Ordinas, A., 1987, Cell surface sialylation of two human tumor cell lines and its correlation with their platelet-activating activity, *Cancer Res.* **47**:1767–1770.
31. Linnemann, D., Raz, A., and Bock, E., 1989, Differential expression of cell adhesion molecules in variants of K1735 melanoma cells differing in metastatic capacity, *Int. J. Cancer* **43**:709–712.
32. Mege, R. M., Matsuzaki, F., Gallin, W. J., Goldberg, J. I., Cunningham, B. A., and Edelman, G. M., 1988, Construction of epithelioid sheets by transfection of mouse sarcoma cells with cDNAs for chicken cell adhesion molecules, *Proc. Natl. Acad. Sci. USA* **85**:7274–7278.
33. Lotan, R., and Raz, A., 1988, Lectins in cancer cells, *Ann. N.Y. Acad. Sci.* **551**:385–396.
34. Tressler, R. J., and Nicolson, G. L., 1988, Cell surface biochemical and metastatic properties of Lens culinaris hemagglutinin-binding variants of a murine large cell lymphoma, *Invasion Metastasis* **8**:351–363.
35. Roossien, F. F., de-Rijk, D., Bikker, A., and Roos, E., 1989, Involvement of LFA-1 in lymphoma invasion and metastasis demonstrated with LFA-1-deficient mutants, *J. Cell Biol.* **108**:1979–1985.
36. Castronovo, V., Colin, C., Parent, B., Foidart, J. M., Lambotte, R., and Mahieu, P., 1989, Possible role of human natural anti-Gal antibodies in the natural antitumor defense system, *J. Natl. Cancer Inst.* **81**:212–216.
37. Dennis, J. W., Laferte, S., and Vanderelst, I., 1989, Asparagine-linked oligosaccharides in malignant tumor growth, *Biochem. Soc. Trans.* **17**:29–31.
38. Johnson, J. P., Stade, B. G., Hupke, U., Holzmann, B., and Riethmuller, G., 1988, The melanoma progression-associated antigen P3.58 is identical to the intercellular adhesion molecule, ICAM-1, *Immunobiology* **178**:275–284.
39. Johnson, J. P., Stade, B. G., Holzmann, B., Schwable, W., and Riethmuller, G., 1989, De novo expression of intercellular-adhesion molecule 1 in melanoma correlates with increased risk of metastasis, *Proc. Natl. Acad. Sci. USA* **86**:641–644.
40. Sher, B. T., Bargatze, R., Holzmann, B., Gallatin, W. M., Matthews, D., Wu, M., Picker, L., Butcher, E. C., and Weissmann, I. L., 1988, Homing receptors and metastasis, *Adv. Cancer Res.* **51**:361–390.
41. Crissman, J. D., Hatfield, J., Schaldenbrand, M., Sloane, B. F., and Honn, K. V., 1985, Arrest and extravasation of B16 amelanotic melanoma in murine lungs. A light and electron microscopic study, *Lab. Invest.* **53**:470–478.
42. Nicolson, G. L., 1982, Metastatic tumor cell attachment and invasion assay utilizing vascular endothelial cell monolayers, *J. Histochem. Cytochem.* **30**:214–220.
43. Crissman, J. D., Hatfield, J. S., Menter, D. G., Sloane, B. F., and Honn, K. V., 1988, Morphological study of

the interaction of intravascular tumor cells with endothelial cells and subendothelial matrix, *Cancer Res.* **48:** 4065–4072.

44. Alby, L., and Auerbach, R., 1984, Differential adhesion of tumor cells to capillary endothelial cells in vitro, *Proc. Natl. Acad. Sci. USA* **81:**5739–5743.
45. Belloni, P. N., and Nicolson, G. L., 1988, Differential expression of cell surface glycoproteins on organ-derived murine vascular endothelia and endothelial cells, *J. Cell Physiol.* **136:**398–410.
46. Streeter, P. R., Berg, E. L., Rouse, B. T., Bargatze, R. F., and Butcher, E. C., 1988, A tissue specific cell molecule involved in lymphocyte homing, *Nature* **331:**41–46.
47. Nakache, M., Berg, E. L., Streeter, P. R., and Butcher, E. C., 1989, The mucosal vascular addressin is a tissue-specific endothelial cell adhesion molecule for circulating lymphocytes, *Nature* **337:**179–181.
48. Cheresh, D. A., 1987, Human endothelial cells synthesize and express an Arg-Gly-Asp-directed adhesion receptor involved in attachment to fibrinogen and von Willebrand factor, *Proc. Natl. Acad. Sci. USA* **84:** 6471–6475.
49. Fitzgerald, L. A., and Phillips, D. R., 1988, Structure of platelet membrane glycoproteins, in: *Platelet Immunology* (T. J. Kumicki and J. George, eds.), Lippincott, Philadelphia.
50. Hynes, R. O., 1987, Integrins: A family of cell surface receptors, *Cell* **48:**549–554.
51. Dejana, E., Colella, S., Conforti, G., Abbadini, M., Gaboli, M., and Marchisio, P. C., 1988, Fibronectin and vitronectin regulate the organization of their respective Arg-Gly-Asp adhesion receptors in cultured human endothelial cells, *J. Cell Biol.* **107:**1215–1223.
52. Muller, W. A., and Gimbrone, M. A., Jr., 1986, Plasmalemmal proteins of cultured vascular endothelial cells exhibit apical–basal polarity: Analysis by surface-selective iodination, *J. Cell Biol.* **103:**2389–2402.
53. Pober, J. S., Lapiere, L. A., Stolpen, A. H., Brock, T. A., Springer, T. A., Fiers, W., Bevilacqua, M. P., Mendrick, D. L., and Gimbrone, M. A., Jr., 1987, Activation of cultured human endothelial cells by recombinant lymphotoxin: Comparison with tumor necrosis factor and interleukin-1 species, *J. Immunol.* **138:**3319–3324.
54. Bevilacqua, M. P., Pober, J. S., Wheeler, M. E., Cotran, R. S., and Gimbrone, M. A., Jr., 1985, Interleukin-1 acts on cultured human vascular endothelium to increase the adhesion of polymorphonuclear leukocytes, monocytes, and related leukocyte cell lines, *J. Clin. Invest.* **76:**2003–2011.
55. Cavender, D., Saegusa, Y., and Ziff, M., 1987, Stimulation of endothelial cell binding of lymphocytes by tumor necrosis factor, *J. Immunol.* **139:**1855–1860.
56. Cotran, R. S., Gimbrone, M. A., Jr., Bevilacqua, M. P., Mendrick, D., L., and Pober, J. S., 1986, Induction and detection of a human endothelial activation antigen in vivo, *J. Exp. Med.* **164:**661–666.
57. Dustin, M. L., and Springer, T. A., 1988, Lymphocyte function-associated antigen-1 (LFA-1) interaction with intercellular adhesion molecule-1 (ICAM-1) is one of at least three mechanisms for lymphocyte adhesion to cultured endothelial cells, *J. Cell Biol.* **107:**321–331.
58. Lauri, D., Bertomeu, M.-C., Orr, F. W., Bastida, E., Sauder, D. N., and Buchanan, M. R., 1989, Differential effects of interleukin-1 and fMLP on chemotaxis and human endothelium adhesivity for A549 tumor cells, *Lab. Invest.* **60:**161–164.
59. Dejana, E., Bertocchi, F., Bortolami, M. C., Fegonesi, A., Tonta, A., Brevario, F., and Giavazzi, R., 1988, Interleukin-1 promotes tumor cell adhesion to cultured human endothelial cells, *J. Clin. Invest.* **82:**1466–1470.
60. Rice, G. E., Gimbrone, M. A., Jr., and Bevilacqua, M. P., 1988, Tumor cell–endothelial interactions. Increased adhesion of human melanoma cells to activated vascular endothelium, *Am. J. Pathol.* **133:** 204–210.
61. Lauri, D., Bertomeu, M.-C., Orr, F. W., Bastida, E., Sauder, D. N., and Buchanan, M. R., 1989, Interleukin-1 increases tumor cell adhesion to endothelial cells through an RGD dependent mechanism: In vitro and in vivo studies, *Clin. Exp. Metastasis* **8:**27–32.
62. Lafrenie, R. M., Podor, T. J., Buchanan, M. R., and Orr, F. W., 1989, Adhesion of A549 tumor cells to interleukin-1alpha stimulated endothelial cells is partially mediated through a vitronectin receptor-like molecule, in: *Proceedings of the First International Conference on Eicosanoids and bioactive lipids in cancer and radiation injury* (K. Honn, ed.) Meeting Abstract P-II-15.
63. Bertomeu, M. C., Whybourne, K., Orr, F. W., and Buchanan, M. R., 1989, Influence of interleukin 1 on 13-HODE synthesis, integrin expression and B16 melanoma cell metastasis in vivo, in: *Proceedings of the First International Conference on Eicosanoids and bioactive lipids in cancer and radiation injury* (K. Honn, ed.) Meeting Abstract II.B-3.

64. Kramer, R. H., Gonzales, R., and Nicolson, G. L., 1980, Metastatic tumor cells adhere preferentially to the extracellular matrix underlying vascular endothelial cells, *Int. J. Cancer* **26**:639–645.
65. Vlodavsky, I., Atzmon, A. R., and Fuks, Z., 1982, Tumor cell attachment to the vascular endothelium and subsequent degradation of the subendothelial extracellular matrix, *Exp. Cell Res.* **140**:149–159.
66. Aumailley, M., and Timpl, R., 1986, Attachment of cells to basement membrane collagen type IV, *J. Cell Biol.* **103**:1569–1575.
67. Rao, C. N., Margulies, I. M. K., Tralka, T. S., Terranova, V. P., Madri, J. A., and Liotta, L. A., 1982, Isolation of a subunit of laminin and its role in molecular structure and tumor cell attachment, *J. Biol. Chem.* **257**:9740–9744.
68. Dejana, E., Colella, S., Languino, L. R., Balconi, G., Corbascio, G. C., and Marchisio, P. C., 1987, Fibrinogen induces adhesion, spreading, and microfilament organization of human endothelial cells in vitro, *J. Cell Biol.* **104**:1403–1411.
69. Ruoslahti, E., and Pierschbacher, M. D., 1987, New perspectives in cell adhesion: RGD and integrins, *Science* **238**:491–497.
70. Fujiwara, S., Wiedmann, H., Timpl, R., Lustig, A., and Engel, J., 1984, Structure and interactions of heparan sulfate proteoglycans from a mouse tumor basement membrane, *Eur. J. Biochem.* **143**:145–157.
71. Robertson, N. P., Starkey, J. R., Hamner, S., and Meadows, G. G., 1989, Tumor cell invasion of three-dimensional matrices of defined composition: Evidence for a specific role for heparan sulfate in rodent cell lines, *Cancer Res.* **49**:1816–1823.
72. Brown, P. J., and Juliano, R. L., 1986, Expression and function of a putative cell surface receptor for fibronectin in hamster and human cell lines, *J. Cell Biol.* **103**:1595–1603.
73. Dedhar, S., Ruoslahti, E., and Pierschbacher, M. D., 1987, A cell surface receptor complex for collagen type I recognizes the Arg-Gly-Asp sequence, *J. Cell Biol.* **104**:585–593.
74. Dedhar, S., Argraves, W. S., Suzuki, S., Ruoslahti, E., and Pierschbacher, M. D., 1987, Human osteosarcoma cells resistant to detachment by an Arg-Gly-Asp-containing peptide overproduces the fibronectin receptor, *J. Cell Biol.* **105**:1175–1182.
75. Liotta, L. A., Horan Hand, P., Rao, C. N., Bryant, G., Barsky, S. H., and Scholm, J., 1985, Monoclonal antibodies to the human laminin receptor recognize structurally distinct sites, *Exp. Cell Res.* **156**:117–126.
76. Pytela, R., Pierschbacher, M. D., and Ruoslahti, E., 1985, Identification and isolation of a 140-Kd cell surface glycoprotein with properties expected of a fibronectin receptor, *Cell* **40**:191–198.
77. Pytela, R., Pierschbacher, M. D., and Ruoslahti, E., 1985, A 125/115-kDa cell surface receptor specific for vitronectin interacts with the arginine-glycine-aspartic acid adhesion sequence derived from fibronectin, *Proc. Natl. Acad. Sci. USA* **82**:5766–5770.
78. Liotta, L. A., 1986, Tumor invasion and membranes. Role of the extracellular matrix, *Cancer Res.* **46**:1–7.
79. McCarthy, J., Hagen, S. T., and Furcht, L. T., 1986, Fibronectin contains distinct adhesion and motility-promoting domains for metastatic melanoma cells, *J. Cell Biol.* **102**:179–188.
80. Lissitzky, J. C., Bouzon, M., Loret, E., Poupor, M. F., and Martin, P. M., 1989, Laminin-mediated adhesion in metastatic rat rhabdomyosarcoma cell lines involves prominent interactions with the laminin E8 fragment, *Clin. Exp. Metastasis* **7**:469–480.
81. Albini, A., Aukerman, S. L., Ogle, R. C., Noonan, D. M., Fridman, R., Martin, G. R., and Fidler, I. J., 1989, The in vitro invasiveness and interactions with laminin of K-1735 melanoma cells. Evidence for different laminin-binding affinities in high and low metastatic variants, *Clin. Exp. Metastasis* **7**:437–451.
82. Ruoslahti, E., and Pierschbacher, M. D., 1986, Arg-Gly-Asp-: A versatile cell recognition signal, *Cell* **44**: 517–529.
83. Hayman, E. G., Pierschbacher, M. D., and Ruoslahti, E., 1985, Detachment of cells from culture substrate by soluble fibronectin peptides, *J. Cell Biol.* **100**:1948–1954.
84. Humphries, M. J., Olden, K., and Yamada, K. M., 1986, A synthetic peptide from fibronectin inhibits experimental metastasis of murine melanoma cells, *Science* **233**:467–470.
85. Humphries, M. J., Yamada, K. M., and Olden, K., 1988, Investigation of the biological effects of anti-cell adhesive synthetic peptides that inhibit experimental metastasis of B16-F10 murine melanoma cells, *J. Clin. Invest.* **81**:782–790.
86. Bretti, S., Neri, P., Lozzi, L., Rustici, M., Comoglio, P., Giancotti, F., and Tarone, G., 1989, Inhibition of experimental metastasis of murine fibrosarcoma cells by oligopeptide analogues to the fibronectin cell-binding site, *Int. J. Cancer* **43**:102–106.
87. Saiki, I., Murata, J., Iida, J., Nishi, N., Sugimura, K., and Azume, I., 1989, The inhibition of murine lung

metastasis by synthetic polypeptides poly(arg-gly-asp) and poly(tyr-ile-gly-ser-arg) with a core sequence of cell adhesion molecules, *Br. J. Cancer* **59**:194–197.

88. Gehlsen, K. R., Argraves, W. S., Pierschbacher, M. D., and Ruoslahti, E., 1989, Inhibition of in vitro tumor cell invasion by Arg-Gly-Asp-containing synthetic peptides, *J. Cell Biol.* **106**:925–930.
89. Kramer, R. H., McDonald, K. A., Crowley, E., Ramos, D. M., and Damsky, C. H., 1989, Melanoma cell adhesion to basement membrane mediated by integrin-related complexes, *Cancer Res.* **49**:393–402.
90. Vollmers, H. P., and Birdmeier, W., 1983, Cell adhesion and metastasis: Monoclonal antibody approach, *Trends Biochem. Sci.* **8**:452–455.
91. Vollmers, H. P., Luihof, B. A., Braun, S., Walker, C. A., Schirrmacher, B., and Burchmeier, W., 1984, Monoclonal antibodies which prevent experimental lung metastasis, *FEBS Lett.* **172**:17–20.
92. Iminura, T., and Nicolson, G. L., 1981, The role of glycoconjugates in metastatic melanoma blood-borne arrest and cell surface properties, *J. Supramol. Struct. Cell Biochem.* **17**:325–336.
93. Lichtner, R. B., Erkell, L. J., Schirrmacher, V., and Nicolson, G. L., 1989, Effects of RA 233 treatment on the adhesive, invasive and metastatic properties of 13762NF rat mammary tumor cells, *Clin. Exp. Metastasis* **7**:175–186.
94. Dennis, J. W., Laferte, S., Waghorne, C., Breitman, J. L., and Kerbel, R. S., 1987, Beta 1-6 branching of ASN-linked oligosaccharides is directly associated with metastasis, *Science* **236**:582–585.
95. Elvin, P., and Evans, C. W., 1984, Cell adhesion and experimental metastasis: A study using the B16 malignant model system, *Eur. J. Cancer Clin. Oncol.* **20**:107–114.
96. Fornabaio, D. M., Alterman, A. L., and Stackpole, C. W., 1988, Metastatic dissemination of B16 melanoma: Evidence that metastases can result from non-specific trapping of disseminated tumor cells, *Invasion Metastasis* **8**:1–16.
97. Sargent, N. S. E., Oestreicher, M., Haidvogl, H., Madnick, H. M., and Burger, M. M., 1988, Growth regulation of cancer metastases by their host organ, *Proc. Natl. Acad. Sci. USA* **85**:7251–7255.
98. Weiss, L., Harlos, J. P., Torhorst, J., Gunthard, B., Hartveit, F., Svendson, E., Huang, W.-L., Grauduiann, E., Eder, M., Zwicknagl, M., Cochrane, E. R., Stock, D., Wright, C., and Horne, C., 1988, Metastatic pattern of renal carcinoma; An analysis of 687 necropsies, *Cancer Res. Clin. Oncol.* **114**:605–612.
99. Weiss, L., 1988, Biomechanical destruction of cancer cells in the heart: A rate-regulator for hematogenous metastasis, *Invasion Metastasis* **8**:228–237.
100. Weiss, L., 1989, Biomechanical destruction of cancer cells in skeletal muscle: A rate-regulator for hematogenous metastasis, *Clin. Exp. Metastasis* **7**:483–491.
101. Weiss, L., 1980, Cancer cell traffic from the lungs to the liver: An example of metastatic inefficiency, *Int. J. Cancer* **25**:385–892.
102. Weiss, L., Ward, P. M., and Holmes, J. C., 1983, Liver-to-lung traffic of cancer cells, *Int. J. Cancer* **32**: 79–83.
103. Weiss, L., 1987, The hemodynamic destruction of circulating cancer cells, *Biorheology* **24**:105–115.
104. Weiss, L., and Schmid-Schonbein, G. W., 1989, Biomechanical interactions of cancer cells with the microvasculature during metastasis, *Cell Biophys.* **14**:187–215.
105. Weiss, L., Harlos, J. P., and Elkin, G., 1989, Mechanism of mechanical trauma to Ehrlich ascites tumor cells in vitro and its relationship to rapid intravascular death during metastasis, *Int. J. Cancer* **44**:143–149.
106. Weiss, L., Harlos, J. P., Elkin, G., and Bixler, B., 1990, Mechanisms for the biomechanical destruction of L1210 leukemia cells: A rate-regulator for metastasis, *Cell Biophys.* **16**:149–159.
107. Gabor, H., and Weiss, L., 1985, Mechanically induced trauma suffered by cancer cells in passing through pores in polycarbonate membranes, *Invasion Metastasis* **5**:71–83.
108. Ingber, D. E., and Folkman, J., 1989, Tension and comprehension as basic determinants of cell form and function. Utilization of a cellular tensegrity mechanism, in: *Cell Shape* (W. D. Stein and F. Brouner, eds.), Academic Press, New York.
109. Halliwell, B., 1989, Superoxide anion, vascular endothelium and reperfusion injury, *Free Radical Res. Commun.* **5**:315–318.
110. Glaves, D., 1986, Intravascular death of disseminated cancer cells mediated by superoxide anion, *Invasion Metastasis* **6**:101–111.
111. Orr, F. W., Adamson, I. Y. R., and Young, L., 1986, Promotion of pulmonary metastasis in mice by bleomycin-induced endothelial injury, *Cancer Res.* **46**:891–897.
112. Vincic, L., Orr, F. W., Warner, D. J. A., Suyama, K. L., and Kay, J. M., 1989, Enhanced cancer metastasis after monocrotaline-induced lung injury, *Toxicol. Appl. Pharmacol.* **100**:259–270.

113. Richters, A., and Richters, V., 1989, Nitrogen dioxide (NO_2) inhalation, formation of microthrombi in lungs and cancer metastasis, *J. Environ. Pathol. Toxicol. Oncol.* **9**(1)**:**45–51.
114. Adamson, I. Y. R., Young, L., and Orr, F. W., 1987, Tumor metastasis after hyperoxic injury and repair of the pulmonary endothelium, *Lab. Invest.* **57:**71–77.
115. Hirata, H., and Tanaka, K., 1984, Artificial metastases and decrease of fibrinolysis in the nude mouse lung after hemithoracic irradiation, *Clin. Exp. Metastasis* **2:**311–319.
116. Dao, T. L., and Yogo, H., 1967, Enhancement of pulmonary metastases by X-irradiation in rats bearing mammary cancer, *Cancer* **20:**2020–2025.
117. Nicolson, G. L., and Custead, S. E., 1985, Effects of chemotherapeutic drugs on platelet and metastatic tumor cell–endothelial cell interactions as a model for assessing vascular endothelial integrity, *Cancer Res.* **45:**331–336.
118. Adamson, I. Y. R., Orr, F. W., and Young, L., 1986, Effects of injury and repair of the pulmonary endothelium on lung metastasis after bleomycin, *J. Pathol.* **150:**279–287.
119. Orr, F. W., Adamson, I. Y. R., Warner, D. J. A., Leroyer, V., Werner, L., Shaughnessy, S., and Young, L., 1988, Brief Review: The effects of oxygen radical-mediated pulmonary endothelial damage on cancer metastasis, *Mol. Cell. Biochem.* **84:**189–198.
120. Shaughnessy, S. G., Warner, D., Buchanan, M. R., Lafrenie, R., and Orr, F. W., 1991, The effects of oxygen radical-mediated pulmonary endothelial damage on cancer metastasis, in: *Proceedings of the First International Conference on Eicosanoids and Bioactive Lipids in Cancer and Radiation Injury* (K. Honn, ed.), Kluwer, Boston, Massachusetts, pp. 147–156.
121. Orr, F. W., Buchanan, M. R., Tron, V. A., et al., 1988, Chemotactic activity of endothelial cell-derived interleukin-1 for human tumor cells, *Cancer Res.* **48:**6758–6763.
122. Zucker, S., Beck, G., DiStefano, J. F., and Lysik, R. M., 1985, Role for different cell proteinases in cancer invasion and cytolysis, *Br. J. Cancer* **52:**223–232.
123. Brizzo, P., Morisset, M., Capony, F., Rougeot, C., and Rochefort, H., 1988, In vitro degradation of extracellular matrix with Mr52,000 cathepsin D secreted by breast cancer cells, *Cancer Res.* **48:**3688–3692.
124. Nakajima, M., Welch, D. R., Belloni, P. N., and Nicolson, G. L., 1987, Degradation of basement membrane type IV collagen and lung subendothelial matrix by rat mammary adenocarcinoma cell clones of differing metastatic potentials, *Cancer Res.* **47:**4869–4876.
125. Masure, S., and Opdenakker, G., 1989, Cytokine-mediated proteolysis in tissue remodelling, *Experientia* **45:**542–549.
126. Pauli, B. U., and Knudson, W., 1988, Tumor invasion: A consequence of destructive and compositional matrix alterations, *Hum. Pathol.* **19:**628–639.
127. Goldfarb, R. H., and Liotta, L. A., 1986, Proteolytic enzymes in cancer invasion and metastasis, *Semin. Thromb. Hemostasis* **12:**294–307.
128. Turpeenniemi-Hujanen, T., Thorgeirsson, U. P., Hart, I. R., Grant, S. S., and Liotta, L. A., 1985, Expression of collagenase IV (basement membrane collagenase) activity in murine tumor cell hybrids that differ in metastatic potential, *J. Natl. Cancer Inst.* **75:**99–103.
129. Leroyer, V., Werner, L., Shaughnessy, S., Goddard, G. J., and Orr, F. W., 1987, Chemiluminescence and oxygen radical generation by Walker carcinosarcoma cells following chemotactic stimulation, *Cancer Res.* **47:**4771–4775.
130. Varani, J., Ginsburg, I., Schuger, L., Gibbs, D. F., Bromberg, J., Johnson, K. J., Ryan, U. S., and Ward, P. A., 1989, Rapid Communication. Endothelial cell killing by neutrophils, *Am. J. Pathol.* **135:**435–438.
131. Marcus, A. J., Silk, S. T., Safier, L. B., and Ullman, H. L., 1977, Superoxide production and reducing activity in human platelets, *J. Clin. Invest.* **59:**149–158.
132. Joseph, M., Capron, A., Tsicopoulos, A., Ameisen, J. C., Martinot, J. B., and Tonnel, A. B., 1987, Platelet activation by IgE and aspirin, *Agents Action* **21**(Suppl.)**:**169–177.
133. Weiss, S. J., Peppin, G., Ortiz, X., Ragsdale, C., and Test, S. T., 1985, Oxidative autoactivation of latent collagenase by human neutrophils, *Science* **227:**747–749.
134. Shah, S. V., Baricos, W. H., and Basci, A., 1987, Degradation of human glomerular basement membrane by stimulated neutrophils: Activation of a metalloproteinase(s) by reactive oxygen metabolites, *J. Clin. Invest.* **79:**25–31.
135. Campbell, E. J., Senior, R. M., and McDonald, J. A., 1982, Proteolysis by neutrophils: Relative importance of cell-surface contact and oxidative inactivation of proteinase inhibitors in vitro, *J. Clin. Invest.* **70:**845–852.
136. Weiss, S. J., and Regiani, S., 1984, Neutrophils degrade subendothelial matrices in the presence of alpha-1-

proteinase inhibitor: Cooperative use of lysosomal proteinases and oxygen metabolites, *J. Clin. Invest.* **73:** 1297–1303.

137. Dean, R. T., Nick, H. P., and Schnebli, H. P., 1989, Free radicals inactivate human neutrophil elastase and its inhibitors with comparable efficiency, *Biochem. Biophys. Res. Commun.* **159:**821–827.
138. Ostrowski, L. E., Ahsan, A., Suther, B. P., Pagast, P., Bain, D. L., Wong, C., Patal, A., and Schultz, R. M., 1986, Selective inhibition of proteolytic enzymes in an in vivo mouse model for experimental metastasis, *Cancer Res.* **46:**4121–4128.
139. Orr, F. W., and Warner, D. J. A., 1987, Effects of neutrophil-mediated pulmonary endothelial injury on the localization and metastasis of circulating Walker carcinosarcoma cells, *Invasion Metastasis* **7:**183–196.
140. Starkey, R., Liggitt, H. D., Jones, W., and Hosick, H. L., 1984, Influence of migratory blood cells on the attachment of tumor cells to vascular endothelium, *Int. J. Cancer* **34:**535–543.
141. Orr, F. W., and Mokashi, S., 1985, Effects of leukocyte activation on the formation of heterotypic tumour-cell aggregates in vitro, *Int. J. Cancer* **35:**101–106.
142. Wood, S., 1958, Pathogenesis of metastasis formation observed in vivo in the rabbit ear chamber, *Arch. Pathol.* **66:**550–568.
143. Shoenfeld, Y., Tal, A., Berliner, S., and Pinkhas, J., 1986, Leukocytosis in non-hematological malignancies—A possible tumour-associated marker, *J. Cancer Res. Clin. Oncol.* **111:**54–58.
144. Ishikawa, M., Koga, Y., Hosokawa, M., and Kobayashi, H., 1986, Augmentation of B16 melanoma lung colony formation in C57BL/6 mice having marked granulocytosis, *Int. J. Cancer* **37:**919–924.
145. Aeed, P. A., Nakajima, M., and Welch, D., 1988, The role of polymorphonuclear leukocytes (PMN) on the growth and metastatic potential of 13762NF mammary adenocarcinoma cells, *Int. J. Cancer* **42:**748–759.
146. Orr, F. W., Adamson, I. Y. R., and Young, L., 1985, Pulmonary inflammation generates chemotactic activity for tumor cells and promotes lung metastasis, *Am. Rev. Respir. Dis.* **131:**607–611.
147. Van Den Brenk, H. A. S., Stone, M., Kelly, H., Orton, C., and Sharpington, C., 1974, Promotion of growth of tumour cells in acutely inflamed tissues, *Br. J. Cancer* **30:**246–260.
148. Orr, F. W., and Warner, D. J. A., 1990, Effects of systemic complement activation and neutrophil-mediated pulmonary injury on the retention and metastasis of circulating cancer cells in mouse lungs, *Lab. Invest.* **62:**331–338.
149. Glaves, D., 1983, Role of polymorphonuclear leukocytes in the pulmonary clearance of arrested cancer cells, *Invasion Metastasis* **3:**160–173.
150. Jones, C. L., Nelson, K. K., Hatfield, J. S., Honn, K. V., and Onoda, J. M., 1988, Regulation of metastasis by polymorphonuclear neutrophils, *Proc. Am. Assoc. Cancer Res.* **29:**62 (Abstract).
151. Gasic, G. J., Gasic, T. B., and Stewart, C. O., 1968, Antimetastatic effect associated with platelet reduction, *Proc. Natl. Acad. Sci. USA* **61:**46–52.
152. Gasic, G., 1984, Role of plasma, platelets, and endothelial cells in tumor metastasis, *Cancer Metastasis Rev.* **3:**99–114.
153. Cavanaugh, P. G., Sloane, B. F., and Honn, K. V., 1988, Role of the coagulation system in tumor-cell-induced platelet aggregation and metastasis, *Haemostasis* **18:**37–46.
154. Bastida, E., and Ordinas, A., 1988, Platelet contribution to the formation of metastatic foci: The role of cancer cell-induced platelet activation, *Haemostasis* **18:**29–36.
155. Tohgo, A., Tanaka, N. G., and Ogawa, H., 1986, Platelet-aggregating activities of metastasizing tumor cells. IV. Effects of cell surface modification on thrombin generation, platelet aggregation and subsequent lung colonization, *Invasion Metastasis* **6:**58–68.
156. Longenecker, G. L., Beyers, B. J., Bowen, R. J., and King, T., 1989, Human rhabdosarcoma cell-induced aggregation of blood platelets, *Cancer Res.* **49:**16–19.
157. Ugen, K. E., Mahalingam, M., Klein, P. A., and Kao, K. J., 1988, Inhibition of tumor cell-induced platelet aggregation and experimental tumor metastasis by the synthetic gly-arg-gly-asp-ser peptide, *J. Natl. Cancer Inst.* **80:**1461–1466.
158. Karpatkin, S., Pearlstein, E., Ambrogio, C., and Coller, B. S., 1988, Role of adhesive proteins in platelet tumor interaction in vitro and metastasis formation in vivo, *J. Clin. Invest.* **81:**1012–1019.
159. Karpatkin, S., Ambrogio, C., and Pearlstein, E., 1988, The role of tumor-induced platelet aggregation, platelet adhesion and adhesive proteins in tumor metastasis, in: *Platelet Membrane Receptors: Molecular Biology, Immunology, Biochemistry, and Pathology* (G. A. Jamieson, ed.), A. R. Liss, New York.
160. Poggi, A., Vicenzi, E., Cioce, V., and Wasteson, A., 1988, Platelet contribution to cancer cell growth and migration: The role of platelet growth factors, *Haemostasis* **18:**18–28.

161. Wallace, A. C., Chew, E.-C., and Jones, D. S., 1977, Arrest and extravasation of cancer cells in the lung, in: *Lung Metastasis* (L. Weiss and M. A. Gilbert, eds.), Hall, Boston, p. 26.
162. Crissman, J., Hatfield, J. S., Menter, D. G., Sloane, B., and Honn, K. V., 1988, Morphological study of interaction of intravascular tumor cells with endothelial cells and subendothelial matrix, *Cancer Res.* **48:** 4065–4072.
163. Weiss, L., 1985, *Principles of Metastasis*, Academic Press, Orlando, Florida, p. 109.
164. Menter, D. G., Steinest, B. W., Sloane, B. F., Taylor, J. D., and Honn, K. V., 1987, A new in vitro model for investigation of tumor cell–platelet–endothelial cell interactions and concomitant eicosanoid biosynthesis, *Cancer Res.* **47:**2425–2432.
165. Kramer, R. H., and Nicolson, G. L., 1979, Interactions of tumor cells with vascular endothelial monolayers: A model for metastatic invasion, *Proc. Natl. Acad. Sci. USA* **76:**5704–5708.
166. Zamora, P. O., Danielson, K. G., and Hosick, H. L., 1980, Invasion of endothelial cell monolayers on collagen gels by cells from mammary tumor spheroids, *Cancer Res.* **40:**4631–4639.
167. Honn, K. V., Grossi, I. M., Diglio, C. A., Wojtukiewicz, M., and Taylor, J. D., 1989, Enhanced tumor cell adhesion to the subendothelial matrix resulting from 12(S)-HETE-induced endothelial cell retraction, *FASEB J.* **3:**2285–2293.
168. Menter, D. G., Steinest, B. W., Sloane, B. F., Gundlach, N., O'Gara, C. Y., Marnett, L. J., Diglio, C., Walz, D., and Honn, K. V., 1987, Role of platelet membrane in enhancement of tumor cell adhesion to endothelial cell extracellular matrix, *Cancer Res.* **47:**6751–6762.
169. Menter, D. G., Sloane, B. F., Steinert, B. W., Onoda, J. M., Craig, R., Hawkins, C., Taylor, J. D., and Honn, K. V., 1987, Platelet enhancement of tumor cell adhesion to subendothelial matrix. Role of platelet cytoskeleton and platelet membrane, *J. Natl. Cancer Inst.* **79:**1077–1098.
170. Skolnik, G., Alpsten, M., and Ivarson, L., 1980, Studies on mechanisms involved in metastasis formation from circulating tumor cells, *J. Cancer Res. Clin. Oncol.* **97:**249–256.
171. Skolnik, G., Ericson, L. E., and Bagge, U., 1983, The effect of thrombocytopenia and antiserotonin treatment on the lodgment of circulating tumor cells, *J. Cancer Res. Clin. Oncol.* **105:**30–37.
172. Glaves, D., and Weiss, L., 1978, Initial tumor cell arrest in animals of defined coagulative states, *Int. J. Cancer* **21:**741–746.
173. Weiss, L., 1985, *Principles of Metastasis*, Academic Press, Orlando, Florida, p. 200.
174. Merriman, R. L., Shacketford, K. A., Tanzer, L. R., Campbell, J. B., and Bernis, J. B., 1989, Drug treatments for metastasis of Lewis lung carcinoma: Lack of correlation between inhibition of lung metastasis and survival, *Cancer Res.* **49:**4509–4511.
175. Hagmar, B., and Boeryd, B., 1969, Disseminating effect of heparin on experimental tumour metastasis, *Pathol. Eur.* **4:**274–282.
176. Sindelar, W. F., Tralka, T. S., and Ketcham, A. S., 1975, Electron microscopic observations on formation of pulmonary metastases, *J. Surg. Res.* **18:**137–161.
177. Machado, E. A., Gerard, D. A., Mitchell, J. R., Lozzio, B. B., and Lozzio, C. B., 1982, Arrest and extravasation of neoplastic cells, *Virchows Arch. A* **396:**73–89.
178. Nakamura, K., Kawaguchi, T., Asahina, S., Sakura, T., Ebina, Y., Yokoya, S., and Morita, M., 1977, Electron microscopic studies on extravasation of tumor cells and early foci of hematogenous metastases, *Gann Monogr.* **1977:**57–71.
179. Lapis, K., Paku, S., and Liotta, L. A., 1988, Endothelialization of embolized tumor cells during metastasis formation, *Clin. Exp. Metastasis* **6:**73–89.
180. D'Amore, P. A., and Thompson, R. W., 1987, Mechanisms of angiogenesis, *Annu. Rev. Physiol.* **49:** 453–464.
181. Folkman, J., 1985, Toward an understanding of angiogenesis: Search and discovery, *Perspect. Biol. Med.* **29:** 10–36.
182. Gross, J. L., Moscatelli, D., and Rifkin, D. B., 1983, Increased capillary endothelial cell protease activity in response to angiogenic stimuli in vitro, *Proc. Natl. Acad. Sci. USA* **80:**2623–2627.
183. Pepper, M. S., Vassalli, J.-D., Moutesano, R., and Orci, L., 1987, Urokinase-type plasminogen activator is induced in migrating capillary endothelial cells, *J. Cell Biol.* **105:**2535–2541.
184. Ausprunk, D. H., and Folkman, J., 1977, Migration and proliferation of endothelial cells in preformed and newly formed blood vessels during tumor angiogenesis, *Microvasc. Res.* **14:**53–65.
185. Furcht, L. T., 1986, Cortical factors controlling angiogenesis: Cell products, cell matrix, and growth factors, *Lab. Invest.* **55:**505–509.

186. Warren, B. A., 1979, Tumor angiogenesis, in: *Tumor Blood Circulation* (H.-I. Peterson, ed.), CRC Press, Boca Raton, p. 49.
187. Vehoeven, D., and Buyssens, N., 1988, Desmin-positive stellate cells associated with angiogenesis in a tumour and non-tumour system, *Virchows Arch. B* **54:**263–272.
188. Thompson, W. D., Shiach, K. J., Fraser, R. A., McIntosh, L. C., and Simpson, J. G., 1987, Tumors acquire their vasculature by vessel incorporation, not vessel ingrowth, *J. Pathol.* **151:**323–332.
189. Folkman, J., and Klagsbrun, M., 1987, Angiogenic factors, *Science* **235:**442–447.
190. Folkman, J., and Ingber, D. E., 1987, Angiostatic steroids, *Ann. Surg.* **206:**374–383.
191. Folkman, J., Weisz, P. B., Joullie, M., Li, W. W., and Ewing, W. R., 1989, Control of angiogenesis with synthetic heparin substitutes, *Science* **243:**1490–1493.
192. D'Amore, P. A., 1988, Antiangiogenesis as a strategy for antimetastasis, *Semin. Thromb. Hemostasis* **14:** 73–78.

28

Possible Relationship between Vascular Permeability Factors, Endothelial Cells, and Peritumoral Brain Edema

A Neurosurgeon's Perspective

Gregory Richard Criscuolo

I. INTRODUCTION

Cerebral edema is a significant cause of the neurological morbidity associated with malignant brain tumors. Patients afflicted with intracranial neoplasms typically present with progressively worsening headaches, vomiting, blurred vision, double vision, and depressed level of consciousness. All of these clinical features are clearly related to elevated intracranial pressure (ICP), which may rapidly result in a patient's demise if not treated promptly and effectively. The brain edema associated with cerebral tumors is an ultrafiltrate of plasma containing water, electrolytes, and plasma proteins, that emanates from the brain tumor microvasculature. Edema fluid typically infiltrates the white matter surrounding an intracerebral tumor in a diffuse manner, while relatively sparing the adjacent cortex. This excessive accumulation of cerebral interstitial tissue fluid contributes to the distortion of normal intracranial structures, and to elevation of the ICP. As a result, cerebral edema is frequently as culpable for a brain tumor patient's symptoms as the primary intracerebral neoplasm. It is now widely accepted that this excessive fluid accumulation results from a flaw in the integrity of the blood–brain barrier. The blood–brain barrier is both an anatomical and a physiological system which normally regulates the entry and egress of substances between the cerebral interstitial and intravascular compartments. The functional components of the blood–brain barrier occur primarily at the level of the vascular endothelial cell and, to a lesser extent, the basement membrane and astrocytic processes which invest the cerebral microvasculature. Ultrastructural examination of normal brain

Gregory Richard Criscuolo • Division of of Neurosurgery, Yale University School of Medicine, New Haven, Connecticut 06510.

Endothelial Cell Dysfunctions, edited by Nicolae Simionescu and Maya Simionescu. Plenum Press, New York, 1992.

microvessels in comparison to brain tumor microvessels supports the time-honored hypothesis that altered tumor vessels are in fact responsible for the abnormal extravasation of fluid and protein into the interstitial space; thus revealing the vasogenic nature of this process. Despite extensive efforts to study this phenomenon, a focused pathophysiological explanation for these anatomical observations has not yet evolved.

It is recognized that treatment of brain tumor patients with high doses of glucocorticosteroids, such as dexamethasone, results in a remarkable resolution of their ICP-related signs and symptoms. Moreover, the onset of this clinical efficacy is readily apparent within a predictable 12 to 48 hr of initiation of this therapy. This dramatic response is of practical significance as it allows the neurosurgeon, patient, and patient's family additional time prior to performing surgery, to discuss issues related to the disease process, surgical therapy, and perioperative care. Furthermore, patients undergoing brain tumor surgery are well known to enjoy a better surgical outcome if their neurologic deficits are minimized preoperatively. Clinical improvement with dexamethasone appears to correlate with partial resolution of vasogenic brain edema as evidenced by computed tomographic (CT) studies. Notably, the clinical responsiveness of brain edema to dexamethasone has been shown to correlate with cytosolic glucocorticoid-receptor concentrations in several types of intracranial tumor. Nevertheless, the pharmacological basis for this remarkable clinical, physiological, and radiographic response is unknown, and the rationale for glucocorticoid use in this setting has been largely empirical. It would appear, therefore, that further study of the extent and specificity of dexamethasone's actions in the setting of neoplastic brain edema would be worthwhile, as it might reveal novel insight into the very mechanism by which certain types of edema evolve.

II. DEFINITION AND CLASSIFICATION OF BRAIN EDEMA

Brain edema is generally defined as a condition whereby disturbed cerebral homeostatic mechanisms result in an abnormal increase in brain tissue volume that is largely attributable to an increase in water content. Many pathological processes are known to result in brain edema formation. Although the pathophysiological mechanisms are not completely understood, it is apparent that several distinct varieties of cerebral edema exist.[4,48,70] Vasogenic brain edema (type I) is associated with marked alterations of the microvascular elements. This disturbance in the blood–brain barrier results in extravasation of plasmalike fluid into the white matter extracellular space (white matter edema). Vasogenic edema is seen clinically in the setting of brain tumor (Fig. 1), abscess, intracerebral hemorrhage, malignant systemic hypertension, and after prolonged seizure activity. Cytotoxic brain edema (type II) is initially associated with damage to cortical cells resulting in impaired cellular membrane homeostasis and permeability. Intracellular accumulation of sodium and water results in cellular swelling (gray matter edema) at the expense of the interstitial space. Cytotoxic brain edema is seen clinically in the setting of cerebral ischemia or infarction, diabetic coma, hepatic encephalopathy, Reye's syndrome, and pseudotumor cerebri. Interstitial brain edema (type III) refers to an infiltration of cerebrospinal fluid (CSF) into the periventricular white matter. The cerebral interstitial fluid normally communicates freely with the CSF space and a bulk diffusion occurs in a centrifugal fashion. Interstitial brain edema occurs in the setting of acute hydrocephalus

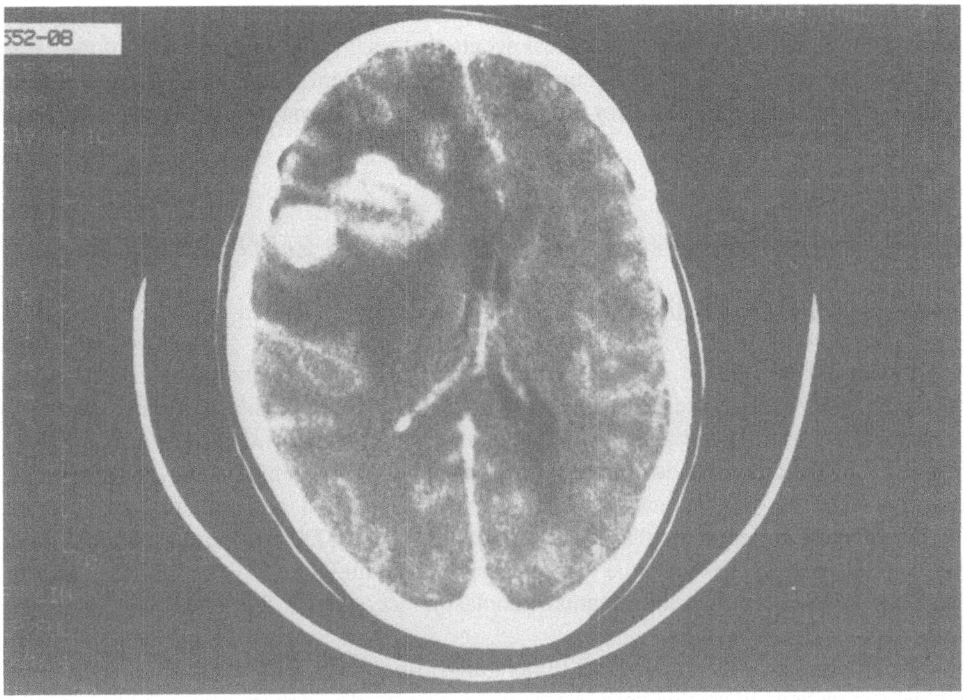

Figure 1. Computer-assisted tomographic brain scan (CT) of a middle-aged person who presented clinically with a 2-week history of progressively worsening headaches, vomiting, double vision, impaired mentation, and left-sided weakness. The multilobular lesion seen within the right frontal lobe was found, at the time of neurosurgical biopsy and excision, to be a highly malignant brain tumor of glial cell origin (glioblastoma multiforme). The brightness of this lesion relates directly to extravasation of an intravenously administered iodinated contrast agent, in areas of frank blood–brain barrier disruption. Note the dark radiolucent region of vasogenic brain edema which arise from within the tumor and characteristically infiltrates the subcortical white matter surrounding the tumor. Cerebral edema appears to be contributing as much to the mass effect and brain distortion as the tumor itself. Within 24 hr of starting high-dose dexamethasone (10 mg i.v. every 4 hr), the patient had returned to normal and was able to undergo an uncomplicated gross total excision of this tumor.

whereby CSF under increased pressure is forced to flow centripetally across the ependymal surface lining the ventricular cavities, and into white matter interstitial space. Blood–brain barrier function is not altered. Osmotic/hydrostatic brain edema (type IV) occurs when a sudden disturbance in Starling's equilibrium results in the development of an osmotic gradient between plasma and cerebral tissue. Initially, the edema consists largely of free water without electrolytes or plasma proteins. The brain eventually loses electrolytes without gaining significant quantities of water. Clinical impairment is more likely related to cellular potassium loss and hyponatremia, as elevated ICP is not a consistent association. Osmotic or hydrostatic brain edema occurs with primary water intoxication, rapid hemodialysis, and the inappropriate secretion of antidiuretic hormone (SIADH). This classification of cerebral edema can be quite sound and useful provided one recognizes its limitations. For instance, it is well recognized that disparate types of brain edema can coexist. For instance, both vasogenic and cytotoxic edema may occur in certain settings depending upon the nature and severity of the initial insult. Furthermore, if one considers the variety of distinct pathophysiological events that may result in expression of the same pattern of edema, a lack of specificity becomes self-evident. Nevertheless, this template rightfully remains the standard for all discussions about edema ultrastructure, pathology, and physiology.

III. BRAIN TUMORS AND TUMOR-ASSOCIATED CEREBRAL EDEMA

A. Survey of Human Brain Tumors

Tumors of the central nervous system (CNS) pose a major challenge to the field of oncology. They continue to be associated with a high degree of morbidity and mortality, despite advances in neurosurgery, radiotherapy, and chemotherapy.[3,8,28,34,40,51,81,83,100] Primary tumors of the CNS account for approximately 10% of all malignancy. Furthermore, cancer is second only to trauma as a cause of death in childhood, and brain tumors represent the second most common childhood malignancy. About 20,000 new primary brain tumors and an additional 15,000 secondary or metastatic brain tumors are diagnosed each year in the United States. In fact, intracranial neoplasms account for 2% of all cancer-related deaths.

Approximately 40% of all CNS tumors are primary tumors derived from neuroectodermal supporting cells (glial cells) such as astrocytes, oligodendrocytes, and ependymal cells. Astrocytomas comprise 60 to 70% of glial cell tumors (gliomas). Several attempts have been made to create a practical grading system for these lesions by correlating histological features with their biological behavior. One of the more commonly employed systems assigns a numerical grade from I through IV. Grade I and grade II astrocytomas are generally considered histologically and biologically benign tumors characterized by variable degrees of hypercellularity, and relatively prolonged survival after treatment with surgical excision or radiation therapy. Grade III (anaplastic astrocytoma) and grade IV (glioblastoma multiforme) astrocytomas display malignant cellular features such as extreme hypercellularity, pleomorphism, increased nucleocytoplasmic ratio, mitotic figures, nuclear hyperchromatism, multinucleation, and bizarre giant cell forms. Malignant stromal changes such as necrosis, pseudopalisading, edema fluid accumulation, and microvascular endothelial proliferation occur almost exclusively in the most highly malignant tumor varieties (glioblastoma multiforme).

Survival in patients with malignant gliomas is poor and ranges from 3 to 6 months untreated. Combined surgical excision, radiation, and chemotherapy have improved survival but mortality still approaches 95 to 100% at 5 years postdiagnosis. A peculiarity of primary brain tumors is the rarity with which they metastasize beyond the CNS. In most instances, malignant brain tumors are locally invasive, and therefore limit their growth to the fixed confines of the intracranial cavity. Recurrences after surgical excision and radiation therapy are most likely to occur at the site of the original tumor. Certain tumors demonstrate a predilection for seeding malignant cells into the subarachnoid space throughout the leptomeninges and spinal cord along CSF pathways (medulloblastoma, ependymoma, pineoblastoma, ependymoblastoma). Spread of a brain tumor in this fashion would naturally pose additional diagnostic and therapeutic challenges.

Brain tumors most likely to be associated with clinical and CT-evident vasogenic cerebral edema include most primary malignant tumors (anaplastic astrocytoma and glioblastoma multiforme), many secondary malignant tumors (metastases from lung, breast, renal, thyroid, and colon cancer, and malignant melanoma), and certain benign tumors (meningioma). Their management is complicated by the presence of peritumoral brain edema, which frequently limits a patient's tolerance for essential brain tumor therapies (surgery, radiotherapy, and chemotherapy) that may themselves initially contribute to cerebral swelling. Furthermore, the risk of a poor outcome after neurosurgical excision of a brain tumor is increased substantially in the setting of severe, symptomatic brain edema. For these reasons, an expanded understanding of the pathophysiology of peritumoral brain edema is essential to the effective treatment of patients with intracranial tumors.

B. The Blood–Brain and Blood–CSF Barriers

The concept of the blood–brain barrier was initially put forth by Paul Ehrlich, a 19th century German bacteriologist who noted that intravascular injection of vital dyes stained all organs except the brain. His interpretation, that this related to a differential organ affinity for the dye, was later proven incorrect when in 1913 a student of his, Edwin E. Goldmann, showed that injection of trypan blue into the CSF readily stained the brain but failed to enter the bloodstream. The blood–brain barrier is now recognized to be a complex, highly selective anatomical and physiological barrier, which regulates the entry and exit of cerebral nutrients and biologically important substances necessary for the maintenance of cerebral metabolism and neuronal activity.[4,13,35,54] Although Goldmann was the first to postulate that brain capillaries were the anatomic basis for the blood–brain barrier, direct confirmation did not occur until the advent of electron microscopy in the 1950s, which demonstrated the continuous tight junctions between adjoining endothelial cells.

Endothelial cells of the cerebral vasculature display several features attesting to their unique participation in blood–brain barrier function. The single most important feature is the high-resistance, pentalaminar tight junctions that fuse endothelial cells together in a continuous layer, and effectively form a physical barrier between circulating molecules and cells, and the cerebral interstitial space. Additional specialized features include: continuous capillary basement membranes, paucity of endothelial micropinocytotic activity, abundance of endothelial cell mitochondria, and absence of endothelial membrane fenestrations.[9–13,20,36,64] Therefore, virtually all macromolecules are excluded by the blood–brain barrier, and diffusion of substances into the brain largely depends upon physical characteristics such as molecular size, electrostatic charge, and lipophilicity. Amino acids, glucose, biogenic amines, and other essential brain nutrients gain entry by a complex system of membrane transporters. The various entry mechanisms may involve active transport (energy-requiring), facilitated transport (not energy-requiring), and enzymatic modification of a molecule prior to entry. The blood–CSF barrier is an analogous but distinct entity whose function is governed by the selective secretory activity of the choroid plexus epithelium. Together, the blood–brain and blood–CSF barriers regulate the composition of the cerebral interstitial fluid and CSF within well-defined limits. Disturbances in any of the components of the blood–brain or blood–CSF barriers, if significant, will result in cerebral edema.

C. Ultrastructure of Brain Edema and the Tumor Microvasculature

Prior reports have described the increased permeability characteristic of the microvasculature within primary and metastatic malignant brain tumors, as well as certain benign tumors.[55–57] Early studies focused primarily on the morphology of cerebral edema, and the altered vascular ultrastructure of tumor-associated and peritumoral blood vessels. Vasogenic brain edema was characterized by infiltration of the white matter interstitial spaces by an ultrafiltrate of plasma.[4,13,48,70] It has been postulated that the specialized features of normal brain endothelial cells result from contact with normal brain tissue.[5,9,10] As the latter consists largely of astrocytes whose cytoplasmic processes enmesh these microvessels, it would follow that the vastly altered milieu inherent in brain tumors, where microvessels develop among abnormal astrocytes (gliomas), or in their absence (metastatic tumors), might account for a dedifferentiation of brain endothelial cells into a less specialized phenotype. One can further speculate that microvessels growing into minimally altered environments might retain some blood–brain barrier features (low-grade astrocytomas associated with minimal or no brain edema), whereas the microvasculature of highly anaplastic gliomas retains little or no semblance to normal brain microvessels (associated with extensive brain edema).

Several features of tumor-associated vascular endothelium, such as widened intracellular junctions, discontinuous tight junctions, membrane fenestrations, noncontiguous basement membranes, active micropinocytosis, and paucity of mitochondria, sharply contrasted with the normal architecture of the blood–brain barrier.[1,12,13,36,55–57,66,94,102,105] However, many of these heterotypical cellular features of brain tumor microvessels typify the endothelium lining normal peripheral vasculature (not involving the blood–brain barrier). While the peripheral vasculature is not inherently permeable to macromolecules, it does appear to be exquisitely sensitive to permeability induction by physiologically occurring substances such as histamine, bradykinin, serotonin, and prostaglandin.[52,62,65] Furthermore, normal cerebral blood vessels appear totally unable to respond to these potent mediators of microvascular extravasation. Therefore, although it appears unlikely that the absence of continuous pentalaminar tight junctions would solely account for the abnormal permeability of brain tumor microvessels, alteration of this critical component of blood–brain barrier may at once render brain tumor microvessels similar to the peripheral microvasculature in anatomical form, and physiological reactivity to permeability-inducing substances.

D. Current Clinical Management of Peritumoral Brain Edema

Patients with brain tumors present most often with clinical manifestations of elevated ICP. They typically complain of headache, nausea, vomiting, drowsiness, and double vision. Seizures and focal neurologic deficits are also commonly present, and their precise pattern, distribution, and severity are dictated largely by the location and size of the tumor and the extent of the surrounding cerebral edema (Fig. 1). Progressive elevation of the ICP will result in clinical progression to stupor and coma. Ultimately, intracranial hypertension leads to cerebral herniation with consequent fatal brain stem compression. In this regard, peritumoral vasogenic cerebral edema deserves reemphasis as a frequent and important accompaniment of intracranial tumors. The combined effects of rapid tumor growth and cerebral swelling within the confines of the cranial vault, dictate the rapidly progressive clinical decline observed in patients with malignant brain tumors. Methods of acutely managing these patients include tumor excision (surgical cytoreduction), fluid restriction, diuresis with intravenous furosemide and mannitol, high-dose intravenous dexamethasone, and ventricular CSF drainage. Tumor excision addresses the tumor mass and its vascular bed directly, and as a consequence, eliminates the source of vasogenic edema fluid. Fluid restriction and diuretics decrease ICP by contracting the intravascular space and facilitating the egress of edema fluid from the cerebral interstitial space and into the vascular compartment. Ventricular CSF drainage lowers ICP by reducing the volume of the cranial contents; however, it is most specific in the setting of hydrocephalus related to tumor obstructing CSF outflow, and may be quite dangerous in the presence of a focal mass lesion with impending brain herniation. Glucocorticoids such as dexamethasone are remarkably effective in reducing the neurological deficits and intracranial hypertension associated with malignant brain tumors.[15,29,42,54,69] Their efficacy is not the result of tumor cytolysis and CT data indicate a direct action upon the peritumoral edema.[33,49,104,106] In fact, myriad studies have clearly shown neoplastic vasogenic brain edema to be the only type of brain edema that responds to glucocorticoid therapy (Fig. 2).[23,38,67] Once a patient has been stabilized, the brain tumor may be more safely neurosurgically excised. The need for additional therapy postoperatively is largely governed by the extent of tumor resection and histopathological diagnosis. Definitive treatment plans are proposed only after consultation between the neurosurgeon, medical oncologist, and radiation therapist. It is not unusual, however, for patients with residual tumor and edema (i.e., those in greatest need of further therapy) to have adjunctive radiation or chemotherapy interrupted as a result of exacerbation of the edemagenic component. Once again, dexamethasone, employed in even higher doses than used initially, may control the brain edema sufficiently to allow resumption of adjunctive therapy.

IV. MEDIATORS OF MICROVASCULAR PERMEABILITY: A CHRONOLOGICAL SYNOPSIS

A. Introduction

The microvasculature of many solid tumors exhibits increased permeability in comparison to normal tissues. Certain pathological reactions associated with neoplastic growth are likely to result from permeability induction by tumor cells. Clinical manifestations of this process include decreases in serum albumin, production of malignant effusions and ascites, tissue edema, and paraneoplastic arthropathy.[78,86,93] Neoplastic vasogenic brain edema likely derives from a physiological alteration in the blood–brain barrier that is manifested by ultrastructural changes in tumor microvessels, and the extravasation of water, salts, and serum proteins into the peritumoral white matter.[13,48,55–57] Several biochemical mediators have been implicated in the pathogenesis of brain edema. Substances such as histamine, serotonin, bradykinin, glutamic acid, polyamines, leukokinins, lymphokines, prostaglandins, thromboxane, prostacyclin, kallidin, lymphocyte permeability factors, kallikrein, and plasminogen activator are potent inducers of microvascular permeability associated with allergic, traumatic, ischemic, infectious, and inflammatory conditions.[6,16,21,24,52,58,59,68,77,91,95,96]

A

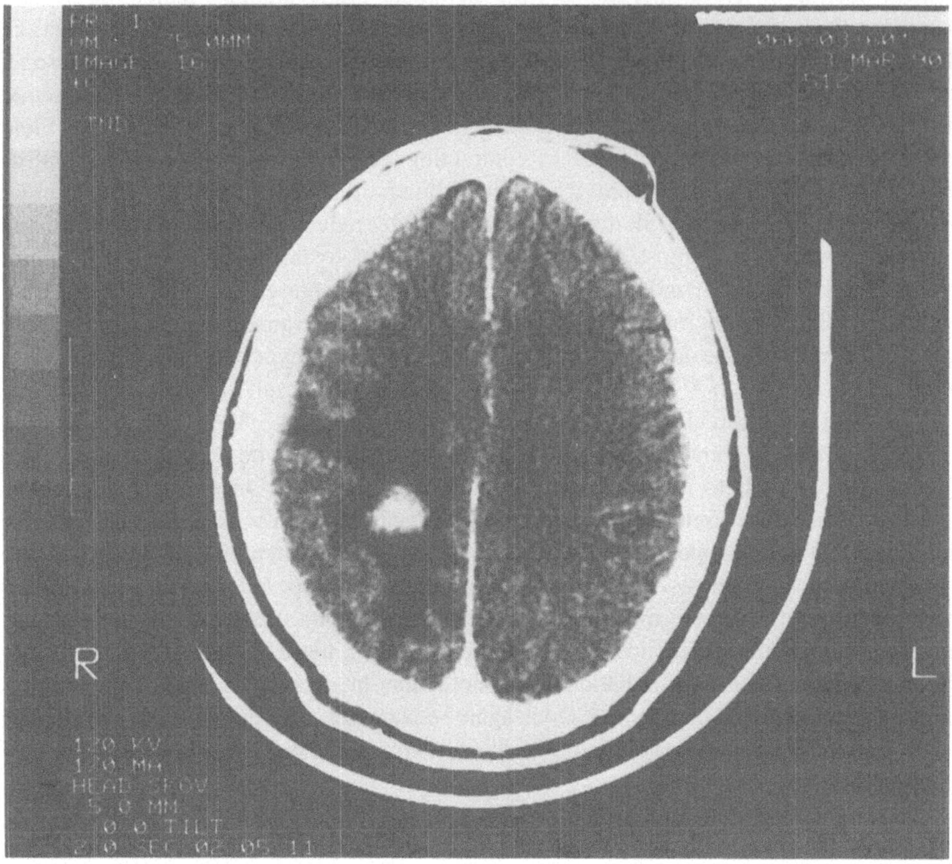

Figure 2. (A) Contrast-enhanced CT scan of a 60-year-old person with lung cancer and a history of progressive left-sided weakness and headache. A small, discrete, brightly enhancing lesion is evident in the right parietal lobe. Equally apparent is the extensive low-attenuation region surrounding this lesion. These findings are quite consistent with a metastatic brain tumor which in this instance is derived from a lung primary. The patient was started on high-dose dexamethasone therapy prior to craniotomy for tumor excision. (B) Contrast-enhanced CT scan performed 72 hr after initiation of dexamethasone therapy. The patient was neurologically normal at this time. The complete absence of contrast enhancement represents a dexamethasone-induced increase in blood–brain barrier integrity. Actual resolution of the edema will occur in a more delayed fashion. The patient underwent a successful total excision of this tumor and was neurologically intact postoperatively.

Furthermore, attention has recently been focused on the potential role of oxygen free radicals in the formation of traumatic and peritumoral brain edema.[43,44]

B. Initial Studies of Tumor-Derived Vascular Permeability Factors

A series of proteinaceous vascular permeability factors (VPFs) have stirred considerable interest in the past decade.[14,17–19,47,50,53,68,76,78–80,86] The existence of VPF was first

B

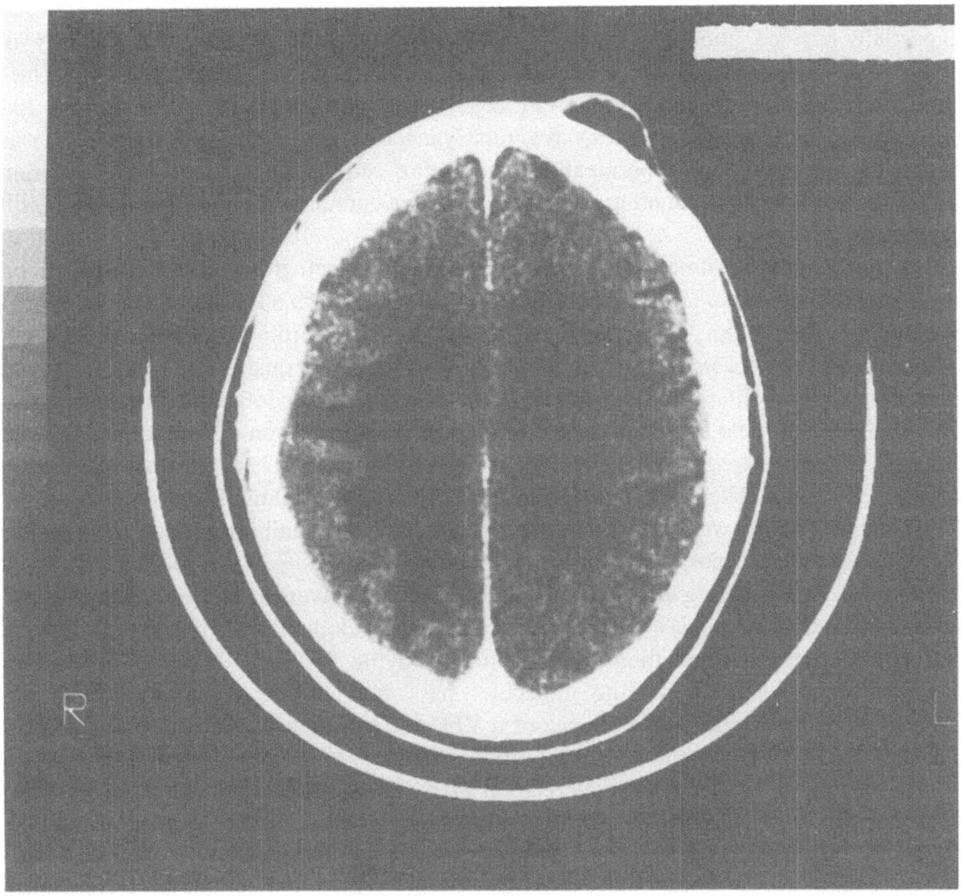

Figure 2. (*continued*)

recognized to be expressed by a guinea pig hepatocarcinoma (line 10) which promoted the accumulation of ascites fluid.[78] This factor (gVPF) was found to be a 34- to 42-kDa basic protein with a strong affinity for immobilized heparin, and a permeability-inducing potential three orders of magnitude greater than histamine (when compared on a molar basis). The presence of gVPF activity has been routinely determined by the Miles assay, a quantitative bioassay for induction of microvascular permeability.[62] Investigators found that gVPF-induced microvascular extravasation was rapid in onset (1 min) and of short duration (20 min), suggesting a direct action upon the microvascular endothelial cell. Furthermore, gVPF activity induced a period of desensitization whereby microvessels previously exposed to this substance became temporarily refractory (for less than 30 min) to further permeability induction. Light and electron microscopy showed that gVPF did not induce endothelial cell damage or mast cell degranulation when injected into the guinea pig subcutaneous microvasculature. Furthermore, intravascular injection of colloidal carbon resulted in the labeling of postcapillary venules, thus supporting an action of gVPF upon these critical microvessels.[58,59,85] Rabbit-derived immunoglobulin (polyclonal IgG) raised to partially purified guinea pig hepatocarcinoma-derived gVPF neutralized essentially all permeability-inducing activity present in tumor ascites fluid and tumor conditioned medium from that cell line. In addition, VPF activity from guinea pig fibrosarcoma (line 104 C1), and Walker rat carcinoma lines was similarly inactivated by anti-gVPF antibody. It was suggested that secretion of permeability-increasing activity (VPF) may be a common feature of tumor cells accounting for the abnormal accumulation of fluid associated with neoplasms.[78,79]

Soon thereafter, a tumor-derived capillary endothelial cell growth factor was identified and purified from rat chondrosarcoma extracellular matrix.[84] This 18-kDa cationic polypeptide had a marked affinity for heparin, and therefore differed from platelet-derived growth factor (PDGF) and epidermal growth factor (EGF). Nanogram quantities of this substance stimulated both proliferation and mobilization of capillary endothelial cells in culture; activities thought to be key components of angiogenesis induction. Other studies showed that while heparin potentiated tumor-induced angiogenesis on the chorioallantoic membrane, angiogenesis was inhibited when both heparin and cortisone were administered simultaneously. Thus, glucocorticoids were once again identified as having an inhibitory action on a specific process essential to the biology of tumor growth and metastasis.[26,27] At this point, investigators interested in factors acting directly upon endothelial cells diverged into two groups: those studying tumor angiogenesis factors (TAFs) and other endothelial, fibroblast, and smooth muscle cell mitogens; and those investigators studying VPFs.[14,18,19,26,27,41,78–80,84,88,101] Interest in these heparin binding molecules has only recently reconverged with the publication of the amino acid sequence of a VPF molecule exhibiting both permeability-inducing and angiogenic mitogen capabilities.[17,47,50]

Partial purification and characterization of a human colon adenocarcinoma-derived VPF (line HT-29) showed it to be a 45-kDa acidic protein which lacked a specific affinity for heparin.[53] Like guinea pig hepatocarcinoma gVPF, HT-29 permeability-inducing activity was not inhibited by antihistamines, antikinins, pepstatin A, or indomethacin. Several other human tumor lines have since been tested for VPF activity. One study demonstrated significant VPF expression by human osteogenic sarcoma, bladder carcinoma, cervical carcinoma, and fibrosarcoma. VPF activity from these tumors comigrated nearly identically with guinea pig line 10 gVPF (34 to 42 kDa) and was neutralized by antibodies raised to line 10 VPF. Furthermore, a VPF expressed by two human tumorigenic cell lines was also characterized by heparin affinity and a size of 38 kDa.[79] Demonstration of the highly conserved nature of VPF molecules across species lines suggested a broader purpose for VPF-mediated permeability induction by tumors (i.e., extravasation of nutrients to support tumor growth in the extracellular matrix).[2,26,27,79]

C. Identification of Human Brain Tumor VPF[14,18,19]

Serum-free conditioned media derived from cultures of primary human malignant brain tumors (anaplastic astrocytoma and glioblastoma multiforme) have been shown to contain a factor that induces microvascular extravasation of fluid and protein in the Miles permeability bioassay.[14] VPF activity was not expressed by cultured human fibroblasts, meningioma cells, or several lines of low-grade astrocytomas (gliomas not usually associated with CT-evident brain edema); however, it was strongly expressed by a number of rapidly growing malignant gliomas (associated with CT-evident edema). Human glioma-derived vascular permeability factor (HG-VPF) appears to differ from commonly recognized mediators of vascular permeability. Early studies did, however, demonstrate several similarities to guinea pig line 10 hepatocarcinoma-derived gVPF. HG-VPF is a 41- to 56-kDa basic cationic polypeptide, with an avid affinity for heparin.[18,19] Like gVPF, the time course of permeability induction by HG-VPF is rapid and transient suggesting a direct action upon endothelial cells. As in previous gVPF studies, prior exposure to HG-VPF results in a period of refractoriness, during which further induction of microvascular extravasation by HG-VPF does not occur. Expression of HG-VPF activity is inhibited by incubation of glioma cell cultures with cycloheximide or dexamethasone. Finally, a polyclonal antibody raised to gVPF completely inactivated HG-VPF activity and further supported homology between these substances.[14]

In order to eliminate variations in HG-VPF expression associated with use of several different tumor lines, serum-free conditioned media derived from low-passage cultures of a single cloned human malignant glioma line (U251) were employed for subsequent HG-VPF studies. A partially purified U251 glioma protein product containing HG-VPF has been shown to induce rapid and reversible elevations in cytosolic calcium in several types of cultured endothelial cells.[19] The HG-VPF-induced intracellular calcium ion changes exhibited dose-responsiveness and were inhibited by nonspecific calcium channel blockers (lithium, cobalt, manganese, and lanthanum ions) suggesting that influx of extracellular calcium ions was responsible for the observed cytosolic calcium transients. Reexposure of endothelial cells to the HG-VPF stimulus failed to produce a second calcium ion transient, suggesting that a period of refractoriness similar to that observed *in vivo* had been induced by the initial HG-VPF exposure. Moreover, HG-VPF-induced calcium changes were completely inhibited in endothelial cells previously exposed to dexamethasone.

D. Purification and Sequencing of Human VPF[17,47,78–80]

A VPF has recently been purified to homogeneity from serum-free conditioned medium of human histiocytic lymphoma cell line U937.[17,47,80] The cDNA sequence of human VPF (hVPF) from this line was shown to code for a 189-amino-acid 38- to 40-kDa polypeptide with two identical 20- to 24-kDa subunits that became evident on reducing gels (SDS–PAGE). Available data suggest that native hVPF is a dimer composed of disulfide-linked subunits, each with the same NH_2-terminal amino acid sequence. The NH_2-terminal region of the predicted amino acid sequence of U937 (human) hVPF was found to be 78% identical to the analogous region of line 10 (guinea pig) VPF. The hVPF molecule also bore some similarities in structure to the B chain of platelet-derived growth factor (PDGF-B),[73] as well as several PDGF/v-*sis* oncogene-related proteins. Similarly, a heparin-binding vascular endothelial cell growth factor (VEGF) has recently been purified from conditioned media derived from bovine pituitary folliculostellate cells and phorbol ester-activated human (HL60) promyelocytic leukemia cells.[50] The cDNA sequence of human VEGF is similar to that of hVPF except for an additional 24 amino acids in the hVPF sequence. VPFs with similar immuno-cross-reactivity have been detected in conditioned media derived from a variety of human and rodent tumor cell lines. In addition to its permeability-inducing activity, hVPF also appears to specifically stimulate endothelial cell growth and angiogenesis.[17,47,50,80] The hVPF molecule did not stimulate [^{3}H]thymidine incorporation or promote growth of other nonendothelial cell types. This feature distinguished it from several other endothelial cell growth factors, such as the heparin-binding fibroblast growth factors (FGFs), which promote growth and replication in nonendothelial, as well as endothelial cell lines. Conversely, several endothelial cell growth factors (EGF, bFGF, aFGF, TGF, IL-1, TNF, and PDGF) failed to exhibit VPF-like permeability-inducing activity when tested in the Miles assay. Other studies have confirmed the permeability-inducing activity of IL-2 but this protein bears no size similarity or sequence homology to the hVPF molecule. [^{125}I]-hVPF was shown to bind specifically and with high affinity to endothelial cells *in vitro*, and could be chemically cross-linked to a high-molecular-weight cell surface receptor, thus suggesting a specific site of interaction between hVPF and the vascular endothelial cell. Complex formation was blocked by excess unlabeled hVPF and anti-hVPF serum, but not by the addition of excess quantities of the aforementioned growth factors. Permeability induction in the Miles assay was observed with as little as 1 nM (8 ng) hVPF.

V. BIOCHEMICAL AND PHYSIOLOGICAL STUDIES OF HUMAN BRAIN-TUMOR VPF

A. HG-VPF Activity, Expression, and Behavior in the Miles Assay[14]

Serum-free conditioned medium from low-passage confluent monolayer cell cultures of human malignant astrocytoma lines evoked macromolecular extravasation (quantitated by measurement of [^{125}I]-BSA accumulation) in the Miles cutaneous microvascular permeability assay.[14] Intradermal injection of this factor results in a permeability change with a rapid onset at 1 to 5 min, a peak at 5 to 15 min, and reversibility by 20 to 30 min. In addition, there is a characteristic period of desensitization or tachyphylaxis whereby injection of HG-VPF into previously exposed sites results in no further induction of permeability. Conditioned medium from benign astrocytoma, meningioma, and fibroblasts demonstrated no significant VPF activity. Fluid aspirated from a cystic glioblastoma contained very high HG-VPF activity, whereas no activity was evident in samples of CSF from a normal volunteer, a patient with a sacral chordoma, or a patient with a malignant cerebral glioma (Table 1). HG-VPF activity increased as the duration of culture incubation was lengthened (Fig. 3).

B. Characterization of HG-VPF[18]

HG-VPF is an acid-stable heat-labile 41- to 56-kDa polypeptide, which is hydrophobic and positively charged under physiological conditions.[18] HG-VPF activity was abolished by treatment with trypsin or pepsin, but was unaffected by ribonuclease A, chondroitinase A,B,C, hyaluronidase, and lipase. Similarly, HG-VPF activity is not inhibited by soybean trypsin inhibitor or hexadimethrine (both known antagonists of tissue plasminogen activator, Hageman factor, and serum kallikrein); or aprotinin (an antagonist of both plasmin and tissue kallikrein); or phenylmethanesulfonyl fluoride [a serine esterase (elastase) inhibitor];

Table 1. Expression of Vascular Permeability Factor (VPF) Activity[a]

Source of test sample	VPF activity
Experiment A	
Control (DMEM)	2.4 ± 4.0
Fibroblast	7.4 ± 4.3
Meningioma 1	5.9 ± 3.2
Meningioma 2	7.7 ± 4.0
Astrocytoma	15.2 ± 5.8
Glioblastoma 1	25.7 ± 3.9
Glioblastoma 2	50.3 ± 12.9
Experiment B	
Control (DPBS)	13.3 ± 7.9
CSF (normal volunteer)	7.2 ± 5.5
CSF (sacral chordoma)	8.6 ± 5.1
CSF (malignant glioma)	5.1 ± 3.9
Cyst fluid (glioblastoma) 1	365.2 ± 80.6
Cyst fluid (glioblastoma) 2	643.3 ± 16.6

[a]The VPF activity for this series of Miles assays is expressed as mean counts per min per mg tissue ± S.E.M. Experiment A: VPF activity in conditioned medium from human tissue lines; experiment B: VPF activity in cerebrospinal fluid (CSF) and cystic glioblastoma fluid. DMEM, Dulbecco's modified Eagle's medium; DPBS, Dulbecco's phosphate-buffered saline. (Reprinted after modification from Bruce *et al.*,[14] with permission.)

or pepstatin A (an acid protease inhibitor which inactivates vascular permeability-inducing leukokinins). Treatment of HG-VPF with dithiothreitol abolished permeability-inducing activity, indicating the presence of at least one essential disulfide bond in this molecule. VPF displays a marked affinity for immobilized heparin (heparin-Sepharose, CL-6B). In addition, we found 90–95% binding of sample activity to hydrophobic resin (phenyl-Sepharose, CL-4B), and hydroxylapatite. While 40–45% of HG-VPF activity was bound

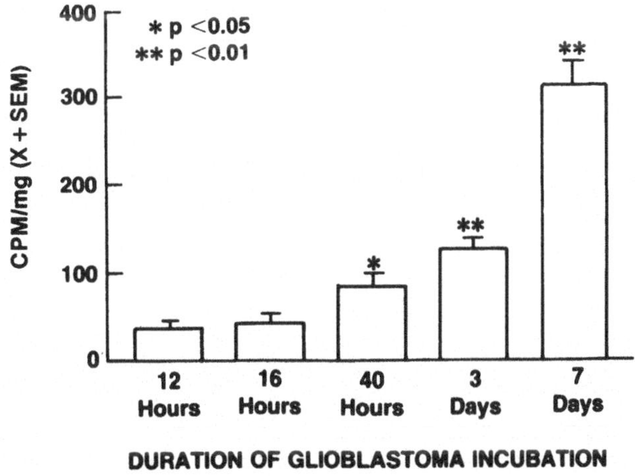

Figure 3. Histogram showing induction of vascular extravasation of ^{125}I-labeled bovine serum albumin, as a function of cell culture incubation time for glioblastoma line 1 samples. Vascular permeability factor (HG-VPF) activity is expressed as mean counts per minute (CPM)/mg + standard error of the mean (SEM). (Reprinted from Bruce *et al.*,[14] with permission.)

by negatively charged resin (carboxymethyl-Sepharose), only 5% of sample activity was bound to positively charged resin (diethylaminoethyl-Sephacel). The latter resin sequestered 80–85% of sample protein. Employing a heparin-Sepharose column and a NaCl gradient generator, it was determined that peak elution of HG-VPF activity occurs at a salt concentration of 0.45 N NaCl. Specific activity was increased 26-fold by this step. When this partially purified sample was applied to an HPLC sizing column, HG-VPF activity eluted in the 41- to 56-kDa fraction. Furthermore, an additional 25-fold increase in specific activity was realized with this step, thus increasing the total approximate purification to 1000-fold (Table 2).

C. Dexamethasone Inhibition of HG-VPF Expression and Activity[18]

Studies to determine how dexamethasone affects HG-VPF expression by cultured cells failed to show a direct toxic effect as measured by cell viability (> 98% by trypan blue exclusion), and final cell counts.[18] Although HG-VPF expression was significantly inhibited by the presence of glucocorticoid, this effect was not associated with a general inhibition of cellular protein synthesis by steroid (Table 3). We had previously determined that treatment of test animals with dexamethasone immediately before performing the permeability assay did not confer protection against HG-VPF-induced extravasation. Coinjection of dexamethasone with HG-VPF also failed to impart a membrane stabilizing protective effect as extravasation was unhindered. In fact, significant inhibition of HG-VPF activity *in vivo* only occurred in test animals given dexamethasone at least 1–2 hr before performing the permeability assay (Fig. 4). To determine whether the inhibitory action of dexamethasone was mediated by *de novo* protein synthesis, a series of *in vivo* studies were performed whereby actinomycin D was administered 2 hr prior to dexamethasone treatment. *In vivo* inhibition of HG-VPF activity by dexamethasone was found to be partially reversed in animals given actinomycin D prior to steroid. This suggested that *de novo* synthesis of a specific polypeptide intermediate is required for the protective effect of dexamethasone to

Table 2. Steps Required to Partially Purify Brain Tumor-Derived Vascular Permeability Factor[a]

Purification step	Volume (ml)	Total activity recovered	Protein activity	Specific purification	Degree of (-fold) recovery (%)	Activity
1. Conditioned serum-free medium	2000	9.78 × 10^6 dpm	200 mg (100%)	49 dpm/μg	—	100
2. Dialysis and lyophilization	30	9.92 × 10^6 dpm	60 mg (30%)	165 dpm/μg	3.4×	101
3. Heparin affinity	250	1.59 × 10^6 dpm	375 μg (2%)	4,244 dpm/μg	87×	16
4. Ultrafiltration	2.5	3.90 × 10^6 dpm	<25 μg (0.01%)	15,600 dpm/μg	318×	4
5. HPLC (TSK2000)	1.0	9.16 × 10^6 dpm	<20 μg (0.01%)	45,800 dpm/μg	935×	9.4

[a]Reprinted after modification from Criscuolo *et al.*,[18] with permission.

Table 3. Effect of Dexamethasone on VPF Expression, Cell Number, Cell Viability, and Protein Synthesis[a,b]

Period of incubation	Dexamethasone	VPF activity[c]	Cell number/well[d]	Protein synthesis[e]
2 days	0 (control)	100%	6.59 ± 0.026	154 ± 6
	10^{-11} M	100%	6.23 ± 0.156	163 ± 7
	10^{-9} M	88%	6.75 ± 0.243	168 ± 13
	10^{-7} M	54%	6.83 ± 0.157	165 ± 7
	10^{-5} M	28%	7.39 ± 0.266	162 ± 6
4 days	0 (control)	100%	6.81 ± 0.620	130 ± 18
	10^{-11} M	70%	7.53 ± 0.610	140 ± 17
	10^{-9} M	69%	7.32 ± 0.264	122 ± 14
	10^{-7} M	68%	8.64 ± 0.164	99 ± 5
	10^{-5} M	54%	7.77 ± 0.303	134 ± 4
7 days	0 (control)	100%	6.56 ± 0.289	126 ± 7
	10^{-11} M	94%	6.62 ± 0.384	131 ± 7
	10^{-9} M	100%	6.55 ± 0.341	138 ± 4
	10^{-7} M	53%	6.96 ± 0.714	129 ± 5
	10^{-5} M	48%	7.55 ± 0.136	118 ± 4

[a]Reprinted after modification from Criscuolo *et al.*,[18] with permission.

[b]Cell viability was > 98% by trypan blue exclusion. Values are means ± S.E.M.

[c]The S.E.M. for VPS activity measurements was determined to be 7.5%. The control VPS activity values (0 dexamethasone) for days 2, 4, and 7 were 1294, 7677, and 11,555 dpm, respectively.

[d]Cell counts were performed in quadruplicate using a Coulter counter and are expressed as multiples of 10^5 cells.

[e]Protein synthesis is expressed as mean picomoles of [^{3}H]leucine incorporated per hour per milligram protein ± S.E.M. for four samples.

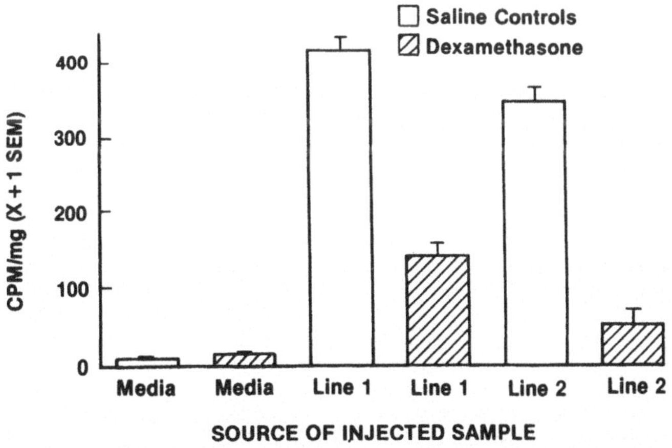

Figure 4. Histogram showing the inhibitory influence of dexamethasone on the *in vivo* activity of vascular permeability factor (HG-VPF) derived from glioblastoma multiforme lines 1 and 2. These findings were determined by the Miles assay. The VPF activity is expressed as mean counts per minute (CPM)/mg + 1 standard error of the mean (SEM). (Reprinted from Bruce *et al.*,[14] with permission.)

occur, thereby strongly arguing against nonspecific steroid-induced membrane stabilization as the key physiological event (Table 4).

D. HG-VPF Effect on Endothelial Cell Cytosolic Calcium Ion Concentration[19]

Recent advances in our understanding of endothelial cell growth requirements have allowed their routine culture for experimental purposes.[5,9,10,20,32,45,46,103] Endothelial cells derived from brain capillaries have received considerable attention because of their unique barrier characteristics.[5,9] Cytosolic free Ca^{2+} plays a pivotal role in endothelial cytoskeletal alterations and subsequent microvascular extravasation.[22,39,52,72,82] Studies of cytosolic calcium changes have been greatly facilitated by the development of a series of novel fluorescent calcium ion probes (quin-2/AM, indo-2/AM, fura-2/AM, fura-3/AM) possessing molecular structures similar to EGTA.[37,60,92] They differ from EGTA in that they contain aromatic rings capable of electrostatic interactions with the functional groups that participate in the chelation of free cytosolic calcium ions $[Ca^{2+}]_i$ Chelation of divalent cations changes the fluorescence and ultraviolet light-absorption properties of these molecules. Alterations in fluorescence are in turn detected and quantitated by a spectrofluorometer. Fura-2/AM has proven particularly useful with its high sensitivity and specificity for calcium ions. Because of its lipophilicity, fura-2/AM is rapidly internalized by endothelial cells. Once internalized, however, deesterification of the acetoxy-methyl group (fura-2/AM → fura-2) converts it to a lipophobic free acid, with only minimal capability for diffusion out of the cell. Cytosolic calcium ion changes in the nanomolar range are readily detectable with this probe. Partially purified HG-VPF has been shown to induce significant calcium ion transients in several varieties of endothelial cells sustained in monolayer cultures.[19] It did not, however, elicit a calcium response in two nonendothelial cell lines (U251 glioma and fibroblasts). The endothelial cytosolic calcium changes were typically rapid or only slightly delayed in onset (15–45 sec), and varied in magnitude depending upon the cell type being studied (Fig. 5). The largest responses were observed in human endothelial cells. It

Table 4. Effect of Actinomycin D on Dexamethasone-Induced Inhibition of Vascular Permeability Factor Activity[a]

Experimental group (20 animals)	$[^{125}I]$-BSA extravasation (mean ± S.E.) Brain tumor VFP	$[^{125}I]$-BSA extravasation (mean ± S.E.) Histamine	% inhibition by dexamethasone Brain tumor VPF	% inhibition by dexamethasone Histamine
Control (n = 5)	22,000 ± 2,427	7977 ± 1005	0%	0%
Dexamethasone (n = 5)	3,160 ± 529	2357 ± 314	86%	71%
Actinomycin D (n = 5)	11,412 ± 860	1623 ± 138	0%	0%
Actinomycin D + dexamethasone (n = 5)	6,976 ± 878	585 ∓ 193	39%	64%

[a]Reprinted after modification from Criscuolo *et al.*,[18] with permission.
$p < 0.0005$ (control versus dexamethasone).
$p = 0.006$ (dexamethasone versus actinomycin D + dexamethasone).
$p = 0.001$ (control versus dexamethasone).
$p = 0.001$ (dexamethosone versus actinomycin D + dexamethasone).

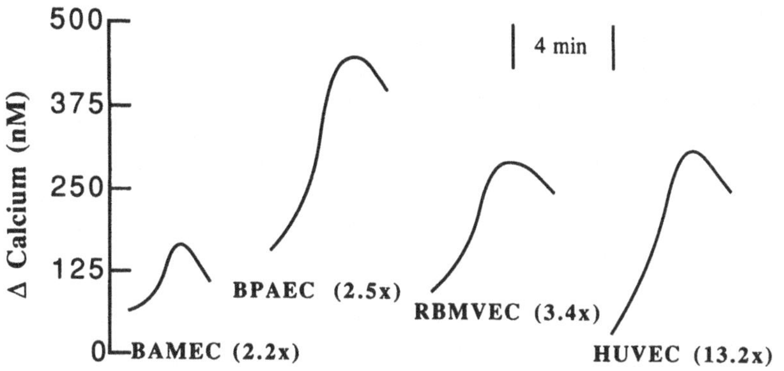

Figure 5. Induction of cytosolic calcium transients in a variety of endothelial cell lines by partially purified human malignant glioma-derived vascular permeability factor (HG-VPF). The relative degree of intracellular calcium change induced by a standard HG-VPF stimulus is indicated in parentheses. All graphic data were obtained by computerized tracing of original hard-copy data after reassignment of *y*-axis values by conversion to absolute Ca^{2+} concentrations. BAMEC, bovine adrenal medullary endothelial cells; BPAEC, bovine pulmonary artery endothelial cells; RBMVEC, rat brain microvessel endothelial cells; HUVEC, human umbilical vein endothelial cells. Note that the largest calcium responses to HG-VPF occurred in human venous endothelial cells. The possible significance of this finding is discussed in the text. (Reprinted from Criscuolo *et al.*,[19] with permission.)

has already been postulated that such variations may relate to some degree of molecular species-specificity despite the highly conserved nature of VPF activity.[79] Peak HG-VPF-induced intracellular calcium ion elevations were observed within 60 sec, exhibited a dose–response phenomenon, and were followed by a sustained elevation above baseline for 5 to 10 min thereafter (Fig. 6). In contrast, exposure of the endothelial cells to the flow-through product of HG-VPF containing glioma-conditioned medium after binding of all HG-VPF activity to a heparin-Sepharose affinity column, did not produce a change in intracellular calcium. The brief time course of the HG-VPF-induced cytosolic calcium changes *in vitro* is compatible with the *in vivo* kinetics, and is supported by other data showing prolonged stimulus–response coupling to occur after only transient intracellular calcium elevations.[72]

Exposure of endothelial cell suspensions to elevated extracellular potassium chloride (120 mM KCl) failed to induce cytosolic calcium transients. Furthermore, verapamil (10 μM) failed to inhibit HG-VPF-induced calcium changes. This finding suggests that HG-VPF increases cytosolic calcium by a route other than verapamil-sensitive voltage-gated calcium channels (Fig. 7). Addition of 2 mM lithium, cobalt, manganese, or lanthanum ions inhibited HV-VPF-induced calcium transients (Fig. 8), as did the absence of extracellular calcium after chelation with EGTA. This indicates that HG-VPF-induced calcium ion changes are largely dependent upon the influx of extracellular calcium $[Ca^{2+}]_e$ through membranous calcium channels. Finally, incubation of endothelial cells for 2 hr with 10 μM dexamethasone before exposure to HG-VPF-containing medium resulted in inhibition of the HG-VPF-induced cytosolic calcium changes (Fig. 9). This inhibition was not observed in similarly treated cells exposed to ATP, nor did it occur when cells were incubated with dexamethasone for less than 1 hr; suggesting that the steroid-induced inhibition was specific for HG-VPF-induced calcium flux, and that dexamethasone may mediate its effect through an intermediary requiring *de novo* synthesis.

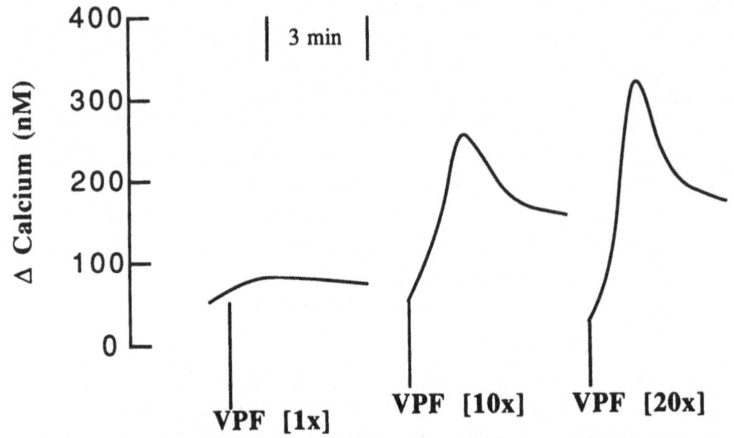

Figure 6. Dose-response relationship between human malignant glioma-derived vascular permeability factor (HG-VPF) stimulus and cytosolic calcium transients induced in human umbilical vein endothelial cells. Fold concentration of HG-VPF is expressed as multiples of a standard solution containing 0.25 mg/ml (1×) of partially purified lyophilized glioma-conditioned medium dissolved in Dulbecco's phosphate-buffered saline (2.5 mg/ml = 10×, 5.0 mg/ml = 20×). (Reprinted from Criscuolo *et al.*,[19] with permission.)

Another polycationic substance (protamine), known to induce changes in vascular permeability,[87,97,98] was also shown to induce physiological changes in endothelial intracellular calcium, when present in concentrations as low as 10 μg/ml. Induction of calcium transients by protamine was rapid in onset, peaked within 30 sec, and sustained a level above baseline for several minutes thereafter. The presence of extracellular divalent cations (2 mM cobalt or manganese) reduced peak $[Ca^{2+}]_i$, but also more completely affected the

Figure 7. Failure of verapamil to inhibit human glioma-derived vascular permeability factor (HG-VPF)-induced calcium transients. Prior exposure of endothelial cells to 10^{-6} M verapamil did not diminish cytosolic calcium changes induced by HG-VPF. Calcium transients were not induced by exposure of endothelial cells to 120 mM KCl (data not shown). This suggests that HG-VPF-induced influx of $[Ca^{2+}]_e$ occurs by a mechanism other than voltage-gated calcium ion channels. (Reprinted from Criscuolo *et al.*,[19] with permission.)

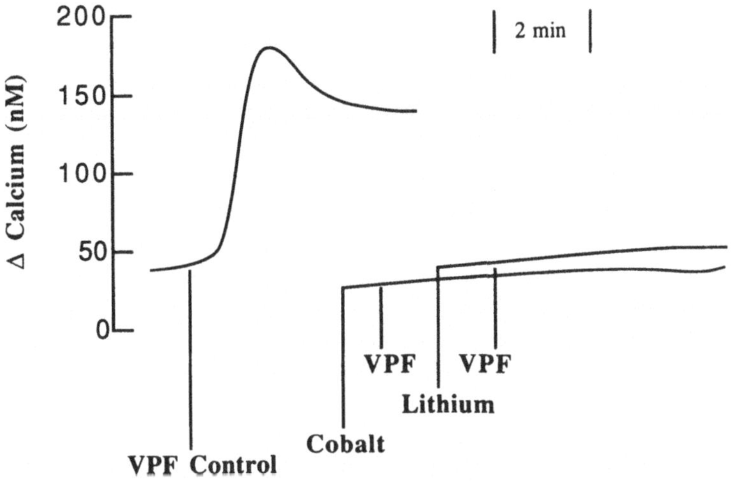

Figure 8. Inhibition of human glioma-derived vascular permeability factor (HG-VPF)-induced calcium ion transients by 2 mM Li^{+} and Co^{2+} cations. Inhibition also occurred with 2 mM concentrations of Mn^{2+} and La^{3+} cations (data not shown). Nonspecific cationic calcium channel blockers appear to inhibit HG-VPF-induced intracellular calcium ion transients. Similarly, chelation of extracellular calcium ions with EGTA partially inhibited induction of calcium transients by HG-VPF (data not shown). This finding suggests that influx of extracellular calcium via non-voltage-gated membranous channels may be the primary event in the HG-VPF-elicited response. There is some indication that mobilization of sequestered intracellular calcium stores $[Ca^{2+}]_{is}$ may also play a role as a small but significant calcium transient was generated despite complete chelation of extracellular calcium by EGTA (data not shown). (Reprinted from Criscuolo *et al.*,[19] with permission.)

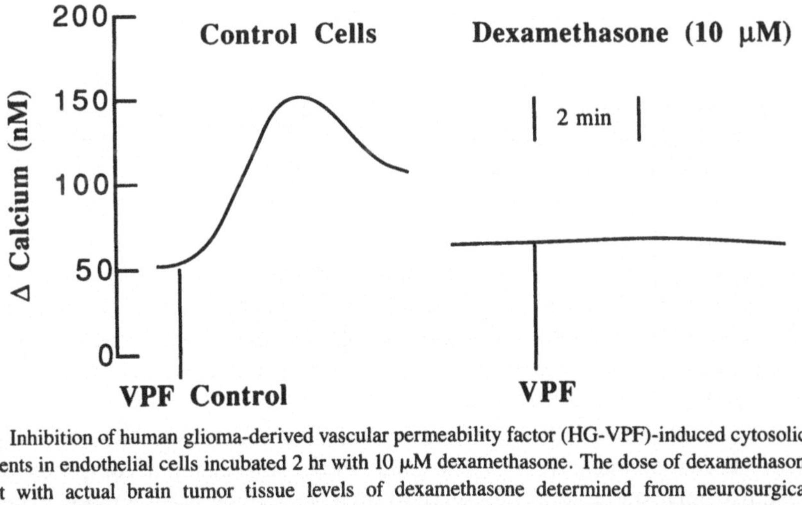

Figure 9. Inhibition of human glioma-derived vascular permeability factor (HG-VPF)-induced cytosolic calcium ion transients in endothelial cells incubated 2 hr with 10 μM dexamethasone. The dose of dexamethasone used is consistent with actual brain tumor tissue levels of dexamethasone determined from neurosurgical biopsy specimens. This finding supports *in vivo* findings that dexamethasone will inhibit microvascular permeability induced by HG-VPF when test animals are given a steroid dose 1–2 hr before injection of HG-VPF samples. (Reprinted from Criscuolo *et al.*,[19] with permission.)

postmaximum levels resulting in a rapid downslope and normalization of intracellular calcium. These findings indicated that combined mobilization of intracellular calcium stores (rapid upslope and peak levels), as well as influx from the extracellular compartment (peak and sustained levels), are involved in the response to HG-VPF and protamine. It also suggests that neutralization of endothelial cell surface charge (anionic) by polycationic substances (VPFs, protamine, angiogenesis factors) may play a role in mediating the actions of these substances. This is supported by ultrastructural evidence that exposure of renal glomeruli to protamine results in disorganization of the epithelial cell foot process architecture (podocytes) investing the microvascular glomeruli, and subsequent leakage of 2protein into the urine. Endothelial surface charge neutralization by polycations may simply eliminate the electrostatic barrier that exists to the vast majority of proteins (negatively charged), allowing them to escape through pores in fenestrated vessels.

VI. SOME THOUGHTS ON THE UNIQUE RELATIONSHIP BETWEEN VPFs, THE BRAIN TUMOR MICROVASCULATURE, DEXAMETHASONE, AND NEOPLASTIC VASOGENIC BRAIN EDEMA

The microvasculature of brain tumors displays several distinctive ultrastructural features that allow ready distinction from normal cerebral microvessels. The presence of widened endothelial cell junctions, discontinuous tight junctions, cellular membrane fenestrations, noncontiguous basement membranes, active micropinocytosis, and paucity of mitochondria, sharply contrast with normal blood–brain barrier architecture.[1,12,36,55–57,64] These features, which also typify endothelium lining the normal peripheral vasculature, have previously been associated with the water and protein extravasation, and consequent cerebral edema that occurs with malignant brain tumors.[13,48,105] Like blood–brain barrier microvessels, however, the normal peripheral vasculature is not inherently permeable to macromolecules. Nevertheless, tumors occurring outside the CNS also exhibit microvascular extravasation that is clinically manifested as malignant effusions, ascites, and tissue edema.[1,78,86,93]

The normal peripheral vasculature is exquisitely sensitive to permeability induction by histamine, bradykinin, serotonin, prostaglandins, and a variety of other physiological substances.[62,65] In contradistinction, normal cerebral microvessels are totally incapable of responding to these substances when tested *in situ*. Similarly, HG-VPF, which clearly is active in the guinea pig and rat cutaneous microvasculature, failed to evoke extravasation of fluorescein–albumin, [^{125}I]-bovine serum albumin, or edema fluid as evidenced by tissue specific gravity determinations, when injected into normal rat brain (unpublished observations of G. R. Criscuolo and E. H. Oldfield). However, injection of HG-VPF into C6 gliomas that had been implanted into rat brains showed brain tumor microvessels to be capable of responding with increased permeability *in situ* (unpublished observations of G. R. Criscuolo and E. H. Oldfield).

Vascular endothelial cells from peripheral and CNS sources are known to possess a complex cytoskeletal architecture.[22,82] The endothelial cytoskeleton is composed of actin, myosin, and tropomyosin proteins that serve to regulate cellular configuration. Therefore, cellular mobility and vascular permeability are processes that are likely to result from changes in intracellular calcium ions that occur in response to angiogenesis factors and VPFs. Stimuli that induce cytosolic calcium transients and change endothelial cell shape, ultimately result in opening of interendothelial junctions at the level of the postcapillary venule, which in turn results in vascular extravasation.[52,63,85,89] Histamine-induced cytosolic calcium changes have been shown to correlate with changes in endothelial F-actin content, and passage of albumin across endothelial monolayers derived from human umbilical veins.[72] In fact, many of the commonly recognized mediators of microvascular permeability have been shown to induce transient elevations in endothelial cytosolic calcium, irrespective of the variety of endothelial cells tested.[19,31,35,52,60,72,89] Significantly, normal brain microvascular endothelial cells are equally capable of responding with calcium ion transients, and, by conjecture, cytoskeletal alterations that would culminate in vascular extravasation. These findings, and the morphological similarities between brain tumor vessels and the normal peripheral vasculature, suggest an alternative hypothesis for the induction of vascular permeability in brain tumors.

Perhaps the reason blood–brain barrier microvessels are incapable of responding to permeability mediators *in situ*, solely relates to their interendothelial junctions being physically joined together by continuous pentalaminar tight junctions. The extreme integrity of this junction is supported by experimental studies of osmotic disruption of the blood–brain barrier by mannitol, wherein investigators showed the interendothelial clefts to be closed, and the tight junctions to be continuous and intact after exposure to that potent physical agent: opening of the barrier by mannitol was associated with an increased transgression of micropinocytotic vesicles between the luminal and abluminal endothelial surfaces.[24] Further support derives from the knowledge that peritumoral brain edema emanates directly from the tumor bed, and is not the result of HG-VPF-induced extravasation in the surrounding blood–brain barrier microvessels.[105] Therefore, one may reasonably put forth the hypothesis that an alteration or absence of such a critical component of blood–brain barrier as the interendothelial tight junction may at once render brain tumor microvessels similar to the peripheral microvasculature, both anatomically and with respect to their ability to respond to permeability mediators. As a result, brain tumor microvessels would be capable of responding to the HG-VPF secreted by surrounding tumor cells, and production of vasogenic cerebral edema would result.

Although the study of VPFs has already yielded valuable insight into a mechanism of tumor-mediated vascular permeability, many questions remain unanswered. Several questions specifically pertinent to neoplastic brain edema immediately arise: Why does peritumoral brain edema not follow an even more malignant course? If brain tumor microvessels remain constantly exposed to HG-VPF, and respond with a constant rate of plasma extravasation, why would a patient not succumb to a rampant increase in ICP within perhaps minutes, hours, or a few days at most? This may be partly explained by known mechanisms of brain edema resolution. Excess extracellular fluid normally travels through the brain interstitium in a centripetal direction (toward the CSF-containing ventricular system) by a passive process referred to as bulk flow. In addition, astrocytic cells are known to be actively involved in the imbibition of excess tissue fluid. This issue may also be partly explained by observations indicating that HG-VPF is capable of inducing a period of unresponsiveness, refractoriness, or tachyphylaxis, whereby consecutively applied stimuli will not result in cytosolic calcium changes or further vascular extravasation.[14,18,19,78,79]

Supporting this notion is the common occurrence of a period of relative unresponsiveness (refractory period) in many physiological cascades, as well as in a variety of physiologically excitable cells (neurons, photoreceptor cells, muscle cells).

The clinical efficacy of dexamethasone in the setting of peritumoral brain edema has been well described and documented.[29,42,48,54,69,104,106] We have been able to demonstrate several levels at which dexamethasone may specifically exert its beneficial effects in this focused pathophysiological setting.[14] Concentrations of dexamethasone that occur in our patients' brain tumors will specifically suppress production of HG-VPF by cultured human malignant glioma cells.[18] This finding alone suggests a powerful mechanism for steroid efficacy. In addition, a distinct and separate mechanism of steroid action may occur at the level of the key biological component for this process; the microvascular endothelial cell. This is supported by the finding of calcium transient suppression by dexamethasone in cultured endothelial cells.[19] Lastly, a glimpse into the mechanism by which dexamethasone affects target cells (glioma cells and endothelial cells) is provided by the following findings: (1) Coinjection of HG-VPF with dexamethasone did not alter the extent of vascular extravasation.[14] (2) Pretreatment of test animals with dexamethasone 1–2 hr before HG-VPF injection substantially inhibited vascular extravasation whereas pretreatment less than 1 hr before resulted in no inhibition of HG-VPF activity.[14,18] (3) Pretreatment of test animals with actinomycin D before dexamethasone exposure, resulted in significant, albeit slightly reduced, HG-VPF-induced extravasation.[18] The implication is that dexamethasone exerts its actions indirectly, by inducing *de novo* synthesis of a polypeptide intermediary, rather than by nonspecific membrane stabilization. This concept has been put forth as an explanation for the protective effect of dexamethasone in rats affected by global cerebral ischemia.[90] Furthermore, a second messenger polypeptide (variably referred to as "macrocortin," "lipocortin," "endocortin," or "renocortin") has been identified, characterized, and found to mediate the effect of glucocorticoids by inhibition of phospholipase A_2.[7] Induction of macrocortin synthesis requires steroid receptor occupation and *de novo* protein synthesis. It therefore appears possible that HG-VPF may act by inhibiting a rate-limiting enzyme in the prostaglandin cascade (phospholipase A_2).[18,19] This is corroborated by the work of others who have shown prostaglandins to be potent inducers of cytosolic calcium transients in endothelial cells.[30,31]

Additional investigations using a completely purified VPF product are clearly warranted, and should better define the kinetics, refractoriness to sequential stimulation, and dose responsiveness of VPF-induced microvascular extravasation. Furthermore, comparison of endothelial cells derived from different anatomical sources (brain, retina, lung, adrenal, umbilical cord), from differing sized vessels (arterial, arteriolar, capillary, venous, venular), and from several species (human, bovine, rodent), should determine the extent, and degree of specificity of HG-VPF activity with regard to these variables.[5,9,10,20,32,45,46,61,74,75,99] Information to date strongly favors a direct action of HG-VPF on the vascular endothelial cell. The rapid and reversible kinetics of HG-VPF activity *in vivo* and *in vitro*, its remarkable affinity for immobilized heparin, and its ability to induce significant changes in endothelial cytosolic calcium lend solid support to, if not affirmation of, this premise.[14,18,19] A well-defined negatively charged glycocalyx, composed primarily of sulfated glycosaminoglycans, is present on capillary endothelial cell surfaces.[2] This cell coat is largely composed of polyanionic heparan sulfate and sialic acid residues. Its primary function is thought to be the selective regulation of endothelial cell binding and transcapillary passage of macromolecules. The cationic nature of the HG-VPF molecule at physiological pH would therefore suggest a means for direct, albeit nonspecific, electrostatic attraction of the HG-VPF molecule to the endothelial cell surface. This in turn would likely facilitate the subsequent binding and interaction between the active site on the HG-VPF molecule and a specific membrane receptor.

The investigations outlined herein support an important role for HG-VPF in the expression of peritumoral brain edema. In addition, an explanation for the efficacy of dexamethasone in the treatment of symptomatic brain edema associated with cerebral neoplasms is strongly suggested. The clinical and experimental findings are also supported by the occurrence of steroid receptors in cerebral tumors,[33,49,104,106] and the lack of steroid efficacy in the setting of ischemic, toxic, hemorrhagic, and traumatic cerebral edema.[23,38,67] Inhibition of HG-VPF activity by glucocorticosteroids such as dexamethasone appears to be related to highly specific actions at the cellular level (i.e., inhibition of tumor cell expression of VPF, and inhibition of VPF-induced endothelial cell calcium responses). Perhaps a better understanding of the subcellular mechanisms involved in steroid-induced HG-VPF inhibition will suggest either worthwhile refinements in our use of these agents, or the means to develop novel methods that are safer and more effective in the treatment of patients with neoplastic vasogenic brain edema.

VII. CLOSING REMARKS

The magnitude of the brain tumor problem has been outlined. Some patients with malignant brain tumors have enjoyed extended functional survival relating largely to aggressive surgical excision, radiotherapy, and chemotherapy. Certain brain tumors (meningiomas) are considered histologically and biologically benign, and therefore ostensibly curable. Nevertheless, a great deal of perioperative neurological morbidity has been associated with excision of these lesions, and largely relates to vasogenic edema production by some meningiomas. One of the most significant therapeutic advances in neuro-oncology in the past two decades has been the recognition of the efficacy of corticosteroids in the symptomatic treatment of patients with malignant brain tumors. Their salutory effects on intracranial hypertension have resulted in better tolerance of both surgical therapy and radiotherapy. However, the extended use of high-dose corticosteroids is not without complications. Consequences of their long-term use include sodium and water retention, hypertension, diabetes mellitus, sepsis related to impaired immune function, connective tissue alterations, gastrointestinal ulceration, aseptic necrosis of the femur head, and altered mentation including mania, psychotic depression, and euphoria. Investigation into the precise mechanism by which steroids act upon peritumoral cerebral edema may eventually provide clinicians with safer and more effective therapeutic alternatives for patients afflicted with brain tumors.

REFERENCES

1. Ackerman, N. B., and Hechmer, P. A., 1978, Studies on the capillary permeability of experimental liver metastases, *Surg. Gynecol. Obstet.* **146**:884–888.

2. Alessandri, G., Raju, K. S., and Gullino, P. M., 1986, Interaction of gangliosides with fibronectin in the mobilization of capillary endothelium. Possible influence on the growth of metastases, *Invasion Metastasis* **6**:145–165.
3. Ammirati, M., Galicich, J. H., Arbit, E., and Liao, Y., 1987, Reoperation in the treatment of recurrent intracranial malignant gliomas, *Neurosurgery* **21**:607–614.
4. Baethmann, A., 1978, Pathophysiological and pathochemical aspects of cerebral edema, *Neurosurg. Rev.* **1**:85–100.
5. Betz, A. L., and Goldstein, G. W., 1980, Transport of hexoses, potassium and neutral amino acids into capillaries isolated from bovine retina, *Exp. Eye Res.* **30**:593–605.
6. Black, K. L., and Hoff, J. T., 1985, Leukotrienes increase blood–brain barrier permeability following intraparenchymal injections in rats, *Ann. Neurol.* **18**:349–351.
7. Blackwell, G. J., Carnuccio, R., Di Rosa, M., Flower, R. J., Parente, L., and Persico, P., 1980, Macrocortin: A polypeptide causing the anti-phospholipase effect of glucocorticoids, *Nature* **287**:147–149.
8. Bleehen, N. M., 1980, Intracranial tumors: Response and resistance to therapeutic endeavors, 1970–1980, *Int. J. Radiat. Oncol. Biol. Phys.* **8**:1083–1133.
9. Bowman, P. D., Betz, A. L., and Goldstein, G. W., 1979, Characteristics of cultured brain capillaries, *J. Cell Biol.* **83**:95a.
10. Bowman, P. D., Betz, A. L., Wolinsky, J. S., Penney, J. B., Shivers, R. R., and Goldstein, G. W., 1981, Primary culture of capillary endothelium from rat brain, *In Vitro* **17**:353–362.
11. Brightman, M. W., and Reese, T. S., 1969, Junctions between intimately apposed cell membranes in the vertebrate brain, *J. Cell Biol.* **40**:648–677.
12. Brightman, M. W., and Reese, T. S., 1970, Types of endothelium in normal and neoplastic brain tissue, *28th Annual Proceedings of the Electron Microscopic Society of America*, pp. 98–99.
13. Brightman, M. W., Klatzo, I., Olsson, Y., and Reese, T. S., 1970, The blood–brain barrier to proteins under normal and pathological conditions, *J. Neurol. Sci.* **10**:215–239.
14. Bruce, J. N., Criscuolo, G. R., Merrill, M. J., Moquin, R. R., Blacklock, J. B., and Oldfield, E. H., 1987, Vascular permeability induced by protein product of malignant brain tumors: Inhibition by dexamethasone, *J. Neurosurg.* **67**:880–884.
15. Casanova, M. F., 1984, Vasogenic edema with intraparenchymal expanding mass lesions: A theory on its pathophysiology and mode of action of hyperventilation and corticosteroids, *Med. Hypotheses* **13**:439–450.
16. Chan, P. H., and Fishman, R. A., 1984, The role of arachidonic acid in vasogenic brain edema, *Fed. Proc.* **43**:210–213.
17. Connolly, D. T., Heuvelman, D. M., Nelson, R., Olander, J. V., Eppley, B. L., Delfino, J. J., Siegel, N. R., Leimgruber, R. M., and Feder, J., 1989, Tumor vascular permeability factor stimulates endothelial cell growth and angiogenesis, *J. Clin. Invest.* **84**:1470–1478.
18. Criscuolo, G. R., Merrill, M. J., and Oldfield, E. H., 1988, Further characterization of malignant glioma-derived vascular permeability factor, *J. Neurosurg.* **69**:254–262.
19. Criscuolo, G. R., Lelkes, P. I., Rotrosen, D., and Oldfield, E. H., 1989, Cytosolic calcium changes in endothelial cells induced by a protein product of human gliomas containing vascular permeability factor activity, *J. Neurosurg.* **71**:884–891.
20. Cunha-Vaz, J. G., 1976, The blood–retinal barrier, *Doc. Ophthalmol.* **41**:237–287.
21. Dempsey, R. J., Roy, M. W., Meyer, K., Tai, H. H., and Olson, J. W., 1985, Polyamine and prostaglandin markers in focal cerebral ischemia, *Neurosurgery* **17**:635–640.
22. Drenckhahn, D., 1983, Cell motility and cytoplasmic filaments in vascular endothelium, *Prog. Appl. Microcirc.* **1**:53–70.
23. Faden, A. I., Jacobs, T. P., Patrick, D. H., and Smith, M. T., 1984, Megadose corticosteroid therapy following experimental traumatic spinal injury, *J. Neurosurg.* **60**:712–717.
24. Farrell, C. L., and Shivers, R. R., 1984, Capillary junctions of the rat are not affected by osmotic opening of the blood–brain barrier, *Acta Neuropathol.* **63**:179–189.
25. Fishman, R. A., and Chan, P. H., 1981, Hypothesis: Membrane phospholipid degradation and polyunsaturated fatty acids play a key role in the pathogenesis of brain edema, *Ann. Neurol.* **10**:75.
26. Folkman, J., Langer, R., Linhardt, R. J., Haudenschild, C., and Taylor, S., 1983, Angiogenesis inhibition and tumor regression caused by heparin or a heparin fragment in the presence of cortisone, *Science* **221**:719.
27. Folkman, J., and Klagsbrun, M., 1987, Angiogenic factors, *Science* **235**:442–447.
28. Frankel, S. A., and Germans, W. J., 1958, Glioblastoma multiforme. Review of 219 cases with regard to natural history, pathology, diagnostic methods, and treatment, *J. Neurosurg.* **15**:489–503.

29. Galicich, J. H., and French, L. A., 1961, Use of dexamethasone in the treatment of cerebral edema resulting from brain tumors and brain surgery, *Am. Pract.* **12**:169–174.
30. Gerritsen, M. E., 1987, Eicosanoid production by the coronary microvascular endothelium, *Fed. Proc.* **46**:47–53.
31. Gerritsen, M. E., Nganele, D. M., and Rodrigues, A. M., 1987, Calcium ionophore (A231887) and arachidonic acid-stimulated prostaglandin release from microvascular endothelial cells: Effects of calcium antagonists and calmodulin inhibitors, *J. Pharmacol. Exp. Ther.* **240**:837–846.
32. Gimbrone, M. A., 1976, Culture of vascular endothelium, *Prog. Hemostasis Thromb.* **3**:1–25.
33. Glick, R. P., Molteni, A., and Fors, E. M., 1983, Hormone binding in brain tumors, *Neurosurgery* **13**: 513–519.
34. Goldsmith, M. A., and Carter, S. K., 1974, Glioblastoma multiforme—A review of therapy, *Cancer Treat. Rev.* **1**:153–165.
35. Greenberg, D. A., 1987, Calcium channels and calcium channel antagonists, *Ann. Neurol.* **21**:317–330.
36. Groothuis, D. R., and Vick, N. A., 1982, Brain tumors and the blood–brain barrier, *Trends Neurosci.* **5**: 232–235.
37. Grynkiewicz, G., Poenie, M., and Tsien, R. Y., 1985, A new generation of Ca^{2+} indicators with greatly improved fluorescence properties, *J. Biol. Chem.* **260**:3440–3450.
38. Gudeman, S. K., Miller, J. D., and Becker, D. P., 1979, Failure of high-dose steroid therapy to influence intracranial pressure in patients with severe head injury, *J. Neurosurg.* **51**:301–306.
39. Hallam, T. J., and Pearson, J. D., 1986, Exogenous ATP raises cytoplasmic free calcium in fura-2 loaded piglet aortic endothelial cells, *FEBS Lett.* **207**:95–99.
40. Harsh, G. R., Levin, V. A., Gutin, P. H., Seager, M., Silver, P., and Wilson, C. B., 1987, Reoperation for glioblastoma and anaplastic astrocytoma, *Neurosurgery* **21**:615–621.
41. Hoshi, H., and McKeehan, W. L., 1984, Brain- and liver cell-derived factors are required for growth of human endothelial cells in serum-free culture, *Proc. Natl. Acad. Sci. USA* **81**:6413–6417.
42. Hossmann, K. A., Hurter, T., and Oschlies, U., 1983, The effect of dexamethasone on serum protein extravasation and edema development in experimental brain tumors of cat, *Acta Neuropathol.* **60**:223–231.
43. Ikeda, Y., Anderson, J. H., and Long, D. M., 1989, Oxygen free radicals in the genesis of traumatic and peritumoral brain edema, *Neurosurgery* **24**:679–685.
44. Ikeda, Y., Ikeda, K., and Long, D. M., 1989, Comparative study of different iron-chelating agents in cold-induced brain edema, *Neurosurgery* **24**:820–824.
45. Jaffe, E. A., Nachman, R. L., Becker, C. G., and Minick, C. R., 1973, Culture of human endothelial cells derived from umbilical veins. Identification by morphologic and immunologic criteria, *J. Clin. Invest.* **52**:2745–2756.
46. Jaffe, E. A., 1977, Endothelial cells and the biology of factor VIII, *N. Engl. J. Med.* **296**:377–383.
47. Keck, P. J., Hauser, S. D., Krivi, G., Sanzo, K., Warren, T., Feder, J., and Connolly, D. T., 1989, Vascular permeability factor, an endothelial cell mitogen related to PDGF, *Science* **246**:1309–1312.
48. Klatzo, I., 1967, Presidential address. Neuropathological aspects of brain edema, *J. Neuropathol. Exp. Neurol.* **26**:1–14.
49. Kornblum, J. A., Bay, J. W., and Gupta, M. K., 1988, Steroid receptors in human brain and spinal cord tumors, *Neurosurgery* **23**:185–188.
50. Leung, D. W., Cachianes, G., Kuang, W. J., Goeddel, D. V., and Ferrara, N., 1989, Vascular endothelial growth factor is a secreted angiogenic mitogen, *Science* **246**:1306–1309.
51. Levin, V. A., 1981, Chemotherapy of recurrent brain tumors, in: *Nitrosoureas* (A. W. Prestayko, ed.), Academic Press, New York, pp. 259–268.
52. Liddell, R. H. A., Scott, A. R. W., and Simpson, G. J., 1981, Histamine-induced changes in the endothelium of post-capillary venules: Effects of chelating agents and cytochalasin B, *Bibl. Anat.* **20**:109–112.
53. Lobb, R. R., Key, M. E., Alderman, E. M., and Fett, J. W., 1985, Partial purification and characterization of a vascular permeability factor secreted by a human colon adenocarcinoma line, *Int. J. Cancer* **36**:473–478.
54. Long, D. M., Hartmann, J. F., and French, L. A., 1966, The response of human cerebral edema to glucosteroid administration. An electron microscopic study, *Neurology* **16**:521–528.
55. Long, D. M., 1970, Capillary ultrastructure and the blood–brain barrier in human malignant brain tumors, *J. Neurosurg.* **32**:127–144.
56. Long, D. M., 1973, Vascular ultrastructure in human meningiomas and schwannomas, *J. Neurosurg.* **38**:409–419.
57. Long, D. M., 1979, Capillary ultrastructure in human metastatic brain tumors, *J. Neurosurg.* **51**:53–58.

58. Majno, G., and Palade, G. E., 1961, Studies on inflammation I. Effect of histamine and serotonin on vascular permeability: An electron microscopic study, *J. Biophys. Biochem. Cytol.* **11**:571–605.
59. Majno, G., Gilmore, V., and Leventhal, M., 1967, On the mechanism of vascular leakage caused by histamine-type mediators, *Circ. Res.* **21**:833–847.
60. Malgaroli, A., Milani, D., Meldolesi, J., and Pozzan, T., 1987, Fura-2 measurement of cytosolic free Ca^{2+} in monolayers and suspensions of various types of animal cells, *J. Cell Biol.* **105**:2145–2155.
61. Meezan, E., Brendel, K., and Carlson, E. C., 1974, Isolation of a purified preparation of metabolically active retinal blood vessels, *Nature* **251**:65–67.
62. Miles, A. A., and Miles, E. M., 1952, Vascular reactions to histamine, histamine liberator and leukotaxine in the skin of guinea pigs, *J. Physiol. (London)* **118**:228–257.
63. Northover, A. M., and Northover, B. J., 1987, Changes in vascular endothelial shape and of membrane potential in response to the ionophore A23187, *Int. J. Microcirc. Clin. Exp.* **6**:137–148.
64. Oldendorf, W. H., Cornford, M. E., and Brown, W. J., 1977, The large apparent work capability of the blood–brain barrier: A study of the mitochondrial content of capillary endothelial cells in brain and other tissues of the rat, *Ann. Neurol.* **5**:409–417.
65. Owen, D. A. A., Poy, E., Woodward, D. F., and Daniel, D., 1980, Evaluation of the role of histamine H_1 and H_2 receptors in cutaneous inflammation in the guinea pig produced by histamine and mast cell degranulation, *Br. J. Pharmacol.* **69**:615–623.
66. Phillipon, J., Foncin, J. F., Grob, R., Srour, A., Poisson, M., and Pertuiset, B. F., 1984, Cerebral edema associated with meningiomas: Possible role of a secretory–excretory phenomenon, *Neurosurgery* **14**: 295–301.
67. Poungvarin, N., Bhoopat, W., Viriyavejakul, A., Rodprasert, P., Buranasiri, P., Sukondhabhant, S., Hensley, M. J., and Strom, B. L., 1987, Effects of dexamethasone in primary supratentorial intracerebral hemorrhage, *N. Engl. J. Med.* **316**:1229–1233.
68. Quindlen, E. A., and Bucher, A. P., 1987, Correlation of tumor plasminogen activator with peritumoral cerebral edema: A CT and biochemical study, *J. Neurosurg.* **66**:729–733.
69. Reichman, H. R., Farrell, C. L., and Del Maestro, R. F., 1986, Effects of steroids and nonsteroid anti-inflammatory agents on vascular permeability in a rat glioma model, *J. Neurosurg.* **65**:233–237.
70. Reulen, H. J., 1976, Vasogenic brain edema. New aspects in its formation, resolution and therapy, *Br. J. Anaesth.* **48**:741–752.
71. Roblin, R., and Young, P. L., 1980, Dexamethasone regulation of plasminogen activator in embryonic and tumor-derived human cells, *Cancer Res.* **40**:2706–2713.
72. Rotrosen, D., and Gallin, J. I., 1986, Histamine type I receptor occupancy increases endothelial cytosolic calcium, reduces F-actin, and promotes albumin diffusion across cultured endothelial monolayers, *J. Cell Biol.* **103**:2379–2387.
73. Rutherford, R. B., and Ross, R., 1976, Platelet factors stimulate fibroblasts and smooth muscle cells quiescent in plasma serum to proliferate, *J. Cell Biol.* **69**:196–203.
74. Ryan, U. S., Clements, E., Habliston, D., and Ryan, J. W., 1978, Isolation and culture of pulmonary artery endothelial cells, *Tissue Cell* **10**:535–554.
75. Ryan, U. S., Mortara, M., and Whitaker, C., 1980, Method of microcarrier culture of bovine pulmonary artery endothelial cells avoiding the use of enzymes, *Tissue Cell* **12**:619–635.
76. Schook, L. B., Otz, U., Lazary, S., De Weck, A. L., Minowadr, J., Odavic, R., Kniep, E. M., and Edy, V., 1981, Lymphokine and monokine activities in supernatants from human lymphoid and myeloid cell lines, *Lymphokines* **2**:1–19.
77. Seiler, N., Knodgen, B., and Bartholeyns, J., 1985, Polyamine metabolism and polyamine excretion in normal and tumor bearing rodents, *Anticancer Res.* **5**:371–378.
78. Senger, D. R., Galli, S. I., Dvorak, A. M., Perruzzi, C. A., Harvey, V. S., and Dvorak, H. F., 1983, Tumor cells secrete a vascular permeability factor that promotes accumulation of ascites fluid, *Science* **219**: 983–985.
79. Senger, D. R., Perruzzi, C. A., Feder, J., and Dvorak, H. F., 1986, A highly conserved vascular permeability factor secreted by a variety of human and rodent tumor cell lines, *Cancer Res.* **46**:5629–5632.
80. Senger, D. R., Connolly, D. T., Van De Water, L., Feder, J., and Dvorak, H. F., 1990, Tumor-secreted vascular permeability factor: Purification and N-terminal amino acid sequence, *Cancer Res.* **50**:1774–1778.
81. Shapiro, W. R., and Byrne, T. N., 1983, Chemotherapy of brain tumors. Basic concepts in oncology of the nervous system, (M. D. Walker, ed.), Nijhoff, The Hague, pp. 65–100.

82. Shasby, D. M., Shasby, S. S., Sullivan, R., and Peach, M. J., 1982, Role of endothelial cell cytoskeleton in control of endothelial permeability, *Circ. Res.* **51**:657–661.
83. Sheline, G. E., 1977, Radiation therapy of brain tumors, *Cancer* **39**:873–881.
84. Shing, Y., Folkman, J., Sullivan, R., Butterfield, C., Murray, J., and Klagsbrun, M., 1984, Heparin affinity: Purification of a tumor-derived capillary endothelial cell growth factor, *Science* **223**:1296–1299.
85. Simionescu, M., Simionescu, N., and Palade, G. E., 1975, Structural basis of permeability in sequential segments of the microvasculature of the diaphragm. II. Pathways followed by microperoxidase across the endothelium, *Microvasc. Res.* **15**:17–36.
86. Stein-Werblowsky, R., 1980, A permeability-enhancing factor produced by tumor. The genesis of malignant effusions, *J. Cancer Res. Clin. Oncol.* **97**:315–320.
87. Strausbaugh, L. J., 1987, Intracarotid infusions of protamine sulfate disrupt the blood–brain barrier of rabbits, *Brain Res.* **409**:221–226.
88. Taylor, S., and Folkman, J., 1982, Protamine is an inhibitor of angiogenesis. *Nature (London)* **297**:307.
89. Thomas, G., 1982, Mechanism of ionophore A23187 induction of plasma leakage and its inhibition by indomethacin, *Eur. J. Pharmacol.* **81**:35–42.
90. Tosaki, A., Koltai, M., Joo, F., Adam, G., Szerdahelyi, P., Lepran, I., Takats, I., Szekeres, L., 1985, Actinomycin D suppresses the protective effect of dexamethasone in rats affected by global cerebral ischemia, *Stroke* **16**:501–505.
91. Tucker, W. S., Kirsch, W. M., Martinez-Hernandez, A., and Fink, L. M., 1978, In vitro plasminogen activator activity in human brain tumors, *Cancer Res.* **38**:297–302.
92. Tsien, R. Y., Rink, T. J., and Poenie, M., 1985, Measurement of cytosolic free Ca^{2+} in individual small cells using fluorescence microscopy with dual excitation wavelengths, *Cell Calcium* **6**:145–157.
93. Ueki, H., Tsunemi, S., and Kubota, Y., 1975, Vascular permeability-increasing action of hypoalbuminemic substance from Ehrlich ascites carcinoma cells, *Gann* **66**:237–243.
94. Underwood, J. C. E., and Carr, I., 1972, The ultrastructure and permeability characteristics of the blood vessels of a transplantable rat sarcoma, *J. Pathol.* **107**:157–166.
95. Unterberg, A., and Baethmann, A. J., 1984, The kallikrein–kinin system as a mediator of vasogenic brain edema. Part 1: Cerebral exposure to bradykinin and plasma, *J. Neurosurg.* **61**:87–96.
96. Vassalli, J. D., Hamilton, J., and Reich, E., 1976, Macrophage plasminogen activator: Modulation of enzyme production by anti-inflammatory steroids, mitotic inhibitors and cyclic nucleotides, *Cell* **8**:271–281.
97. Vehaskari, V. M., Root, E. R., Germuth, F. G., and Robson, A. M., 1982, Glomerular charge and urinary protein excretion: Effects of systemic and intrarenal polycation infusion in the rat, *Kidney Int.* **22**:127–135.
98. Vehaskari, V. M., Chang, C. T. C., Stevens, J. K., and Robson, A. M., 1984, The effects of polycations on vascular permeability in the rat, *J. Clin. Invest.* **73**:1053–1061.
99. Voyta, J. C., Via, D. P., Butterfield, C. E., and Zetter, B. R., 1984, Identification and isolation of endothelial cells based on their increased uptake of acetylated-low density lipoprotein, *J. Cell Biol.* **99**:2034–2040.
100. Walker, M., Alexander, E., Hunt, W., MacCarty, C., Mahaley, M. S., Mealy, J., Norrell, H., Owens, G., Ransohoff, J., Wilson, C. B., Gehan, E., and Strike, T., 1978, Evaluation of BCNU and or radiotherapy in the treatment of anaplastic gliomas. A cooperative clinical trial, *J. Neurosurg.* **49**:333–343.
101. Wall, R. T., Harker, L. A., Quadracci, L. J., and Striker, G. E., 1978, Factors influencing endothelial cell proliferation, *J. Cell. Physiol.* **96**:203–214.
102. Ward, J. D., Hadfield, M. G., Becker, D. P., and Lovings, E. T., 1974, Endothelial fenestrations and other vascular alterations in primary melanoma of the central nervous system, *Cancer* **34**:1982–1991.
103. Weibel, E. R., and Palade, G. E., 1964, New cytoplasmic components in arterial endothelia, *J. Cell Biol.* **23**:101–112.
104. Yamada, K., Bremer, A. M., and West, C. R., 1979, Effects of dexamethasone on tumor-induced brain edema and its distribution in the brain of monkeys, *J. Neurosurg.* **50**:361–367.
105. Yamada, K., Ushio, Y., Hayakawa, T., Kato, A., Yamada, N., and Mogami, H., 1982, Quantitative autoradiographic measurements of blood–brain barrier permeability in the rat glioma model, *J. Neurosurg.* **57**:394–398.
106. Yu, Z. Y., Wrange, O., Boethius, J., Hatam, A., Granholm, L., and Gustafsson, J. A., 1981, A study of glucocorticoid receptors in intracranial tumors, *J. Neurosurg.* **55**:575–760.

VII

Endothelial Cell in Other Disturbances

29

Effects of Polyunsaturated Fatty Acids in Endothelium

Arthur A. Spector and Steven A. Moore

I. INTRODUCTION

The omega-6 polyunsaturated fatty acids are necessary for proper bodily function and health. This class consists of a series of six fatty acids that can be interconverted through elongation, desaturation, and retroconversion.[1] The metabolic pathway and the structures of the major components, linoleic acid (18:2) and arachidonic acid (20:4), are illustrated in Fig. 1. Linoleic acid, the main dietary component, can be converted to the other members of the omega-6 series. Prostaglandins and lipoxygenase products are formed from arachidonic acid in animal tissues, including the endothelium. These eicosanoids are especially important for endothelial function. PGI_2, the main eicosanoid produced by arterial endothelium, inhibits platelet aggregation and causes relaxation of vascular smooth muscle.[2] Several other eicosanoids produced by endothelium, including PGE_2,[3,4] hydroxyeicosatetraenoic acids (HETEs),[3] and dihydroxyeicosatetraenoic acids (diHETEs),[5,6] probably also play a role in endothelial function. Furthermore, linoleic acid can be converted to a hydroxylated metabolite, hydroxyoctadecadienoic acid (HODE),[7–9] which can influence the antithrombogenic properties of the endothelial surface as well as modulate eicosanoid formation and degradation.[7,9]

The metabolic pathway and structures of the main components of the omega-3 class of polyunsaturated fatty acids, linolenic acid (18:3), EPA (20:5), and DHA (22:6), also are shown in Fig. 1. The omega-3 fatty acids are structural analogues of the corresponding members of the omega-6 class; they contain one extra double bond three carbons distant from the methyl terminus of the fatty acyl chain. Like the omega-6 class, the omega-3 fatty acids cannot be completely synthesized and are ultimately derived from the diet. EPA and DHA are modulators of arachidonic acid metabolism and reduce its conversion to pros-

Arthur A. Spector • Department of Biochemistry, University of Iowa College of Medicine, Iowa City, Iowa 52242. *Steven A. Moore* • Department of Pathology, University of Iowa College of Medicine, Iowa City, Iowa 52242.

Endothelial Cell Dysfunctions, edited by Nicolae Simionescu and Maya Simionescu. Plenum Press, New York, 1992.

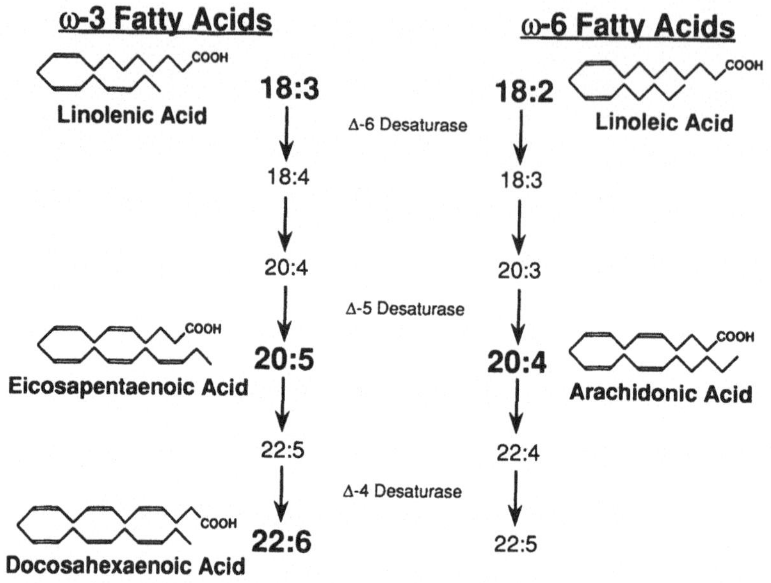

Figure 1. Metabolic pathways for the omega-6 and omega-3 polyunsaturated fatty acids. The structures of the main components of each class are shown.

taglandins.[10–12] In addition, cerebral microvascular endothelium can convert substantial amounts of EPA to prostaglandins and lipoxygenase products,[13] suggesting that omega-3 fatty acids may have an important functional role in the cerebral circulation and blood–brain barrier.

II. LARGE VESSEL ENDOTHELIUM

Polyunsaturated fatty acid metabolism and function have been studied in endothelial cultures derived from human umbilical veins,[10,14–16] bovine pulmonary artery,[17] and bovine aorta.[11,12,18,19] Generally similar results were obtained, suggesting that polyunsaturated fatty acids are handled similarly in the endothelium of many different large blood vessels.

A. Incorporation into Endothelial Lipids

Linoleic acid,[14,17] arachidonic acid,[15,16] docosatetraenoic acid (22:4),[20] linolenic acid,[10] EPA,[10,11] and DHA[12] are readily taken up and incorporated into endothelial lipids. Whether movement into the cell occurs by diffusion across the plasma membrane or is a carrier-mediated process has not been resolved.[21] In this regard, microvessel endothelial cells contain a surface membrane protein that binds albumin.[22] It is possible that this protein acts as a receptor for albumin and mediates the transfer of fatty acids from albumin to the endothelial cell in the microvasculature.

Polyunsaturated fatty acids are incorporated into endothelial phospholipids, particularly the choline phosphoglycerides (PC). Arachidonic acid also rapidly enters the inositol phospholipids (PI),[16] and DHA accumulates in the ethanolamine phosphoglycerides (PE), including ethanolamine plasmalogens.[12] Pulse–chase experiments indicate that some of the EPA initially incorporated into PC is subsequently transferred to PE.[12] Likewise, arachidonic acid initially incorporated into PC is transferred to PE and ethanolamine plasmalogen.[23] These transfers occur spontaneously; activation by an agonist is not required.[23] Some of the EPA and arachidonic acid probably are elongated to 22-carbon products prior to transfer to the ethanolamine fraction. In addition, even when only small amounts are available, a sizable fraction of the arachidonic acid uptake is incorporated into triglycerides.[18]

The availability of saturated or monounsaturated fatty acids does not reduce the incorporation of arachidonic acid,[15] EPA,[11] or DHA[12] into endothelial phospholipids, but these polyunsaturated fatty acids compete with each other. The most likely mechanism is competition at either the acylCoA synthase or acyltransferase reactions.

1. *Polarity of Uptake*

Endothelial cells have been cultured on micropore filters, and the filters suspended in tissue culture wells so that the polarity of uptake and release processes can be studied.[24] Smooth muscle cells can be grown in the basolateral compartment of these chambers, converting this into a coculture system mimicking the vessel wall.[25] Bovine aortic endothelial monolayers grown on these filters take up palmitic acid (16:0) and oleic acid (18:1) to a greater extent from the apical fluid, whereas all of the polyunsaturated fatty acids, including arachidonic acid, are taken up equally well from either the apical or basolateral medium. The distribution of the newly incorporated arachidonic acid in endothelial lipids was similar when uptake occurred from either surface.[26] The findings with this model system suggest that endothelium has the capacity to take up polyunsaturated fatty acids when they are available either in the circulation or within the vascular wall.

2. *Phospholipid Molecular Species*

Human umbilical vein endothelial cultures contain 36 molecular species of PC and 34 species of PE.[27] When the cells are incubated with [^{3}H]arachidonic acid, molecular species having two unsaturated fatty acids (16:1–20:4, 18:1–20:4, 18:2–20:4, 20:4–20:4) have higher specific radioactivities than 16:0–20:4 or 18:0–20:4. This indicates that the diunsaturated molecular species containing arachidonic acid have higher turnover rates than molecular species containing a saturated fatty acid.

B. *Elongation and Desaturation and Retroconversion*

Because arachidonic acid is the main substrate for endothelial eicosanoid synthesis, one would think that these cells might have a large capacity to convert linoleic acid to arachidonic acid. This does not occur when endothelium from large vessels is cultured under the usual conditions. Only small amounts of linoleic acid are converted to arachidonic acid[15,17]; instead, some linoleic is elongated to docosadienoic acid (20:2). By contrast, eicosatrienoic acid (20:3) is converted to 20:4, and 20:4 to 22:4.[15] Therefore, the elongation steps and the 5-desaturase are intact, but the 6-desaturase does not operate effectively in endothelial cells cultured from large blood vessels.

When the serum content of the medium is reduced, 6-desaturase activity increases and more linoleic acid is converted to arachidonic acid.[28] Since the luminal surface of large vessels is continuously exposed to plasma lipoproteins and free fatty acid, the conversion of linoleic to arachidonic acid probably is suppressed in the vascular endothelium. Therefore, large vessel endothelium probably obtains most of its arachidonic acid preformed from the circulation.

Linolenic acid also is not extensively converted to either EPA or DHA when large vessel endothelial cells are cultured under ordinary conditions.[10] EPA is elongated to 22:5, but it is not desaturated to 22:6.[10,11]

Human umbilical vein endothelial cells can form 24- and 26-carbon polyunsaturates from arachidonic acid.[29] The fatty acids formed include 24:4, 24:5, 24:6, and 26:5. These fatty acids are not retained in the cell and are spontaneously released into the medium. Furthermore, bovine aortic endothelial cells can retroconvert 22:4 to arachidonic acid.[20] Therefore, one function of 22:4 probably is to act as a storage form of arachidonic acid. In addition, DHA can be retroconverted to EPA.[12]

C. Modification of Fatty Acid Composition

The fatty acid composition of endothelial lipids varies to some degree in response to changes in the type of fatty acid available in the extracellular fluid.[30] In particular, the arachidonic acid content of the endothelial phospholipids will change in response to changes in the amount of arachidonic acid that is available. Under certain conditions, the arachidonic acid content of PC can double, and substantial increases in the arachidonic acid content of PI and PE also can occur.

When the endothelial lipids become highly enriched in arachidonic acid, 22:4 begins to accumulate.[20] The endothelial phospholipids also can become enriched in linoleic acid,[14,28] but there is little or no accumulation of arachidonic acid under these conditions.[14] In fact, the arachidonic acid content of the endothelial phospholipids decreases when large amounts of linoleic acid accumulate.[30] Endothelial phospholipids also can accumulate omega-3 fatty acids if substantial amounts of either EPA or DHA are available.[10–12] When EPA accumulates, the arachidonic acid content of the phospholipids decreases,[10,11] indicating that these two fatty acids compete for entry into the cellular phospholipid pools.

D. Free Fatty Acid Release

Cultured endothelial cells are able to spontaneously release stored fatty acid when the fatty acid content of the medium decreases.[19] This process, which is particularly pronounced in the case of arachidonic acid,[19] also occurs with DHA and oleic acid.[12,19] The arachidonic acid released under basal conditions is derived primarily from endothelial triglycerides.[19] Albumin must be present in the medium for release to occur, presumably because an extracellular acceptor for the free fatty acid is required.[21] These findings suggest that the endothelium can serve as a storage site for essential polyunsaturated fatty acids and subsequently release them for use by surrounding tissues. Arachidonic acid is released in both the apical and basolateral direction.[26] Therefore, the endothelium can transfer polyunsaturated fatty acids, particularly arachidonic acid, from the blood into the tissues, or from the tissues into the blood, with the endothelial cell temporarily retaining any excess in the form of triglycerides. This is indicated schematically in Fig. 2, which illustrates the known pathways of endothelial arachidonic acid metabolism.

Fatty acid released from the endothelial cells can be taken up by adjacent smooth muscle cells,[25] and the smooth muscle cells can convert arachidonic acid obtained in this way to PGE_2. Furthermore, fatty acid released from smooth muscle cells can be taken up by endothelial cells.[25] Similar fatty acid interchange probably also can occur between the endothelium and other cells present either in the circulation or in the vascular wall.

E. Microvessel Endothelium

Endothelium cultured from foreskin or coronary microvessels produces prostaglandins,[31,32] and brain microvessel endothelial cells also have quite active pathways for the production of eicosanoids.[4,13,33,34] In addition, thc clongation and desaturation of polyunsaturated fatty acid precursors readily occurs in microvessel endothelium,[35,36] a finding that differs from what occurs in large vessel endothelium.

1. Elongation and Desaturation

As demonstrated for large vessel endothelial cells,[15,16] murine cerebromicrovascular endothelial cells take up preformed arachidonic acid or EPA when they are available in the extracellular fluid.[4,13,36] However, in contrast to minimal desaturation of precursors of these essential fatty acids in large vessel endothelium,[15,28] the endothelial cells from the brain microvasculature extensively elongate and desaturate linoleic and linolenic acids.[36] This also occurs in cultured human omental microvascular endothelium.[35] As shown in Fig. 3, the major elongation and desaturation product in the omega-6 pathway is arachidonic acid, that of the omega-3 pathway is EPA. The cerebromicrovascular endothelial cells elongate and desaturate considerable amounts of linoleic and linolenic acids even in the presence of large amounts of serum. As opposed to the large vessel cells,[28] the production of arachidonic acid and EPA increases as the amount of serum in the medium is increased from

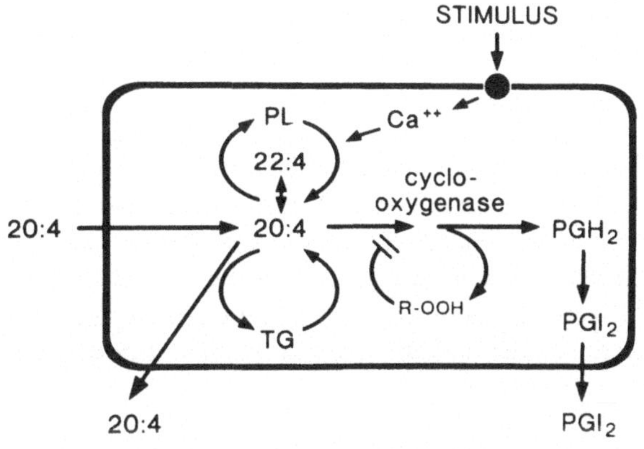

Figure 2. Schematic representation of arachidonic acid metabolism in vascular endothelial cells. Reprinted from *Semin. Thromb. Hemostasis* **14**:196–201, 1988.

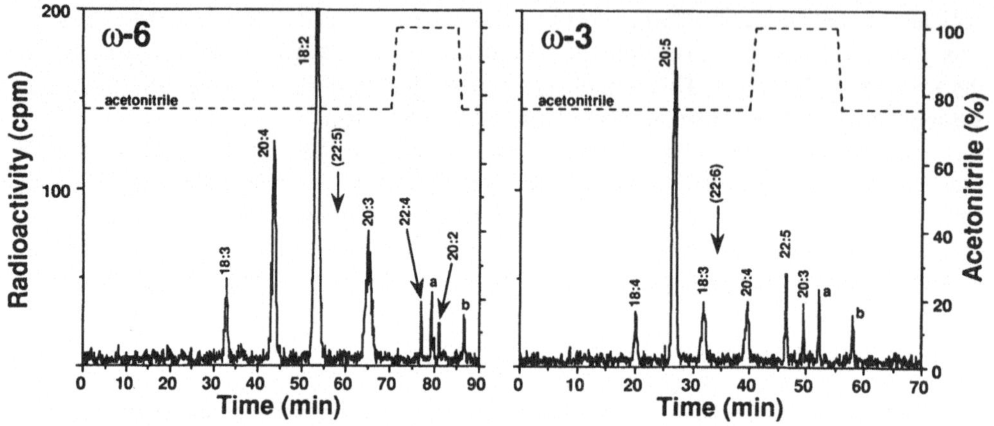

Figure 3. Elongation and desaturation of linoleic and linolenic acids by mouse brain microvessel endothelial cultures. The cells were incubated for 24 hr with radioactive linoleic (left side) or linolenic (right side) acids, and the labeled products were separated by HPLC with an acetonitrile:water gradient.[36]

1 to 20%.[36] The failure of relatively high serum concentrations to suppress these conversions suggests that, in contrast to large vessel endothelium, these microvascular endothelial cells perform similar desaturation and elongation functions *in vivo*.

The products of these elongations and desaturations can be either incorporated into endothelial lipids or released into the medium. Therefore, the microvessel endothelium may be a major site of processing of omega-6 and omega-3 fatty acids. The liver presently is thought to be the only important site for the production of circulating 20- and 22-carbon essential fatty acids formed from the dietary 18-carbon precursors.[37] It is possible that by releasing their elongation/desaturation products into the blood, the microvascular endothelia could collectively also be the source of a substantial amount of these highly unsaturated fatty acids. In addition to the potential of supplying these fatty acids to the blood and tissues, this intracellular elongation/desaturation pathway probably is an important source of the arachidonic acid and EPA needed for microvascular eicosanoid production.

The microvessel endothelial cell elongation and desaturation pathway exhibits a preference for omega-3 fatty acid substrates.[35,36] This may have added significance with respect to cerebromicrovascular endothelium because of the large amount of omega-3 fatty acid found in the brain.[38,39]

2. 4-Desaturase Activity

The desaturation and elongation of linoleic and linolenic acids appear to follow the pathways shown in Fig. 1.[37,40] Of note, however, is the absence of 4-desaturase activity in the cerebral endothelial cultures.[36] These cells have undergone several passages, and 4-desaturase activity often is lost during cell culture.[41] However, this enzyme is active in Y79 retinoblastoma cells even after repeated passage.[42] Primary cultures of human omental microvascular endothelium also possess 4-desaturase activity and synthesize small quantities of DHA.[35]

III. EICOSANOID METABOLISM

Figure 4 shows the radioactive eicosanoids produced by microvessel and large vessel endothelial cells incubated with [^{3}H]arachidonic acid, as determined by high-performance liquid chromatography (HPLC). In both cases, the main product formed is PGI_2, detected as the stable inactivation product, 6-keto-prostaglandin $F_{1\alpha}$ (6-keto-$PGF_{1\alpha}$).[4,16] The microvessel endothelial cells also produce substantial amounts of PGE_2 and in some cases, their PGE_2 output can exceed PGI_2.[4] The widely held view,[43] however, that microvascular endothelium does not produce appreciable amounts of PGI_2 is not consistent with our findings, as shown in Fig. 4. Additional products observed are prostaglandin $F_{2\alpha}$ ($PGF_{2\alpha}$), 12-hydroxyheptadecatrienoic acid (HHT), and HETEs. The formation of lipoxygenase products also has been detected in calf aortic and bovine coronary artery endothelial cultures.[44,45]

Similar radiolabeled eicosanoid products are observed when endothelial cultures are activated by thrombin or ionophore A23187. When bovine aortic endothelial cells grown on micropore filters are exposed to 2 μM A23187,[26] these eicosanoid products are released into both the apical and basolateral medium. Microvessel endothelium grown on filters also releases prostaglandins in both directions and as shown in Fig. 5, the release is greater in the basolateral direction. If such polarity actually occurs *in vivo*, the eicosanoids released by endothelial cells should be accessible to underlying smooth muscle or parenchymal cells, in addition to blood cells.

Human umbilical vein endothelial cells can convert docosatetraenoic acid to dihomo-prostaglandins.[46] Bovine aortic endothelial cells, however, do not make measurable amounts of these dihomo-hydroxylated products.

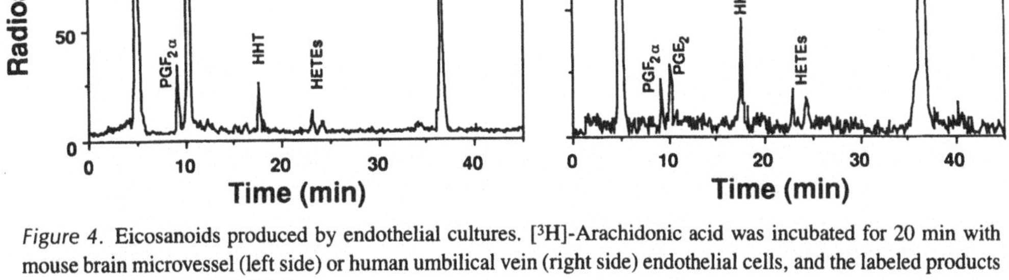

Figure 4. Eicosanoids produced by endothelial cultures. [^{3}H]-Arachidonic acid was incubated for 20 min with mouse brain microvessel (left side) or human umbilical vein (right side) endothelial cells, and the labeled products were determined by HPLC.[4,49]

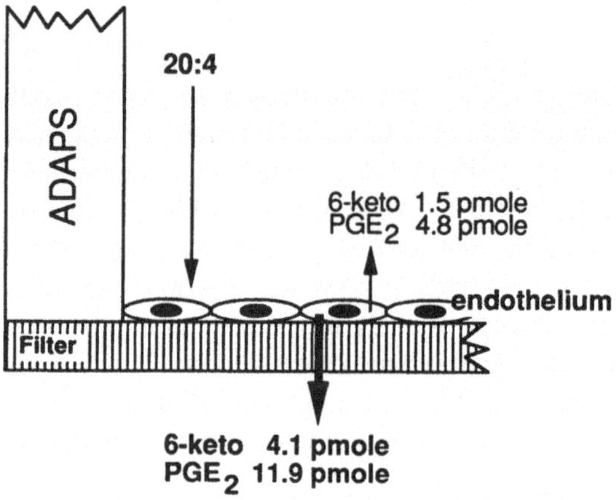

Figure 5. Representation of prostaglandin release from endothelial cultures grown on micropore filters. Less PGI_2 (6-keto) and PGE_2 are released into the apical solution than the basolateral fluid.[4]

A. Sources of Arachidonic Acid

The intracellular arachidonic acid utilized for eicosanoid production is released from phospholipids. Endothelial cells contain both phospholipases A_2 and C, and there is evidence that the arachidonic acid is derived from both PC and PI.[47] Since phosphatidylinositol-4,5-bisphosphate (PIP_2) and PC are hydrolyzed simultaneously, it seems unlikely that the inositol trisphosphate (IP_3) released from PIP_2 by phospholipase C mediates the calcium-activated release of arachidonic acid from PC.[48]

Arachidonic acid supplied directly from the extracellular fluid also can serve as a substrate for PGI_2 formation,[16] as indicated in Fig. 2. Most of the PGI_2 is formed during the first 15 min of incubation; subsequently, the cells become refractory. The cultures recover the ability to synthesize PGI_2 in 24 hr, but recovery is blocked by cycloheximide. The inhibition probably involves inactivation of cyclooxygenase due to the formation of excessive hydroperoxyeicosatetraenoic acid (HPETE),[16] as illustrated in Fig. 2. However, if the endothelial lipids become enriched in arachidonic acid and cyclooxygenase inactivation is prevented, the cells will release more PGI_2 upon subsequent exposure to thrombin or A23187.[16] Presumably, when the phospholipid storage pools become enriched in arachidonic acid, they release more arachidonic acid, providing more substrate for prostaglandin formation.

Plasma Lipoproteins

Endothelial cells release PGI_2 when they are exposed to plasma lipoproteins.[49] Initially, this was thought to be a special property of high-density lipoproteins (HDL),[50] possibly contributing to the protection afforded by HDL against coronary heart disease. However, low-density lipoproteins (LDL) also stimulate PGI_2 release,[51,52] and there is no evidence that this is due to a cytotoxic effect of the LDL on the endothelial cells.[52] These lipoproteins probably increase PGI_2 formation by directly providing arachidonic acid to the endothelium,[56] as illustrated in Fig. 2.

B. Effects of Omega-3 Fatty Acids

EPA suppresses PGI_2 formation in large vessel endothelial cells.[10,11] Two mechanisms appear to be involved. One is incorporation of EPA into the phospholipid substrate pools, replacing some of the arachidonic acid usually present in these pools.[10] When the cell is subsequently activated, less arachidonic acid is available for release, providing less substrate to the cyclooxygenase. In addition, EPA competes with arachidonic acid for access to cyclooxygenase and thereby directly reduces PGI_2 formation.[11] DHA also can suppress PGI_2 formation.[12] In human umbilical vein endothelial cultures, however, DHA is a less effective inhibitor than EPA,[10] while linolenic acid produces little or no suppression of PGI_2 synthesis.[10]

Only a relatively small amount of trienoic prostaglandins are produced from EPA by human umbilical vein endothelial cultures,[10] and little or none is produced by bovine aortic endothelial cells.[11] Therefore, under the usual conditions, the main effect of EPA in large vessel endothelium is to reduce the formation of dienoic prostaglandins without quantitative replacement by trienoic prostaglandins. The formation of PGI_3 is increased if omega-6 polyunsaturates are added together with EPA, apparently because the omega-6 hydroperoxides that are formed activate cyclooxygenase,[53] but the amounts produced are still relatively small.

When humans are fed EPA, gas chromatography combined with mass spectrometry (GC/MS) analysis of urinary metabolites indicates that large amounts of PGI_3 are formed.[54,55] The endothelium is thought to be the source of the PGI_3. Furthermore, as assessed from the measurements of urinary metabolites, there is no reduction in PGI_2 formation.[54,55] These findings are inconsistent with what has been observed with endothelial cells cultured from large vessels.[10–12,56] However, with respect to PGI_3 formation, they agree with what is observed with cerebral microvascular endothelial cultures and isolated cerebral microvessels.

1. Microvessel Endothelium and Isolated Microvessels

Cerebromicrovascular endothelial cultures convert EPA to PGI_3 and PGE_3,[13] as shown in Fig. 6 (left side). Small amounts of these prostaglandins, as well as $PGF_{3\alpha}$, are produced by isolated cerebral microvessels (Fig. 6, right side). These findings may have important implications regarding potential effects of omega-3 fatty acid supplementation on the microcirculation. Dietary EPA reportedly has a beneficial effect on tissues subjected to ischemia followed by reperfusion.[57] If eicosanoids derived from EPA are produced in the microcirculation *in vivo* to the extent observed *in vitro*, they may mediate some of the protective actions through antiplatelet activity or direct vasoactive effects.

With respect to suppression of dienoic prostaglandin formation by EPA, however, the microvessel results are similar to those reported with large vessel cultures,[10,11] and not with the GC/MS urinary metabolite results obtained in the clinical studies.[54,55] Exposure to EPA substantially decreases PGI_2 and PGE_2 formation.[13] The largest reductions occur in microvascular endothelial cells stimulated with A23187 or acetylcholine, where intracellular arachidonic acid is the substrate for prostaglandin formation.

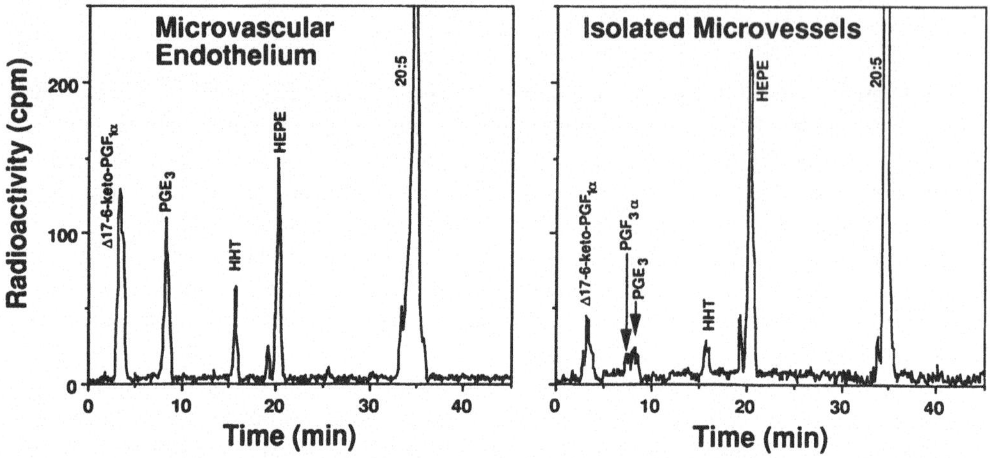

Figure 6. Eicosanoids produced from EPA by cerebral microvessel endothelium. Results with endothelial cultures are shown on the left; those with isolated cerebral microvessels on the right. These preparations were incubated with radioactive EPA for 30 min, and the labeled products were determined by HPLC.[13]

2. Growth Factors

Endothelial cells produce growth factors, including a platelet-derived growth factor-like protein ($PDGF_c$).[58] Exposure of bovine aortic endothelial cells to MaxEPA, a fish oil preparation containing EPA and DHA, inhibits $PDGF_c$ formation.[59] Antioxidants reduce the inhibitory effect of MaxEPA, indicating that a free radical mechanism is involved in the protective action. This suppression of endothelium-mediated smooth muscle proliferation may be the mechanism whereby the omega-3 fatty acids produce their antiatherogenic effect.

C. Lipoxygenase Products

Calf aortic endothelial cells convert extracellular arachidonic acid to 12-HETE and a metabolite that has the chromatographic properties of a diHETE.[44] Bovine aortic endothelial cell homogenates produce 5-, 11-, and 15-HETE from arachidonic acid.[3] However, the total production of these HETEs is only 18% as much as the amount of prostaglandins formed by the homogenates. Likewise, bovine coronary artery endothelial cells produce 11-, 12-, and 15-HETE when they are incubated with arachidonic acid,[45] but the amounts formed are small relative to PGI_2. Human umbilical vein endothelial cells incubated with arachidonic acid also form 15-HETE.[5,53]

While these results suggest that large vessel endothelial cells contain lipoxygenase activity, this conclusion is equivocal. First, endothelial 11- and 15-HETE can be produced by cyclooxygenase.[60] A cytochrome P-450 monooxygenase that utilizes arachidonic acid also is present in endothelium.[61] This pathway produces arachidonic acid epoxides,[62] which can give rise to HETEs or diHETEs. In addition, endothelial cells take up lipoxygenase products and convert them to other metabolites. For example, 15-HPETE is converted to 8,15-diHETE,[5] and leukotriene (LT) A_4 is converted to LTC_4 and LTD_4.[47] Therefore, release of lipoxygenase products does not necessarily imply the presence of a lipoxygenase in the endothelial cell.

The evidence that microvessel endothelial cells contain lipoxygenase activity is stronger. A major arachidonic acid metabolite of cerebral microvessels is 12-HETE,[4,63] as shown in Fig. 7. This appears to be derived from the endothelium,[34] even though the ability of the cerebromicrovascular endothelium to produce large amounts of 12-HETE is lost when the cells are isolated and cultured. As shown in Fig. 6, both the isolated cerebral microvessels (right side) and cultured cerebromicrovascular endothelial cells produce a major hydroxylated product, presumed to be a hydroxyeicosapentaenoic acid (HEPE), from radioactive EPA.

IV. HYDROXYEICOSATETRAENOIC ACID (HETE) METABOLISM

As in the case of 15-HPETE and LTA_4,[5,64] large vessel endothelial cells take up 5-HETE,[65,66] 12-HETE,[66] and 15-HETE.[67] Uptake of HETEs by cells grown on micropore filters can occur from either the apical or basolateral surface.[67] Microvessel endothelial cells also take up 12-HETE.[33] All of the HETEs are incorporated into endothelial phospholipids, 5- and 12-HETE primarily into PC and 15-HETE primarily into PI.[65–67] This may perturb the usual tight packing of the fatty acyl chains in certain regions of the lipid bilayer and thereby affect certain membrane functions,[68] as illustrated in Fig. 8. In addition, 15-HETE may affect membrane signal transduction since it is targeted to PI.

Exposure to HETEs reduces the capacity of large vessel endothelial cells to produce PGI_2 when the cells are incubated subsequently with arachidonic acid, thrombin, or A23187.[69] 12-HETE also reduces prostaglandin formation in microvessel endothelial cells.[33] Furthermore, rat aortic endothelial cell monolayers retract if they are exposed to 12(*S*)-HETE,[70] the 12-HETE isomer that is released by activated platelets. The retraction is

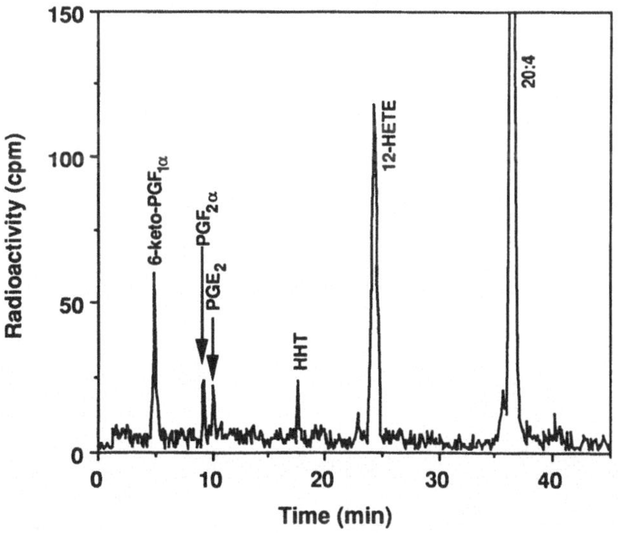

Figure 7. Eicosanoids produced from arachidonic acid by cerebral microvessels.[34]

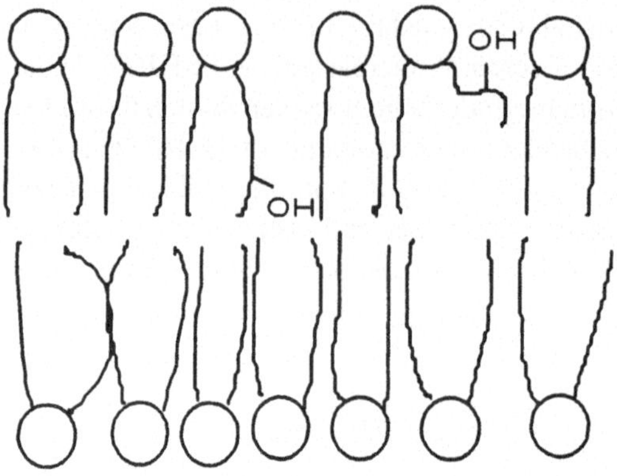

Figure 8. Schematic representation of the perturbations that may occur in the membrane lipid bilayer as a result of HETE incorporation into phospholipids. Reprinted from *Prog. Lipid Res.* **27**:271–323, 1988.

time-dependent and reversible, requiring from 15 min to 1 hr to become manifest and reverting back to normal within 36 hr. Spaces between endothelial cells open as a result of the retraction, allowing tumor cells to adhere to the subendothelial matrix. These findings suggest a mechanism whereby cells that release HETEs, such as activated platelets and leukocytes, may compromise endothelial function by reducing the capacity of the endothelium to form prostaglandins and allowing the penetration of blood cells or neoplastic cells into the vascular wall.

A. Oxidative Catabolism

Large vessel and microvessel endothelial cells can catabolize 12- and 15-HETE through oxidation.[33,68] As shown in Fig. 9, one major and several minor labeled metabolites are detected by reverse-phase HPLC when microvessel endothelial cells are incubated with

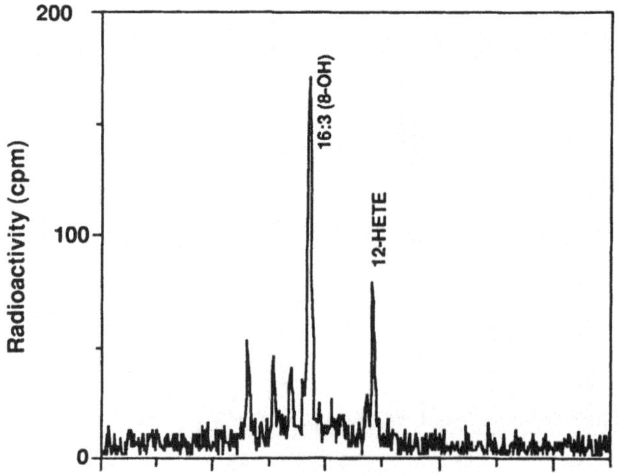

Figure 9. Metabolic products formed from 12-HETE by cerebral microvessel endothelial cultures. The cells were incubated with radioactive 12-HETE, and the labeled products were separated by reverse-phase HPLC.[33]

radioactive 12-HETE. Analysis of the main metabolite by GC/MS, as shown in Fig. 10, demonstrates that the major metabolite of 12-HETE is 8-hydroxyhexadecatrienoic acid [16:3(8-OH)].[33] A corresponding product, 16:3(11-OH) is formed from 15-HETE.[67] It is likely that these metabolites are formed by two beta-oxidation cycles, removing 4-carbons from the carboxyl-terminus of the HETE.[33,67] Evidence obtained from skin fibroblasts suggests that the beta-oxidation occurs in peroxisomes.[71] The chain-shortened oxidation products accumulate in the extracellular fluid and relative to HETEs, are only poorly incorporated into endothelial lipids.[67]

The beta-oxidation process probably is protective, enabling the endothelial cells to catabolize the 12- and 15-HETE and thereby clear them from intracellular lipids. In this regard, 12-HETE activates smooth muscle migration under certain conditions.[72] Therefore, one of the functions of the endothelium may be to protect against atherogenesis by taking up and inactivating any HETE released in the arterial intima. Since HETEs also reduce

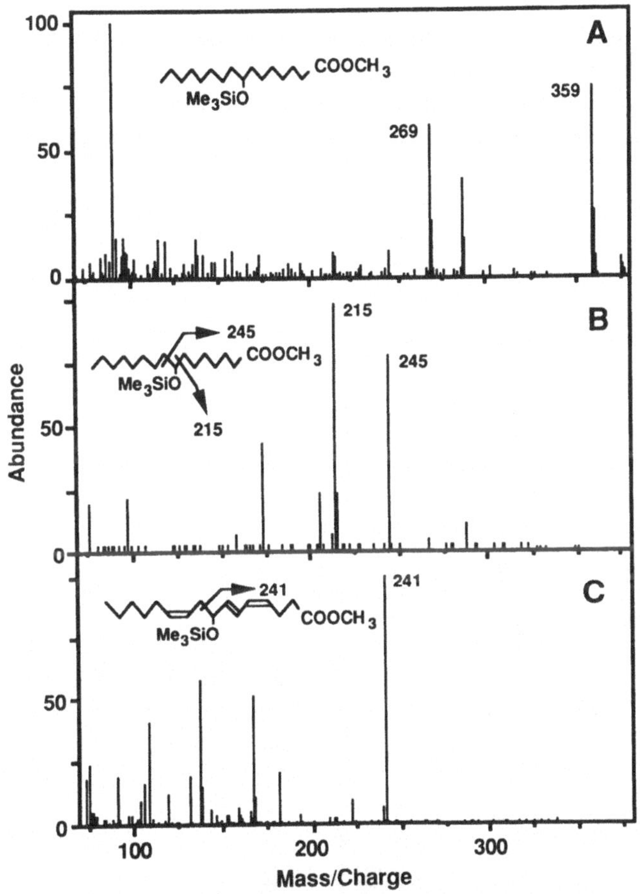

Figure 10. Mass spectra of the main metabolite formed from 12-HETE. The metabolite was methylated and silylated. (A) EI mode, reduced metabolite; (B) CI mode in NH_3, reduced metabolite; (C) EI mode, metabolite. Reprinted from *J. Cell. Physiol.* **137**:75–85, 1988.

endothelial prostaglandin formation and the tightness of cellular contacts,[33,69,70] another function of this oxidative process may be to restore integrity of the monolayer and the complete antithrombotic and vasodilator capacity of the endothelium following an influx of HETE. In addition, it is possible that the products formed, 16:3(8-OH) and 16:3(11-OH), may themselves have functional effects. The concept of HETE-induced vascular injury and the putative protective action of the endothelium are summarized in Fig. 11.

V. HYDROXYOCTADECADIENOIC ACID (HODE) FORMATION

Primary human umbilical vein endothelial cultures can convert linoleic acid to a hydroxylated derivative, HODE.[7–9] Two isomeric forms of HODE have been detected, 13-HODE[7,8] and 9-HODE.[9] Some evidence suggests that 13-HODE is a 15-lipoxygenase product,[7,8] whereas other data indicate that both forms of HODE are cyclooxygenase products.[9] The HODE that is formed remains associated with the cells under some conditions,[7,8] while in other cases, it is entirely released into the medium.[9] More 9- than 13-HODE is recovered in the extracellular fluid. Of the various cells so far tested, only the human umbilical vein endothelium produces appreciable amounts of HODE.[7] The reason for this specificity is difficult to understand, particularly if HODE is produced by either 15-lipoxygenase or cyclooxygenase, for these enzymes are present and active in other types of endothelium.

Although the role of HODE has not been conclusively established, several potentially important actions have been observed. 13-HODE prevents adhesion of blood cells to the

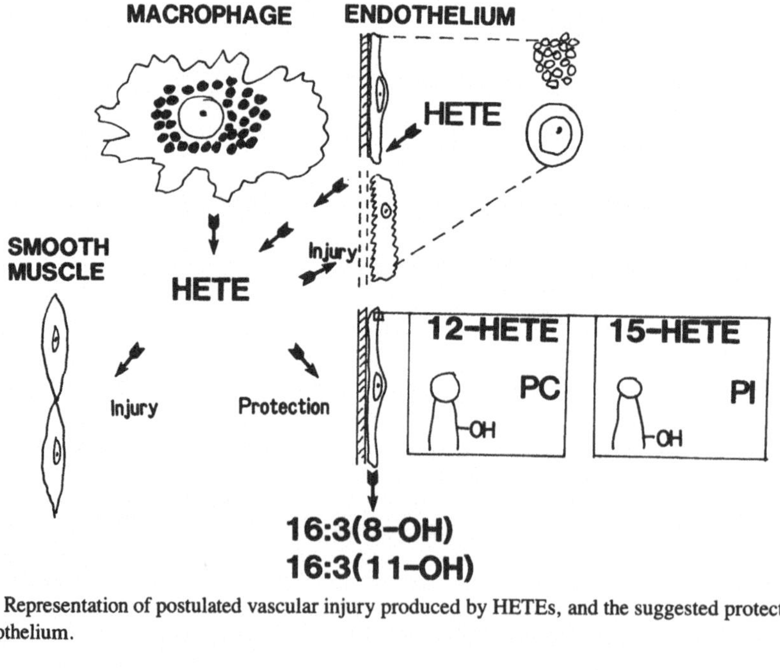

Figure 11. Representation of postulated vascular injury produced by HETEs, and the suggested protective action of the endothelium.

endothelial surface.[7] In addition, both 9- and 13-HODE reduce PGI_2 formation and HETE oxidation by the endothelial cells.[9] These findings suggest that HODE may be an autocrine mediator of endothelial function.

ACKNOWLEDGMENTS. These studies were supported by SCOR/A grant HL 14230 from the National Heart, Lung and Blood Institute, a Program Project Grant NS24261 from the National Institute of Neurological Disorders and Strokes, and a Clinical Investigator Development Award NS01096 from the National Institute of Neurological Disorders and Stroke, National Institutes of Health.

REFERENCES

1. Sprecher, H., 1981, Biochemistry of essential fatty acids, *Prog. Lipid Res.* **20**:13–22.
2. Moncada, S., 1982, Prostacyclin and arterial wall biology, *Arteriosclerosis* **2**:193–207.
3. Mayer, B., Moser, R., Gleispach, H., and Kukovetz, W. R., 1986, Possible inhibitory function of endogenous 15-hydroxyeicosatetraenoic acid on prostacyclin formation in bovine aortic endothelial cells, *Biochim. Biophys. Acta* **875**:641–653.
4. Moore, S. A., Spector, A. A., and Hart, M. N., 1988, Eicosanoid metabolism in cerebromicrovascular endothelium, *Am. J. Physiol.* **254**:C37–C44.
5. Hopkins, N. K., Oglesby, T. D., Bundy, G. L., and Gorman, R. R., 1984, Biosynthesis and metabolism of 15-hydroperoxy-5,8,11,13-eicosatetraenoic acid by human umbilical vein endothelial cells, *J. Biol. Chem.* **259**: 14048–14053.
6. Johnson, A. R., Revtyak, G., and Campbell, W. B., 1985, Arachidonic acid metabolites and endothelial injury: Studies with cultures of human endothelial cells, *Fed. Proc.* **44**:19–24.
7. Buchanan, M. R., Haas, T. A., Lagarde, M., and Guichardant, M., 1985, 13-Hydroxyoctadecadienoic acid is the vessel wall chemorepellant factor, LOX, *J. Biol. Chem.* **260**:16056–16059.
8. Takayama, H., Gimbrone, M. A., Jr., and Schaefer, A. I., 1987, Vascular lipoxygenase activity. Synthesis of 15-hydroxyeicosatetraenoic acid from arachidonic acid by blood vessel and cultured vascular endothelial cells, *Thromb. Res.* **45**:803–816.
9. Kaduce, T. L., Figard, P. H., Leifur, R., and Spector, A. A., 1989, Formation of 9-hydroxyoctadecadienoic acid from linoleic acid in endothelial cells, *J. Biol. Chem.* **264**:6823–6830.
10. Spector, A. A., Kaduce, T. L., Figard, P. H., Norton, K. C., Hoak, J. C., and Czervionke, R. L., 1983, Eicosapentaenoic acid and prostacyclin production by cultured human endothelial cells, *J. Lipid Res.* **24**: 1595–1604.
11. Hadjiagapiou, C., Kaduce, T. L., and Spector, A. A., 1986, Eicosapentaenoic acid utilization by bovine aortic endothelial cells: Effects on prostacyclin production, *Biochim. Biophys. Acta* **875**:369–381.
12. Hadjiagapiou, C., and Spector, A. A., 1987, Docosahexaenoic acid metabolism and effect on prostacyclin production in endothelial cells, *Arch. Biochem. Biophys.* **253**:1–12.
13. Yerram, N. R., Moore, S. A., and Spector, A. A., 1989, Eicosapentaenoic acid metabolism in brain microvessel endothelium: Effect on prostaglandin formation, *J. Lipid Res.* **30**:1747–1757.
14. Spector, A. A., Hoak, J. C., Fry, G. L., Denning, G. M., Stoll, L. L., and Smith, J. B., 1980, Effect of fatty acid modification on prostacyclin production by cultured human endothelial cells, *J. Clin. Invest.* **65**:1003–1012.
15. Spector, A. A., Kaduce, T. L., Hoak, J. C., and Fry, G. L., 1981, Utilization of arachidonic and linoleic acids by cultured human endothelial cells, *J. Clin. Invest.* **68**:1003–1011.
16. Spector, A. A., Kaduce, T. L., Hoak, J. C., and Czervionke, R. L., 1983, Arachidonic acid availability and prostacyclin production by cultured human endothelial cells, *Arteriosclerosis* **3**:323–331.
17. Kaduce, T. L., Spector, A. A., and Bar, R. S., 1982, Linoleic acid metabolism and prostaglandin production by cultured bovine pulmonary artery endothelial cells, *Arteriosclerosis* **2**:380–389.
18. Denning, G. M., Figard, P. H., Kaduce, T. L., and Spector, A. A., 1983, Role of triglycerides in endothelial arachidonic acid metabolism, *J. Lipid Res.* **24**:993–1001.

19. Figard, P. H., Hejlik, D. P., Kaduce, T. L., Stoll, L. L., and Spector, A. A., 1986, Free fatty acid release from endothelial cells, *J. Lipid Res.* **27:**771–780.
20. Mann, C. J., Kaduce, T. L., Figard, P. H., and Spector, A. A., 1986, Docosatetraenoic acid in endothelial cells: Formation, retroconversion to arachidonic acid, and effect on prostacyclin production, *Arch. Biochem. Biophys.* **244:**813–823.
21. Spector, A. A., 1986, Plasma albumin as a lipoprotein, in: *Biochemistry and Biology of Plasma Lipoproteins* (A. M. Scanu and A. A. Spector, eds.), Dekker, New York, pp. 247–279.
22. Ghinea N., Eskenasy, M., Simionescu, M., and Simionescu, N., 1989, Endothelial albumin binding proteins are membrane-associated components exposed on the cell surface, *J. Biol. Chem.* **264:**4755–4758.
23. Wey, H. E., Jakubowski, J. A., and Deykin, D., 1986, Incorporation and redistribution of arachidonic acid in diacyl and ether phospholipids of bovine aortic endothelial cells, *Biochim. Biophys. Acta* **878:**380–386.
24. Hennig, B., Shasby, D. M., Fulton, A. B., and Spector, A. A., 1984, Exposure to free fatty acid increases the transfer of albumin across cultured endothelial monolayers, *Arteriosclerosis* **4:**489–497.
25. Stoll, L. L., and Spector, A. A., 1987, Lipid transfer between endothelial and smooth muscle cells in coculture, *J. Cell. Physiol.* **133:**103–110.
26. Shasby, D. M., Stoll, L. L., and Spector, A. A., 1987, Polarity of arachidonic acid metabolism by bovine aortic endothelial cell monolayers, *Am. J. Physiol.* **253:**H1177–H1183.
27. Blank, M. L., Spector, A. A., Kaduce, T. L., and Snyder, F., 1986, Composition and incorporation of [^{3}H]arachidonic acid into molecular species of phospholipid classes by cultured human endothelial cells, *Biochim. Biophys. Acta* **877:**211–215.
28. Rosenthal, M. D., and Whitehurst, C., 1983, Fatty acyl $\Delta 6$ desaturation activity of cultured human endothelial cells. Modulation by fetal bovine serum, *Biochim. Biophys. Acta* **750:**490–496.
29. Rosenthal, M. D., and Hill, J. R., 1984, Human vascular endothelial cells synthesize and release 24- and 26-carbon polyunsaturated fatty acids, *Biochim. Biophys. Acta* **795:**171–178.
30. Spector, A. A., Hoak, J. C., Fry, G. L., Stoll, L. L., Tanke, C. T., and Kaduce, T. L., 1981, Essential fatty acid availability and prostacyclin production by cultured human endothelial cells, *Prog. Lipid Res.* **20:**471–477.
31. Charo, I. F., Shak, M. A., Karasek, P. M., Davison, P. M., and Goldstein, I. M., 1984, Prostaglandin I_2 is not a major metabolite of arachidonic acid in cultured endothelial cells from human foreskin microvessels, *J. Clin. Invest.* **74:**914–919.
32. Gerritsen, M. D., and Cheli, C. D., 1983, Arachidonic acid and prostaglandin endoperoxide metabolism in isolated rabbit coronary microvessels and cultivated coronary microvessel endothelial cells, *J. Clin. Invest.* **72:**1658–1671.
33. Moore, S. A., Prokuski, L. J., Figard, P. H., Spector, A. A., and Hart, M. N., 1988, Murine cultured microvascular endothelium incorporate and metabolize 12-hydroxyeicosatetraenoic acid, *J. Cell. Physiol.* **254:**C37–C44.
34. Moore, S. A., Figard, P. H., and Spector, A. A., 1989, Brain microvessels produce 12-hydroxyeicosatetraenoic acid, *J. Neurochem.* **53:**376–382.
35. Moore, S. A., Yoder, E., and Spector, A. A., 1989, Human microvascular endothelium produce arachidonate (20:4ω-6), eicosapentaenoate (20:5ω-3), and docosahexaenoate (22:6ω-3) from essential fatty acid precursors, *FASEB J.* **3:**A705.
36. Moore, S. A., Yoder, E., and Spector, A. A., 1990, Role of the blood–brain barrier in the formation of long-chain ω-3 and ω-6 fatty acids from essential fatty acid precursors, *J. Neurochem.* **55:**391–402.
37. Naughton, J. M., 1981, Supply of polyenoic fatty acids to the mammalian brain: The ease of conversion of the short-chain essential fatty acids to their longer chain polyunsaturates in liver, brain, placenta and blood, *Int. J. Biochem.* **13:**21–32.
38. Neuringer, M., Anderson, G. J., and Connor, W. E., 1988, The essentiality of N-3 fatty acids for the development and function of the retina and brain, *Annu. Rev. Nutr.* **8:**517–541.
39. Salem, N., Jr., Kim, H.-Y., and Yergey, J. A., 1986, Docosahexaenoic acid: Membrane function and metabolism, in: *Health Effects of Polyunsaturated Fatty Acids in Seafoods* (A. P. Simopoulos, ed.), Academic Press, New York, pp. 263–317.
40. Holman, R. T., 1986, Nutritional and biochemical evidence of acyl interaction with respect to essential polyunsaturated fatty acids, *Prog. Lipid Res.* **25:**29–39.
41. Spector, A. A., Mathur, S. N., Kaduce, T. L., and Hyman, B. T., 1981, Lipid nutrition and metabolism of cultured mammalian cells, *Prog. Lipid Res.* **19:**155–186.

42. Hyman, B. T., and Spector, A. A., 1981, Accumulation of N-3 polyunsaturated fatty acids by cultured human Y79 retinoblastoma cells, *J. Neurochem.* **37**:60–69.
43. Smith, W. L., 1986, Prostaglandin biosynthesis and its compartmentation in vascular smooth muscle and endothelial cells, *Annu. Rev. Physiol.* **48**:251–262.
44. Kuhn, H., Ponicke, K., Halle, W., Weisner, R., Schewe, T., and Forster, W., 1985, Metabolism of [1-^{14}C]-arachidonic acid by cultured calf aortic endothelial cells: Evidence for the presence of a lipoxygenase pathway, *Prostaglandins Leukotrienes Med.* **17**:291–303.
45. Revtyak, G. E., Johnson, A. R., and Campbell, W. B., 1988, Cultured bovine coronary artery endothelial cells synthesize HETEs and prostacyclin, *Am. J. Physiol.* **254**:C8–C19.
46. Campbell, W. B., Falck, J. R., Okita, J. R., Johnson, A. R., and Callahan, K. S., 1985, Synthesis of dihomoprostaglandins from adrenic acid (7,10,13,16-docosatetraenoic acid) by human endothelial cells, *Biochim. Biophys. Acta* **837**:67–76.
47. Hong, S. L., and Deykin, D., 1982, Activation of phospholipases A2 and C in pig aortic endothelial cells synthesizing prostacyclin, *J. Biol. Chem.* **257**:7151–7154.
48. Kaya, H., Patton, G. M., and Hong, S. L., 1989, Bradykinin-induced activation of phospholipase A2 is independent of activation of polyphosphoinositide-hydrolyzing phospholipase C, *J. Biol. Chem.* **264**:4972–4977.
49. Spector, A. A., 1988, Lipid and lipoprotein effects on endothelial eicosanoid formation, *Semin. Thromb. Hemostasis* **14**:196–201.
50. Fleisher, L. N., Tall, A. R., Witte, L. D., Miller, R. W., and Cannon, P. J., 1982, Stimulation of arterial endothelial cell prostacyclin synthesis by high density lipoproteins, *J. Biol. Chem.* **257**:6653–6655.
51. Spector, A. A., Scanu, A. M., Kaduce, T. L., Figard, P. H., Fless, G. M., and Czervionke, R. L., 1985, Effect of human plasma lipoproteins on prostacyclin production by cultured endothelial cells, *J. Lipid Res.* **26**: 288–297.
52. Holland, J. A., Pritchard, K. A., Rogers, N. J., and Stemerman, M. B., 1988, Perturbation of human endothelial cells by atherogenic levels of low density lipoprotein, *Am. J. Pathol.* **132**:474–478.
53. Bordet, J.-C., Guichardant, M., and Lagarde, M., 1988, Hydroperoxides produced by n-6 lipoxygenation of arachidonic and linoleic acids potentiate synthesis of prostacyclin related compounds, *Biochim. Biophys. Acta* **958**:460–468.
54. Fisher, S., and Weber, P. C., 1984, Prostaglandin I_3 is formed in vivo in man after dietary eicosapentaenoic acid, *Nature* **307**:165–168.
55. Knapp, H. R., Reilly, I. A. G., Alessandrini, P., and Fitzgerald, G. A., 1986, In vivo indexes of platelet and vascular function during fish-oil administration in patients with atherosclerosis, *N. Engl. J. Med.* **314**: 937–942.
56. Brox, J. H., and Nordoy, A., 1983, The effect of polyunsaturated fatty acids on endothelial cells and their production of prostacyclin, thromboxane and platelet inhibitory activity, *Thromb. Hemostasis* **50**:762–767.
57. Black, K. L., Hoff, J. T., Radin, N. S., and Deshmukh, G. D., 1984, Eicosapentaenoic acid: Effect on brain prostaglandins, cerebral blood flow and edema in ischemic gerbils, *Stroke* **15**:65–69.
58. DiCorleto, P. E., and Bowen-Pope, D. F., 1983, Cultured endothelial cells produce a platelet-derived growth factor-like protein, *Proc. Natl. Acad. Sci. USA* **80**:1919–1923.
59. Fox, P. L., and DiCorleto, P. E., 1988, Fish oils inhibit endothelial cell production of platelet-derived growth factor-like protein, *Science* **241**:453–456.
60. Yamaja Setty, B. N., Stuart, M. J., and Walenga, R. W., 1985, Formation of 11-hydroxyeicosatetraenoic acid and 15-hydroxyeicosatetraenoic acid in human umbilical arteries is catalyzed by cyclooxygenase, *Biochim. Biophys. Acta* **833**:484–494.
61. Pinto, A., Abraham, N. G., and Mullane, K. M., 1986, Cytochrome P-450-dependent monoxygenase activity and endothelial-dependent relaxations induced by arachidonic acid, *J. Pharmacol. Exp. Ther.* **236**:445–451.
62. Falck, J. R., Schueler, V. J., Jacobson, H. R., Siddhanta, A. K., Pramanik, B., and Capdevila, J., 1987, Arachidonate epoxygenase: Identification of epoxyeicosatrienoic acids in rabbit kidney, *J. Lipid Res.* **28**: 840–846.
63. Hambrecht, G. S., Adesuyi, S. A., Holt, S., and Ellis, E. F., 1987, Brain 12-HETE formation in different species, brain regions, and in brain microvessels, *Neurochem. Res.* **12**:1029–1033.
64. Feinmark, S. J., and Cannon, P. J., 1986, Endothelial cell leukotriene C_4 synthesis results from intercellular transfer of leukotriene A_4 synthesized by polymorphonuclear leukocytes, *J. Biol. Chem.* **261**:16466–16472.

65. Richards, C. F., Johnson, A. R., and Campbell, W. B., 1986, Specific incorporation of 5-hydroxy-6,8,11,14-eicosatetraenoic acid into phosphatidylcholine in human endothelial cells, *Biochim. Biophys. Acta* **875:** 569–581.
66. Schafer, A. I., Takayama, H., Farrell, S., and Gimbrone, M. A., Jr., 1986, Incorporation of platelet and leukocyte lipoxygenase metabolites by cultured vascular cells, *Blood* **67:**373–378.
67. Shen, X.-Y., Figard, P. H., Kaduce, T. L., and Spector, A. A., 1988, Conversion of 15-hydroxyeicosatetraenoic acid to 11-hydroxyhexadecatrienoic acid by endothelial cells, *Biochemistry* **27:**996–1004.
68. Spector, A. A., Gordon, J. A., and Moore, S. A., 1988, Hydroxyeicosatetraenoic acids (HETEs), *Prog. Lipid Res.* **27:**271–323.
69. Hadjiagapiou, C., and Spector, A. A., 1986, 12-Hydroxyeicosatetraenoic acid reduces prostacyclin production by endothelial cells, *Prostaglandins* **31:**1135–1144.
70. Honn, K. V., Grossi, I. M., Diglio, C. A., Wojtukiewicz, M., and Taylor, J. D., 1989, Enhanced tumor cell adhesion to the subendothelial matrix resulting from 12(S)-HETE-induced endothelial cell retraction, *FASEB J.* **3:**2285–2293.
71. Gordon, J. A., Figard, P. H., and Spector, A. A., 1990, HETE metabolism in cultured human skin fibroblasts: Evidence for peroxisomal β-oxidation, *J. Clin. Invest.* **85:**1173–1181.
72. Nakao, J., Ito, H., Koshihara, Y., and Murota, S., 1984, Age-related increase in the migration of aortic smooth muscle cells induced by 12L-hydroxy-5,8,10,14-eicosatetraenoic acid, *Atherosclerosis* **51:**179–187.

30

Modulation of Endothelial Coagulant Properties and Barrier Function by Factors in the Vascular Microenvironment

H. Gerlach, M. Clauss, S. Ogawa, and D. M. Stern

I. INTRODUCTION

As the cells forming the luminal vascular surface, endothelium regulates both barrier and hemostatic functions, comprising pro- and anti-coagulant reactions.[1–3] Endothelial cells carry out these functions by controlling the expression of cell surface molecules, such as receptors which regulate the hemostatic balance, those which affect permeability across the endothelial monolayers, and those which mediate leukocyte adhesion. The schematic representation in Fig. 1 emphasizes that endothelial regulation of these processes occurs in response to environmental stimuli, and that these stimuli include both blood-borne mediators, as well as factors in the vessel wall. In response to these perturbants, barrier function of the endothelial monolayer is affected and coagulant function can be shifted to favor clot formation. Barrier function of the monolayer and cell surface coagulant properties are closely linked since activation of coagulation leads to formation of proteases and fibrin, which can, in turn, increase endothelial permeability. Conversely, diminished barrier function and ingress of plasma proteins into the subendothelium promotes clotting.

In this chapter, recent observations from our laboratory which may contribute toward understanding why many pathologic processes become localized to particular parts of the vascular tree will be described. We have examined mechanisms that can lead to localized clot formation and increased vascular permeability on the endothelialized vessel surface. These include factors modulating the endothelial response to the cytokine tumor necrosis factor/cachectin (TNF), and hypoxia-induced perturbation of endothelial function.

H. Gerlach, M. Clauss, S. Ogawa, and D. M. Stern • Rover Research Laboratory, Department of Physiology and Cellular Biophysics, College of Physicians and Surgeons, Columbia University, New York, New York 10032.

Endothelial Cell Dysfunctions, edited by Nicolae Simionescu and Maya Simionescu. Plenum Press, New York, 1992.

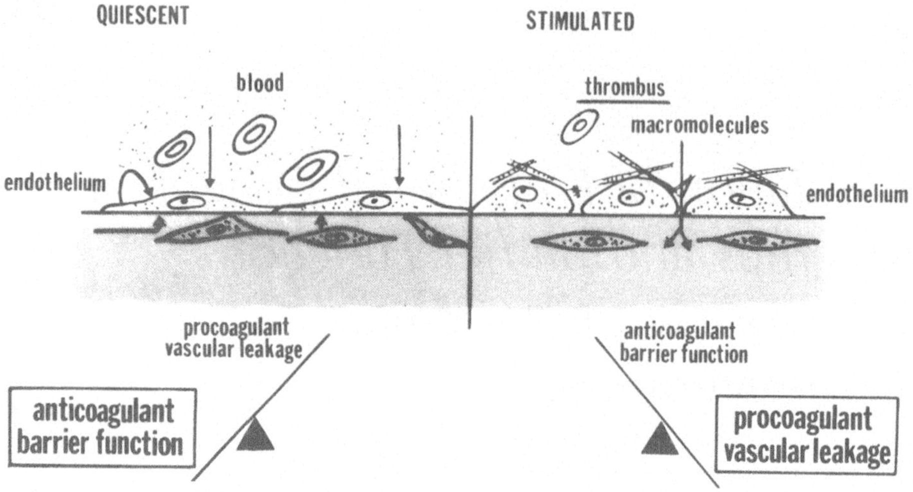

Figure 1. Schematic depiction of endothelial coagulant and barrier properties in the quiescent and stimulated state.

II. A TUMOR-DERIVED MEDIATOR WHICH ALTERS THE ENDOTHELIAL RESPONSE TO TNF

Necrosis of tumors sensitive to the cytokine TNF has been closely linked to changes in the tumor vasculature: since the earliest reports, vascular collapse in the tumor, hemorrhage, and thrombosis in tumor vessels have been observed.[4–6] Furthermore, there has not been a close correlation between the cytotoxic effect of TNF on tumor cells *in vitro* and TNF-induced tumor regression *in vivo*, suggesting that TNF might act, at least in part, through host-induced mechanisms, such as a vascular response.[4–6] To investigate this hypothesis, mice bearing Meth A fibrosarcomas were treated with TNF.[7] First, the mice were infused with radiolabeled fibrinogen, then they were infused with TNF or inactivated TNF (or saline alone), and deposition of radioactivity in the tissues was measured. In mice infused with TNF there was a prompt rise in radioactivity in the tumor, compared with controls (Fig. 2), but deposition of radioactivity in other organs in response to TNF was not as striking. To determine if accumulation of radioactivity in the tumor tissue was accompanied by fibrin formation, the tumor tissue was extracted and the solubilized material subjected to SDS–PAGE and Western blot analysis with a monoclonal antibody which recognizes a fibrin-specific epitope (Fig. 2, inset).[8] Only in tumor-bearing mice treated with TNF was there a dense band on the immunoblots. These results indicated that activation of coagulation with fibrin formation was occurring in the tumor vasculature and led to the question: where is this activation of coagulation occurring? Does it begin in an intact, minimally perturbed vascular bed, or in one which is visibly damaged?

A morphologic study of the tumor vasculature in control and TNF-treated animals[7] showed that tumor vessels in control animals (i.e., not infused with TNF) were patent, without evidence of adherent cells or fibrin (Fig. 3A). In contrast, after the infusion of TNF,

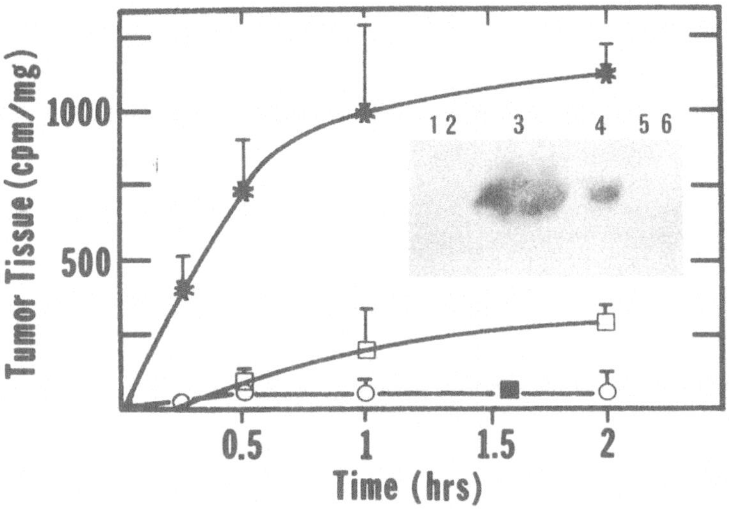

Figure 2. Effect of TNF on the incorporation of fibrin/fibrinogen into Meth A sarcomas. Mice bearing Meth A sarcomas were infused with radioiodinated fibrinogen alone (○) or in the presence of TNF (*). At the indicated times, tumor tissue was excised, weighed, and counted. Other mice were treated either with 3(α-acetonylbenzyl)-4-hydroxycoumarin for 4 days before the TNF infusion (□) or were infused with heat-inactivated TNF (■). (Inset) Mice were infused with unlabeled human fibrinogen in the presence or absence of TNF. Two hours after the TNF infusion, samples of tumor tissue were processed for reduced SDS–PAGE and Western blotting. Blots were reacted with monoclonal antibody to a fibrin-specific epitope. Lane 1, tumor tissue from a mouse not infused with TNF or human fibrinogen; lane 2, tumor tissue from a mouse infused with TNF but not human fibrinogen; lane 3, tumor tissue from a mouse infused with TNF and human fibrinogen; lane 4, tumor tissue from a mouse treated with the coumarin derivative and subsequently infused with TNF and human fibrinogen; lane 5, tumor tissue from a mouse infused with heat-treated TNF and human fibrinogen; lane 6, tumor tissue from a mouse infused with human fibrinogen alone.

fibrin was evident in close association with the inner aspect of the vessel wall (Fig. 3B). Fibrin was identified immunologically and ultrastructurally (based on the characteristic 21-nm periodicity of the strands).[9] Fibrin deposition appears to occur on a morphologically intact/minimally perturbed endothelial layer, as indicated in the scanning electron micrograph (Fig. 3B). After several hours, fibrin deposits progressed to occlusive thrombosis and, by 3 hr subsequent to the infusion of TNF, perfusion in the tumor vasculature was reduced ≈80%. The most remarkable feature of this fibrin deposition was its extreme selectivity for the tumor vasculature. Neither fibrin deposits nor platelet aggregates were evident in other organs examined, and within the tumors, visible fibrin was restricted, at least initially, to the intravascular space. These observations suggested that study of the vascular response had led to the recognition of a thrombosis model in which localization of clot formation was due to the interaction of TNF with tumor vasculature. No external physical maneuvers to induce clot formation, such as denudation of the endothelial monolayer, were required.

In order to explain the localization of clotting to the tumor vasculature, our attention was focused on endothelium in the tumor and its perturbation by environmental and growth/regulatory factors. These include tumor-derived mediators and changes in the growth and/or

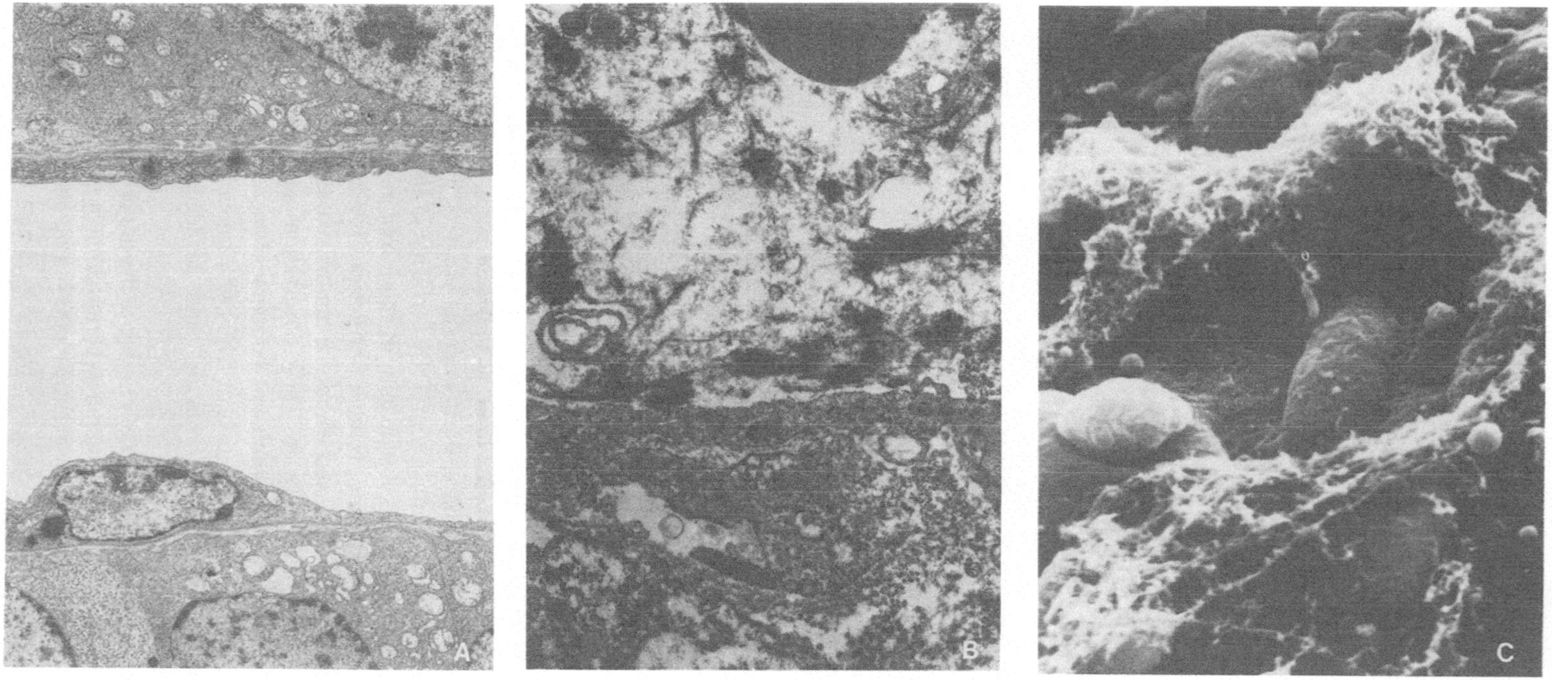

Figure 3. TNF-induced changes in tumor vasculature. Mice bearing Meth A sarcomas were infused with TNF and subjected to perfusion fixation at the indicated times. Transmission (A) electron microscopy of tumors from saline-injected controls showing patency of tumor vessels, which are delimited by an intact luminal endothelium. Scanning (B) and transmission (C) electron microscopy 30 min after the injection of TNF. There is evidence of fibrin accumulation at the endothelial cell surface. A, B, ×5040; C, ×3360.

migratory state of the endothelium, which could potentially alter the cellular response to TNF. Another important environmental factor we have examined is hypoxia, since hypoxemia is often ubiquitous in tumor vasculature.

In the quiescent state, endothelium promotes blood fluidity through the coordinated expression of multiple anticoagulant mechanisms, and the absence of procoagulant activity which initiates activation of coagulation, i.e., lack of tissue factor expression.[1–3] After exposure to the cytokine TNF, the usual balance of endothelial homeostatic mechanisms is shifted to promote clot formation by suppression of anticoagulant mechanisms, such as the protein C/protein S and fibrinolytic pathways, and the induction of tissue factor.[10–13] These changes in cellular coagulant properties are not due to toxic effects of TNF on endothelium, but the result of specific receptor-activated processes.[14] For example, TNF induces the transcription, translation, and insertion into the plasma membrane of tissue factor. Tissue factor activity in endothelium is maximal 3–8 hr after exposure of cultures to TNF and falls off, reaching the baseline by 24 hr.[10,11] These results could potentially explain TNF-induced generalized activation of coagulation in response to the infusion of high concentrations of TNF. However, they do not account for the selective fibrin deposition observed in the tumor vasculature after the infusion of lower concentrations of TNF (as in the experiments depicted in Figs. 2 and 3). If the induction of tissue factor in endothelium were enhanced and localized to the tumor bed, then this could begin to explain thrombosis in the tumor microvasculature. This led us to ask whether the tumor cells were priming the endothelium to promote clotting and to increase their responsiveness to the cytokine.[14]

When conditioned medium from Meth A fibrosarcoma tumor cells was incubated with endothelium, induction of procoagulant activity, characterized as tissue factor, was observed. In the presence of a low concentration of TNF, which by itself had only a small effect, the conditioned medium enhanced the procoagulant response compared with that observed with either agent alone.[7] These observations led us to characterize the nature of the activity responsible for this induction of endothelial procoagulant activity (i.e., tissue factor), which appeared to be synergistic with the procoagulant effect of TNF on endothelium.[14] Starting with serum-free medium from Meth A tumor cells, a purification procedure was developed. By sequential ammonium sulfate precipitation, ion exchange, and preparative SDS–PAGE, a polypeptide was purified to homogeneity (about 500,000-fold) from the culture supernatants.[14] This polypeptide, termed Meth A factor because of its production by Meth A fibrosarcoma cells, migrated as a single band, $M_r \approx 44$ kDa on nonreduced SDS–PAGE, and, with reduction, its mobility slowed somewhat, suggesting the presence of intrachain disulfide bonds (Fig. 4). When the gel was sliced and electroeluted, a peak of tissue factor-inducing activity was observed in the region corresponding to this band (Fig. 4). Exposure of Meth A factor to protease K destroyed its activity and prevented the appearance of the band on the gel. However, a variety of harsh treatments such as heat, acid, base, high-salt urea, and other agents had little effect.

Meth A factor, by immunologic and functional criteria, is distinct and different from multiple other cytokines and growth factors, including interleukins 1, 2, and 6, transforming growth factors α/β, tumor necrosis factors α/β, fibroblast growth factor, epidermal growth factor, platelet-derived growth factor, etc. The molecule would thus appear to be a novel protein.[14] However, definitive conclusions about the uniqueness of Meth A factor, and its relation to tumor-produced permeability enhancing factor described by others,[53] will have to await sequencing/cloning and further study. Consistent with its potential biologic

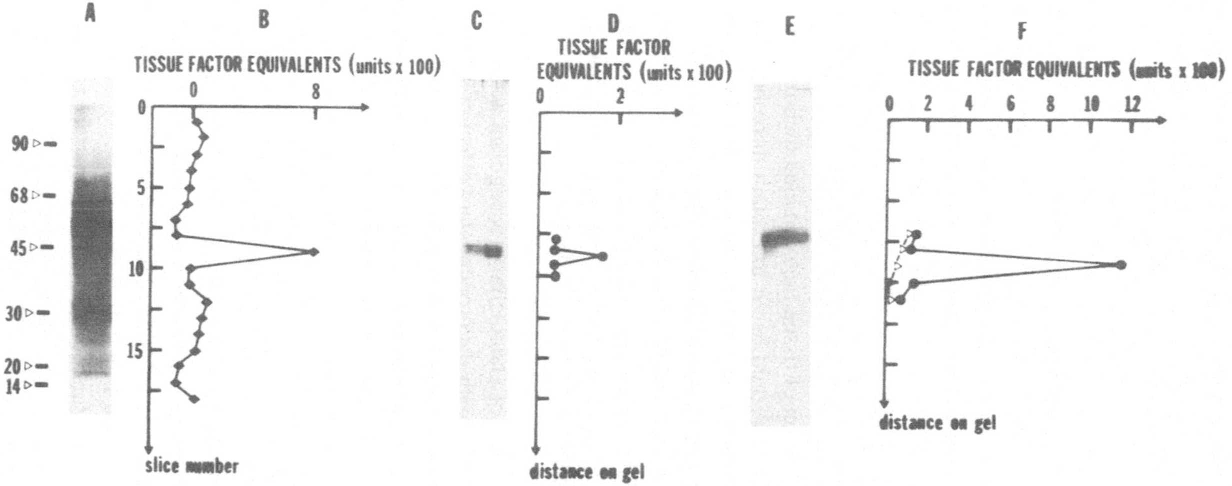

Figure 4. SDS–PAGE and electroelution of Meth A factor. Lane A, nonreduced SDS–PAGE (12.5%) of partially purified factor visualized by silver staining. Lane B, activity profile of material eluted from the indicated slice in lane A. A nonstained lane on the gel, otherwise identical to lane A, was sliced and subjected to electroelution. Tissue factor inducing activity is seen to be in the slice corresponding to ≈ 45 kDa. Lane C, the electroeluted material with peak activity in lane B was again subjected to nonreduced SDS–PAGE (12.5%) and material on the gel was visualized by silver staining. Lane D, activity profile of the material in lane C. Lane E, the same electroeluted sample as in lane C, but run under reduced conditions with material on the gel visualized by silver staining. Lane F, protease K treatment of electroeluted Meth A factor. Meth A factor eluted from nonreduced SDS–PAGE was incubated with buffer alone (●) or in the presence of protease K (Δ). material was then tested for its ability to induce endothelial cell tissue factor.

importance, Meth A factor-like activity is produced not only by murine Meth A fibrosarcomas, but also by a diverse group of human and rodent tumors.

Purified Meth A factor induced tissue factor activity in endothelium in the absence of TNF in a time-dependent manner, being maximal by 6–12 hr into the incubation period.[14] In the presence of TNF, the tissue factor response was enhanced. Expression of tissue factor following exposure of endothelial cultures to Meth A factor was also dependent on the concentration of mediator (Fig. 5). Tissue factor induction was saturable, being half-maximal at ≈ 3–4 pM. Comparing TNF-mediated tissue factor induction with the effect of Meth A factor on endothelial tissue factor indicates that these two agents are quite similar in this regard: maximal tissue factor induction comparable, as are the concentrations of mediator required for a half-maximal response. This suggests that a locally produced agent, such as Meth A factor, can potentially have a selective and profound effect on thrombogenicity in the tumor vascular bed.

The effect of Meth A factor on endothelial tissue factor was due to the induction of *de novo* transcription and translation of the tissue factor gene. When endothelial cells were incubated with low concentrations of TNF or Meth A factor alone, there was an increase in the level of mRNA detectable by hybridization to a cDNA probe for tissue factor (Fig. 6). Both agents together demonstrated a striking increase in the level of tissue factor mRNA (Fig. 6). As expected, cycloheximide blocked expression of tissue factor antigen/functional activity, indicating that Meth A factor required integrity of cellular biosynthetic mechanisms to exert its effect on endothelial tissue factor.

Although our initial studies have focused on the effect of Meth A factor on endothelial tissue factor, recent pilot studies have demonstrated that this mediator alters a range of cellular functions: assembly of the cytoskeleton and cell shape are altered, barrier function of the monolayer is perturbed, and apparently new proteins are inserted into the cell membrane. Further studies will be required to characterize Meth A factor and assess its

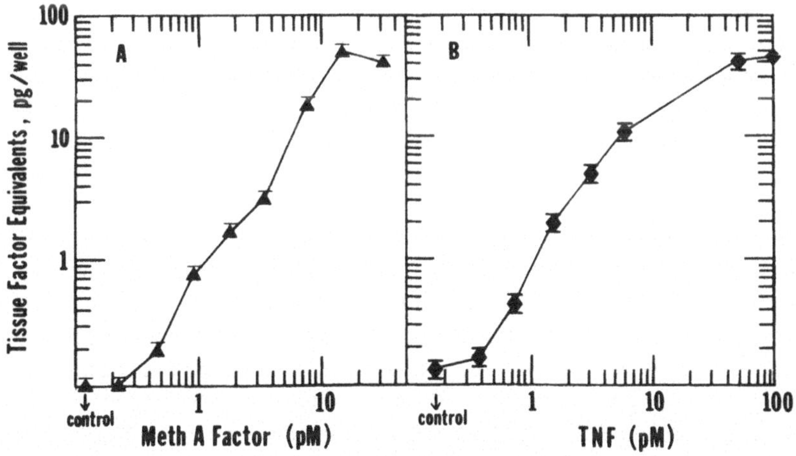

Figure 5. Induction of endothelial cell tissue factor by Meth A factor: comparison with the effect of TNF on endothelial tissue factor. Endothelial cells were incubated for 6 hr with the indicated concentrations of Meth A factor (A) or TNF (B). Then tissue factor activity of the monolayers was assessed.

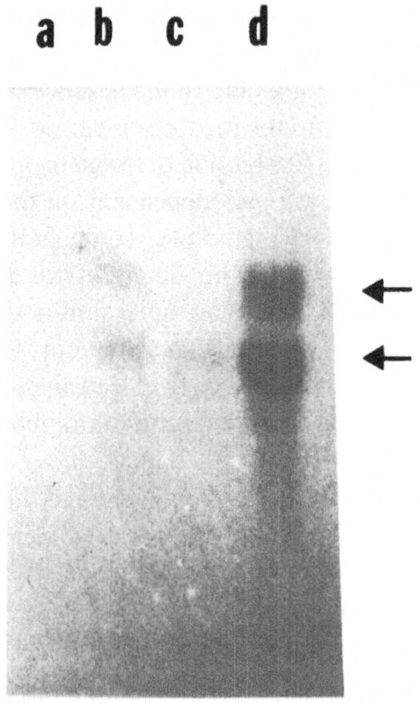

Figure 6. Northern blots of endothelial cell tissue factor mRNA: effect of Meth A factor and TNF. Endothelial cells were incubated with either buffer alone (a), TNF (30 ng; b), or Meth A factor (5 ng; c) or both factors together at the same concentrations (d). RNA was isolated and 10 μg of each sample was run on a 1% agarose gel, transferred to nitrocellulose, and incubated with radiolabeled cDNA for tissue factor (the cDNA probe for tissue factor was generously provided by Dr. E. Sadler, Washington University School of Medicine, St. Louis, Mo.). Arrows point to bands corresponding to ≈ 3.5 and 2.3 kb.

ability to alter properties of normal and tumor vasculature (in the presence and absence of TNF).

III. THE EFFECT OF THE GROWTH/MOTILE STATE ON THE RESPONSIVENESS OF ENDOTHELIUM TO TNF/CACHECTIN

To explain the striking localization of TNF effects to the tumor vasculature, we reasoned that several mechanisms would probably be working together. One distinguishing feature of tumor neovasculature is the increased number of endothelial cells in the growth/migratory configuration, as compared with quiescent, contact-inhibited endothelium in normal vessels. As the studies to be described in this section will demonstrate, the growth state, and the configuration which accompanies it, can modify the endothelial response to TNF.

Thrombomodulin is a cell surface anticoagulant cofactor of endothelium which has a central role in the protein C/protein S anticoagulant pathway.[15] This cofactor functions as a receptor which binds thrombin with high affinity and promotes effective activation of the plasma zymogen protein C (free thrombin is not an effective activator of protein C). The product, activated protein C, is an anticoagulant enzyme. Thrombomodulin is expressed constitutively by quiescent endothelium. By immunofluorescence, unstimulated endothelial cells show a relatively diffuse pattern of thrombomodulin on the cell surface (Fig. 7A). After TNF, some thrombomodulin is rapidly internalized and degraded,[16] the remainder redistributing into discrete patches (Fig. 7B). With time, these patches coalesce into larger aggregates (Fig. 7C), which are shed from the cell surface. TNF is also reported to depress

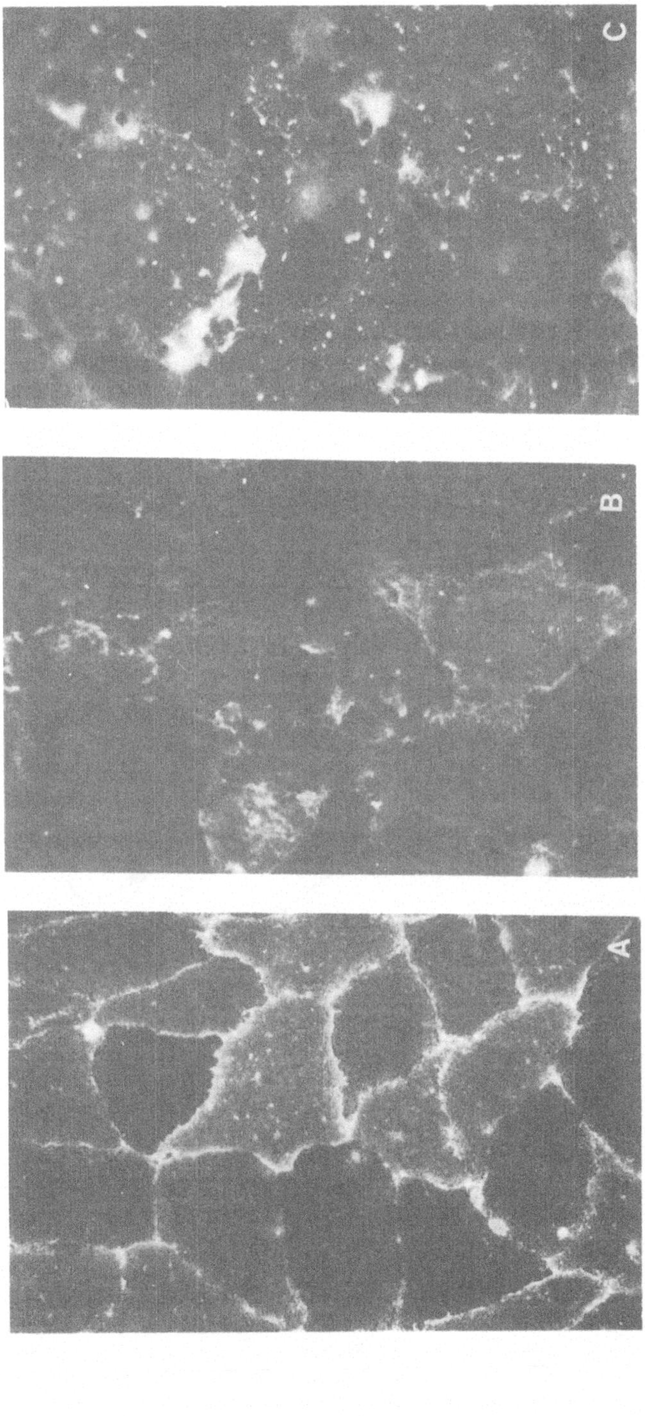

Figure 7. Immunofluorescence study of the effect of TNF on endothelial thrombomodulin. Endothelial monolayers were incubated in the absence (A) or presence of TNF (3 nM) for 2 hr (B) and 4 hr (C). Thrombomodulin was visualized by indirect immunofluorescence using a monospecific antibody to thrombomodulin. ×650.

levels of thrombomodulin mRNA.[17] By these means, TNF induces a decrease in both cell surface and total thrombomodulin, and thereby downregulates a protective antithrombotic mechanism of the vessel surface.

When the effect of TNF on endothelial thrombomodulin was compared in subconfluent, just confluent, and postconfluent cultures (cultures were characterized based on the labeling index, cell density, and time in culture, see Fig. 8 legend),[18] the following pattern of cellular response emerged (Fig. 8): the concentration of TNF resulting in half-maximal thrombomodulin suppression was ≈ 0.1 nM on the subconfluent cultures, ≈ 1 nM on the confluent cultures, and ≈ 2 nM on the postconfluent cultures. Thus, there was about a 20-fold decrease in the sensitivity of endothelium to TNF from the subconfluent to the postconfluent state, in terms of modulation of the expression of thrombomodulin. To determine if the latter results could be extended to other endothelial functions which TNF perturbs, induction of tissue factor and alteration of barrier function were studied. Induction of endothelial tissue factor also occurred at lower concentrations of TNF on subconfluent than on postconfluent cultures. Recent studies have indicated that TNF can increase the permeability of endothelial monolayers to solutes.[19–21] When TNF-mediated perturbation of

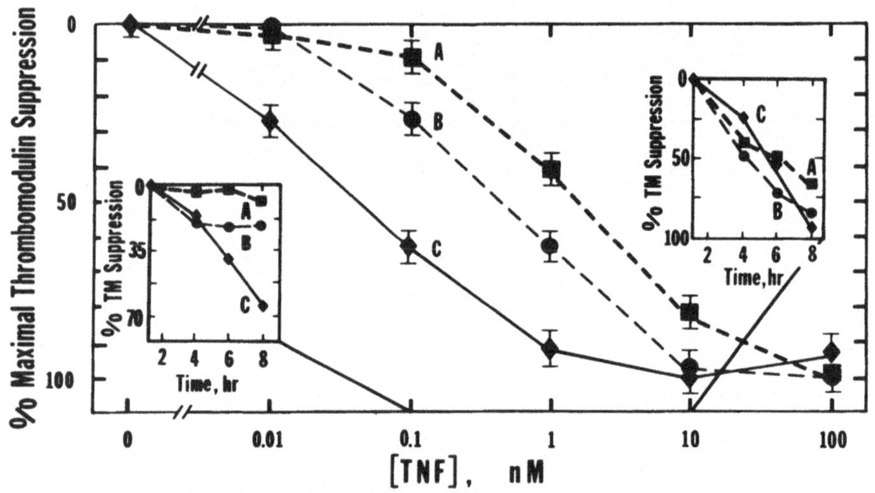

Figure 8. Effect of the growth state of endothelium on TNF-induced suppression of thrombomodulin. Endothelial monolayers (subconfluent, C, diamond; confluent, B, circle; late postconfluent, A, square) were incubated with the indicated concentration of TNF, and, after 8 hr, thrombomodulin activity, thrombin-mediated activated protein C formation, was assessed. Cultures were classified based on the number of days after plating, the number of cells per cm^2, and the labeling index: subconfluent (2 days after plating, 3.6×10^4 cells/cm^2, labeling index 30%), confluent (4 days after plating, 4.8×10^4 cells/cm^2), early postconfluent (7 days after plating, 1.3×10^5 cells/cm^2, labeling index 1–2%), and late postconfluent (14 days after plating, 2×10^5 cells/cm^2, labeling index $< 1\%$). Results are expressed as percent maximal suppression of thrombomodulin activity. Inset lower left: Time course of TNF-induced suppression of thrombomodulin with subconfluent (C, diamond), confluent (B, circle), and late postconfluent (A, square) endothelial cultures. Cultures were incubated with TNF (0.1 nM) for the indicated times and thrombin-mediated protein C activation was assessed. Results are expressed as percent maximal suppression of thrombomodulin activity as above. Inset upper right: Time course of TNF-induced suppression of thrombomodulin as in lower left inset except that the TNF concentration was 10 nM.

monolayer barrier function was compared in cultures which had just achieved confluence (and just established barrier function), lower concentrations of cytokine were required to increase permeability than with postconfluent cultures.[18]

Taken together, these results were consistent with the hypothesis that subconfluent endothelium, which is in the growth/migratory state, was more sensitive to TNF. To understand the basis for these changes in endothelial responsiveness to TNF, we examined expression of TNF binding sites on endothelium as a function of the growth state (Fig. 9).[18] Equilibrium binding studies employing radioiodinated TNF and endothelial cultures in the indicated growth were performed. Subconfluent cells displayed high-affinity binding sites with $K_d \approx 0.2$ nM, whereas when the cells arrived at confluence, the affinity of TNF binding was decreased. Postconfluent endothelial cultures displayed a progressively lower affinity for TNF. The affinity of TNF–endothelial binding appeared to decrease by 10- to 20-fold from the subconfluent to the postconfluent state. This parallels the results of the function studies. In fact, on an intact aortic vessel segment or an umbilical cord, only lower-affinity TNF binding sites could be detected. Consistent with these data, subconfluent cultures briefly exposed to TNF showed more intense staining, by indirect immunofluorescence with anti-TNF antibody, than postconfluent cultures (Fig. 10). These results were confirmed by studies with gold-labeled TNF: TNF–gold conjugates were only seen on the surface of subconfluent cultures at low concentrations of the cytokine.

The mechanisms underlying this change in the affinity of endothelium for TNF were studied under four experimentally defined conditions.[18] Cross-linking experiments with radioiodinated TNF and endothelium showed additional bands corresponding to ≈ 66 and

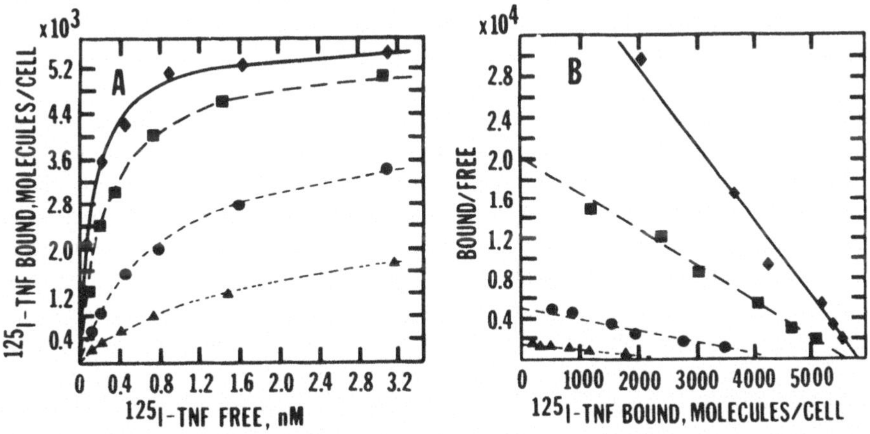

Figure 9. Binding of [^{125}I]-TNF to endothelium: comparison of subconfluent (diamond), confluent (square), early (circle), and late postconfluent cultures (diamond). (A) Endothelial monolayers were incubated for 2 hr at 4°C with the indicated concentrations of [^{125}I]-TNF alone (total binding) or [^{125}I]-TNF in the presence of excess unlabeled TNF (nonspecific binding). Specific binding (total minus nonspecific binding) is plotted versus free [^{125}I]-TNF. Data were analyzed by nonlinear least-squares analysis and the curve indicates the best fit line. (B) Scatchard analysis of the data shown in A. Parameters of binding, K_d, and the number of molecules bound per cell, follow: subconfluent (2 days after plating, ≈ 0.12 nM, 6×10^3 molecules/cell), confluent (4 days after plating, ≈ 0.27 nM, 5.4×10^3 molecules/cell), early postconfluent (7 days after plating, ≈ 0.85 nM, 4.3×10^3 molecules/cell), and late postconfluent (14 days after plating, ≈ 1.9 nM, 2.7×10^3 molecules/cell).

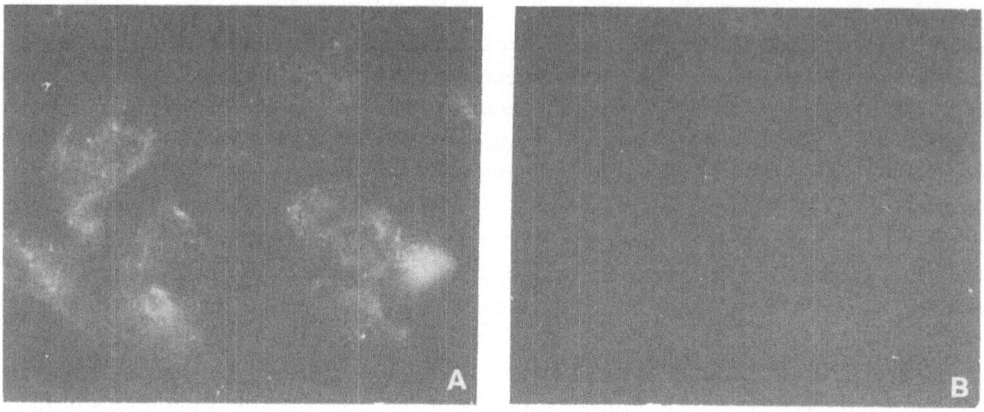

Figure 10. Visualization of TNF bound to subconfluent (A) and late postconfluent (B) endothelial monolayers by immunofluorescence. Endothelial cultures were incubated with TNF (0.3 nM) for 1 hr at 4°C, washed, and then, after 5 min at room temperature, fixation was carried out. TNF was visualized by adding rabbit anti-TNF antibody and FITC-conjugated anti-rabbit IgG. ×650.

≈ 84 kDa with subconfluent cultures which were not observed with postconfluent cultures. Experiments with X-irradiated endothelium, whose growth but not motility was blocked, showed that proliferation was not required for induction of high-affinity sites. Postconfluent endothelium, triggered to enter the proliferative cycle by microtubule poisons, expressed high-affinity TNF binding sites together with changes in cell shape/cytoskeleton prior to their entry into S phase. These observations suggested that alterations in cell shape/cytoskeleton which accompany migration/growth of endothelium have a central role in regulating the affinity of TNF binding sites and the cellular response to TNF.

Based on these findings, we considered that one mechanism for localizing TNF action within the vasculature would be local stimulation of endothelial growth or motility, which could enhance expression of high-affinity TNF receptors and thereby magnify the TNF response. To test this, we performed *in vitro* wounding studies in which postconfluent monolayers were perturbed by physically removing part of the culture and comparing the TNF response of cells close to the wound (cells at the wound edge and those which have migrated into the wound acquire a motile configuration and undergo mitosis) with those farther removed (the latter cells remain in the quiescent state throughout the response to the wound). First, wounded monolayers were allowed to recover for 48 hr, and then a low concentration of TNF was added. Localization of cytokine to the wound edge and cells which migrated into the wound was evident compared with cells farther removed from the wound. Similarly, induction of endothelial tissue factor and suppression of thrombomodulin were evident in proximity to the wound in the presence of low concentrations of TNF compared with cells distal to the wound, and control cultures wounded in the same way but not exposed to TNF. These results are consistent with the hypothesis that on subconfluent endothelial cells the receptor for TNF may be modulated to a form with higher affinity. This suggests a mechanism for targeting the action of TNF to growing/migratory endothelium, which is found in situations of angiogenesis, and injury and repair.

IV. THE EFFECT OF HYPOXIA ON ENDOTHELIAL CELL FUNCTION

Another important environmental factor prevalent in tumor vasculature which could potentially perturb endothelium is hypoxemia. Among the states which affect central properties of the endothelium, surely the most important, pathogenetically, are the consequences of hypoxemia, usually related to disturbances of blood flow. In the context of tumors, tumor vasculature is associated with increased permeability and microthromboses, both of which are closely related to the pathologic picture seen in severe hypoxemia.[22–29] Venous thrombosis is also associated with hypoxemia.[27,28] These considerations have led us to examine in detail the effect of hypoxia on endothelial barrier and coagulant properties.[29] Our hypoxia studies have employed a closed model system in which endothelium is exposed to low concentrations of oxygen, in the range of those observed in pathologic states *in vivo*.[27,30] An experiment can be carried out in its entirety in the chamber in a controlled atmosphere, to produce the effects of hypoxia, without reoxygenation, on endothelium.

A hypoxic environment is not toxic to endothelial cells; the cells remain viable by all criteria: they exclude trypan blue, remain adherent, continue to synthesize protein, and do not release LDH, consistent with previous findings.[31] However, on a more subtle level, hypoxia induces multiple reversible changes in endothelial physiology. When bovine aortic endothelium was grown to confluence in an ambient atmosphere and transferred to a hypoxic environment ($pO_2 \approx 14$ mm Hg), the endothelial cells were larger and flatter than their normoxic counterparts, and rearrangement of the actin-based cytoskeleton was observed (Fig. 11). A part of the change in cell shape included formation of small gaps between contiguous cells which are seen in Fig. 11, and were evident on scanning electron

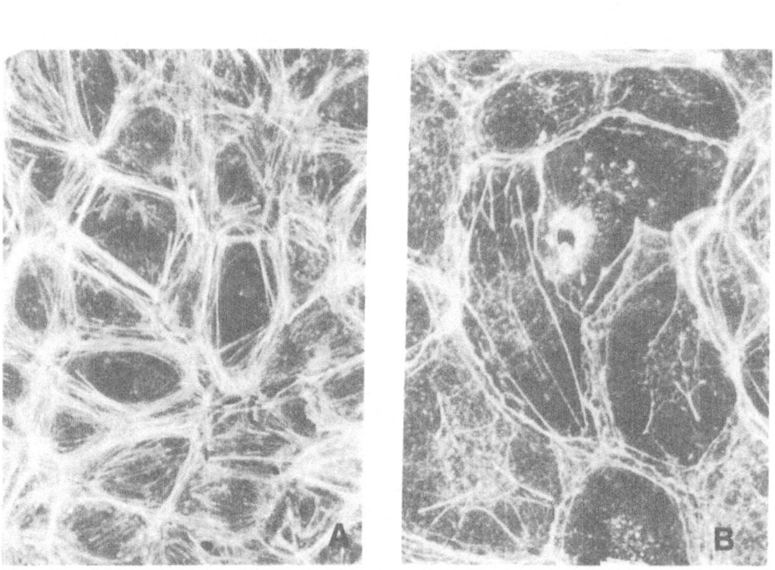

Figure 11. Hypoxia-induced perturbation of the actin-based cytoskeleton and cell shape in endothelial cell cultures. Endothelial monolayers were grown to confluence in an ambient air atmosphere (A), and transferred to hypoxia ($pO_2 \approx 14$ mm Hg) for about 24 hr (B). The actin-based cytoskeleton was visualized by rhodamine phalloidin staining. ×650.

micrographs of hypoxic cultures. These observations suggested that barrier function of the monolayer would be altered under hypoxia.

To study barrier properties of hypoxic endothelial cells, cultures were grown on filters, and passage of macromolecules and solutes of lower molecular weight was measured. In this model, endothelial cells achieve barrier function and the monolayers reproducibly display increased permeability to a range of stimuli which also increase vascular permeability *in vivo*, such as thrombin, histamine, and TNF.[20,21,32–38] However, the baseline permeability of this system is greater than that of endothelium *in vivo*,[36] indicating that only relative measurements can be made *in vitro*, and the results must be compared with *in vivo* studies using the same perturbants. Using this system, hypoxia induced a decrease in barrier function to radiolabeled tracers of a range of molecular sizes—albumin, inulin, and sorbitol (Fig. 12). The steady increase in diffusion of the three different tracers across the monolayer occurred in a time-dependent manner, increasing steadily up to 72 hr after placing cultures in hypoxia. These changes were reversible, as the bars marked R for recovery in Fig. 12A indicate. Hypoxia-induced increase in monolayer permeability is also dependent on the oxygen concentration (Fig. 12B), being most evident at the lower oxygen tensions.

The results of these studies suggest that the basis for the increased vascular permeability in hypoxia involves a decrease in barrier function of the endothelium. Increased permeability of the endothelial monolayer in hypoxia provides a mechanism for the escape of critical plasma factors into the subendothelium. This is one route by which hypoxia can enhance procoagulant events: contact of hemostatic components with the matrix. In addition, hypoxia directly modulates molecules on the endothelial surface which regulate coagulation. Our studies have examined the effect of hypoxia on thrombomodulin and endothelial cell procoagulant activity.[30]

When endothelial cultures are exposed to hypoxia, cell surface thrombomodulin activity and total cellular thrombomodulin antigen decline in parallel (Fig. 13). By 48 hr of hypoxia, levels of thrombomodulin have declined 70–80% compared with normoxic controls (Fig. 13A). This contrasts with the general decrease in endothelial protein synthesis induced by hypoxia, which decreased by about 25–35% under these conditions. Following replacement of cultures back into an ambient air atmosphere, thrombomodulin levels increased and protein synthesis also rebounded. Both actually overshot the original normoxic baseline (defined as 100%) early in the recovery period. Suppression of thrombomodulin was dependent on the oxygen concentration in the atmosphere, and was most evident at lower oxygen concentrations. At this point it was important to determine whether suppression of hypoxia reflected a general decrease in endothelial function, a hypoxia-induced "dormancy," or if it was due to specific mechanisms activated in an atmosphere with lower oxygen concentrations. To examine the selectivity of hypoxia for modulation of thrombomodulin, production of this protein and elaboration of fibronectin by endothelial cultures were compared. In contrast to thrombomodulin, hypoxia did not decrease production of fibronectin (the latter is consistent with the findings of Lynch *et al.*[39]). Furthermore, Northern blots of endothelial RNA hybridized with cDNA probes for thrombomodulin and fibronectin demonstrated no reduction in fibronectin message, but loss of the band corresponding to thrombomodulin message (Fig. 14). Although mechanisms by which hypoxia suppresses thrombomodulin are not yet completely clear, based on these data, a selective decrease in its synthesis appears to be involved.

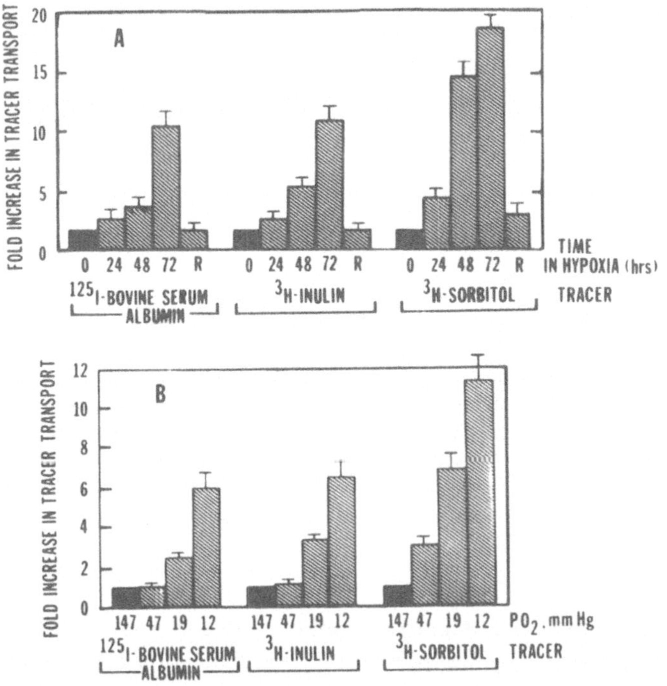

Figure 12. Permeability of endothelial monolayers exposed to hypoxia: effect of incubation time and oxygen concentration. (A) Dependence of permeability on incubation time in hypoxia. Confluent endothelial monolayers grown on filters were incubated in an environment with normal oxygen content (solid bars) or with $pO_2 \approx 14$ mm Hg (cross-hatched bars), and permeability assays were carried out at the indicated times (24, 48, 72 hr) by adding tracers ([^{125}I]albumin, [^{3}H]inulin; [^{3}H]sorbitol) to the compartment above the monolayer. R denotes cultures placed in an ambient air atmosphere (recovery) for 48 hr after 72 hr in hypoxia. (B) Dependence of permeability in oxygen concentration. Monolayers on filters were incubated for 48 hr in an atmosphere containing the indicated concentration of oxygen, and then permeability assays were carried out by adding tracers ([^{125}I]albumin; [^{3}H]inulin; [^{3}H]sorbitol) to the compartment above the monolayer. In each case (A, B), transfer of tracer to the compartment below the monolayer was assessed after a 4-hr incubation period of endothelium with the tracers, and is shown as the ratio of transfer of tracer in hypoxic cultures to transfer of tracer in normoxic cultures (fold increase in hypoxia compared with normoxic controls). Transfer across normoxic cultures for each tracer in 4 hr was arbitrarily defined as 1.

Previous studies have demonstrated that venous valve pockets can be subject to extremely low oxygen tensions approaching 0 mm Hg. In addition, under these conditions fibrin forming on the endothelial surface of the valve cusp has been observed in pathologic studies. This led Malone[28] to speculate, 23 years ago, that hypoxia might modulate properties of the vessel surface to promote clot formation. The predominant feature of endothelial coagulant properties in the quiescent state is expression of anticoagulant properties. So far, our results indicated that thrombomodulin, a central part of the protein C/protein S pathway, was suppressed. Data from another group have demonstrated a decline in net fibrinolytic activity elaborated by hypoxic endothelium.[40] Taken together, these alterations in coagulant properties would render the vessel wall less resistant to clot formation. To effectively shift the balance of coagulant properties on the endothelial surface

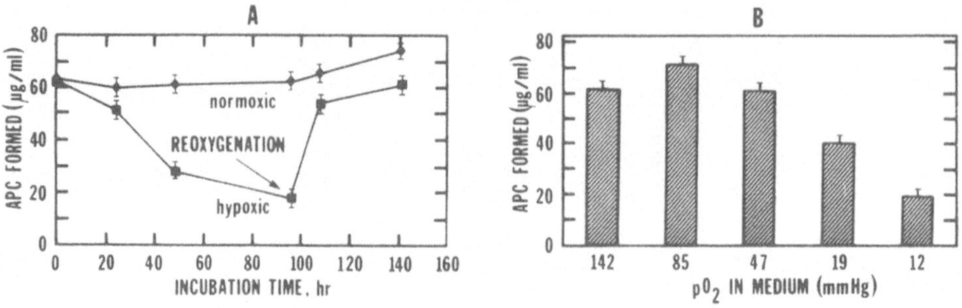

Figure 13. Hypoxia-induced suppression of endothelial thrombomodulin. (A) Effect of incubation time. Confluent endothelial monolayers were incubated in an atmosphere with $pO_2 \approx 14$ mm Hg and thrombomodulin function (squares) was assessed at the indicated times. Functional assays of endothelium-dependent activated protein C (APC) formation were carried out on intact monolayers in a hypoxic environment by incubating cultures with thrombin and protein C. The diamonds represent APC formation on normoxic endothelial cultures. (B) Effect of oxygen concentration. Confluent endothelial monolayers were incubated for 48 hr in an atmosphere with the indicated oxygen concentration and thrombomodulin functional activity was assessed. Data are shown as percent APC formation in normoxic controls/cell, the latter arbitrarily designated as maximal APC formation, 100%.

to clot formation, suppression of anticoagulant mechanisms would have to be coordinated with an induction of procoagulant properties, i.e., expression of a mechanism(s) which could initiate clot formation. According to this formulation, studies were performed to assess the ability of hypoxic endothelial cells to activate coagulation.[30] As cultures were incubated under hypoxic conditions, their ability to shorten the clotting time of plasma was increased, compared with normoxic controls (Fig. 15A). Induction of procoagulant activity also depended on the oxygen concentration, being most striking at lower pO_2 values (Fig. 15B).

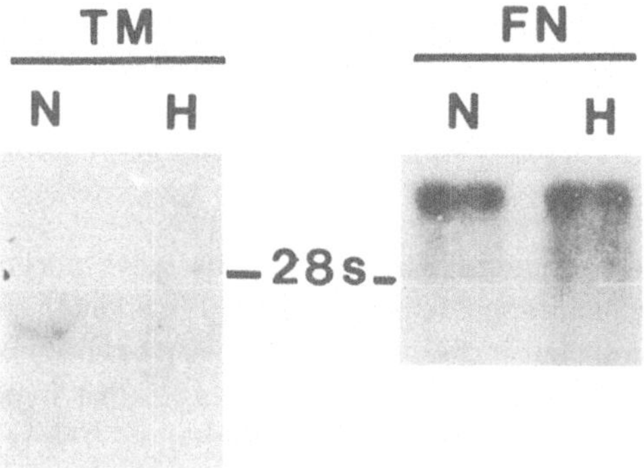

Figure 14. Northern blot of hypoxic endothelial cell RNA hybridized with cDNA probes for thrombomodulin and fibronectin. Confluent endothelial cultures were incubated in an ambient air (N) or hypoxic (H) environment for 18 hr and then RNA was extracted. Samples containing ≈ 10 µg of total RNA were applied to each lane of the gel, transferred to nitrocellulose, and then hybridized with a cDNA probe to either thrombomodulin (TM) or fibronectin (FN) (the cDNA probe for thrombomodulin was generously provided by Dr. E. Sadler and the probe for fibronectin was generously provided by Dr. R. Hynes, MIT, Cambridge, Mass.). The thrombomodulin message is between 3.7 and 3.8 kb and the fibronectin message is ≈ 8.2 kb.

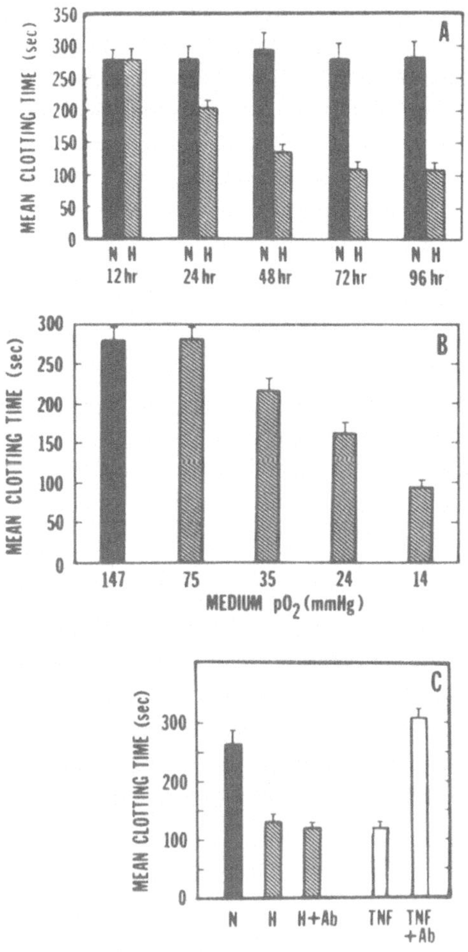

Figure 15. Hypoxia-induced expression of procoagulant activity by endothelium. (A) Incubation time. Confluent endothelial monolayers were grown to confluence and then transferred to a hypoxic atmosphere ($pO_2 \approx 14$ mm Hg) for the indicated times (H). (B) Oxygen concentration. Confluent endothelial monolayers were grown to confluence and then transferred to an atmosphere with the indicated oxygen concentration for 48 hr. In A and B, procoagulant activity was assessed based on shortening of the clotting time of recalcified plasma. N denotes normoxic controls. (C) Effect of blocking tissue factor monoclonal antibody on the procoagulant activity of hypoxic and tumor necrosis factor (TNF)-treated endothelial cultures. Confluent endothelial cultures were exposed to hypoxia ($pO_2 \approx 14$ mm Hg) for 72 hr and then their ability to shorten the clotting time was tested after preincubation for 30 min in the presence of anti-tissue factor antibody (H + Ab) or nonimmune mouse IgG (H). Control cultures were maintained in normoxia throughout the experiment (N).

Most of the time when procoagulant activity is induced in endothelium it is due to the expression of tissue factor. For example, in an earlier section of this review, data indicating that TNF induced synthesis of endothelial tissue factor were discussed. The procoagulant activity of TNF-treated endothelium is due to expression of tissue factor, which can be neutralized by the addition of a blocking monoclonal antibody to tissue factor (Fig. 15C). In contrast, anti-tissue factor antibody had no effect on the procoagulant activity of hypoxic endothelium. Since hypoxia did not appear to induce tissue factor in endothelium, this led us to characterize further the nature of the procoagulant activity responsible for shortening the clotting time of hypoxic cultures. Excluding a contribution of the extrinsic pathway based on the latter data, it seemed reasonable that hypoxia might induce endothelium-mediated direct activation of factors IX, X, and/or prothrombin. To test this, hypoxic cultures were incubated with each of these purified coagulation factors. No activation of factor IX or prothrombin was detectable. Hypoxic endothelium did activate factor X. SDS–PAGE of radiolabeled factor X showed formation of the heavy chains of factor Xaα and Xaβ in parallel with detection of factor Xa activity.[41] Induction of factor X activating capacity in endothelium by hypoxia depended on the incubation time in hypoxia (Fig. 16A), and was reversible on replacement of cultures in normoxia. In addition, factor X activation by hypoxic endothelium was most evident in cultures exposed to lower oxygen concentrations (Fig. 16B). Formation of factor Xa was not affected by the presence of blocking antibodies to factors IX, VIII, VII, and tissue factor, consistent with the interpretation that it occurred independent of the classical intrinsic and extrinsic pathways of coagulation.

Additional studies have been carried out to examine properties of the factor X activator of hypoxic endothelium in order to further characterize it and assess its possible relevance to the prothrombotic tendency observed in hypoxemic vasculature.[30] The factor X activator was expressed on the cell surface of intact hypoxic cultures, and subcellular fractionation experiments demonstrated that it was concentrated in cell membranes. In contrast, no subcellular fraction from normoxic endothelium activated factor X. Kinetic studies have shown that the factor X activator was saturable with respect to the substrate, and Michaelis–Menten analysis revealed the K_m to be ≈ 25 μg/ml at pH 7.4. Although our kinetic studies have indicated that the hypoxia-induced factor X activator is less effective than the extrinsic or intrinsic systems in the activation of factor X, alterations in the cellular milieu in hypoxia may enhance its efficacy. For example, acidotic conditions enhance its activity. Furthermore, hypoxic conditions, which are often accompanied by a severe fall in blood flow, would be associated with decreased delivery of protease inhibitors which neutralize factor X to the site of its formation on the hypoxic vessel wall. In order for hypoxic endothelium to express the factor X activator, the integrity of protein synthesis must be maintained. Thus, addition of cycloheximide to cultures simultaneous with hypoxia blocked appearance of the factor X activator. Once the factor X activator was expressed, its activity could be blocked by mercury chloride, an inhibitor of cysteine proteases, but not by phenylmethylsulfonyl fluoride, an inhibitor of serine proteases. These data suggested that a sulfhydryl group was

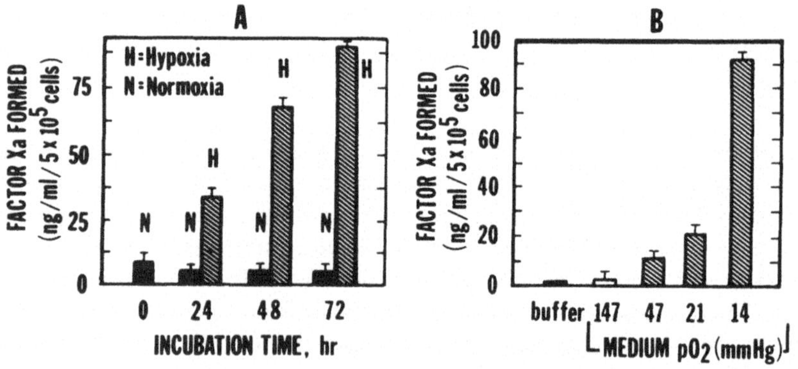

Figure 16. Factor X activation by hypoxic endothelium. (A) Effect of incubation time on the capacity of hypoxic endothelium to activate factor X. Confluent endothelial monolayers were exposed to hypoxia (pO_2 ≈ 14 mm Hg) for the indicated time, washed with EDTA-containing buffer, and then their ability to directly activate purified factor X was assessed. (B) Effect of oxygen concentration on the capacity of hypoxic endothelium to activate factor X. Confluent endothelial monolayers were incubated for 72 hr in an atmosphere with the indicated oxygen concentration, washed with EDTA-containing buffer, and then their ability to directly activate purified factor X was assessed. The bar labeled "buffer" refers to a reaction mixture containing factor X incubated for 1 hr with buffer alone in the absence of endothelial cells.

important for activity of this enzyme, raising the possibility that it could be related to a previously described factor X activator expressed selectively by malignant tissues.[42,43]

To begin to bridge the gap between these observations in cell culture and properties of hypoxemic vasculature *in vivo*, experiments were performed with the native endothelium of aortic vessel segments.[30] Vessel segments were excised and incubated in an ambient atmosphere or hypoxia. Then, factor X activation and thrombomodulin function (endothelium-dependent, thrombin-mediated formation of activated protein C) was studied. Only the hypoxic organ cultures promoted activation of factor X. In addition, activated protein C formation was selectively decreased on the hypoxic cultures. This indicated that native aortic endothelium responded to hypoxia in the same ways as our endothelium in culture.

Taken together, the results of these studies of hypoxic endothelial cells indicate that an environment with low concentrations of oxygen alters cellular functions in a reversible and selective manner. These changes result, at least in part, from a redirection of endothelial biosynthetic capacity with suppression of certain proteins and induction of others. In this context, previous studies indicated that hypoxia induced synthesis of oxygen-regulated proteins, the production of α_1-adrenergic receptors, and mitogens/inhibitors for smooth muscle cell growth.[44–52] Thus, one can expect that further studies of hypoxic endothelium should elucidate the structure and function of novel activities which may contribute to the pathogenesis of vascular dysfunction in hypoxemia.

V. CONCLUSION

The results presented in this review indicate that regulation of endothelial properties occurs in a dynamic manner, modulated by local mediators, such as the tumor-derived Meth A factor, and environmental perturbations leading to induction of the growth/migratory state of endothelium or hypoxia. In response to these alterations in the local environment, the endothelialized vessels' surface allows increased permeability to solutes and promotes a shift in the balance of cellular coagulant properties to favor clot formation. Working in concert, these stimuli can potentially create thrombogenic foci within the vasculature. Future studies will be required to test this hypothesis *in vivo*.

ACKNOWLEDGMENTS. We thank Mr. Samuel Rover for his generous contribution and Ms. Carmen Viana for her generous efforts on behalf of our laboratory. We are grateful to Dr. Gabriel Godman (Department of Pathology) for helpful suggestions during the course of this work and for assistance in the preparation of this manuscript. This work was supported by grants from the PHS (HL34625, HL42833, HL42507, CA43902), New York Lung Association (Research Award), and Council for Tobacco Research (1971 and 2101R1). D.M.S. completed this work during the tenure of a Genentech-Established Investigator Award from the American Heart Association.

REFERENCES

1. Gimbrone, M. (ed.), 1986, *Vascular Endothelium in Hemostasis and Thrombosis*, Churchill & Livingstone, Edinburgh.

2. Simionescu, N., and Simionescu, M. (eds.), 1988, *Endothelial Cell Biology in Health and Disease*, Plenum Press, New York.
3. Stern, D., and Nawroth, P. (eds.), 1987, *Vessel Wall, Semin. Thromb. Hemostasis* **13**.
4. Old, L., 1986, Tumor necrosis factor, *Science* **230:**630–634.
5. Asher, A., Mule, J., Reichert, C., Shiloni, E., and Rosenberg, S., 1987, Studies of the anti-tumor efficacy of systemically administered recombinant tumor necrosis factor against several murine tumors in vivo, *J. Immunol.* **138:**963–975.
6. Carswell, E., Old, L., Kassel, R., Green, S., Fiore, N., and Williamson, B., 1975, An endotoxin induced serum factor that causes necrosis of tumors, *Proc. Natl. Acad. Sci. USA* **72:**3666–3670.
7. Nawroth, P., Handley, D., Matsueda, G., De Waal, R., Gerlach, H., Blohm, D., and Stern, D., 1988, Tumor necrosis factor/cachectin-induced intravascular fibrin formation in meth A fibrosarcomas, *J. Exp. Med.* **168:**637–647.
8. Hui, K., Haber, E., and Matsueda, G., 1983, Monoclonal antibodies to a synthetic fibrin-like peptide bind to human fibrin but not fibrinogen, *Science* **222:**1129–1132.
9. Bale, M., and Mosher, D., 1986, Effects of thrombospondin on fibrin polymerization and structure, *J. Biol. Chem.* **261:**862–870.
10. Nawroth, P., and Stern, D., 1986, Modulation of endothelial cell hemostatic properties by TNF, *J. Exp. Med.* **164:**740–745.
11. Bevilacqua, M., Pober, J., Majeau, G., Fiers, W., Cotran, R., and Gimbrone, M., 1986, Recombinant TNF induces procoagulant activity in endothelium, *Proc. Natl. Acad. Sci. USA* **83:**4533–4537.
12. Schleef, R., Bevilacqua, M., Sawdey, M., Gimbrone, M., and Loskutoff, D., 1986, Cytokine activation of vascular endothelium. Effects on tissue-type plasminogen activator and type I plasminogen inhibitor, *J. Biol. Chem.* **263:**5797–5804.
13. Nawroth, P., Bank, I., Handley, D., Cassimeris, J., Chess, L., and Stern, D. J., 1989, Enhanced responsiveness of endothelium in the growing/motile state to tumor necrosis factor/cachectin, *Exp. Med.* **163:**1363–1375.
14. Clauss, M., Murray, C., Vianna, M., De Waal, R., Thurston, G., Nawroth, P., Gerlach, H., Gerlach, M., Bach, R., Familletti, P., and Stern, D., 1990, A polypeptide factor produced by fibrosarcoma cells that induces endothelial tissue factor and enhances the procoagulant response to tumor necrosis factor/cachectin, *J. Biol. Chem.* **265:**7078.
15. Esmon, C., 1987, The regulation of natural anticoagulant pathways, *Science* **235:**1348–1352.
16. Moore, K., Esmon, C., and Esmon, N., 1989, Tumor necrosis factor leads to the internalization and degradation of thrombomodulin from the surface of bovine aortic endothelial cells in culture, *Blood* **73:** 159–165.
17. Conway, E., and Rosenberg, R., 1988, Tumor necrosis factor suppresses transcription of the thrombomodulin gene in endothelial cells, *Circ.* **78**(Suppl. II)**:**0462a.
18. Gerlach, H., Liebermann, H., Brett, J., Bach, R., Godman, G., and Stern, D., 1989, Enhanced responsiveness of endothelium in the growing/motile state to tumor necrosis factor/cachectin, *J. Exp. Med.* **170:**913–931.
19. Henning, B., Goldblum, S., and McClain, C., 1987, Interleukin I and tumor necrosis factor/cachectin increase endothelial permeability in vitro, *J. Leukocyte Biol.* **42:**551a.
20. Clark, M., Chen, M.-J., Crooke, S., and Bomalaski, J., 1988, Tumor necrosis factor induces phospholipase A_2-activating protein in endothelial cells, *Biochem. J.* **250;**125–132.
21. Brett, J., Gerlach, H., Nawroth, P., Steinberg, S., Godman, G., and Stern, D., 1989, Tumor necrosis factor/cachectin increases permeability of endothelial cell monolayers by a mechanism involving regulatory G proteins, *J. Exp. Med.* **169:**1977–1991.
22. Stelzner, T., O'Brien, R., Sato, K., and Weil, J., 1988, Hypoxia-induced increases in pulmonary transvascular protein escape in rats, *J. Clin. Invest.* **82:**1840–1847.
23. Hultgren, J., 1978, High altitude edema, in: *Lung Water and Solute Exchange* (N. Staub, ed.), Dekker, New York, pp. 437–469.
24. Kinasewitz, G., Groome, L., Marshall, R., Leslie, W., and Diana, H., 1986, Effect of hypoxia on permeability of pulmonary endothelium of canine visceral pleura, *J. Appl. Physiol.* **61:**554–560.
25. Olesen, S.-P., 1986, Rapid increase in blood brain barrier permeability during severe hypoxia and metabolic inhibition, *Brain Res.* **368:**24–29.
26. Sevitt, S., 1967, The acutely swollen leg and deep vein thrombosis, *Br. J. Surg.* **54:**886–890.
27. Hamer, J., Malone, P., and Silver, I., 1981, The pO_2 in venous valve pockets: Its possible bearing on thrombogenesis, *Br. J. Surg.* **68:**166–170.

28. Malone, P., 1977, A hypothesis concerning the aetiology of venous thrombosis, *Med. Hypotheses* **5**:189–201.
29. Ogawa, S., Gerlach, H., Esposito, C., Pasagian-Macaulay, A., Brett, J., and Stern, D., 1990, Hypoxia modulates the barrier and coagulant function of cultured bovine endothelium, *J. Clin. Invest.* **85**:1090–1097.
30. Roca, J., Hogan, M., Story, D., Bebout, D., Haab, P., Gonzalez, R., Ueno, O., and Wagner, P., 1989, Evidence for tissue diffusion limitation of VO_2max in normal humans, *J. Appl. Physiol.* **67**:291–299.
31. Lee, S.-L., and Fanburg, B., 1987, Glycolytic activity and enhancement of serotonin uptake by endothelial cells exposed to hypoxia/anoxia, *Circ. Res.* **60**:653–688.
32. Bottaro, D., Shepro, D., Peterson, S., and Hechtman, H., 1986, Serotonin, norepinephrine, and histamine mediation of endothelial cell barrier function in vitro, *J. Cell. Physiol.* **128**:189–197.
33. Del Vecchio, P., Siflinger-Birnboim, A., Shepard, J., Bizios, R., Cooper, J., and Malik, A., 1987, Endothelial monolayer permeability to macromolecules, *Fed. Proc.* **46**:2511–2515.
34. Meyrick, B., Perkett, E., Harris, T., and Brigham, K., 1987, Correlation of permeability with the structure of the endothelial layer of pulmonary artery intimal explants, *Fed. Proc.* **46**:2516–2520.
35. Shasby, D., and Roberts, R., 1987, Transendothelial transfer of macromolecules in vitro, *Fed. Proc.* **46**:2506–2510.
36. Albelda, S., Sampson, P., Haselton, F., McNiff, J., Mueller, S., Williams, S., Fishman, A., and Levine, E., 1988, Permeability characteristics of cultured endothelial cell monolayers, *J. Appl. Physiol.* **64**:308–317.
37. Navab, M., Hough, G., Van Lenten, B., Berliner, J., and Fogelman, A., 1988, Low density lipoproteins transfer lipopolysaccharides across endothelial monolayers in a biological active form, *J. Clin. Invest.* **81**: 601–612.
38. Retrosen, D., and Gallin, J., 1986, Histamine type I receptor occupancy increases endothelial cytosolic calcium, reduces F-actin, and promotes albumin diffusion across cultured endothelial monolayers, *J. Cell Biol.* **103**:2379–2380.
39. Lynch, D., Ansel, P., and Levene, R., 1989, Effects of anoxia on gene expression in human endothelial cells, *J. Cell Biol.* **107**:581a.
40. Wojta, J., Jones, R., Binder, B., and Hoover, R., 1988, Reduction in PO_2 decreases the fibrinolytic potential of cultured bovine endothelial cells derived from pulmonary arteries and lung microvasculature, *Blood* **71**:1703–1706.
41. Fujikawa, K., Coan, M., Legaz, M., and Davie, E., 1974, The mechanism of activation of bovine factor X by intrinsic and extrinsic pathways, *Biochemistry* **13**:5290–5299.
42. Falanga, A., and Gordon, S., 1985, Isolation and characterization of cancer procoagulant: A cysteine proteinase from malignant tissue, *Biochemistry* **24**:5558–5567.
43. Falanga, A., Alessio, M., Donati, M., and Barbui, T., 1988, A new procoagulant in leukemia, *Blood* **71**: 870–875.
44. Subjeck, J., and Thung-Tai, S., 1986, Stress protein synthesis of mammalian cells, *Am. J. Physiol.* **250**:C1–C17.
45. Anderson, G., Stoler, D., and Scarcello, L., 1989, Normal fibroblasts responding to anoxia exhibit the features of the malignant phenotype, *J. Biol. Chem.* **264**:14885–14892.
46. Anderson, G., Stoler, D., and Scarcello, L., 1989, Retrotransposon-like VL30 elements are efficiently induced in anoxic rat fibroblasts, *J. Mol. Biol.* **205**:765–769.
47. Sciandra, J., Subjeck, J., and Hughes, H., 1984, Induction of glucose-regulated proteins during anaerobic exposure and of heat-shock proteins after reoxygenation, *Proc. Natl. Acad. Sci. USA* **81**:4843–4847.
48. Wilson, R., and Sutherland, R., 1989, Enhanced synthesis of specific proteins, RNA, and DNA caused by hypoxia and reoxygenation, *J. Radiat. Oncol. Biol. Phys.* **16**:957–961.
49. Hassoun, P., Pasricha, P., Teofel, E., Lee, S., and Fanburg, B., 1989, Hypoxia stimulates the release by pulmonary artery endothelial cells of an inhibitor of pulmonary artery smooth muscle cell growth, *Am. J. Respir. Cell Mol. Biol.* **1**:377–384.
50. Bounelis, P., Magargal, W., King, S., Booyse, F., Oparil, S., and Miller, D., 1989, Hypoxia stimulates platelet-derived growth factor gene expression by pulmonary artery endothelial cells, *Clin. Res.* **36**:503a.
51. King, S., Booyse, F., Lin, P.-H., Traylor, M., Narkates, A., and Oparil, S., 1989, Hypoxia stimulates endothelial cell angiotensin converting enzyme antigen synthesis, *Am. J. Physiol.* **256**:C1231–C1238.
52. Krulewitz, A., and Fanburg, B., 1984, The effect of oxygen tension on the in vitro production and release of angiotensin-converting enzyme by bovine pulmonary artery endothelial cells, *Am. Rev. Respir. Dis.* **130**:866–869.
53. Senger, D., Galli, S., Dvorak, A., Perruzzi, C., Harvey, V., and Dvorak, H., 1983, Tumor cells secrete a vascular permeability factor that promotes accumulation of ascites fluid, *Science* **219**:983–986.

31

Parathyroid Endothelium

A New Look at the Pathophysiology of Idiopathic Hypoparathyroidism

Maria Luisa Brandi and Andrea Fattorossi

I. INTRODUCTION

Idiopathic hypoparathyroidism (IHP) is a metabolic disorder characterized by deficient secretion of parathyroid hormone (PTH) and consequent hypocalcemia. Although most cases are sporadic, familial occurrence of IHP has been reported as well, with no preference for age or sex. The rarity and heterogeneity of the disease gave rise to different classifications of the various forms of IHP. This is also the explanation given for the relative paucity of information on the mechanisms that underlie PTH deficiency. Autoimmune mechanisms have been suggested to be pathogenetically instrumental in IHP, even though the disease seems to be a very rare condition in comparison with other autoimmune endocrine failures. An autoimmune basis for IHP is supported by several clinical, histological, serological, and experimental observations: (1) IHP has been described in association with Addison's disease, primary hypothyroidism, type I diabetes, malabsorption syndromes, primary hypogonadism, vitiligo, pernicious anemia, and alopecia. All these entities are believed to be of autoimmune origin or associated with autoimmune phenomena. The presence of two of the triad of Addison's disease, hypoparathyroidism, and chronic mucocutaneous candidiasis is recognized as an autosomal recessive trait, the type I polyglandular autoimmune syndrome (Type I PGA).[1] (2) Histological examination in IHP has shown atrophy of the parathyroid glands with lymphocytic infiltration.[2] (3) Antiparathyroid antibodies occurred in 38% of pediatric patients with IHP using an indirect immunofluorescence assay.[3] (4) Inoculation of homologous parathyroid tissue induced isoimmune hypoparathyroidism.[4] (5) Antibodies directed against cell surface antigens inhibited PTH secretion in IHP.[5] (6) Complement-dependent cytotoxicity was shown using sera of patients with IHP and bovine parathyroid cells in long-term culture.[6]

Maria Luisa Brandi • Department of Physiopathology, University of Florence Medical School, Florence 50139, Italy. *Andrea Fattorossi* • Immunology Unit, Medical Department, DASRS, Rome, Italy.

Endothelial Cell Dysfunctions, edited by Nicolae Simionescu and Maya Simionescu. Plenum Press, New York, 1992.

Although IHP is almost certainly immunologically mediated so that IHP is nowadays largely synonymous with autoimmune hypoparathyroidism, neither the precise mechanism(s) producing parathyroid cell damage nor the target(s) of the autoimmune reaction have been defined. The parathyroid chief cell appears to be the best candidate. However, as Table 1 indicates, there are fragmentary and conflicting results regarding both the nature of the primary target antigen(s) and the functional effects of circulating autoantibodies in IHP. Blizzard *et al.* found antibodies to parathyroid tissue in patients with IHP.[3] No information on the antigen location within the parathyroid tissue was obtainable from these studies. Successively, antibodies of the IgG class to parathyroid oxyphil cells were described in the serum of a patient with IHP.[7] Moreover, using an indirect immunofluorescence technique, others have found that IHP patient sera which were reactive against human parathyroid cells showed a similar positive reactivity against other human cells rich in mitochondria.[8] This reactivity could be abolished by absorption with human mitochondria and the antigen recognized by the IHP sera was a protein different from previously described mitochondrial autoantigens. More recently, Posillico *et al.* were able to demonstrate a functional role of parathyroid autoantibodies in IHP using human dispersed parathyroid cells.[5] They demonstrated, in a selected number of cases, that antiparathyroid antibodies reacting with cell surface antigens caused a marked inhibition of PTH secretion.[5]

However, the pathogenesis of IHP remains obscure and although IHP is conventionally included in the list of destructive organ-specific autoimmune disorders, it has proved difficult to demonstrate cytotoxic autoantibodies in this disorder. Our studies on the pathogenesis of IHP have taken the form of *in vitro* models consisting of long-term cultures of bovine parathyroid (PT-b) cells as targets, and sera from IHP patients as effectors.[6,9] The end point has been some index of cell damage, such as a decrease in thymidine incorporation and loss of cytoplasmic membrane integrity utilizing the chromium release technique. The results of our studies have been consistent with the hypothesis that in IHP sera there is an immunoglobulin component capable of damaging parathyroid cells.

Recently, certain facts suggested the possibility that our culture system contained epithelial as well as endothelial cells. Indeed, bovine parathyroid glands, our source of cells, represent highly vascularized endocrine organs, and approximately one third of the cell population is estimated to be endothelial in nature. Moreover, bovine parathyroid cells, although maintained in long-term culture for many passages, could not be cloned,

Table 1. Circulating Autoantibodies Reactive with Parathyroid Cells

Substrate	Detection	Target	Effect	Ref.
Human parathyroid tissue	Immunofluorescence	—	—	3
Human parathyroid tissue	Immunofluorescence	Mitochondria	—	8
Human parathyroid tissue and cells	Immunofluorescence; PTH secretion	Cell surface	Inhibition of PTH secretion	5
Bovine parathyroid cells	Immunofluorescence; Cr release	Cell surface	Cytotoxicity	6
Bovine parathyroid endothelial cells	Immunofluorescence	Cell surface	—	15

suggesting the possibility that interactions among different cell types in the culture facilitate long-term growth. Favorable growth of bovine parathyroid cells was fostered in a medium containing bovine pituitary and hypothalamic extracts, which are extremely rich in mitogens for endothelial cells. These observations prompted us to clone selectively epithelial and endothelial cells. The epithelial cells have been cultured and cloned from rat parathyroid tissue,[10] instead, cloned endothelial cells have been obtained from bovine parathyroid tissue.[11]

Our earlier studies using the PT-b cells indicated the possibility that autoimmunity may have an important role in the etiology or progression of hypoparathyroidism. We have further investigated this possibility, taking into account another component of the parathyroid tissue, the endothelium.

II. EXPERIMENTAL PROCEDURES

A. Patients

Serum samples were obtained from 42 patients with the following diagnoses: 6, IHP; 7, autoimmune thyroid disease; 4, Addison's disease; 3, idiopathic pulmonary fibrosis; 4, primary biliary cirrhosis; 3, rheumatoid arthritis; 8, systemic lupus erythematosus; 7, normal subjects. Sera for antibodies were analyzed after storage at $-20°$ for up to 3 years. Complement was inactivated by incubating sera at 56°C for 30 min. Clinical protocols were carried out after informed consent was obtained from all patients.

B. Cell Culture

PT-b cells were prepared and cultured as previously described in detail.[12] These cells were cultured in serum-free Coon's modified Ham's F12 medium that contained antibiotics, 5 μg of bovine pituitary extract, 150 μg of bovine hypothalamic extract, 100 ng of epidermal growth factor, 10 μg of insulin, 5 μg of transferrin, 5 ng of selenous acid, 3.5 ng of hydrocortisone, 25 pg of triiodothyronine, 15 ng of retinoic acid, 200 μg of D-galactose, and 30 μg of putrescine per milliliter, as well as 0.3 mM calcium and 0.5 mM magnesium. PT-b cells were studied between the second and tenth passage. Cloned rat parathyroid epithelial (PT-r) cells were cultured as described previously in detail.[10] The cells were grown in a mixture (1:1) of Coon's modified Ham's F12 and Dulbecco's modified Eagle's minimal essential (DME) medium containing 5% calf serum, 1% Nutridoma-SP™, 1 mM calcium, and 0.5 mM magnesium. Endothelial cells from bovine parathyroid tissue (BPE) were cloned by limiting dilution and cultured in Coon's modified Ham's F12 medium that contained antibiotics, 10% Nu-Serum™, 2% Ultroser-G™, 200 μg/ml of D-galactose, 1.0 mM calcium, and 0.5 mM magnesium.[11] Bovine adrenal medullary endothelial (BAME) cells were grown in minimal essential medium (MEM) supplemented with antibiotics and 10% heat-inactivated fetal calf serum (FCS) as described.[13] Pulmonary artery endothelial (BPAE) cells (ATCC No. CCL 209) were maintained in MEM with antibiotics and 20% FCS. Normal human skin fibroblasts (HSF) (ATCC No. CRL 1229) were maintained in DME medium supplemented with antibiotics and 20% FCS.

C. Flow Cytometry

Low-density lipoprotein uptake, a hallmark of endothelial cell phenotype, was measured using the procedures of Voyta *et al.*,[14] with modification. Cells to be labeled were incubated with 10 μg/ml of 1,1′-dioctadecyl-3,3,3′,3′-tetramethyl-indocarbocyanine perchlorate acetylated low-density lipoprotein (DiI-Ac-LDL) for 3 hr at 37°C. At the end of the incubation time, cells were washed once with serum-free culture medium and treated with trypsin to yield a single cell suspension. Suspended cells were washed with cold culture medium, centrifuged, resuspended at a concentration of 10^6/ml, and finally analyzed by flow cytometry.

Sera were tested in an indirect immunofluorescence assay against a series of cultured cells using flow cytometry. For flow cytometry experiments, cells were released from culture plates using trypsin solution, resuspended in cold phosphate-buffered saline (PBS), and incubated for 45 min in an ice bath with a 1:20 dilution of serum from patients and normal subjects. After repeated washing with PBS, cells were incubated for an additional 45 min in an ice bath with a 1:40 dilution of a monospecific, affinity-purified, fluorescein-conjugated goat anti-human IgG, IgM, or IgA. The cells were washed again with PBS and used for flow cytometry.

D. Indirect Immunofluorescence in Attached Cells

Cells were cultured on glass Lab-Tek™ 8-chamber slides, washed with PBS, and incubated for 45 min in ice with a 1:20 dilution of serum from patients or normal subjects. After repeated washing with PBS, cells were incubated for 45 min in ice with a 1:40 dilution of a monospecific, affinity-purified, fluorescein-conjugated goat anti-human IgG, IgM, or IgA, and examined with a fluorescence microscope.

E. Immunohistologic Staining

Immunofluorescence was performed on cold 1:1 methanol/acetone-fixed frozen sections obtained from bovine parathyroid, adrenal, pituitary, kidney, and liver tissues. Human parathyroid adenoma tissues were also used. Sections were sequentially incubated with a 1:20 dilution of sera from IHP patients or normal subjects. Indirect immunofluorescence was developed with 1:20 fluorescein-conjugated goat anti-human IgM as second antibody. Double fluorescence staining was also performed to compare the distribution of vascular endothelium and antigen(s) recognized by IHP sera. Factor VIII-related antigen (factor VIII-R Ag) was chosen as marker for the endothelium. Sections were sequentially incubated with a 1:20 dilution of IHP serum and with a 1:30 dilution of rabbit antiserum for factor VIII-R Ag. Sections were washed with PBS and incubated with a mixture of a 1:20 dilution of fluorescein-conjugated goat anti-human IgM and a 1:30 dilution of rhodamine-conjugated goal anti-rabbit IgG, washed, and scanned by fluorescence microscopy.

F. Adsorption Studies

Different cell types (10×10^6) and their tenfold concentrated conditioned media were incubated with a 1:10 final dilution of IHP serum for 4 hr at 4°C. The supernatants from the cells and the concentrated conditioned media were then immediately used for indirect immunofluorescence studies.

Adsorption of IHP sera was also performed with 1 to 5 mg/ml concentrations of purified collagen type I, III, IV, and V, and of laminin and fibronectin. Supernatants were clarified by centrifugation and used for immunofluorescence analysis.

III. RESULTS

A. Clinical Features

All IHP patients were clinically hypoparathyroid. As previously described, these IHP sera mediated specific lysis of PT-b cells.[6] Patient 1 showed borderline-detectable adrenal antibodies. Patient 2 displayed seropositive rheumatoid arthritis. Patients 3–6 expressed mucocutaneous candidiasis, patients 5 and 6 were alopecic, and patients 4 and 6 displayed adrenal insufficiency. Patients 1 and 2 are twin sisters, patients 3–5 are siblings in another family. There were no other affected members in the family of patient 6.

B. Culture Characteristics

BPE cells exhibited a profound ability to take up and package exogenous DiI-Ac-LDL in punctate storage sites in the perinuclear cytosol.[11] BPE cells from replicate cultures not exposed to DiI-Ac-LDL showed no fluorescence. Cultured PT-b cells contained brightly fluorescent colonies. Flow cytometry analysis showed a strong signal with BPE cells, while PT-r cells gave a weak fluorescence signal (not shown). The data from these analyses were used to set scatter and fluorescence sampling for analysis and sorting of PT-b cells. This made it possible to quantify the endothelial cell component in PT-b cells. The profile of primary culture and tenth passage of PT-b cells maintained in the usual medium is shown in Fig. 1. The fluorescence shift of PT-b cells at the tenth passage was similar to that of primary cultures, accounting for approximately 8% of the whole PT-b population. However, PT-b cells from tenth passage had a much greater fluorescence intensity and most importantly, took up a high amount of DiI-Ac-LDL (Fig. 1). PT-b cells switched from the usual medium at the first passage to BPE growth medium showed after ten passages a substantially larger population of DiI-Ac-LDL positive cells (Fig. 1).

C. Immunofluorescence in Cultured Cells

Immunofluorescence analysis of trypsinized resuspended PT-b, BPE, BPAE, or BAME cells revealed that sera from IHP patients reacted with the outer cell membrane of the entire cell population (Fig. 2). Normal sera and sera from patients with autoimmune diseases other than IHP did not react (not shown). Tests with monospecific anti-human immunoglobulins indicated that only IgM antibodies reacted with the bovine cells. PT-r cells and HSF were completely nonreactive (not shown). A similar extracellular distribution of the antigen in punctate storage sites was found in adherent BPE cells freshly stained with IHP sera (Fig. 3). Flow cytometry analysis confirmed the fluorescence microscopy observations and showed a single peak of fluorescence with PT-b, BPE, BAME and BPAE

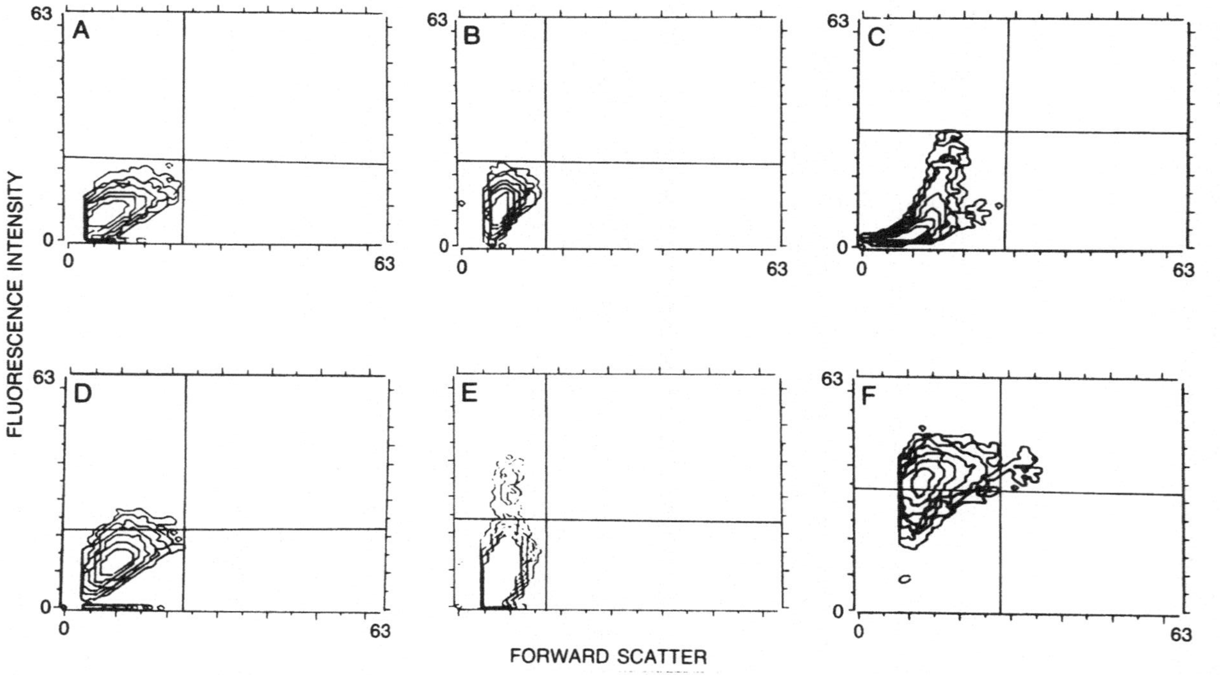

Figure 1. Contour plots for uptake of DiI-Ac-LDL with PT-b cells at different passages. Control plots for PT-b cells in primary culture, tenth passage in the usual growth medium, and tenth passage in medium used to culture BPE cells are depicted in A, B, and C, respectively. Profiles for the corresponding DiI-Ac-LDL labeled cells are represented in the lower panels (D, E, and F).

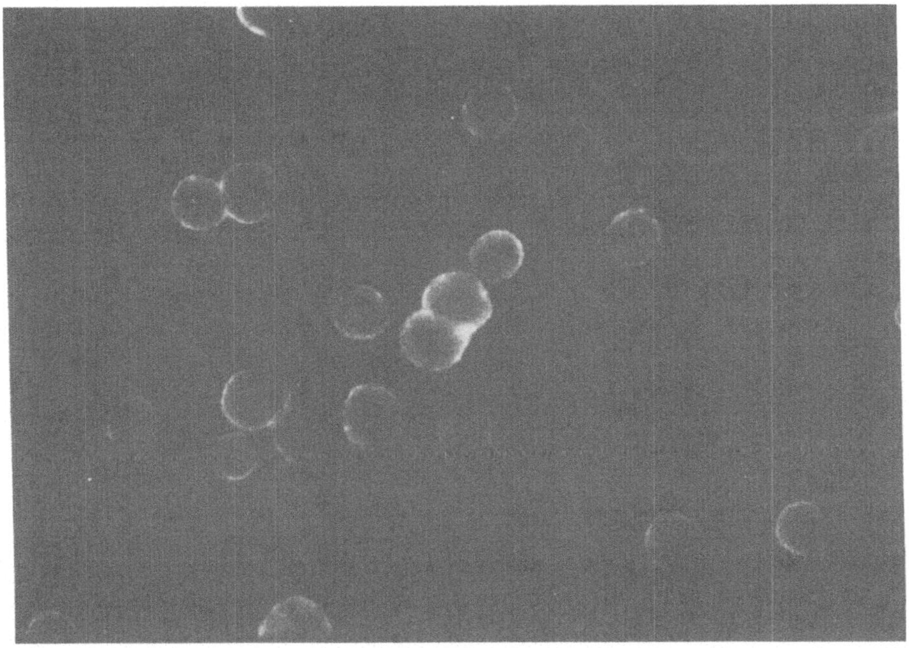

Figure 2. Staining pattern of BPE cells in suspension after incubation with IHP serum and fluorescein-conjugated anti-human IgM.

cells (Fig. 4). The intensity of staining varied among the bovine endothelial cells (Fig. 4) and among IHP patients (Fig. 5), suggesting a different antigen expression and different antibody titer, respectively. The fluorescence intensity in BPE, BAME, and BPAE cells incubated with IHP sera was significantly higher than in PT-r cells and HSF (Fig. 5). Among the non-IHP patients, serum from only one, an addisonian, contained IgM-reactive antibodies.[15] Sera from normal subjects produced a fluorescence reactivity significantly lower than that obtained with sera from IHP sera (Fig. 5). Reactivity with endothelial cells was abolished after adsorption with PT-b, BPE, BAME, and BPAE cells, while PT-r cells and HSF were inert (Table 2). Sera from IHP patients adsorbed with conditioned medium from PT-b, BPE, BAME, and BPAE cells significantly reduced the reactivity, as calculated by the shift in the mean peak of fluorescence of endothelial cells toward the control value (Table 2). Adsorption experiments with conditioned medium from PT-r cells and HSF were without effect (not shown). Incubation with IHP sera with collagen type I, III, IV, and V, laminin, and fibronectin, all components of the extracellular matrix, did not change the fluorescence reactivity of endothelial cells (Table 2).

D. Immunohistology

Consistent with data obtained with cultured cells, all sera from IHP and the non-IHP addisonian patients contained IgM antibodies that reacted with determinants in bovine

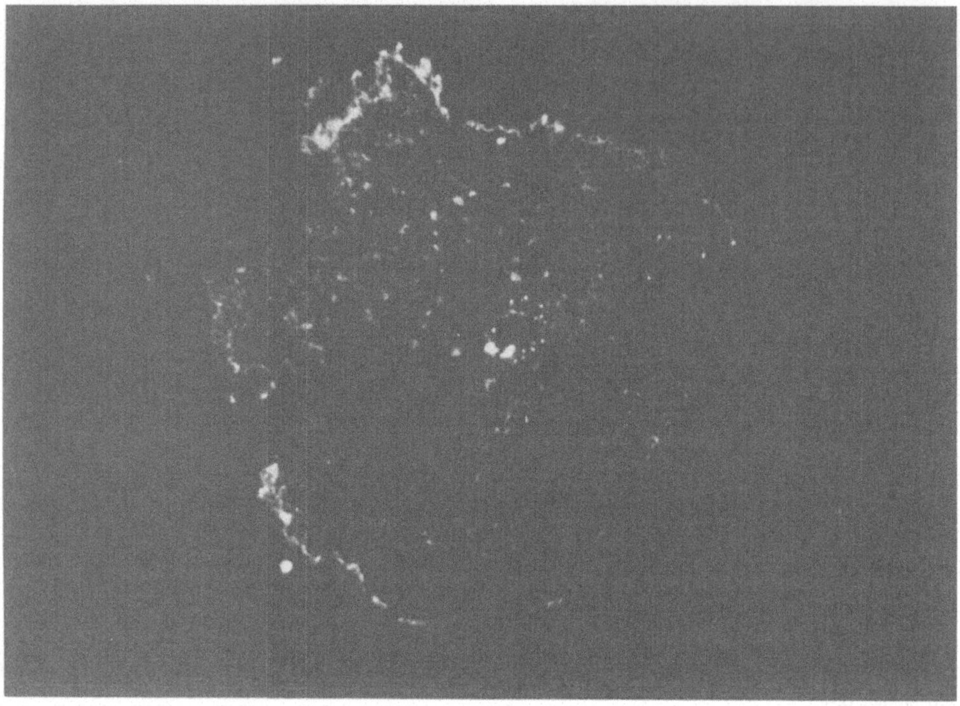

Figure 3. Indirect immunofluorescence staining of adherent BPE cells after incubation with IHP serum and fluorescein-conjugated anti-human IgM.

parathyroid tissue closely related to the endothelium and in close apposition with the epithelial cell membrane. A similar endothelium-related reactivity was observed in bovine adrenal, pituitary, and kidney, but the intensity of the staining was much lower and the antigen(s) were not as evenly distributed as in the parathyroid tissue (not shown). Sera from normal subjects and patients with non-IHP autoimmune patients did not show such endothelium-related pattern in any tissue examined (not shown). Adsorption of IHP sera with PT-b, BPE, BAME, and BPAE cells or their conditioned media reduced the reactivity of bovine parathyroid tissue, whereas adsorption on PT-r cells or proteins of the extracellular matrix produced no detectable changes (Table 2).

Sera from IHP patients gave less consistent results when tested on human parathyroid adenoma tissues.[15] One serum was positive with all three human specimens analyzed; two other sera were positive with one specimen each; and the remaining three sera showed no reactivity with any of the adenomas.[15] Adsorption of positive sera with PT-b, BPE, and BPAE cells dramatically reduced the fluorescence intensity related to the vascular endothelium, whereas adsorption with PT-r cells or the proteins of the extracellular matrix had no effect (Table 2).

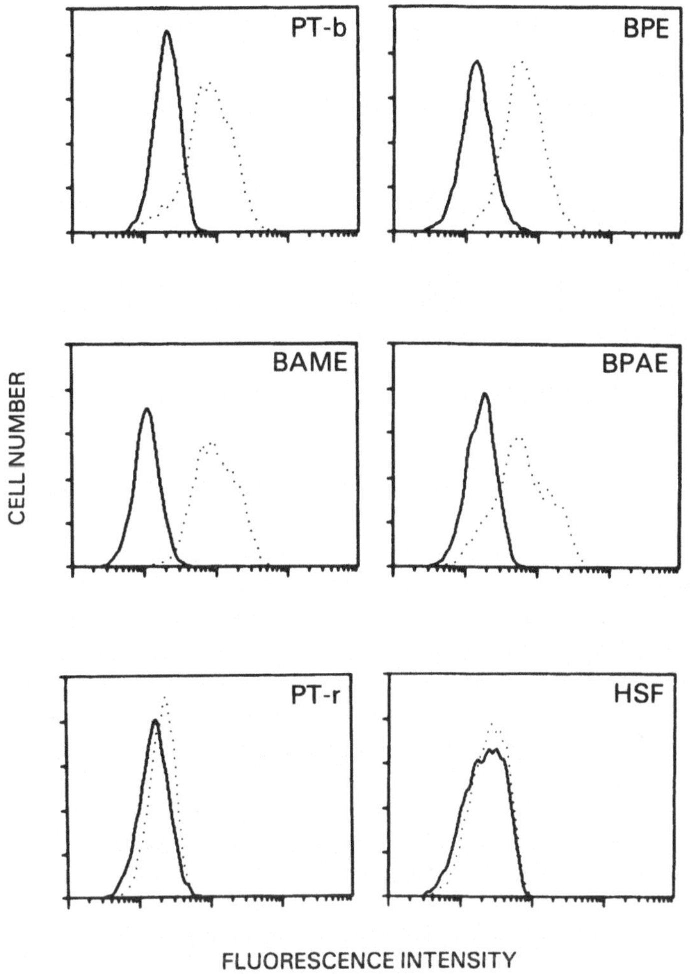

Figure 4. Flow cytometry profiles of PT-b, BPE, BAME, BPAE, PT-r, and HSF incubated with serum from IHP patient and fluorescein-conjugated anti-human IgM.

IV. DISCUSSION

An epithelium-directed autoimmunity would affect parathyroid through two major mechanisms: (1) a slow destructive process, by which the normal parenchyma is attacked by antibodies and loss of physiological function occurs; and (2) a blocking process, in which antibodies play a major role by competing with agonists for binding to cell surface receptors and by producing gradual atrophy of the parathyroid tissue. Earlier we reported that use of a long-term parathyroid cell culture system, the so-called PT-b cells, allowed detection of cytotoxic autoantibodies of the IgM class in IHP.[6,9] Accordingly, atrophy of parathyroid tissue in IHP would be the result of a destructive process mediated by cytotoxic antibodies. Although the direct immune hypothesis on the epithelial cells is attractive,

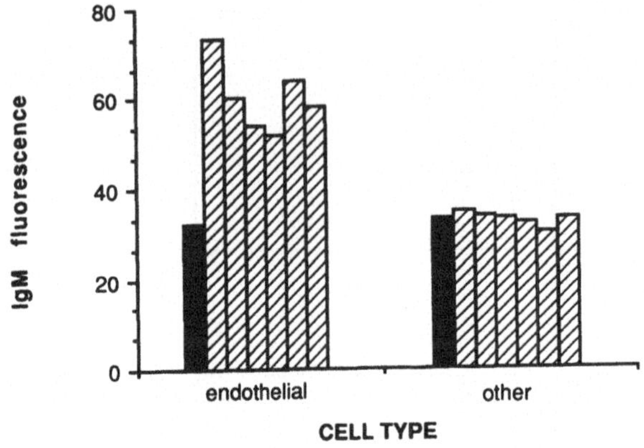

Figure 5. Reactivity of serum from subjects with six patients affected by IHP (textured bars) and from healthy subjects (solid bars) on bovine endothelial cells (BPE, BAME, BPAE) and on nonendothelial cells (PT-r, HSF).

another mechanism may be operative in the pathogenesis of IHP. At present, we think that anti-endothelial cell antibodies of the IgM class may be the expression of IHP.[15] The vascular hypothesis suggests that the endothelial cell attack might be a primary event in the initiation of the disease.

The prospect that autoantibodies to the endothelial cell membrane in IHP may be initiating cellular damage, either directly or indirectly, indicates possible new directions for the study and classification of IHP. The endothelium has become the focus in a number of studies as a potential target for cell-specific autoantibody binding. Antibodies directed against endothelial cells have been demonstrated in a variety of disease situations, including

Table 2. Effect of Adsorption of IHP Serum with Different Cell Lines, Concentrated Conditioned Medium from Endothelial Cells, and Proteins of the Extracellular Matrix on the Immunofluorescence Reactivity of Cultured Endothelial Cells and of Parathyroid Tissue

	Fluorescence[a] intensity			Tissue[b]	
Substrate	BPE	BAME	BPAE	Bovine	Human[c]
None	74	62	65	+++	+++
PT-b	32	33	31	+	+
BPE	38	43	37	+	+
BAME	40	39	39	+	nd
BPAE	36	41	40	+	+
PT-r	70	61	67	+++	+++
HSF	72	58	60	nd	nd
Conditioned medium[d]	37–44	36–44	40–44	++	nd
Proteins of the extracellular matrix[e]	69–71	62–64	61–55	+++	+++

[a]Mean peak of fluorescence in units of channel number. For each data point, numbers represent the mean value of at least three separate measurements. The average of four healthy donors assayed in parallel was 38 ± 4.

[b]+++, strong reaction; ++, moderate reaction; +, weak to negative reaction.

[c]One parathyroid adenoma.

[d]Conditioned media from PT-b, BPE, BAME, and BPAE cells.

[e]Collagen type I, III, IV, and V, laminin, and fibronectin.

systemic lupus erythematosus, rheumatoid arthritis, scleroderma, thrombotic thrombocytopenic purpura, and multiple sclerosis. The conventional concepts of endothelial physiology, prevalent decades ago, were simple and convenient. The understanding of endothelial cell ultrastructure and the improvements in culture techniques greatly contributed to the recognition of a marked variability of endothelial cells from tissue to tissue. Alterations of the endothelial cells offer an attractive mechanism by which some manifestation of specific diseases could be mediated. It should also be noted that the presence of antibodies against endothelium with the described specificity might explain the involvement of several organs in type I PGA.

The significance of antiendothelium antibodies in the pathogenesis of IHP cannot be conclusively established. The reactivity with the outer cell membrane in culture and with accessible determinants in whole tissue fulfills one of the essential prerequisites for a pathogenic antibody. A constant blood flow is necessary for the maintenance of tissue integrity. Therefore, any condition affecting the function of endothelium can have devastating consequences for the physiology of the whole organ. However, we have not yet been able to demonstrate any complement-mediated cytotoxicity in parathyroid endothelial cells *in vitro* using IHP serum (M. L. Brandi, personal communication). These observations do not rule out the possibility that IHP antiendothelial antibodies may become cytotoxic *in vivo*. Moreover, endothelial cells may still have a pivotal role in the pathogenesis of parathyroid atrophy in IHP. It is hypothesized that in IHP the endothelial antigen(s) is transferred to epithelial cells, where the cytotoxic effect would take place. Support for this view comes from the following observations: (1) clonal parathyroid epithelial cells do not react with IHP sera; (2) in flow cytometry experiments, PT-b cells, which are comprised of both epithelial and endothelial components, gave a single peak of fluorescence with IHP sera, indicating a uniform distribution of reacting antibodies in the mixed cell population; (3) concentrated conditioned medium from parathyroid endothelial cells can adsorb the IHP antibodies; (4) the target for IgM antibodies in tissue sections was often found in close apposition to the epithelial cells. All these observations suggest the possibility that the IHP antigen(s) is shed from endothelial cells and adsorbed by the surrounding epithelial cells, where several factors may contribute to the expression of a cytotoxic effect (Fig. 6).

One of the major questions raised from this proposed model comes from the observation that the antigenic determinants are not restricted to the parathyroid endothelium, as consistent reactivity was also found in endothelial cells derived from other tissues. However, the etiology of autoimmune diseases is clearly multifactorial and therefore, several intrinsic and extrinsic factors are critical to the development of a clinically overt autoimmune damage. This may also explain the diversity in organ involvement among patients affected by IHP.

Although not organ-specific, the antiendothelial antibodies circulating in IHP patients appear to be disease-specific. In only one patient without clinical evidence of parathyroid dysfunction were characteristic IgM antibodies detectable. This patient displayed Addison's disease. Adrenocortical insufficiency is an extremely frequent finding in IHP patients with type I PGA and a 20% incidence of parathyroid antibodies in patients with isolated idiopathic adrenocortical insufficiency has been reported. The disease specificity of these antiendothelial IgM antibodies may represent a reliable system for the diagnosis and follow-up of IHP patients. In fact, inconsistencies in detecting antiparathyroid antibodies by using human parathyroid adenomas in IHP have already been noted.

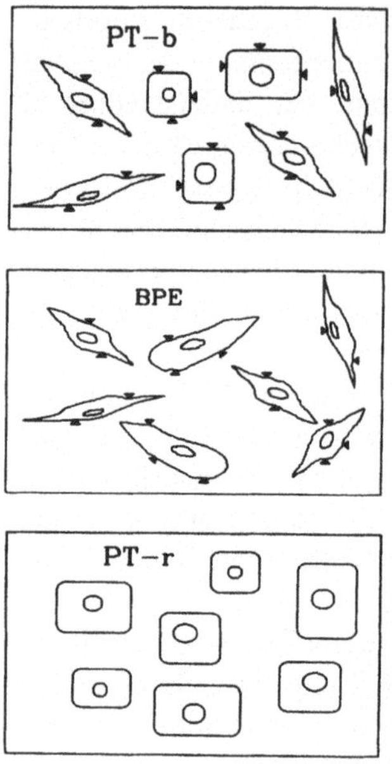

Figure 6. Schematic representation of PT-b, BPE, and PT-r cells. Solid triangles depict antigenic determinants recognized by IHP antibodies.

The presence of antibodies of the IgM class directed against endothelial antigen(s) offers an attractive explanation for some manifestations of IHP. These antibodies in combination with other factors may contribute to the development of parathyroid tissue damage. Future studies should make it possible to clarify the role of endothelium in parathyroid pathophysiology.

V. PERSPECTIVES

The biological and clinical significance of circulating anti-endothelial cell antibodies in IHP at present remains unknown. Further studies are needed to ascertain their clinical relevance, their usefulness in predicting the course of the disease and possibly directing the therapeutic management. Assuming that antiendothelial antibodies play a role in the tissue damage of IHP, it is critical to determine the chronological relationship of them by carrying out prospective studies of newly diagnosed patients and their relatives.

The BPE cells and the studies initiated here should make it possible to characterize further the nature of cell-specific antibodies in IHP and perhaps to identify the antigen(s) involved. Two major bands of 200 and 130 kDa in BPE cell membranes have been recognized by immunoblotting techniques.[15] A molecular approach, isolating cDNA(s) for the autoantigen(s) identified in the parathyroid tissue, will allow comparison of the corresponding proteins for homology.

According to our theory the destruction of parathyroid tissue is induced by organ-specific hypersensitivity of the humoral type. However, a discrete infiltration of the parathyroid parenchyma by mononuclear cells in IHP patients suggests the presence of an organ-specific hypersensitivity of the cellular type. Presumably, if immunological mechanisms are of importance in the etiology of IHP, they operate through a synergism between delayed hypersensitivity and circulating antibodies. A complete evaluation of the immunology of IHP accordingly must include an estimation of cell-mediated autoreactivity.

Finally, assuming that only cross-reactive antibodies have the ability to cause multiple organ-specific associations, one must explain why clinical evidence for damage of all the corresponding target organs is not present in all patients. Local tissue factors such as homing receptors, T cell infiltration, major histocompatibility complex expression, and antigen availability, as well as variable affinity of a single antibody for different tissue epitopes, may determine whether or not disease occurs.

ACKNOWLEDGMENTS. The authors thank Drs. Carolyn Bondy, Penelope Feuillan, Lorraine Fitzpatrick, Stephen Marx, Merily Poth, and Allen Spiegel for referral of their patients at the National Institutes of Health, Bethesda, Md.

REFERENCES

1. Doniach, D., and Bottazzo, G. F., 1981, Polyendocrine autoimmunity, in: *Clinical Immunology Update* (E. C. Franklin, ed.), Elsevier, Amsterdam, p. 95.
2. Craig, J. M., Schiff, L. H., and Boone, J. E., 1955, Chronic moniliasis associated with Addison's disease, *Am. J. Dis. Child.* **89**:669–684.
3. Blizzard, R. M., Chee, D., and Davis, W., 1966, The incidence of parathyroid and other antibodies in the sera of patients with idiopathic hypoparathyroidism, *Clin. Exp. Immunol.* **1**:119–128.
4. Lupulescu, A., Potorac, E., Pop, A., Heitmanck, S., Merculiev, E., Chisiu, N., Oprisan, R., and Neacsu, C., 1968, Experimental investigations on immunology of the parathyroid gland, *Immunology* **14**:475–482.
5. Posillico, J. T., Wortsman, J., Srikanta, S., Eisenbarth, G. S., Mallette, L. E., and Brown, E. M., 1986, Parathyroid cell surface autoantibodies that inhibit parathyroid hormone secretion from dispersed human parathyroid cells, *J. Bone Mineral Res.* **1**:475–483.
6. Brandi, M. L., Aurbach, D. G., Fattorossi, A., Fattorossi, A., Quarto, R., Marx, S. J., and Fitzpatrick, L. A., 1966, Antibodies cytotoxic to bovine parathyroid cells in autoimmune hypoparathyroidism, *Proc. Natl. Acad. Sci. USA* **83**:8366–8369.
7. Irvine, W. J., and Scarth, L., 1969, Antibody to the oxyphil cells of the human parathyroid in idiopathic hypoparathyroidism, *Clin. Exp. Immunol.* **4**:505–510.
8. Betterle, C., Caretto, A., Zeviani, M., Pedini, B., and Salvianti, C., 1985, Demonstration and characterization of anti-human mitochondria autoantibodies in idiopathic hypoparathyroidism and in other conditions, *Clin. Exp. Immunol.* **62**:353–360.
9. Quarto, R., Brandi, M. L., Fitzpatrick, L. A., Fattorossi, A., and Aurbach, G. D., 1986, Antibodies to parathyroid cells in autoimmune hypoparathyroidism, in: *Current Advances in Skeletogenesis* (S. Hurwitz and J. Sela, eds.), Heiliger, Jerusalem, p. 230.
10. Sakaguchi, K., Santora, A., Zimering, M., Curcio, F., Aurbach, G. D., and Brandi, M. L., 1987, Functional epithelial cell line cloned from rat parathyroid glands, *Proc. Natl. Acad. Sci. USA* **84**:3269–3273.
11. Brandi, M. L., Ornberg, R. L., Sakaguchi, K., Curcio, F., Fattorossi, A., Lelkes, P. I., Matsui, T., Zimering, M., and Aurbach, G. D., 1990, Establishment and characterization of a clonal line of parathyroid endothelial cells, *FASEB J.* **4**:3152–3158.
12. Brandi, M. L., Fitzpatrick, L. A., Coon, H. G., and Aurbach, G. D., 1986, Bovine parathyroid cells: Cultures maintained for more than 140 population doublings, *Proc. Natl. Acad. Sci. USA* **83**:1709–1713.
13. Banerjee, D. K., Ornberg, R. L., Youdim, M. B. H., Heldman, E., and Pollard, H. B., 1985, Endothelial cells from bovine adrenal medulla develop capillary-like growth patterns in culture, *Proc. Natl. Acad. Sci. USA* **82**:4702–4706.
14. Voyta, J. C., Via, D. P., Butterfield, C. E., and Zetter, B. R., 1984, Identification and isolation of endothelial cells based on their increased uptake of acetylated-low density lipoprotein, *J. Cell Biol.* **99**:2034–2042.
15. Fattorossi, A., Aurbach, G. D., Sakaguchi, K., Cama, A., Marx, S. J., Streeten, E. A., Fitzpatrick, L. A., and Brandi, M. L., 1988, Anti-endothelial cell antibodies: Detection and characterization in sera from patients with autoimmune hypoparathyroidism, *Proc. Natl. Acad. Sci. USA* **85**:4015–4019.

Index

SPRINGER NATURE

GPSR Compliance

The European Union's (EU) General Product Safety Regulation (GPSR) is a set of rules that requires consumer products to be safe and our obligations to ensure this.

If you have any concerns about our products, you can contact us on ProductSafety@springernature.com

In case Publisher is established outside the EU, the EU authorized representative is:

Springer Nature Customer Service Center GmbH
Europaplatz 3
69115 Heidelberg, Germany

The manufacturer's authorised representative in the EU is Springer Nature Customer Service Centre GmbH, Europaplatz 3, 69115 Heidelberg, Germany. If you have any concerns regarding our products, please contact ProductSafety@springernature.com

Printed and bound by CPI Group (UK) Ltd, Croydon, CR0 4YY
14/10/2026
02233117-0020